VOLUME IX

In Preparation: CHEMICAL EXPERIMENTATION UNDER EXTREME CONDITIONS

Edited by Bryant W. Rossiter

VOLUME X

APPLICATIONS OF BIOCHEMICAL SYSTEMS IN ORGANIC CHEMISTRY, in Two Parts

Edited by J. Bryan Jones, Charles J. Sih, and D. Perlman

VOLUME XI

CONTEMPORARY LIQUID CHROMATOGRAPHY

R. P. W. Scott

VOLUME XII

SEPARATION AND PURIFICATION, Third Edition

Edited by Edmond S. Perry and Arnold Weissberger

VOLUME XIII

LABORATORY ENGINEERING AND MANIPULATIONS, Third Edition

Edited by Edmond S. Perry and Arnold Weissberger

VOLUME XIV

THIN-LAYER CHROMATOGRAPHY, Second Edition

Justus G. Kirchner

VOLUME XV

THEORY AND APPLICATIONS OF ELECTRON SPIN RESONANCE

Walter Gordy

TECHNIQUES OF CHEMISTRY

ARNOLD WEISSBERGER, *Editor*

VOLUME XV

THEORY AND APPLICATIONS OF ELECTRON SPIN RESONANCE

To Eileen and Terry

INTRODUCTION TO THE SERIES

Techniques of Chemistry is the successor to the Technique of Organic Chemistry Series and its companion — Technique of Inorganic Chemistry. Because many of the methods are employed in all branches of chemical science, the division into techniques for organic and inorganic chemistry has become increasingly artificial. Accordingly, the new series reflects the wider application of techniques, and the component volumes for the most part provide complete treatments of the methods covered. Volumes in which limited areas of application are discussed can be easily recognized by their titles.

Like its predecessors, the series is devoted to a comprehensive presentation of the respective techniques. The authors give the theoretical background for an understanding of the various methods and operations and describe the techniques and tools, their modifications, their merits and limitations, and their handling. It is hoped that the series will contribute to a better understanding and a more rational and effective application of the respective techniques.

The present volume, *Theory and Applications of Electron Spin Resonance* by Walter Gordy, completes *Chemical Applications of Spectroscopy,* edited by W. West and published as Volume IX of the series Technique of Organic Chemistry.

It is a pleasure to recognize Dr. West's never-failing interest in a task he undertook more than twenty years ago. His devotion is matched by the author's willingness to invest time and effort in a treatise commensurate to the great importance of the field.

Authors and editors hope that readers will find the volumes in this series useful and will communicate to them any criticisms and suggestions for improvements.

ARNOLD WEISSBERGER

Research Laboratories
Eastman Kodak Company
Rochester, New York 14650

of the order of 3570 G must be applied. If one wishes to observe it with 1-cm waves (30,000 MHz), a field of the order of 10,700 G is required. Since these fields are easily obtainable in the laboratory, most of the spectra are observed in the microwave region of 1 to 3 cm. Excellent commercial epr spectrometers are now available for observations in this range.

Most of the stable molecules, excluding those containing transition elements, have all their electrons neatly paired so that their spin moments point in opposite directions and mutually cancel. The pairing of electrons was recognized as an important phenomenon in chemical bonding even before it was realized that the electron has a spin moment. Now it is known from the Pauli exclusion principle that two electrons can occupy the same atomic or molecular orbital only when their spin moments are opposed. Thus the two electrons of a fully occupied bonding molecular orbital or of a covalent chemical bond have no resultant electron-spin moment and can have no esr. Only molecules or radicals having an orbital or orbitals occupied by a single (unpaired) electron can exhibit esr. However, one can produce unpaired electrons by breaking chemical bonds or by lifting electrons from paired, singlet-state orbitals into excited triplet states and can then use esr to examine the resulting species.

If the electron spin were sufficiently free to give the simple resonance frequency of (1.1), we could learn that unpaired spins were present and could learn their concentrations, their rates of buildup or decay—but little else. In practice, however, there are always magnetic fields within the paramagnetic substances that interact with the electron-spin moment to perturb the resonance. These internal fields may split the esr into a complex multiplet or may simply alter its width or frequency. Among the most informative of the internal perturbing fields are those arising from nuclear magnetic moments. Internal magnetic fields also arise from orbital motions of the electrons. Although the orbital magnetic moments in molecular free radicals are largely quenched by crystalline electric fields of solids or by strong electric fields along chemical bonds, weak internal fields can be induced by the applied field through admixture of the orbital states. These induced fields displace the resonance slightly from its free-spin value. This displacement is generally anisotropic so that the g factor of (1.1) is most often not a simple constant but a tensor. By evaluating the principal elements of the anisotropic **g** tensor, one can often learn the symmetry of the internal electric fields and the orientation of certain chemical bonds and orbitals in the free radical or excited molecule.

In the absence of an externally applied magnetic field or of internal interactions, the spin moments will have no preferred direction, and the spins will be randomly oriented. If a uniform magnetic field is imposed on the sample consisting of a magnetic species with $S = 1/2$, the components of spin angular momentum will be resolved along the field, corresponding to the two allowed states, $M_S = 1/2$ (spin up) and $M_S = -1/2$ (spin down). Within an extremely small fraction of a second after the field is on—the time it takes for the spins to come to thermal equilibrium—slightly more than half the spins are found to be in the state of lower energy, $M_S = -1/2$. For example, such a paramagnetic sample at room temperature in a magnetic field of 3000 G will have approximately 50.05% of its spins in the lower state and 49.95% of its spins in the upper state. This slight difference in population has significant consequences. If there were no difference in population of energy states, spin resonance could not be detected, since a radiation field at the resonant frequency induces transitions between two states in either direction with equal probability. If the population of the two states were equal, as much energy would be emitted as absorbed, and no net effect would be observed. A net absorption occurs only when the population of the lower state is greater.

1. MAGNETIC RESONANCE OF FREE ATOMS

1.a. The Landé *g* Factor

Although we are concerned here primarily with the magnetic resonance of unpaired electrons in molecular free radicals, or crystals, we can advantageously consider briefly the gross paramagnetic resonance of free atoms, first neglecting nuclear spin effects. Since the electrons in completely filled subvalence shells contribute nothing to the magnetism of the atom, they can be ignored. Each electron has a spin momentum **s**, a spin magnetic moment μ_s, and (except when in an *s* orbital) an orbital momentum $\boldsymbol{l}$ and an orbital magnetic moment μ_ℓ. The angular momentum and spin moments are coupled in various ways in different atoms. The most common coupling scheme is the Russell-Saunders coupling, in which the various vectors l_1, l_2, and so on of the different electrons combine to form a resultant **L**, and the various spin vectors $\mathbf{s}_1$, $\mathbf{s}_2$, and so on combine to form a resultant **S**; **L** and **S** then combine to form a resultant **J**. The total quantum number J may have integral or half-integral values differing by integers and ranging from $|L + S|$ to $|L - S|$. When there is only one electron outside a closed shell, there is only the coupling between the spin and the orbital motion of that electron, which can combine to give only $j = l + 1/2$, or $j = l - 1/2$. The components of the magnetic moments μ_L and μ_S combine to give a resultant μ_J along **J**, which can be expressed by

$$\mu_J = -g_J \beta \, \mathbf{J} \tag{1.4}$$

The classical interaction of μ_J with an applied magnetic field **H** is

$$\mathcal{H} = -\mu_J \cdot \mathbf{H} = g_J \beta \, \mathbf{J} \cdot \mathbf{H} = g_J \beta H |\mathbf{J}| \cos\theta \,, \tag{1.5}$$

where θ is the angle between **J** and **H**. Quantum mechanically, **J** is resolved along

PREFACE

At this advanced stage in the development of electron paramagnetic resonance (epr) an attempt to contain in one volume a comprehensive treatment of it would be foolhardy. No such attempt is made here, but the selection of the limited material to be included has been one of the more difficult tasks involved in the preparation of this volume. Electron magnetic resonance spectroscoppy has spread like a forest fire throughout solid-state physics, organic and inorganic chemistry, biochemisry, and biological physics and is now spreading rapidly in pure and applied biology. The power it gives to observe with precision unpaired electrons within condensed matter—liquids and solids—has made it a tool for all scientists, but most particularly for chemists.

Our limited goal has been the explanation of the theory essential for the understanding and interpretation of electron-spin resonance (esr) spectra in condensed matter and its applications to paramagnetic species induced in nonmagnetic or diamagnetic substances by various physical or chemical means. We treat doublet-spin-state spectra of free radicals produced by bond dissociation or molecular ionization and the triplet-state spectra of molecules, of charge-transfer complexes, and of exchange-coupled radical pairs. We include the spectra of ordered species in single crystals; of randomly oriented species in powders, polymers, or glassy solutions; and of tumbling free radicals in liquid solutions. Numerous spectra of various types are interpreted, and the derived parameters tabulated. To a very limited degree, the volume is a source book of selected information that has been gained from esr in the past.

By selection of *electron-spin resonance* as a title rather than the more general term, *electron paramagnetic resonance,* we sought to remove from consideration all problems of gaseous atoms or molecules, in which the orbital angular momentum joins in the resonance. Nevertheless, partly to show what the subject is not, we found it desirable in the introductory chapter to give a brief discussion of the magnetic resonance of free atoms. Although the orbital moments do not directly participate in the resonance motions in the condensed systems that we treat, they do influence to a degree the spin-resonance frequencies through residual spin-orbit couplings. Thus one cannot entirely forget the orbitals in any treatment of esr. The most pronounced effects of orbital moments on electron magnetic resonance occur for the salts of the transition elements, especially for the rare-earth salts. These naturally occurring magnetic species were the first to be investigated with esr and they have been the most thoroughly and extensively studied. Hence we decided that the most significant containment of our subject material could be made by omission of the transition ions. Their inclusion is rendered superfluous by the lengthy treatise, *Paramagnetic Resonance of the Transition Ions,* by Abragam and Bleaney (Clarendon, Oxford; 1970), probably the two most qualified to write on this topic. To further contain the subject material, descriptions of experimental techniques and instrumentation are not included. Regretably, we had to omit the description of electron-electron nuclear double resonance (ELDOR) and saturation-transfer spectroscopy, which are subjects suitable for individual monographs. Although the basic theory for electron-nuclear double resonance (ENDOR) is developed, its application to the spectra is not specifically illustrated. A recent book, *Electron Spin Double Resonance Spectroscopy,* by Kevan and Kisper (Wiley-Interscience, New York, 1976), is devoted entirely to ENDOR spectroscopy. Saturation-transfer esr spectroscopy, a new technique developed by Larry R. Dalton and his associates, is most useful for observation of the large spin-labeled biological polymers in viscous solutions, a subject of great importance in biophysical and biochemical areas. The powerful electron spin-labeling technique introduced by H. M. McConnell and his associates has developed into an extensive subfield of esr spectroscopy, which we also regret having to exclude from this limited volume.

This book could not have been written without the diligent and effective work of my wife, Vida Miller Gordy, who typed and edited the entire first draft and read and marked the final copy. Mrs. Jean Luffman typed the final manuscript with accuracy and clarity. Many of the illustrations were drawn by Mrs. Dorothy Bailey. The former Duke graduate students and postdoctoral fellows who worked in esr spectroscopy contributed directly or indirectly to this volume through their earlier collaborations with me. I am particularly indebted to Drs. Ichiro Miyagawa, Donald Chesnut, Peter Smith, Howard Shields, Joseph Hadley, William Nelson, F. M. Atwater, and Louis Dimmey, who kindly read parts of this manuscript and made helpful suggestions. During the considerable time required for completion of this volume I have benefited from the kindly encouragement and exceptional patience of Dr. William West. For these courtesies, and for his careful reading of the manuscript, I am sincerely grateful.

WALTER GORDY

Durham, North Carolina
May 1979

CONTENTS

Chapter I

INTRODUCTION

Electronic paramagnetic resonance, first detected by the Russian scientist Zavoisky [1] in 1945, represents a powerful method for the study of many properties of matter of interest to chemists, physicists, and biologists. The theoretical and experimental methods were developed mostly at Oxford, England by a theoretical group led by M. H. L. Pryce and A. Abragam and an experimental team led by B. Bleaney. The foundation of this field of spectroscopy was provided by the earlier theoretical work on magnetism, notably by J. H. Van Vleck of Harvard, and the experimental measurements on magnetic dispersion, notably by C. J. Gorter at Leyden. General reviews describing these early developments are listed at the end of the chapter [2-10].

Classically, this resonance consists of the Larmor precession of a free electron-spin moment, an atomic or a molecular moment about an externally applied magnetic field. Quantum mechanically, the resonance is described as a transition between Zeeman levels separated by the applied field. The method is not limited to the naturally occurring paramagnetic substances, since electrons can always be unpaired and paramagnetism produced by irradiation or other physical or chemical treatment of nonmagnetic substances. Paramagnetic resonance is commonly observed in solids, liquids, and gases and also in both liquid and solid solutions. Probably no method is more effective for the study of radiation chemistry of the solid state than is electron paramagnetic resonance

or electron spin resonance, commonly designated as epr and esr. Microwave radiation used for observation of the spectra readily penetrates most nonmetallic solids. Increasingly, the epr or esr method is being used to detect trace radicals in organic and biological substances and to study transient radicals in chemical reactions. Essentially all organic chemicals and biochemicals can be studied with this method.

In its purest form, esr corresponds to a flipping over of the free electron-spin vector—along with its associated spin magnetic moment—in an imposed magnetic field. The classical energy of a magnetic dipole moment $\boldsymbol{\mu}$ in a magnetic field $\mathbf{H}$ is $-\boldsymbol{\mu} \cdot \mathbf{H}$, or $-\mu_H H$, where μ_H is the component of the dipole moment along the field. The electron-spin moment can have only two components along the field, $\mu_H = \beta$ and $-\beta$, where β is the Bohr magneton. These correspond to the two allowed values of the electron-spin quantum number, $M_S = 1/2$ and $M_S = -1/2$. Thus there are two spin states corresponding to energies βH and $-\beta H$, with energy difference $2\beta H$. More precisely, this energy difference is $g_s\beta H$, where $g_s = 2.0023$ for the completely free electron-spin moment. With the Bohr rule, $h\nu = E_{M=1/2} - E_{M=-1/2} = g\beta H$, one finds the esr frequency to be

$$\nu = \frac{g_s\beta H}{h} \tag{1.1}$$

where h is Planck's constant. In the ideal case, where there are no internal interactions but only interactions with the externally applied field, $\mathbf{H}$, the resonance occurs at a frequency of

$$\nu(\text{ in megahertz}) = 2.8024\ \text{H (in gauss)} \tag{1.2}$$

It is interesting that this frequency corresponds to the classical Larmor precessional frequency of the spin moment about the field direction.

In actual observation of the resonance one usually leaves the microwave frequency fixed and sweeps through the resonance by varying the strength of the magnetic field imposed. Thus the esr curves traced by the spectrometer represent a plot of intensity of absorption against magnetic field strength at a fixed radiation frequency. At a constant frequency of ν_0, the resonant field strength is

$$H_0 = \frac{h\nu_0}{g_s\beta} \tag{1.3}$$

The ideal case of the free electron spin is not completely achieved but is closely approached in almost all resonances in molecular free radicals. Thus if one expects to observe the resonance with 3-cm microwaves (10,000 MHz), a field

Library of Congress Cataloging in Publication Data

Gordy, Walter, 1909-
Theory and applications of electron spin resonance.

(Techniques of chemistry ; v. 15)
"A Wiley-Interscience publication."
"Completes Chemical applications of spectroscopy."
Includes index.
1. Electron paramagnetic resonance spectroscopy.
I. Chemical applications of spectroscopy.
II. Title.

QD61.T4 vol. 15 [QD96.E4] 542'.08]
ISBN 0-471-93162-4 79-12377

Printed in the United States of America

10 9 8 7 6 5 4 3 2 1

TECHNIQUES OF CHEMISTRY

VOLUME XV—W. West, *Editor*

THEORY AND APPLICATIONS OF ELECTRON SPIN RESONANCE

WALTER GORDY

Duke University

A WILEY-INTERSCIENCE PUBLICATION

JOHN WILEY & SONS

New York · Chichester · Brisbane · Toronto

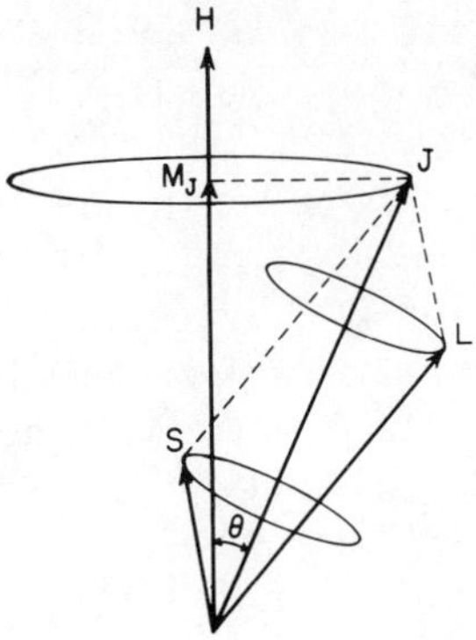

Fig. 1.1 Vector model of an atom in an applied field **H** when $g_J\beta\mathbf{J} \cdot \mathbf{H} << \lambda\, \mathbf{L} \cdot \mathbf{S}$ and there is no nuclear coupling.

the applied field with quantized components that in units of ħ are $M_J = J$, $J - 1, \ldots, -J$. From the vector model of Fig. 1.1, $\cos\theta = M_J/|\mathbf{J}|$. Substitution of this value in (1.5) gives the allowed Zeeman energies as

$$E_J = g_J \beta M_J H \tag{1.6}$$

The magnetic resonance frequency corresponding to the transition $M_J \rightarrow M_J \pm 1$ is

$$\nu = \frac{g_J \beta H}{h} \tag{1.7}$$

where $\beta = 5.05038 \times 10^{-24}$ erg/G is the Bohr magneton and g_J is the gyromagnetic ratio or Landé g factor. This well-known formula can be derived easily from the vector model of Fig. 1.1. For example,

$$\mu_J = \mu_L \cos(\mathbf{L}, \mathbf{J}) + \mu_S \cos(\mathbf{S}, \mathbf{J}) \tag{1.8}$$

with $\mu_J = g_J\beta\mathbf{J}$, $\mu_L = \beta\mathbf{L}$, and $\mu_S = 2\beta\mathbf{S}$. This gives

$$g_J = \frac{\beta[\mathbf{L}\cos(\mathbf{L}, \mathbf{J}) + 2\mathbf{S}\cos(\mathbf{S}, \mathbf{J})]}{\mathbf{J}} \tag{1.9}$$

From the law of the cosine, $\cos(\mathbf{L}, \mathbf{J}) = (\mathbf{J}^2 + \mathbf{L}^2 - \mathbf{S}^2)/2\mathbf{LJ}$ and $\cos(\mathbf{S},\mathbf{J}) = (\mathbf{J}^2 + \mathbf{S}^2 - \mathbf{L}^2)/2\mathbf{SJ}$. Substitution of the eigenvalues $\mathbf{J}^2 = J(J+1)$, $\mathbf{L}^2 = L(L+1)$, $\mathbf{S}^2 = S(S+1)$ in the cosines, with some rearrangement, yields the Landé g factor for the atom:

$$g_J = 1 + \frac{J(J+1) + S(S+1) - L(L+1)}{2J(J+1)} \tag{1.10}$$

A case of special interest is that of S states for which $L = 0$, $J = S$, and $g_J = g_S = 2$.

In the preceding Hamiltonian we did not include the spin-orbit interaction $\lambda\mathbf{L} \cdot \mathbf{S}$ because this interaction is so strong that it is not broken down or measurably perturbed by applicable magnetic strengths. A change in $\lambda\mathbf{L} \cdot \mathbf{S}$ associated with a change in the eigenvalues of $J = L + S, L + S - 1, L + S - 2, \ldots, -(L + S)$ represents a very large energy change as compared with the Zeeman splitting and corresponds to a change in the electronic state of the atom. Thus in observation of epr spectra of atoms, one can consider **J** as well as **L** and **S** as constants of the motion. Only changes in the resolvable components M_J need be considered.

The conventional labeling of the atomic states signifies the particular values of J, L, and S for the state. For example, L of 0,1,2,3 corresponds to S,P,D,F states. A superscript to the left is added to signify the spin degeneracy $2S + 1$, and a subscript on the right is added to indicate the value of J. For example, the ground state of nitrogen, $^4S_{3/2}$, has $L = 0$, $S = 3/2$, $J = S = 3/2$, and hence a Landé g factor $g_J = 2$, from (1.10). Atomic oxygen has a $^3P_2$ ground state, $L = 1$, $S = 1$, $J = S + L = 2$, and $g_J = 3/2$. Atomic fluorine has a $^2P_{3/2}$ ground state with $S = 1/2$, $L = 1$, $J = 3/2$, and hence $g_J = 4/3$.

1.b. Nuclear-magnetic Interactions

When the nucleus of an atom has a spin $I \neq 0$, it will have an associated magnetic moment $\mu_I = g_I\beta_I\mathbf{I}$, where g_I is the nuclear g factor and $\beta_I = eh/4\pi Mc$ is the nuclear magneton. This μ_I interacts with the averaged atomic field at the nucleus $\mathbf{H}_J$ that is in the direction of **J**. Because interaction with field components perpendicular to **J** is canceled by the precessional motions, the nuclear magnetic interaction when no field is applied can be expressed as

$$\mathcal{H}_N = \mu_I \cdot \mathbf{H}_J = A\mathbf{I} \cdot \mathbf{J} \tag{1.11}$$

where A is the nuclear coupling constant between **I** and **J**. The last expression follows because $\mu_I \sim \mathbf{I}$ and $\mathbf{H}_J \sim \mathbf{J}$. Thus when **J** is a constant of the motion,the nuclear interaction is an isotropic $\mathbf{I} \cdot \mathbf{J}$ interaction, in contrast to the tensor interaction in solids, where the spin-orbit coupling is broken down (see Chapter II). For spin values, $I > 1/2$, the nucleus may also have an electric quadrupole moment Q that gives additional coupling to **J**. As this interaction is treated in Chapter II, it is omitted here.

In the absence of an external field, **I** and **J** are thus coupled to form a resultant designated $\mathbf{F} = \mathbf{J} + \mathbf{I}$. Since **J** and **I** commute, $\mathbf{J} \cdot \mathbf{I} = \mathbf{I} \cdot \mathbf{J}$ and

$$\mathbf{F}^2 = (\mathbf{J} + \mathbf{I})^2 = \mathbf{J}^2 + \mathbf{I}^2 + 2\mathbf{I} \cdot \mathbf{J} \tag{1.12}$$

and hence

$$\mathbf{I} \cdot \mathbf{J} = (1/2)\ (\mathbf{F}^2 - \mathbf{J}^2 - \mathbf{I}^2) \tag{1.13}$$

Quantum mechanically, the square of the total angular momentum $\mathbf{F}^2$ has the eigenvalues $F(F+1)$, and the resolvable components of $\mathbf{F}$ have eigenvalues M_F. In this coupling model, $\mathbf{J}^2$ and $\mathbf{I}^2$ are also quantized with $\mathbf{J}^2 = J(J+1)$ and $\mathbf{I}^2 = I(I+1)$. Thus $\mathcal{H}_N$ is diagonal in the representation $|F, M_F, J, I\rangle$, and the diagonal element represents the specific energy values

$$E_N = (1/2)A\,[F(F+1) - J(J+1) - I(I+1)] \tag{1.14}$$

where the F quantum numbers have the values

$$F = J + I,\ J + I - 1,\ J + I - 2,\ \ldots,\ -(J + I)$$

If a weak magnetic field is imposed, one in which the Zeeman energies are much smaller than $A\mathbf{I} \cdot \mathbf{J}$, a resultant $\mathbf{F}$ is still formed by $\mathbf{J}$ and $\mathbf{I}$. In the vector model of Fig. 1.2, $\mathbf{J}$ and $\mathbf{I}$ precess about $\mathbf{F}$, and $\mathbf{F}$ precesses more slowly about the direction of the applied field $\mathbf{H}$. The components of the total angular momentum that are resolvable along a fixed direction in space have the quantized values

$$M_F = F,\ F - 1,\ F - 2,\ \ldots,\ -F$$

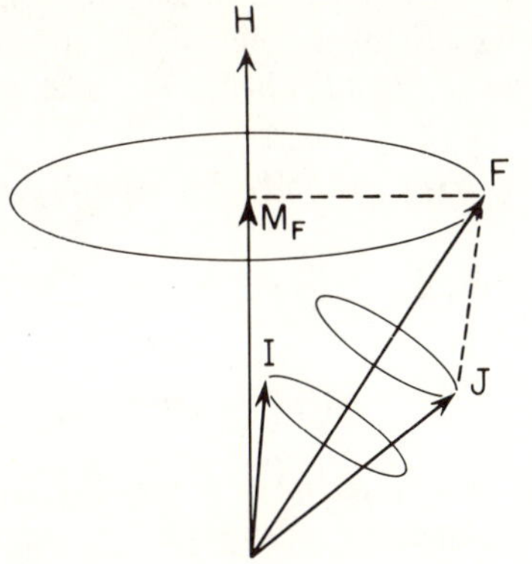

Fig. 1.2 Vector model of an atom in an applied field **H** when there is nuclear coupling and $g_I\beta_I\mathbf{I} \cdot \mathbf{H} \ll A\mathbf{I} \cdot \mathbf{J}$ (weak-field case).

The Zeeman Hamiltonian for this weak-field case is

$$\mathcal{H}_M = -\mu_F \cdot \mathbf{H} \tag{1.15}$$

where

$$\mu_F = \mu_J \cos(\mathbf{J}, \mathbf{F}) - \mu_I \cos(\mathbf{I}, \mathbf{F}) \ . \tag{1.16}$$

The components of μ_J and μ_I perpendicular to **F** average to zero by the precession of **J** and **I** about **F**. The sign of the last term is negative because μ_J and μ_I are opposed in sign when μ_I is positive. Let us define g_F such that $\mu_F = g_F\beta\mathbf{F}$. Then with $\mu_J = g_J\beta\mathbf{J}$ and $\mu_I = g_I\beta_\mathrm{I}\mathbf{I}$, (1.16) gives

$$g_F = g_J \cos(\mathbf{J}, \mathbf{F}) - g_\mathrm{I}\frac{\beta_\mathrm{I}}{\beta} \cos(\mathbf{I}, \mathbf{F}) \tag{1.17}$$

With cos (**J**,**F**) and cos(**I**,**F**) determined from the cosine law and with $\mathbf{F}^2 = F(F + 1)$, $\mathbf{J}^2 = J(J + 1)$, $\mathbf{I}^2 = I(I + 1)$, (1.17) for g_F becomes

$$g_F = g_J \frac{F(F+1) + J(J+1) - I(I+1)}{2F(F+1)} - g_I\left(\frac{\beta_I}{\beta}\right)\frac{F(F+1) + I(I+1) - J(J+1)}{2F(F+1)} \tag{1.18}$$

where g_J is given by (1.9).

By substitution of $g_F\beta\mathbf{F}$ for μ_F, the Zeeman Hamiltonian, (1.15), can be expressed as

$$\mathcal{H}_M = g_F\beta\mathbf{F} \cdot \mathbf{H} = g_F\beta(F_xH_y + F_yH_y + F_zH_z) \tag{1.19}$$

Since in this weak-field case $\mathcal{H}_M$ is small as compared with $\mathcal{H}_N$, it can be treated as a perturbation of $\mathcal{H}_N$. The first-order perturbation is simply an average of $\mathcal{H}_M$ over the wave functions $|F,M_F,J,I)$ of $\mathcal{H}_N$. This average is

$$E_{M_F} = (F,M_F,J,I|\mathcal{H}_M|F,M_F,J,I) \tag{1.20}$$

Since the diagonal elements of F_z are M_F and those of F_x and F_y vanish, this average is

$$E_{M_F} = g_F\beta HM_F \tag{1.21}$$

where g_F is given by (1.18) and $H = H_z$. Alternately, one can express $\mathcal{H}_M$ as $g_F|\mathbf{F}|H \cos(\mathbf{F},\mathbf{H})$ and then with $\cos(\mathbf{F},\mathbf{H}) = (F_z/|\mathbf{F}|) = (M_F/|\mathbf{F}|)$ can derive the same formula. The vector-model treatment is equivalent to the first-order perturbation theory. This weak-field Zeeman splitting is superimposed on each of

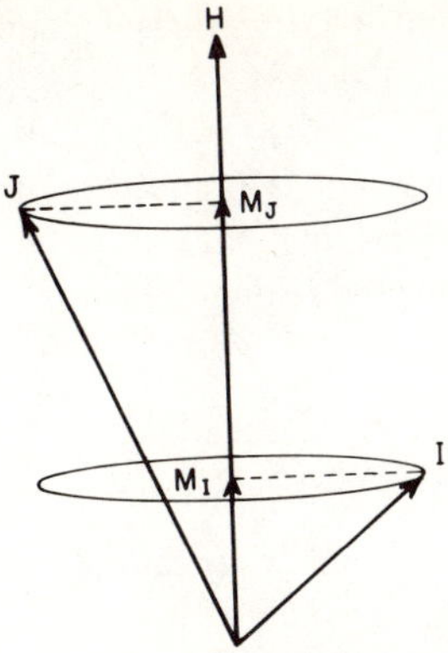

Fig. 1.3 Vector model of an atom in an applied field **H** when $g_I\beta_I\mathbf{I}\cdot\mathbf{H} >> A\mathbf{I}\cdot\mathbf{J}$(strong-field case).

the nuclear hyperfine levels (corresponding to each value of F) as given by (1.14). Because $(\beta_I/\beta) = 5.446 \times 10^{-4}$, the last term in the g_F of (1.18) can be omitted in approximate treatments.

A second important case occurs when the nuclear $\mathbf{I} \cdot \mathbf{J}$ interaction is much smaller than the $\mathbf{J} \cdot \mathbf{H}$ interaction. In this strong-field or Paschen-Back case, **I** and **J** do not form a resultant, but each is individually resolved along (or in the vector model of Fig. 1.3, precesses about) the direction of the applied field. The Hamiltonian for the magnetic interaction is

$$\begin{aligned}\mathcal{H}_M &= -\mu_J \cdot \mathbf{H} - \mu_I \cdot \mathbf{H} + A\mathbf{I} \cdot \mathbf{J} \\ &= g_J\beta\mathbf{J} \cdot \mathbf{H} - g_I\beta_I\mathbf{I} \cdot \mathbf{H} + A\mathbf{I} \cdot \mathbf{J}\end{aligned} \tag{1.22}$$

The axis of quantization of both **J** and **I** is the direction of **H** and is chosen as the z axis. In the representation $|J,M_J,I,M_I)$, in which the first two terms are diagonal, the diagonal matrix elements of the last term are AM_JM_I. This follows from the fact that diagonal matrix elements of J_z are M_J and those of I_z are M_I, whereas the diagonal elements of J_x, J_y, I_x, and I_y are zero. Thus the energy values to first order, $(J,M_J,I,M_I|\mathcal{H}_M|J,M_J,I,M_I)$, are

$$E_M = g_J\beta HM_J - g_I\beta_I HM_I + AM_JM_I \tag{1.23}$$

where g_J is given by (1.9) and A has the same value as in the weakfield case, (1.14). A more precise treatment, which takes into account the incomplete resolution of the **J** and **I** along the applied field, has been given by Breit and Rabi [11] for $J = 1/2$ with any value of I. Similar corrections are derived in

Chapter III, Sections 4.b and 4.k for nuclear couplings of doublet-state free radicals in solids.

The selection rules for epr are $\Delta M_J = \pm 1$ and $\Delta M_I = 0$. These rules mean, in effect, that the electronic moment is induced by the microwaves to flip over in the applied field **H**, whereas the nuclear moment is not. These selection rules give the epr frequency for the strong-field case as

$$\nu = \frac{g_J \beta H}{h} + \frac{A}{h} M_I \tag{1.24}$$

The nuclear magnetic quantum number M_I can take the values

$$M_I = I, I - 1, I - 2, \ldots, -I$$

where I is the nuclear spin value. Thus there are $(2I + 1)$ hyperfine components of the epr transition. These have equal spacing and are of equal intensity when observed under normal conditions of thermal equilibrium. The spacing of consecutive components gives the coupling constant A/h directly. At a constant frequency the "resonant field" values are

$$H_0 = \frac{h\nu_0}{g_J \beta} - \frac{A}{g_J \beta} M_J \tag{1.25}$$

where A is in energy units (ergs). For convenience, A is often expressed in frequency units. When this is done, the last term of (1.25) must be multiplied by h because $A(\text{ergs}) = hA(\text{hertz})$. Thus in magnetic field units the spacing of the hyperfine components is $hA/g_J\beta$, where A is in frequency units (hertz) rather than in ergs.

Figure 1.4 illustrates the energy-level diagrams for the zero-field, weak-field, and strong-field cases for an atom in a $^2S_{1/2}$ state with nuclear spin $I = 1/2$. Although the nuclear magnetic hyperfine splitting differs in the weak- and strong-field cases, the same value of the coupling constant A is obtained from measurements of either of the two types of spectra. Approximately the same value of the nuclear coupling constant is obtained from analysis of esr spectra of atoms in solids where the $\mathbf{L} \cdot \mathbf{S}$ coupling is also broken down, and where neither J nor F is a good quantum number. See Chapter VIII, Section 3 for a treatment of atoms trapped in solids. Theoretical formulas for calculation of A are given in Chapter II. Values of some coupling constants measured from atomic spectra are given in Table A.1. These atomic coupling constants are used in the interpretation of esr hyperfine structure in condensed matter.

This brief discussion of magnetic interactions in free atoms is given as an aid to the understanding of the epr or esr of condensed matter where the para-

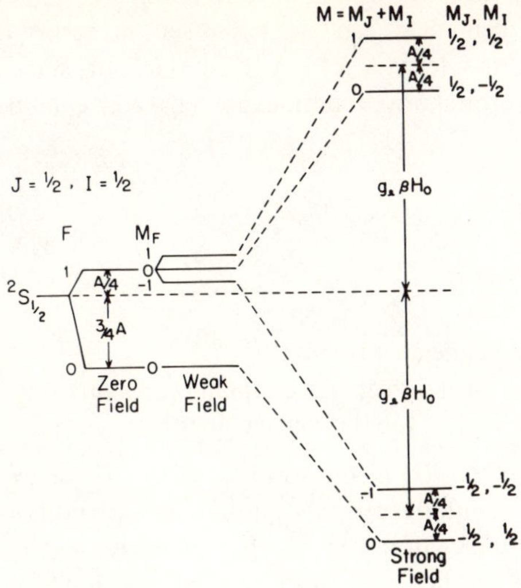

Fig. 1.4 Energy-level diagrams of an atom in an $^2S_{½}$ state ($L = 0, J = S = ½$) with a coupling nucleus having $I = ½$, when no field is applied and when weak and strong magnetic fields are applied.

magnetic elements involve bonded or trapped atoms. Those concerned with the study of magnetic resonance of free atoms should consult texts on atomic spectra and atomic beam resonance [12-15].

2. DESCRIPTION OF ATOMIC ORBITALS

In addition to the spin functions, electronic and nuclear, one is concerned primarily with the orientation-dependent parts of the atomic orbital wave functions in treating esr spectra. We deal primarily with *s* and *p* orbitals and to a lesser extent with *d* orbitals. The *s* orbital is spherically symmetric and thus is independent of orientation ($l = 0$). The orientation of the *p* orbitals ($l = 1$), of which there are three (corresponding to $m_l = 1, 0, -1$), can be expressed in terms of the *x*, *y*, and *z* coordinates. For most esr applications, the *p*-orbital wave functions are most conveniently expressed as

$$p_x = xf(r), \qquad p_y = yf(r), \qquad p_z = zf(r) \tag{1.26}$$

where $f(r)$ is independent of the orientation. The p_x, p_y, and p_z orbitals are oriented along the respective coordinate axes.

There are five independent d-orbital wave functions ($l = 2, m_l = 2,1,0,-1,-2$). Their orientation dependence can be expressed in terms of xy/r^2, yz/r^2, zx/r^2, $[(x^2 - y^2)/2]$, and $[(2z^2 - x^2 - y^2)/2\sqrt{3}]/r^2$; or, since $(1/r^2)$ can be included in the $f(r)$, the orbital wave functions are most conveniently expressed for esr as

$$d_{xy} = xyf(r), \qquad d_{yz} = yzf(r), \qquad d_{zx} = zxf(r) \tag{1.27}$$

$$d_{x^2-y^2} = (1/2)(x^2-y^2)f(r), d_{2z^2-x^2-y^2} = (1/2\sqrt{3})(2z^2-x^2-y^2)f(r) \tag{1.28}$$

where $f(r)$ is again independent of orientation.

In evaluations of nuclear hyperfine coupling constants one must know the averaged quantity $\left(1/r_l^3\right)_{\text{av}}$ for the particular orbital considered. This average, of course, depends on the $f(r)$ in the wave function. There are, however, good theoretically or experimentally derived values of this quantity already available for the common coupling atoms in the free radicals treated in this volume. Therefore, we are not particularly concerned here with the radial part of the atomic wave functions.

With the radial part of the wave function eliminated from the operation, it is sufficient and convenient to label the functions with the orbital and spin quantum numbers only. The orbital wave function might be labeled simply ψ_{lm_l} or $|lm_l\rangle$. When the electron spin is included, the function would be indicated by ψ_{l,m_l,m_s} or $|lm_l,m_s\rangle$. In a partially filled shell in which the individual l vectors are coupled to form the resultant **L** and the individual spins are coupled to form the resultant **S**, the functions would be signified by $|L,M_L,S,M_S\rangle$. In operations in which **L** and **S** remain constant, the functions would be adequately labeled by $|M_L,M_S\rangle$. In the operations to follow, however, it is important that the spin and orbital parts of the function are independent and separable.

3. NATURE OF RESONANCE IN CONDENSED MATTER

One might expect paramagnetic resonance spectra of condensed matter to be much more complicated than that of isolated atoms or molecules in the gaseous state. Actually, in most instances the opposite is true. Certain factors act to simplify the resonance in condensed matter, particularly that for molecular free radicals. The greatest simplification results from the quenching of the orbital moments. In the free atoms of a gas, the orbit of an electron coupled with the electron spin may precess about an applied magnetic field and may change orientation as a consequence of a paramagnetic transition. In contrast, the orbits of electrons in molecular free radicals are fixed in the molecular frame by the strong Coulomb interactions of the chemical bonds. Nevertheless, in the gaseous state the molecular frame to which the resultant orbital moment is fixed can itself precess about the applied field. In condensed matter this precession is generally prohibited by intermolecular interactions. Even in trapped atoms or ionically bonded paramagnetic ions in salts of the transition elements, the orbital orientations are controlled by the strong crystalline electrical fields, and, quite generally, their orbital moments are quenched to a high approximation. Thus only the electron spin moment is free to precess about an applied magnetic field in condensed systems. The observed resonance is approximately that of an atom in an S state. The orbital moments are not entirely quenched, however; higher-order interactions can lead to complications in the spectra and can give valuable information about the chemical bonds and crystalline fields.

3.a. Effects of Chemical Bonds

The quenching of the p-orbital momentum is illustrated with a common type of organic free radical in which the unpaired electron density is concentrated in a p orbital of a bonded carbon. Normally, in saturated organic compounds the s and p valence orbitals of carbon are hybridized into equivalent sp_3 orbitals that form four tetrahedral covalent bonds. When one of these bonds is broken to form a free radical, the hybridization changes; the three remaining bonding orbitals (unless restricted by steric factors) become planar sp_2 hybrids in most instances, and the unpaired electron goes into a pure p orbital perpendicular to the sp_2 plane. The most common free radical of this type is one in which the unpaired electron is in a delocalized π orbital, and only a fraction of its density is in the p orbital of a particular carbon that is perpendicular to the sp_2 plane. For simplicity, let us consider a radical like CH_3 in which the odd electron is localized on a single carbon. A diagram of such a radical is shown in Fig. 1.5 with p_x as the orbital of the unpaired electron and with the sp_2 hybrid in the yz plane. It is readily apparent that this p_x orbital could not undergo classical precession when the yz plane is fixed in space by the three chemical bonds.

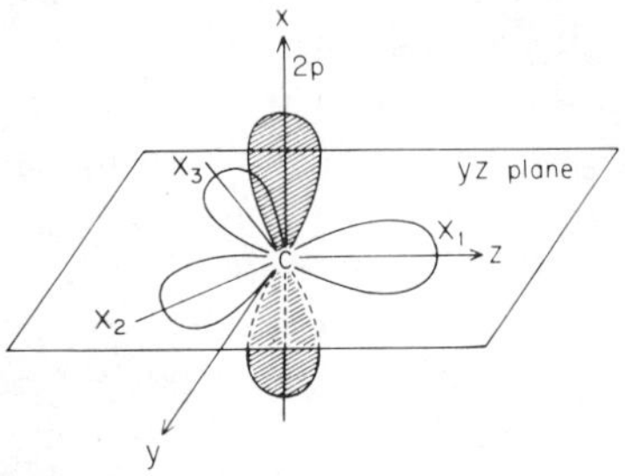

Fig. 1.5 Diagram illustrating the quenching of the orbital angular momentum of an unpaired $2p_x$ electron of carbon by the chemical bonds in the molecular yz plane. The orbital of the unpaired electron is not free to precess or to be reoriented in an applied field **H** as it is in a free atom, but its spin moment **S** is free to do so.

Quantum mechanically, the orbital angular momentum of a p electron changes direction when the unpaired electron switches orbitals, from p_x to p_y or p_z. In a radical like that of Fig. 1.5, this switch could occur in two ways: (1) an electron from a σ-bonding orbital could switch to the p_x orbital, leaving an unpaired electron in one of the σ-bonding orbitals; or (2) the unpaired p_x electron could switch to an antibonding σ orbital in the xy plane, leaving the p_x orbital empty. In the first instance one electron of a bonding pair is removed; in the second an antibonding one is added. In either case, the energy of the system is raised appreciably. Consequently, the $2p$ shell of the carbon is split into an orbital singlet that has the unpaired electron in the p_x orbital and two excited states of much higher energy in which the unpaired electron has an orbital in the yz plane. The important fact for esr is that the ground state of the radical is an orbital singlet and a spin doublet.

It is customary to define an effective or reduced orbital quantum number L' for the bonded atom such that $2L' + 1$ is the orbital degeneracy of the ground state. Therefore, in the radical of Fig. 1.5, $2L' + 1 = 1$, and $L' = 0$. This means, in effect, that here the bonded C on which the unpaired electron is concentrated can be considered as being in an S state. The spin of the one unpaired electron is free to become oriented in an applied field with resolvable components, $m_s = 1/2$ or $-1/2$. Except for differences in nuclear hyperfine structure and slight anisotropy in g due to very small admixing of the ground state with the higher orbital states, the esr of such a free radical having a single unpaired electron is like that of the free hydrogen atom.

The d orbitals are important in paramagnetic resonance primarily because of the iron-group elements that have unpaired electrons in unfilled d shells, although some unpaired d-orbital electron density may occur on atoms such as P, S, or Cl in certain molecular free radicals. The iron-group salts were the first paramagnetic species to be investigated extensively with epr, and their spectra have been treated in numerous reviews and comprehensive monographs [4-10]. Paramagnetic resonance of the transition ions [10] is too extensive in scope for coverage in the present volume, which is limited for the most part to molecular free radicals involving electron-spin density in s and p orbitals. It is sufficient to say that the crystalline field in the iron-group salts splits the d-orbital degeneracies and usually leaves an orbital singlet as the lowest level, which produces an effective singlet orbital group state $L' = 0$, similar to that described earlier for the p orbitals in bonded atoms. In first order this lowest orbital singlet may be considered as an S state with the spin free to precess about an imposed magnetic field and produce an esr. However, the separation of the lowest orbital singlet from the higher orbital levels by the crystalline field is not as great—nor is the orbital quenching as complete—as that produced by chemical bonds. Therefore, the deviation of the g factor from that of the free electron spin is generally greater for the iron-group elements than for the molecular free radicals in which the orbital degeneracies are lifted by chemical bonds.

3.b. Orbital Substates and Effective Orbital Quantum Numbers

In summary, the orbital degeneracy of free radicals or ions in solids is lifted by strong crystalline fields or chemical bonds, leaving orbital substates that are widely separated as compared with normal Zeeman splittings. Certain of these substates may, however; have residual orbital degeneracies, depending on the symmetry of the fields. In treatments of esr, it is advantageous to assign a reduced, effective orbital angular momentum quantum number L' for the orbital substate such that $2L' + 1$ is equivalent to its residual orbital degeneracy. Generally, the lower the symmetry of the internal fields, the more complete the splitting (i.e., the fewer the remaining orbital degeneracies). For free radicals in organic solids and most inorganic salts, the lowest orbital substate, the ground state of the paramagnetic element, is an orbital singlet with an effective orbital quantum number such that $2L' + 1 = 1$. Hence $L' = 0$, $M_L = 0$, and the orbital function of the ground state is designated $|0\rangle$. The orbital functions of substates of higher orbitals, which may or may not have residual orbital degeneracies, are designated $|n\rangle$.

These conditions greatly simplify the epr calculations in solids. They make possible the development of a general spin Hamiltonian applicable to diverse types of paramagnetic species in solids and liquids. The development of this spin Hamiltonian is described in the next chapter. Its solution for the energy states and spin resonance frequencies is described in Chapter III, and various applications are given in later chapters.

3.c. Other Treatments of esr

The literature on electron magnetic resonance is now so voluminous that it is impossible to discuss here all phases of the subject or even to survey all the excellent work that has been done. This book is not intended to serve as a review but as an aid to an understanding and interpretation of the spectra of molecular free radicals. The comprehensive treatise by Abragam and Bleaney on paramagnetic resonance of transition ions [10] provides the most complete and most rigorous treatment available on this phase of epr. A later volume by Poole and Farach [16] provides a selective but extensive bibliography of reviews, conference proceedings, and books on epr that appeared before 1972. A textbook treatment by Wertz and Bolton [17] was also published in 1972. Atherton's book on esr [18], which appeared in 1973, treats both molecular radicals and transition elements, covering theoretical aspects as well as applications. Kevan and Kispert have contributed a monograph on ENDOR spectroscopy [19]. Applications of esr and ENDOR in studies of radiation effects are covered in a recent monograph (1977) by Box [20]. A second edition of the well-known book by Slichter, *Principles of Magnetic Resonance* [21], has just appeared (1978).

References

1. E. Zavoisky, *J. Phys. USSR,* **9**, 211, 245 (1945).
2. J. H. Van Vleck, *Theory of Electric and Magnetic Susceptibilities,* Clarendon, Oxford, 1932.
3. C. J. Gorter, *Paramagnetic Relaxation,* Elsevier, New York, 1947.
4. A. Abragam and M. H. L. Pryce, *Proc. Roy. Soc. (Lond.),* **A205**, 135-153 (1951).
5. B. Bleaney and D. J. E. Ingram, *Proc. Roy. Soc. (Lond.),* **A205**, 336-356 (1951).
6. B. Bleaney and K. W. H. Stevens, *Rep. Progr. Phys. (Lond. Phys. Soc.),* **16**, 108-159 (1953).
7. K. D. Bowers and J. Owen, *Rep. Progr. Phys. (Lond. Phys. Soc.),* **18**, 304-375 (1955).
8. D. M. S. Bagguley and J. Owen, *Rep. Progress in Physics (Lond. Phys. Soc),* **20**, 304-378 (1957).
9. William Low, *Paramagnetic Resonance in Solids,* Academic, New York, 1960.
10. A. Abragam and B. Bleaney, *Electron Paramagnetic Resonance of Transition Ions,* Clarendon, Oxford, 1970.
11. G. Breit and I. I. Rabi, *Phys. Rev.,* **38**, 2082 (1931).
12. H. Kopfermann, *Nuclear Moments* (English version prepared by E. E. Schneider from 2nd German ed.), Academic, New York, 1958.
13. H. E. White, *Introduction to Atomic Spectra,* McGraw-Hill, New York, 1934.
14. E. U. Condon and G. H. Shortly, *The Theory of Atomic Spectra,* Cambridge U. P., Cambridge, England, 1959.
15. N. F. Ramsey, *Nuclear Moments,* Wiley, New York, 1953.
16. C. P. Poole, Jr., and H. A. Farach, *The Theory of Magnetic Resonance,* Wiley-Interscience, New York, 1972.
17. J. E. Wertz and J. H. Bolton, *Electron Spin Resonance: Elementary Theory and Practical Applications,* McGraw-Hill, New York, 1972.
18. N. M. Atherton, *Electron Spin Resonance: Theory and Applications,* Wiley, New York, 1973.
19. L. Kevan and L. D. Kispert, *Electron Spin Double Resonance Spectroscopy,* Wiley, New York, 1976.
20. H. C. Box, *Radiation Effects: ESR and ENDOR Analysis,* Academic, New York, 1977.
21. C. P. Slichter, *Principles of Magnetic Resonance,* 2nd ed., Springer-Verlag, New York, 1978.

Chapter **II**

THE SPIN HAMILTONIAN

The theory of paramagnetic resonance was first developed for the ions of the transition elements [1]. Although this volume is mainly concerned with molecular free radicals, essentially the same theory is applicable to these species. The principal difference is that the unpaired electrons giving rise to the paramagnetism in the salts of transition elements are mostly concentrated in atomic orbitals of the paramagnetic ion, whereas the unpaired electron in the molecular free radical is often concentrated in a molecular orbital that spreads over more than one atom. This difference causes no great difficulty in adaptation of the theory because, as in many other applications, the molecular orbital of the free radical can be expressed as a linear combination of the atomic orbitals of the atoms composing it. To a good approximation, the cross terms involving products of wave functions of different atoms may be neglected. When this is done, the procedure is reduced to solution of the separate atomic problems and combination of the results through addition of the tensor elements, as is illustrated in later sections. This reduction of the molecular problem assumes a knowledge of the unpaired electron density on each atom of the radical or, more specifically, of the electron-spin density in particular orbitals of the different atoms. The spin density does not necessarily coincide with the electronic charge or electronic mass density because of the possibility of negative spin density [2], to be discussed later. From a fitting of the theory to the observed spectra one can obtain a knowledge of the distribution of spin densities on the different atoms.

As explained in Chapter I, normal Zeeman interaction of a free atom with an external magnetic field involves the resultant **J** of the electronic orbital moment **L** and the electron spin moment **S**. In condensed matter, however, the orbital momentum vector is generally restrained by the chemical bonds or the crystalline fields from formation of a resultant **J** with **S** or from alignment with an applied field. In the first-order approximation the orbital magnetism does not contribute to the Hamiltonian for the interaction of the paramagnetic species with an externally applied field. In higher-order approximations, however, effects of orbital moments do not vanish.

The treatment by Pryce and Abragam [3,4] was designed to take into account the effects of the orbital momentum separately and in advance and thus to derive a reduced Hamiltonian that includes only electron-spin and nuclear-spin operators. This procedure is possible because the orbital and spin operators commute and hence are readily separated. The reduced Hamiltonian from which the **L** operators are removed is commonly called the *spin Hamiltonian.* Its specific form depends on the symmetry and nature of the internal fields, but a generalized spin Hamiltonian applicable to essentially all classes of free radicals observed in the solid or liquid state can be expressed compactly through use of tensor formalism.

The development of such a generalized spin Hamiltonian has brought a unity and order into the enormously diverse and complex magnetic resonance spectra of condensed matter. Most of the present monograph is concerned with the tailoring of this spin Hamiltonian to specific paramagnetic species and the derivation of its eigenvalues for interpretation of observed epr spectra. First we give a description of the origin of different terms in the Hamiltonian and describe methods for its solution, which gives the paramagnetic energies and frequencies.

1. NATURE OF THE INTERACTIONS

The total energy of a paramagnetic atom or ion situated within a molecule or crystal can be broken down into parts of different forms and magnitudes. The largest of these parts are the kinetic energy and the Coulomb interaction of the electrons with the nucleus and with each other. These terms are

$$W_F = \sum_k \left(\frac{P_k^2}{2m} - \frac{Ze^2}{r_k} \right) + \sum_{jk} \frac{e^2}{r_{jk}} \tag{2.1}$$

Because it remains fixed in the transformations considered here, we are not particularly concerned with the specific form of W_F.

Next in magnitude is the interaction of the electron orbitals with the crystalline electric fields of the crystal or the chemical bond of the molecules. The form and magnitude of this interaction, designated V_c, depends on the strength and symmetry of the internal fields. Since V_c remains fixed and is generally much greater than the magnetic interactions that give rise to esr, its specific form is not required for the present applications.

To a very good approximation, Russell-Saunders coupling applies in elements of the iron group and for the atoms that constitute most free radicals. When there is more than one unpaired electron on the atom or ion, the orbital momenta $l_1, l_2, l_3, \ldots$, of the different electrons in the ion couple to form a resultant angular momentum represented by the vector **L**. Likewise, the spin moments $s_1, s_2, s_3, \ldots$, of the individual electrons form a resultant spin momentum represented by the vector **S**. Associated with each of these vectors is a magnetic moment. That along **L** is

$$\mu_L = -\beta \mathbf{L} \tag{2.2}$$

and that along **S** is

$$\mu_S = -g_s \beta \mathbf{S} \tag{2.3}$$

in which β is the Bohr magneton (0.92731×10^{-20} erg/G) and g_s is the gyromagnetic ratio of the electron spin. For a completely free electron spin, $g_s = 2(1 + \alpha/2\pi \cdots) = 2.0023$, where α is the fine-structure constant. Through interaction of their associated magnetic moments, **L** and **S** are coupled to form a resultant angular momentum designated by the vector **J**. The coupling energy of **L** and **S** is

$$\mathcal{H}_{LS} = \lambda \mathbf{L} \cdot \mathbf{S} \tag{2.4}$$

where λ is the spin-orbit coupling constant. For many elements, values of λ are known from atomic spectroscopy.

The orbital and spin magnetic moments $-\beta\mathbf{L}$ and $-g_s\beta\mathbf{S}$ have an interaction energy

$$\mathcal{H}_H = -(\mu_L + \mu_S) \cdot \mathbf{H} = \beta(\mathbf{L} + g_s\mathbf{S}) \cdot \mathbf{H} \tag{2.5}$$

with an applied magnetic field **H**.

Before treating electron spin-spin and nuclear interactions, we next describe how the L-dependent factors are eliminated from the preceding terms.

2. SPIN HAMILTONIAN WITHOUT NUCLEAR INTERACTIONS

The Hamiltonian that we consider first is

$$\mathcal{H} = \mathcal{H}_F + \mathcal{H}_V + \mathcal{H}_{LS} + \mathcal{H}_H \tag{2.6}$$

We designate $\mathcal{H}^0 = \mathcal{H}_F + \mathcal{H}_V$ and consider

$$\mathcal{H}' = \mathcal{H}_{LS} + \mathcal{H}_H = \lambda\mathbf{L} \cdot \mathbf{S} + \beta\mathbf{H} \cdot (\mathbf{L}+g_s\mathbf{S})$$

as a perturbation on the ground state of $\mathcal{H}^0$, for which we designate the energy value by E_0, with the energies of higher orbital states by $E_1, ..., E_n$ and the corresponding eigenfunctions by $\psi_0, \psi_1, ..., \psi_n$, or by the kets $|0), |1), ..., |n)$. The lowest orbital state $|0)$, for reasons already given, is assumed to be an orbital singlet.

The first-order perturbation energy is simply the average of the perturbing Hamiltonian over the wave function of the unperturbed state:

$$\begin{aligned} E_0^{(1)} &= \int \psi_0 \mathcal{H}' \psi_0 d\tau = (0|\mathcal{H}'|0) \\ &= \lambda(0|\mathbf{L} \cdot \mathbf{S}|0) + \beta(0|\mathbf{L} \cdot \mathbf{H}|0) \\ &\quad + g_s\beta(0|\mathbf{S} \cdot \mathbf{H}|0) \end{aligned} \tag{2.7}$$

Because the space coordinates of **L** are independent of those of the spin **S**, the operations can be carried out separately, one at a time. To separate the coordinates, one must express the scalar products of $\mathcal{H}'$ in terms of their components. With z chosen as the axis of quantization, the result is

$$\mathcal{H}_S^{(1)} = \lambda\,(0|L_z|0)\;S_z + \beta H_z\;(0|L_z|0)\;S_z + (0|0)\;g_s\beta\mathbf{H} \cdot \mathbf{S} \tag{2.8}$$

With z chosen as the axis of quantization, L_z alone has possible nonvanishing diagonal elements. Therefore, terms in $(0|L_x|0)$ and $(0|L_y|0)$ are zero and are not included. Because of the normalization, $(0|0) = 1$. The justification for dropping the terms in $(0|L_z|0)$ follows. In the free atom the states in which L and M_L are good quantum numbers, characterized by different M_L values, have the same energy when no field is applied; hence the ground state has a $2L + 1$ orbital degeneracy. For reasons already given in Chapter 1, we assume that for the paramagnetic species considered here this degeneracy is lifted and the ground state E_0 is an orbital singlet. The orbital states of paramagnetic elements of condensed systems are considered as having a reduced effective orbital angular momentum L' corresponding to the residual orbital degeneracy $2L' + 1$. For the singlet orbital ground state, $2L' + 1 = 1$, which gives $L' = 0$ and hence $M_{L'} = 0$. Thus for the ground state, $(0|L_z|0) = (0|L_z'|0) = M_{L'} = 0$. With this simplification and with the normalization $(0|0) = 1$, the first-order interaction (2.8) reduces to

$$\mathcal{H}_S^{(1)} = g_s \beta \mathbf{H} \cdot \mathbf{S} \tag{2.9}$$

We now carry out the evaluation of the L-dependent parts of $\mathcal{H}'$ to second order. From this, we see how the interactions with the higher orbital states, $|1) \ldots |n)$, lead to an anisotropy in the $\mathbf{g}$ for the ground state. The second-order perturbation energy is given by

$$\begin{aligned}\mathcal{H}^{(2)} &= -\sum_{n \neq 0} \frac{(0|\mathcal{H}'|n)(n|\mathcal{H}'|0)}{E_n - E_0} \\ &= -\sum_{n \neq 0} \frac{(0|\lambda \mathbf{L} \cdot \mathbf{S} + \beta \mathbf{H} \cdot (\mathbf{L} + g_s \mathbf{S})|n)(n|\lambda \mathbf{L} \cdot \mathbf{S} + \beta \mathbf{H} \cdot (\mathbf{L} + g_s \mathbf{S})|0)}{E_n - E_0}\end{aligned} \tag{2.10}$$

Because $\mathcal{H}'$ does not involve r explicitly, only the orientation-dependent factor in the orbital wave function need be applied. The radial part of the wave function can be considered as factored out and normalized to unity. The orbital wave functions are orthogonal and thus $(0|n)$ equals zero. Hence the parts that do not involve the orbital variable, such as $(0|g_s \beta \mathbf{H} \cdot \mathbf{S}|n) = (0|n)\,(g_s \beta \mathbf{H} \cdot \mathbf{S})$, are zero. Therefore,

$$\begin{aligned}\mathcal{H}_S^{(2)} &= -\sum_{n \neq 0} \frac{(0|\lambda \mathbf{L} \cdot \mathbf{S} + \beta \mathbf{H} \cdot \mathbf{L}|n)(n|\lambda \mathbf{L} \cdot \mathbf{S} + \beta \mathbf{H} \cdot \mathbf{L}|0)}{E_n - E_0} \\ &= -\sum_{i,j = x,y,z} \sum_{n \neq 0} \frac{[(0|L_i|n)(\beta H_i + \lambda S_i)]\,[(n|L_j|0)(\beta H_j + \lambda S_j)]}{E_n - E_0} \\ &= -\sum_{i,j = x,y,z} \Lambda_{ij}\,(\beta H_i + \lambda S_i)(\beta H_j + \lambda S_j)\end{aligned} \tag{2.11}$$

where

$$\Lambda_{ij} = \sum_{n \neq 0} \frac{(0|L_i|n)\ (n|L_j|0)}{E_n - E_0} \tag{2.12}$$

By carrying out the indicated multiplications, one obtains

$$\mathcal{H}_S^{(2)} = -2\beta\lambda \sum_{i,j=x,y,z} \Lambda_{ij} H_i S_j - \lambda^2 \sum_{i,j=x,y,z} \Lambda_{ij} S_i S_j - \beta^2 \sum_{i,j=x,y,z} \Lambda_{ij} H_i H_j \tag{2.13}$$

The last term, that in $H_i H_j$, represents a diamagnetic interaction that is very small and is usually omitted. The term in $S_i S_j$ represents an anisotropic, spin-

spin interaction introduced via the spin-orbit coupling. This term is small; it vanishes when there is only one unpaired electron. It will be enveloped in a larger, spin-spin tensor derived in Section 5. The term H_iS_j causes an anisotropy in the g which is of significance in most paramagnetic substances. To obtain the total interaction of the electronic spin with **H**, this second-order term is added to the first-order interaction of (2.9). Thus with the omission of the S_iS_j and H_iH_j interactions,

$$\begin{aligned}\mathcal{H}_S &= g_s\beta\mathbf{H}\cdot\mathbf{S} - 2\beta\lambda \sum_{i,j=x,y,z} \Lambda_{ij}H_iS_j \\ &= \beta \sum_{i,j=x,y,z} g_{ij}H_iS_j\end{aligned} \tag{2.14}$$

where

$$g_{ij} = g_s\,\delta_{ij} - 2\lambda\Lambda_{ij} \tag{2.15}$$

and $\delta_{ij} = 1$ when $i = j$ and 0 when $i \neq j$. We see that the **g** factor is a tensor because of the components $2\lambda\Lambda_{ij}$. The quantities Λ_{ij} cannot be evaluated until the particular orbital wave function and the energy eigenvalues E_n involved in a particular application are known. Illustrations of the evaluation are given in Chapter VII. This spin Hamiltonian for the interaction of **S** with the applied field **H** is most compactly expressed as

$$\mathcal{H}_S = \beta\mathbf{H}\cdot\mathbf{g}\cdot\mathbf{S} \tag{2.16}$$

This important expression was originally obtained by Pryce [3].

3. NUCLEAR MAGNETIC INTERACTIONS

3.a. Interactions of Nuclei with an Applied Field

The direct interaction of the nuclear magnetic moment with the applied field **H** gives rise to the well-known energies

$$\mathcal{H}_{\mathbf{H}\cdot\mathbf{I}} = -\mu_I\cdot\mathbf{H} = -g_I\beta_I\mathbf{I}\cdot\mathbf{H} \tag{2.17}$$

commonly observed in nuclear magnetic resonance (nmr) spectroscopy. However, in normal esr transitions for which the selection rules are $M_S \to M_S \pm 1$, $M_I \to M_I$, the nuclear orientation does not change, and the $\mathbf{I}\cdot\mathbf{H}$ term, which displaces equally all spin levels, causes no detectable effects. In second-order esr transitions and in ENDOR spectra, the nuclear spin vector changes orientation when the electron spin flips. Then, effects of the $\mathbf{I}\cdot\mathbf{H}$ term are detectable and significant.

3.b. Nuclear Interactions with Unpaired Electrons in Molecular Orbitals

In molecular free radicals each individual nucleus exerts its separate influence independently on the electron moment of the free radical. Interactions between the different nuclear spins, generally not detectable in esr spectroscopy, can be neglected. The effects of the different nuclei are additive, and their total effect is to produce a composite, often complex, hyperfine structure from which one can hope to identify and measure the splitting resulting from the individual nuclei.

The theoretical problem can be reduced to an essentially atomic one by expression of the molecular orbitals of the free radical as a linear combination of atomic orbitals. For illustration, let us here neglect configuration interaction and assume that the wave function of an unpaired electron can be represented by the linear combination of the atomic wave functions ψ_a and ψ_b of atoms A and B:

$$\psi_{\text{mol}} = a\psi_a + b\psi_b \tag{2.18}$$

If at some point in the orbital the interaction of the electron with the nucleus of A is $\mathcal{H}_a$ and if that with the nucleus of B is $\mathcal{H}_b$, the energy values could be obtained from the average of

$$\mathcal{H}_N = \mathcal{H}_a + \mathcal{H}_b \tag{2.19}$$

over the orbital of the electron. At this point we need not consider the specific forms of $\mathcal{H}_a$ and $\mathcal{H}_b$, which are treated in the next section. The average is expressed as

$$\begin{aligned} \int \psi_{\text{mol}}\mathcal{H}_N\psi^*_{\text{mol}}\,d\tau &= \int (a\psi_a+b\psi_b)(\mathcal{H}_a+\mathcal{H}_b)(a\psi_a^*+b\psi_b^*)d\tau \\ &= a^2 \int \psi_a\mathcal{H}_a\psi_a^*\,d\tau + 2ab \int \psi_a\mathcal{H}_a\psi_b^*\,d\tau + b^2 \int \psi_b\mathcal{H}_a\psi_b^*\,d\tau \\ &\quad + b^2 \int \psi_b\mathcal{H}_b\psi_b^*\,d\tau + 2ab \int \psi_a\mathcal{H}_b\psi_b^*\,d\tau + a^2 \int \psi_a\mathcal{H}_b\psi_a^*\,d\tau \end{aligned} \tag{2.20}$$

when ψ_{mol} is normalized by

$$\begin{aligned} \int \psi_{\text{mol}}\psi^*_{\text{mol}}\,d\tau &= a^2 \int \psi_a\psi_a^*\,d\tau + 2ab \int \psi_a\psi_b^*\,d\tau + b^2 \int \psi_b\psi_b^*\,dt \\ &= a^2 + 2ab \int \psi_a\psi_b^*\,d\tau + b^2 = 1 \end{aligned} \tag{2.21}$$

Because the operators $\mathcal{H}_a$ and $\mathcal{H}_b$ are Hermitian, integrals such as $\int \psi_a\mathcal{H}_a\psi_b^*\,d\tau$ and $\int \psi_b^*\mathcal{H}_a\psi_a\,d\tau$ are equivalent and are combined in the preceding expression.

A great simplification, which costs little in accuracy, can be achieved by omission of the cross terms in the expression for the energy, (2.20), and in the normalization, (2.21). The coupling energy is then expressed as

$$E_N = E_{N_1} + E_{N_2} = a^2 \int \psi_a \mathcal{H}_a \psi_a^* \, d\tau + b^2 \int \psi_b \mathcal{H}_a \psi_b^* \, d\tau + b^2 \int \psi_b \mathcal{H}_b \psi_b^* \, d\tau + a^2 \int \psi_a \mathcal{H}_b \psi_a^* \, d\tau \quad (2.22)$$

with the orbitals normalized by $(a^2 + b^2) = 1$. Justification for this procedure is that the distortion effects of the bond orbital overlap have little effect on the nuclear coupling energy because of its inverse cube variation with distance from the nucleus. Actually, the *s*-orbital interaction (Fermi contact interaction) occurs entirely at the nucleus. The normalization, in effect, distributes the overlap density $2ab \int \psi_a \psi_b^* \, d\tau$ of (2.21) equally between the two atomic orbitals ψ_a and ψ_b. For identical atoms, a would equal b, and the density for the two atoms would be equal. When there is ionic character in the bond, a and b differ; their relative values depend on the degree of the ionic character.

Further simplication can be achieved by omission of the terms $a^2 \int \psi_a \mathcal{H}_b \psi_a^* \, d\tau$ and $b^2 \int \psi_b \mathcal{H}_a \psi_b^* \, d\tau$, which represent the interaction of the unpaired electron density on atom A with the nucleus of B, and vice versa. Except for the short bonds involving hydrogen, these interactions are usually negligible. Calculation of these terms for hydrogen is described in Chapter VI. The example can be easily generalized to couplings of a π orbital that spreads over n atoms. With the above simplifications, the coupling energy of the π electron to the ith nucleus would be

$$E_{N_i} \approx a_i^2 \int \psi_i \mathcal{H}_i \psi_i^* \, d\tau = a_i^2 (i|\mathcal{H}_i|i) \quad (2.23)$$

when the expanded molecular wave function

$$\psi_{\text{mol}} = \sum_{i=1}^{n} a_i \psi_i \quad (2.24)$$

is normalized by $\sum_{i=1}^{n} a_i^2 = 1$. Thus for most applications, the analysis of nuclear coupling in molecular free radicals reduces to one of evaluation of atomic couplings and the weighting coefficients of the coupling electron density on the different atoms.

For simplicity in the preceding discussion we have considered a single unpaired electron and have neglected its interaction with other electrons of the radical. In this "one-electron" orbital the "unpaired-electron" density on the different atoms as measured by a_i^2 is equivalent to the population density of the orbitals. As a general rule, the unpaired electron interacts to some extent with other electrons of the radical (configuration interaction) to induce measurable nuclear interactions with other "paired-electron" orbitals of the radical. Thus nuclear coupling is, strictly speaking, a plural-electron problem even in doublet-state radicals. In a complete theory one must consider the total wave function for all electrons of the molecule or radical, but in practice only the valence-shell electrons need be included. When all electrons in the valence shells are taken into account, β as well as α spins must be included in the calculations. Since α and β spins of the same orbital have coupling equal in magnitude but opposite in sign, their resultant nuclear coupling is proportional to the orbital spin density defined by

$$\rho_j = P^{\alpha}_{\psi_j} - P^{\beta}_{\psi_j} \tag{2.25}$$

where $P_{\psi_j}{}^{\alpha}$ is the probability that ψ_j will have an electron with an α spin and $P_{\psi_j}{}^{\beta}$ is the probability that it will have one with a β spin. The more general form of (2.23) for the nuclear coupling energy of the ith atom is, therefore,

$$E_{N_i} = \sum_j \rho_{ij} \, (j|\mathcal{H}_{ij}|j) \tag{2.26}$$

where ρ_{ij} is the spin density of the jth orbital of the ith atom. For a doublet-state free radical, $\sum_i \sum_j \rho_{ij} = 1$. Note that when both α and β spins are considered, the spin density may differ from the orbital population density or the charge density, $\rho_{\psi_j}{}^{\alpha} + \rho_{\psi_j}{}^{\beta}$. A discussion of spin densities in free radicals is given in Chapter VI.

In the following section we deal with the evaluation of orbital-dependent factors in the integral of (2.23) or (2.26), leaving the spin-dependent factors in $\mathcal{H}_i$ to be evaluated separately, thus reducing $\mathcal{H}_i$ to a spin-only Hamiltonian. This procedure is possible because the orbital and spin operations commute. An additional simplification results from the fact that the orientation-dependent factor of an orbital can be separated from the factors that depend only on separation of the electron from the coupling nucleus. In Cartesian coordinates, the orientation-dependent factors are expressed as functions of x, y, and z in (1.27) and (1.28).

In spherical coordinates, r θ, ϕ the orbital part of the "hydrogen-like" atomic wave function is expressed as

$$\psi_{n,l,m_l} = C_N R_{n,l}(r) P^{|m_l|}(\cos\theta) \exp\,(\pm\, i\, m_l\, \phi) \tag{2.27}$$

where R is the radial function that is independent of orientation, $P^{|m_l|}(\cos\theta)$ is the associated Legendre polynomial that depends on the angle θ, and $\exp \pm im_l\phi$ is the function that depends on the orbital orientation ϕ. Of the three orbital quantum numbers n, l, and m_l, only m_l depends on the orbital orientation. The electron-spin functions indicated by $\alpha(m_s = +1/2)$ and $\beta(m_s = -1/2)$ are completely independent of the coordinate functions and signify only the two allowed spin orientations.

3.c. Nuclear Coupling Tensor for Bonded Atoms

Nuclear magnetic interactions with electron moments are of two kinds, the Fermi contact interaction and the dipole-dipole interaction. The Fermi contact interaction is completely isotropic and arises only with electron-spin density at the nucleus. Since only *s* orbitals have nonvanishing density $\psi_0\psi_0{}^*$at the nucleus, only *s*-orbital density can give the Fermi contact interaction. Because of its spherical symmetry, the interaction of electron spin with nuclear spin for the *s*-orbital density outside the nucleus averages to zero. Although the contact interaction is independent of the orientation of the applied field, it is an interaction between the magnetic moments of the nuclear spin and the electron spin and thus depends on the relative orientation of these moments; that is, it is an $\mathbf{S} \cdot \mathbf{I}$ interaction shown by Fermi [5] to be

$$\mathcal{H}_S = A_S \mathbf{S} \cdot \mathbf{I} \tag{2.28}$$

where the coupling of a particular *s* electron is

$$\begin{aligned} A_S &= (8/3)\,\beta\beta_I g_s g_I (\psi_s \psi_s^*)_0 \\ &= \frac{8}{3} \frac{hcg_I R\,\alpha^2\, Z_{\text{eff}}^3}{n^2} \end{aligned} \tag{2.29}$$

in which β is the Bohr magneton, β_I is the nuclear magneton, g_I is the g factor of the nucleus, h is Planck's constant, c is the velocity of light, R is the Rydberg constant, α is the fine-structure constant, Z_{eff} is the effective nuclear charge, and n is the total quantum number. For atomic hydrogen in the ground state A_S = 1420 MHz or 507 G, which is the doublet splitting for the strong-field, Paschen-Back case.

The orbitals for which $l \neq 0, p, d,$ and so on have no Fermi contact interaction because their density vanishes at the nucleus, $\psi_0\psi_0^* = 0$. Their nuclear interaction is anisotropic and arises from the magnetic field at the nucleus due to the orbital field and to the electron-spin moment.

Classically, the orbital field at the nucleus arises from the motion of the electronic charge around the nucleus. From electromagnetic theory the field at N due to the motion of the electronic charge, $-e$, with velocity $\mathbf{v}$ at a distance r from N is

$$\mathbf{H} = \frac{e\mathbf{v} \times \mathbf{r}}{cr^3} = -\left(\frac{e}{mc}\right) \frac{\mathbf{r} \times \mathbf{p}}{r^3} \tag{2.30}$$

In units of ħ, the angular momentum of the electron about N is $l = (\mathbf{r} \times \mathbf{p}/\hbar)$. With this substitution in (2.30) one obtains

$$\mathbf{H}_l = \frac{-2\beta l}{r^3} \tag{2.31}$$

for the field at N due to the orbital motion. In this equation, $\beta = (e\hbar)/(2mc)$ is the Bohr magneton. The interaction of this field with the nuclear moment $\mu_I = g_I\beta_I\mathbf{I}$ leads to the energy:

$$W_l = -\mu_I \cdot \mathbf{H}_l = 2g_I\beta\beta_I l \cdot \frac{\mathbf{I}}{r_3} \tag{2.32}$$

The field at N due to a dipole moment μ at a distance r is

$$\begin{aligned} \mathbf{H}_\mu &= \frac{-\text{grad.}\,(\mu \cdot \mathbf{r})}{r^3} \\ &= \frac{-\mu}{r^3} + \frac{3\,r(\mu \cdot \mathbf{r})}{r^5} \end{aligned} \tag{2.33}$$

Thus the field at N due to the electron-spin moment $\mu_s = -2\beta\mathbf{s}$ at a distance r is

$$\mathbf{H}_s = -2\beta\left[\frac{-\mathbf{s}}{r^3} + \frac{3r(\mathbf{s} \cdot \mathbf{r})}{r^5}\right] \tag{2.34}$$

The energy of interaction of $\mathbf{H}_s$ with the nuclear moment is

$$W_s = -\mu_I \cdot \mathbf{H}_s = 2g_I\beta\beta_I\left[-\frac{\mathbf{s} \cdot \mathbf{I}}{r^3} + 3\,\frac{(\mathbf{s} \cdot \mathbf{r})\mathbf{I} \cdot \mathbf{r}}{r^5}\right] \tag{2.35}$$

To obtain the quantized interaction energies, one must average W_l and W_s over the orbitals of all coupling electrons for which $l \neq 0$ and sum the result. Electrons in closed shells can be neglected because their interactions mutually cancel. The simplest method for obtaining this average is to convert the preceding expression into an equivalent Hamiltonian operator involving only orbital angular momentum and spin operators and then to apply perturbation theory. This is the method employed by Abragam and Pryce in their classic paper [4], but the details of the development are not given by them. Some of the required transformations are described in the review by Bleaney and Stevens [6], and a thorough treatment is given in the monograph by Abragam and Bleaney [1].

The orbital term W_l is simple. Except for the radial factor $1/r^3$, which can be averaged separately, it involves only constants and the orbital and nuclear spin operators. The Hamiltonian operator for this part is

$$\mathcal{H}_l = 2g_I\beta\beta_I\left(\frac{1}{r^3}\right)_{av} l \cdot \mathbf{I} \tag{2.36}$$

If there is more than one electron in the same L shell (same value of l), the $(1/r^3)_{av}$ factor will be the same for each one; when $\mathbf{L}$ (= $\sum_k l_k$) is a constant of the motion, the resultant Hamiltonian is

$$\mathcal{H}_L = 2g_I\beta\beta_I\left(\frac{1}{r_l^3}\right)_{av} \mathbf{L} \cdot \mathbf{I} \tag{2.37}$$

Transformation of the spin term W_s is more difficult than that for W_l because it involves the vector operators $\mathbf{r}$. A procedure for converting the coordinate operators of $\mathbf{r}$ to angular momentum operators is described by Stevens [7], who shows by comparisons with the results from group theory that for obtaining the matrix elements of the Hamiltonian the components of the unit vector $\mathbf{r}/|r|$ can be replaced by those of the respective angular momentum components l_x/l, l_y/l, and l_z/l times a proportionality constant that is the same for all the components. Because the components of r commute whereas those of l do not, the substituted l components must be properly symmetrized. The appropriate substitutions are, for example, $(xy)/r^2 = (xy + yx)/(2r^2) = \tau(l_xl_y + l_yl_x)/(2l^2)$, where τ is a constant. These substitutions can be justified by the simpler (if less rigorous) argument that the averaged value of $\mathbf{r}$ and that of the averaged classical angular momentum l have the same directions because the components of both $\mathbf{r}$ and l, which are perpendicular to the symmetry lobes of the orbitals, will cancel in the averaging process. Since l^2 can be replaced by its eigenvalue $l(l + 1)$, the l^2 in the denominator is usually included in the proportionality constant which is designated by $-\xi$. Because the individual components of l commute with themselves, $[(\mathbf{r} \cdot \mathbf{r})/r^2] = \xi l \cdot l = \xi l(l + 1)$. With these various substitutions, the spin-dipole Hamiltonian operator equivalent to (2.35) can be expressed as

$$\mathcal{H}_s = 2g_I\beta\beta_I\left(\frac{1}{r^3}\right)_{av} \xi\{l(l + 1)\mathbf{s} \cdot \mathbf{I} - \frac{3}{2}[(l \cdot \mathbf{s})(l \cdot \mathbf{I}) + (l \cdot \mathbf{I})l \cdot \mathbf{s}]\} \tag{2.38}$$

The constant ξ can be evaluated from a particular case; for a single electron it has the value

$$\xi = \frac{2}{(2l-1)(2l+3)} \tag{2.39}$$

When there is more than one unpaired electron within the same l shell, the Russell-Saunders coupling can usually be assumed; that is, **L** and **S** are constants of the motion. The corresponding operators L_x, L_y, and L_z can be substituted for l_x, l_y, and l_z and S_x, S_y, and S_z for the components s_x, s_y, and s_z, but the value of the proportionality constant is different because of the coupling of the different spins and different orbits. The spin-dipole Hamiltonian is then expressed as

$$\mathcal{H}_S = 2g_I\beta\beta_I\left(\frac{1}{r_l^3}\right)_{av} \xi\{L(L+1)\mathbf{S}\cdot\mathbf{I} - \frac{3}{2}[(\mathbf{L}\cdot\mathbf{S})(\mathbf{L}\cdot\mathbf{I}) + (\mathbf{L}\cdot\mathbf{I})(\mathbf{L}\cdot\mathbf{S})]\} \tag{2.40}$$

where [4]

$$\xi = \frac{(2l+1) - 4S}{S(2l-1)(2l+3)(2L-1)} \tag{2.41}$$

It is seen that this expression for ξ reduces to that of (2.39) when $L = l$ and $S = 1/2$.

The factor $\left(\frac{1}{r_l^3}\right)_{av}$, which depends on the radial functions $\psi_n{}^l$, is independent of the orientation of the orbital and has the same value for all electrons of a particular L shell. Approximate values of it have been calculated for many atoms, and more precise values for certain ones have been derived from experimental measurements in atomic spectra, especially from atomic-beam resonance experiments. Tabulated values of $\left(\frac{1}{r_l^3}\right)_{av}$ for different atoms are available [8]. For hydrogen-like orbitals, (2.48) can be used for approximate calculations of the quantity.

As was done for (2.6), L-dependent factors in the Hamiltonian are evaluated first, and the operations involving **S** and **I** are postponed; thereby, a spin Hamiltonian independent of the **L** operator is obtained. This is possible because **L**, **S**, and **I** are commuting operators; their coordinates are independent. The

L-dependent terms in the nuclear interactions are generally small and can be treated with first-order perturbation theory.

As in the derivation of the **g** tensor, the orbital ground state is assumed here to be a singlet signified by $|M'_L) = |0)$; higher orbital states are signified by $|n)$. Those interested in the rarer case of an orbital-degenerate ground state should consult the original paper by Abragam and Pryce [4].

The first-order treatment is simply an averaging of the $\mathcal{H}_N$ over the orbital ground state. The $\mathcal{H}_L$ and $\mathcal{H}_S$ parts can be averaged separately:

$$\mathcal{H}_N' = (0|\mathcal{H}_N|0) = (0|\mathcal{H}_L|0) + (0|\mathcal{H}_S|0) \tag{2.42}$$

Because there are no diagonal matrix elements of L_x and L_y and $(0|L_z|0) = 0$, the $\mathcal{H}_L$ contribution vanishes in first order:

$$\begin{aligned}(0|\mathcal{H}_L|0) &= 2g_I\beta\beta_I\left(\frac{1}{r_l^3}\right)_{av} (0|\mathbf{L}\cdot\mathbf{S}|0) \\ &= 2g_I\beta\beta_I\left(\frac{1}{r_l^3}\right)_{av} [(0|L_x|0)S_x \\ &\quad + (0|L_y|0)S_y + (0|L_x|0)S_z] = 0\end{aligned} \tag{2.43}$$

The spin-dipole part, $\mathcal{H}_S$, is thus responsible for the magnetic coupling. This part is reduced as follows:

$$\begin{aligned}(0|\mathcal{H}_S|0) &= 2g_I\beta\beta_I\left(\frac{1}{r_l^3}\right)_{av} \xi\{L(L+1)(0|\mathbf{S}\cdot\mathbf{I}|0) \\ &\quad -\frac{3}{2}[(0|(\mathbf{L}\cdot\mathbf{S})(\mathbf{L}\cdot\mathbf{I})|0) + (0|(\mathbf{L}\cdot\mathbf{I})(\mathbf{L}\cdot\mathbf{S})|0)]\}\end{aligned} \tag{2.44}$$

The first term inside the brace does not contain the orbital operator and is unaffected by the averaging. The orbitals are normalized so that $(0|0)$, and hence $(0|\mathbf{S}\cdot\mathbf{I}|0) = (0|0)\ \mathbf{S}\cdot\mathbf{I} = \mathbf{S}\cdot\mathbf{I}$, For evaluation, the terms within brackets must be expressed in terms of their component products. The indicated multiplication yields terms of the form $L_xL_yS_xI_y$, $L_yL_xS_xI_y$, and so forth. Since S_x and I_y commute, S_xI_y is equal to I_yS_x. Furthermore, the components of **I** and **S** commute with those of **L**. The terms in the brackets can thus be written in the compact form

$$\sum_{i,j=x,y,z} (0|L_iS_iL_jI_j|0) = \sum_{i,j=x,y,z} (0|L_iL_j|0)\, S_iI_j$$

$$= \sum_{i,j=x,y,z} \sum_n (0|L_i|n)(n|L_j|0)S_iI_j \qquad (2.45)$$

The summation includes all excited orbital states, all M_L values, of the L being manipulated. The matrix elements in the summation are the same as those in the numerator of Λ_{ij} in (2.12). They cannot be evaluated until the specific orbital involved is known. Here we simply indicate them. To calculate the total nuclear spin Hamiltonian, one must add the Fermi contact term for s orbitals, (2.28), to (2.45). The result can be written

$$\mathcal{H}_N = \sum_{i,j=x,y,z} \{A_S\delta_{ij} + 2g_I\beta\beta_I\left(\frac{1}{r_l^3}\right)_{av} \xi[L(L+1)\delta_{ij} - 3\sum_n (0|L_i|n)(n|L_j|0)]\} \ S_iI_j \qquad (2.46)$$

where δ_{ij} equals 1 when $i = j$ and 0 when $i \neq j$. The term in the brackets represents the total nuclear coupling and is designated by

$$A_{ij} = A_S\delta_{ij} + 2g_I\beta\beta_I\left(\frac{1}{r_l^3}\right)_{av} \xi[L(L+1)\delta_{ij} - 3\sum_n (0|L_i|n)(n|L_j|0)] \qquad (2.47)$$

The s-orbital coupling is given by (2.29); ξ is given by (2.41). When it is not known from atomic spectra, $\left(\frac{1}{r_l^3}\right)_{av}$ can be calculated approximately with the theoretical formula [9]

$$\left(\frac{1}{r_l^3}\right)_{av} = \frac{Z^3_{\text{eff}}}{a_0^3n^3\,(l+1)(l+1/2)l} \qquad (2.48)$$

which assumes hydrogen-like wave functions.

The electron spin-nuclear spin Hamiltonian can now be expressed in the following compact form:

$$\mathcal{H}_N = \sum_{i,j} A_{ij}S_iI_j = \mathbf{S} \cdot \mathbf{A} \cdot \mathbf{I} \qquad (2.49)$$

in which $\mathbf{A}$ is the nuclear coupling tensor given by (2.47). Methods for derivation of the principal elements of A are described in Chapters III and V.

When the electron-spin density concentrated on one atom, A, interacts with the nucleus of another atom, B, the classical form of the interaction is that of (2.35) for two dipoles. (The most important illustration of this is the ⟩CH fragment.) The resulting spin Hamiltonian is a tensor product like that of (2.49). Thus the Hamiltonian of (2.49) can be used for the general analysis of hyperfine structure caused by dipole-dipole interaction, regardless of the distribution of the electronic spins.

4. NUCLEAR QUADRUPOLE INTERACTIONS

Any nuclear isotope with a spin I greater than 1/2 has an electric quadrupole moment that can interact with electric field gradients in an atom, molecule, or crystal. This interaction gives rise to pure quadrupole resonance spectroscopy in solids and to nuclear quadrupole hyperfine structure in the rotational spectra of "nonmagnetic" molecules in singlet ground states. Much information has been gained about chemical bonding from these sources. As a result, the value of the nuclear quadrupole moment as a chemical probe to gain information about molecules is recognized by most chemists.

In esr spectroscopy as commonly observed, the nuclear quadrupole interaction is not detected because the orientation of the nuclear spin does not change; selection rules are $\Delta M_S = \pm 1, \Delta M_I = 0$. For a particular nuclear orientation with constant M_I value, the nuclear quadrupole displacement is the same for all spin levels. There is no electric moment associated with the electron spin, and hence no interaction is possible between **S** and the electric quadrupole moment of the nucleus. Thus there is no change in the quadrupole interaction energy when the electronic spin flips unless there is an associated change in the nuclear orientation. Through the coupling of the magnetic dipole of **S** and **I** the nuclear spin can be induced to flip with the electron spin, thus causing detectable changes in the nuclear quadrupole energy. These transitions, called *second-order transitions,* are less probable than the normal or first-order transitions but are observable in many paramagnetic crystals with sensitive esr spectrometers. In ENDOR experiments changes in the nuclear spin orientation, $\Delta M = \pm 1$, are induced by an nmr radiofrequency imposed simultaneously with the esr microwave frequency. Thus the nuclear quadrupole interaction, as well as the $\mathbf{I} \cdot \mathbf{H}$, interaction is observable in these experiments.

The nuclear quadrupole Hamiltonian for a free atom is developed in a classic paper by Casimir [10]. In deriving the spin Hamiltonian for the nuclear quadrupole interaction in solids, one can start with the expression for the interaction in a free atom or ion and can eliminate the L-dependent factors by perturbation theory with the assumption that the orbital moments are quenched in first order. The procedure used for the nuclear magnetic interaction is

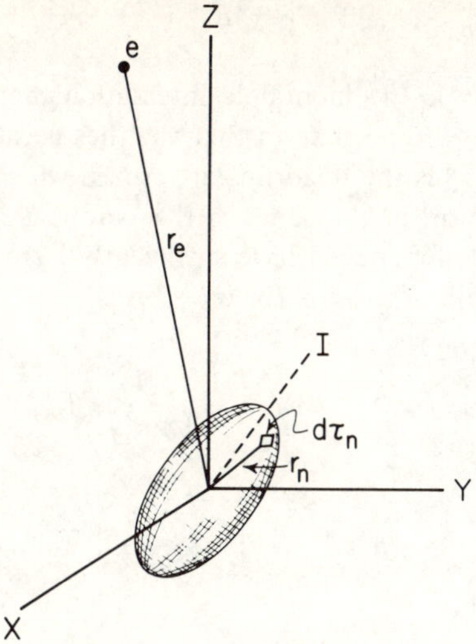

Fig. 2.1 Reference system for expressing interaction of the nuclear quadrupole moment with the extranuclear charges.

described in Section 3. It is simpler, however, to derive the Hamiltonian for the solid state directly, with the assumption that the electric field gradient with which the quadrupole moment interacts is fixed in space. This assumption envelops the one that the orbital angular momentum is quenched. However, this gradient arises from all charges outside the nucleus, except those spherically distributed about it, and not simply those of unpaired electrons.

Although the charges of a quadrupolar nucleus are not spherically symmetric, they have an axially symmetric distribution, either prolate (positive quadrupole moment Q) or oblate (negative Q). The axis of symmetry is the spin axis, and the axial symmetry may be considered as arising from a spinning motion. Although the spin axis is not defined quantum mechanically, but only its resolvable component along a space-fixed axis, let us first consider it as a classical ellipsoid with symmetry axis z along **I**, as indicated in Fig. 2.1. The classical energy of interaction of the nuclear charge with a potential $V(x,y,z)$ is

$$E_n = \int \rho_n(x,y,z)\, V(x,y,z) d\tau \tag{2.50}$$

where ρ_n is the nuclear charge density in the elemental volume $d\tau = dx\, dy\, dz$, where x,y,z is the body-fixed system with the origin at the nuclear center, and where z is along **I** (Fig. 2.1). The integration is taken over the nuclear volume. By expansion of V in a Taylor series about the origin, (2.50) may be expressed as

$$E_n = E_0 + E_1 + E_2 + \cdots \tag{2.51}$$

where $E_0 = V_0 \int \rho_n \, d\tau$ is the monopole interaction that is independent of nuclear orientation, E_2 is a dipolar term that vanishes because the nucleus has no dipole moment, and E_2 is the quadrupolar term in which we are interested. The next nonvanishing term in the series is the octopolar term, which we do not consider here because few nuclei have sufficiently large octopole moments for detection. The complete expression for $E_2 = E_Q$ is:

$$\begin{aligned} E_Q = {} & \frac{1}{2}\left(\frac{\partial^2 V}{\partial x^2}\right)_0 \int \rho_n x_n^2 \, d\tau_n + \frac{1}{2}\left(\frac{\partial^2 V}{\partial y^2}\right)_0 \int \rho_n y_n^2 \, d\tau_n \\ & + \frac{1}{2}\left(\frac{\partial^2 V}{\partial z^2}\right)_0 \int \rho_n z_n^2 \, d\tau_n + \left(\frac{\partial^2 V}{\partial x \partial y}\right)_0 \int \rho_n x_n y_n \, d\tau_n \\ & + \left(\frac{\partial^2 V}{\partial x \, \partial y}\right)_0 \int \rho_n x_n z_n \, d\tau_n + \left(\frac{\partial^2 V}{\partial y \, \partial z}\right)_0 \int \rho_n y_n z_n \, d\tau_n \end{aligned} \tag{2.52}$$

where x_n, y_n, and z_n are the coordinates at the point of the elemental nuclear volume $d\tau_n$. Because of the nuclear symmetry,

$$\int \rho_n x_n y_n \, d\tau_n = 0, \quad \int \rho_n x_n z_n \, d\tau_n = 0, \quad \int \rho_n y_n z_n \, d\tau_n = 0$$

and

$$\int \rho_n x_n^2 \, d\tau_n = \int \rho_n y_n^2 \, d\tau_n = \frac{1}{2} \int \rho_n (x_n^2 + y_n^2) d\tau_n = \frac{1}{2} \int \rho_n (r_n^2 - z_n^2) d\tau_n$$

With these substitutions, (2.52) becomes

$$E_Q = \frac{1}{4}\left[\left(\frac{\partial^2 V}{\partial x^2}\right)_0 + \left(\frac{\partial^2 V}{\partial y^2}\right)_0\right] \int \rho_n (r_n^2 - z_n^2) d\tau_n + \frac{1}{2}\left(\frac{\partial^2 V}{\partial z^2}\right)_0 \int \rho_n z_n^2 \, d\tau_n \tag{2.53}$$

Since the charges giving rise to V are outside the nucleus, Laplace's equation, $\Delta^2 V = 0$, holds

$$\left(\frac{\partial^2 V}{\partial x^2}\right)_0 + \left(\frac{\partial^2 V}{\partial y^2}\right)_0 = -\left(\frac{\partial^2 V}{\partial z^2}\right)_0 \tag{2.54}$$

and (2.52) reduces further to

$$E_Q = \frac{1}{4}\left(\frac{\partial^2 V}{\partial z^2}\right)_0 eQ^* \tag{2.55}$$

where

$$Q^* \equiv \frac{1}{e}\int \rho_n(3z_n^2 - r_n^2)d\tau_n \tag{2.56}$$

is defined as the intrinsic quadrupole moment of the nucleus.

The intrinsic quadrupole moment is not directly observable, and the quadrupole moment Q, as usually defined, is the component of Q^* that is resolvable along a space-fixed axis about which **I** has its maximum component $M_I = I$. In the vector model, **I** will precess about the axis of quantization Z, and it is evident that Q will be axially symmetric about Z. If a space-fixed system X, Y, Z is now chosen with Z as the axis of symmetry of Q, a derivation similar to the preceding one yields

$$E_Q = \frac{1}{4}\left(\frac{\partial^2 V}{\partial Z^2}\right)_0 eQ \tag{2.57}$$

where

$$Q = \frac{1}{e}\int \rho_n(3Z_n^2 - R_n^2)d\tau_n \tag{2.58}$$

where ρ_n and $d\tau_n$ are the nuclear density $\rho_n(X_n Y_n Z_n)$ and the elemental value $d\tau_n = dX_n\, dY_n\, dZ_n$ within the space coordinates of Q. One can easily find the relationship between Q and Q^* by resolution of $(\partial^2 V/\partial z^2)_0$ along Z. If θ is the angle between Z and z (Z and I in Fig. 2.1), the transformation is

$$\begin{aligned}\left(\frac{\partial^2 V}{\partial z^2}\right)_0 &= \frac{1}{2}\left(\frac{\partial^2 V}{\partial Z^2}\right)_0 (3\cos^2\theta - 1)\\ &= \frac{1}{2I(I+1)}\left(\frac{\partial^2 V}{\partial Z^2}\right)_0 [(3I_Z^2 - I(I+1)]\end{aligned} \tag{2.59}$$

since $\cos\theta = I_Z/|I| = M_I/[I(I+1)]^{1/2}$

With this transformation, (2.55) becomes

$$E_Q = \frac{1}{8}\frac{eQ^*}{I(I+1)}\left(\frac{\partial^2 V}{\partial Z^2}\right)_0 [3I_Z^2 - I(I+1)] \tag{2.60}$$

When $M_I = I$, this energy is equivalent to that of (2.57). Thus

$$\frac{1}{4} eQ \left(\frac{\partial^2 V}{\partial Z^2}\right)_0 = \frac{1}{8} \frac{eQ^*}{I(I+1)} \left(\frac{\partial^2 V}{\partial Z^2}\right)_0 [3I^2 - I(I+1)] \tag{2.61}$$

which gives

$$Q^* = \frac{2(I+1)}{2I - 1} Q \tag{2.62}$$

Substitution of this value of Q^* in (2.55) gives

$$E_Q = \frac{(I + 1)}{2(2I - 1)} eQ \left(\frac{\partial^2 V}{\partial z^2}\right)_0 \tag{2.63}$$

Now let us assume that X,Y,Z, the orientation of the space-fixed reference system, is such that the coordinates are not principal axes of the field gradient. The transformation of $(\partial^2 V/\partial z^2)_0$ to this system will then have the form

$$\begin{aligned}\left(\frac{\partial^2 V}{\partial z^2}\right)_0 = {} & \left(\frac{\partial^2 V}{\partial X^2}\right)_0 \left(\frac{\partial X}{\partial z}\right)^2 + \left(\frac{\partial^2 V}{\partial Y^2}\right)_0 \left(\frac{\partial Y}{\partial z}\right)^2 + \left(\frac{\partial^2 V}{\partial Z^2}\right)_0 \left(\frac{\partial Z}{\partial z}\right)^2 \\ & + \left(\frac{\partial^2 V}{\partial X \partial Y}\right)_0 \left(\frac{\partial X}{\partial z}\right) \left(\frac{\partial Y}{\partial z}\right) + \left(\frac{\partial^2 V}{\partial Y \partial X}\right)_0 \left(\frac{\partial Y}{\partial z}\right) \left(\frac{\partial X}{\partial z}\right) \\ & + \left(\frac{\partial^2 V}{\partial Y \partial Z}\right)_0 \left(\frac{\partial Y}{\partial z}\right) \left(\frac{\partial Z}{\partial z}\right) + \left(\frac{\partial^2 V}{\partial Z \partial Y}\right)_0 \left(\frac{\partial Z}{\partial z}\right) \left(\frac{\partial Y}{\partial z}\right) \\ & + \left(\frac{\partial^2 V}{\partial Z \partial X}\right)_0 \left(\frac{\partial Z}{\partial z}\right) \left(\frac{\partial X}{\partial z}\right) + \left(\frac{\partial^2 V}{\partial X \partial Z}\right)_0 \left(\frac{\partial X}{\partial z}\right) \left(\frac{\partial Z}{\partial z}\right)\end{aligned} \tag{2.64}$$

Substitution of this value for $(\partial^2 V/\partial z^2)_0$ into (2.63) yields the classical Hamiltonian in the X,Y,Z system. The operator equivalent can be obtained easily by the substitution

$$\frac{\partial X}{\partial z} = \cos\theta_{Xz} = \frac{I_X}{|I|}, \quad \frac{\partial Y}{\partial z} = \cos\theta_{Yz} = \frac{I_Y}{|I|}, \quad \frac{\partial Z}{\partial z} = \cos\theta_{Zz} = \frac{I_Z}{|I|}$$

For simplicity, we also make the substitutions

$$\chi_{XX} = eQ\left(\frac{\partial^2 V}{\partial X^2}\right)_0, \quad \chi_{XY} = eQ\left(\frac{\partial^2 V}{\partial X\,\partial Y}\right)_0$$

and so on. With these substitutions and $|I|^2 = I(I+1)$, the general form of the Hamiltonian operator for the quadrupole interaction is found to be

$$\begin{aligned}\mathcal{H}_Q = \left(\frac{1}{2I(2I-1)}\right)[&\chi_{XX}I_X^2 + \chi_{YY}I_Y^2 + \chi_{ZZ}I_Z^2 \\ &+ \chi_{XY}I_XI_Y + \chi_{YX}I_YI_X + \chi_{YZ}I_YI_Z \\ &+ \chi_{ZY}I_ZI_Y + \chi_{ZX}I_ZI_X + \chi_{XZ}I_XI_Z] = \left(\frac{1}{2I(2I-1)}\right)\mathbf{I}\cdot\chi\cdot\mathbf{I}\end{aligned} \tag{2.65}$$

Since $(\partial^2 V)/(\partial X\,\partial Y) = (\partial^2 V)/(\partial Y\,\partial X)$, and so on, χ is a symmetrical tensor, and in the classical Hamiltonian there are only six independent terms because $(\partial V/\partial X)(\partial V/\partial Y) = (\partial V/\partial Y)(\partial V/\partial X)$, and so forth. In the operator equivalent, however, the associated terms are not equal because the components of **I** do not commute, $I_XI_Y \neq I_YI_X$.

When X, Y, and Z are principal axes of the field gradient, the cross terms vanish and

$$\mathcal{H}_Q = \frac{1}{2I(2I-1)}[\chi_{XX}I_X^2 + \chi_{YY}I_Y^2 + \chi_{ZZ}I_Z^2] \tag{2.66}$$

In the event of axial symmetry about Z, $\mathcal{H}_Q$ reduces further to

$$\mathcal{H}_Q = \frac{\chi_{ZZ}}{4I(2I-1)}(3I_Z^2 - I^2) \tag{2.67}$$

because $\chi_{XX} = \chi_{YY}$ and, from Laplace's equation, $\chi_{XX} + \chi_{YY} + \chi_{ZZ} = 0$.

The preceding expressions of the nuclear quadrupole Hamiltonian have the same form as those employed in pure nuclear quadrupole resonance spectroscopy of solids. The elements χ_{XX}, χ_{YY}, and so on, are the quadrupole coupling constants as usually defined in nuclear quadrupole spectroscopy or in gaseous spectroscopy. In esr spectroscopy the factor $1/[2I(2I-1)]$ is usually included in the definition of the coupling elements, and the coupling tensor is designated by **P**, where

$$\mathbf{P} = \frac{1}{2I(2I-1)}\chi \tag{2.68}$$

The Hamiltonian can then be expressed in the most convenient form:

$$\mathcal{H}_Q = \mathbf{I} \cdot \mathbf{P} \cdot \mathbf{I} \tag{2.69}$$

Simplified methods for relating the coupling constants χ to properties of the chemical bond are described in a companion volume in the series, *Microwave Molecular Spectra* [11]. The coupling constants usually depend on the nature of the filling of the valence p shell of the coupling atom. They are generally independent of the type of spectroscopy used to evaluate them; however, because of intermolecular interaction in the solid, those of molecules in the solid state are only approximately equal to those of the same molecules in the gaseous state.

Although the solid-state $\mathcal{H}_Q$ employed in esr, ENDOR, and nuclear quadrupole resonance (nqr) has the same form, its application may be quite different for esr or ENDOR and more complicated than for nqr. In magnetic resonance spectroscopy the internal magnetic fields and the externally applied fields compete with the electric field gradients for alignment of the nuclear spin, whereas in nqr, as commonly observed in diamagnetic substances, there is no such competition.

5. ELECTRON SPIN-SPIN INTERACTIONS

Most paramagnetic species treated in this volume are isolated free radicals having only one unpaired electron. There is, of course, no electron spin-spin interaction in these free radicals. However, molecules in triplet states and in coupled free radicals or biradicals are being observed increasingly. In them, the spin-spin coupling produces gross effects on the esr and can yield valuable information about the substances.

In the triplet state the two unpaired electrons must have different orbitals because of the Pauli exclusion principle. The parallel alignment of their spins results from exchange interaction that in molecular free radicals is usually isotropic or very nearly so. This interaction is the basis for Hund's rule, which specifies that the resultant spin **S** in atoms will have maximum value consistent with the exclusion principle. The isotropic exchange energy is expressed as

$$\mathcal{H}_J = J_{1,2}\, \mathbf{s}_1 \cdot \mathbf{s}_2 \tag{2.70}$$

where $J_{1,2}$ is the exchange coupling constant. When there is anisotropy in **g**, an anisotropic component in the exchange interaction can occur that adds a tensor term to the $\mathcal{H}_J$. A treatment of different types of exchange coupling is given in the treatise by Abragam and Bleaney [1].

In most triplet states the isotropic exchange energy is much greater than the spin resonance energy $\beta\mathbf{H} \cdot \mathbf{g} \cdot \mathbf{S}$. This large isotropic exchange energy displaces equally all the Zeeman levels and causes no noticeable effects on the esr spectrum. One observes $\Delta M_S = \pm 1$ among the three orientations of **S** corresponding to $M_S = 1, 0, -1$ rather than between the two orientations, $+1/2$ and $-1/2$, of an individual spin. In weakly coupled biradicals, however, the exchange energy can be of the same magnitude as the spin resonance energy. In these cases the resonance may be complicated by a partial breakdown of the exchange. Furthermore, it is complicated by anisotropic dipole-dipole interaction of the two spins that, in triplet states of molecules or biradicals, does not average out as it does for S states of atoms or ions. In triplet states of molecules or coupled biradicals, discussed in Chapter X, the dipole-dipole interaction causes the three spin levels $M_S = 1, 0, -1$ to be unequally separated and leads to an anisotropic doublet splitting of the esr. The larger isotropic exchange term usually ensures that the two spins maintain their parallel alignment so that M_S is still a good quantum number.

The spin dipole-dipole Hamiltonian may be obtained from the classical expression, (2.33), by substitution of $\mu_1 = -g\beta\mathbf{s}_1$ and $\mu_2 = -g\beta\mathbf{s}_2$ and by designation of the separation vector of the two electrons by $\mathbf{r}_{1,2}$. The result is

$$\mathcal{H}_d = g^2\beta^2 \left[\frac{\mathbf{s}_1 \cdot \mathbf{s}_2}{r_{1,2}^3} - \frac{3(\mathbf{r}_{1,2} \cdot \mathbf{s}_1)(\mathbf{r}_{1,2} \cdot \mathbf{s}_2)}{r_{1,2}^5} \right] \tag{2.71}$$

Although this Hamiltonian has the same form as that for the electron spin-nuclear spin interaction, (2.35), its reduction to a purely spin Hamiltonian is more difficult. When the two electrons are confined to a single atom, one can express $\mathbf{r}_{1,2}$ in terms of the position coordinates of $\mathbf{s}_1$ and $\mathbf{s}_2$ with reference to a coordinate system centered at the nucleus. Then one can use the principle of equivalent operators to replace the intercept coordinates that represent direction cosines of $\mathbf{r}_1$ and $\mathbf{r}_2$ with conjugate l_1 and l_2 operators, as was done for the nuclear interaction (2.35). When the two electrons are in the same atomic shell and Russell-Saunders coupling prevails, the Hamiltonian can be expressed as [4]

$$\mathcal{H}_d = -\rho[\mathbf{L} \cdot \mathbf{S} + \frac{1}{2}(\mathbf{L} \cdot \mathbf{S}) - \frac{1}{3}L(L + 1)S(S + 1)] \tag{2.72}$$

where ρ is a constant. With the assumption of an effective singlet orbital ground state $L' = 0$, the L operators in this Hamiltonian can be evaluated with perturbation theory as was done for (2.40); this evaluation would leave a reduced Hamiltonian involving only the electron-spin operators.

For calculation of the electron spin-spin interaction of a triplet state in a molecule or coupled biradical, the model of the atomic orbital is not as useful as it is for calculation of nuclear interactions since the interacting spins which give rise to $\mathcal{H}_d$ are mostly on different atoms, whereas the electron spin-nuclear interactions occur mostly on the same atom. For example, the electron spin-spin interaction of the triplet state of O_2 can be calculated closely with the assumption that the interacting electron densities are on the two opposite atoms, whereas the ^{17}O nuclear coupling of $^{16}O^{17}O$ can be calculated to a good approximation by consideration of the unpaired electron density on ^{17}O only.

For evaluation of the spin-spin interaction energies, $\mathcal{H}_d$ must be averaged over the electron orbitals, and the correlation of the electron spins must be taken into account. This process is exceedingly complicated in the general case, but it is not difficult to derive the general form of the spin-spin Hamiltonian. One can use this $\mathcal{H}_{SS}$ for fitting the experimental data from which the specific form of the spin-spin coupling tensor can be obtained. From the latter, information about the electronic structure of the species can be derived.

To simplify the derivation, let us specify that $\mathbf{r}_{1,2} = \mathbf{r}_2 - \mathbf{r}_1 = \mathbf{r}$, and let x, y, and z be the intercept coordinates $x_2 - x_1$, $y_2 - y_1$, and $z_2 - z_1$. With these substitutions and expansions of the dot products, (2.71) can be expressed as

$$\begin{aligned}\mathcal{H}_d = g^2\beta^2\left(\frac{1}{r^3}\right)\{ & s_{1x}s_{2x}\left[1 - 3\left(\frac{x}{r}\right)^2\right] + s_{1y}s_{2y}\left[1 - 3\left(\frac{y}{r}\right)^2\right] \\ & + s_{1z}s_{2z}\left[1 - 3\left(\frac{r}{z}\right)^2\right] - 3(s_{1x}s_{2y} + s_{1y}s_{2x})\frac{xy}{r^2} \\ & -3(s_{1y}s_{2z} + s_{1z}s_{2y})\frac{yz}{r^2} - 3(s_{1z}s_{2x} + s_{1x}s_{2z})\frac{xz}{r^2}\}\end{aligned} \tag{2.73}$$

Since the spins are assumed to be coupled by exchange interaction to form a resultant $\mathbf{S} = \mathbf{s}_1 + \mathbf{s}_2$, the $\mathcal{H}_d$ can be expressed in terms of the $\mathbf{S}$ operator. In making the substitution one should remember that the components of $\mathbf{s}_1$ and $\mathbf{s}_2$ commute (i.e., $s_{1x}s_{2y} - s_{2y}\,s_{1x} = 0$, etc.), but that the components of the individual spins do not (i.e., $s_{1x}s_{1y} - s_{1y}s_{1x} = is_{1z}$). The components of $\mathbf{S}$ obey the same commutation rules as those of the individual spins (i.e., $S_xS_y - S_yS_x = iS_z$, etc.). To find the first three terms, we write

$$\mathbf{S}\cdot\mathbf{S} = (\mathbf{s}_1 + \mathbf{s}_2)\cdot(\mathbf{s}_1 + \mathbf{s}_2) = \mathbf{s}_1\cdot\mathbf{s}_1 + 2\mathbf{s}_1\cdot\mathbf{s}_2 + \mathbf{s}_2\cdot\mathbf{s}_2 \tag{2.74}$$

With $\mathbf{s}_1\cdot\mathbf{s}_1 = \mathbf{s}_2\cdot\mathbf{s}_2 = 3/4$, this can be written as

$$s_{1x}s_{2x} + s_{1y}s_{2y} + s_{1z}s_{2z} = S_x^2 + S_y^2 + S_z^2 - 3/4 \tag{2.75}$$

from which

$$s_{1x}s_{2x} = S_x^2 - 1/4, \quad s_{1y}s_{2y} = S_y^2 - 1/4, \quad s_{1z}s_{2z} = S_z^2 - 1/4$$

By setting $S_xS_y + S_yS_x = (s_{1x} + s_{2x})(s_{1y} + s_{2y}) + (s_{1y} + s_{2y})(s_{1x} + s_{2x})$ and by applying the commutation rules, one finds that

$$s_{1x}s_{2y} + s_{2x}s_{1y} = \frac{1}{2}(S_xS_y + S_yS_x) \tag{2.76}$$

By substituting these expressions into (2.73) and dropping the numerical constants that displace all levels equally, one obtains

$$\begin{aligned}\mathcal{H}_d = \frac{1}{2} g^2\beta^2 \frac{1}{r^3}\Bigg\{ & S_x^2\left[1 - 3\left(\frac{x}{r}\right)^2\right] + S_y^2\left[1 - 3\left(\frac{y}{r}\right)^2\right] \\ & + S_z^2\left[1 - 3\left(\frac{z}{r}\right)^2\right] - 3(S_xS_y + S_yS_x)\frac{xy}{r^2} \\ & - 3(S_yS_z + S_zS_y)\frac{yz}{r^2} - 3(S_zS_x + S_xS_z)\frac{zx}{r^2}\Bigg\}\end{aligned} \tag{2.77}$$

Since the spin operators commute with any space operators, the direction cosines and $1/r^3$ can be averaged separately over the orbital wave functions when enough is known about a particular species. At this point we simply indicate this averaging and express $\mathcal{H}_d$ as a spin Hamiltonian

$$\begin{aligned}\mathcal{H}_{SS} = \mathbf{S}\cdot\mathbf{D}\cdot\mathbf{S} = & D_{xx}S_x^2 + D_{yy}S_y^2 + D_{zz}S_z^2 \\ & + D_{xy}S_xS_y + D_{yx}S_xS_x + D_{yz}S_yS_z \\ & + D_{zy}S_yS_z + D_{zx}S_zS_x + D_{xz}S_xS_z\end{aligned} \tag{2.78}$$

where $\mathbf{D}$ is a symmetrical tensor with elements $D_{xx} = k\left\langle 1 - 3\left(\frac{x}{r}\right)^2\right\rangle$, $Dyy = k\left\langle 1 - 3\left(\frac{y}{r}\right)^2\right\rangle$, $D_{zz} = k\left\langle 1 - 3\left(\frac{z}{r}\right)^2\right\rangle$, $D_{xy} = D_{yx} = -3\,k\left\langle\frac{xy}{r^2}\right\rangle$, $Dyz = D_{zy} = -3\,k\left\langle\frac{zy}{r^2}\right\rangle$, and $D_{zx} = D_{xz} = -3\,k\left\langle\frac{xz}{r^2}\right\rangle$, where $k = \frac{1}{2}g^2\beta^2\langle r^{-3}\rangle$ is a constant for a particular radical species.

In general, the problem of obtaining these averages is exceedingly complex. For many important triplet state molecules or biradicals, however, the problem is simplified by symmetry or by concentration of the interacting electron-spin densities on well separated atoms so that the problem becomes solvable, at least to useful approximations. Applications of the Hamiltonian are given in Chapter X.

6. COMPOSITE SPIN HAMILTONIAN

The general form of the spin Hamiltonian, which includes all the terms developed previously in this chapter, is expressed most compactly by

$$\mathcal{H}_S = \beta \mathbf{S} \cdot \mathbf{g} \cdot \mathbf{H} + \mathbf{S} \cdot \mathbf{D} \cdot \mathbf{S} + \mathbf{S} \cdot \mathbf{A} \cdot \mathbf{I} + \mathbf{I} \cdot \mathbf{P} \cdot \mathbf{I} - g_I \beta_I \mathbf{H} \cdot \mathbf{I} - \mathbf{H} \cdot \Lambda \cdot \mathbf{H} \quad (2.79)$$

This Hamiltonian is adequate for interpretation of all types of esr spectra treated in this volume. The last term contains no spin operators, but it can influence some of the other terms through cross interaction. Seldom, if ever, will all the terms be applicable to one type of radical. In its application to the spectra of specific radicals some terms can be deleted in advance from a pre-knowledge of the type of radical or from the nature of the observed spectra. The remaining terms can often be simplified from such knowledge or from comparison of preliminary calculations with the observed spectra. The specific form of the final $\mathcal{H}_S$ is determined by a precise fitting of the predicted spectrum (often by high-speed computers) to the spectrum that is observed. Thus the spin Hamiltonian is tailored to the specific paramagnetic species observed, partly by theoretical knowledge and partly by empirical fitting. Although only one, or at most a few, of the terms in (2.79) may be left in the final form of $\mathcal{H}_S$ for a specific radical, one needs to be aware of (if not familiar with) the terms and forms that are rejected in the tailoring process. For efficient and rapid application, one also needs to be familiar with the various simplified forms and solutions applicable to special cases. In Chapter III we describe solutions for the more common specialized cases as well as the general solution.

References

1. A. Abragam and B. Bleaney, *Electron Paramagnetic Resonance of Transition Ions,* Clarendon, Oxford, 1970.
2. H. M. McConnell, *J. Chem. Phys.,* **28**, 1188 (1958).
3. M. H. L. Pryce, *Proc. Phys. Soc. (Lond.),* **A63**, 25 (1950).
4. A. Abragam and M. H. L. Pryce, *Proc. Roy. Soc. (Lond.),* **A205**, 135 (1951).
5. E. Fermi, *Z. Phys.,* **60**, 320 (1930).
6. B. Bleaney and K. W. H. Stevens, *Rep. Prog. Phys. (Lond. Phys. Soc.),* **16**, 108 (1953).
7. K. W. H. Stevens, *Proc. Phys. Soc. (Lond.),* **A65**, 209 (1952).
8. R. G. Barnes and W. V. Smith, *Phys. Rev.,* **93**, 95 (1954).

9. H. Kopfermann, *Nuclear Moments* (English version prepared by E. E. Schneider from 2nd German ed.), Academic, New York, 1958.

10. H. B. G. Casimir, *On the Interaction between Atomic Nuclei and Electrons,* Teyler's Tweede Genootschap, E. F. Bohn, Haarlem, 1936.

11. W. Gordy and R. L. Cook, *Microwave Molecular Spectra,* Wiley-Interscience, New York, 1970, Chapter 14.

Chapter **III**

CHARACTERISTIC ENERGIES AND FREQUENCIES

1. METHOD FOR FINDING ENERGY VALUES AND EIGENFUNCTIONS

If ψ represents the eigenfunction of the Hamiltonian operator $\mathcal{H}$, the corresponding eigenvalues E of $\mathcal{H}$ are given by solution of the Schrödinger equation

$$\mathcal{H}\psi = E\psi \tag{3.1}$$

However, one generally does not know the correct eigenfunctions, and it is customary to express them in terms of a complete, normalized, orthogonal set designated by Φ

$$\psi = \sum_n C_n \Phi_n \tag{3.2}$$

Substitution of this expression in (3.1) yields

$$\sum_n C_n \mathcal{H} \Phi_n = E \sum_n C_n \Phi_n \tag{3.3}$$

Multiplication of (3.3) by Φ_m^* and integration over all coordinate space gives

$$\sum_n C_n \int \Phi_m^* \mathcal{H} \Phi_n \, d\tau = E \sum_n C_n \int \Phi_m^* \Phi_n \, d\tau \tag{3.4}$$

Because the Φ functions are orthonormal

$$\int \Phi_m^* \Phi_n \, d\tau = \delta_{mn} \tag{3.5}$$

where $\delta_{mn} = 1$ when $m = n$ and 0 when $m \neq n$. With this substitution and with the more convenient designation, $\mathcal{H}_{nm} = \int \Phi_m^* \mathcal{H}\Phi_n \, dr = (m|\mathcal{H}|n)$, (3.4) can be expressed as

$$\sum_n C_n [\mathcal{H}_{mn} - E\,\delta_{mn}] = 0 \tag{3.6}$$

This summation represents a set of n homogeneous equations in n unknown coefficients that has a nontrivial solution only if the determinants of the coefficients vanish. This leads to the secular equation

$$\begin{bmatrix} \mathcal{H}_{11} - E & \mathcal{H}_{12} & \mathcal{H}_{13} & \cdots & \mathcal{H}_{1n} \\ \mathcal{H}_{21} & \mathcal{H}_{22} - E & \mathcal{H}_{23} & \cdots & \mathcal{H}_{2n} \\ \mathcal{H}_{31} & \mathcal{H}_{32} & \mathcal{H}_{33} - E & \cdots & \mathcal{H}_{3n} \\ \cdots & \cdots & \cdots & \cdots & \cdots \\ \mathcal{H}_{n1} & \mathcal{H}_{n2} & \mathcal{H}_{n3} & \cdots & \mathcal{H}_{nn} - E \end{bmatrix} = 0 \tag{3.7}$$

the solution of which gives n values for E that represent the energy eigenvalues of the Hamiltonian. For simplicity, only one subscript is used in the expansion (3.2) to label the functions Φ, but often more than one quantum number or subscript is required to specify the particular numbers of the set represented by Φ. The usage of plural subscripts where needed in later discussions should cause no confusion. Since the energy operator $\mathcal{H}$ is Hermitian, the determinant of (3.7) is symmetric, that is, $\mathcal{H}_{mn} = \mathcal{H}_{nm}$.

In principle, one can choose any complete set of orthonormal functions for the expansion of (3.2), but in practice one must choose a set for which the matrix elements of $\mathcal{H}$ are known or can be evaluated. For the spin Hamiltonian, the obvious choice is the electron spin function together with the nuclear spin functions, for which the matrix elements of the spin operators are known. If $\mathcal{H}_S$ contains only electron-spin operators, the appropriate representation is S, M_S. These are the numbers that distinguish, or label, the particular wave function Φ that may be designated in this representation by Φ_{S,M_S}, or in Dirac ket notation by $|S,M_S)$. If both electronic and nuclear spin operators are contained in $\mathcal{H}_S$, the appropriate representation would be S,M_S,I,M_I; in the ket form the basis functions would be $|S,M_S,I,M_I)$. In any particular problem where S and I are the same throughout, these subscripts may be dropped, and the simpler designation $|M_S,M_I)$ can be used without confusion. Spin Hamiltonians containing only the nuclear spin operators are encountered later. For dealing with them, the appropriate representation is $|I,M_I)$, or simply $|M_I)$.

Note that the eigenfunctions ψ of $\mathcal{H}$ are not required for finding the energy eigenvalues E, but only the matrix elements of the operator $\mathcal{H}$ in the chosen representation. When needed, these eigenfunctions can be obtained by an evaluation of the coefficients in the expansion of (3.2) by solution of the set of (3.6) for the C_n values (called *eigenvectors*) after the matrix elements $(m|\mathcal{H}|n)$ are obtained and the secular (3.7) is solved for the eigenvalues of E. Solution of this set, which consists of n equations with n unknown coefficients, is inadequate for a complete solution, however. The additional equation required is provided by normalization of ψ and is

$$\int \psi\psi^* \, d\tau = \sum_n C_n C_n^* = 1 \tag{3.8}$$

which follows from the fact that the functions Φ_n form a normalized, orthogonal set.

The solution of the secular equation (3.7) is equivalent to the diagonalization of the energy matrix. This follows from the fact that the energy eigenvalues, which are the roots E of (3.7), also correspond to the elements of the diagonal energy matrix. If, for example, $|k)$ represents an eigenfunction of $\mathcal{H}$, (3.1) may be expressed as

$$\mathcal{H}|k) = E_k|k) \tag{3.9}$$

Operation on both sides of this expression by $|k)$ with $(k|k) = 1$ gives the eigenvalues E_k in terms of the elements

$$E_k = (k|\mathcal{H}|k) \tag{3.10}$$

of the diagonal matrix. The off-diagonal elements vanish when the matrix is expressed in the eigenfunctions of $\mathcal{H}$. In matrix mechanics, diagonalization of the energy matrix corresponds to a transformation of the arbitrary functions in which the energy matrix is originally expressed to the eigenfunctions of $\mathcal{H}$.

2. MATRIX ELEMENTS OF SPIN OPERATORS

The matrix elements that will be needed in dealing with esr spectra are those of the electron-spin operators S_x, S_y, and S_z and the corresponding nuclear spin operators I_x, I_y, and I_z. These are of the same form, and their derivations [1] are well known. We give only the results here. For simplicity, we omit from the ket brackets the quantum numbers S or I that do not change in the operators. It must be remembered that there are $2S + 1$ values for M_S, varying by integral steps from S to $-S$. Similarly, there are $2I + 1$ values of M_I, ranging by unit steps from I to $-I$. The nonzero matrix elements of S_z, S_x, and S_y are

$$(M_S|S_z|M_S) = M_S \tag{3.11}$$

$$(M_S|S_x|M_S \pm 1) = \frac{1}{2}[S(S+1) - M_S(M_S \pm 1)]^{1/2} \tag{3.12}$$

$$(M_S|S_y|M_S \pm 1) = \pm\frac{i}{2}[S(S+1) - M_S(M_S \pm 1)]^{1/2} \tag{3.13}$$

In some problems it is more convenient to use the matrix elements of $S_x \pm iS_y$. For brevity, we designate $S_\pm = S_x \pm iS_y$. The upper signs are to be used together and the lower together. These are often called *raising* or *lowering* operators since

$$S_+|M_S) = [S(S+1) - M_S(M_S + 1)]^{1/2}|M_S + 1) \tag{3.14}$$

$$S_-|M_S) = [S(S+1) - M_S(M_S - 1)]^{1/2}|M_S - 1) \tag{3.15}$$

The nonvanishing matrix elements are thus

$$\begin{aligned}(M_S + 1|S_+|M_S) &= (M_S|S_-|(M_S + 1)\\ &= [S(S+1) - M_S(M_S + 1)]^{1/2}\end{aligned} \tag{3.16}$$

In most problems encountered $S = 1/2$, for which the specific elements are

$$\begin{aligned} (\pm 1/2|S_x|\mp 1/2) &= 1/2 \\ (\mp 1/2|S_y|\pm 1/2) &= \pm i/2 \\ (\pm 1/2|S_z|\pm 1/2) &= \pm 1/2 \end{aligned} \tag{3.17}$$

where the upper signs are to be applied together and likewise the lower. Also

$$(1/2|S_+|-1/2) = (-1/2|S_-|1/2) = 1 \tag{3.18}$$

The matrix elements of the nuclear spin operators have the same form as those for S and can be obtained by substitution of I for S and of M_I for M_S in these formulas.

The matrix elements of the products such as S_xS_y or of the operators S_x^2, S_y^2, and so on, can be found from the preceding formulas by the application of the matrix-product rule

$$(M_S|S_xS_y|M_S') = \sum_{M_S''} (M_S|S_x|M_S'')(M_S''|S_y|M_S') \tag{3.19}$$

and the Hermitian property of the matrices. The nonvanishing elements of S_z^2, S_x^2, and S_y^2 are

$$(M_S|S_z^2|M_S) = M_S^2 \tag{3.20}$$

$$(M_S|S_x^2|M_S) = (M_S|S_y^2|M_S) = 1/2[S(S+1) - M_S^2] \tag{3.21}$$

$$(M_S|S_x^2|M_S \pm 2) = -(M_S|S_y^2|M_S \pm 2)$$

$$= 1/4[S(S+1) - M_S(M_S\pm 1)]^{1/2} \; [S(S+1) - (M_S\pm 1)(M_S\pm 2)]^{1/2} \tag{3.22}$$

Obviously, in the last expression $|M'| = |M\pm 2| \leqslant S$.

3. SPECTRA FOR DOUBLET SPIN STATES ($S = 1/2$) WITH NO NUCLEAR EFFECTS

3.a. Isotropic g Factor

In this case, representing the simplest of all esr spectra, the spin Hamiltonian reduces to the one term $g\beta\mathbf{H} \cdot \mathbf{S}$. As explained in Chapter I, the resulting spectrum consists of a single line of frequency $g\beta H/h$.

3.b Anisotropic g: Principal Axes

Let us now consider the $\mathcal{H}_S$ with an anisotropic **g** tensor expressed in terms of its principal axis, still without nuclear interactions. In the principal axes the cross terms vanish, and the spin Hamiltonian of (2.16) becomes

$$\mathcal{H}_S = \beta(g_x H_x S_z + g_y H_y S_y + g_z H_z S_z) \tag{3.23}$$

in which we have dropped the unnecessary double subscripts. Suppose now that the magnetic field is applied in some arbitrary direction to the uniformly oriented radicals of a single crystal. If θ_x, θ_y, and θ_z represent the angles of H with the respective principal axes, we have

$$H_x = H\cos\theta_x, \quad H_y = H\cos\theta_y, \quad H_z = H\cos\theta_z \tag{3.24}$$

and

$$\mathcal{H}_S = \beta H(g_x S_x \cos\theta_x + g_y S_y \cos\theta_y + g_z S_z \cos\theta_y + g_z S_z \cos\theta_z) \tag{3.25}$$

The matrix elements of this $\mathcal{H}_S$ can be readily found since it contains only electron-spin operators and constants. From the matrix elements of the spin operators given in (3.17) the nonvanishing matrices of the $\mathcal{H}_S$ of (3.25) when $S = 1/2$ are

$$(1/2|\mathcal{H}_S|1/2) = (1/2)\,\beta H g_z \cos\theta_z \tag{3.26}$$

$$(-1/2|\mathcal{H}_S|1/2) = (1/2)\,\beta H(g_x \cos\theta_x + i g_y \cos\theta_y) \tag{3.27}$$

$$(1/2|\mathcal{H}_S|-1/2) = (1/2)\,\beta H(g_x \cos\theta_x - i g_y \cos\theta_y) \tag{3.28}$$

$$(-1/2|\mathcal{H}_S|-1/2) = (-1/2)\,\beta H g_z \cos\theta_z) \tag{3.29}$$

The secular equation is

$$\begin{vmatrix} (1/2)\,\beta H g_z \cos\theta_z - E & (1/2)\beta H(g_x \cos\theta_x + i g_y \cos\theta_y) \\ (1/2)\,\beta H(g_x \cos\theta_x - i g_y \cos\theta_y) & -(1/2)\beta H g_z \cos\theta_z - E \end{vmatrix} = 0 \tag{3.30}$$

The solution of this equation is

$$E_{\pm} = \pm(1/2)\,\beta H(g_x^2 \cos^2 \theta_x + g_y^2 \cos^2 \theta_y + g_z^2 \cos^2 \theta_z)^{1/2}$$
$$= \pm(1/2)\, g\beta H \qquad (3.31)$$

where

$$g = (g_x^2 \cos^2 \theta_x + g_y^2 \cos^2 \theta_y + g_z^2 \cos^2 \theta_z)^{1/2} \qquad (3.32)$$

is the effective g value for the given direction. The esr frequency is given by the Bohr relation $h\nu = W_+ - W_-$, as

$$\nu = \frac{g\beta H}{h} \quad \text{or} \quad H = \frac{h\nu_0}{g\beta} \qquad (3.33)$$

The expression is of the same form as (1.1), but the g factor here is not isotropic.

For many crystals the g factor, although not isotropic, is axially symmetric. Let us assume that z is the symmetry axis and designate $g_z = g_{\parallel}$. Then $g_x = g_y = g_{\perp}$. Also, $\cos^2 \theta_x + \cos^2 \theta_y = 1 - \cos^2 \theta_z = \sin^2 \theta_z$. By the use of these relations, the effective g factor for the axially symmetric case can be expressed as

$$g = (g_{\parallel}^2 \cos^2 \theta + g_{\perp}^2 \sin^2 \theta)^{1/2} \qquad (3.34)$$

where θ is the angle between the applied field and the symmetry axes.

The preceding solution is for $S = 1/2$, but the same effective g is obtained for the triplet state $S = 1$ if the spin-spin interaction is neglected. Even though spin-spin interaction is seldom negligible in organic free radicals, it is often sufficiently small that it can be treated with perturbation theory, with the preceding solution representing the unperturbed ground state. Solutions for the triplet state are given in Chapter X.

Sometimes the principal axes can be surmised from known symmetry properties of the single crystal or free-radical species and can be confirmed by preliminary measurements. If all the paramagnetic elements or free radicals have the same orientations relative to the crystal surfaces, a magnetic axis of summetry is also a principal axis of **g**. In most cases, however, the principal axes must be found the hard way, by diagonalization of a **g** tensor obtained from measurements of the effective g relative to a system arbitrarily oriented with respect to the principal axes. The reference system for the measurements is defined with respect to distinguishable features of the crystal surface. Methods for dealing with this more general problem are described in Chapter V.

3.c Anisotropic g: Arbitrary Axes

When x, y, and z are not the principal axes, the cross terms in $\mathcal{H}_S$ do not vanish. To find the energy eigenvalues, one can solve the secular equation as was done in Section 3.b, but the cross terms make this solution rather cumbersome. The same solution can be found in a simpler way with the vector model treatment described in the paragraph that follows. This treatment also gives some physical insight into the resonance process. It will be of advantage later in calculations of nuclear interactions.

The tensor interaction of the spin **S** with an applied field **H** can be reduced to a simple dot product of two vectors in the following way. In vector notation, the tensor interaction is

$$\begin{aligned}\beta\mathbf{H}\cdot\mathbf{g}\cdot\mathbf{S} = \beta(iH_x + jH_y + kH_z)\cdot(iig_{xx} + ijg_{xy}\\ + ikg_{xz} + jig_{yx} + jjg_{yy} + jkg_{yz}\\ + kig_{zx} + kjg_{zy} + kkg_{zz})\cdot(iS_x + jS_y + kS_z)\end{aligned} \tag{3.35}$$

where i, j, and k are the unit vectors along the x, y, and z axes. By carrying out the operation on the left only, one obtains

$$\beta\mathbf{H}\cdot\mathbf{g}\cdot\mathbf{S} = g_e\beta\mathbf{H}_e\cdot\mathbf{S} \tag{3.36}$$

where

$$\begin{aligned}\mathbf{H}_e &= \left(\frac{1}{g_e}\right)\mathbf{H}\cdot\mathbf{g}\\ &= \left(\frac{1}{g_e}\right)[i(H_xg_{xx} + H_yg_{yx} + H_zg_{zx}) + j(H_xg_{xy} + H_yg_{yy} + H_zg_{zy})\\ &\quad + k(H_xg_{xz} + H_yg_{yz} + H_zg_{zz})]\\ &= i(H_e)_x + j(H_e)_y + k(H_e)_z\end{aligned} \tag{3.37}$$

is the effective field with which the spin moment interacts. Physically, $\mathbf{H}_e$ is the vector sum of the applied field and the induced internal field $\mathbf{H}_e = \mathbf{H} + \mathbf{H}_i$. Since the $\mathcal{H}$ is not isotropic, $\mathbf{H}_i$ and hence $\mathbf{H}_e$ will not have the same direction as **H**.

In the vector model (Fig. 3.1) the electron-spin vector **S** precesses about the effective $\mathbf{H}_e$; that is, it is quantized along $\mathbf{H}_e$ with component values M_S. The energy values of $\mathcal{H}_S$ are easily found from this model. Since

$$\beta\mathbf{H}_e\cdot\mathbf{S} = \beta|\mathbf{H}_e||\mathbf{S}|\cos(\mathbf{H}_e,\mathbf{S}) \tag{3.38}$$

and

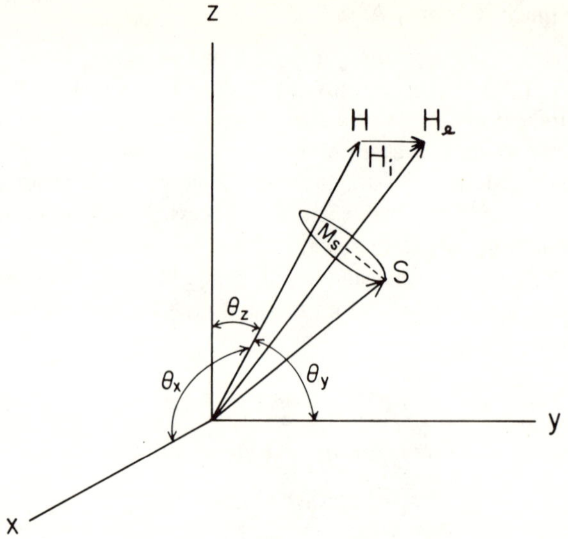

Fig. 3.1 Vector model of the electron-spin precession when there is **g** anistotropy in the paramagnetic specimen. The field $\mathbf{H}_e$ is the resultant of the applied field **H** and the internally induced field $\mathbf{H}_i$.

$$\cos(\mathbf{H}_e,\mathbf{S}) = \frac{M_S}{|\mathbf{S}|} \tag{3.39}$$

it is evident that

$$E_H = g_e\beta|\mathbf{H}_e|M_S \tag{3.40}$$

where

$$|H_e| = (\mathbf{H}_e \cdot \mathbf{H}_e)^{1/2}$$
$$= \left(\frac{1}{g_e}\right)[(H_x g_{xx} + H_y g_{yx} + H_z g_{zx})^2 + (H_x g_{yx} + H_y g_{yy} + H_z g_{yz})^2 + (H_x g_{zx} + H_y g_{zy} + H_z g_{zz})^2]^{1/2} \tag{3.41}$$

Suppose now that the magnetic field **H** is applied at angles θ_x, θ_y, and θ_z from the respective axes x, y, and z. Let us designate the direction cosines as $\cos\theta_x = l_x$, $\cos\theta_y = l_y$, $\cos\theta_z = l_z$ so that $H_x = Hl_x$, $H_y = Hl_y$, $H_z = Hl_z$. With these substitutions in $\mathbf{H}_e$, (3.40) reduces to

$$E_H = g\beta HM_S \tag{3.42}$$

and the resonant field values are

$$H_0 = \frac{h\nu_0}{g\beta} \tag{3.43}$$

where

$$g = [(g_{xx}l_x + g_{yx}l_y + g_{zx}l_z)^2 + (g_{xy}l_x + g_{yy}l_y + g_{zy}l_z)^2 + (g_{xz}l_x + g_{yz}l_y + g_{zz}l_z)^2]^{1/2} \tag{3.44}$$

This general expression for the effective g is the same as that obtained from a solution of the secular equation. Simpler, specialized formulas can be readily obtained from it. When x, y, and z are the principal axes of g, the cross terms involving g_{xy}, g_{yz}, and so on vanish, and the expression for g reduces to that of (3.32) obtained from a solution of the secular equation.

In finding the principal g values one usually makes measurements with **H** applied along the different axes x, y, and z and with **H** in the coordinate planes xy, xz, and zy. One thus obtains numerical values of g for the different orientations and a set of algebraic equations that can be solved for the tensor elements g_{xx}, g_{xy}, and so on. Since the **g** tensor is symmetric ($g_{xy} = g_{yx}$, etc.), there are only six independent elements to be evaluated. With **H** applied along one of the axes, along x, for example, $l_x = 1$, $l_y = 0$, $l_z = 0$, and

$$g_x = [g_{xx}{}^2 + g_{xy}{}^2 + g_{xz}{}^2]^{1/2} \tag{3.45}$$

When **H** is applied in one of the coordinate planes, in the xy plane, for example, $l_z = 0$, $l_x = \cos\theta$, and $l_y = \sin\theta$, where θ is the angle of **H** with the x axis, (3.44) reduces to

$$g_\theta = [(g_{xx}\cos\theta + g_{yx}\sin\theta)^2 + (g_{xy}\cos\theta + g_{yy}\sin\theta)^2 + (g_{xz}\cos\theta + g_{yz}\sin\theta)^2]^{1/2} \tag{3.46}$$

Methods for using these formulas with the experimental measurements to obtain the principal g values are described in Chapter V, Section 1.

4. NUCLEAR HYPERFINE STRUCTURE

The general spin Hamiltonian,

$$\mathcal{H}_S = \beta\mathbf{H}\cdot\mathbf{g}\cdot\mathbf{S} + \mathbf{S}\cdot\mathbf{D}\cdot\mathbf{S} + \mathbf{S}\cdot\mathbf{A}\cdot\mathbf{I} + \mathbf{I}\cdot\mathbf{P}\cdot\mathbf{I} - g_I\beta_I\mathbf{I}\cdot\mathbf{H}, \tag{3.47}$$

described in Chapter II, includes different types of electron spin-nuclear spin interactions, some of which do not influence the most commonly observed esr hyperfine spectra treated in this section. The spin-spin term $\mathbf{S} \cdot \mathbf{D} \cdot \mathbf{S}$, important for the triplet state $S = 1$, vanishes for the more common doublet state $S = 1/2$ that is considered here. Neither the nuclear quadrupole term $\mathbf{I} \cdot \mathbf{P} \cdot \mathbf{I}$ nor the nuclear magnetic term $g_I\beta_I\mathbf{I} \cdot \mathbf{H}$ necessarily vanishes when $S = 1/2$, but the energies corresponding to these terms do not change in the normal esr transitions for which the selection rules are $\Delta M_S = 1$, $\Delta M_I = 0$. These terms, important for ENDOR spectra, influence the normal esr frequencies only in higher-order approximations not generally detectable. Therefore, the $\mathcal{H}_S$ adequate for treating most esr having hyperfine structure is

$$\mathcal{H}_S = \beta\mathbf{H} \cdot \mathbf{g} \cdot \mathbf{S} + \mathbf{S} \cdot \mathbf{A} \cdot \mathbf{I} \tag{3.48}$$

In essentially all esr spectra observed at microwave frequencies, the first term in $\mathcal{H}_S$, $\beta\mathbf{H} \cdot \mathbf{g} \cdot \mathbf{S}$, is much larger than the nuclear interaction, $\mathbf{S} \cdot \mathbf{A} \cdot \mathbf{I}$. With common H values of 10 kG, the levels due to the first term have separations of the order of 20,000 MHz, whereas the hyperfine splittings due to the second term are usually less than 300 MHz. For this reason, the simplest and most common treatment is independent diagonalization of the energy matrix of the term $\beta\mathbf{H} \cdot \mathbf{g} \cdot \mathbf{S}$, as was done in the previous section, followed by treatment of the nuclear term as a perturbation on the electron-spin states $|S,M_S)$. When, as is most often the case, $\mathbf{S} \cdot \mathbf{A} \cdot \mathbf{I} \ll \beta\mathbf{H} \cdot \mathbf{g} \cdot \mathbf{S}$, first-order perturbation theory is adequate for prediction of the observed hyperfine splitting to the accuracy of the measurements. This is equivalent to taking the average of $\mathbf{S} \cdot \mathbf{A} \cdot \mathbf{I}$ over the representation $|S,M_S)$ that diagonalizes $\beta\mathbf{H} \cdot \mathbf{g} \cdot \mathbf{S}$. When first-order theory is inadequate, the matrix of the nuclear term $\mathbf{S} \cdot \mathbf{A} \cdot \mathbf{I}$ must be diagonalized or second-order perturbation theory applied. Diagonalization of the nuclear energy matrix is achieved by solution of the secular equation, as will be illustrated. Hyperfine structure of different types is discussed in the following sections, with simpler cases considered first.

4.a. Isotropic Interactions: First-order Treatment

First let us consider the case in which the nuclear coupling is completely isotropic and the anisotropy in g is negligible. This condition occurs in liquid solutions of free radicals. Since the couplings of the different nuclei of a given free radical are mutually independent, the hyperfine structure of each is calculated separately, and the multiplets of the different nuclei are summed. Completely isotropic interactions arise only from s-orbital density of the unpaired electron. Orbital spin density on one atom, however, may give rise to detectable anisotropic coupling to nuclei of neighboring atoms. Orbitals with angular momentum ($l \neq 0$), such as p or d orbitals, have no density at the nucleus and no Fermi contact interaction. In radicals that are tumbling rapidly, such as the molecules of a liquid, the anisotropic nuclear couplings of non-*s* orbitals average to zero. Such motions leave the isotropic *s*-orbital coupling undiminished. Isotropic *s*-orbital coupling to H nuclei frequently occurs through hyperconjugation.

For isotropic **g** and **A**, the $\mathcal{H}_S$ for $S = 1/2$, excluding quadrupole interactions, is

$$\mathcal{H}_S = g\beta \mathbf{H} \cdot \mathbf{S} + A\,\mathbf{I} \cdot \mathbf{S} - g_I\beta_I \mathbf{H} \cdot \mathbf{I} \tag{3.49}$$

Most isotropic splitting observed in esr is sufficiently small that at microwave frequencies the first-order treatment is adequate. The first-order nuclear energies are

$$(M_S M_I | A_f \mathbf{I} \cdot \mathbf{S} - g_I\beta_I \mathbf{H} \cdot \mathbf{I} | M_S M_I) = A_f M_S M_I - g_I \beta_I H M_I \tag{3.50}$$

With the $g\beta\mathbf{S} \cdot \mathbf{H}$ evaluated in Section 3.a the total energy is

$$E_{M_S, M_I} = g\beta H M_S + A_f M_S M_I - g_I\beta_I H M_I \tag{3.51}$$

The selection rules, $\Delta M_S = 1$, $\Delta M_I = 0$ (see Chapter IV), yield for the resonant frequencies

$$h\nu_0 = g\beta H + A_f M_I \tag{3.52}$$

where $M_I = I - 1, I - 2, \ldots, -I$ and A_f is the Fermi contact coupling. It is seen that the nuclear magnetic term $g_I\beta_I\mathbf{H} \cdot \mathbf{I}$ has no effect on the esr frequencies to first order. For a single coupling nucleus, there are $2I + 1$ hyperfine components corresponding to the different values of M_I.

The isotropic coupling gives a measure of the *s*-orbital density on the atom having the coupling nucleus. If A_f represents the observed isotropic Fermi coupling and A_{ns} represents the atomic coupling for the *s* orbital that causes the interaction, the spin density in this *s* orbital of the coupling atom in the radical can be obtained from the relation

$$\rho_s = \frac{A_f}{A_{ns}} \tag{3.53}$$

The atomic coupling A_{ns} can be calculated from (2.29), but for many atoms it is known from measurements of atomic spectra. Table A.1 gives values of A_{ns} for the coupling atoms most commonly found in organic free radicals. Hydrogen, with its 1*s*-bonding orbitals, is often found to cause isotropic hyperfine structure in organic free radicals. Sometimes several coupling hydrogens are observed in the same free radical.

4.b Isotropic Interactions: Precise Treatment

When the esr measurements are not made at sufficiently high frequencies to achieve the strong-field condition $A\mathbf{S} \cdot \mathbf{I} \ll g\beta\mathbf{H} \cdot \mathbf{S}$, an admixture of the electron-spin and nuclear spin states causes measurable perturbation in the esr spectra predicted by the first-order formula of (3.52). Adequate corrections may be made with second-order perturbation theory or in simpler cases by solution of the secular equation for the complete $\mathcal{H}_S$, as described in the following paragraph.

For a coupling nucleus having spin $I = 1/2$ with isotropic coupling in a free radical with one unpaired electron ($S = 1/2$) having isotropic g, the secular equation for the complete spin Hamiltonian is easily solved. Nuclei with $I = 1/2$ have no quadrupole interaction, and radicals with $S = 1/2$ have no spin-spin interaction. Therefore, the complete $\mathcal{H}_S$ for this case is

$$\mathcal{H}_S = g\beta\,\mathbf{H} \cdot \mathbf{S} + A\,\mathbf{I} \cdot \mathbf{S} - g_I\beta_I\mathbf{H} \cdot \mathbf{I} \tag{3.54}$$

With the isotropic g and A, the spins will be aligned by the applied field regardless of its direction. Hence, without loss of generality, we can choose the axis of quantization z as the direction of the applied field. Thus with $H_z = H$, $H_x = 0$, and $H_y = 0$, the expanded form of the Hamiltonian is then

$$\mathcal{H}_S = (g\beta S_z - g_I\beta_I I_z)H + A(I_xS_x + I_yS_y + I_zS_z) \tag{3.55}$$

The matrix elements for S_x, S_y, and S_z for $S = 1/2$ are given by (3.17), and those for the corresponding components for $I = 1/2$ have the same values. Because the components of the **I** and **S** operators commute, the elements for the products I_xS_x and so on are obtained by simple multiplication of those for the respective components. Thus the nonvanishing matrix elements $(M_SM_I|\mathcal{H}_S|M_S'M_I')$ of $\mathcal{H}_S$ in the representation M_S, M_I are obtained easily. The secular equation is formed simply by subtraction of E from the diagonal terms of the energy matrix and by treatment of the resulting array as a secular determinant. The elements are arranged in the sequence corresponding to decreasing values of $M = M_S + M_I$, 1,0,0,−1. The secular equation is

$$\begin{array}{c|cccc|c}
\begin{matrix} M'_S M'_I \\ \\ M_S M_I \end{matrix} & 1/2, 1/2 & 1/2, -1/2 & -1/2, 1/2 & -1/2, -1/2 & \\
\hline
1/2, 1/2 & \frac{a+A}{4} - E & 0 & 0 & 0 & \\
1/2, -1/2 & 0 & \frac{b-A}{4} - E & \frac{A}{2} & 0 & \\
 & & & & & = 0 \\
-1/2, 1/2 & 0 & \frac{A}{2} & -\frac{b+A}{4} - E & 0 & \\
-1/2, -1/2 & 0 & 0 & 0 & -\frac{a-A}{4} - E &
\end{array} \tag{3.56}$$

where $a = (1/2)(g\beta - g_I\beta_I)H$ and $b = (1/2)(g\beta + g_I\beta_I)H$. This determinant is seen to factorize into easily solved minors that yield the energy values

$$E_{M=1} = \frac{1}{2}(g\beta H - g_I\beta_I H) + \frac{A}{4} \tag{3.57}$$

$$E_{M=-1} = -\frac{1}{2}(g\beta H - g_I\beta_I H) + \frac{A}{4} \tag{3.58}$$

$$E^{\pm}_{M=0} = \pm\frac{1}{2}[(g\beta H + g_I\beta_I H)^2 + A^2]^{1/2} - \frac{A}{4} \tag{3.59}$$

Note that the two formulas corresponding to $M = 1(M_S = 1/2, M_I = 1/2)$ and $M = -1\ (M_S = -1/2, M_I = -1/2)$ are the same as those for the first-order energies. Only the states having the same total angular momentum $|1/2, -1/2\rangle$ and $|-1/2, 1/2\rangle$ are intermixed.

For microwave esr where the strong-field case is always closely approached, (3.59) can be simplified thus:

$$E^{\pm}_{M=0} = \pm\frac{1}{2}(g\beta H + g_I\beta_I H)\left\{1 + \frac{A^2}{2(g\beta H + g_I\beta_I H)^2} + \cdots\right\} - \frac{A}{4} \tag{3.60}$$

This expression, with (3.57) and (3.58), yields the frequencies corresponding to the allowed transitions $M = 0 \rightarrow 1$ and $-1 \rightarrow 0$ to second order as

$$\nu^{\pm} = \frac{g\beta H_0}{h} \pm \frac{A}{2h} + \frac{A^2}{4h(g\beta + g_I\beta_I)H_0} \tag{3.61}$$

where $g\beta H_0/h = \nu_0$ is the hypothetical unperturbed esr frequency that would be observed if there were no hyperfine splitting. It is seen that the doublet splitting

$$\Delta\nu = \nu^+ - \nu^- = \frac{A}{h} \tag{3.62}$$

is independent of the second-order correction. However, the doublet center $(\nu^+ + \nu^-)/2$ differs from the unperturbed frequency

$$\nu_0 = \frac{\nu^+ + \nu^-}{2} - \frac{(\Delta\nu)^2}{4(\nu_0 + \nu_{\text{nmr}})} \tag{3.63}$$

Thus the effect of the higher-order correction is to shift the doublet pattern to higher frequency without changing its spacing. The doublet splitting, in gauss, is

$$\Delta H = H^+ - H^- = \frac{h\Delta\nu}{g\beta} \tag{3.64}$$

where $\Delta\nu$ is in hertz = cps units.

The nuclear resonance frequency $\nu_{\text{nmr}} = g_I\beta_I H_0/h$ is of the order of 10^{-3} smaller than ν_0 and may be omitted from the denominator of the correction term without significant loss of accuracy. Thus the term $g_I\beta_I\mathbf{H}\cdot\mathbf{I}$ has no significant effect on the esr spectra, even in the higher-order treatment when the nuclear coupling is isotropic. This is not necessarily true when the coupling is anisotropic.

As an illustration of the higher-order perturbations, let us consider atomic hydrogen for which the coupling constant A/h = 1420.4 MHz. For esr measurements at a constant field H_0 such that $\nu_0 = g\beta H_0/h$ = 30,000 MHz, the first-order equation (3.52) predicts the hyperfine components at $\nu^+ = \nu_0 + A/2h$ = 30,710.2 MHz, and $\nu_0 - A/2h$ = 29,289.8 MHz, whereas the corresponding frequencies predicted by (3.57) to (3.59) are ν^+ = 30,726.9 MHz and ν^- = 29,306.5 MHz. When ν_0 = 10,000 MHz, the first-order predictions are ν^+ = 10,710.2 MHz and 9,289.8 MHz, whereas the precise theory predicts ν^+ = 10,760.3 MHz and ν^- = 9,339.9 MHz. Note that both components ν^+ and ν^- occur at higher frequencies in the refined treatment but that the doublet separation $\Delta\nu = (\nu^+ - \nu^-)$ = 1,420.4 MHz remains unchanged. Thus one can obtain an accurate value of the coupling constant A from the doublet separation by application of the first-order theory. However, one cannot take the halfway position between the doublets as the position of the undisplaced ν_0 frequency in measurements of g, as indicated by first-order theory. The center of the doublet is shifted 16.7 MHz at 30,000 MHz and 50.1 MHz at 10,000 MHz. For common free radicals where the doublet splitting is of the order of 100 MHz, the corresponding shift of the doublet center is only 0.25 MHz at 10,000 MHz. The H atom represents an upper limit to the hydrogen splitting.

For isotropic g and $S = 1/2$, the secular determinant also factorizes into solvable subdeterminants for all values of I. Thus for the important case of the free radical having one unpaired electron, the secular equation can be solved, and the exact energy values for any I can be obtained whenever the nuclear quadrupole coupling is zero or has negligible effects. When $S = 1/2$, the possible value of $M = M_I + M_S$ are

$$M = I + 1/2, \quad (I + 1/2) - 1, \quad (I + 1/2) - 2, \quad \ldots, \quad -(I + 1/2)$$

A solution of the secular equation [2] corresponding to the $\mathcal{H}_S$ of (3.55) yields the following closed formulas for the energies.

For maximum and minimum values of $M = \pm\,(I + 1/2)$,

$$E_M^{\pm} = \pm\,\frac{1}{2}(g\beta H - 2\,g_I\beta_I I\,H) + \frac{AI}{2} \tag{3.65}$$

and for M values between $I + 1/2$ and $-(I + 1/2)$,

$$E_M^{\pm} = \pm\,\frac{1}{2}\,[(g\beta H + g_I\beta_I H)^2 + 2\,A\,M(g\beta H + g_I\beta_I H) + (I + 1/2)^2\,A^2]^{1/2} - g_I\beta_I H\,M - \frac{A}{4} \tag{3.66}$$

It is evident that these formulas reduce to those of (3.57) to (3.59) when I=1/2.

Solutions of the intermediate-field case for atomic S states were first obtained by Breit and Rabi [2], and the higher-order corrections to the strong-field solution of (3.51) are often referred to as *Breit-Rabi corrections.* Their formulas, (3.57) to (3.59) and (3.65) and (3.66), hold rigorously for S states of atoms for which the quadrupole coupling vanishes because of spherical symmetry. In the pseudo-S states of free radicals, however, the charge cloud about the nuclei seldom has spherical symmetry. The electron spin is free, but the nuclear spin is not necessarily so when $I > 1/2$. Because of possible quadrupole coupling, the axis of quantization of I may differ from that of S. The energy matrix is then more complicated, and the precise solution is more difficult. The quadrupole effects treated in Section 6 are simplest when the field is applied along a common principal axis of the **A** and **P** tensors. Fortunately, in most free radicals, the quadrupole coupling is sufficiently small that these higher-order effects on the esr frequencies can be neglected.

When these formulas are applied with the selection rule $\Delta M = 1$ for calculation of normal esr spectra, the terms involving the nmr factor $g_I\beta_I H$ can be omitted except for unusually precise measurements. For ENDOR measurements, however, these terms must be retained.

4.c. Isotropic g and Anisotropic A: Principal Axes

We consider next an isotropic **g** tensor combined with an anisotropic nuclear coupling. The treatment also applies to slightly anisotropic **g** tensors, those for which effects of the **g** anisotropy on the nuclear coupling are negligible. We omit here the nmr term, $g_I\beta_I\mathbf{H}\cdot\mathbf{I}$, and possible quadrupole interactions that can influence esr spectra only in higher-order approximations, and we consider the Hamiltonian

$$\mathcal{H}_S = g\beta\mathbf{H}\cdot\mathbf{S} + \mathbf{S}\cdot\mathbf{A}\cdot\mathbf{I} \tag{3.67}$$

In evaluating this $\mathcal{H}_S$, we can assume the axis of quantization of **S** to be the direction of the applied field **H**. Although the nuclear field effective on the electron will not be in the direction of **H**, its magnitude, ~100 G, is so small in comparison to the usual **H**, ~10,000 G, that its effects on the direction of quantization of **S** may be neglected. However, one may not neglect the field of the electron spin on the nucleus, which is about 1500 times greater than the field of **I** effective on **S**, or about 150,000 G. Thus the field of the electron-spin moment at the nucleus is, in general, much greater than the applied field **H** and is the dominant factor in determining the axis of qunatization of **I**. Because of these conditions, we can evaluate the $g\beta\mathbf{H}\cdot\mathbf{S}$ term independently of the nuclear interaction and in the evaluation of the nuclear interaction can assume **S** to be aligned along **H**.

With **S** quantized along **H**, the $g\beta\,\mathbf{S}\cdot\mathbf{H}$ term has the values $g\beta HM_S$. Expressed in the principal axes x, y, and z of the **A** tensor, the nuclear term $\mathbf{S}\cdot\mathbf{A}\cdot\mathbf{I}$ is

$$\mathcal{H}_N = A_xS_xI_x + A_yS_yI_y + A_zS_zI_z \tag{3.68}$$

Since **S** is quantized along **H**, one can set $S_x = l_xS_H$, $S_y = l_yS_H$, and $S_z = l_zS_H$, where l_x, l_y, and l_z are the direction cosines between H and the principal axes of A. By substitution of these values with $S_H = M_S$, one obtains a reduced $\mathcal{H}_N$ with only nuclear spin operators

$$\mathcal{H}_N = (A_xl_xI_x + A_yl_yI_y + A_zl_zI_z)M_S \tag{3.69}$$

As an example, we solve (3.69) for $S = 1/2$ and $I = 1$. Hyperfine structure of ^{14}N or deuterium, for which $I = 1$, is often observed in organic free radicals. When $I = 1$, the M_I values are 1,0,−1, and the matrix elements of $\mathcal{H}_N$ for the upper states $M_S = 1/2$ are

$$(1|\mathcal{H}_N|1) = \frac{1}{2}A_z l_z = \text{a}$$

$$(1|\mathcal{H}_N|0) = \frac{\sqrt{2}}{4}(A_x l_x - iA_y l_y) = b^*$$

$$(0|\mathcal{H}_N|1) = \frac{\sqrt{2}}{4}(A_x l_x + iA_y l_y) = b$$

$$(0|\mathcal{H}_N|0) = 0 \tag{3.70}$$

$$(0|\mathcal{H}_N|-1) = \frac{\sqrt{2}}{4}(A_x l_x - iA_y l_y) = b^*$$

$$(-1|\mathcal{H}_N|0) = \frac{\sqrt{2}}{4}(A_x l_x + iA_y l_y) = b$$

$$(-1|\mathcal{H}_N|-1) = -\frac{1}{2}A_z l_z = -a$$

Therefore, the secular equation is

$$\begin{vmatrix} a-E_N & b^* & 0 \\ b & E_N & b^* \\ 0 & b & -a-E_N \end{vmatrix} = 0 \tag{3.71}$$

which reduces to

$$E_N(a^2 - E^2) + 2E_N(bb^*) = 0 \tag{3.72}$$

$$E_N\ [(1/4)A_z^2 l_z^2 - E_N^2] + (1/4)E_N(A_x^2 l_x^2 + A_y^2 l_y^2) = 0 \tag{3.73}$$

The solutions are

$$E_N = 0, \pm (1/2)A \tag{3.74}$$

where

$$A = (A_x^2 l_x^2 + A_y^2 l_y^2 + A_z^2 l_z^2)^{1/2} \tag{3.75}$$

The solutions for $\mathcal{H}_N$ for the lower spin state, $M_S = -1/2$, are similar.

When $I = 1$, there are three nuclear sublevels for each electronic spin state, one of which is not displaced. There are three hyperfine components with resonant field strengths at constant observation frequency ν_0 of

$$H_1 = \frac{h\nu_0}{g\beta} + \frac{A}{g\beta} \tag{3.76}$$

$$H_0 = \frac{h\nu_0}{g\beta} \tag{3.77}$$

$$H_{-1} = \frac{h\nu_0}{g\beta} - \frac{A}{g\beta} \tag{3.78}$$

4.d. Strong-field, Paschen-Back Case

An isotropic component A_f in the coupling gives rise to a magnetic field $(\beta/B_I)(A_f/g\beta) = (A_f/g\beta_I)$ at the nucleus that is always parallel to the axis of quantization of **S** and hence in the direction of **H** when **S** is aligned with **H**, as assumed earlier. Thus when the anisotropic component $\mathbf{A}_\mu$ is smaller than A_f or the applied field is sufficiently strong that $(A_\mu/g\beta_I) \ll (H + A_f/g\beta_I)$, the nuclear spin **I** as well as **S** will be aligned with the field, and $I_x = l_x I_H$, $I_y = l_y I_H$, and $I_z = l_z I_H$, where $(M_I|I_H|M_I) = M_I$. Substitution of these values into (3.69) gives the first-order energies for the strongfield case as

$$E = AM_S M_I \tag{3.79}$$

where the effective coupling constant is

$$A = A_x l_x^2 + A_y l_y^2 + A_z l_z^2 \tag{3.80}$$

With the substitutions $A_x = A_f + (A_\mu)_x$, $A_y = A_f + (A_\mu)_y$, $A_z = A_f + (A_\mu)_z$, and $l_x^2 + l_y^2 + l_z^2 = 1$, (3.80) can be expressed as

$$A = A_f + (A_\mu)_x l_x^2 + (A_\mu)_y l_y^2 + (A_\mu)_z l_z^2 \tag{3.81}$$

Equation (3.80) or (3.81) for the effective coupling A is more convenient to apply than (3.75), but the strong-field conditions for which it holds are less frequently encountered. Note, however, that when **H** is applied along a principal axis of **A**, the formulas give identical results. For example, when **H** is along the x axis, $l_x = 1 = l_x^2$, $l_y = 0 = l_y^2$, $l_z = 0 = l_z^2$, and $A = A_x$ from (3.75) or (3.80). Thus the strong-field condition (parallel alignment of **S** and **I**) always holds when **H** is imposed along a common principal axis of **g** and **A**.

4.e. Anisotropic g and A: Arbitrary Axes

The spin Hamiltonian that we consider here is

$$\mathcal{H}_S = \beta\mathbf{H}\cdot\mathbf{g}\cdot\mathbf{S} + \mathbf{S}\cdot\mathbf{A}\cdot\mathbf{I} - g_I\beta_I\mathbf{H}\cdot\mathbf{I} \tag{3.82}$$

It is assumed that $\mathbf{S} \cdot \mathbf{A} \cdot \mathbf{I} \ll \beta \mathbf{S} \cdot \mathbf{g} \cdot \mathbf{H}$ so that effects of the nuclear field on the alignment of the electron spin are negligible. The separate solution of $\beta \mathbf{H} \cdot \mathbf{g} \cdot \mathbf{S}$ already obtained (Section 2) can be used. A vector-operator treatment similar to that used for $\beta \mathbf{H} \cdot \mathbf{g} \cdot \mathbf{S}$ is employed in the solution of

$$\mathcal{H}_N = \mathbf{S} \cdot \mathbf{A} \cdot \mathbf{I} - g_I \beta_I \mathbf{H} \cdot \mathbf{I} = (\mathbf{S} \cdot \mathbf{A} - g_I \beta_I \mathbf{H}) \cdot \mathbf{I} \tag{3.83}$$

for the nuclear sublevels.

If $\mathbf{h}_e$ is the resultant magnetic field at the nucleus, the nuclear magnetic interaction may also be expressed as

$$\mathcal{H}_N = -g_I \beta_I \mathbf{h}_e \cdot \mathbf{I} = g_I \beta_I (\mathbf{h} - \mathbf{H}_e) \cdot \mathbf{I} \tag{3.84}$$

which is equivalent to

$$\mathcal{H}_N = g_I \beta_I |\mathbf{h}||\mathbf{I}| \cos(\mathbf{h}, \mathbf{I}) - g_I \beta_I |\mathbf{H}_e||\mathbf{I}| \cos(\mathbf{H}_e, \mathbf{I}) \tag{3.85}$$

where $H_e \approx H$ is given by (3.37). The last term is evaluated in Section 5. In

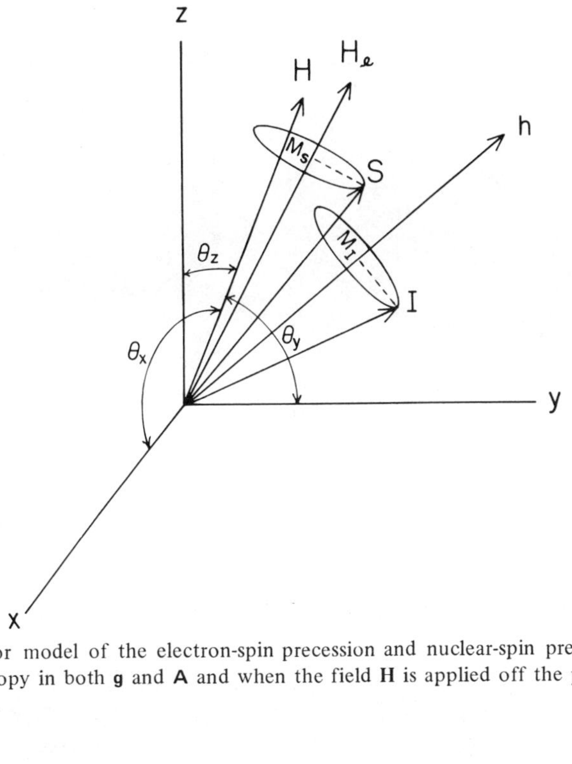

Fig. 3.2 Vector model of the electron-spin precession and nuclear-spin precession when there is anisotropy in both $\mathbf{g}$ and $\mathbf{A}$ and when the field $\mathbf{H}$ is applied off the principal axes of both tensors.

first order it does not affect the esr hyperfine structure because of the selection rule $\Delta M_I = 0$. Here we evaluate the first term on the right, which represents the magnetic hyperfine interaction of **I** with **S**. To do this, we assume $H_e \ll h$ and take **h** to be the axis of quantization of **I** as indicated in the vector model of Fig. 3.2. Then

$$\cos(\mathbf{h},\mathbf{I}) = \frac{I_h}{|\mathbf{I}|} = \frac{M_I}{|\mathbf{I}|} \tag{3.86}$$

and the first term on the right side of (3.85) transforms to

$$\mathcal{H}_N^{(h)} = g_I \beta_I |\mathbf{h}| M_I \tag{3.87}$$

The problem remaining is evaluation of $|\mathbf{h}|$.

Comparison of the equivalent expressions for $\mathcal{H}_N$ in (3.83) and (3.84) shows that with neglect of the small $(H_e - H)$ term

$$\mathbf{h} = \frac{\mathbf{S} \cdot \mathbf{A}}{g_I \beta_I} \tag{3.88}$$

Now **S** is aligned by the dominant term $\beta \mathbf{H} \cdot \mathbf{g} \cdot \mathbf{S}$ along the effective field $\mathbf{H}_e$ (3.37) and hence we can set

$$\mathbf{S} = \frac{\mathbf{H}_e}{|\mathbf{H}_e|} S_{H_e} = \frac{\mathbf{H} \cdot \mathbf{g}}{|\mathbf{H} \cdot \mathbf{g}|} M_S \tag{3.89}$$

where $\mathbf{H}_e/|\mathbf{H}_e|$ is the unit vector along the axis of quantization of **S**. Substitution of **S** from (3.89) into (3.88), together with substitution of the magnitude of the resulting expression for **h** into (3.87), shows the magnetic hyperfine energy to be

$$E_{M_S,M_I} = \frac{|\mathbf{H} \cdot \mathbf{g} \cdot \mathbf{A}|}{|\mathbf{H} \cdot \mathbf{g}|} M_S M_I \tag{3.90}$$

If l_x, l_y, and l_z are the direction cosines of the applied field with the reference axes x, y, and z, then $\mathbf{H} = H(\mathbf{i}l_x + \mathbf{j}l_y + \mathbf{k}l_z)$ and thus

$$\mathbf{H} \cdot \mathbf{g} = H(\mathbf{i}g_a + \mathbf{j}g_b + \mathbf{k}g_z) \tag{3.91}$$

where

$$
\begin{aligned}
g_a &= g_{xx}l_x + g_{yx}l_y + g_{zx}l_z \\
g_b &= g_{xy}l_x + g_{yy}l_y + g_{zy}l_z \\
g_c &= g_{xz}l_x + g_{yz}l_y + g_{zz}l_z
\end{aligned}
\tag{3.92}
$$

Therefore,

$$
\begin{aligned}
\mathbf{H} \cdot \mathbf{g} \cdot \mathbf{A} &= H(\mathbf{i}g_a + \mathbf{j}g_b + \mathbf{k}g_c) \cdot \mathbf{A} \\
&= H[\mathbf{i}(g_aA_{xx} + g_bA_{yx} + g_cA_{zx}) + \mathbf{j}(g_aA_{xy} \\
&\quad + g_bA_{yy} + g_cA_{zy}) + \mathbf{k}(g_aA_{xz} + g_bA_{yz} + g_cA_{zz})]
\end{aligned}
\tag{3.93}
$$

By taking the magnitudes of the vectors of (3.91) and (3.93) and substituting the results into (3.90) one obtains

$$E_{M_S,M_I} = A\, M_SM_I \tag{3.94}$$

where A is the effective coupling constant given by

$$
\begin{aligned}
A = \frac{1}{g}\,[(g_aA_{xx} &+ g_bA_{yx} + g_cA_{zx})^2 + (g_aA_{xy} + g_bA_{yy} \\
&+ g_cA_{zy})^2 + (g_aA_{xz} + g_bA_{yz} + g_cA_{zz})^2]^{1/2}
\end{aligned}
\tag{3.95}
$$

and

$$g = (g_a^2 + g_b^2 + g_c^2)^{1/2} \tag{3.96}$$

with g_a, g_b, and g_c defined by (3.92). The total electron-spin and nuclear-spin energy is

$$E_{M_S,M_I} = g\beta H\, M_S + A\, M_SM_I \tag{3.97}$$

where g is again given by (3.96) combined with (3.92), and A is given by (3.95).

Equation (3.95) holds for any value of S and I except for complications that may arise from spin-spin interaction when $S > 1/2$ and from nuclear quadrupole interaction when $I > 1/2$. These complications are treated later.

Since the x,y,z reference system is arbitrary, the same reference system applies for the **g** and the **A** tensors, even though the principal axes of these tensors may not coincide. Although the general expression for **A** in (3.95), which includes off-diagonal elements of **g** as well as of **A**, is rather complicated to apply directly, it has the advantage that various specialized forms can be quickly obtained from it.

4.f. Anisotropic g and A: Principal Axes

When the principal axes of **g** and **A** are different and both are unknown, the simplest solution of the problem would be to first find the principal elements of **g** independently of **A**. For this purpose, measurements of the center positions of the hyperfine patterns for different orientations of the radicals would be used. The theory developed in Section 3.b for no hyperfine splitting is applicable to these measurements. When the principal axes of **g** are thus determined, one can then choose these axes as the reference system for the still unknown **A** tensor. With the x,y,z system as the principal axes of **g** but not of **A**, (3.95) may be expressed as

$$A = \frac{1}{g}\,[(g_x l_x A_{xx} + g_y l_y A_{yx} + g_z l_z A_{zx})^2 + (g_x l_x A_{xy} + g_y l_y A_{yy} + g_z l_z A_{zy})^2 + (g_x l_x A_{xz} + g_y l_y A_{yz} + g_z l_z A_{zz})^2]^{1/2} \quad (3.98)$$

where

$$g = [(g_x l_x)^2 + (g_y l_y)^2 + (g_z l_z)^2]^{1/2} \quad (3.99)$$

and g_x, g_y, and g_z are the principal elements of **g**. When the reference axes x,y, and z are the principal axes of **A** as well as of **g**, the formula for A becomes

$$A = \frac{1}{g}\,[(g_x l_x A_x)^2 + (g_y l_y A_y)^2 + (g_z l_z A_z)^2]^{1/2} \quad (3.100)$$

where A_x, A_y, A_z are the principal values of **A** and where **g** is given by (3.99).

4.g. Isotropic g and Anisotropic A: Arbitrary Axes

When the effects of g anisotropy on the nuclear coupling are negligible, one can set $g_x = g_y = g_z = g$ in (3.98) and obtain

$$A = [(l_x A_{xx} + l_y A_{yx} + l_z A_{zx})^2 + (l_x A_{xy} + l_y A_{yy} + l_z A_{zy})^2 + (l_x A_{xz} + l_y A_{yz} + l_z A_{zz})^2]^{1/2} \quad (3.101)$$

In the principal axis of **A** the off-diagonal elements vanish, and the formula for the effective A reduces to

$$A = [(l_x A_x)^2 + (l_y A_y)^2 + (l_z A_z)^2]^{1/2} \tag{3.102}$$

which is the expression obtained before, (3.75).

For experimental determination of the **A** tensor, one generally measures the hyperfine splitting with the field imposed along the x, y, and z axes and at various angles in the coordinate planes, xy, xz, and zy. Sufficient independent information can be obtained in this way for evaluation of all the tensor elements of **A**. With **H** imposed in the xy plane θ^0 from x, $l_x = \cos\theta$, $l_y = \sin\theta$, $l_z = 0$, and (3.101) is expressed as

$$A = [(A_{xx}\cos\theta + A_{yx}\sin\theta)^2 + (A_{xy}\cos\theta + A_{yy}\sin\theta)^2 + (A_{xz}\cos\theta + A_{yz}\sin\theta)^2]^{1/2} \tag{3.103}$$

When **H** is imposed along the x axis, $\cos\theta = 1$, $\sin\theta = 0$, and (3.103) reduces to

$$A = (A_{xx}^2 + A_{xy}^2 + A_{xz}^2)^{1/2} \tag{3.104}$$

The values for the other coordinate planes are similarly obtained from (3.101).

4.h. Axially Symmetric Coupling: *p* Orbitals

The anisotropic coupling most often observed in molecular free radicals arises from p-orbital spin density on the coupling atom and is axially symmetric about this orbital. Usually, such coupling is superimposed on an isotropic component caused by s-orbital spin density on the same atom, but the axis of symmetry is determined by the p-orbital component. Here we treat only axially symmetric dipole-dipole interactions such as that caused by p-orbital interaction alone. The next topics deal with combined s- and p-orbital coupling.

Free radicals with unpaired electron density in p orbitals usually have g anisotropy so small that it has a negligible effect on the nuclear coupling. Therefore, to simplify the discussion, we neglect the small effects of g anisotropy and apply (3.102), in which the reference axes are the principal axes of **A**. Let us choose z as the symmetry axis and designate $A_z = A_{\|}$, and $A_x = A_y = A_{\perp}$, with θ as the angle between z and the applied field **H**. Then $(A_z l_z)^2 = A_{\|}^2 \cos^2\theta$ $(A_x l_x)^2 + (A_y l_y)^2 = A_{\perp}^2(l_x^2 + l_y^2) = A_{\perp}^2(1 - \cos^2\theta) = A_{\perp}^2 \sin^2\theta$. With these substitutions, (3.103) becomes

$$A = (A_{\|}^2 \cos^2\theta + A_{\perp}^2 \sin^2\theta)^{1/2} \tag{3.105}$$

When **H** is imposed along the symmetry axis, $\theta = 0$ and $A = \pm A_{\|}$; when $\theta = 90°$, $A = \pm A_{\perp}$.

Pure dipole-dipole interaction (no isotropic, Fermi-contact component) averages to zero over all directions of space. The principal elements of a dipole-dipole coupling tensor $\mathbf{A}_\mu$ satisfy the condition

$$(A_\mu)_x + (A_\mu)_y + (A_\mu)_z = 0 \tag{3.106}$$

For axially symmetric dipole-dipole interaction with $(A_\mu)_z = (A_\mu)_\parallel$ and $(A_\mu)_x = (A_\mu)_y = (A_\mu)_\perp$,

$$(A_\mu)_\parallel = -2(A_\mu)_\perp \tag{3.107}$$

When the $\mathbf{g}$ tensor is also axially symmetric and has the same axis of symmetry as the $\mathbf{A}$ tensor, (3.105) can be simply corrected for any significant effects of $\mathbf{g}$ anisotropy by multiplication of $A_\parallel$ by $g_\parallel/g$ and $A_\perp$ by $g_\perp/g$.

4.i. Combined Isotropic and Anisotropic Coupling

It should be noted that the various formulas derived for the effective coupling A holds when A has an isotropic as well as an anisotropic component. In this case the principal elements are: $A_x = A_f + (A_\mu)_x$, $A_y = A_f + (A_\mu)_y$, and $A_z = A_f + (A_\mu)_z$, where A_f is the isotropic component and $(A_\mu)_x$, $(A_\mu)_y$, $(A_\mu)_z$ are the principal elements of the anisotropic dipole-dipole tensor $\mathbf{A}_\mu$. In practice, it is usually simpler first to determine the values of A_x, A_y, and A_z by diagonalization of the $\mathbf{A}$ tensor and then to separate the isotropic and anisotropic parts with the help of (3.106). An example of combined A_f and A_μ coupling is the mixed s and p coupling described in the following section.

4.j. Mixed s- and p-Orbital Coupling

Most anisotropic couplings observed for molecular species are superimposed on an isotropic component that is often a large fraction of the total coupling observed. When there is no s hybridization in the coupling orbital, configuration interaction leading to spin polarization of paired orbitals having s character usually gives a measurable isotropic s component in the coupling. Thus the mixed s and p coupling considered here is probably the most common type of coupling observed in molecular free radicals. For salts of the transition elements where the anisotropic component is due to d- or f-orbital density, configuration interaction may also lead to an isotropic component.

Because of the relationship expressed by (3.107), it is evident that an axially symmetric, pure dipole-dipole interaction (with no isotropic component) can be represented completely by a single coupling constant. It is customary to specify this constant by B such that

$$B = (1/2)(A_\mu)_\parallel = -(A_\mu)_\perp \tag{3.108}$$

Also, A is commonly used to specify the isotropic s-orbital coupling. We adopt these conventions here, except that we use the subscript f on A to signify that the term represents only the isotropic, Fermi contact component of the general **A** tensor.

With the preceding designations, the principal elements of an axially symmetric coupling made up of s- and p-orbital contributions are

$$\begin{aligned} A_{\parallel} &= A_f + 2B \\ A_{\perp} &= A_f - B \end{aligned} \tag{3.109}$$

With these substitutions, (3.105) for axially symmetric coupling is expressed as

$$A = [(A_f + 2B)^2 \cos^2 \theta + (A_f - B)^2 \sin^2 \theta]^{1/2} \tag{3.110}$$

In magnetic field units, the separations of the successive hyperfine components are

$$\Delta H = \frac{1}{g\beta} [(A_f + 2B)^2 \cos^2 \theta + (A_f - B)^2 \sin^2 \theta]^{1/2} \tag{3.111}$$

By measurement of $\Delta H_{\parallel}$ and $\Delta H_{\perp}$, $\theta = 0$ and $90°$, one obtains

$$\Delta H_{\parallel} = \frac{1}{g\beta} |A_f + 2B| \tag{3.112}$$

$$\Delta H_{\perp} = \frac{1}{g\beta} |A_f - \mathrm{B}|$$

When, as is most often true, A_f and B have the same sign, these equations yield

$$A_f = \frac{g\beta}{3}(\Delta H_{\parallel} + 2\Delta_{\perp}) \tag{3.113}$$

$$B = \frac{g\beta}{3}(\Delta H_{\parallel} - \Delta H_{\perp}) \tag{3.114}$$

From the coupling values determined by the experimental measurements and the known atomic coupling constants, one can find the s- and p-orbital spin densities on the particular coupling atom of the radical. The tabulated coupling values, A_{ns} and A_{np}, (Table A.1) are those for an unpaired electron in the corresponding orbital of the free atom; the A_{np} value is that for the perpendicular orientation of the orbital. The relationships of the observed free-radical coupling to this constant are

$$A_f = \sum_n \rho_{ns} A_{ns} \tag{3.115}$$

$$B = \sum_{\substack{n \\ i = x,y,z}} \rho_{np_i} A_{np_i} \tag{3.116}$$

where ρ_{ns} and ρ_{np} represent the spin densities in the particular orbitals ns and np that are responsible for the coupling and where A_{ns} and A_{np} correspond to the atomic coupling constants. The n subscript refers to the radial quantum number that specifies the column of the atom in the atomic table.

The summation is included in (3.115) because sometimes more than one s orbital contributes to A_f. Because of the spherical symmetry in the s-orbital coupling, there is no way to distinguish the contributions from different s electrons. The valence s orbital usually has much the greater spin density, but slight spin densities of subvalence s electrons can cause significant contribution to A_f because of the $1/n^3$ variation of A_{ns}. Such slight spin densities can arise from configuration interactions on the coupling atom. In contrast, the coupling p orbital can be identified from the axis of symmetry in the coupling, and the B coupling constant usually comes from spin density of a particular valence p electron but sometimes from p_x, p_y, and p_z densities. See Chapter VI.

4.k. Second-order Corrections

The foregoing treatment of anisotropic interactions did not take into account the effects of intermixture of electron-spin and nuclear-spin states (Breit-Rabi corrections), as was done for the isotropic interactions. The most precise treatment is diagonalization of the energy matrix corresponding to the complete $\mathcal{H}_S$, but this approach is unnecessarily complicated. When the nuclear interactions are so large that the first-order treatment is inadequate, second-order perturbation theory is usually found to be accurate within the experimental measurements. As can be seen from the illustrations to be given, these corrections are of the order of $(\Delta\nu)^2/4\nu_0$, where $\Delta\nu$ is the hyperfine component separation and $\nu_0 = (g\beta H_0/h)$ is the undisplaced frequency that would be observed if there were no nuclear interactions. In addition to this Breit-Rabi type of correction, which applies to isotropic as well as anisotropic interactions, important "forbidden" transitions can result from interactions between the term $g_I\beta_I\mathbf{H} \cdot \mathbf{I}$ and off-diagonal components of an anisotropic $\mathbf{A}$ tensor when $\mathbf{S} \cdot \mathbf{A} \cdot \mathbf{I}$ and $g_I\beta_I\mathbf{H} \cdot \mathbf{I}$ are of comparable magnitude. These transitions are described in later chapters. Also, when the nuclear quadrupole interaction is comparable in magnitude to the nuclear magnetic interaction, perturbations of the esr magnetic hyperfine structure can occur, especially when the $\mathbf{A}$ and $\mathbf{P}$ tensors have different principal axes (see Section 6). We first treat the more common perturbation that arises from an admixture of electron-spin and nuclear spin states.

First, let us consider anisotropic $\mathbf{g}$ and $\mathbf{A}$ tensors that have common principal axes. With perturbation theory, we derive the higher-order corrections for the doublet spin states, $S = 1/2$, for any value of I; but to simplify the problem, we consider an $\mathcal{H}_S$ with no quadrupole interactions and with the magnetic field imposed along a principal axis or in a principal plane. With the principal axis as the reference system and with **H** chosen along z, this $\mathcal{H}_S$ is expressed by

$$\mathcal{H}_S = \beta H g_z S_z + A_z S_z I_z + A_x S_x I_x + A_y S_y I_Y - g_I \beta_I H I_z \tag{3.117}$$

The first-order energies are the diagonal elements of $\mathcal{H}_S$:

$$E^{(1)} = (M_S M_I | \mathcal{H}_S | M_S M_I) = g_z \beta H M_S + A_z M_S M_I - g_I \beta_I H M_I \tag{3.118}$$

The second-order terms are

$$E^{(2)} = \sum_{M'_S, M'_I \neq M_S, M_I} \frac{|(M_S M_I | \mathcal{H}_S | M'_S M'_I)|^2}{E_{M_S M_I} - E_{M'_S M'_I}} \tag{3.119}$$

which here are

$$E^{(2)}_{M_S M_I} = \Sigma' \frac{|(M_S M_I | A_x S_x I_x + A_y S_y I_y | M'_S M'_I)|^2}{E_{M_S, M_I} - E_{M'_S, M'_I}} \tag{3.120}$$

The required matrix elements are given by (3.12) and (3.13); those for M_I are obtained by the substitutions of I for S, and thus M_I for M_S. Because $g_I \beta_I H <<$ $g\beta H$, to a good approximation $E_{M_S, M_I} - E_{M'_S, M'_I} = g_z \beta H$, and the second-order energies for the $M_S = \pm 1/2$ spin levels are found to be

$$E^{(2)}_{\pm 1/2, M_I} = \pm \frac{1}{16 g_z \beta H} \{(A_x \pm A_y)^2 [I(I+1) - M_I(M_I + 1)] + (A_x \mp A_y)^2 [I(I+1) - M_I(M_I - 1)]\} \tag{3.121}$$

where the upper signs apply for the $M_S = 1/2$ state. The total energies $E = E^{(1)} + E^{(2)}$ are obtained by combination of (3.118) and (3.121). By application of the selection rules $M_S \leftrightarrow M_S + 1$ and $M_I \leftrightarrow M_I$, the esr frequencies are found to be

$$h\nu = g_z \beta H + A_z M_I + \frac{1}{8 g_z \beta H} [(A_x + A_y)^2 + (A_x - A_y)^2] [I(I+1) - M_I^2] \tag{3.122}$$

When there is axial symmetry about z, $A_z = A_{||}$, $A_x = A_y = A_\perp$, $g_z = g_{||}$, and this expression becomes

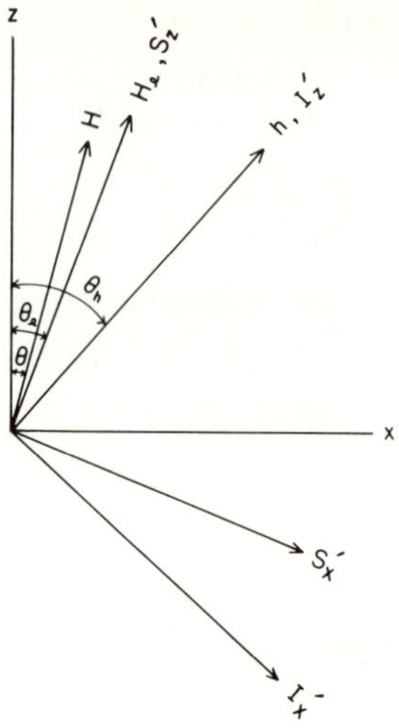

Fig. 3.3 Coordinate system for rotation of the components of **S** and **I** when the field **H** is applied in the *zx* plane. Compare with Fig. 3.2.

$$h\nu = g_{\parallel}\beta H + A_{\parallel}M_I + \frac{A_{\perp}^2}{2g_{\parallel}\beta H}\,[I(I+1) - M_I^2] \tag{3.123}$$

The problem becomes more complicated when the magnetic field is not oriented along a principal axis. However, if the off-axis orientations are restricted to a principal plane, it is possible to derive a reasonably simple, closed formula for the second-order corrections. Let us assume that **H** is in the zx principal plane at angle θ from z, as indicated in Fig. 3.3. Then $H_z = H\cos\theta$, $H_x = H\sin\theta$, $H_y = 0$, and $\mathcal{H}_S$ expressed in the common principal axes of **g** and **A** becomes

$$\mathcal{H}_S = \beta H(g_x S_X \sin\theta + g_z S_z \cos\theta) + A_x S_x I_x + A_y S_y I_y + A_z S_z I_z \tag{3.124}$$

when the $g_I\beta_I\mathbf{H}\cdot\mathbf{I}$ and nuclear quadrupole interactions are omitted. Effects of $g_I\beta_I\mathbf{H}\cdot\mathbf{I}$ are negligible because $g_I\beta_I \ll g\beta$. Nuclear quadrupole effects are treated in Section 6.

The first-order solutions of (3.124) are contained in the general first-order solutions already obtained: with $l_x = \sin\theta$, $l_y = 0$, $l_z = \cos\theta$, (3.32) becomes, for the present case,

$$g = (g_x{}^2 \sin^2 \theta + g_z \cos^2 \theta)^{1/2} \tag{3.125}$$

and (3.100) becomes

$$A = \frac{1}{g} [(g_x A_x)^2 \sin^2 \theta + (g_z A_z)^2 \cos^2 \theta]^{1/2} \tag{3.126}$$

The calculation of the second-order corrections for the $\mathcal{H}_S$ of (3.124) is complicated by the fact that the principal axis z is not the axis of quantization for **S** or **I**. As shown earlier, **S** is quantized along $\mathbf{H}_e = \mathbf{g} \cdot \mathbf{H}$, and **I** is quantized along $\mathbf{h} = \mathbf{H} \cdot \mathbf{g} \cdot \mathbf{A}$. To evaluate the second-order corrections, we express $A_x S_x I_x + A_y S_y I_y + A_z S_z I_z$ in the rotated (primed) system in which **S** is along $\mathbf{H}_e$ and **I** along **h** (Fig. 3.3). From Fig. 3.3 it is evident that

$$\begin{aligned} S_x &= S'_z \sin \theta_e + S'_x \cos \theta_e \\ S_y &= S'_y \\ S_z &= S'_z \cos \theta_e - S'_x \sin \theta_e \end{aligned} \tag{3.127}$$

and

$$\begin{aligned} I_x &= I'_z \sin \theta_h + I'_x \cos \theta_h \\ I_y &= I'_y \\ I_z &= I'_z \cos \theta_h - I'_x \sin \theta_h \end{aligned} \tag{3.128}$$

The angles can be evaluated from the relation $(\mathbf{H}_e/|\mathbf{H}_e|) = (\mathbf{H} \cdot \mathbf{g}/|\mathbf{gH}|)$, which, in the principal coordinates, is

$$\frac{i(H_e)_x + j(H_e)_y + k(H_e)_z}{H_e} = \frac{i g_x H_x + j g_y H_y + k g_z H_z}{gH} \tag{3.129}$$

Since $\cos \theta_e = (H_e)_z/H_e$, $\sin \theta_e = (H_e)_x/H_e$, $\cos \theta = H_z/H$, and $\sin \theta = H_x/H$, it is evident from (3.129) that

$$\cos \theta_e = \frac{g_z}{g} \cos \theta \tag{3.130}$$

$$\sin \theta_e = \frac{g_x}{g} \sin \theta \tag{3.131}$$

where

$$g = (g_x{}^2 \sin^2 \theta + g_z{}^2 \cos^2 \theta)^{1/2} \tag{3.132}$$

From combination of (3.88) and (3.89) it is seen that the unit vector in the direction of **H** is

$$\mathbf{h}/|\mathbf{h}| = \frac{\mathbf{H} \cdot \mathbf{g} \cdot \mathbf{A}}{|\mathbf{H} \cdot \mathbf{g} \cdot \mathbf{A}|} \tag{3.133}$$

from which

$$\cos \theta_h = \frac{g_z A_z}{gA} \cos \theta \tag{3.134}$$

$$\sin \theta_h = \frac{g_x A_x}{gA} \sin \theta \tag{3.135}$$

where g is given by (3.132) and where

$$A = \frac{1}{g} (g_x{}^2 A_x{}^2 \sin^2 \theta + g_z{}^2 A_z{}^2 \cos^2 \theta)^{1/2} \tag{3.136}$$

By the use of the transformations of (3.127) and (3.128) with the values of (3.130), (3.131), (3.134), and (3.135), one finds that

$$\begin{aligned} \mathcal{H}_N &= A_z S_z I_z + A_x S_x I_x + A_y S_y I_y \\ &= A S'_z I'_z + \frac{A_x A_z}{A} S'_x I'_x + \left[\frac{g_x g_z}{A g^2} (A_x{}^2 - A_z{}^2) \sin \theta \cos \theta \right] S'_x I'_z \\ &\quad + A_y S'_y I'_y \end{aligned} \tag{3.137}$$

where θ is the angle of **H** with the z axis. Note that an off-diagonal term $S'_x I'_z$ is introduced by the transformation. The coefficient of the corresponding $S'_z I'_x$ term vanishes. The first term on the right is diagonal in the transformed coordinates and has the value $A M_S M_I$, with the value of A given by (3.136) as already found in the first-order treatment.

The second-order corrections are obtained by application of the second-order formula, (3.119), to the terms in $S'_xI'_x$, $S'_yI'_y$, and $S'_xI'_z$. The required matrix elements are given by (3.12) and (3.13). The resulting energies for the most useful case, $S = 1/2$, with any value of I are

$$E^{(2)}_{\pm 1/2, M_I} = \pm \frac{1}{16A^2 g\beta H_0} \{(A_xA_z \pm A_yA)^2 \; [I(I+1) - M_I(M_I+1)] + (A_xA_z \mp A_yA)^2 \; [I(I+1) - M_I(M_I-1)] + 4\left(\frac{g_xg_z}{g^2}\right)^2 (A_z{}^2 - A_x{}^2)^2 M_I{}^2 \sin^2\theta \cos^2\theta\} \quad (3.138)$$

If $E^{(2)}$ is added to the first-order energies and the selection rules $M_S \to M_S + 1$ and $M_I \to M_I$ are applied, the transition frequencies are found to be

$$h\nu = g\beta H_0 AM_I + \frac{1}{8A^2 g\beta H_0} \{[(A_xA_z + A_yA)^2 + (A_xA_z - A_yA)^2][I(I+1) - M_I{}^2] + \left(\frac{g_xg_z}{g^2}\right)^2 (A_z{}^2 - A_x{}^2)^2 \; M_I{}^2 \sin^2 2\theta\} \quad (3.139)$$

where **g** is given by (3.125) and **A** by (3.126).

When **g** and **A** have a common axis of symmetry, $g_z = g_\parallel$, $g_x = g_y = g_\perp$, $A_z = A_\parallel$, $A_x = A_y = A_\perp$, and (3.139) reduces to

$$h\nu = g\beta H_0 + AM_I + \frac{1}{8A^2 g\beta H_0} \{[2A_\perp{}^2 (A_\parallel{}^2 + A^2)] \; [I(I+1) - M_I{}^2] + \left(\frac{g_\perp g_\parallel}{g^2}\right)^2 (A_\parallel{}^2 - A_\perp{}^2)^2 M_I{}^2 \sin^2 2\theta\} \quad (3.140)$$

where $g^2 = g_\parallel{}^2 \cos^2\theta + g_\perp{}^2 \sin^2\theta$ and $g^2 A^2 = (g_\parallel A_\parallel)^2 \cos^2\theta + (g_\perp A_\perp)^2 \sin^2\theta$.

5. NUCLEAR INTERACTION WITH THE APPLIED FIELD

In these discussions we have for the most part neglected the nuclear interaction with the applied field because it has no first-order effect on the esr spectra for which $\Delta M_I = 0$ and only a very slight effect on the second-order interaction. However, for ENDOR spectra where the nuclear-spin vector is flipped by a radio-frequency field, $\Delta M_I = \pm 1$, and in "forbidden" transitions corresponding to changes in M_I, the **I** · **H** interaction is detected as a first-order effect. For these reasons we consider it here in more detail.

The first-order $\mathbf{H} \cdot \mathbf{I}$ interaction is quite simple when $\mathbf{H}$ is applied along a principal axis of the $\mathbf{A}$ tensor for then the axis of quantization of $\mathbf{I}$ is along $\mathbf{H}$ and $\mathbf{H} \cdot \mathbf{I}$ is diagonal. In this case

$$E_{M_I}^{(H)} = -G_I M_I \tag{3.141}$$

where $G_I = g_I \beta_I H$.

When the applied field is not along a principal axis of $\mathbf{A}$, the $\mathbf{H} \cdot \mathbf{I}$ interaction is complicated by the fact that $\mathbf{H}$ is not along an axis of quantization of $\mathbf{I}$. As already described, $\mathbf{I}$ is aligned by the effective field $\mathbf{h}_e = \mathbf{h} - \mathbf{H}$ where $\mathbf{h}$, the field at the nucleus due to the electron-spin moment, is usually the dominant component. When $\mathbf{H}$ is along a principal axis of $\mathbf{A}$, $\mathbf{h}$ has the same direction as $\mathbf{H}$, but otherwise its direction is quite different from that of $\mathbf{H}$. Because in most applications $\mathbf{H} \ll \mathbf{h}$, we assume $\mathbf{I}$ to be quantized along $\mathbf{h}$ even when $\mathbf{H}$ is not along a principal axis. A direct approach to the problem is to resolve $\mathbf{H}$ along the axis of quantization $\mathbf{h}$, designated z' in Fig. 3.3. Because the angle between $\mathbf{H}$ and $\mathbf{h}$ is not known explicitly, a simpler approach is to resolve both $\mathbf{H}$ and $\mathbf{I}$ along the principal axes of $\mathbf{A}$ and evaluate their component interactions. In a first-order treatment we resolve only the component of I_h that is quantized along $\mathbf{h}$. Thus

$$\mathbf{I} = I_h\,[i(l_n)_x + j(l_n)_y + k(l_n)_z] \tag{3.142}$$

and

$$\mathbf{H} = H(il_x + jl_y + kl_z) \tag{3.143}$$

where l_n signifies the direction cosines of $\mathbf{h}$ and l signifies those of $\mathbf{H}$ with the x, y, and z axes (see Fig. 3.3). From (3.133) it is evident that

$$(l_n)_x = \frac{l_x g_x A_x}{gA}, \qquad (l_n)_y = \frac{l_y g_y A_y}{gA}, \qquad (l_n)_z = \frac{l_z g_z A_z}{gA} \tag{3.144}$$

With these values and with $I_h = M_I$, the first-order energies are seen to be

$$E_{M_I}^{(H)} = \frac{G_I M_I}{gA}\,(l_x{}^2 g_x A_x + l_y{}^2 g_y A_y + l_z{}^2 g_z A_z) \tag{3.145}$$

where g_I is assumed to be isotropic and $G_I = g_I \beta_I H$, $gA = [(l_x g_x A_x)^2 + (l_y g_y A_y)^2 + (l_z g_z A_z)^2]^{1/2}$ and where l_x, l_y, and l_z are the direction cosines of $\mathbf{H}$ with the reference system.

When **H** is applied in the xz plane at angle θ with z, $l_z = \cos\theta$, $l_x = \sin\theta$, $l_y = 0$, and

$$E_{M_I}^{(H)} = -\frac{G_I M_I}{gA}(g_x A_x \sin^2\theta + g_z A_z \cos^2\theta). \tag{3.146}$$

When x, y, and z are not the principal axes of **g** and **A**, we can set

$$\mathbf{I} = \frac{\mathbf{H}\cdot\mathbf{g}\cdot\mathbf{A}}{|\mathbf{H}\cdot\mathbf{g}\cdot\mathbf{A}|}M_I = \frac{l\cdot\mathbf{g}\cdot\mathbf{A}}{gA}M_I \tag{3.147}$$

and obtain the more general expression

$$E_{M_I}^{(H)} = -\frac{G_I M_I}{gA}\, l\cdot\mathbf{g}\cdot\mathbf{A}\cdot 1 \tag{3.148}$$

where g is given by (3.96) and A by (3.95) and where $l = \mathbf{i}l_x + \mathbf{j}l_y + \mathbf{k}l_z$, with l_x, l_y, and l_z, the direction cosines of **H** with the reference axes.

6. NUCLEAR QUADRUPOLE INTERACTIONS

Nuclear quadrupole interactions have no first-order effects on the normal esr spectrum for which the selection rules $M_S \to M_S \pm 1$ and $M_I \to M_I$ apply. In second order, however, the quadrupole interactions can perturb the esr transitions slightly, and off-diagonal elements in the interactions can lead to weak, "forbidden" transitions corresponding to a flipping of the nuclear spin, $M_I \to M_I \pm 1$. More correctly, these forbidden transitions correspond to transitions of $\Delta M = 0$ or 2 in the total quantum number $M = M_S + M_I$ of the admixed electronic and nuclear spin states. In ENDOR experiments the $M_I \to M_I \pm 1$ transitions are deliberately induced by the imposition of a radiofrequency resonant field simultaneous with that of the microwave field that induces the electron-spin transitions $M_S \to M_S \pm 1$. Here we are concerned mainly with derivation of the nuclear quadrupole energies that contribute to the displacements of the nuclear sublevels of the esr states and their possible perturbations of the normal esr transitions.

The nuclear quadrupole Hamiltonian

$$\mathcal{H}_Q = \mathbf{I}\cdot\mathbf{P}\cdot\mathbf{I} = \sum_{i,j=x,y,z} P_{ij} I_i I_j \tag{3.149}$$

derived in Chapter II does not involve the electron-spin operator and has the same form as that for nonmagnetic solids commonly observed in pure nqr. Under certain conditions, notably when the **g**, **A**, and **P** tensors all have common principal axes and when the magnetic field **H** is imposed along one of the common principal axes, the solution of the $\mathcal{H}_Q$ and the expression of the first-order quadrupole energies is the same as that of pure nqr. These energies are simply superimposed on the nuclear magnetic splitting. When, however, the magnetic field is not imposed along a common principal axis, the axis of quantization of **I** is shifted away from the principal axis of the nuclear quadrupole interaction. Thus the imposition of the strong magnetic field required for esr can significantly alter the quadrupole interaction, even though the quadrupole interaction has only a slight, usually negligible, effect on the normal esr frequencies.

6.a. Axially Symmetric Coupling: H along Symmetry Axis

Let us first consider the case in which **H** is imposed along a common axis of symmetry of **g**, **A**, and **P**. The axis, which we designate as z, will then be the axis of quantization of both **S** and **I**. Except for the Breit-Rabi interaction, which is negligible if **H** is strong, the nuclear quadrupole energies can be derived separately and added to those already found for the **g** and **A** interactions. The simplest case is that for an axially symmetric interaction (see Chapter II, Section 4) for which

$$\mathcal{H}_Q = \frac{P_z}{2}\,[3I_z{}^2 - I(I+1)] \tag{3.150}$$

and the resulting energies are

$$E_Q = \frac{P_z}{2}\,[3M_I{}^2 - I(I+1)] \tag{3.151}$$

The quadrupole coupling constant, as usually designated and expressed in hertz, is related to P_z in energy units by

$$\chi_{zz} = eQq_z = \frac{P_z 2I(2I-1)}{h} \tag{3.152}$$

In some treatments, the right-hand side of (3.150) is written as $P_{\|}[I_z{}^2 - (1/3)I(I+1)]$, where $P_{\|} = (3/2)P_z$; in other treatments, as $P_{\|}[3I_z{}^2 - I(I+1)]$, where then $P_{\|} = P_z/2 = h\chi_{zz}/[4I(2I-1)]$. Also the coupling constant, $eQq_z = eQ(\partial^2 V/\partial z^2)$, is sometimes written eQq_z and sometimes e^2Qq_z, depending on the units used for q_z.

For a transition in which both the electron spin and the nuclear spin are flipped, $M_S \to M_S \pm 1$, $M_I \to M_I \pm 1$, as in ENDOR, the increment in the radiofrequency caused by the quadrupole interaction is

$$(\Delta\nu)_Q = \frac{3\chi_{zz}}{4I(I+1)}(2M_I - 1) \tag{3.153}$$

in which the coupling constant χ_{zz} is expressed in frequency units. Note that the $(\Delta\nu)_Q$, unlike the magnetic hyperfine displacement, is independent of the electron-spin state and is also independent of the sign of M_I. It is evident from (3.151) that $(\Delta\nu)_Q = 0$ when $\Delta M_I = 0$, as in normal esr spectra.

6.b. Asymmetric Coupling: H along Common Principal Axis of A and P

When $\mathcal{H}_Q$ is not axially symmetric but is expressed in the principal axes, (3.149) reduces to the form

$$\mathcal{H}_Q = P_x I_x^2 + P_y I_y^2 + P_z I_z^2 \tag{3.154}$$

Because the constants P are proportional to field gradients that obey Laplace's equation, $\nabla^2 V = 0$,

$$P_x + P_y + P_z = 0 \tag{3.155}$$

and hence there are only two independent coupling constants. These can be expressed in terms of P_z and the asymmetry parameter as

$$\eta = \frac{P_x - P_y}{P_z} = \frac{\chi_x - \chi_y}{\chi_z} \tag{3.156}$$

With these substitutions and with $P_z = -(P_x + P_y)$, (3.154) can be expressed in the form

$$\mathcal{H}_Q = \frac{P_z}{2}[3I_z^2 - I(I+1)] + \frac{\eta P_z}{2}(I_x^2 - I_y^2) \tag{3.157}$$

The nonvanishing matrix elements of I_x^2, I_y^2, and I_z^2 can be obtained from (3.20) to (3.22) by substitution of I and its components for S and its components. With these elements substituted in (3.157), the nonvanishing matrix elements of $\mathcal{H}_Q$ are seen as

$$(I,M_I|\mathcal{H}_Q|I,M_I) = \frac{P_z}{2}[3M_I^2 - I(I+1)] \tag{3.158}$$

$$(I,M_I|\mathcal{H}_Q|I,M_I \pm 2) = \frac{\eta P_z}{4}\{[I(I+1) - M_I(M_I \pm 1)][I(I+1) - (M_I \pm 1)(M_I \pm 2)]\}^{1/2} \tag{3.159}$$

where the upper signs are applied together and the lower signs together and where $|M'| = |M_I \pm 2| \leqslant I$. Thus there are diagonal elements and nondiagonal elements off by ±2 only.

For the most common spin values, $I = 1, 3/2$, the secular equation for the nuclear interactions, including the quadrupole term, can be solved explicitly and the energies expressed in closed formulas. For higher spin values, second-order perturbation treatment of the off-diagonal terms is usually adequate when $\eta \leqslant 0.1$. Accurate computer solution of the energy matrix for any value of η or I is possible.

As an illustration, we solve the problem for $I = 1$ with the magnetic field chosen along the principal axis z that is assumed to be also a principal axis of **g** and **A**. We further assume the common condition that $A \ll g\beta H$ so that effects of the off-diagonal elements $A_x S_x I_x$ and $A_y S_y I_y$ are negligible. Since we are considering possible $\Delta M_I = \pm 1$ as well as $\Delta M_I = 0$ transitions, the nmr term $-g_I \beta_I \mathbf{H} \cdot \mathbf{I}$ must be included. For these conditions,

$$\mathcal{H}_S = G_z S_z + A_z S_z I_z - G_I I_z + \mathcal{H}_Q \tag{3.160}$$

where $G_z = g_z \beta H$ and $G_I = g_I \beta_I H$ and where $\mathcal{H}_Q$ is given by (3.157). For $I = 1$, the off-diagonal matrix elements of $\mathcal{H}_Q$ are, from (3.159), $(\mp 1|\mathcal{H}_Q|\pm 1) = (\eta P_z/2)$. Since S_z is diagonal, the energy matrix factors into 3 × 3 submatrices corresponding to each value of M_S. The submatrix for the particular M_S value is

M_I \ M_I'	1	0	−1
1	$(1\|\mathcal{H}_S\|1)$	0	$\frac{\eta P_z}{2}$
0	0	$(0\|\mathcal{H}_S\|0)$	0
−1	$\frac{\eta P_z}{2}$	0	$(-1\|\mathcal{H}_S\|-1)$

where $(0|\mathcal{H}_S|0) = G_z M_S - P_z$ and $(\pm 1|\mathcal{H}_S|\pm 1) = (G_z M_S \pm A_z M_S \mp G_I + P_z/2)$. Solution of the corresponding secular equations when $M_S = 1/2$ and $M_S = -1/2$ gives the energies for the doublet spin state

$$E^0_{\pm 1/2} = \pm \frac{1}{2} G_z - P_z \tag{3.161}$$

$$E^{\pm}_{1/2} = \frac{1}{2} G_z + \frac{P_z}{2} \mp \left\{ \left[\frac{A_z}{2} - G_I\right]^2 + \left(\frac{\eta P_z}{2}\right)^2 \right\}^{1/2} \tag{3.162}$$

$$E^{\pm}_{-1/2} = -\frac{1}{2} G_z + \frac{P_z}{2} \mp \left\{ \left[\frac{A_z}{2} + G_I \right]^2 + \left(\frac{\eta P_z}{2} \right)^2 \right\}^{1/2} \tag{3.163}$$

The normal esr transitions that correspond to $1/2 \leftrightarrow 1/2$, $+ \leftrightarrow +$, $- \leftrightarrow -$, and $0 \leftrightarrow 0$ are

$$h\nu_0 = G_z \tag{3.164}$$

$$h\nu_+ = G_z + \left\{ \left[\frac{A_z}{2} - G_I \right]^2 + \left(\frac{\eta P_z}{2} \right)^2 \right\}^{1/2} + \left\{ \left[\frac{A_z}{2} + G_I \right]^2 + \left(\frac{\eta P_z}{2} \right)^2 \right\}^{1/2} \tag{3.165}$$

$$h\nu_- = G_z - \left\{ \left[\frac{A_z}{2} - G_I \right]^2 + \left(\frac{\eta P_z}{2} \right)^2 \right\}^{1/2} - \left\{ \left[\frac{A_z}{2} + G_I \right]^2 + \left(\frac{\eta P_z}{2} \right)^2 \right\}^{1/2} \tag{3.166}$$

where $G_z = g_z \beta H$ and $G_I = g_I \beta_I H$. Although the $M_I = \pm 1$ states are intermixed by the off-diagonal quadrupole terms, the $+ \leftrightarrow +$ corresponds in the limit of $\eta \to 0$ to $M_I = 1 \to 1$ and $- \leftrightarrow -$ to $M_I = -1 \to -1$ for the axially symmetric case. It is of interest to note that the off-diagonal quadrupole elements displace the upper component ν_+ to higher frequencies and the lower component ν_- to lower frequencies by an equal amount, so that the average,

$$\frac{\nu_+ + \nu_-}{2} = \nu_0 = \frac{G_z}{h} = \frac{g_z \beta H}{\mathrm{h}} \tag{3.167}$$

is the undisplaced center frequency. Thus the effect of the η term is to increase the hyperfine splitting of the esr multiplet without displacing the multiplet center, whereas the Breit-Rabi effect shifts the center without changing the splitting.

Solutions of the secular equations for the $\mathcal{H}_S$ of (3.160) for $1 < I$ are rather complicated, and we derive the energies for $S = 1/2$ and $I = 3/2$ with perturbation theory. The first-order energies are simply the diagonal elements of the $\mathcal{H}_S$ and are

$$E^{(1)}_{\pm 1/2, M_I} = \pm \frac{1}{2} G_z \pm \frac{1}{2} A_z M_I - G_I M_I + \frac{1}{2} P_z \left(3M_I^2 - \frac{15}{4}\right) \tag{3.168}$$

where $M_I = 3/2, 1/2, -1/2, -3/2$. The second-order terms are obtained by substitution of the off-diagonal matrix elements of (3.159) with the corresponding energy differences into the second-order formula, (3.119). As an example, the off-diagonal matrix element for $M_I = 3/2$ is $(3/2|\mathcal{H}_Q|-1/2) = (\sqrt{3}/2)\,\eta P_z$, the corresponding energy difference for the $M_S = 1/2$ state is $E_{1/2,3/2} - E_{1/2,-1/2} = A_z - 2G_I + 3P_z$, and the resulting second-order energy from (3.119) is $E^{(2)}_{1/2,3/2} = (3/4)\eta^2 P_z^2/(A_z - 2G_I + 3P_z)]$. With the second-order terms for the other levels similarly obtained and with the first-order terms from (3.168), the energy levels to second order, $E_{M_S,M_I} = E^{(1)} + E^{(2)}$, for $S = 1/2$ and $I = 3/2$ are found to be

$$E_{\pm 1/2,3/2} = \pm\frac{1}{2}G_z \pm \frac{3}{4}A_z - \frac{3}{2}G_I + \frac{3}{2}P_z + \frac{3}{4}\left(\frac{\eta^2 P_z{}^2}{\pm A_z - 2G_I - 3P_z}\right) \quad (3.169)$$

$$E_{\pm 1/2,1/2} = \pm\frac{1}{2}G_z \pm \frac{1}{4}A_z - \frac{1}{2}G_I - \frac{3}{2}P_z + \frac{3}{4}\left(\frac{\eta^2 P_z{}^2}{\mp A_z + 2G_I + 3P_z}\right) \quad (3.170)$$

$$E_{\pm 1/2,-1/2} = \pm\frac{1}{2}G_z \mp \frac{1}{4}A_z + \frac{1}{2}G_I - \frac{3}{2}P_z + \frac{3}{4}\,\frac{\eta^2 P_z{}^2}{\pm A_z + 2G_I + 3P_z} \quad (3.171)$$

$$E_{\pm 1/2,-1/2} = \pm\frac{1}{2}G_z \mp \frac{3}{4}A_z + \frac{3}{4}G_I + \frac{3}{2}P_z + \frac{3}{4}\left(\frac{\eta^2 P_z{}^2}{\pm A_z + 2G_I + 3P_Z}\right) \quad (3.172)$$

where the upper signs apply for the $M_S = 1/2$ state, the lower for the $M_S = -1/2$ state, and $G_z = g\beta H$, $G_I = g_I\beta_I H$.

The microwave esr quanta correspond to $E_{+1/2,M_I} - E_{-1/2,M_I}$, whereas the radiofrequency quanta measured in ENDOR correspond to $E_{M_S,M_I} - E_{M_S,M_I\pm 1}$. Frequencies corresponding to either type of transition are readily obtained from the energy formulas derived in this section. However, one should beware of using the second-order formula unless $(\eta P_z)^2 \ll A_z$.

6.c. Magnetic Field Applied off the Axis

When the applied magnetic field **H** is not along a common principal axis of the **A** and **P** tensors, complications result from the fact that the axis of quantization of **I** is shifted away from the principal axis z partly by the **H** · **I** coupling, but mostly by the direction of the much stronger coupling **h** · **I** where **h** is the magnetic field at the nucleus due to the electron-spin moment. Although the direction of **h** is determined by the alignment of **S** by **H**, it is usually not in the direction of either **S** or **H**. The direction of **h** relative to **H**, already derived in Section 5, is given by (3.144) (see Fig. 3.3). In most paramagnetic species the field **h** is still the dominant factor in the alignment of the nuclear spin, even when there is quadrupole interaction, because for most species the nuclear magnetic interaction is much larger than the quadrupole interaction. When these two interactions are of comparable magnitude, the secular equation for the combined tensor interaction $\mathcal{H}_N = \mathbf{S} \cdot \mathbf{A} \cdot \mathbf{I} + \mathbf{I} \cdot \mathbf{P} \cdot \mathbf{I}$ must be solved with off-diagonal elements of both tensors included.

In this treatment we assume $\mathbf{I} \cdot \mathbf{P} \cdot \mathbf{I} \ll \mathbf{S} \cdot \mathbf{A} \cdot \mathbf{I} \ll \mathbf{H} \cdot \mathbf{g} \cdot \mathbf{S}$ so that, in effect, $\mathbf{H_e}$ (3.37) determines the alignment of **S**, and $\mathbf{S} \cdot \mathbf{A}$ and hence $\mathbf{H} \cdot \mathbf{g} \cdot \mathbf{A}$ determines the alignment of **I**. We neglect effects of the much smaller $\mathbf{H} \cdot \mathbf{I}$ interaction in determining the direction of **I**. When the assumed conditions prevail, the solutions already obtained for the magnetic interactions $\mathbf{H} \cdot \mathbf{g} \cdot \mathbf{S}$ and $\mathbf{S} \cdot \mathbf{A} \cdot \mathbf{I}$ can be used, and the quadrupole energies can be derived with usable accuracy from perturbation theory.

With the assumed conditions, **I** is aligned by the magnetic forces along **h**, as described elsewhere in Sections 4,e and 4.k and diagrammed in Fig. 3.3. In the primed system x',y',z' of Fig. 3.3 with z' along **h**, the effective quadrupole Hamiltonian $\mathcal{H}'_Q$ may be expressed as

$$\mathcal{H}'_Q = P'_{zz} I'^2_z + P'_{xx} I'^2_x + P'_{yy} I'^2_y \tag{3.173}$$

where P'_{zz}, P'_{xx}, and P'_{yy} are the diagonal elements of **P** with reference to the x',y',z' system. Although this is not the principal system for the quadrupole interaction, the cross terms such as $P'_{xz} I'_x I'_y$ are averaged out by the rapid precession of **I** about z' when the $\mathbf{h} \cdot \mathbf{I}$ interaction is large in comparison with $\mathbf{I} \cdot \mathbf{P} \cdot \mathbf{I}$. This results from the changing of the signs of I'_x and I'_y over a complete cycle of the precession. Contributions from the I'^2_x and I'^2_y terms will not be averaged out in this way, but these terms mutually cancel in the first-order treatment.

With an asymmetry parameter for the primed system defined by

$$\eta' = \frac{P'_{xx} - P'_{yy}}{P'_{zz}} \tag{3.174}$$

and with use of the relation, $P'_{xx} + P'_{yy} + P'_{zz} = 0$, (3.173) can be expressed in the form

$$\mathcal{H}'_Q = \frac{P'_{zz}}{2} [3I'_z - I(I+1)] + \frac{\eta' P'_{zz}}{2} (I'^2_x - I'^2_y) \tag{3.175}$$

Because (3.175) has the same form as (3.157), one can use the solutions already obtained for finding the energy values of $\mathcal{H}'_Q$. However, the first-order evaluation of $\mathcal{H}'_Q$ is easily obtained and easily applied with ENDOR for the finding of values of P'_{zz} as a function of the orientation of **H**, from which good approximate values of η as well as P_z can be obtained.

With z' as the axis of quantization, the diagonal elements of the $\mathcal{H}'_Q$ of (3.175) that correspond to the first-order energies are

$$E_Q^{(1)} = (P'_{zz}/2)[3M_I^2 - I(I+1)] \tag{3.176}$$

The last term of (3.175) vanishes in first order because the diagonal matrix elements of I'^2_x and I'^2_y are equal, and thus their difference is zero. This first-order formula obviously holds for any value of I, also for any value of S or M_S since it is independent of M_S.

The effecitve values P_{zz}' are related to the tensor elements of **P** by the following transformation. The direction of z', which is the direction **h** [see (3.133)] is given by

$$\frac{\mathbf{h}}{|\mathbf{h}|} = \frac{\mathbf{H}\cdot\mathbf{g}\cdot\mathbf{A}}{|HgA|} = \frac{\mathbf{l}\cdot\mathbf{g}\cdot\mathbf{A}}{|gA|} \tag{3.177}$$

where $\mathbf{l}$ is the unit vector in the direction of the applied field, that is, $\mathbf{H} = |\mathbf{H}|\mathbf{l} = |\mathbf{H}|(il_x + jl_y + kl_z)$. The transformation of the **P** tensor to the z' axis gives P_{zz}'

$$P_{zz}' = \frac{\mathbf{h}\cdot\mathbf{P}\cdot\mathbf{h}}{h^2} = \frac{(\mathbf{l}\cdot\mathbf{g}\cdot\mathbf{A})\cdot\mathbf{P}\cdot(\mathbf{l}\cdot\mathbf{g}\cdot\mathbf{A})}{(gA)^2} \tag{3.178}$$

Therefore, the first-order quadrupole energies for any orientation of the field when $P \ll A$ may be expressed by substitution of (3.178) in (3.176), as

$$E_Q^{(1)} = \frac{1}{2}[3M_I^2 - I(I+1)]\,\frac{(\mathbf{l}\cdot\mathbf{g}\cdot\mathbf{A})\cdot\mathbf{P}\cdot(\mathbf{l}\cdot\mathbf{g}\cdot\mathbf{A})}{(gA)^2} \tag{3.179}$$

In the general case, g is given by (3.96) and A, by (3.95). When x, y, and z are the principal axes of **g** and **A** but not of **P**, the transformation may be expressed as

$$P'_{zz} = \frac{1}{(gA)^2}[l_xg_xA_x,\ l_yg_yA_y,\ l_zg_zA_z]\begin{vmatrix} P_{xx} & P_{xy} & P_{xz} \\ P_{yx} & P_{yy} & P_{yz} \\ P_{xz} & P_{zy} & P_{zz}\end{vmatrix}\begin{Bmatrix} l_xg_xA_x \\ l_yg_yA_y \\ l_zg_zA_z\end{Bmatrix}$$

$$= \frac{1}{(gA)^2}[l_xg_xA_x)^2 P_{xx} + (l_yg_yA_y)^2 P_{yy} + (l_zg_zA_z)^2 P_{zz}$$

$$+\ 2(l_x g_x A_x)(l_y g_y A_y)P_{xy} + 2(l_x g_x A_x)(l_z g_z A_z)P_{xz} + 2(l_y g_y A_y)(l_z g_z A_z)P_{yz}] \qquad (3.180)$$

where $(gA)^2 = (l_x g_x A_x)^2 + (l_y g_y A_y)^2 + (l_z g_z A_z)^2$. The combining of the off-diagonal elements is possible because the **P** tensor is symmetric (i.e., $P_{yx} = P_{xy}$, etc.).

If x, y, and z are principal axes of **P** as well as of **g** and **A**, the cross terms in (3.180) drop out, and the transformation formula becomes

$$P'_{zz} = \frac{1}{(gA)^2}[(l_x g_x A_x)^2 P_x + (l_y g_y A_y)^2 P_y + (l_z g_z A_z)^2 P_z] \qquad (3.181)$$

which can be expressed in terms of the two independent parameters P_z and η as

$$P'_{zz} = \frac{P_z}{2(gA)^2}\{[3(l_z g_z A_z)^2 - (gA)^2] + \eta[(l_x g_x A_x)^2 - (l_y g_y A_y)^2]\} \qquad (3.182)$$

where η is the asymmetry parameter defined by (3.156). In these expressions, l_x, l_y, and l_z are the direction cosines of the applied field with the reference axes here assumed to be the principal axes of **A**. When the quadrupole coupling is axially symmetric, that is, when $\eta = 0$, the first-order formula for off-axis directions of **H**, (3.176), with P'_{zz} from (3.182) becomes

$$E_Q^{(1)} = \left(\frac{P_z}{4}\right)[3M_I^2 - I(I+1)]\left[\frac{3(l_z g_z A_z)^2}{(gA)^2} - 1\right] \qquad (3.183)$$

It is evident from (3.182) that the asymmetry parameter η as well as P_z can be evaluated from application of the first-order theory, even though the asymmetry parameter η' for the rotated system does not appear in the first-order formula, (3.176). This results from the fact that P_x and P_y as well as P_z contribute to the observable P'_{zz} values.

With the assumptions made, the various interaction energies are additive:

$$E_{M_S,M_I} = g\beta HM_S + AM_SM_I + E_{M_I}^{(H)} + E_Q^{(1)} \qquad (3.184)$$

where the general forms of **g** and **A** are given by (3.96) and (3.95), respectively, and where $E_{M_I}^{(H)}$ is given by (3.148).

As for measurements along the principal axes, there are no first-order effects of quadrupole interaction on the esr frequencies for any orientation of the crystal because of the $M_I \leftrightarrow M_I$ selection rule. However, the off-diagonal elements of the last term of (3.180) do not vanish but give rise to slight displacements of the esr lines and to weak, second-order esr transitions for off-axis orientations (see Chapter V, Section 4.b). The most important applications of the first-order formula, (3.179), are those to ENDOR spectra for which the radiofrequency measurements correspond to $M_I \leftrightarrow M_I \pm 1$ and the quadrupole term of (3.176) gives rise to frequency displacements of

$$(\Delta\nu)_Q = (3/2)P'_{zz}\,(2M_I \pm 1) \tag{3.185}$$

which is superimposed on that caused by the nuclear magnetic displacements.

In the experimental evaluation of the elements of the P tensor one can first evaluate the principal elements and directions of the principal axes of **g** and **A** from measurements of the normal esr transitions, $M_S \leftrightarrow M_S \pm 1, M_I \leftrightarrow M_I$, and can then apply (3.180) to measurements of $M_I \leftrightarrow M_I \pm 1$ transitions with ENDOR to obtain the tensor elements of **P**. The resulting **P** tensor is diagonalized in the usual manner to give the principal elements of the quadrupole coupling. If the principal axis of P_z is found to coincide with that of **g** and **A**, the more precise formulas of (3.161) to (3.163) or (3.169) to (3.172) can then be applied to measurements with **H** along this axis for improvement of the value of P_z.

By application of second-order theory to the $\mathcal{H}'_Q$ of (3.175), as was done in the $I = 3/2$ case for the similar $\mathcal{H}_Q$ of (3.157), or by solution of the secular equation, as was done for $I = 1$ in the preceding section, energy formulas involving η' as well as P'_{zz} can be obtained. These permit an evaluation of η' as well as P'_{zz} from the experimental data and an improved value for P'_{zz} from which P_z and η are derived. Since $\mathcal{H}'_Q$ has the identical form of $\mathcal{H}_Q$, one can use for this purpose the solutions already obtained for the z principal axis in the preceding section. The increased accuracy can, however, be more easily achieved in most instances by measurements along the principal axes after these axes are located approximately by application of first-order theory. For the axially symmetric $\mathcal{H}_Q$, Bleaney [3] gives an expression of the energies to second order for off-axis orientations of the applied field that holds when the nuclear magnetic interaction is much larger than the quadrupole interaction.

7. INTERATOMIC HYPERFINE INTERACTIONS

The magnetic dipole-dipole interaction of the nucleus of an atom X with the electron-spin density of a second atom Y has the same mathematical form, (2.35), as the dipole-dipole interaction already treated. The corresponding Hamiltonian operator has the same tensor form, $\mathcal{H}_\mu = \mathbf{S} \cdot \mathbf{A} \cdot \mathbf{I}$, as that already solved in Section 4, and the solutions obtained there may be applied in the analysis of observed hyperfine magnetic structure and in the experimental evaluation of the coupling constants. Only in the interpretation of the observed coupling constants does the analysis differ. The theoretical formulas for calculation of atomic coupling constants cannot, of course, be used for this type of coupling, nor can the tabulated atomic coupling constants.

When the **g** anisotropy is negligible, the effective coupling expressed in the principal axes of the **A** tensor is

$$A = (A_x{}^2 \cos^2 \theta_x + A_y{}^2 \cos^2 \theta_y + A_z{}^2 \cos^2 \theta_z)^{1/2} \tag{3.186}$$

where

$$\begin{aligned} A_x &= A_f + (A_\mu)_x \\ A_y &= A_f + (A_\mu)_y \\ A_z &= A_f + (A_\mu)_z \end{aligned} \tag{3.187}$$

and θ_x, θ_y and θ_z are the angles of the applied field **H** with the x, y, z principal axes of the dipole-dipole tensor $\mathbf{A}_\mu$. The isotropic component A_f always arises from s-orbital spin density on the coupling atom, but the anisotropic component A_μ may result from spin density on either the coupling atom, neighboring atoms, or both. Thus, after the principal elements $(A_\mu)_x$, $(A_\mu)_y$, and $(A_\mu)_z$ are experimentally evaluated, there remains the problem of determining their source. However, when the coupling atom is H, one can safely assume that the $\mathbf{A}_\mu$ components arise entirely from interaction with spin density on neighboring atoms, primarily with spin density on the atom to which the H is bonded. There is not likely to be enough $2p$ density on the H to give detectable anisotropic coupling. For other commonly observed coupling nuclei, the $\mathbf{A}_\mu$ term most generally arises from p orbital density on the coupling atom with a very small, usually negligible, contribution from spin density on other atoms. The relative contributions of these sources to the observed $\mathbf{A}_\mu$ tensor must be calculated or be estimated theoretically. This is easiest when the intra- and interatomic contributions to $\mathbf{A}_\mu$ have common principal axes, as is likely when both arise from spin density of the same π orbital.

Rigorous theoretical calculation of interatomic coupling, although straightforward, is exceedingly complicated. It involves the averaging of the dipole-dipole Hamiltonian of (2.35) over the orbital of the electron when the nucleus is not at the coordinate center of the orbital. With some simplifying assumptions, McConnell and Strathdee [4] have calculated the interaction for the important CH fragment. The formulas that they derive can be applied with adjusted parameters to certain related radicals such as the CF or CCl fragment. Simpler methods for calculation of approximate values are described in Chapter VI, Section 1.d.

8. PLURAL NUCLEAR COUPLING

The hyperfine structure of esr as calculated in Section 4 is for a single coupling nucleus. In many paramagnetic species, electron-spin density occurs on more than one atom and interacts with more than one nucleus. Unlike hyperfine interactions in molecular spectra of gases, plural nuclear interactions cause no complexity in the theory of hyperfine interactions in the esr of condensed matter. The interactions by the different nuclei are independent within the accuracy of resolution, and their displacements or splittings of the esr levels are simply additive. Thus the theory developed for individual nuclei is applicable and adequate for plural nuclear coupling. The only complications arise from the resolution and identification of individual components in the very complex hyperfine patterns that frequently occur when the multiplets of different nuclei are superimposed.

For simplicity, we omit in the present discussion the nuclear quadrupole and nmr interactions because in first order they do not influence the esr hyperfine spectra. However, these interactions are also independent and additive for the different nuclei. The magnetic hyperfine structure of a given electron-spin state M_S when there are n coupling nuclei may be expressed as

$$E_{M_S,M_{I_i}} = g\beta H M_S + M_S \sum_{i=1}^{n} A_i M_{I_i} \tag{3.188}$$

where I_i is the spin and A_i is the effective coupling constant of the ith nucleus. The frequencies of the esr transitions, $M_S \to M_S \pm 1, M_{I_i} \to M_{I_i}$, are

$$\nu = \nu_0 + \frac{1}{h} \sum_{i=1}^{n} A_i M_{I_i} \tag{3.189}$$

where $\nu_0 = (g\beta H_0/h)$. At constant frequency ν_0

$$H = H_0 + \frac{1}{g\beta} \sum_{i=1}^{n} A_i M_{I_i} \tag{3.190}$$

where $H_0 = (h\nu_0/g\beta)$, and for the ith nucleus M_{I_i} can have the values

$$M_{I_i} = I_i, I_i - 1, I_i - 2, \ldots, -I_i \tag{3.191}$$

The summation is taken over all coupling nuclei with a particular M_{I_i} value for each. The resonant frequencies, or resonant field values, are obtained from different combinations of particular M_{I_i} values of the different nuclei.

If the n coupling nuclei have identical coupling A, one can set

$$T = \sum_{i=1}^{n} I_i \tag{3.192}$$

and

$$M_T = T, T - 1, T - 2, \ldots, -T \tag{3.193}$$

Then the resonant field values at constant ν_0 are simply

$$H = H_0 + (\Delta H) M_T \tag{3.194}$$

where $\Delta H = (A/g\beta)$.

When there is no degeneracy and all components are separated, the number of hyperfine components in the esr spectrum will be the product

$$N = \prod_{i=1}^{n} (2I_i + 1) \tag{3.195}$$

When the n nuclei have the same spin value I and there is no degeneracy, the number of hyperfine components will be

$$N = (2I + 1)^n \tag{3.196}$$

When the different nuclei have equivalent couplings, the number of hyperfine components will be

$$N = 2T + 1 \tag{3.197}$$

where T is the sum of the spin values. If, in addition, the n equally coupling nuclei have the same spin value I, the number of hyperfine components will be

$$N = 2nI + 1 \tag{3.198}$$

For example, four nuclei with spins of 1/2, 1, 3/2, and 5/2, respectively, with inequivalent coupling can give rise to an esr hyperfine pattern having a maximum of 144 components. When all four nuclei have equal coupling, the number of components is only 12. Six protons, $I = 1/2$, with inequivalent coupling, can give rise to 64 components; when all six have equivalent coupling, there are only seven components. In general, n nuclei with $I = 1/2$ and with equivalent couplings give rise to a hyperfine structure of $n + 1$ components.

Figure 4.2 illustrates the energy-level diagram and esr frequency spectrum obtained when there are two nuclei with $I = 1$ having different coupling ratios, A_1/A_2. This example shows how qualitatively different esr patterns are obtained simply by alteration of the relative coupling of two nuclei having the same spin. Under conditions of thermal equilibrium, the intensities of the components due to a particular nucleus have equal strength within the accuracy of the measurements. The differences in line strength illustrated in Fig. 4.2 arise from superposition of hyperfine components of the different nuclei. Line intensities are treated theoretically in Chapter IV. Composite hyperfine structure is illustrated by Figs. 4.3 and 4.4 and by many applications in later chapters.

References

1. E. Feenberg and G. E. Pake, *Notes on the Quantum Theory of Angular Momentum,* Addison-Wesley, Cambridge, Mass., 1953.
2. G. Breit and I. Rabi, *Phys. Rev.,* **38**, 2082 (1931).
3. B. Bleaney, *Phil. Mag.,* **42**, 441 (1951).
4. H. M. McConnell and J. Strathdee, *Molec. Phys.,* **2**, 129 (1959).

Chapter **IV**

LINE STRENGTHS, LINE SHAPES, AND RELAXATION PHENOMENA

The time-dependent phenomena considered in this chapter are treated partly with classical mechanics, partly with quantum mechanics, and in some aspects with a combination of the two. The quantum-mechanical approach is most advantageous for derivation of selection rules, transition probabilities, and line strengths, whereas classical or semiclassical methods are most advantageous in description of the resonance process and in treatment of relaxation phenomena.

The interactions that we have treated so far have been those between spin magnetic moments and a static magnetic field. We have derived characteristic energies of stationary states of the spin moments in applied static fields and calculated transition frequencies between these stationary states. However, we have not discussed the causes or mechanisms of such transitions. These form the subject of this chapter.

As is true for all other spectroscopy, transitions between spin resonance states are induced by electromagnetic radiation having the frequency required by the Bohr relation, but there are differences, although not basic ones. Only radiation-field components that are perpendicular to the applied static field can induce the normal magnetic resonance transitions, whereas the field components inducing transitions between intrinsic (zero-field) states of a gas are independent of space-fixed directions. Most, but not all, spectral transitions of gaseous molecules are induced by a coupling of the electric dipole moment with the electric component of radiation, whereas transitions in esr must always be induced by magnetic dipole coupling since the electron spin has no associated electric moment. Electron spin resonance transitions are induced only by alternating-current (ac) magnetic field components that are normal to the space-fixed direction of the static magnetic field.

1. QUANTUM-MECHANICAL TRANSITION PROBABILITIES

1.a. Electron-spin Transitions

We first consider probabilities of transitions between two electron spin states without hyperfine structure. After the overall transition strength has been calculated, the relative intensities of hyperfine components are separately derived.

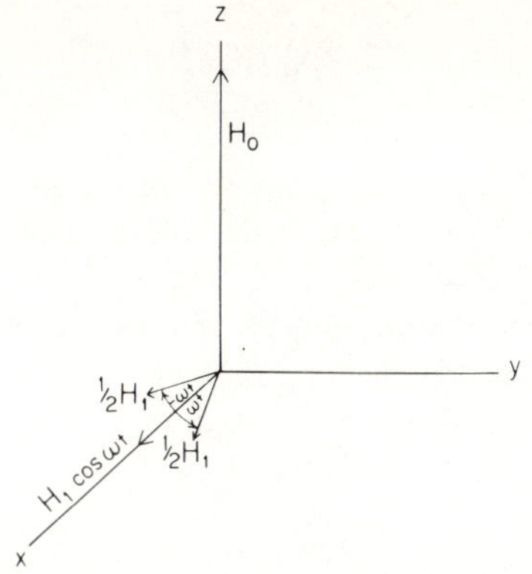

Fig. 4.1 Diagram showing relative orientations of the static magnetic field and the alternating components of the radiation field imposed to induce magnetic resonance. The ac field, H_1 cos ωt, along the x axis is equivalent to the sum of two components of amplitude (½) H_1 rotating in opposite sense in the xy plane, only one of which is effective in inducing transitions.

Let us assume that a static magnetic field $\mathbf{H}_0$ imposed on an electron-spin system is along the z axis and that a plane-polarized radiation field is imposed in the xy plane, with the alternating magnetic component along the x axis, as indicated in Fig. 4.1. There is no loss of generality in this choice of coordinates since the only requirement is that the effective field component must be perpendicular to $\mathbf{H}_0$. The spin Hamiltonian for this system is

$$\mathcal{H}_S = \mathcal{H}_S{}^0 + \mathcal{H}_S{}^1(t) \tag{4.1}$$

where $\mathcal{H}_S{}^0$ is the static-field Hamiltonian already evaluated in Chapter III and $\mathcal{H}_S{}^1(t)$ represents the time-dependent interaction of the radiation field with the spin system in the static field. The alternating magnetic-field component that gives rise to $\mathcal{H}_S{}^1(t)$ will normally depend on the type of resonant cavity or the nature of the radiation mode in which the sample is placed, but for plane-polarized radiation with the magnetic component along x, it may be written as

$$H = H_x = H_1 \cos \omega t = \frac{1}{2} H_1 (e^{i\omega t} + e^{-i\omega t}) \tag{4.2}$$

where H_1 is the amplitude of the magnetic component assumed to be a constant in these calculations and where $\omega = 2\pi\nu$ is the angular frequency of the radiation. With the assumed conditions,

$$\mathcal{H}_S{}^1(t) = \mu_x H_x = \frac{1}{2} g\beta S_x H_1 \, (e^{i\omega t} + e^{-i\omega t}) \tag{4.3}$$

where $\mu_x = g\beta S_x$. For the present purpose, we can neglect the usually small **g** anisotropy. Normally, $H_1 \ll H_0$, and hence $\mathcal{H}_S{}^1(t) \ll \mathcal{H}_0$; therefore, $\mathcal{H}_S{}^1(t)$ can be treated as a first-order perturbation on the eigenstates of $\mathcal{H}_0$. This time-dependent perturbation induces transitions between the eigenstates of $\mathcal{H}_S{}^0$, as described in the paragraphs that follow.

A spin having the state m with wave function ψ_m at $t = 0$ has, as a result of the perturbation $\mathcal{H}_S{}^1(t)$, the probability

$$p_{mn}(t) = a_n(t)\, a_n{}^*(t) \tag{4.4}$$

of undergoing a transition to the state ψ_n where a_n is given by first-order, time-dependent perturbation theory [1] to be

$$a_n(t) = -\frac{i}{\hbar} \int_\sigma \int_0^t \Psi_m \mathcal{H}_S{}^1(t)\, \Psi_n{}^*\, d\sigma\, dt \tag{4.5}$$

With

$$\Psi_m = \psi_m(x,y,z) \exp \frac{-iE_m t}{\hbar}, \quad \Psi_n = \psi_n(x,y,z) \exp \frac{-iE_n t}{\hbar} \tag{4.6}$$

$$\omega_{mn} = \frac{E_n - E_m}{\hbar} \tag{4.7}$$

and with $\mathcal{H}_S{}^1(t)$ given by (4.3), the time and space coordinates are easily separated and (4.5) expressed as

$$a_n(t) = -\frac{i}{\hbar} \frac{g\beta H_1}{2} (m|S_x|n) \int_0^t \left| \exp[i(\omega + \omega_{mn})t] + \exp[-i(\omega - \omega_{mn})t] \right| dt \tag{4.8}$$

where $(m|S_x|n) = \int_\sigma \psi_m S_x \psi_n{}^*\, d\sigma$. On integration, (4.8) becomes

$$a_n(t) = \frac{g\beta H_1}{2\hbar} (m|S_x|n) \left\{ \frac{1 - \exp[i(\omega + \omega_{mn})t]}{\omega + \omega_{mn}} + \frac{1 - \exp[-i(\omega - \omega_{mn})t]}{\omega - \omega_{mn}} \right\} \tag{4.9}$$

It is evident that the numerator of each of the terms within the brace of (4.9) cannot be greater than 2. Hence neither term is significantly large unless its denominator is small. When ω and ω_{mn} are alike in sign, the first term is negligible and the second is significant; when they are unlike in sign, the first term becomes important and the second, negligible. However, only the component that has the same rotational sense as the resonance (like signs of ω and ω_{mn}) is effective in inducing magnetic resonance transitions (see Sections 7 and 8). Thus we can omit the first term and set

$$a_n(t) = \frac{g\beta H_1}{2\hbar}(m|S_x|n)\,\frac{1-\exp[-i(\omega-\omega_{mn})t]}{\omega-\omega_{mn}} \tag{4.10}$$

With the designation of $\theta = (\omega - \omega_{mn})t$, this can be written as

$$a_n(t) = \frac{g\beta H_1}{2\hbar}(m|S_x|n)t\,\frac{1-\cos\theta + i\sin\theta}{\theta} \tag{4.11}$$

and hence

$$a_n(t)a_n{}^*(t) = \frac{(g\beta H_1)^2}{4\hbar^2}\,|m|S_x|n)|^2t^2\left[\frac{\sin^2\frac{1}{2}\theta}{\frac{1}{2}\theta^2}\right] \tag{4.12}$$

Equation (4.12) gives the probability of an induced transition in time t when the inducing radiation has a particular frequency $\nu = \omega/2\pi$. To get the total or integrated probability of transition when the sample is exposed to radiation of uniform intensity at all frequencies, one must integrate this expression over all frequencies from $-\infty$ to $+\infty$ with H_1 held constant. To do this, note that $d\nu = (1/\pi t)d(1/2)\theta)$ and that

$$\int_{-\infty}^{\infty}\frac{\sin^2\frac{1}{2}\theta}{\left(\frac{1}{2}\theta\right)^2}\,d\nu = \frac{2}{\pi t}\int_0^{\infty}\frac{\sin^2\frac{1}{2}\theta}{\left(\frac{1}{2}\theta\right)^2}\,d\left(\frac{1}{2}\theta\right) = \frac{1}{t} \tag{4.13}$$

The last integral is a well known form that has the value $\pi/2$. Since the other factors of (4.12) are independent of frequency, the integrated probability for the transition in t seconds is

$$p_{mn}(t) = \frac{\pi^2}{h^2}(g\beta H_1)^2\,|(m|S_x|n)|^2t \tag{4.14}$$

where h is Planck's constant.

The probability of an induced transition in unit time is often expressed as the Einstein B coefficient times the inducing radiation density ρ_{mn}. Thus it is of interest to express H_1^2 in terms of the energy denisty. With plane-polarized radiation having the magnetic component entirely along x, as we have assumed, the radiation density $\rho_{mn} = H_1^2/8\pi$. Therefore,

$$p_{mn}(t{=}1) = B_{mn}\rho_{mn} = \frac{8\pi^3}{h^2}\rho_{mn}(g\beta)^2\,|(m|S_x|n)|^2 \tag{4.15}$$

which gives the Einstein coefficient

$$B_{mn} = \frac{8\pi^3}{h^2}(g\beta)^2\,|(m|S_x|n)|^2 \tag{4.16}$$

Equation (4.16) is consistent with the usual expression for the Einstein B coefficient for isotropic absorbers or emitters such as field-free gaseous molecules in isotropic radiation, which is

$$B_{mn} = B_{nm} = \frac{8\pi^3}{3h^2}\left[|(m|\mu_x|n)|^2 + |(m|\mu_y|n)|^2 + |(m|\mu_z|n)|^2\right] \tag{4.17}$$

where μ_x, μ_y, and μ_z are the matrix elements of the dipole-moment operator μ resolved along space-fixed axes. For the isotropic conditions, only a third of the radiation would be effective in inducing transitions by the μ_x operator, whereas in deriving (4.16) we have assumed that the polarization is such that all the radiation is effective along x.

Because the dipole-moment operator is Hermitian, $|m|\mu_x|n)|^2 = |(n|\mu_x|m)|^2$. Likewise $|(m|S_x|n)|^2 = |(n|S_x|m)|^2$ and (4.16) or (4.17) shows that the coefficients of stimulated absorption and stimulated emission between the same levels are exactly equal, that is, $B_{mn} = B_{nm}$, as originally postulated by Einstein. This equivalence has important consequences for all spectroscopy because it signifies that there can be no net change in the energy of the radiation field from stimulated transitions except when there is a difference in population of the two levels m and n. Thus the relaxation processes (Section 9) that tend to maintain a difference in population of the levels are essential to the observation of spectral lines.

It is easily shown from Planck's radiation law and Boltzmann's law (see, for example, Ref. 2, Chapter 3) that the Einstein coefficient of spontaneous emission A_{nm} is related to the B coefficient by

$$A_{nm} = \frac{8\pi h\nu_{mn}{}^3}{c^3}B_{nm} = \frac{8\pi h}{\lambda^3}B_{nm} \tag{4.18}$$

The probability of spontaneous emission, unlike that of stimulated emission, does not depend on the radiation to which the sample is exposed. For this reason and because of the inverse cube variation with wavelength of the emitted radiation, spontaneous emission is a negligible factor in microwave esr spectroscopy.

1.b. Nuclear-spin Transitions

The probability of an induced nuclear-spin transition is calculated in the same manner as that for an electron-spin transition. With the same conditions assumed, (4.15) can be used, with a substitution of the nuclear dipole moment $\mu_I = g_I\beta_I I_x$. Thus the B coefficients for nmr may be expressed as

$$B_{M_I',M_I} = B_{M_I,M_I'} = \frac{8\pi^3}{h^2} (g_I\beta_I)^2 (M_I|I_x|M_I')^2 \tag{4.19}$$

Numerically, for most nuclei $g_I \approx g_S$, and the B coefficient for nuclear-spin transitions is a factor of $(\beta_I/\beta)^2 = 3 \times 10^{-7}$, smaller than that for the esr transitions. Nevertheless, nmr transitions are detectable! They are detectable partly because this transition probability is concentrated in a small frequency band (sharp lines) and partly because the $\mathbf{H}_1$ field or power density can be made very large in the radiofrequency region, thereby enhancing the probability of transitions. In normal esr, where the esr and nmr states are not intermixed by off-diagonal elements in the nuclear coupling, the probability of simultaneous flipping of the nuclear spin and the electron spin is negligibly small. However, the nuclear-spin transitions are important in ENDOR experiments, and certainly in nmr.

2. SELECTION RULES

The transition probabilities derived in Section 1 depend on the spin-dipole matrix elements connecting the states between which the transitions occur, and the vanishing or nonvanishing of these elements is the basis of the selection rules for magnetic resonance transitions. Since the magnetic dipole moments are proportional to the spin moments, the selection rules are determined by the matrix elements of the spin components. In magnetic resonance experiments, the spin ($\mathbf{S}$ or $\mathbf{I}$) is quantized along a space-fixed axis defined by the imposed static field $\mathbf{H}_0$, which is customarily chosen as z or Z. The nonvanishing elements of S_z or of I_z are thus diagonal and connect the same states. The spin-dipole matrix elements that give rise to transitions between states of different energy are the off-diagonal ones. When z is the axis of quantization, the transition must result from the nonvanishing elements of the x or y components. The only nonvanishing elements of these components are off-diagonal by ± 1. These give rise to the selection rules for electron-spin transitions,

$$M_S \leftrightarrow M_S \pm 1, \quad M_S \nleftrightarrow M_S \tag{4.20}$$

and, similarly, for nuclear-spin transitions,

$$M_I \leftrightarrow M_I \pm 1, \quad M_I \nleftrightarrow M_I \tag{4.21}$$

When straight esr is observed, only the $\Delta M_S = \pm 1$ rule is important since $M_S \leftrightarrow M_S$ corresponds to zero energy change. Similarly, when straight nmr is observed, only the rule $\Delta M_I = \pm 1$ is significant. However, when the spin Hamiltonian involves both **S** and **I**, a combination of changes in M_S and M_I are involved, and $\Delta M_I = 0$ becomes significant. In conventional esr with hyperfine structure, the appropriate selection rules are

$$M_S \leftrightarrow M_S \pm 1, \quad M_I \leftrightarrow M_I \tag{4.22}$$

The $\Delta M_I = 0$ results from the very low probability of a flipping of the nuclear spin in comparison with that of the electron spin because $\mu_I << \mu_S$, as explained earlier.

When a strong radiation field is applied at the nuclear hyperfine frequency simultaneously with the microwave field at the esr frequency, as in ENDOR, both nuclear-spin and electron-spin transitions are induced. In terms of the combined quantum numbers $M = M_S + M_I$, the selection rules are

$$M \leftrightarrow M \pm 2, \quad M \leftrightarrow M \tag{4.23}$$

The selection rules of (4.23) also correspond to "forbidden" second-order transitions of esr where the electron-spin and nuclear-spin states are intermixed. The second-order interactions that give rise to forbidden transitions are described in later sections.

3. LINE STRENGTHS

In Section 2 we calculated the probability p_{mn} that a spin **S**, originally in magnetic state m would undergo a transition to state n in unit time when exposed to resonance radiation. If the number of such spins in the lower state m is N_m, the total number of spins expected to undergo such transitions per second is $N_m p_{mn}$, each of which absorbs an energy quantum $h\nu_{mn}$. Thus the total energy absorbed per second, the power absorption, is

$$P_{mn} = N_m p_{mn} h\nu_{mn} \tag{4.24}$$

If N_n is the number of such spins in the upper spin state n, the probable number undergoing the opposite transition, $n \rightarrow m$, per second is $N_n p_{nm}$, and the number undergoing spontaneous emission per second will be $N_n A_{mn}$, where A_{mn} is given by (4.18). Thus the total power emitted by the spin in state n is expected to be

$$P_{nm} = N_n p_{mn} h\nu_{mn} + N_n A_{mn} h\nu_{mn} \tag{4.25}$$

The stimulated emission will be in phase with the stimulating radiation and hence with the absorbed power. However, the radiation spontaneously emitted will have random phase and be undetectably small in the microwave region. Therefore, we can neglect the last term of (4.25); since $p_{nm} = p_{mn}$ (see Section 1), the net power change in the radiation field is predicted to be

$$\Delta P_{mn} = P_{nm} - P_{mn} = (N_n - N_m) p_{mn} h\nu_{mn} \tag{4.26}$$

where p_{mn} is given by (4.14) or (4.15). From (4.26) it is evident that there will be no net change of power in the radiation field when $N_m = N_n$. Thus observable absorption or emission of power requires a difference in population of the two levels. When population of the lower level is the greater, $N_n < N_m$, there will be a net loss of power (absorption); when $N_m < N_n$, there will be a net emission of power (ΔP positive).

Since esr spectroscopy is generally observed in absorption, we deal for the most part with systems in which the lower level has the greater population. Absorption spectroscopy is the more common and the more convenient type to observe because thermal processes tend to maintain a greater number of particles in the lower state. Nevertheless, various methods are now known to achieve a greater population in the upper state and thus make possible stimulated emission spectroscopy for selected systems. Under conditions of thermal equilibrium, the difference in population of nondegenerate states n and m is given by the well-known Boltzmann law:

$$\frac{N_n}{N_m} = \exp \frac{-h\nu_{mn}}{kT} \tag{4.27}$$

where k is Boltzmann's constant and T is the absolute temperature. With this relation and (4.26), the net absorbed power may be expressed as

$$P_{\text{abs}} = -\Delta P = N_m \left(1 - \exp \frac{-h\nu_{mn}}{kT}\right) p_{mn} h\nu_{mn} \tag{4.28}$$

Equation (4.28) must be further refined before it can be usefully applied in esr absorption spectroscopy. For various reasons to be discussed in Section 4, the electronic spins of a paramagnetic sample absorb over a band of frequencies that is usually much greater than the frequency spread of the source power employed in esr experiments. In fact, the frequency band of the source power is usually so narrow that it can be considered as a monochromatic source in comparison with the much wider band absorbed by the spin system. To calculate the power absorbed at a particular frequency, one must multiply (4.28) by a normalized shape function that gives the relative absorption at different frequencies. Since the total or integrated absorption by all the spins will not be changed by the spreading of the transition frequencies into a band, the shape function $f(\nu - \nu_0)$ must be normalized by

$$\int_0^\infty f(\nu - \nu_0)d\nu = 1 \tag{4.29}$$

Here ν is the variable frequency and ν_0 is the frequency of maximum resonance. The shape function may have different forms for different spin systems, as described in Section 4.

When plane-polarized source power is imposed normal to the static field $\mathbf{H}_0$, as described in Section 4, the power absorption of frequency ν by a spin system maintained at thermal equilibrium may be expressed by

$$P_{\text{abs}}(\nu) = \frac{\pi^2}{h}(g\beta H_1)^2(m|S_x|n)^2 N_m\left[1 - \exp\frac{-h\nu_{mn}}{kT}\right]\nu f(\nu-\nu_0) \tag{4.30}$$

This expression is obtained from (4.28) by substitution of the value of p_{mn} from (4.14) and multiplication by the shape function. The matrix elements of S_x are given by (3.12), and expressions for $f(\nu - \nu_0)$ and N_m are given in Sections 4 and 6. For most purposes we can use the approximations

$$N_m \approx \frac{N}{2S + 1} \tag{4.31}$$

and

$$1 - \exp\frac{-h\nu}{kT} \approx \frac{h\nu}{kT} \tag{4.32}$$

Then, for the $M_S \leftrightarrow M_S - 1$ transition when thermal equilibrium is maintained,

$$P_{\text{abs}}(\nu) = \frac{N\pi^2(g\beta H_1)^2\,\nu^2\,[S(S+1) - M_S(M_S - 1)]}{4kT\,(2S+1)}\,f(\nu - \nu_0) \tag{4.33}$$

where N represents the total number of spins in the paramagnetic sample.

For unsaturated, homogeneously broadened lines, the Lorentzian shape function, (4.40), is usually a good approximation to $f(\nu - \nu_0)$. This function has the peak value of $1/(\pi \Delta\nu_{1/2})$, (4.41). Under these conditions, the power absorption at the peak resonant frequency is given by

$$P_{\max}(\nu_0) = \frac{N\pi(g\beta H_1)^2 \nu_0^2 [S(S+1) - M_S(M_S-1)]}{4kT(2S+1)(\Delta\nu)_{1/2}} \tag{4.34}$$

where $\Delta\nu_{1/2}$ is the line half-width.

One of the important uses of the power absorption equation is for measurement of the concentration of paramagnetic elements in various materials. To make absolute measurements, one must express the amplitude H_1 in terms of the measurable oscillator power. The latter problem depends on experimental details such as the type of resonant cavity employed in the spectrometer, its operating mode and Q value. Such information may be found in other treatments that concentrate on experimental methods [3,4].

Measurement of the ratio of signal strength of a sample containing an unknown quantity of spins to the signal strength of a calibrated sample containing a known quantity of spins is commonly used for finding the value of N. Unless the two signals are identical in width and shape, as is unlikely, the integrated signals $\int P_{abs}(\nu)d\nu$ must be compared. This is done by comparison of the areas under the two absorption curves obtained with known relative gain and sensitivity of the spectrometer. With spectrometer sensitivity held constant and with a calibrated gain control, one can adjust the gain control so that the signals are of equal or comparable strength. For many substances this relative method is not only simpler, but also more accurate than are those methods that depend on measurement of absolute power absorption for evaluation of N. Calibrated samples of stable free radicals containing known numbers of free radicals, such as 1,1-diphenyl-2-picrylhydrazyl (DPPH) can be chemically prepared. Physical methods for measurement of the concentration of free radicals in irradiated samples are described by Köhnlein and Müller [5].

3.a. Relative Intensities of Hyperfine Components

Except for a small, usually negligible, difference caused by the Boltzmann factor, the $2I + 1$ spin states of a given nucleus are equally populated under conditions of thermal equilibrium (see Section 6). Thus when there is only one coupling nucleus in the paramagnetic element, the hyperfine components of the esr transition $\Delta M_S = \pm 1$, $\Delta M_I = 0$ are expected to be of equal intensity. For example, ^{14}N with a spin of unity would split the esr resonance into a multiplet of three equally intense components. The Mn nucleus with a spin of 5/2 would give rise to $2(5/2) + 1 = 6$ equally intense components. Each component has strength $1/(2I + 1)$ that of the total esr transition predicted by (4.34).

When the paramagnetic element or free radical has more than one coupling nucleus, the relative intensities of the components are not necessarily equal because of the possibility that the components arising from different nuclei may fall at the same frequency. For instance, two equally coupling hydrogen nuclei (with $I = 1/2$) will give a triplet having intensities in the ratio 1:2:1, whereas two unequally coupling hydrogens will give four equally strong components. Two equally coupling ^{14}N nuclei (with $I = 1$) will give a quintet of intensity ratio 1:2:3:2:1, whereas two unequally coupling ^{14}N nuclei with one coupling appreciably greater than the other will give nine equal components, a triplet with a triplet substructure. Some coupling combinations are illustrated by diagrams in Fig. 4.2. If there are n coupling nuclei and no degeneracy or coinciding components, each hyperfine component will have strength $1/(2I_1 + 1)(2I_2 + 1) \cdots (2I_n + 1)$ that of the total integrated intensity of the esr transition.

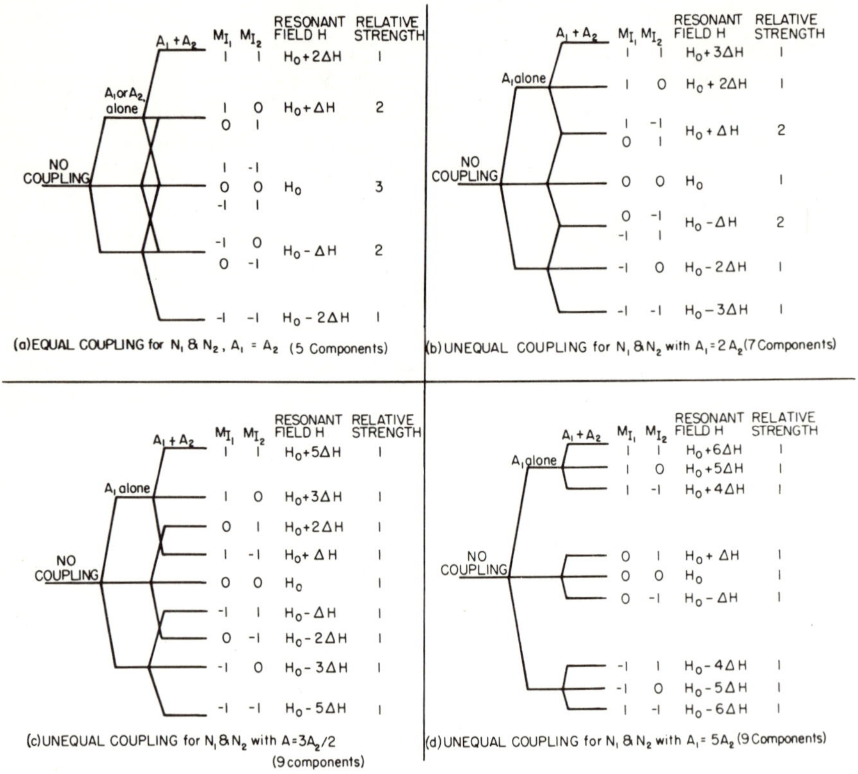

Fig. 4.2 Diagrams of the composite esr hyperfine structure of two nuclei, N_1 and N_2, both with spin $I = 1$ but with different magnetic coupling constants, A_1 and A_2. These diagrams show how the number of hyperfine components as well as their relative strengths and spacings depends on the ratios of the separate splittings by the nuclei.

One often encounters equal coupling by three, and sometimes by several, protons ($I = 1/2$) in organic free radicals. It is easy to calculate the relative intensities of such a multiplet or submultiplet from probability theory. When I equals 1/2, each nucleus must point up (when $M_I = 1/2$) or down (when $M_I = -1/2$) in the static magnetic field. The outside components of the multiplet are given by the combination of the n nuclei pointing in the same direction, all up or all down. There is only one combination for "all up," and one for "all down." Thus the outside components are assigned unit intensity. The component next to these will have all of the nuclei except one pointing in the same direction. Of the n nuclei, $r = n - 1$ must point in the same direction in this case. The number of ways this can happen is simply the number of possible combinations of n things taken $n - 1$ at a time. For the next component, $r = n - 2$ of the nuclei must point in the same direction, and so on. Generally, the number of combinations of n things taken r at a time is

$$ {}_nC_r = \frac{n!}{r!(n-r)!} \tag{4.35} $$

As an illustration, consider six equally coupling hydrogens:

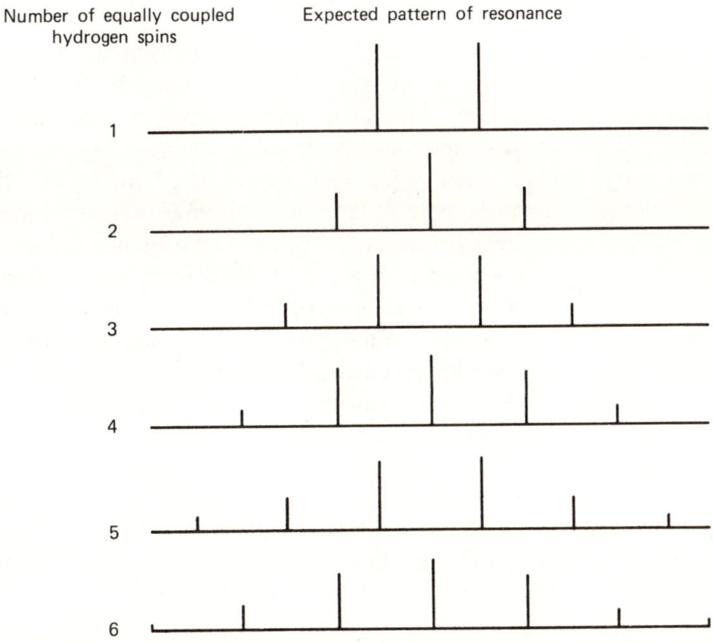

Fig. 4.3 Types of composite hyperfine structure predicted for different numbers of equally coupling nuclei with spin $I = ½$.

$${}_6C_6 = 1, \qquad {}_6C_5 = \frac{6!}{5!} = 6$$

$${}_6C_4 = \frac{6!}{4!\,2!} = 15, \qquad {}_6C_3 = \frac{6!}{3!\,3!} = 20$$

Thus there will be a septet with relative intensities 1:6:15:20:15:6:1. Note that the intensity ratio of the first two components is 1:6. In general, the intensity ratio of the first two components is $1{:}n$ for n equally coupling nuclei with $I = 1/2$. The unit intensity of the outside components corresponds to $1/(2I + 1)^6 = 1/64$ of the summed intensity of all components of the multiplet. Figure 4.3 shows the hyperfine structure expected for various configurations of equally coupling nuclei, all with $I = 1/2$.

For combinations of nuclei with different couplings or spins, one can easily find the relative intensities by making simple diagrams of the relative displacements by different nuclei and counting the components that coincide. This method is illustrated for nuclei with $I = 1$ in Fig. 4.2. In a similar way, the relative intensities were obtained for different coupling combinations of nuclei with spin 1/2, shown in Fig. 4.4.

4. LINE WIDTHS AND SHAPES

Electron-spin-resonance contours vary in form and width because of a number of factors. Some of these factors are inherent in the physical or chemical state of the paramagnetic substance, and some are instrumental distortions. The resonances of powdered samples are considerably distorted by anisotropy in **g** or in nuclear coupling. These effects are treated in Chapter VIII. Lines of oriented radicals in single crystals are relatively free of these anisotropic effects but not completely because of the imperfect alignment of the radicals or because of impurity centers in the crystal. Unresolved hyperfine structure is a common contributor to resonance shape and width for solids, both powders and single crystals. In samples for which the paramagnetic elements are concentrated, the resonances are broadened by dipole-dipole interaction between the spin moments of neighboring elements but may be sharpened by exchange interaction between the spins [6]. Paramagnetic impurities that differ from those being directly observed also can cause distortion, particularly through alteration of the spin-lattice relaxation time. Instrumental distortion may result from inhomogeneities in the static magnetic field $\mathbf{H}_0$, saturation of the spin system with excessive power, nonlinearity of the amplifier or detector, also from various modulation schemes used to enhance detection.

4.a. Homogeneous and Inhomogeneous Broadening

It is important, particularly for an understanding of power saturation effects,

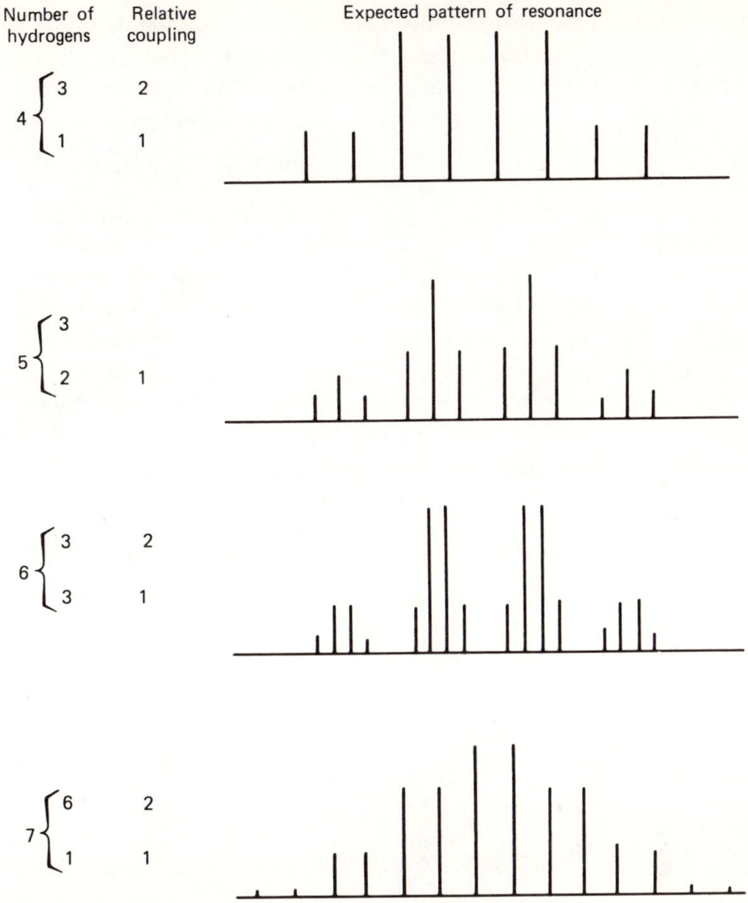

Fig. 4.4 Illustrative patterns for two sets of $I = \frac{1}{2}$ nuclei. All nuclei of each set have the same coupling, but the couplings of the two sets differ as indicated. The marked differences in the patterns arise from degeneracy of certain components, as demonstrated for $I = 1$ nuclei in Fig. 4.2.

to distinguish between homogeneous and inhomogeneous broadening. In a homogeneously broadened line, all spins in the sample respond to, and may absorb, radiation quanta of any frequency within the frequency-response band of the total spin system; in other words, the spins of the sample respond as a unit. When a line is inhomogeneously broadened, only certain groups of spins or spin packets in the sample can directly absorb quanta of a particular frequency within the band of frequencies absorbed by the total sample. Portis [7] gives a theoretical treatment of the two classes of broadening and lists the following sources that give rise to each. Sources of homogeneous broadening include: (1)

dipolar interaction between like spins, (2) spin-lattice relaxation, (3) interaction with the radiation field (saturation broadening), (4) motion of spins in the microwave fields, (5) diffusion of spin excitation through the sample, and (6) motionally narrowed fluctuations in the local field. Sources of inhomogeneous broadening include: (1) hyperfine interaction, (2) anisotropic broadening (broadening due to anisotropy in **g** or **A** tensors), (3) dipolar interaction between spins with different Larmor frequencies, and (4) inhomogeneities in applied magnetic fields.

The homogeneously broadened line has a symmetrical, approximately Lorentzian shape [see (4.40)], when not broadened by excessive power (Section 5). The inhomogeneously broadened line may be considered as comprised of many separate, homogeneously broadened lines of spin packets that have their peak absorptions at slightly different frequencies because of differences in local field values within the sample. An inhomogeneously broadened line, when not caused by anisotropic interactions, more often approaches the Gaussian line shape [see (4.57)].

4.b. Dependence of Line Breadth on State Lifetime: Natural Line Width

When the distortion and broadening due to unresolved structure, **g** and **A** anisotropy, and the various instrumental effects are eliminated, the line shape and width are determined essentially by the lifetime in the states involved in in the transitions. The lifetime in the state and the spread of the transition energy are related by the uncertainty principle

$$\Delta \mathrm{t} \cdot \Delta \mathrm{E} \approx \hbar \tag{4.36}$$

The associated frequency spread, or line width, is

$$\Delta\nu = \frac{\Delta E}{h} \approx \frac{1}{2\pi(\Delta t)} \tag{4.37}$$

Equation (4.37) provides a basis for theoretical prediction of the "natural line width." If a spin system in a state n were so nearly isolated that the lifetime in the state was limited mainly by spontaneous emission, the mean lifetime of the state would be approximately

$$\Delta t \approx \frac{1}{A_{nm}} \tag{4.38}$$

where A_{nm} is the Einstein coefficient of spontaneous emission given by (4.18). If m were the only lower state to which the system could fall, the approximate width due to spontaneous emission would be

$$\Delta\nu = \frac{1}{2\pi} A_{nm} = \frac{32\pi^3 \nu_{mn}^3}{3hc^3} |(m|\mu|n)|^2 \tag{4.39}$$

For a doublet resonance state $S = 1/2$, the squared dipole-moment matrix elements are $(1/2)(g\beta)^2 = 2\beta^2$. In this case the natural line width is $\Delta\nu \approx 6.4 \times 10^{-42}$ ν_{mn}^3. For an esr frequency in the 3-cm wave region where $\nu = 10^{10}$ Hz, the natural line width is $\Delta\nu \approx 6.4 \times 10^{-12}$ Hz. It is evident that the natural line width (*i.e.,* the line spread due to spontaneous emission) is unobservably small at microwave and radiofrequencies.

Although spontaneous emission can be safely neglected, other processes that terminate lifetime in magnetic resonance states are important factors in determining the frequency spread of the resonance. In microwave spectroscopy of gases, the lifetime in a state is limited mainly by collisions between molecules. The collision-broadened line has an essentially Lorentzian shape [8] or, more precisely, a VanVleck-Weisskopf line shape [9] and a width determined by the collision time. In esr of solids there are no such collisions, but the lifetime in the esr states is determined mainly by thermal motions and atomic oscillations or vibrations, which have a statistical distribution similar to that of the thermal motions in the gas. These thermal motions are coupled through residual spin-orbit coupling or spin-dipole coupling with the electron-spin moment and can thus cause transitions that interrupt the spin states. There will be a distribution of state lifetimes depending on the distribution of atomic vibrational or phonon frequencies and the associated frequency distribution of the varying internal magnetic fields effective on the different spins. Interruptions of the states result from induced transitions by components of the fluctuating local fields. However, the resonance can also be broadened by a spread in the static, or slowly varying, internal fields that causes the resonant frequencies of spins of the system to differ without causing transitions. Obviously, the various factors that determine the line shapes and widths in magnetic resonance of solids and liquids are very different from those that produce the collision-broadened or Doppler-broadened lines of gases. Nevertheless, when free of the anisotropic effects of **g** and **A**, these lines are often found to have resonances approximately Lorentzian in shape (like collision broadening) and sometimes approximately Gaussian in shape (similar to the Doppler-broadened lines of gases) and sometimes a shape that is a composite of the two (like a Doppler- and pressure-broadened gaseous line).

4.c. Lorentzian Line Shape

In frequency units, the Lorentzian line-shape function has the form [8]

$$f(\nu - \nu_0) = \frac{1}{\pi} \frac{\Delta\nu_{1/2}}{(\nu-\nu_0)^2 + (\Delta\nu_{1/2})^2} \tag{4.40}$$

which is normalized by (4.29). Here ν_0 is the frequency for the peak response (maximum absorption or emission) and $\Delta\nu_{1/2} = 1/(2\pi\ \tau)$ is the mean lifetime of the state. Also, $\Delta\nu_{1/2}$ may be defined as one-half the width of the lines as measured between half-intensity points. For magnetic resonance lines not distorted by power saturation, this shape function is obtained from a solution of the Bloch equations [see (4.120)]. At the peak of the resonance, $\nu = \nu_0$, the Lorentzian function, has the value

$$f_{max} = \frac{1}{\pi(\Delta\nu_{1/2})} \tag{4.41}$$

Note that (4.40) has the same form as the shape function (4.120) derived from the Bloch theory (see Section 8).

Magnetic resonance is usually observed at a constant radiation frequency, and the resonance contour is swept out as the magnetic field imposed along z is varied. If the field value for peak absorption at the constant frequency is designated as H_0, the Lorentzian shape function in terms of H, when normalized by

$$\int_0^\infty f(H - H_0)dH = 1 \tag{4.42}$$

is

$$f(H - H_0) = \frac{1}{\pi}\ \frac{\Delta H_{1/2}}{(H - H_0)^2 + (\Delta H_{1/2})^2} \tag{4.43}$$

The value of this function at the peak resonant field value $H = H_0$ is

$$f_{max} = \frac{1}{\pi\Delta H_{1/2}} \tag{4.44}$$

where $\Delta H_{1/2}$ is one-half the line width (in guass) between half-intensity points. If T_{eff} is the effective relaxation time, then

$$\Delta H_{1/2} = \frac{h}{g\beta}\ \frac{1}{T_{eff}} \tag{4.45}$$

For homogeneously broadened lines, T_{eff} is approximately the spin-spin relaxation time T_1 [see (4.140)].

For greater sensitivity, most spectrometers record the first or second derivatives of the line shape rather than the true line shape. The first derivative of the

Lorentzian shape function, (4.43), is

$$f'(H - H_0) = -\frac{2\Delta H_{1/2}}{\pi}\frac{H - H_0}{[(H-H_0)^2 + (\Delta H_{1/2})^2]^2} \tag{4.46}$$

At the peak resonance field, $H = H_0$ and $f'(H - H_0) = 0$. The maximum and minimum values of $f'(H - H_0)$ occur at field values

$$H_{\pm} = H_0 \pm \frac{\Delta H_{1/2}}{\sqrt{3}} \tag{4.47}$$

which are found by solution of $f''(H - H_0) = 0$. A convenient measure of the line width is the separation, $\Delta H_{pp} = H_- - H_+$, of the extrema of the first-derivative curve, called the *peak - peak width.* From (4.47) it is evident that

$$\Delta H_{1/2} = \frac{\sqrt{3}}{2}\Delta H_{pp} \tag{4.48}$$

These widths are indicated in the diagrams of Fig. 4.5.

By substitution of the $H_{\pm}$ values from (4.47) into (4.46) one obtains the difference between the positive and negative peak values:

$$f'_{\max} - f'_{\min} = \frac{3\sqrt{3}}{4\pi(\Delta H_{1/2})^2} \tag{4.49}$$

Equation (4.49) gives a measure of the sensitivity of the first-derivative detection. The ratio of the maximum range in first-derivative signal to that of the undistorted signal is

$$\frac{f'_{\max} - f'_{\min}}{f_{\max}} = \frac{1.30}{\Delta H_{1/2}} \tag{4.50}$$

Thus the peak-to-peak signal range for first-derivative detection relative to that of normal line-shape detection increases with decrease of line width. The most important advantage of first-derivative detection is realized with its use in phase-locked amplifiers.

Also frequently used with phase-lock-in detection is the second derivative of the actual line-shape function. The second derivative is detected when small-amplitude modulation is used and the phase-lock-in amplifier is tuned to the second harmonic of the modulation frequency. The second derivative of the Lorentzian function of (4.43) is

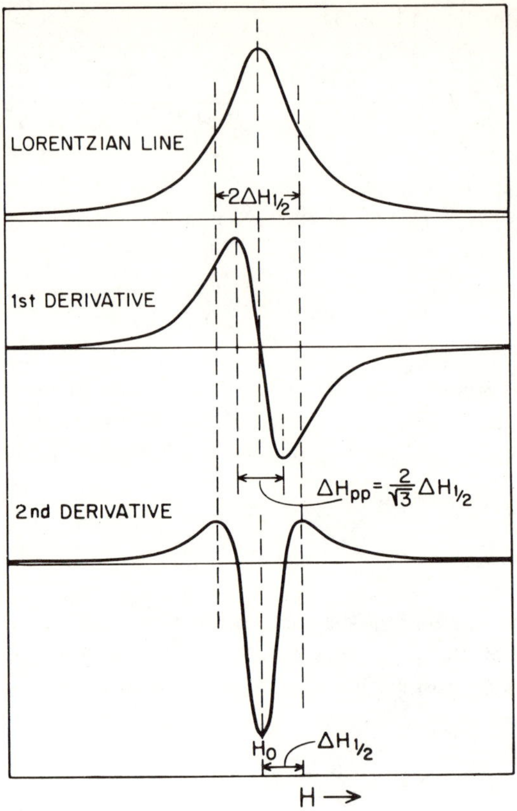

Fig. 4.5 The Lorentzian line shape with its first- and second-derivative contours.

$$f''(H - H_0) = \left(\frac{2\Delta H_{1/2}}{\pi}\right) \frac{3(H - H_0)^2 - (\Delta H_{1/2})^2}{[(H - H_0)^2 + (\Delta H_{1/2})^2]^3} \tag{4.51}$$

The function f'' has a minimum value at the resonant field value $H = H_0$ of

$$f''_{\text{min}} = -\frac{2}{\pi(\Delta H_{1/2})^3} \tag{4.52}$$

By setting $f'''(H - H_0) = 0$ and solving for H, one finds that $f''(H - H_0)$ has a maximum on either side of H_0 at

$$H = H_0 \pm \Delta H_{1/2} \tag{4.53}$$

and at these points $f''(H - H_0)$ has the maximum value

$$f''_{max} = \frac{1}{2\pi(\Delta H_{1/2})^3} \tag{4.54}$$

The separation of these two peaks, $2\Delta H_{1/2}$, provides a convenient method of measurement of $\Delta H_{1/2}$ and hence of the relaxation time with (4.45). The magnitude of the total signal range obtained from the difference between the positive and negative peaks gives a measure of the detectability. The ratio of this difference to f_{max} of the undistorted signal shape is

$$\frac{f''_{max} - f''_{min}}{f_{max}} = \frac{2.5}{(\Delta H_{1/2})^2} \tag{4.55}$$

Comparison of this expression with (4.50) shows that

$$\frac{f''_{max} - f''_{min}}{f'_{max} - f'_{min}} = \frac{1.92}{\Delta H_{1/2}} \tag{4.56}$$

which reveals that the sensitivity of second-derivative detection relative to first-derivative detection varies inversely as line half-width.

Figure 4.5 shows graphically the Lorentzian line shape, together with its first and second derivatives. The peak values at $H = H_0$ in the undistorted line and in the second-derivative curve are of opposite sign. To conform to the usual custom, we have plotted the true line shape as positive.

The distinguishable features of the first- and second-derivative curves are useful for location of the resonant peaks and half-power points and hence for measurement of **g** and **A**, of peak power absorption, of the number of spins in the sample, of $\Delta H_{1/2}$, and thus of the effective relaxation times. To make use of these features, however, the observer must be sure that the frequency modulation employed in the detection is sufficiently small in comparison with the line width to give the true derivative curves. Also, to apply the relations to find the relaxation times and other constants, he must know the nature or causes of the line broadening.

4.d. Gaussian Line Shape

The Gaussian shape of magnetic resonance lines has the same mathematical form as the Doppler line shape of gases [2]. At constant magnetic field H_0, the frequency distribution is

$$f(\nu - \nu_0) = f_{max} \exp\left\{-\text{natlog } 2 \left[\frac{\nu - \nu_0}{\Delta\nu_{1/2}}\right]^2\right\} \tag{4.57}$$

where $\Delta\nu_{1/2}$ is the half-width of the line and where f_{max}, determined by the normalization,

$$\int_0^\infty f(\nu - \nu_0)d\nu = 1 \tag{4.58}$$

occurs at $\nu = \nu_0$ and has the value $f_{max} = (\text{natlog } 2/\pi)^{1/2}/\Delta\nu_{1/2}$. In magnetic field units at constant frequency, the corresponding expression is

$$f(H - H_0) = f_{max} \exp\left\{-\text{natlog } 2 \left[\frac{H - H_0}{\Delta H_{1/2}}\right]^2\right\} \tag{4.59}$$

where

$$f_{max} = \frac{(\text{natlog } 2/\pi)^{1/2}}{\Delta H_{1/2}} \tag{4.60}$$

and $\Delta H_{1/2}$ again represents half the line width measured between half-intensity points.

The true Gaussian shape is not likely to be observed in esr, but approximate Gaussian shapes may be observed when there is symmetrical inhomogeneous broadening. More often, the lines can be represented as a combination of Lorentzian and Gaussian shapes. The Gaussian and Lorentzian line shapes are compared in Fig. 4.6.

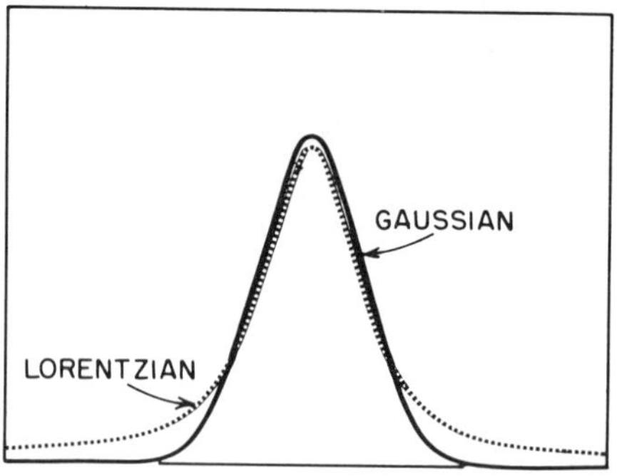

Fig. 4.6 Comparison of the Lorentzian and Gaussian line shapes.

4.e. Anisotropic Distortions of Line Shape

Anisotropic internal interactions cause asymmetric shapes in the esr signals of powdered samples, glasses, or polycrystalline materials in which the paramagnetic species have random orientations. The degree of the signal spread and the type of asymmetry depend on the nature of the various anisotropic interactions. Therefore, no general line-shape formula can be given, such as the Lorentzian shape for symmetric lines. Even so, useful information can be obtained from the asymmetries caused in the line shape by these distortions.

When the various anisotropic interaction constants are known from measurements on single crystals, one can predict by statistical analysis the signal shape to be expected for the polycrystalline form of the substance. More important is the inverse problem of derivation of the anisotropic interaction constants from asymmetric esr curves of materials for which data on single crystals cannot be obtained. This is done most effectively by computer simulation of the signal with various assumed anisotropies in the interaction constants and with assumed line shapes and widths for the corresponding components of the uniformly oriented radicals. The various parameters assumed are adjusted until the observed esr curve is fitted. Lorentzian line shapes are usually assumed in this process. The magnitude of the principal values of **g** and **A** can be derived to a useful accuracy simply from the bend points in the asymmetric signals. Methods for doing this are described in Chapter VIII.

5. POWER SATURATION EFFECTS

Realization of the line shape of Fig. 4.5 requires, among other things, that the inducing power level be sufficiently low that thermal relaxation processes can, to a good approximation, maintain the Boltzmann equilibrium between the spin levels. When the power level exceeds that amount, the homogeneously broadened line broadens further and becomes flattened or rounded in the region of peak absorption (see Fig. 4.7). This phenomenon is called the *saturation effect,* or *saturation broadening.* In magnetic resonance, it depends on the relaxation times T_1 and T_2 (Section 8) as well as on the power level.

Effects of power saturation on the shapes of homogeneously broadened magnetic resonance lines may be derived classically from the Bloch equations (see Section 8). The resulting line-shape function is

$$f(\omega - \omega_0) = \frac{2T_2}{1 + T_2(\omega - \omega_0)^2 + \gamma^2 H_1^2 T_1 T_2} \tag{4.61}$$

where H_1 is the amplitude of the magnetic component of the absorbed power, T_1 is the spin-lattice relaxation time, and T_2 is the spin-spin relaxation time.

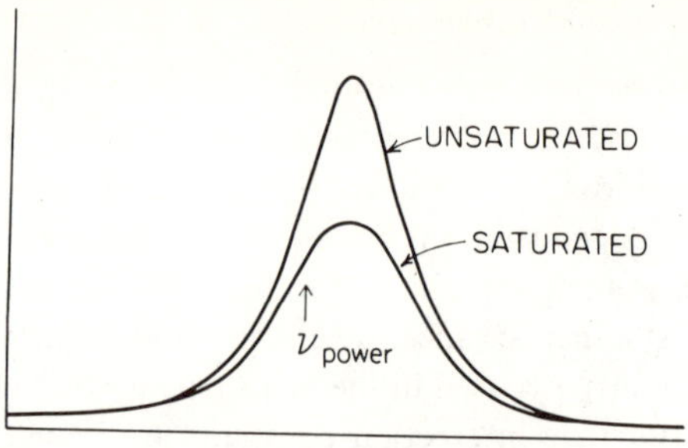

Fig. 4.7 Illuatration of power-saturation effects on the homogeneously broadened line. Saturating power may be applied at any frequency within the absorption band, but the amount required increases with deviation of the power frequency from the center of the resonance.

Note that when the power level is sufficiently small that $\gamma^2 H_1{}^2 T_1 T_2 \ll 1$ and $T_2 = T_{\text{eff}} = (1/\Delta\omega_{1/2})$, this expression reduces to the Lorentzian shape function, (4.40). An expression similar to (4.61) has been derived for gaseous absorption lines from quantum mechanics by Karplus and Schwinger [10].

The effect of saturation on an inhomogeneously broadened line is notably different from that on a homogeneously broadened line. Indeed, this difference can be used to distinguish the two types of broadening. If saturating power is applied at a particular frequency within the inhomogeneously broadened band, only the spin packet responsible for the absorption is saturated. The absorbed power cannot be distributed to the other spin packets, at least not in a time that is short in comparison to the spin-lattice relaxation time T_1. If the saturating power is then turned off and the entire absorption contour swept out in the usual

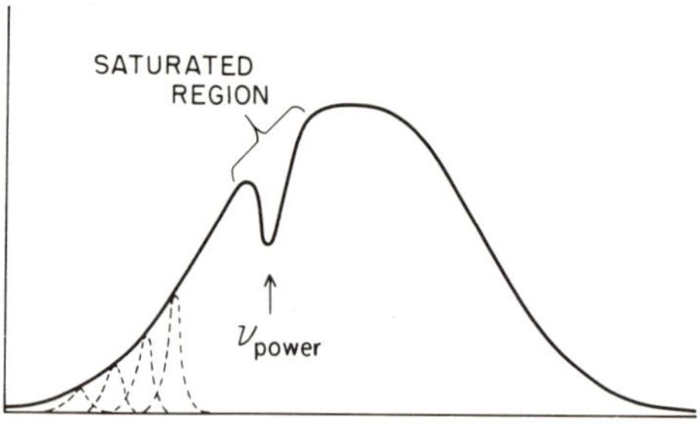

Fig. 4.8 Illustration of power-saturation effects on the inhomogeneously broadened line. The inhomogeneously broadened line represents the contour of unresolvable homogeneous components arising from spin packets of differing resonant frequencies, as indicated by the dotted curves. The rate of spin exchange between the different packets is so slow that power applied at a given frequency saturates only the packets absorbing at the frequency. This selective saturation indicated by the dip in the curve is commonly called *hole burning.*

way but in a time that is short compared with T_1, a hole appears in the contour at the frequency of the previously applied saturating power. This phenomenon, illustrated in Fig. 4.8, is called *hole burning.* It does not occur for the homogeneously broadened lines because the saturating power is diffused through spin exchange to the other spins throughout the sample in a time that is much shorter than the spin-lattice relaxation time.

If saturating power is applied simultaneously at all frequencies over the inhomogeneously broadened line or if the entire contour is swept over in a time that is short in comparison with the spin-lattice relaxation time, the various spin packets will be uniformly saturated, and there will be no change in the line shape, in contrast to the saturation-broadening effects observable under similar conditions for the homogeneously broadened resonance. Although the effects on their line shapes are different, the percentage of power absorption for both types of lines is decreased as the degree of power saturation increases. The rate of energy absorption in both cases is limited by the rate at which the energy absorbed by the spin systems can be transferred to the lattice motions.

The relative rate of absorption and dissipation of energy by the spin system can be expressed in terms of the rate of change of the population difference $n = (N_n - N_m)$. The total rate of change of n is

$$\left(\frac{dn}{dt}\right)_{\text{total}} = \left(\frac{dn}{dt}\right)_{P_\nu} + \left(\frac{dn}{dt}\right)_{\text{S-L}} \tag{4.62}$$

where the first term on the right results from absorption of the source power and the last term, from spin-lattice relaxation. For continuous absorption at a constant power level, equilibrium will be established, $(dn/dt)_{\text{total}} = 0$, and n is constant. Then

$$-\left(\frac{dn}{dt}\right)_{P_\nu} = \left(\frac{dn}{dt}\right)_{\text{S-L}} = \left(\frac{n_0 - n_c}{T_1}\right) \tag{4.63}$$

where n_c is the population difference at the constant power level and n_0, that for thermal equilibrium. The last form follows from Block's relation, (4.124). The signs of the derivatives are opposite because power absorption tends to decrease the population difference and the spin-lattice relaxation tends to increase it. At the limiting value $n = 0$, complete saturation, $(dn/dt)_{\text{S-L}}$ has its maximum value of n_0/T_1, and power absorption is at a maximum. However, the percentage of total power absorbed at saturation is less than that for incomplete saturation. When $n_c = n_0$, thermal equilibrium, $(dn/dt)_{\text{S-L}} = 0$, and likewise $(dn/dt)_{P_\nu} = 0$. Hence there can be no absorption of power without alteration of thermal equilibrium to some degree. This condition seems to contradict our assumption of thermal equilibrium in the derivation of the power absorption [see (4.33)]. The justification of this assumption is that T_1 is so small that the operating value of n_c can be very close to n_0 whereas $(dn/dt)_{\text{S-L}}$ is still significantly large. Although the absolute power absorption becomes smaller as thermal equilibrium is approached, the percentage absorption increases. Because of this and the fact that the most sensitive detectors operate at low power levels, the highest spectrometer sensitivity is usually achieved with $n \approx n_0$. However, the most sensitive operating power level depends on the value of T_1. When T_1 is large, saturation is reached with very low power levels, and the maximum rate of continuous absorption n_0/T_1 is small, becoming vanishingly small as $T_1 \to \infty$.

6. POPULATION OF SPIN STATES

Let us consider nondegenerate states of a system that is in thermal equilibrium. If E_0 is the energy of the lowest spin state and N_0 is the number of spins of the sample in this state, the populations of the higher states will be

$$N_1 = N_0 \exp\frac{-(E_1 - E_0)}{kT}, \qquad N_2 = N_0 \exp\frac{-(E_2 - E_0)}{kT}, \qquad \ldots, N_n = N_0 \exp\frac{-(E_n - E_0)}{kT} \tag{4.64}$$

where $N_1, N_2, \ldots, N_n$ are the population numbers and $E_1, E_2, \ldots, E_n$ are the energies of the higher spin states. The total number of spins in the sample may be expressed as

$$N = N_1 + N_2 + N_3 + \cdots + N_n = N_0 \sum_n \exp\frac{-(E_n - E_0)}{kT} \tag{4.65}$$

The number in a particular state m is thus

$$N_m = N \frac{\exp[-(E_m - E_0)/kT]}{\sum_n \exp[-(E_m - E_0)/kT]} \tag{4.66}$$

If the m level has degeneracy d_m and the different n levels have degeneracy d_n, (4.66) is modified to become

$$N_m = N \frac{d_m \exp[-(E_m - E_0)/kT]}{Q} \tag{4.67}$$

where

$$Q = \sum_n d_n \exp \frac{-(E_n - E_0)}{kT} \tag{4.68}$$

is the partition function. When all the levels have the same degeneracy, the d factors cancel from the numerator and the denominator, and the state population is given by (4.66).

For esr, $E_1 - E_0 = h\nu = g\beta H$ and $E_n - E_0 = nh\nu = ng\beta H$, where n has the $2S + 1$ integral values,

$$n = 0,1,2,3, \ldots, 2S$$

and m is a particular value of n. When the states are nondegenerate,

$$Q = \sum_n \exp \frac{-ng\beta H}{kT} \approx (2S + 1) \frac{1 - Sg\beta H}{kT} \tag{4.69}$$

and

$$N_m \approx N \frac{(1 - mg\beta H/kT)}{(2S + 1)(1 - Sg\beta H/kT)} \tag{4.70}$$

For the important spin-doublet state, $S = 1/2$,

$$Q = 1 + \exp \frac{-g\beta H}{kT} \approx 2\left(1 - \frac{1}{2}\frac{g\beta H}{kT}\right) \tag{4.71}$$

When there is thermal equilibrium, the population of the lowest state, $M_S = -1/2$ $(m = 0)$, is

$$N_{-1/2} = N \frac{1}{1 + \exp(-g\beta H/kT)} \approx \left(\frac{N}{2}\right)\left(1 + \frac{1}{2}\frac{g\beta H}{kT}\right) \tag{4.72}$$

where N is the total number of spins in the sample. The corresponding population of the upper state, $M_S = 1/2$ $(m = 1)$, is

$$N_{1/2} = \frac{N}{2}\left(1 - \frac{1}{2}\frac{g\beta H}{kT}\right) \tag{4.73}$$

and the population difference is

$$n = N_{-1/2} - N_{1/2} \approx \frac{Ng\beta H}{2kT} \tag{4.74}$$

For convenience in numerical analysis,

$$\frac{g\beta H}{kT} = \frac{h\nu}{kT} = \frac{1.439\ \nu(\mathrm{cm}^{-1})}{T} = \frac{48.0 \times 10^{-6}\ \nu\,(\mathrm{MHz})}{T} \tag{4.75}$$

where ν is the frequency of the observation and T is the absolute temperature. At $T = 300^\circ$ (room temperature) and an observation frequency of 30,000 MHz (K band), $g\beta H/kT$ is 48×10^{-3}. Thus for esr measurements in the centimeter-wave region at room temperature, $Q \approx 2S + 1$, and one can use the approximation $N_m \approx N/(2S + 1)]$, (4.31), for calculation of N_m in (4.30). However, for low-temperature measurements, the more precise expression must usually be employed.

6.a. Nuclear Substates

All the electron-spin states of a given spin system have the same number of nuclear-spin substates. When these substates are degenerate or unresolved, the nuclear degeneracy cancels from (4.67) and has no effect on the relative state populations. When nuclear hyperfine structure is resolved, one needs to consider the populations of the nuclear-spin substates in calculation of the intensities of the hyperfine components.

The calculation of the population of the nuclear-spin substate in a given electronic state is carried out in a manner similar to that described earlier for the electron-spin states. Because the separations of the nuclear sublevels are orders of magnitude smaller than kT, one can neglect the small differences caused by the Boltzmann factor and, for purposes of calculating relative intensities of esr hyperfine components, assume the substates of a given nucleus to be equally populated. If the spin of a particular nucleus of a free radical is I_i it will give rise to $2I_i + 1$ sublevels for each electron-spin state m, each with population given to a good approximation by

$$N_{m,M_{I_i}} = \frac{N_m}{2I_i + 1} \tag{4.76}$$

where N_m is the population of the electron-spin state m given by (4.66) or (4.70).

7. CLASSICAL DESCRIPTION: THE LARMOR PRECESSION

The classical description of magnetic resonance (electron or nuclear spin) as a Larmor precession of a macroscopic spin moment or magnetization vector provides an idealized model that is very helpful in the understanding of certain of its aspects. When modified to take account of the time variation of the components of the magnetization vector, as in the Bloch theory (Section 8), it is capable of presenting a satisfactory description of such phenomena associated with resonance as power absorption, line shapes, and relaxation effects.

In an applied static magnetic field under conditions of thermal equilibrium, a free-spin system will have a macroscopic magnetic moment in the direction of the field **H** because of the difference in population of the spin states. For $S = 1/2$, there are only two levels; the macroscopic moment M_z (per gram of sample) in the direction of **H**, defined as the molar magnetization, may be easily calculated from the population difference given by (4.74). Each spin has an associated spin magnetic moment along the field direction z of $g\beta M_S = (1/2)g\beta$. The difference between the number of spins pointing down and those pointing up, (4.74), is $Ng\beta H/2kT$. Hence, the resultant macroscopic moment M_z, or magnetization per gram, is

$$M_z = \frac{Ng^2\beta^2 H}{4kT} \tag{4.77}$$

where N represents the number of spins per gram. For spin values greater than 1/2, the corresponding expression is somewhat more complicated in derivation because the upward and downward moments must be summed for all the different M_S values. The corresponding expression for any value of S is

$$M_z = N\,\frac{g^2\beta^2 S(S + 1)H}{3kT} \tag{4.78}$$

The static molar magnetic susceptibility defined by

$$\chi_0 = \frac{M_0}{H_0} \tag{4.79}$$

called the *static Curie susceptibility,* is accordingly

$$\chi_0 = \frac{Ng^2\beta^2\, S(S + 1)}{3kT} \tag{4.80}$$

Associated with the macroscopic magnetic moment is a macroscopic mechanical moment or resultant spin moment $\mathbf{S}_R$ that is due to the orientations of the individual spin moments by the field. Since the magnetic and spin moments of the individual spins **S** are proportional, $\mu_S = -g\beta\mathbf{S}$, one can assume also that the resultant spin moment will be proportional to **M**, the macroscopic magnetic moment. Thus

$$\mathbf{M} = -\gamma \mathbf{S}_R \tag{4.81}$$

where **M** is defined such that its M_z component is given by (4.78) and γ is the magnitude of the classical macroscopic gyromagnetic ratio or magnetomechanical ratio, which is negative for electrons because μ_S and **S** are opposite in direction.

First, let us assume the magnetization to be represented by a single vector **M**, of constant magnitude, with associated angular momentum $\mathbf{S}_R$. Classically, the interaction of the magnetic field **H** with **M** will put a torque, $\mathbf{M} \times \mathbf{H}$, on $\mathbf{S}_R$ that will cause it to precess about **H**. According to Newtonian mechanics, the rate of change in angular momentum caused by the torque is

$$\frac{d\mathbf{S}_R}{dt} = \mathbf{M} \times \mathbf{H} \tag{4.82}$$

Substitution of $\mathbf{S}_R = -\mathbf{M}/\gamma$ from (4.81) into this expression yields

$$\frac{d\mathbf{M}}{dt} = -\gamma \mathbf{M} \times \mathbf{H} \tag{4.83}$$

Thus, classically, the magnetic moment **M** precesses about **H** to give the measurable component M_z indicated by (4.78). It is interesting that the classical precessional frequency of **M** is the same as the quantum-mechanical resonance frequency. This frequency, called the *Larmor precessional frequency,* may be found easily by solution of (4.83), as shown in the following paragraph.

When z is chosen along **H**, then $H_x = 0$, $H_y = 0$, and $H_z = H$. The component equations from (4.83) are:

$$\frac{dM_x}{dt} = -\gamma M_y H \tag{4.84}$$

$$\frac{dM_y}{dt} = \gamma M_x H \tag{4.85}$$

$$\frac{dM_z}{dt} = 0 \tag{4.86}$$

Solution of the last of these is simple and shows that M_z is constant. Solution of the first two requires separation of the x and y variables. This is achieved by differentiation of the first and substitution of the resulting value of dM_y/dt into the second. The resulting equation for M_x is

$$\frac{d^2 M_x}{dt^2} = -\gamma^2 H^2 M_x \tag{4.87}$$

Similarly,

$$\frac{d^2 M_y}{dt^2} = -\gamma^2 H^2 M_y \tag{4.88}$$

The solutions of (4.87) and (4.88) yield

$$M_x = A \cos \omega t + B \sin \omega t \tag{4.89}$$

and

$$M_y = A' \cos \omega t + B' \sin \omega t \tag{4.90}$$

where

$$\omega = \omega_L = \gamma H \tag{4.91}$$

clearly has the dimensions of angular velocity or frequency. It is the well-known Larmor precessional frequency. Since M_z is a constant component of **M**, it can be expressed as

$$M_z = M_0 = M \cos \alpha = \text{constant} \tag{4.92}$$

where M is the magnitude of **M** and α is the angle of precession of **M** about the field. Thus with **H** held constant, after thermal equilibrium is established, it is obvious that **M** is constant and that α, the angle of precession, is constant as well as ω_L.

The various integration constants are not independent. By use of (4.92) and the relation $M_x{}^2 + M_y{}^2 + M_z{}^2 = M^2$ with choice of the particular solution $M_x = A \cos \omega t$, values of M_x and M_y can be expressed as

$$M_x = M \sin \alpha \cos \omega t \tag{4.93}$$

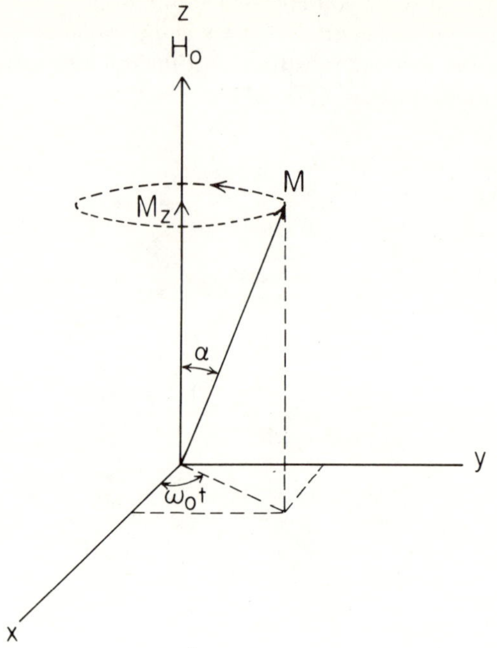

Fig. 4.9 Vector diagram of the Larmor precession of the magnetization **M** about the direction of an applied dc magnetic field $\mathbf{H}_0$.

and

$$M_y = M \sin \alpha \sin \omega t \tag{4.94}$$

A vector model illustrating the classical precession of **M** is shown in Fig. 4.9.

In the classical treatment of induced absorption of radiation, it is convenient to consider the magnetic vector of the radiation as rotating in the xy plane, perpendicular to the z axis along which the static field $\mathbf{H}_0$ is imposed. Experimentally, the ac field may be plane polarized with the magnetic vector along x, as assumed in Section 1, but this is equivalent to two circularly polarized components rotating in opposite sense, as is evident from the equivalence

$$2H_1 \cos \omega t = H_1 e^{j\omega t} + H_1 e^{-j\omega t} \tag{4.95}$$

In practice, only one of these components will rotate in the same sense as the Larmor precession of **M**, and hence only one will be effective in inducing transitions. The sense of the rotation that induces esr absorption of radiation is positive; the effective ac component can be represented by

$$\mathbf{H}_1 = \mathbf{i}H_1 \cos \omega t + \mathbf{j}H_1 \sin \omega t \tag{4.96}$$

where **i** and **j** are unit vectors along the x and y axes. If $\mathbf{H}_0$ represents the static component along z, the resultant field is

$$\mathbf{H} = \mathbf{H}_1 + \mathbf{H}_0 \tag{4.97}$$

The simplest way to treat the resonance problem when the ac field is small compared with $\mathbf{H}_0$ is to transform (4.83) to a coordinate system that rotates with $\mathbf{H}_1$ in the xy plane about $\mathbf{H}_0$ [11]. This well-known transformation may be expressed as

$$\frac{d\mathbf{M}}{dt} = \frac{d\mathbf{M}'}{dt} + \omega \times \mathbf{M} \tag{4.98}$$

where the prime signifies the rate of change of **M** relative to the rotating coordinates and ω signifies the angular velocity of the coordinate rotation. Under the assumed condition, ω also corresponds to the angular velocity of $\mathbf{H}_1$ about z, and its magnitude is 2π times the radiation frequency. With the value $d\mathbf{M}/dt$ from (4.83), and with the equivalence $\omega \times \mathbf{M} = -\mathbf{M} \times \omega$, (4.98) is readily transformed to

$$\frac{d\mathbf{M}'}{dt} = -\gamma \mathbf{M} \times \frac{\mathbf{H} - \omega}{\gamma} \tag{4.99}$$

which may be expressed as

$$\frac{d\mathbf{M}'}{dt} = -\gamma \mathbf{M} \times \mathbf{H}_{\text{eff}} \tag{4.100}$$

where

$$\mathbf{H}_{\text{eff}} = \frac{\mathbf{H} - \omega}{\gamma} = \frac{\mathbf{H}_1 + \mathbf{H}_0 - \omega}{\gamma} \tag{4.101}$$

Both $\mathbf{H}_0$ and $-\omega/\gamma$ are along the z axis but are oppositely directed because γ as used in this treatment signifies the magnitude only of the electron gyromagnetic ratio and is hence a positive number. If ω_0 is the Larmor precessional velocity about $\mathbf{H}_0$ alone, we can set $\mathbf{H}_0 = (\omega_0/\gamma)$ and express (4.101) by

$$\mathbf{H}_{\text{eff}} = \mathbf{H}_1 + \frac{1}{\gamma}(\omega_0 - \omega) = \mathbf{H}_1 + \mathbf{H}' \tag{4.102}$$

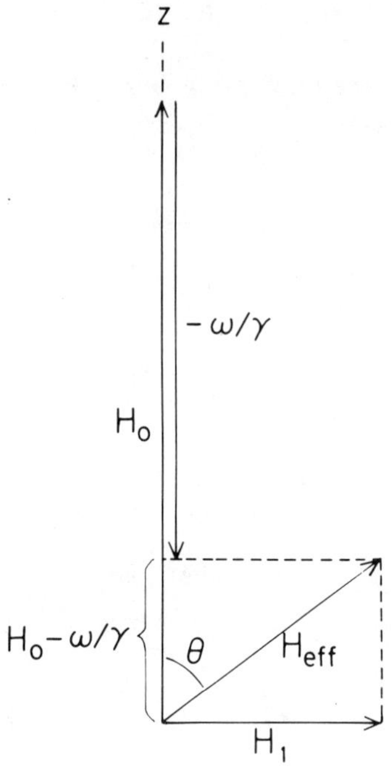

Fig. 4.10 Diagram showing the resultant z field $\mathbf{H}' = \mathbf{H}_0 - (\omega/\gamma)$ and the effective magnetic field $\mathbf{H}_{\text{eff}}$ relative to a coordinate system rotating with $\mathbf{H}_1$ about z with velocity ω.

where $\mathbf{H}' = (1/\gamma)(\omega_0 - \omega)$ is the resultant z field relative to the rotating system. The vector diagram of Fig. 4.10 shows the relationship of these various components as measured in the rotating system. It is evident from the diagram that

$$\theta = \arctan \frac{H_1}{H_0 - \gamma\omega} \tag{4.103}$$

Equation (4.100) has the same form as (4.83), and its solution obtained in a similar way yields the precessional frequency of the magnetization vector relative to the rotating (primed) system

$$\omega' = \gamma \mathbf{H}_{\text{eff}} \tag{4.104}$$

with magnitude

$$\omega' = [\gamma^2 H_1^2 + (\omega_0 - \omega)^2]^{1/2} \tag{4.105}$$

Usually $H_1 << H_0$ and, except when ω equals or closely approaches ω_0, $\gamma^2 H_1^2 << (\omega_0 - \omega)^2$, and

$$\omega' \approx \omega_0 - \omega \tag{4.106}$$

With reference to the fixed laboratory frame, $\mathbf{M}$ precesses about $\mathbf{H}_{\text{eff}}$ with frequency ω' while $\mathbf{H}_{\text{eff}}$ precesses about $\mathbf{H}_0$ at frequency ω, as indicated by the vector diagram of Fig. 4.11. When $\omega = \omega_0$, it follows that $\theta = 0$, $\mathbf{H}_{\text{eff}} = \mathbf{H}_1$, and the magnetization vector precesses about $\mathbf{H}_1$ in the rotating frame at frequency $\omega' = \gamma H_1$, as indicated in Fig. 4.12.

It is evident from Fig. 4.12 that under resonant radiation $\mathbf{M}$ would precess about $\mathbf{H}_1$ and that the z component of the magnetization would in time average to zero, regardless of how small $\mathbf{H}_1$ may be. This incorrect deduction results from our having neglected relaxation times and having treated the magnitudes M and S_R as constant, classical quantities. The assumption of a constant M value implies an infinite spin-lattice relaxation time. When, however, the relaxation time is very long as compared with $1/\omega'$, the idealized model of Fig. 4.11 or 4.12 is closely approached. The model is very helpful in the understanding of experiments involving pulse techniques, negative spin temperatures, spin echoes,

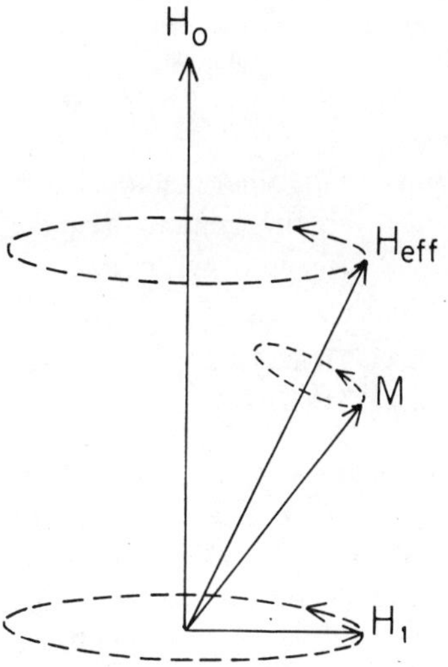

Fig. 4.11 Vector diagram of the precession of the magnetization $\mathbf{M}$ relative to the fixed coordinate system when a rotating field component $\mathbf{H}_1$ is applied normal to $\mathbf{H}_0$; $\mathbf{M}$ precesses about the $\mathbf{H}_{\text{eff}}$ of Fig. 4.10 with angular velocity $\gamma\mathbf{H}_{\text{eff}}$ while $\mathbf{H}_{\text{eff}}$ turns about $\mathbf{H}_0$ with the angular velocity ω of $\mathbf{H}_1$.

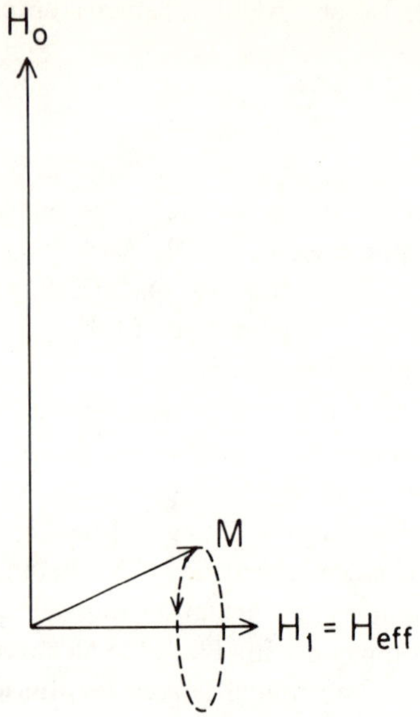

Fig. 4.12 Vector diagram of the precession of **M** relative to the $\mathbf{H}_1$ when it is rotating at the Larmor frequency $\omega_0 = \gamma H_0$. In this case the $\mathbf{H}_0 - (\omega/\gamma)$ of Fig. 4.10 is zero and $\mathbf{H}_{\text{eff}} = \mathbf{H}_1$. As seen from the laboratory frame, **M** precesses about $\mathbf{H}_1$ at velocity $\gamma\mathbf{H}_1$ while $\mathbf{H}_1 = \mathbf{H}_{\text{eff}}$ turns about $\mathbf{H}_0$ at ω_0. Since $\mathbf{H}_1$ is normal to $\mathbf{H}_0$, there is for this resonant condition no static or time-averaged component of **M** along $\mathbf{H}_0$, but there is a varying M_z component that oscillates with frequency $(\gamma\mathbf{H}_1/2\pi)$.

adiabatic fast passage, and certain other phenomena. To give a complete description of magnetic resonance, this theory must be modified to take into account relaxation phenomena, as described in Section 8.

8. MODIFIED CLASSICAL TREATMENT: THE BLOCH PHENOMENOLOGICAL THEORY

The classical theory of the Larmor precession of the macroscopic magnetization vector **M** does not provide a complete description of magnetic resonance, mainly because it does not properly take into account the statistical nature of **M**. By postulating separate decay mechanisms and relaxation times for the horizontal and vertical components of the magnetization, Bloch [11] obtained a modified classical theory that provides a satisfactory and useful description of the resonance. Although this theory was derived for, and first applied to, nmr, it is applicable also to esr.

The usual resonance experiments involve the imposition of a static field $\mathbf{H}_0$ along z and a much weaker rotating ac field perpendicular to $\mathbf{H}_0$. Thus the components of **H** are $H_z = 0$, $H_x = H_1 \cos \omega t$ and $H_y = H_1 \sin \omega t$. To the component equations, Bloch added terms to take into account separate decay rates for the components of magnetization both parallel and perpendicular to the applied field. The component equations of (4.83) are thus modified to

$$\frac{dM_x}{dt} = -\gamma M_y H_0 + \gamma M_z H_1 \sin \omega t - \frac{M_x}{T_2} \tag{4.107}$$

$$\frac{dM_y}{dt} = \gamma M_x H_0 - \gamma M_z H_1 \cos \omega t - \frac{M_y}{T_2} \tag{4.108}$$

$$\frac{dM_z}{dt} = -\gamma(M_x H_1 \sin \omega t - M_y H_1 \cos \omega t) + \frac{M_0 - M_z}{T_1} \tag{4.109}$$

Solution of these equations is most easily accomplished by transformation of them to the coordinate system rotating about $\mathbf{H}_0$ with velocity ω, as was done for the simpler equations in Section 7. These solutions may then be transformed back to the fixed-reference frame. The somewhat tedious process is described elsewhere [11-13]. In the fixed-reference frame, the resulting component values are

$$M_x = \frac{1}{2} M_0 \gamma T_2 \frac{T_2(\omega_0 - \omega)\, 2H_1 \cos \omega t + 2H_1 \sin \omega t}{1 + T_2(\omega_0 - \omega)^2 + \gamma^2 H_1^2 T_1 T_2} \tag{4.110}$$

$$M_y = \frac{1}{2} M_0 \gamma T_2 \frac{2H_1 \cos \omega t - T_2(\omega_0 - \omega) 2H_1 \sin \omega t}{1 + T_2(\omega_0 - \omega)^2 + \gamma^2 H_1^2 T_1 T_2} \tag{4.111}$$

$$M_z = M_0 \frac{1 + T_2^2 (\omega_0 - \omega)^2}{1 + T_2(\omega_0 - \omega)^2 + \gamma^2 H_1^2 T_1 T_2} \tag{4.112}$$

where $\omega_0 = \gamma H_0$ and γ is the magnitude of the electron gyromagnetic ratio.

8.a. Power Absorption and Line-shape Functions

From the preceding solutions it is easy to obtain the power absorbed from the ac source. The total energy of the interaction of **H** with the moment **M** per unit volume is $W = -\mathbf{M} \cdot \mathbf{H}$, and its rate of change is

$$\frac{dW}{dt} = -\mathbf{M} \cdot \frac{d\mathbf{H}}{dt} - \mathbf{H} \cdot \frac{d\mathbf{M}}{dt} \tag{4.113}$$

The first term on the right represents the power absorbed from the varying magnetic field and the second, that from the varying **M** caused by spin-lattice relaxation. Under steady-state operation there is no net change of energy in the spin system, and the power absorbed from the source is exactly equal to that given up to the thermal motions of the lattice. Thus for the steady-state operation,

$$-\mathbf{M} \cdot \frac{d\mathbf{H}}{dt} = \mathbf{H} \cdot \frac{d\mathbf{M}}{dt} \tag{4.114}$$

In component form this is

$$\mathbf{i}M_x \frac{dH_x}{dt} + \mathbf{j}M_y \frac{dH_y}{dt} + \mathbf{k}M_z \frac{dH_z}{dt} = -\left(\mathbf{i}H_x \frac{dM_x}{dt} + \mathbf{j}H_y \frac{dM_y}{dt} + \mathbf{k}H_z \frac{dM_z}{dt}\right) \tag{4.115}$$

Now $H_z = H_0$, and hence $dH_z/dt = 0$. We assume here that the component of the rotating ac field is $H_x = H_1 \cos \omega t$ and that $H_y = H_1 \sin \omega t$. With these conditions, the average power absorption can be expressed as

$$P_{\text{abs}} = -\left(M_x \frac{dH_x}{dt} + M_y \frac{dH_y}{dt}\right)_{\text{av}} \tag{4.116}$$

By substitution of the values of M_x and M_y from (4.110) and (4.111) and of $dH_x/dt = -\omega H_1 \sin \omega t$ and $dH_y/dt = \omega H_1 \cos \omega t$, with the averages $(\cos^2 \omega t)_{\text{av}} = 1/2$, $(\sin^2 \omega t)_{\text{av}} = 1/2$, one finds the power absorption to be

$$P_{\text{abs}} = \frac{\gamma\, \omega H_1^2 M_0 T_2}{1 + T_2(\omega - \omega_0)^2 + \gamma^2 H_1^2 T_1 T_2} \tag{4.117}$$

With $M_0 = \chi_0 H_0$ and $\omega - \omega_0 = \gamma H_0$, this can be expressed in the form

$$P_{\text{abs}} = \frac{1}{2} \chi_0\, \omega^2 H_1^2 f(\omega - \omega_0) \tag{4.118}$$

where

$$f(\omega - \omega_0) = \frac{2T_2}{1 + T_2(\omega - \omega_0)^2 + \gamma^2 H_1^2 T_1 T_2} \tag{4.119}$$

is the resonance shape function. When H_1 is not sufficiently large to produce saturation broadening, that is, when $\gamma^2 H_1^2 T_1 T_2 \ll 1$, the last term in the denominator is negligible, and the line has the Lorentzian shape

$$f(\omega - \omega_0) = \frac{2T_2}{1 + T_2^2 (\omega - \omega_0)^2} \tag{4.120}$$

With $\omega = 2\pi\nu$ and $T_2 = 1/\Delta\omega_{1/2} = 1/(2\pi\Delta\nu_{1/2})$, this expression may be converted to the form of (4.40).

It should be noted that a linearly polarized field, $2H_1 \cos \omega t$, is required to produce the rotating component assumed in the preceding solutions. Therefore, the predicted power absorption that depends on the square of H_1 is 4 times the one that would occur if a linearly polarized field of $H_1 \cos \omega t$ were assumed, as is often done. It should be remembered that the linearly polarized field is equivalent to two fields rotating in the opposite sense, only one of which is effectively absorbed by magnetic resonance because the component rotating in opposition to the Larmor precession is effectively out of phase with the rotating components of **M**. The preceding derivation used only one of the rotating components equivalent to the linear field $2 H_1 \cos \omega t$. In the quantum-mechanical derivation of the intensity formula (4.33), we assumed a linearly polarized component of $H_1 \cos \omega t$. The classical expression for the power absorption expected for this field component is, therefore, only one-fourth that expressed by (4.118), or

$$P_{\text{abs}} = \frac{1}{8} \chi_0 \omega^2 H_1^2 f(\omega - \omega_0) \tag{4.121}$$

By setting $\chi_0 = N(g\beta)^2 S(S+1)/3kT$ from (4.80), with $\omega = 2\pi\nu$ and $f(\omega - \omega_0) = f(\nu - \nu_0)$, we can transform (4.121) to

$$P_{\text{abs}} = \frac{N\pi^2\nu^2(g\beta H_1)^2 S(S+1) f(\nu - \nu_0)}{6\,kT} \tag{4.122}$$

To transform the quantum mechanical expression of Eq. (4.33) to the classical form, one must sum the power absorption over all the transitions. If one substitutes the value of the summation

$$\sum_{M_S = -(S-1)}^{M_S = S} [S(S+1) - M_S(M_S - 1)] = \frac{2}{3} S(S+1)(2S+1) \tag{4.123}$$

for $[S(S+1) - M_S(M_S - 1)]$ into (4.33), it transforms to (4.122).

Because it describes the resonance of all like spins in terms of the components of a single precessing vector, the Bloch theory applies only to homogeneously broadened lines. Thus it is not surprising that the line-shape function that it predicts is that for homogeneously broadened lines. However, an inhomogeneously broadened resonance can often be treated as a summation of the homogeneous absorption by independent spin packets in the sample. In this way, the Bloch theory may be applied to inhomogeneously broadened resonances.

8.b. Relaxation Times T_1 and T_2

The rate of change of M_z after $\mathbf{H}_0$ is extinguished is given by the Bloch equation

$$\frac{dM_z}{dt} = \frac{M_0 - M_z}{T_1} \tag{4.124}$$

This equation provides an analytical definition of the longitudinal relaxation time T_1, which is equivalent to the spin-lattice relaxation time (Section 9). Integration of the equation with the evaluation of the integration constants determined by choice of $M_z = 0$ at $t = 0$ yields

$$M_z = M_0 \,(1 - e^{-t/T_1}) \; . \tag{4.125}$$

Quantitatively, T_1 is the time required for M_z to decay to $1/e$ of its equilibrium value M_0 after the field is turned off, or to grow to $1/e$ of its equilibrium value after the field is turned on. This measure of T_1 requires that the static field be turned off or on in a time that is short as compared with T_1, an impossible feat when T_1 is very short. Without turning off the static field, however, one can reduce M_z to zero, or very nearly zero, through a 90° pulse of resonance radiation (see Section 10), which can be turned off in a time that is short enough for measurement of very short relaxation times. If M_z is reduced to zero in this way, it will again grow to $1/e$ of its value in time T_1 after the saturating power is turned off.

Figure 4.13 shows the buildup of the magnetization M_z to M_0 after it has been reduced to zero by a saturating pulse of ac power. This curve also expresses the rate of dissipation of the energy $M_0 H_0$ absorbed by the spin system from the ac source. Therefore, (4.124) expresses the rate of exchange of energy between the spin system and the thermal energy of the lattice, and T_1 is a measure of this rate of exchange.

In a fixed, laboratory frame, there are no time-averaged components of magnetization perpendicular to the static field $\mathbf{H}_0$ comparable to the M_z component. The components of M_x and M_y of (4.110) and (4.111) average to zero over a complete cycle. The Bloch T_2 is the decay time of the amplitudes

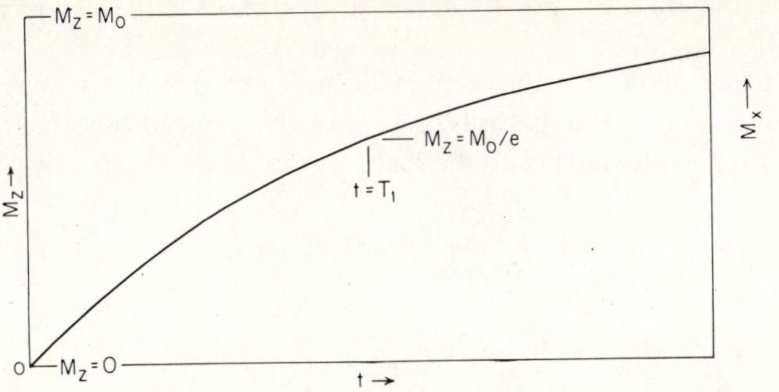

Fig. 4.13 Exponential growth of the M_z magnetization after a saturating pulse of power is removed.

of any alternating components of magnetization transverse to the static field $\mathbf{H}_0$. For this reason, it is commonly called the *transverse relaxation time.*

Analytically, T_2 is defined by the Bloch equations that, expressed in the frame rotating at ω_L with $\mathbf{H}_1$ turned off, are

$$\frac{dM_x'}{dt} = -\frac{M_x'}{T_2} \tag{4.126}$$

and

$$\frac{dM_y'}{dt} = -\frac{M_y'}{T_2} \tag{4.127}$$

If $(M_x')_0$ and $(M_y')_0$ are the values of the magnetization along x' and y' at $t =$ 0 when $\mathbf{H}_1$ is turned off, their values at a later time t will be

$$M_x' = (M_x')_0 e^{-t/T_2} \tag{4.128}$$

and

$$M_y' = (M_y')_0\, e^{-t/T_2} \tag{4.129}$$

The resultant transverse magnetization in the rotating frame at time t is

$$M_1 = (M_x'^2 + M_y'^2)^{1/2} = (M_1)_0 e^{-t/T_2} \tag{4.130}$$

where $(M_1)_0$ is the component along $\mathbf{H}_1$ at time t_0. Quantitatively, T_2

represents the time for M_1 to decay to $1/e$ of its value $(M_1)_0$ after $\mathbf{H}_1$ is turned off.

If the phase of the rotating system is chosen so that M_1 is along the space-fixed x axis at $t = 0$, it is easily seen that the perpendicular components of magnetization with reference to the space-fixed axis are

$$M_x = (M_1)_0 e^{-t/T_2} \cos \omega t \tag{4.131}$$

and

$$M_y = (M_1)_0 e^{-t/T_2} \sin \omega t \tag{4.132}$$

The corresponding z component is given by (4.125). Figure 4.14 shows diagramatically the decay of the M_x component after the field $\mathbf{H}_1$ is extinguished. The nature of T_1 and T_2 and the processes on which they depend are described further in the section to follow.

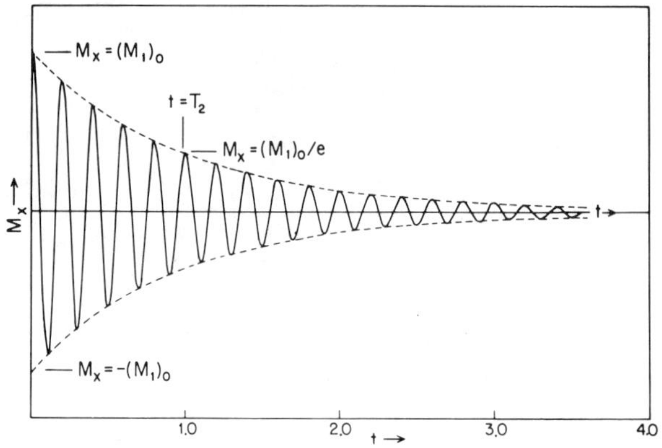

Fig. 4.14 Exponential decay of the M_x magnetization after a saturating pulse of power is removed.

9. RELAXATION PROCESSES

9.a. Spin-Lattice Relaxation

Exchange of energy between the thermal motions of the solid and the magnetic spin states, usually referred to as *spin-lattice relaxation,* is essential to the observation of magnetic resonance. Without such an exchange, a continuous absorption of microwave resonant energy would not be possible.

Although the Bloch phenomenological theory correctly takes into account the functional time variation in the components of macroscopic magnetization, it includes no description of the mechanism of this variation. To understand the causes for the time variation in these components, one must consider the many separate spin moments that together constitute the gross classical moment. Quantum mechanically, the macroscopic magnetization must be considered a statistical quantity, an expectation value that depends on the difference in population of the magnetic spin states.

At equilibrium in zero field, the spins have random orientations in space; their resultant moment is zero. When a static magnetic field $\mathbf{H}_0$ is applied, the spins become aligned parallel or antiparallel to the field, with slightly more spins having the antiparallel alignment, to give the population difference and observable magnetization M_z, as already described. A finite time is required, however, for the spins to become reoriented in the field and to achieve the population difference and the $M_0 = \chi_0 H_0$ value for thermal equilibrium. The time required for the randomly oriented spins to achieve their thermal equilibrium orientations in a suddenly applied field is the same as that required for the spins to become again randomly oriented when the field is suddenly extinguished. This reorientation time is measured by T_1 in the Bloch equations. Since the alignment energy, $M_0 H_0$, is taken from, or given back to, the thermal motions of the lattice during this process, T_1 is also a measure of the spin-lattice relaxation time.

Despite much theoretical and experimental work on the subject, the various processes contributing to spin-lattice relaxation are not yet completely understood. The observed and predicted values of T_1 seldom agree closely. In most paramagnetic species, multiple processes are involved. The simplest, but not the most significant, mechanism for spin-lattice relaxation is a direct exchange of esr quanta with lattice vibrations (phonons) of the same frequency. This method was first advanced by Waller [14], who proposed that the resonance exchange is induced by phonon modulation of the magnetic dipole field of a given spin acting on its neighboring spins. Unfortunately, there are certain difficulties that prevent this beautifully simple method from being very significant. The first is that only a small fraction of the normal distribution of the thermal energy is concentrated in vibrational frequencies as low as the microwave spin-resonance frequency. Therefore, the vibrational quanta, phonons, at the resonance frequency are normally scarce. The second is the so-called phonon bottleneck. Phonons induced by the spin system must be dissipated. Unless these phonons can escape from the sample or be converted to other lattice frequencies or other forms of energy as fast as they are produced by the spin system, they build up to an excessive concentration that blocks further relaxation of the spin system by this process. Comparison of the theoretically predicted T_1 values with those experimentally observed shows that the direct process is orders of magnitude too slow to account for the relaxation rates normally observed at room temperature [15]. Because the rate of relaxation by this process decreases much less rapidly with temperature than that by other processes, it becomes relatively more important at low temperature.

Waller, who carried out his pioneering theoretical work in 1932, long before experimental measurements of magnetic resonance or relaxation times were made, also proposed the Raman process of relaxation in which the lattice phonons are "scattered" by the spins, with the esr frequency added to or subtracted from the frequency of the "scattered" phonons, as in the Raman process. For this method to be effective, there must be lattice modes having difference frequencies equal to the esr frequency. Such vibrations, however, can be in the higher-frequency region, more densely populated at ordinary temperatures. Furthermore, many mode pairs can participate in this process since the only requirement is that their difference frequency be equal to the esr frequency. It is evident that the effectiveness of this process would decrease rapidly with decrease of temperature since vibrations at higher frequencies would be frozen out as the temperature is lowered.

Orbach has proposed an indirect process of relaxation that involves a transition to an excited orbital level of the paramagnetic element from one of the ground magnetic doublet states, with a return to the other [16]. Like the Raman process, this one involves two optical frequencies having a difference equal to the magnetic resonance frequency. It is more restricted than the Raman process because two specific optical transitions are involved. However, this process does not require simultaneous absorption and emission of the two optical quanta, as does the Raman process. The probability of such transitions depends on the Boltzmann factor, $\exp(-\Delta_n/kT)$, where Δ_n is the difference in energy between the doublet magnetic ground state and the excited orbital level.

For paramagnetic ions of free radicals having an odd number of electrons (Kramers' doublets), the temperature and magnetic-field dependence of the spin-lattice relaxation by the processes described in the preceding paragraphs is expressed by [15]

$$\frac{1}{T_1} = \overbrace{AH^4T}^{\text{Direct process}} + \overbrace{BH^2T^7 + B'T^9}^{\text{Raman process}} + \overbrace{C\exp\frac{-\Delta_n}{kT}}^{\text{Orbach process}}$$

where the coefficents, A, B, B', and C, are constant for a particular species. At liquid helium temperature, only the direct process (first-power dependence on temperature) is usually significant, whereas the Raman and Orbach terms become dominant at higher temperatures. The weighting coefficients and the relative contributions of the terms at intermediate temperatures are difficult to predict theoretically.

Equal in importance with the available vibrational modes for spin-lattice relaxation is the mechanism for the coupling of these modes with the magnetic spin transitions. The probability of conversion of an esr quantum to a phonon, or vice versa, depends on the strength of this coupling. Mechanical vibration modes (phonons) do not interact directly with the magnetic spin states but must do so indirectly by modulation of the magnetic field components effective on the electron spin. Since the effect of the dipole field of one spin on another depends on the separation of the spins, it is evident that a phonon that varies this distance at the phonon frequency would impose an alternating magnetic field component on the spin that might induce spin flipping, as does the magnetic field component of microwave radiation. This was the coupling mechanism assumed by Waller for his early calculations. Because of the rapid decrease (inverse cube variation) of a dipole field with distance, this form of coupling mechanism is effective only for rather concentrated spins.

As a result of theoretical work by Heitler and Teller [17], Fierz [18], Krönig [19], and especially by J. H. Van Vleck [20, 21], as well as improved methods of measurement of relaxation time, it is now recognized that the most common and effective coupling of the lattice vibrations with magnetic spin states occurs indirectly through residual spin-orbit coupling. As explained in Chapter I, the crystalline electric fields in ionic salts or chemical bonds in a molecular free radical lift the degeneracy of the orbital states, usually leaving an orbital singlet as a ground state. To first order, this orbital singlet behaves like an atomic S state. Because of a second-order admixture of the singlet orbital ground state with higher orbital states (see Chapter II), the orbital magnetic field is not entirely quenched, and thus there is a slight orbital magnetic field acting on the spin moment. The orbital moments are strongly coupled to the lattice of crystals or to the chemical bonds of free radicals through strong internal electric fields (Stark splitting). Hence the residual orbital magnetic field acting on the electron spin is effectively modulated by all vibrational modes or phonons. The probability that a magnetic spin transition will be induced by such modulation at the resonant frequency is proportional to the square of the residual orbital field components that are transverse to $\mathbf{H}_0$. In general, one expects and finds the rate of relaxation to increase, $\mathbf{T}_1$ to decrease, with the strength of residual spin orbital coupling. When the orbital motions are effectively quenched (splitting of the orbital states, Δ_n, large) and when the spin-spin interaction is small (diluted samples), one expects T_1 and also T_2 to be long and, consequently, the esr lines to be sharp. Since the residual spin-orbit coupling is also responsible for the $\mathbf{g}$ anisotropy (see Chapters II and III), T_1 tends to be longer and the lines sharper as the $\mathbf{g}$ anisotropy decreases. Because of the strong quenching of the orbital moments by the chemical bonds in organic free radicals, their $\mathbf{g}$ anisotropy is generally small and their esr lines sharp. This is a very favorable circumstance for easy and accurate measurements of the esr frequencies of such radicals.

Pulse techniques for observation of spin-lattice relaxation are discussed briefly in Section 10. Detailed description of experimental techniques for measurement of spin-lattice relaxation times and theoretical methods for prediction of them would require too much space for the scope of this volume. Comprehensive treatments of these subjects are available in specialized monographs. Rather thorough and lucid discussions may be found in volumes by Abragam and Bleaney [22], Orbach and Stapleton [23], Poole and Farach [24], and Standley and Vaughan [15].

9.b. Transverse Relaxation and Phase Memory

To account mechanistically for the amplitude variation of the transverse components of magnetization, one considers spins of the system as precessing independently with different phase relaxations among their components rotating in the *xy* plane. The transverse components of two spins precessing in the same sense with phase difference of 180° cancel, but their longitudinal *z* components are additive. The spin phases may have random differences so that their resultant transverse magnetization is very small, or zero, even when the M_z component is large. Furthermore, the phase relations of those spins having slightly different precessional frequencies will be constantly changing. The transverse magnetization thus depends on the phase coherence of the precessing spins.

If all the spins of a homogeneous system having the same Larmor frequency were to have the same phases at a given time, they would gradually lose their phase coherence because of spin flips that randomly alter the phases of the precessional spins. If the various spin packets that give rise to an inhomogeneous line were to have the same phase at a given time, they would lose their phase coherence, not only because of spin flips, but also—and usually much faster—because of their differences in precessional frequency. Either of these processes of destruction of phase coherence contributes to the decay of the transverse magnetization, but the two are inherently different processes and can be readily distinguished by the spin echo technique. If the inhomogeneous fields are static, the phase coherence lost in the latter process can be regained or reconstructed by the application of successive pulses to produce echo signals, as described in Section 10. The characteristic that allows this recovery of phase coherence is generally called *phase memory.* The loss of phase coherence that is not recoverable is said to result from loss of phase memory. Although spin flips are the most prevalent cause for loss of phase memory, they are not the sole cause. Randomly varying precessional frequencies caused by fluctuating internal fields can cause irrecoverable loss of phase coherence without inducing spin flips or without producing an inhomogeneity in the resonance. Thus the phase memory time designated as T_M is not necessarily the same as the spin-spin relaxation time, but it is strictly equivalent to the Block relaxation time T_2 for homogeneous resonance. When there are no fluctuating internal fields, T_M is also equal to the spin-spin relaxation time. In most applications, these three time rates are numerically equivalent and are usually indiscriminately designated by T_2.

9.c. Spin Temperature

When the spin-lattice relaxation time is long, the spin system is thermally rather well isolated from the lattice motions, and during magnetic resonance experiments the spin system is seldom in thermal equilibrium with the lattice motions. Consequently, it is advantageous to consider the spin system as having its own temperature, distinct from that of the lattice. The term *spin temperature* has become common in the literature on magnetic resonance. When T_1 is long, it is easy to raise the spin temperature t_s many orders of magnitude above that of the lattice by exposure of the system to resonant radiation. The rate of return of t_s to the lattice temperature after the resonant radiation is cut off depends, of course, on the value of T_1. When the spins of a system are similar and give rise to a homogeneously broadened resonance, the spin temperature can usually be considered the same throughout the system. However, for systems having spin packets with resonance frequencies different from those of other spin packets, such as those having inhomogeneously broadened lines, it is possible to raise the spin temperature of a particular spin packet well above that of other spins of the system by exposing the system to radiation at the resonant frequency of that particular packet. The rate at which the spin temperature is equalized throughout the spin system after the heating pulse is extinguished is determined by the spin-spin relaxation T_2.

Quantitatively, the spin temperature is defined in terms of the population distribution of the magnetic spin states. For a two-level spin system ($S = 1/2$), in which all the spins have the same level spacing, $(E_{1/2} - E_{-1/2}) = h\nu$, a quantitative definition is quite simple. If the spins of the system are in thermal equilibrium (but with the spin system not necessarily in equilibrium with the lattice), the ratio of the populations of the upper and lower states is given by the Boltzmann relation

$$\frac{N_{1/2}}{N_{-1/2}} = \exp\frac{-h\nu}{kt_s} \tag{4.133}$$

where t_s is the spin temperature in Kelvin units. This equation is easily solved, to give

$$t_s = \frac{h\nu}{k \log_e\left(\dfrac{N_{-1/2}}{N_{1/2}}\right)} \tag{4.134}$$

When the spin system is in equilibrium with the lattice, the population ratio $N_{-1/2}/N_{1/2}$ will be such as to give t_s the same value as the lattice temperature. Absorption of resonant radiation will tend to equalize the state population and thus increase t_s above the lattice temperature T. As $N_{-1/2}/N_{1/2}$ approaches unity, the spin temperature approaches infinity. When T_1 is long, only moderate absorption of resonant radiation can make $N_{-1/2}/N_{1/2}$ closely approach unity and t_s extremely high as compared to the lattice temperature.

Note that when the upper state population is the greater, $N_{-1/2} < N_{1/2}$, (4.134) indicates the spin temperature as negative. Although the absorption of resonant radiation can make $N_{-1/2}$ approach $N_{1/2}$ very closely, it cannot make $N_{-1/2}$ exceed $N_{1/2}$. However, there are now a number of well known methods such as adiabatic fast passage or optical pumping that can bring about a greater population in the upper state, $N_{-1/2} < N_{1/2}$, and thus achieve t_s negative. Consequently, the term *negative spin temperature* is commonly used in the description of magnetic resonance phenomena, and the concept of negative temperature is very useful in the theory of masers and lasers.

9.d. Spin-Spin Relaxation

Normally a system of like spins in a constant magnetic field has the same spin temperature throughout the system. Let us assume that such a system is subjected to a radiation pulse that raises the spin temperature of one group of spins above that of others. The process of equalization of the spin temperature or the restoration of thermal equilibrium within the perturbed system requires the exchange of energy between different spins, but not necessarily between the spins and the lattice. The spin exchange that restores thermal equilibrium within the spin system itself is called *spin-spin relaxation.*

Although the mutual flipping of two equivalent spins, one up and one down, often called a *flip-flop*, does not change the total energy of the spin system, it does transfer energy from one spin to another, and it does alter the phase relations of the spins and the transverse magnetization. Consequently, for spin systems that give homogeneously broadened resonances, T_2 of the Block equations gives a measure of the rate of exchange of energy between the spins of the system (spin-spin relaxation), and T_2 is commonly used to designate the spin-spin relaxation time. The decay of the transverse magnetization may be regarded as the classical analogue of the spin-spin relaxation.

If a spin flips without an accompanying spin flop, there will be a change in both the horizonal and transverse magnetization. Thus unaccompanied spin flips influence both T_1 and T_2, whereas spin flip-flops influence only T_2. For this reason, T_1 is an upper limit to the value of T_2, and T_2 can be (and often is) much shorter than T_1. In nuclear resonance T_2 is almost always very much shorter than T_1, which is relatively long because nuclear spins are not effectively coupled to the lattice motions. In esr, T_1 can be short and comparable in value to T_2 because of residual spin-orbital coupling. For free radicals in which this residual coupling is very small or for samples at low temperature where the lattice motions are frozen out, T_1 is generally much greater than T_2.

Spin-spin relaxation is the principal determinant of T_2 for homogeneously broadened lines of samples in which the spins are concentrated. In diluted samples in which the spins are widely separated, spin-spin interaction is relatively unimportant, spin flipping results mostly from spin-lattice relaxation, and T_2 approaches T_1 in value.

9.e. Inhomogeneous Relaxation and Spin Diffusion

For inhomogeneously broadened lines, different spins or spin-packets in the sample precess at slightly differing frequencies because of either inhomogeneities in the applied magnetic field or a difference in the internal field effective at the various sites. It is evident that the phase relations of transverse components rotating at different angular velocities would be constantly changing, even if no spin flip were to occur. Thus the transverse relaxation in such samples cannot be ascribed entirely to mutual spin flipping although the difference in the local fields may result from spin-spin interactions. Inhomogeneous broadening is due in part to a spread of resonant frequencies of various spin packets, and not entirely to relaxation processes. Therefore, the spin-spin relaxation time is not correctly measured by the inverse width of inhomogeneously broadened lines. Nevertheless, it is useful to define a pseudorelaxation time $T_2{}^*$ by [15]

$$T_2{}^* = \frac{1}{\Delta\omega_{1/2}} = \frac{1}{\gamma \Delta H_{1/2}} \tag{4.135}$$

where $\Delta\omega_{1/2}$ or $\Delta H_{1/2}$ is half the line width between half-power points of the inhomogeneously broadened line. This definition is loosely justified by the fact that spins precessing in phase with a Larmor frequency difference of $\Delta\omega$ will get out of phase in a time approximately equal to $1/(\Delta\omega)$. Equation (4.135) is analogous to the expression for T_2 of the Lorentzian line shape, (4.120), and T_2^* approaches T_2 as the inhomogeneity in the broadening decreases. In liquid samples the differences in the local fields are usually averaged out by the rapid motions of the spins so that the Lorentzian shape is observed, and the inverse line width gives a true value of the spin-spin relaxation T_2 when $T_2 << T_1$.

As indicated in Fig. 4.8, the inhomogeneous line is a composite of an overlapping sequence of more narrow homogeneous lines of spin packets having slightly different peak frequencies. Spin flip-flop exchange can occur between spins of packets having adjacent, overlapping absorptions. In this way, a spin-flip excitation can move from packet to packet in random movement throughout the entire sample, provided that the excitation is not too quickly dissipated by spin-lattice relaxation. This process of spin-spin relaxation in inhomogeneous systems is called *spin diffusion.* It is generally too slow to prevent the "hole burning" shown in Fig. 4.8 and yet must be fast enough to precede the spin-lattice relaxation. In many species it is a significant relaxation process, however, one that can be readily recognized from the change in absorption signal as detected with a nonsaturating spectrometer set on one part of the band when staturating power is applied in another. This is known as *electron-spin electron-spin double resonance* (ELDOR). [25, 26].

In a fluid or semifluid medium, spin diffusion can occur through movement of the paramagnetic elements to other sites in the sample. Measurement of saturation transfer within inhomogeneous resonances resulting from motional spin diffusion can give information about the rate of movement of free radicals and spin-labeled molecules in liquid systems. Although the resonances of nonviscous liquids are generally homogeneous, motional spin diffusion within liquid systems can be studied by deliberate introduction of inhomogeneity in the applied field.

9.f. Saturation Transfer by Spectral Diffusion

A very important type of relaxation occurs in slowly rotating radicals that have anisotropy in their **g** tensors or nuclear hyperfine tensors, or both. Let us assume that one such radical with a doublet spin state has the correct orientation to absorb a quantum of frequency ν_0 and after doing so turns slowly to another angle without undergoing a reverse transition. While in the excited state, its transition frequency has been changed because the internal field of the radical effective on the spin has changed with the orientation. However, it cannot absorb a quantum at the new frequency because it is still in the upper spin state. A dispersal of the transition frequencies in this way, without change of spectral state, is called *spectral diffusion.*

If the various radicals of a magnetically anisotropic polycrystalline sample are rotating in a random manner at rates too slow to induce transitions but sufficiently rapid to cause frequency dispersal before appreciable spin-lattice relaxation occurs, power saturation imposed at a particular frequency ν_0 will thus be spread to other frequencies within the anisotropic band. Saturation transfer by this process can be observed with the double resonance techniques mentioned earlier [25, 26], but more sensitive modulation techniques have been developed for the purpose [27, 28]. Observations of such saturation transfer provide the most powerful method for measurement of the very slow rotations of large organic radicals in viscous media. The method has proved to be of special importance for study of the motions of spin-labeled biological polymers in membranes [29, 30].

To achieve the adiabatic dispersal of the frequencies required for this kind of saturation transfer, the rotations must be slow as compared with the Larmor frequency of the spins. Furthermore, they must be too slow to average out the anisotropy of the resonances as do the rapidly tumbling motions of radicals in nonviscous liquids. The normal sweep spectrometer would not detect the effects of such slow rotation and would sweep out the anisotropic signals expected for polyoriented, static radicals.

9.g. Cross Relaxation

In samples observed with esr there are often more than one paramagnetic species. When two such species have overlapping resonances—if only in the wings of their absorption contours—and when their spins are not too widely separated in space, a flip-flop spin exchange can occur in which a spin of one species A emits a quantum that is absorbed by a neighboring spin of the second species B. If the spin-spin coupling between species A and B is sufficiently strong that the exchange is rapid in comparison with the individual relaxation times of A and B, it will tend to reduce the longer relaxation time to the shorter.

Let us assume, for example, that $T_1{}^A$ (for species A) is very long and that $T_1{}^B$ (for species B) is very short. If a quantum absorbed by a spin of type A is given to a spin of type B through a spin exchange, the quantum will have a high probability of being lost to the lattice by the B spin before it is transferred back to the A spin through a reverse exchange. The spin-lattice relaxation time of the A spin system is thus effectively shortened by the cross relaxation.

Cross relaxation commonly occurs in irradiated solids where more than one stable free radical species is produced by the irradiation. Figure 4.15 illustrates the effects of cross relaxation on the intensity of the lower-field component of the H-atom doublet produced by γ-irradiation of H_2O at 4.2°K [31]. In the absence of preferential cross relaxation, the two H-atom hyperfine components are expected to be of equal intensity and, in the absence of saturation broadening or site splitting, very sharp. Power absorption by either component is severely limited by the inability of the spins to dispose of the absorbed energy. Any

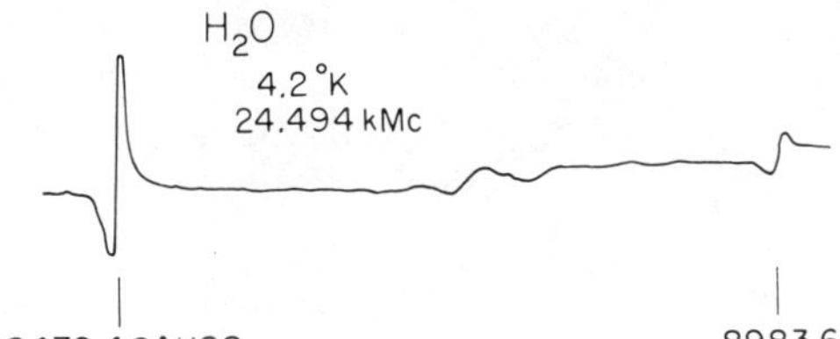

Fig. 4.15 Illustration of the effects of cross relaxation on the relative intensities of the H-atom doublet in the esr of γ-irradiated H_2O at 4.3°K. The two outer components are the lines of the H doublet. Normally, the two components have equal strength. Here the lower-field component is much the stronger because its spin lattice time is reduced through cross relaxation with OH and possibly other radicals, the absorptions of which are spread to the lower-field side by **g** anisotropy. From H. N. Rexroad and W. Gordy, *Phys. Rev.*, **125**, 242 (1962).

process that assists the H atoms to dispose of the quantum absorbed by either component will increase the power absorption and strengthen the signal observed for that component. Note that the lower field component of the H-doublet in Fig. 4.15 is orders of magnitude stronger than the higher field component. This difference is believed to be caused by the overlapping of the lower field component by absorption of either OH radicals, H_2O^+ radical ions, or both [31], which are produced along with the H atoms by the irradiation of the frozen H_2O. Both OH and H_2O^+ radicals have **g** anisotropy that, at the observation frequency, causes the wings of their absorption to spread over the lower- but not the upper-field component of the H-atom doublet. The **g** anisotropy also indicates the presence of residual spin-orbital coupling that would reduce T_1 for these radicals significantly below that for the trapped H atoms.

As was shown by Bloembergen et al. [32], who did pioneering work on cross relaxation, the two spin systems need not have the same peak resonance frequency for cross relaxation to occur between them. They need only to overlap in the wings of their absorption contours. Energy absorbed at any frequency within the band of a homogeneously broadened line is spread quickly to other spins of the system through rapid, spin-spin relaxation. Furthermore, the individual spins engaging in a cross-relation, flip-flop exchange need not have the same magnetic energy. The difference in energy of the two transitions is supplied by, or dissipated in, the spin systems through the process of spin diffusion.

Cross relaxation in irradiated solids where two kinds of radical species are often created simultaneously by bond dissociation or molecular ionization is not uncommon. In frozen solids the pairs simultaneously created usually are trapped close together and hence experience strong spin-spin interaction that is conducive to rapid cross relaxation. This pair-type relation in irradiated ice is illustrated in Fig. 4.15. Cross relaxation in irradiated solids often has an important influence on ENDOR signal strength.

Slight paramagnetic impurities can also give rise to marked decrease in the spin-lattice relaxation of a spin system through cross relaxation. Let us assume, for example, that the spin system A has a very long spin-lattice relaxation time, whereas the impurity species B is closely coupled to the lattice, that is, has a very short T_1. Quanta absorbed by spins of species A will migrate to the impurity elements B through spin exchange in species A and will be transferred to spins of B through spin exchange (flip-flop) between neighboring A and B spins. Quanta thus absorbed by B will be quickly converted to thermal motions of the lattice.

Quantitative treatments of cross relaxation may be found in the original paper by Bloembergen et al. [32], and in the book on electron-spin relaxation by Standley and Vaughan [15].

4.10 TECHNIQUES FOR OBSERVATION OF RELAXATION TIMES

10.a. Pulse Techniques

From Section 7 and Fig. 4.12 it is evident that resonant radiation $\omega = \omega_0$ applied normal to $\mathbf{H}_0$ will cause $\mathbf{M}$ to precess about $\mathbf{H}_1$ with angular velocity $\omega_p = \gamma H_1$, or with a period

$$T_p = \frac{1}{\nu_p} = \frac{2\pi}{\gamma H_1} \tag{4.136}$$

Let us assume that H_1 is sufficiently large that $T_p \ll T_1, T_2$. Then, for short pulses of radiation, the model of Fig. 4.12 holds very well. A pulse of duration $(1/4)T_p$ (90° pulse) will leave $\mathbf{M}$ rotating in the xy plane with frequency ω_0. Thus at the instant such a pulse is turned off, M_z will be zero, but the M_z component will gradually return to its thermal equilibrium value M_0 at a rate determined by T_1, as is indicated in (4.124) and shown graphically in Fig. 4.13. At the end of the 90° pulse, the magnetization in the xy plane will decay at a rate determined by T_2. In the rotating frame of reference, this is measured by the exponential decay of M_1 of (4.130) and in the laboratory frame by the decay of the alternating M_x or M_y component of (4.131) and (4.132). Figure 4.14 shows graphically the decay of the M_x component after the quarter-wave pulse is turned off.

The magnetization rotating in the xy plane will induce signals at frequency ω_0 in a receiver coil placed in this plane. Such an arrangement can be used for detection of the M_x decay signal of Fig. 4.14 and for measurement of T_2 when the loading effects of the coil are taken into account or when the coupling to the coil is sufficiently weak that these effects are negligible. This method of detection of M_x is similar to the Bloch induction method for detection of magnetic resonance. To detect a continuous signal, one would apply a cw radiation component at ω_0 along x and the pickup coil along y. The spin system then acts simply as a transformer to transfer energy at ω_0 from the primary coil (source power) to the secondary coil (receiver).

If the length of the applied pulse is $t = (1/2)T_p = \pi/(\gamma H_1)$ (180° pulse), the z component of the magnetization M_0 will be negative at the end of the pulse. This will be evident from Fig. 4.12, which shows that as $\mathbf{M}$ precesses about $\mathbf{H}_1$ the z component can be represented by $M_z = M_0 \cos \gamma H_1 t$. After the 180° pulse is extinguished, M_z will return to its former positive value at a rate that is determined by the spin-lattice relaxation time T_1. Quantum mechanically, a negative M_0 means that the populations of the upper and lower levels are inverted, a condition that corresponds to a negative spin temperature (see Section 9).

10.b Spin Echoes

The spin-echo technique contributed to magnetic resonance by Hahn [33] is especially useful for measuring relaxation times, particularly phase memory, and for distinguishing between T_2 and T_2*. Among other applications, it is also useful for study of spin diffusion, whether due to indirect spin flip-flop exchanges or to migration as in liquids of the paramagnetic elements. The early experiments with spin echoes were in nmr, but applications in esr are numerous and are increasing. Reviews of these experiments are given in various monographs on relaxation [15, 23, 24]. A particularly lucid and thorough treatment of spin echoes is given by Mims [34]. To give the nonspecialist in relaxation phenomena an introduction to this ingenious technique, we describe briefly the basic method for generation of spin echoes as originally introduced by Hahn. The classical model developed in Section 7 and the pulse technique described in the preceding section are prerequisites to an understanding of the process.

Spin echoes are made possible by inhomogeneous broadening that may result from inhomogeneity either in the internal fields or in the externally applied field $\mathbf{H}_0$. As a result of this inhomogeneity, individual spin packets of the sample have slightly differing resonant frequencies. Let us assume that a strong 90° pulse of radiation with frequency ω_0 at the center of an inhomogeneously broadened resonance is applied with $\mathbf{H}_1$ along x'. If $H_1 >> \Delta H$ where ΔH is the total spread in the inhomogeneous field, the pulse will cause all components of $\mathbf{M}$ that contribute to $\mathbf{M}_0$ to rotate smoothly about x' so that at the end of the 90° pulse $\mathbf{M}_0$ will lie along y'. Immediately after the pulse, the component vectors that comprise $\mathbf{M}_0$ will, because of their slightly differing precessional frequencies, begin to fan out in the $x'y'$ plane. Relative to the fixed coordinates, those with ω slightly less than ω_0 will drop behind y', whereas those with ω greater than ω_0 will lead y'. Relative to y', assumed to rotate counterclockwise from y at ω_0, the components with $\omega < \omega_0$ will rotate in a clockwise sense and those with $\omega > \omega_0$, in a counterclockwise sense. After a given time, the various components will have so spread out in the $x'y'$ plane that $\mathbf{M}_0$ will vanish and also the induction signal picked up in the receiver coil. This fanning out of the component vectors after time τ is indicated in Fig. 4.16 where the frequency differences in the spin packets are assumed, for simplicity, to be uniform. At time τ after the first pulse when the vectors are thus fanned out, a second pulse is applied, but with twice the duration (180° pulse). This second pulse will cause all the fanned-out components of magnetization to rotate through 180° about x'. Relative to the fixed frame, this rotation will reverse the positions of the slower and faster components, putting ahead those rotating more slowly, so that a later convergence of the components will occur. In time τ after the 180° pulse, the magnetization vector will be reconstructed along the y' axis, but in the opposite direction ($-y'$) because of the 180° rotation of all the $\mathbf{M}$ components, and with a smaller amplitude than that of the original $\mathbf{M}_0$ because of the

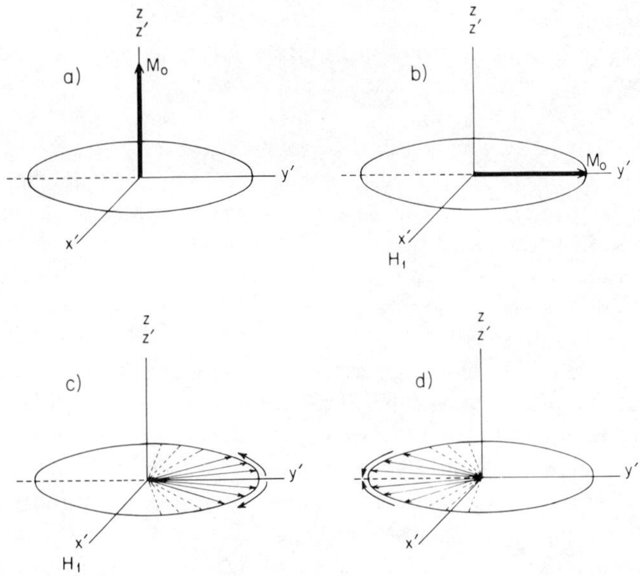

Fig. 4.16 Diagrams illustrating the mechanism of the pulse generation of spin echoes. Before a pulse is applied, the magnetization $\mathbf{M}_0$ is along the common z or z' axis, as shown in (a). A 90° pulse applied with $\mathbf{H}_1$ along the x' axis rotates $\mathbf{M}_0$ about x' to the y' axis of the rotating frame, as in (b). Because of inhomogeneous broadening some components of $\mathbf{M}_0$ rotate faster and some slower than y', which itself rotates at the frequency ω_θ of the applied power pulse assumed to be the center Larmor frequency of the resonance. As a result, components of $\mathbf{M}_0$ begin to fan out in the $x'y'$ plane immediately after the quarter-wave pulse is applied, as indicated in (c). Note that relative to y' the faster components rotate clockwise and the slower rotate counterclockwise. After a time τ when the components have fanned out sufficiently to cancel the $x'y'$ magnetization, an 180° pulse is applied with $\mathbf{H}_1$ again along x' to rotate all the component vectors through 180° about x'. The effect of this rotation is to cause the vectors again to converge on y', as indicated in (d). In time τ after the 180° pulse (or 2τ after the original 90° pulse), the $x'y'$ magnetization will again be along the y' axis, but in the negative direction. The reconstructed magnetization will produce an induction signal (an echo) that can be observed by a pickup coil in the xy plane. After their convergence, the components will again fan out, and the echo signal will disappear. If a second 180° pulse is applied in time 2τ after the first 180° pulse (or 3τ after the 90° pulse), the magnetization will again be reconstructed along the y' axis, and a second echo may be observed. This process can be repeated, but gradually the reconstructed magnetization, and consequently the echo signal strength, becomes weaker because of loss of phase memory, as explained in the text and illustrated in Fig. 4.17.

loss of phase memory of the spins. The loss of phase memory arises from random spin flips resulting from spin-spin interactions (measured by T_2) and from spin-lattice relaxation (measured by T_1), whereas the orderly dispersion of the components in the $x'y'$ plane due to slight differences in precessional frequencies of various spin packets causes loss of phase coherence but not of phase memory.

Since the individual vectors continue to rotate at the same rate *after* the 180° pulse as *before* it, the time required for them to converge is τ, the same as that required for them to disperse. Thus in time τ after the 180° pulse, an induction signal may again be observed in the receiver of the pickup coil imposed in the xy plane. This signal, called the *spin echo*, will be 180° out of phase with the original induction signal pulse observable immediately after the 90° pulse. If at time τ after the first echo—when the vectors are again dispersed—a second 180° pulse is applied exactly as before, a second echo will be observed, provided that the phase memory has not been lost. This echo will be in phase with the original signal because the second rotation of 180° again reverses the direction of the magnetization. By similarly repeated 180° pulses, a sequence of echoes can be observed until the phase memory is entirely lost through random spin flips. Such a sequence is indicated in Fig. 4.17.

The exponential decay of the echo signals (Fig. 4.17) results from spin-spin and spin-lattice relaxations. The characteristics of this decay curve give information about T_1 and T_2. When T_1 is known from measurements of the M_z growth curve (Fig. 4.13), T_2 can be obtained. When T_1 is known to be very long, as compared with T_2, one might attribute the decay entirely to spin-spin relaxation; in this instance T_2 can be determined directly from the spin-echo decay curve.

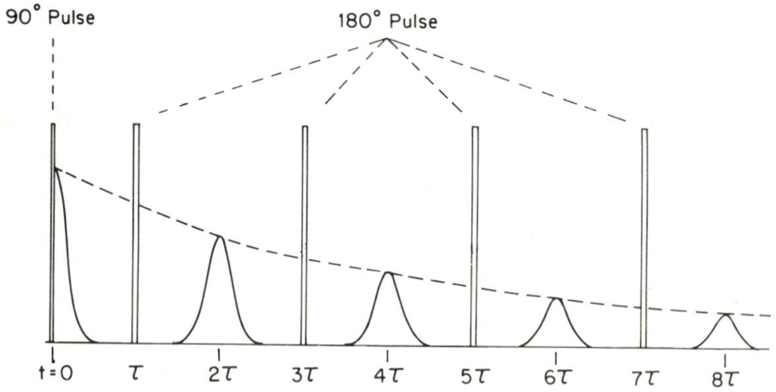

Fig. 4.17 Illustration of exponential decay of a spin-echo sequence resulting from loss of phase memory in the spin system. For further description, see legend for Fig. 4.16.

10.c. Adiabatic Fast Passage

Another commonly used method for reversing the populations of a doublet-spin state (achieving negative magnetization $\mathbf{M}_0$) is simply to sweep the frequency of the radiation through the spin resonance in a time that is very short as compared with T_1 and T_2 but slow as compared with the precessional frequency of $\mathbf{M}$ about the effective field $\mathbf{H}_{eff}$ of Fig. 4.11. Under these conditions, the magnitude of $\mathbf{M}$ will not change significantly during the sweep over the resonance, but its M_z component will decrease from $+M_0$ to $-M_0$. Equation (4.102) or (4.103) and Fig. 4.10 show that the orientation θ of the effective field $\mathbf{H}_{eff}$ relative to the z axis depends on the value of $(\omega_0 - \omega)$ and changes from positive to negative when ω becomes greater than ω_0. If the orientation θ is not changed too rapidly, $\mathbf{M}$ will continue to precess about $\mathbf{H}_{eff}$ as θ changes from positive to negative. Usually, ω is left constant, and $(\omega_0 - \omega)$ is changed as ω_0 is swept through ω. This is accomplished by a sweep of the z-field H. The Ehrenfest conditions for adiabatic perturbation [35] require that the change of orientation be slow, as compared with the rate of precession of $\mathbf{M}$ about $\mathbf{H}_{eff}$, that is,

$$\frac{d\theta}{dt} = \gamma H_{eff} \tag{4.137}$$

From Fig. 4.10 it is evident that $\Delta\theta = \Delta H_{eff}/H_{eff}$ and hence that

$$\frac{d\theta}{dt} = \frac{1}{H_{eff}}\left(\frac{dH_{eff}}{dt}\right) << \gamma H_{eff} \tag{4.138}$$

When the resonance is swept out in the usual way by variation of the z field, $\mathbf{H}_0$, with ω constant and with $\mathbf{H}_1$ also constant in the rotating reference frame, one sees from differentiation of (4.102) that $dH_{eff}/dt = dH_0/dt$ and that the requirement of (4.138) for adiabatic passage may be expressed as

$$\frac{1}{H_{eff}}\left(\frac{dH_0}{dt}\right) << \gamma H_{eff} \tag{4.139}$$

Since the smallest H_{eff} is H_1, this condition limits the maximum rate of sweep of the z field to

$$\frac{1}{H_1}\left(\frac{dH_0}{dt}\right) << \gamma H_1 \tag{4.140}$$

for adiabatic passage. This maximum allowable sweep rate must be sufficient for the passage to be achieved before significant relaxation of the spins has occurred. This requirement can be expressed as

$$\frac{1}{H_1}\left(\frac{dH_0}{dt}\right) >> \gamma\Delta H_{1/2} \approx \left(\frac{1}{T_1}+\frac{1}{T_2}\right) \tag{4.141}$$

where $\Delta H_{1/2}$ is the line half-width. The combined requirements limiting the choices of the power level and sweep rate may be expressed as

$$\gamma\Delta H_{1/2} << \frac{1}{H_1}\left(\frac{dH_0}{dt}\right) << \gamma H_1 \tag{4.142}$$

Obviously, these conditions are more easily met when $\Delta H_{1/2}$ is small, that is, when the relaxation times are long.

REFERENCES

1. L. Pauling and E. B. Wilson, Jr., *Introduction to Quantum Mechanics,* McGraw-Hill, New York, 1935.
2. W. Gordy and R. L. Cook, *Microwave Molecular Spectra,* Wiley-Interscience, New York, 1970.
3. C. P. Poole, Jr., *Electron Spin Resonance: A Comprehensive Treatise on Experimental Techniques,* Interscience, New York, 1967.
4. R. S. Alger, *Electron Paramagnetic Resonance: Techniques and Applications,* Interscience, New York, 1968.
5. W. Köhnlein and A. Müller, *Phys. Med. Biol.,* **6**, 599 (1962); *Int. J. Radiat. Biol.,* **8**, 141 (1964).
6. J. H. Van Vleck, *Phys. Rev.,* **74**, 1168 (1948).
7. A. M. Portis, *Phys. Rev.,* **91**, 1071 (1953).
8. H. A. Lorentz, *Proc. Amsterdam Acad.,* **8**, 591 (1906).
9. J. H. Van Vleck and V. F. Weisskopf, *Rev. Mod. Phys.,* **17**, 227 (1945).
10. R. Karplus and J. Schwinger, *Phys. Rev.,* **73**, 1020 (1948).
11. F. Bloch, *Phys. Rev.,* **70**, 460 (1946).
12. G. E. Pake and T. L. Estle, *The Physical Principles of Electron Paramagnetic Resonance,* 2nd ed., W. A. Benjamin, Reading, Mass., 1973.
13. H. G. Hecht, *Magnetic Resonance Spectroscopy,* Wiley, New York, 1967.
14. I. Waller, *Z. Physik,* **79**, 370 (1932).
15. K. J. Standley and R. A. Vaughan, *Electron Spin Relaxation Phenomena in Solids,* Adam Hilger, London, 1969.
16. R. Orbach, *Proc. Roy. Soc.,* **A264**, 458 (1961).
17. W. Heitler and E. Teller, *Proc. Roy. Soc.,* **A155**, 629 (1936).

18. M. Fierz, *Physica,* **5**, 433 (1938).
19. R. de L. Krönig, *Physica,* **6**, 33 (1939).
20. J. H. Van Vleck, *J. Chem. Phys.,* **7**, 72 (1939).
21. J. H. Van Vleck, *Phys. Rev.,* **57**, 426 (1940).
22. A. Abragam and B. Bleaney, *Electron Paramagnetic Resonance of Transition Ions,* Clarendon, Oxford, 1970.
23. R. Orbach and H. J. Stapleton, "Electron Spin-Lattice Relaxation," in *Electron Paramagnetic Resonance,* S. Geschwind, Ed., Plenum, New York, 1972, Chapter 2.
24. C. P. Poole, Jr. and H. A. Farach, *Relaxation in Magnetic Resonance: Dielectric and Mossbauer Applications,* Academic, New York, 1971.
25. J. S. Hyde, J. C. W. Chien, and J. H. Freed, *J. Chem. Phys.,* **48**, 4211 (1968).
26. J. S. Hyde, R. C. Sneed, Jr. and G. H. Rist, *J. Chem. Phys.,* **51**, 1404 (1969).
27. J. S. Hyde and L. Dalton, *Chem. Phys. Lett.,* **16**, 568 (1972).
28. L. R. Dalton, B. H. Robinson, L. A. Dalton, and P. Coffey, in *Advances in Magnetic Resonance,* Vol. 8, J. S. Waugh, Ed., Academic, New York, 1976, pp. 149-259.
29. T. Stone, T. Buckman, P. Nordio, and H. M. McConnell, *Proc. Nat. Acad. Sci.* (USA), **54**, 1010 (1965).
30. H. M. McConnell and B. G. McFarland, *Quart. Rev. Biophys.,* **3**, 91 (1970).
31. H. N. Rexroad and W. Gordy, *Phys. Rev.,* **125**, 242 (1962).
32. N. Bloembergen, S. Shapiro, P. S. Pershan, and J. O. Artman, *Phys. Rev.,* **114**, 445 (1959).
33. E. L. Hahn, *Phys. Rev.,* **80**, 580 (1950).
34. W. B. Mims, "Electron Spin Echoes," in *Electron Paramagnetic Resonance,* S. Geschwind, Ed., Plenum, New York, 1972, Chapter 4.
35. P. Ehrenfest, *Proc. Amsterdam Acad.,* **19**, 576 (1916).

Chapter V

ANALYSIS OF SPECTRA IN SINGLE CRYSTALS

Solutions are obtained in Chapter III for the anisotropic spin Hamiltonian expressed in terms of the principal axes of **g** and **A**, as well as in arbitrarily chosen orthogonal axes fixed in space. In studies of single crystals, one usually does not know in advance the principal axes, and the problem is to find them. The first step in the solution of this problem is to evaluate the tensor elements of **g** and **A** relative to an orthogonal system that can be identified by obvious structural features of the crystal. The second step is to diagonalize the **g** and **A** tensors to find the principal elements and direction cosines relative to the predefined axes.

The seven crystal systems identified by crystallographers from measurement of angles between the crystal faces without regard to internal structure are indicated by Fig. 5.1 and Table 5.1 [1]. These classifications are used by esr spectroscopists in choosing a reference system for measurement and analysis of esr spectra of single crystals. When not already known or obvious, the crystal type can be readily identified by measurement of the angles between the crystal faces. If the crystal is orthorhombic, the logical choice of reference axes is the *a*, *b*, and *c* crystallographic axes. For monoclinic crystals, one might select the

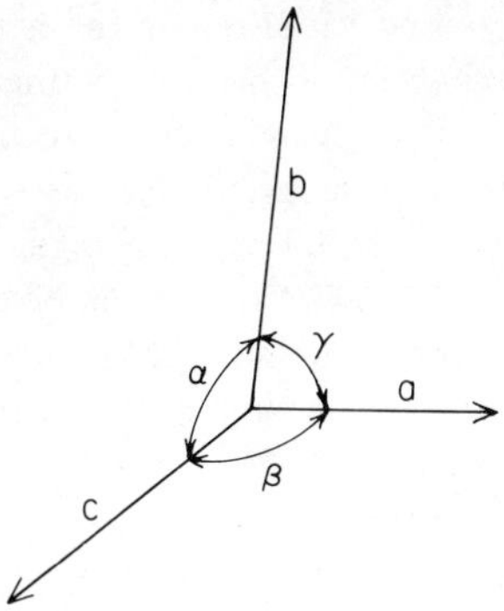

Fig. 5.1 Conventional labeling of crystal axes and angles. Compare with Table 5.1.

Table 5.1 Distinguishing Dimensions of the Conventional Crystal Systems[a]

System	Number of Lattices	Axes	Angles
Triclinic	1	$a \neq b \neq c$	$\alpha \neq \beta \neq \gamma$
Monoclinic	2	$a \neq b \neq c$	$\alpha = \gamma = 90^\circ \neq \beta$
Orthorhombic	4	$a \neq b \neq c$	$\alpha = \beta = \gamma = 90^\circ$
Tetragonal	2	$a = b \neq c$	$\alpha = \beta = \gamma = 90^\circ$
Cubic	3	$a = b = c$	$\alpha = \beta = \gamma = 90^\circ$
Trigonal	1	$a = b = c$	$\alpha = \beta = \gamma < 120^\circ, \neq 90^\circ$
Hexagonal	1	$a = b \neq c$	$\alpha = \beta = 90^\circ, \gamma = 120^\circ$

[a] Compare with assignment of sides and angles in Fig. 5.1. For a further description of the crystal systems, consult Kittel [1].

two orthogonal crystallographic axes a and b and then designate the third axis c' that is perpendicular to the ab plane. For triclinic crystals, for which none of the crystallographic axes is orthogonal, one might choose the crystallographic b axis along an edge of the crystal and a second axis a' perpendicular to it and in a face of the crystal; the third axis c' is then chosen orthogonal to the $a'b$ plane. When the internal structure of the crystal is known, the measured esr parameters can be related directly to this structure. However, much useful information is obtainable from esr of crystals of unknown internal structure, particularly of irradiated crystals.

Measurements of esr spectra are usually made with the paramagnetic crystal mounted in a high-Q, resonant cavity at a fixed observing frequency ν_0 that is the peak resonant frequency of the cavity. The resonant frequency or frequencies of the oriented paramagnetic species in the crystal are then swept through the frequency ν_0 by slow variation of the applied field **H**. The field value H_0 corresponding to the peak absorption at the chosen frequency ν_0 is recorded for each component of the particular esr spectrum. Because of anisotropies in **g** and possible hyperfine interactions, the resonant field values will depend on the orientation of the crystal relative to the applied field. In typical measurements, the crystal is mounted so that it can be rotated about one of the reference axes that is normal to the applied field. For example, if this axis is z, the applied field would then be in the xy plane, and the measurements of the resonant field values made with **H** at different angles relative to x or y are said to be made in the xy plane. One then makes plots of the observed spectrum as a function of orientation of the crystal in each of the coordinate planes and seeks to fit the theoretical formulas to the observed curves by adjustment of the parameters of the formulas, usually by computer simulation. In some crystal types, free radicals of the same kind have more than one orientation relative to spaced-fixed axes. Often, two or four distinct orientations are observed, depending on the crystal symmetry or the number of molecules in the unit cell. The experimenter must unscramble the patterns of these differently oriented radicals before a fitting of the experimental and theoretical curves can be made. When the free radicals are produced by irradiation of the crystal, more than one free-radical species is often produced. The patterns of these species must also be unscrambled before an analysis can be made. Different isotopic species of the same radical type can give rise to qualitatively different spectra when the isotopic species do not have the same nuclear spins or couplings. These various complications are sometimes frustrating to the observer, but they provide him insurance against boredom!

1. EVALUATION OF THE g TENSOR

As explained in Chapter III, the effects of nuclear fields on **g** anisotropy are usually negligibly small. For this reason, it is possible to abstract the elements of the **g** tensor from experimental data before an analysis of the hyperfine structure is made. To first order, one can obtain g-only experimental curves by taking measurements at the center of the symmetrical hyperfine multiplets. Second-order effects can, however, skew the center of the hyperfine patterns when the nuclear splitting, ΔH, is large, that is, unless $\Delta H \ll g\beta H$. These second-order effects are discussed in Section 3. Here we describe the application of the first-order theory for finding the principal g values.

For applications in the three principal planes, the formulas of Chapter III, Section 3 can be expressed in more convenient forms. If measurements are taken in the xy plane where **H** forms an angle θ with the x axis, the effective $g_\theta^{\,2}$ for various θ values is given by (3.46).

$$g_\theta^2 = (g_{xx}\cos\theta + g_{yx}\sin\theta)^2 + (g_{xy}\cos\theta + g_{yy}\sin\theta)^2 + (g_{xz}\cos\theta + g_{yz}\sin\theta)^2$$

$$= g_x^2\cos^2\theta + g_y^2\sin^2\theta + 2(g_{xx}g_{yx} + g_{xy}g_{yy} + g_{xz}g_{yz})\sin\theta\cos\theta \qquad (5.1)$$

where

$$g_x^2 = g_{xx}^2 + g_{xy}^2 + g_{xz}^2 \qquad (5.2)$$

$$g_y^2 = g_{yy}^2 + g_{yx}^2 + g_{yz}^2 \qquad (5.3)$$

and where g_x and g_y are the effective g values along the x and y axes. By use of standard trigonometric relations, (5.1) can be put in the more convenient form for plotting:

$$g_\theta^2 = K_1(x,y) + K_2(x,y)\cos 2\theta + K_3(x,y)\sin 2\theta \qquad (5.4)$$

where

$$K_1(x,y) = \frac{1}{2}(g_x^2 + g_y^2) \qquad (5.5)$$

$$K_2(x,y) = \frac{1}{2}(g_x^2 - g_y^2) \qquad (5.6)$$

$$K_3(x,y) = g_{xx}g_{yx} + g_{xy}g_{yy} + g_{xz}g_{yz} \qquad (5.7)$$

Expressions for the other coordinate planes can be obtained by obvious exchange of the coordinates. It is sometimes convenient to use the maximum and minimum values of **g** observed in the coordinate planes. These are obtained when $dg_\theta/d\theta$ is set equal to zero. The result is

$$\tan 2\theta_m = \frac{K_3}{K_2} \qquad (5.8)$$

where θ_m is the value of θ for which the extreme values of g_θ in the xy plane are observed.

Measurements along the three axes give the three effective values g_x, g_y, and g_z, and measurements in the three planes away from the axes at any θ can give only three additional constants. Because the tensor is symmetric, $g_{xy} = g_{yx}$, and so on, there are only six unknown tensor elements to be evaluated, and there are six independent equations linking these unknowns to the measurable constants. Thus all elements of the **g** tensor are determinable.

The effective g_θ values to be used in (5.4) may be obtained from the relation

$$g_\theta = \frac{\nu_0}{\beta H_\theta} \tag{5.9}$$

where H_θ are the resonant field values at the fixed frequency ν_0 for the different orientations θ. However, for the small **g** anisotropies normally observed in free radicals, it is possible (and more convenient) to express the relationship of (5.4) in terms of the displacements, ΔH, in resonant field values caused by the anisotropy. For example, let us write

$$g_\theta = g_0 + \Delta g \tag{5.10}$$

and

$$H_\theta = H_0 + \Delta H \tag{5.11}$$

where g_0 and H_0 are the effective g and H for resonance of an isotropic reference signal at the observing frequency ν_0. Theoretically, g_0 might best be chosen as the free spin value 2.0023, but the nearly isotropic DPPH signal with g = 2.0036 near the free-spin value is often chosen because of its convenience as a reference standard. By substitution of these expressions into (5.9) one obtains

$$g_0 + \Delta g = \frac{\nu_0}{\beta(H_0 + \Delta H)} = \frac{\nu_0}{\beta H_0}\left[1 - \frac{\Delta H}{H_0} + \left(\frac{\Delta H}{H_0}\right)^2 - \cdots\right] \tag{5.12}$$

For small **g** anisotropies, the second-order and higher terms can be neglected. With $g_0 = (\nu_0/\beta H_0)$, we then obtain

$$\frac{\Delta g}{g_0} = -\frac{\Delta H}{H_0} \tag{5.13}$$

With the omission of the second-order terms $(\Delta g)^2$, one can set $g_\theta{}^2 = g_0{}^2 + 2g_0\,\Delta g = g_0{}^2 - 2g_0{}^2(\Delta H/H_0)$ into (5.4) and, after rearrangement, obtain

$$\Delta H = K_1' + K_2' \cos 2\theta + K_3' \sin 2\theta \tag{5.14}$$

where

$$K_1' = \frac{H_0(g_0^2 - K_1)}{2g_0^2} \tag{5.15}$$

$$K_2' = \frac{-H_0K_2}{2g_0^2} \tag{5.16}$$

$$K_3' = \frac{-H_0K_3}{2g_0^2} \tag{5.17}$$

When the matrix elements of **g** expressed in the arbitrary system x,y,z are found as described in the preceding paragraphs, the resulting tensor can be diagonalized by a rotational transformation of the tensor to the principal axes. Let l_{ix}, l_{iy}, and l_{iz} be the direction cosines between the principal axis i and the reference axes x, y, and z, respectively, and let g_i be the corresponding principal value. Since the off-diagonal elements vanish in the principal system, the transformation may be expressed as

$$\begin{bmatrix} g_{xx} & g_{xy} & g_{xz} \\ g_{yx} & g_{yy} & g_{yz} \\ g_{zx} & g_{zy} & g_{zz} \end{bmatrix} \begin{bmatrix} l_{ix} \\ l_{iy} \\ l_{iz} \end{bmatrix} = g_i \begin{bmatrix} l_{ix} \\ l_{iy} \\ l_{iz} \end{bmatrix} \tag{5.18}$$

By performing the indicated matrix multiplications, equating the corresponding elements, and transforming the terms on the right, one can obtain the following equations:

$$(g_{xx} - g_i)l_{ix} + g_{xy}l_{iy} + g_{xz}l_{iz} = 0 \tag{5.19}$$

$$g_{yx}l_{ix} + (g_{yy} - g_i)l_{iy} + g_{yz}l_{iz} = 0 \tag{5.20}$$

$$g_{zx}l_{ix} + g_{zy}l_{iy} + (g_{zz} - g_i)l_{iz} = 0 \tag{5.21}$$

These three simultaneous equations in the three unknown cosines, l_{ix}, l_{iy}, and l_{iz}, have nontrivial solutions only when the determinant of the coefficients vanishes, that is, when

$$\begin{vmatrix} (g_{xx} - g_i) & g_{xy} & g_{xz} \\ g_{yx} & (g_{yy} - g_i) & g_{yz} \\ g_{zx} & g_{zy} & (g_{zz} - g_i) \end{vmatrix} = 0 \tag{5.22}$$

Solution of this cubic equation yields three values of g_i, which are the principal values of **g**. We use the designations g_u, g_v, and g_w, where g_u is the smallest, g_v is the intermediate, and g_w is the largest principal element of **g** and where u, v, and w signify the principal axes. Substitution of each of the principal values in turn for g_i into (5.19) to (5.21) yields a set of three equations involving the three direction cosines for the particular principal axis. These equations with the auxiliary equation $l_{ix}^2 + l_{iy}^2 + l_{iz}^2 = 1$ may be solved for the direction cosines for the particular principal value $i = u,v,w$ that is substituted.

As a method of procedure, it is usually simpler to obtain the elements $(g^2)_{ij}$ of the squared g matrix $(\mathbf{g}^2)_i$ from (5.4) or (5.14) than those of the unsquared (**g**). One can then diagonalize the squared matrix and take the square root of the diagonal elements to obtain the principal g values. Since the g values are always positive, this procedure causes no ambiguity in sign. The relationship of the squared to the unsquared elements is easily found from the matrix-product rule:

$$(g^2)_{ij} = (i|g^2|j) = \sum_{k=x,y,z} (i|g|k)(k|g|j) = \sum_k g_{ik}g_{ki} \tag{5.23}$$

The resulting expressions are simplified because the g matrix is symmetric, that is, $g_{ij} = g_{ji}$. For example,

$$(g^2)_{xx} = g_{xx}g_{xx} + g_{xy}g_{yx} + g_{xz}g_{zx} = g_{xx}^2 + g_{xy}^2 + g_{xz}^2 = g_x{}^2 \tag{5.24}$$

Similarly, $(g^2)_{yy} = g_y{}^2$, $(g^2)_{zz} = g_z{}^2$, and

$$(g^2)_{xy} = g_{xx}g_{xy} + g_{xy}g_{yy} + g_{xz}g_{zy} = (g^2)_{yx} = K_3(x,y) = K_3(y,x) \tag{5.25}$$

With the corresponding relationship for the other coordinates, $(\mathbf{g}^2)$ is expressed in terms of the observed constants by

$$\begin{vmatrix} (g^2)_{xx} & (g^2)_{xy} & (g^2)_{xz} \\ (g^2)_{yx} & (g^2)_{yy} & (g^2)_{yz} \\ (g^2)_{zx} & (g^2)_{zy} & (g^2)_{zz} \end{vmatrix} = \begin{vmatrix} g_x{}^2 & K_3(x,y) & K_3(x,z) \\ K_3(y,x) & g_y{}^2 & K_3(y,z) \\ K_3(z,x) & K_3(z,y) & g_z{}^2 \end{vmatrix} \tag{5.26}$$

where the g_x^2, g_y^2, and g_z^2 values along the respective axes are related to the K constants by

$$g_x^2 = K_1(x,y) + K_2(x,y) = K_1(z,x) - K_2(z,x) \tag{5.27}$$

$$g_y^2 = K_1(y,x) + K_2(y,x) = K_1(x,y) - K_2(x,y) \tag{5.28}$$

$$g_z^2 = K_1(z,y) + K_2(z,y) = K_1(y,z) - K_2(y,z) \tag{5.29}$$

The squared principal elements g_i^2 may then be obtained from solution of the cubic secular equation,

$$\begin{vmatrix} [K_1(x,y)+K_2(x,y)]-g_i^2 & K_3(x,y) & K_3(x,z) \\ K_3(y,x) & [K_1(y,x)+K_2(y,x)]-g_i^2 & K_3(y,z) \\ K_3(z,x) & K_3(z,y) & [K_1(z,y)+K_2(z,y)]-g_i^2 \end{vmatrix} = 0 \tag{5.30}$$

for the three values of g_i^2 that are the squared principal elements, g_u^2, g_v^2, and g_w^2, of the **g** tensor.

The direction cosines, or eigenvectors, for the principal elements are, of course, the same as those for the squared principal elements and hence may be obtained from solution of the set of equations

$$[K_1(x,y) + K_2(x,y) - g_i^2]l_{ix} + K_3(x,y)l_{iy} + K_3(x,z)l_{iz} = 0 \tag{5.31}$$

$$K_3(y,x)l_{ix} + [K_1(y,x) + K_2(y,x) - g_i^2]l_{iy} + K_3(y,z)l_{iz} = 0 \tag{5.32}$$

$$K_3(z,x)l_{ix} + K_3(z,y)l_{iy} + [K_1(z,x) + K_2(z,y) - g_i^2]l_{iz} = 0 \tag{5.33}$$

where $i = u,v,w$ and where the values g_u^2, g_v^2, and g_w^2 obtained from solution of (5.30) are in turn substituted for g_i^2. The solution of these equations yields only the ratios of the direction cosines; the additional relation $l_{ix}^2 + l_{iy}^2 + l_{iz}^2 = 1$ is required to give their specific value.

Figure 5.2 illustrates the variation of ΔH due to **g** anisotropy for the three orthogonal planes by sulfur-centered free radicals produced by irradiation of single crystals of cystine dihydrochloride at room temperature [2]. These crystals are monoclinic, with C_2 symmetry and have two molecules in the unit cell [3]. Two of the reference axes are identified with the orthogonal crystallographic axes *a* and *b*; and the third, *c'*, which is perpendicular to *a* and *b*, is not identified with any surface feature. There are two distinguishable orientations for the radicals that give rise to the two crossing patterns in the *bc'* plane, but all the radicals have equal orientations in the other two planes. The doublet

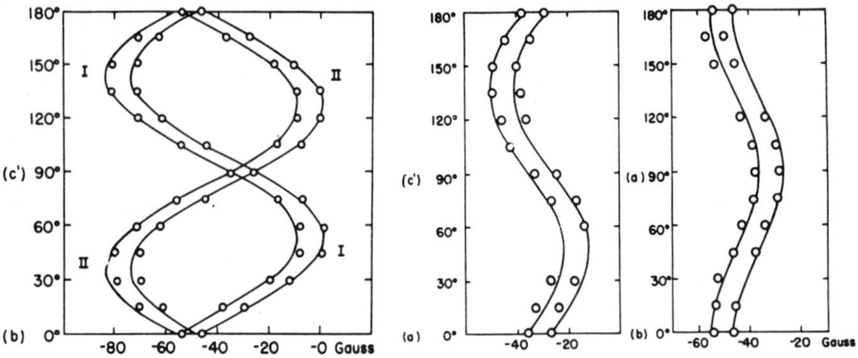

Fig. 5.2 Example of variations due to **g** anisotropy of esr line positions as a function of crystal orientation in the magnetic field for three orthogonal crystal planes. The resonance is that of a sulfur-centered radical in a γ-irradiated single crystal of L-cystine dihydrochloride. The nearly isotropic doublet splitting is due to proton nuclear coupling; the two sets of lines in the bc' plane are due to two magnetically distinguishable orientations of the radicals. All measurements were made at room temperature and at a constant observation frequency of 9 kMHz. A diagram of the crystal form (Fig. 7.8) and an interpretation of the **g** tensor of these radicals are given in Chapter VII, Section 6.a. From Kurita and Gordy [2].

curves separated by about 10 G result from essentially isotropic coupling of a proton. Thus in derivation of the **g** tensor from these curves, one would adjust the parameters of the theoretical formula such as (5.14) to fit a curve drawn halfway between these doublets. The **g** anisotropy of these sulfur-centered radicals, the principal g values of which were found to be 2.003, 2.029, and 2.052 [2], is somewhat larger than that for most other organic free radicals. The spectra and structures of these and other sulfur-centered radicals are treated in Chapter VII, Section 6.

Figure 5.3 gives plots of the effective g values as a function of orientation in three orthogonal planes of a sulfur-centered radical, the **g** anisotropy of which is exceptionally large for an organic free radical. These plots were obtained by Saxebøl and Herskedal [4] for an irradiated single crystal of N-acetyl-L-cysteine, which forms a triclinic crystal with only one molecule in the unit cell. Note that the observed patterns indicate that all the radicals have a common orientation for each reference plane. Figure 5.4 indicates the chosen reference axes relative to the crystal surfaces. The observed resonance has a four-component, approximately isotropic, hyperfine pattern, and the effective g values are determined from the field values corresponding to the centers of the hyperfine pattern at the different angles. The principal g values obtained from analysis of these curves are 1.990, 2.006, and 2.214. An even larger **g** anisotropy, 1.99 to 2.29, was observed earlier by Akasaka in irradiated single crystals of cysteine HCl at 77°K [5].

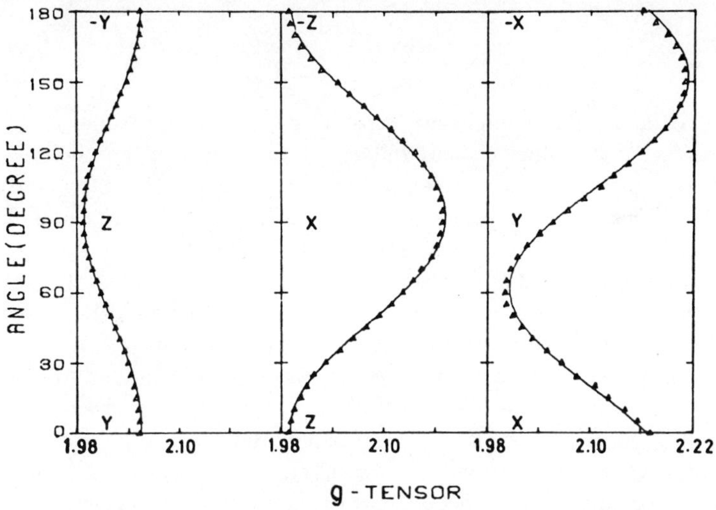

Fig. 5.3 Observed and calculated *g* variations in three orthogonal crystal planes for a sulfur-centered radical in an irradiated crystal of *N*-acetyl-L-cysteine. The crystal form and reference planes are given in Fig. 5.4. From Saxebøl and Herskedal [4].

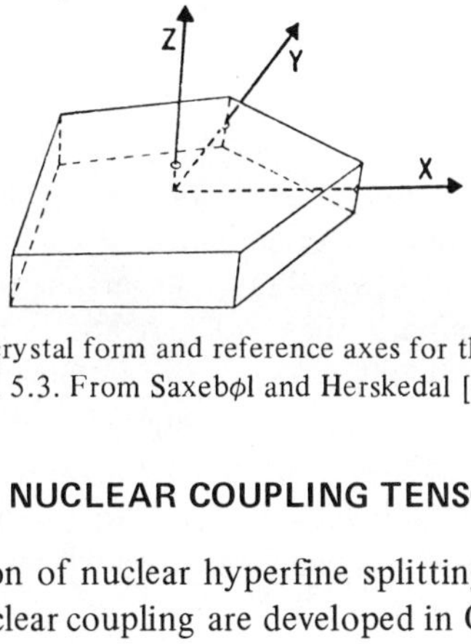

Fig. 5.4 Diagram of the crystal form and reference axes for the triclinic *N*-acetyl-L-cysteine crystal. Compare with Fig. 5.3. From Saxebøl and Herskedal [4].

2. EVALUATION OF NUCLEAR COUPLING TENSORS

Formulas for calculation of nuclear hyperfine splitting of esr spectra for various kinds and degrees of nuclear coupling are developed in Chapter III. In this section we describe how these formulas are applied to measurements of single crystals to give the principal elements of anisotropic coupling tensors. Only magnetic coupling is treated here since it is the only type of nuclear coupling detected in normal esr transitions for which the quantum number of the nuclear spin does not change.

2.a. Effective Couplings and Units of Measurement

The effective magnetic coupling of a nucleus n for any orientation of the crystal in the magnetic field may be expressed as

$$(A_{\text{eff}})_{\text{n}} = \left(\frac{g_{\text{eff}}\beta}{h}\right)\Delta H_{\text{n}} \tag{5.34}$$

where ΔH_{n} is the component separation of the hyperfine structure or substructure due to that nucleus and g_{eff} is the g value of the observed radical effective at the particular orientation. Here, A_{eff} is expressed in frequency units, and ΔH is measured in magnetic field units at constant frequency. Thus by measurement of the component spacing or line splitting due to a particular nucleus, one can find its effective coupling for any orientation of the crystal. When there is only one coupling nucleus, there is no problem of assignment. One simply measures the separation of the adjacent components, or measures the separation of the outside components and divides by the number of spacings. In a composite pattern where the submultiplet structure of a given nucleus n is completely separated, there is again no problem; the ΔH_{n} is obtained by measurement of the component spacings in the submultiplet. However, the substructure due to a particular nucleus in a composite pattern is not always easy to identify. The description given in Chapter 4, Section 3.a for the composite hyperfine patterns resulting from nuclei with various combinations and couplings will be helpful for this purpose.

Before describing the procedures for analysis of the hyperfine measurements, it is desirable to clarify the relationship of the units commonly used for measurement and analysis of the nuclear coupling constants and those commonly used in theoretical treatments. In the development of the theory (Chapters II and III), the nuclear coupling tensor was expressed in energy units (erg), whereas in the application of this theory here and in later chapters these constants are expressed in frequency units (megahertz) or in magnetic field units (gauss). This inconsistency, also found in most other treatments of magnetic resonance, is amply justified by the simplification of the formulations that it allows. The spin Hamiltonian and its eigenvalues are energy expressions, and these expressions are simplified if the coupling constants are expressed in energy units. To express them in frequency units would require that each coupling tensor or its elements in the many theoretical equations be multiplied by Planck's constant h. On the other hand, since the hyperfine structure from which these constants are derived is measured in frequency units of megahertz (MHz) or, more often, in guass (G), it is more convenient to express the observed constants in these units. Despite the fact that hyperfine splitting is almost always measured in gauss at a constant frequency, it is generally more desirable to express the couplings in frequency units, which are independent of the value of g. Nevertheless, for organic radicals where the observed g is nearly isotropic and is close to the free-spin value, the coupling is commonly expressed directly in the magnetic-field units used for their measurement. For comparison of the coupling of the same nucleus in radicals for which the g values are different or for analysis of $\mathbf{A}$

tensors when **g** is significantly anisotropic, it is advisable to convert the splitting observed in gauss to coupling constants in frequency units that are independent of g.

The following relations are convenient for conversion of the units:

$$A\ (\mathrm{Hz}) = \frac{1}{h} A\ (\mathrm{erg}) = \frac{g\beta}{h} A\ (\mathrm{G}) \tag{5.35}$$

where h is Planck's constant. Usually the coupling elements are measured in gauss and then converted to megahertz. Numerically,

$$A_{\mathrm{eff}}\ (\mathrm{MHz}) = 1.3996\, g_{\mathrm{eff}} A_{\mathrm{eff}}\ (\mathrm{G}) \tag{5.36}$$

where $A_{\mathrm{eff}}(\mathrm{G}) = \Delta H(\mathrm{G})$ = the corresponding component separation. When g has the isotropic, free-spin value 2.0023, the relationship is simply

$$A\ (\mathrm{MHz}) = 2.8024\, A\ (\mathrm{G}) \tag{5.37}$$

2.b. Evaluation of the A Tensor When g Anisotropy is Negligible

In most organic free radicals the anisotropy in **g** is sufficiently small that its effect on the nuclear coupling tensor is negligible. For these radicals we can use the nuclear coupling formulas of Chapter III, Section 4.g. Since the splitting by nuclei with different coupling can be measured and analyzed separately, we need only treat a single coupling tensor. The procedure for finding the principal elements is the same as that already described for the **g** tensor in Chapter V, Section 1. The same orthogonal reference system may be used for both.

When effects of **g** anisotropy are neglected, the square of the effective coupling in the xy reference plane as obtained from (4.105) may be expressed in the form

$$A_\theta^2 = A_x^2 \cos^2\theta + A_y^2 \sin^2\theta + 2(A_{xx}A_{yx} + A_{xy}A_{yy} + A_{xz}A_{yz}) \sin\theta \cos\theta \tag{5.38}$$

where A_x and A_y are the effective coupling along the x and y axes, which are expressed in terms of the squared tensor elements by

$$A_x^2 = A_{xx}^2 + A_{xy}^2 + A_{zx}^2 \tag{5.39}$$

$$A_y^2 = A_{yy}^2 + A_{yx}^2 + A_{yz}^2 \tag{5.40}$$

Note that these equations have the same form as those for the **g** tensor, (5.1) to (5.3), and hence may be expressed as

$$A_\theta^2(x,y) = C_1(x,y) + C_2(x,y)\cos 2\theta + C_3(x,y)\sin 2\theta \tag{5.41}$$

where θ is the angle between the x axis and the applied field, and where

$$C_1(x,y) = \frac{1}{2}(A_x^2 + A_y^2) \tag{5.42}$$

$$C_2(x,y) = \frac{1}{2}(A_x^2 - A_y^2) \tag{5.43}$$

$$C_3(x,y) = A_{xx}A_{yx} + A_{xy}A_{yy} + A_{xz}A_{yz} \tag{5.44}$$

Similar expressions for the other coordinate planes may be obtained by obvious changes of the coordinates.

By measuring the hyperfine splittings in the three coordinate planes and by squaring the resulting values, one can evaluate the six independent C constants as described for the $(\mathbf{g}^2)$ tensor in Section 1. In this case, when effects of **g** anisotropy are negligible, A_θ in gauss is equivalent to the component spacings ΔH for all values of θ.

The elements of the squared tensor $(\mathbf{A}^2)$ may be obtained from these measurable constants by use of the relations:

$$(A^2)_{xx} = A_x^2 = C_1(x,y) + C_2(x,y) = C_1(z,x) - C_2(z,x) \tag{5.45}$$

$$(A^2)_{yy} = A_y^2 = C_1(y,x) + C_2(y,x) = C_1(x,y) - C_2(x,y) \tag{5.46}$$

$$(A^2)_{zz} = A_z^2 = C_1(z,y) + C_2(z,y) = C_1(y,z) - C_2(y,z) \tag{5.47}$$

$$(A^2)_{xy} = (A^2)_{yx} = C_3(x,y) = C_3(y,x) \tag{5.48}$$

$$(A^2)_{xz} = (A^2)_{zy} = C_3(x,z) = C_3(z,x) \tag{5.49}$$

$$(A^2)_{yz} = (A^2)_{zy} = C_3(y,z) = C_3(z,y) \tag{5.50}$$

which are similar to those for $(\mathbf{g}^2)$. One substitutes the numerical values of $(A^2)_{ij}$ thus obtained into the cubic secular equation:

$$\begin{vmatrix} (A^2)_{xx} - A_i^2 & (A^2)_{xy} & (A^2)_{xz} \\ (A^2)_{yx} & (A^2)_{yy} - A_i^2 & (A^2)_{yz} \\ (A^2)_{zx} & (A^2)_{zy} & (A^2)_{zz} - A_i^2 \end{vmatrix} = 0 \qquad (5.51)$$

and solves for the three values $A_i^2 \doteq A_u^2, A_v^2, A_w^2$, which are the squared principal elements. The direction cosines for each of the principal axes $i = u,v,w$, may then be found by solution of the simultaneous equations

$$[(A^2)_{xx} - A_i^2]l_{ix} + (A^2)_{xy}l_{iy} + (A^2)_{xz}l_{iz} = 0 \qquad (5.52)$$

$$(A^2)_{yx}l_{ix} + [(A^2)_{yy} - A_i^2]l_{iy} + (A^2)_{yz}l_{iz} = 0 \qquad (5.53)$$

$$(A^2)_{zx}l_{ix} + (A^2)_{zy}l_{iy} + (A_{zz}{}^2 - A_i^2)l_{iz} = 0 \qquad (5.54)$$

with the square of each principal value substituted in turn for A_i^2 and with the use of the additional relation, $l_{ix}^2 + l_{iy}^2 + l_{iz}^2 = 1$.

Solution of the preceding secular equation yields the squared principal elements A_u^2, A_v^2, and A_w^2, from which the magnitudes of the principal coupling elements are obtained by simple extraction of the square roots of these numbers. With the A_θ values measured in gauss, the resulting A_u, A_v, and A_w values will likewise be in gauss. They can be converted to megahertz with (5.36) and the known isotropic g value.

The signs of the couplings are not obtained from the preceding analysis nor are they found from straight esr measurement of the symmetrical, first-order hyperfine splitting by any method of analysis. They are usually derived theoretically from the known signs of the nuclear moments, but they may also be obtained from such types of measurements as second-order transitions of ENDOR. If the esr hyperfine splitting does not become zero for any orientation, all three principal values of the coupling must have the same sign. This experimentally observable fact, together with theoretical considerations, often allows one to deduce the signs of all the coupling elements. Although the elements of an anisotropic dipole-dipole coupling such as the p-orbital coupling are not all of the same sign, this is often masked by a dominant isotropic component that causes all the observed principal values to have like signs. Because of the significant isotropic component, the three principal values of the proton coupling in the CH fragment are all of the same sign despite the rather large dipole-dipole contribution.

Although the direction cosines of the principal axes of the nuclear coupling tensor are often the same as those of the **g** tensor, they are not necessarily the same, nor are they necessarily the same for different coupling nuclei of the same radical. As a rule, the direction cosines of the principal axes for each nucleus with distinguishable coupling must be separately evaluated by the process described earlier.

There is a simple procedure that often may be used to give the principal elements of the **A** tensor for the more common types of organic free radicals for which the anisotropy in **A** is due to axially symmetric p-orbital coupling. In such radicals, the minimum g value, g_u, is usually in the direction of aligned p-orbital components of a π-molecular orbital, in which the electron-spin density is concentrated. The nuclear coupling will then be axially symmetric about g_u, the direction of which is presumably known from a preanalysis of **g**. For these radicals, one can easily find the principal elements of **A** by measuring the hyperfine splitting with the magnetic field along g_u and in a plane perpendicular to g_u. As a rule, it is a time-saving procedure to measure first the hyperfine splitting of each nucleus in the plane perpendicular to g_u to ascertain whether the coupling has such axial symmetry. When the **g** is so nearly isotropic that the direction of g_u cannot be measured with reliable accuracy, this method cannot be used. One must then use a general procedure such as that described earlier when the axes of symmetry cannot be easily discovered by preliminary observations. In such radicals, the **A** tensor provides a much more reliable measurement of the direction of the p orbitals than does the **g** tensor.

2.c. Evaluation of the A Tensor When the g Anisotropy is Large

A large **g** anisotropy can so influence the effective **A** values that the formulas such as (5.34) do not give accurate values for the principal elements of **A**. Formulas for calculation of effective nuclear coupling that take into account the effects of **g** anisotropy are expressed by (3.95), (3.98), and (3.100). Here, we outline a procedure for applying these formulas to find the eigenvalues and eigenvectors of **A**.

Probably the simplest procedure is to first find the principal axes of **g** by the method described in Section 1 and then to use these axes as the reference system in which to measure the hyperfine splitting. As before, measurements are made in three orthogonal planes, but now the planes are the principal planes of the **g** tensor. This procedure has the advantage of simplifying the equations that must be applied, and it often simplifies the problem further by quickly revealing whether the coupling is axially symmetric about a principal axis of **g**, as is frequently true when the spin density is in a π orbital of the radical.

When reference axes x, y, and z are the principal axes of **g**, the effective coupling A is given by (3.98). For measurements in the xy plane, the square of the effective coupling obtained from (3.98) may be expressed in the form

$$A_\theta^2 = \frac{1}{g_\theta^2}\,[(g_x A_x)^2 \cos^2\theta + (g_y A_y)^2 \sin^2\theta + 2\,g_x g_y (A_{xx} A_{yx}$$

$$+ A_{xy}A_{yy} + A_{xz}A_{yz})\sin\theta\cos\theta] \qquad (5.55)$$

where θ is the angle of **H** with the x axis and

$$g_\theta^2 = (g_x^2\cos^2\theta + g_y^2\sin^2\theta) \qquad (5.56)$$

where g_θ is the effective g value in the xy principal plane. The quantities

$$A_x^2 = A_{xx}^2 + A_{xy}^2 + A_{xz}^2 \qquad (5.57)$$

$$A_y^2 = A_{yy}^2 + A_{yx}^2 + A_{yz}^2 \qquad (5.58)$$

are the square of the effective coupling along the x and y axes, respectively. Equation (5.55) may be expressed [cf. with (5.4)] in the form

$$A_\theta^2 = \frac{1}{g_\theta^2}[C_1'(x,y) + C_2'(x,y)\cos 2\theta + C_3'(x,y)\sin 2\theta] \qquad (5.59)$$

where

$$C_1'(x,y) = \frac{1}{2}(g_x^2A_x^2 + g_y^2A_y^2) \qquad (5.60)$$

$$C_2'(x,y) = \frac{1}{2}(g_x^2A_x^2 - g_y^2A_y^2) \qquad (5.61)$$

$$C_3'(x,y) = g_xg_y(A_{xx}A_{yx} + A_{xy}A_{yy} + A_{xz}A_{yz}) \qquad (5.62)$$

The corresponding relations for the other two coordinate planes may be obtained by obvious substitution of the coordinates in these expressions.

Again, it is simpler to solve these equations for the elements $(A^2)_{ij}$ of the squared tensor rather than for the elements. The method for obtaining the squared elements is the same as that described for the **g** tensor in Section 1. Here, the elements of the squared tensor are related to the observable constants by

$$(A^2)_{xx} = A_x^2 = \frac{1}{g_x^2}[C_1'(x,y) + C_2'(x,y)] \qquad (5.63)$$

Similarly

$$(A^2)_{yy} = A_y^2 = \frac{1}{g_y^2}[C_1'(x,y) - C_2'(x,y)] \qquad (5.64)$$

$$(A^2)_{zz} = A_z^2 = \frac{1}{g_z^2}\,[C_1'(x,z) - C_2'(x,z)] \tag{5.65}$$

$$(A^2)_{yx} = (A^2)_{xy} = \frac{1}{g_x g_y}\,C_3'(x,y) \tag{5.66}$$

$$(A^2)_{xz} = (A^2)_{zx} = \frac{1}{g_z g_x}\,C_3'(z,x) \tag{5.67}$$

$$(A^2)_{yz} = (A^2)_{zy} = \frac{1}{g_y g_z}\,C_3'(y,z) \tag{5.68}$$

The **g** anisotropy does not influence the effective A_i values when the applied field is along a principal axis of g. One can obtain the $(A^2)_{xx}$, $(A^2)_{yy}$, and $(A^2)_{zz}$ values simply by squaring the splitting as observed along the principal g axes, x, y, z. Solutions for the C_3' constants require additional observations in the principal planes between the coordinates. Measurements of the component splitting, as in the preceding section, give directly the effective A_θ, in gauss, which is squared and substituted into (5.59). The required g_θ^2 values in this equation are calculated from (5.56) with the known g_x, g_y, and g_z values and the particular θ for which the measurement is made.

Thus from the measurements of the hyperfine splitting with the preobserved g_x, g_y, and g_z values, all the elements $(A^2)_{ij}$ can be obtained from (5.63) to (5.68). These elements are then substituted into (5.51) to obtain the secular equation that is solved for the squared principal values and into (5.52) to (5.54), which are solved for the direction cosines.

When **g** and **A** have common principal axes, the effective nuclear coupling as expressed relative to these axes (Chapter III, Section 4.f) is

$$A = \frac{1}{g}\,[(g_x A_x l_x)^2 + (g_y A_y l_y)^2 + (g_z A_z l_z)^2]^{1/2} \tag{5.69}$$

where $g = [(g_x l_x)^2 + (g_y l_y)^2 + (g_z l_z)^2]^{1/2}$ and l_x, l_y, and l_z are the direction cosines of **H** relative to the principal axes x, y, z. When the axes are known from analysis of the **g** tensor, the principal values A_x, A_y, A_z can then be found simply by measurement of the hyperfine splitting with **H** along each axis. Although one conceivably might have to diagonalize the **A** tensor to be sure that its principal axes are the same as those of **g**, this can be guessed in most cases from a consideration of the probable structure of the radical or be discovered by preliminary observations that reveal an axis of symmetry in the coupling about one of the principal axes of **g**.

Because of obscuring signals from other radical species, the hyperfine splitting cannot always be measured for all orientations of the crystal, nor even along all three principal axes. The obscuring signals may come from other radical species, from the same radicals with different orientations in the crystal, or from different isotopic species of the same radicals as those under study. For these reasons, one often must derive the principal values of **A** from measurement over a limited range of crystal orientation in which the **g**-anisotropy effects are not negligible.

2.d. Illustrative Examples

^{14}N Coupling in Irradiated Single Crystals of $(KSO_3)_2$ NOH

The esr of $(KSO_3)_2NO$ radicals observed by Hamrick, Shields, and Gangwer [6] for irradiated single crystals of $(KSO_3)_2NOH$ provides a relatively uncomplicated illustration of the evaluation of nuclear coupling constants from plots of the hyperfine splitting in three orthogonal reference planes. The spin density of the radical is concentrated on the ^{14}N that has spin $I = 1$, and thus the hyperfine structure is a triplet with components of equal intensity. The crystal is monoclinic with unit cell dimensions: $a = 9.74$ Å, $b = 6.51$ Å, and $c = 14.13$ Å; $\alpha = 90°$, $\beta = 105°50'$, and $\gamma = 90°$. There are four molecules in the unit cell

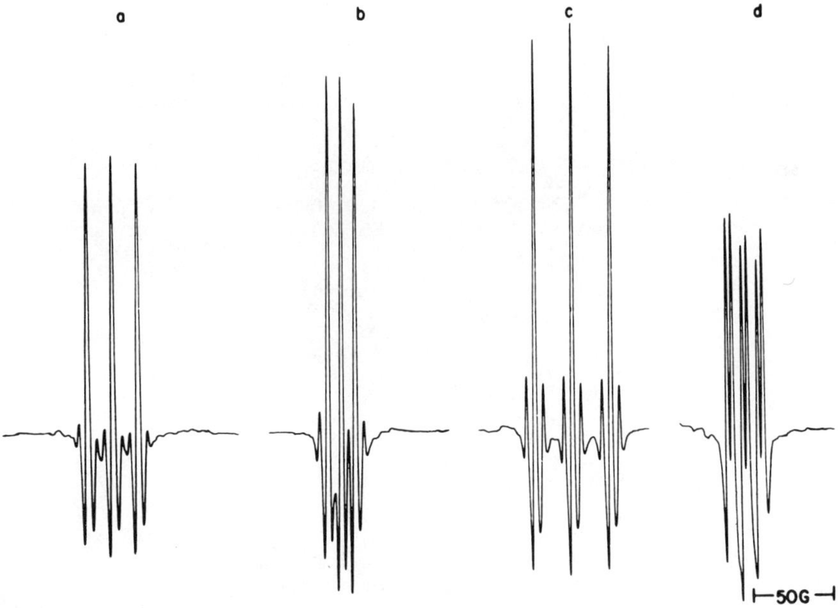

Fig. 5.5 The ^{14}N splitting of the nitrosyldisulfonate ion with the magnetic field oriented along the respective crystal axes *a*, *b*, and *c*; and with *d* 45° from *a* in the *ac** reference plane. Plots of the splitting are given in Fig. 5.6. From Hamrick, Shields, and Gangwer [6].

that give rise to radicals with two distinguishable orientations for some directions of the applied field. The orthogonal crystallographic axes *a* and *b* were designated as two of the reference axes; the third, perpendicular to them, was designated as c^*. Figure 5.5 shows the spectra observed with **H** along each of the reference axes and at 45° from *a* in the ac^* plane. The lines were recorded with an *X*-band Varian spectrometer.

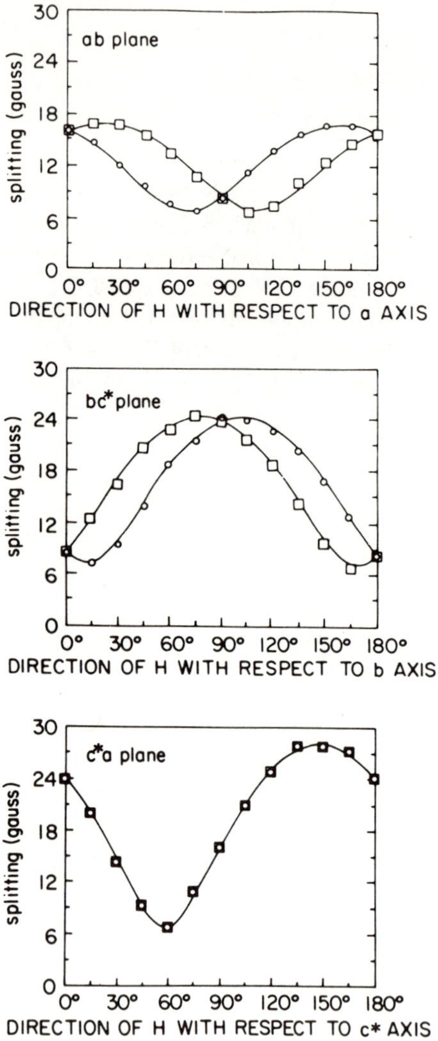

Fig. 5.6 Plots of the ^{14}N splitting of the nitrosyldisulfonate ion in three orthogonal reference planes of the crystal in the magnetic field. The squares and dots represent splitting for the radicals having two distinguishable orientations in the *ab* and ac^* planes. See the observed spectra in Fig. 5.5. From Hamrick, Shields, and Gangwer [6].

Because the radicals have common orientations relative to the reference axes, only one ^{14}N triplet is observed with **H** along these axes. Note, however, that there are weak satellite lines, strongest for the c^* direction, on both sides of each component. These satellites are due to the flipping of environmental protons, as explained in Section 4.c. Figure 5.6 shows plots of the observed ^{14}N splitting in the three orthogonal planes; the theoretical curves (solid lines) are calculated with the derived coupling constants. The two curves for the *ab* and bc^* planes arise from the two distinguishable radical sites for these planes. The principal values of the ^{14}N magnetic coupling derived from these curves are 77, 21.6, and 15.4, in megahertz, with an isotropic component of 38 MHz. Principal g values derived in the same study are 2.0026, 2.0055, and 2.0094.

Irradiated Crystal of Malonic Acid

The esr of the $\dot{C}H(COOH)_2$ radicals in irradiated single crystals of malonic acid, $CH_2(COOH)_2$, were measured and analyzed by McConnell, Heller, Cole, and Fessenden [7] to provide the first reliable values for the principal elements of the proton coupling in the important CH fragment. Although their paper is rather complicated, mostly because of their discussion of weak signals due to forbidden transitions (see Section 4) and to secondary radicals, their analysis of the first-order spectrum of the $H\dot{C}(COOH)_2$ radical is basically simple. This simplicity was made possible by the known structure of the unit cell in this crystal and by their correct hypothesis of the structure and orientation of the principal radical formed by the irradiation.

Figure 5.7 shows the unit cell structure [8]. McConnell et al. [7] correctly assumed that the radical $H\dot{C}(COOH)_2$ would be formed by the loss of an H and that the CH bond would become oriented in the plane of the three carbons in such direction as to bisect the angle between the CC bonds. Because of

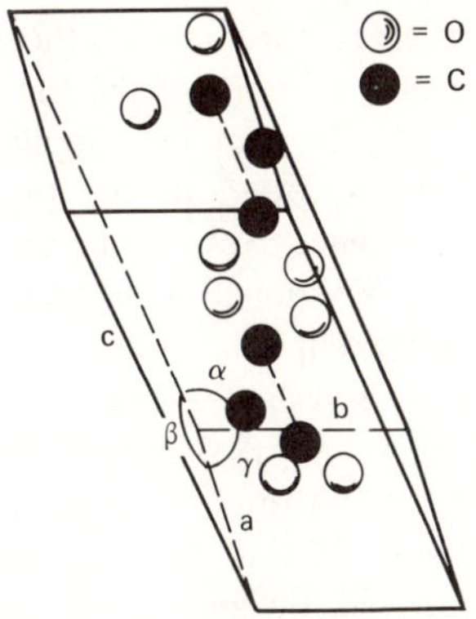

Fig. 5.7 Sketch of the unit cell of malonic acid. From McConnell, Heller, Cole, and Fessenden [7].

symmetry, one principal axis of the proton coupling should then be along this bisector; another should be perpendicular to this direction and also in the CCC plane; the third should be perpendicular to the CCC plane. These directions, easily identifiable from the known crystal structure, were chosen as their x,y,z, reference system, as indicated in Fig. 5.8. As anticipated, these axes

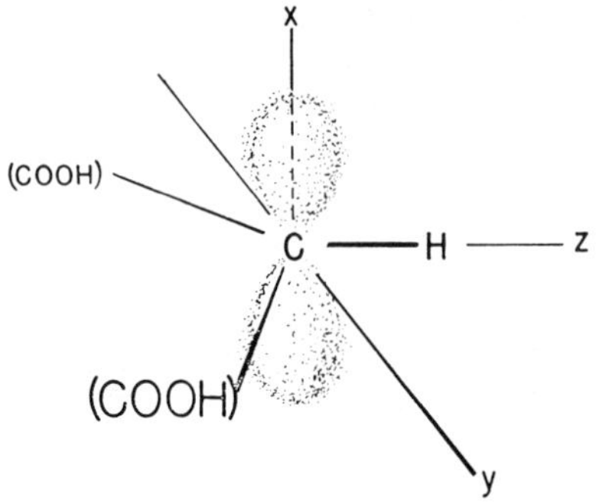

Fig. 5.8 Sketch of the π radical $CH(COOH)_2$. The dotted area represents the distribution of the π-electron spin density. From T. Cole, C. Heller, and H. M. McConnell, *Proc. Nat. Acad. Sci. (USA)*, **45**, 525 (1959).

proved to be principal axes for the observed proton doublet splitting as well as the **g** tensor. Figure 5.9 reproduces the proton doublet pattern observed with **H** imposed along the three axes. The smallest splitting is observed for the x direction, which is along the p orbital; the largest is observed for the y direction, which is perpendicular to the p orbital and to the CH bond, as indicated in Fig. 5.8. From the crystal structure, the radicals are expected to have a common orientation for any direction of the applied magnetic field, and this proved to be true.

Figure 5.10 shows the observed variation of the splitting of the doublets (outer plots) for the different orientations of the field in the zx and zy planes. The fact that these curves can be fitted by the coupling formula, $A = (A_x{}^2 l_x{}^2 + A_y{}^2 l_y{}^2 + A_z{}^2 l_z{}^2)^{1/2}$, (3.79), expressed in the principal axes verifies their assumption that the chosen reference axes are the principal axes of the proton coupling. In their calculations, McConnel et al. used polar coordinates rather than the Cartesian coordinates used in this book.

Once the assumed principal axis system was verified as described above, it was a simple problem to find the principal values of the proton coupling. These values are given directly by the splitting of the three doublets, shown in Fig. 5.9. The fact that the observed splitting does not become zero in any of the principal planes, as demonstrated by Fig. 5.10, proves that all the principal elements are of the same sign. The magnitudes of the principal values

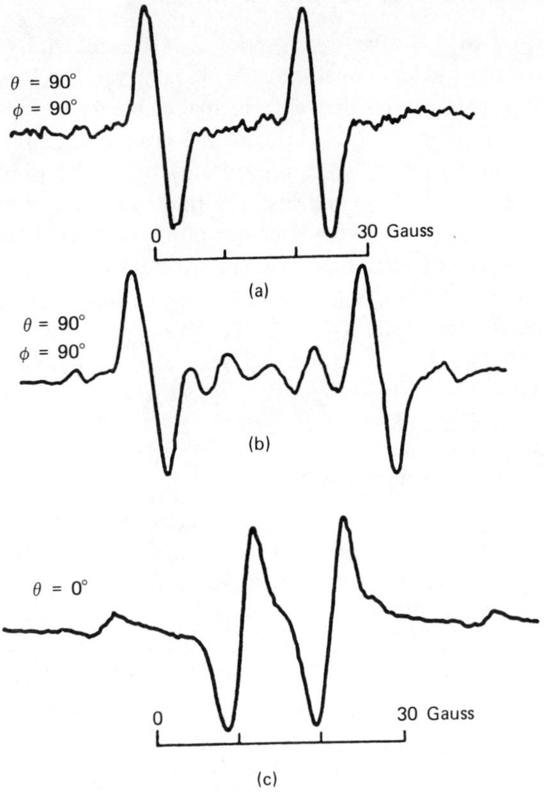

Fig. 5.9 Splitting by the CH proton in the esr of the $CH(COOH)_2$ radical with the magnetic field along the three principal axes of the proton-coupling tensor. From McConnell, Heller, Cole, and Fessenden [7].

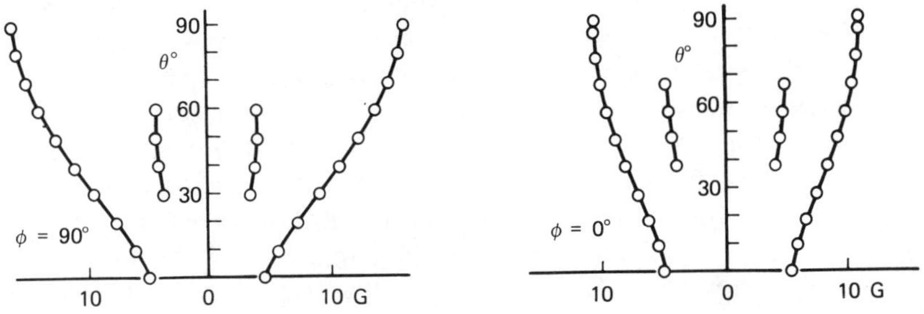

Fig. 5.10 Variation of the CH proton splitting of the $CH(COOH)_2$ esr as a function of the orientation of the magnetic field in the two principal planes of the coupling tensor. The inside plots represent second-order transitions. From McConnell, Heller, Cole, and Fessendon [7];

obtained are: $|A_x| = 61 \pm 2$ MHz, $|A_y| = 91 \pm 2$ MHz, and $|A_z| = 29 \pm 2$ MHz. Theoretical arguments by McConnell et al. [7], later verified by measurement of the ^{13}C coupling [9], proved that all principal elements are negative in sign. A discussion of the coupling of the $C_\alpha H$ fragment is given in Chapter VI.

The chosen reference axes, x, y, z, proved also to be the principal axes of the **g** tensor for the $H\dot{C}(COOH)_2$ radicals. The principal g values are then easily evaluated by a fitting of (3.103) to the center position of the doublet. Because the doublet splitting is relatively small, second-order shifts of the doublet center described in Section 3 are not significant. In such hydrocarbon radicals, the **g** tensor is very nearly isotropic. The principal values were found to be: $g_x = 2.0026$, $g_y = 2.0035$, and $g_z = 2.0033$, all very close to the value of the free electron spin. The inside plots in Fig. 5.10 represent weak second-order transitions, which are explained in Section 4.

An Irradiated Trifluoroacetamide Crystal

Trifluoroacetamide provides a somewhat more complex example in which the radicals formed by the irradiation have two coupling nuclei and two magnetically distinguishable orientations in a crystal of unknown unit-cell structure. The esr has been measured and analyzed by Lontz and Gordy [10]. The spectrum is clean in that signals of only one radical species are observed, but second-order transitions (see Section 4) are detected at both X- and K-band frequencies. The crystals are approximately monoclinic in form with the angles $\alpha \approx \gamma \approx 90°$ and $\beta = 68°$. The crystal form with the orthogonal reference system used for description of the spectrum in the original work is shown in Fig. 5.11. To conform to the conventional notation of Table 5.1, the designations of the b and c edges of the crystal must be reversed.

Figure 5.12 shows the hyperfine pattern observed with **H** at selected orientations in the $a'b$ reference plane. The top curve superficially appears to be the spectral pattern of a single radical with four equally coupling nuclei, but this is not true. Measurements show that the spacing of the two outside components from the center line is more than twice that of those closer in. This suggests that the quintet is actually two superimposed triplet patterns, one with slightly more than twice the spacing of the other, with only their central components superimposed. Observations at other orientations quickly confirmed this interpretation. Note that for the 30° orientation the central components are not quite superimposed, and that there are six observable lines. Here the relative positions of the triplets are slightly different because of **g** anisotropy. At the 56° orientation, where the two triplets have the same splitting and the same g, they become completely superimposed. This suggests that the two triplets arise from two radicals of the same chemical form, with two F nuclei having the same coupling, and that the two radicals have magnetically equivalent orientations at 56° and inequivalent orientations for other directions of **H** in this plane. Further obser-

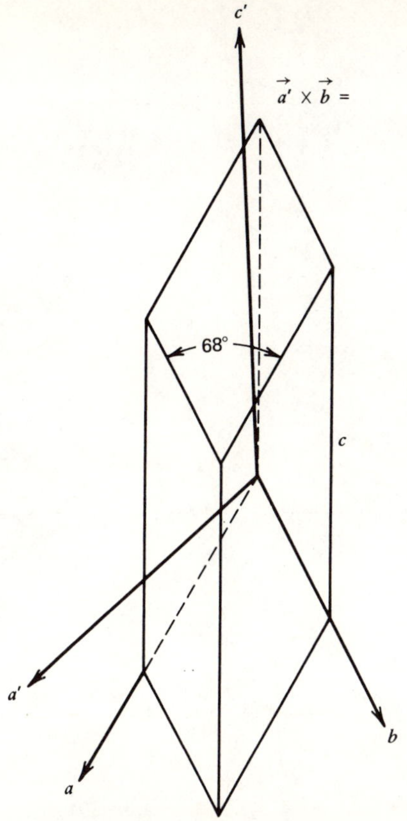

Fig. 5.11 Crystal form of trifluoroacetamide and the reference axes used in evaluation of the **g** and **A** tensors of the free radicals produced by γ-irradiation. From Lontz and Gordy [10].

vations, which need not be described, confirmed this interpretation.

Please note the very weak, closely spaced doublet between the stronger triplet components of the top curve of Fig. 5.12 and the similar, but somewhat stronger, doublets in the bottom curve. These are second-order transitions forbidden in the first-order theory. They are explained in Section 4.

Plots of the positions of the outside components of the two triplets (labeled I and II) in two of the orthogonal reference planes are shown in Fig. 5.13. Calculation of the principal coupling elements from both patterns gave the same axially symmetric principal values, but with differently oriented axes. One symmetry axis was found to be perpendicular to the *ac* plane, and the other was found to be perpendicular to the *bc* plane. Confirming evidence for this

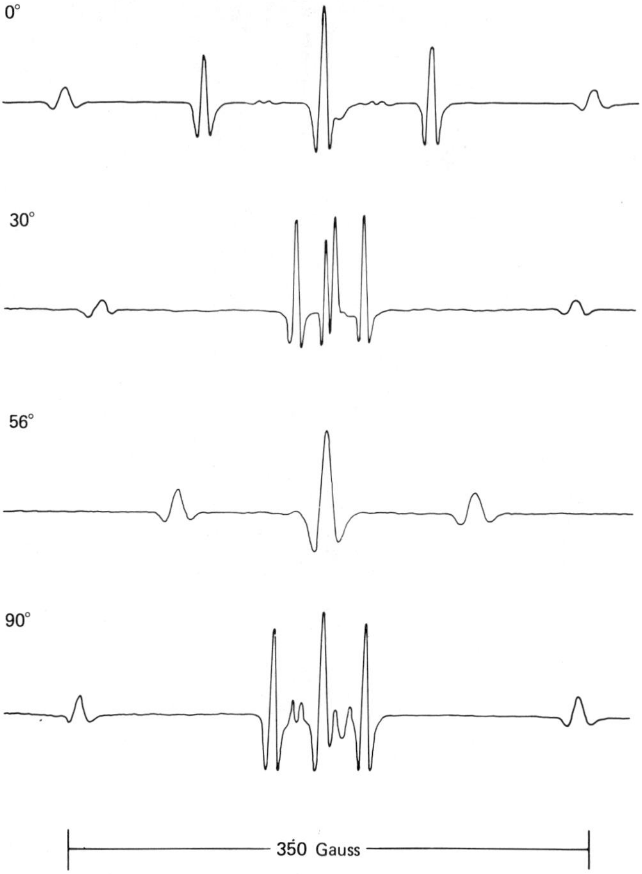

Fig. 5.12 Hyperfine patterns due to ^{19}F observed for selected crystal orientations in the $a'b$ plane (see Fig. 5.11) of a γ-irradiated single crystal of trifluoroacetamide. From Lontz and Gordy [10].

analysis was then obtained from measurements made with **H** rotated alternately in each of these planes. Figure 5.14 shows plots of the positions of the outside components of both triplets with **H** rotated in the *bc* plane. Obviously, the two equivalent fluorines that give rise to pattern I have axially symmetric coupling about the normal to this plane. Similar patterns obtained for the *ab* plane verify the conclusion that radical II has its coupling axis perpendicular to the *ab* plane. Otherwise, the two radicals are magnetically identical.

The only resonable interpretation of these features is that the two equivalent patterns arise from the identical radicals in the two crystalline planes, as shown in Fig. 5.15. For the coupling of the two ^{19}F nuclei of each radical to be equivalent for all orientations, as observed, their coupling p orbitals must be

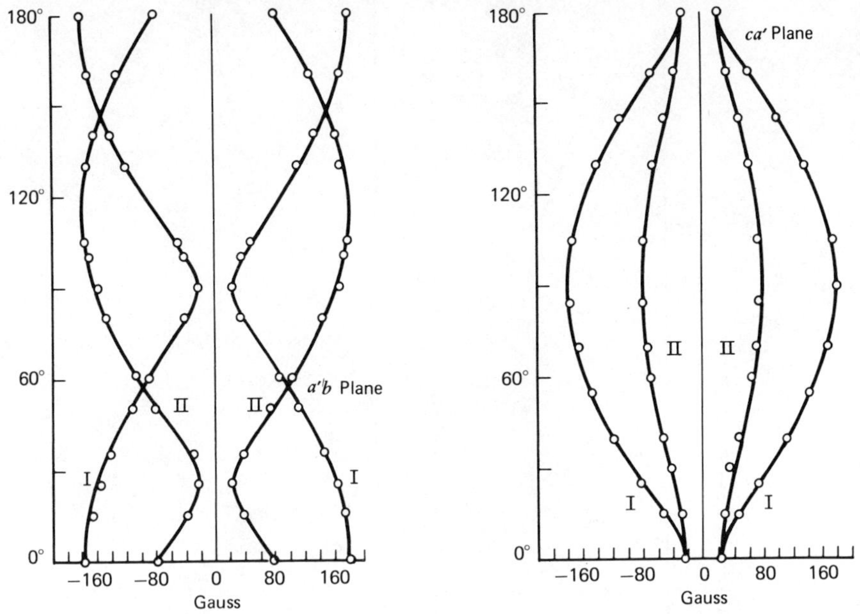

Fig. 5.13 Plots of the positions of the outside components of the two triplets, I and II, as shown in Fig. 5.12, for various orientations of the magnetic field in the *a′b* and *ca′* planes. From Lontz and Gordy [10].

strictly parallel, and both must have the same spin density. These features indicate the planar form of the radicals, shown in Fig. 5.15. The unpaired electron has a π orbital perpendicular to the molecular plane. The principal elements of **g**, as well as those of the ^{19}F coupling, conform to this interpretation. The derived principal elements are $g_{\parallel} = 2.0025$, $g_{\perp} = 2.0045$, $A_{\parallel} = 498$ MHz, and $A_{\perp} = 67$ MHz, where the $\parallel$ and $\perp$ subscripts refer to either the *ac* or *bc* plane.

The preceding examples not only indicate how the spectra are analyzed, but also suggest how esr can give information about unknown crystal structure, the structure of free radicals, and the nature of radiation damage. More detailed analysis and interpretation of observed parameters are given in Chapter VI. Although these examples should be helpful to a beginner in learning to observe and analyze esr of single crystals, each irradiated crystal is a new problem, not exactly like any of those described in the literature. This is what makes the pursuit interesting, at times even exciting.

3. SECOND-ORDER SHIFTS OF HYPERFINE COMPONENTS

The first-order theory described in Section 2 is adequate for analysis of most hyperfine structure observed in single crystals. However, second-order shifts of

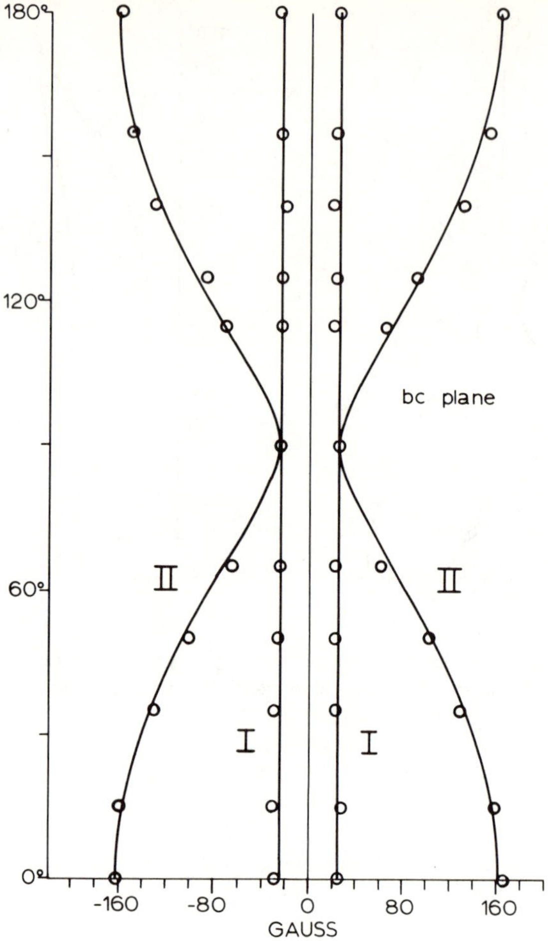

Fig. 5.14 Plots of the outside components of the two triplets, I and II, for various orientations of the magnetic field in the *bc* plane. The plots show that the ^{19}F couplings for radical I have axial symmetry about the perpendicular to this plane. Similar plots show that the ^{19}F couplings of radical II have axial symmetry about the perpendicular to the *ac* plane. The radical structures and their relative orientations deduced from the analysis are shown in Fig. 5.15. From Lontz and Gordy [10].

the components can be observed for certain coupling combinations or applied-field values. These deviations, which are usually of the order of 1% or less, can be calculated to the accuracy of the measurement with second-order perturbation theory, as described in Chapter III, Section 4.k. Except when the hyperfine splitting is very great (several hundred megahertz), significant line shifts can be avoided by an increase in the observation frequency ν_0 and hence in the applied

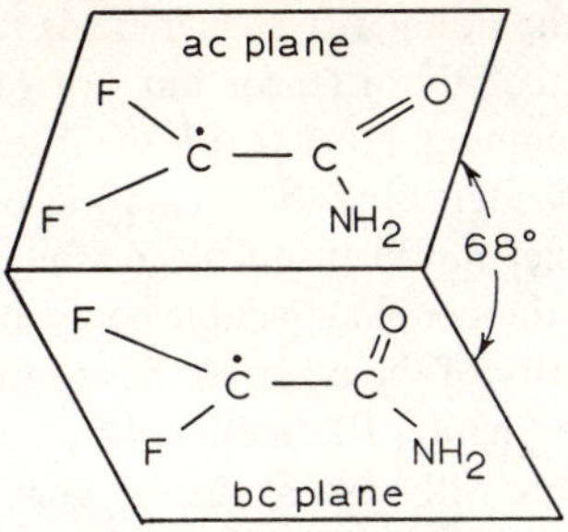

Fig. 5.15 Structure and orientation of the free radicals formed by γ-irradiation of a single crystal of trifluoroacetamide, as deduced from the ^{19}F splitting of their esr spectra. From Lontz and Gordy [10].

field values to ensure that the separation, $g\beta H$, of the electron spin states is very large as compared with that of the nuclear substates. When this is not feasible, the recommended procedure is to find and diagonalize the first-order tensors, as already described, and then to apply the second-order corrections to the principal elements, or in the principal planes.

Large anisotropic coupling in free radicals is due mostly to the p orbital and is axially symmetric about the coupling p-orbital axis. Second-order corrections for axially symmetric magnetic coupling are given by (3.140). At a constant observation frequency ν_0, the components for $S = 1/2$ radicals occur at field values

$$H = H_0 + \left(\frac{h}{g_\theta \beta}\right) A_\theta M_I + \frac{h^2}{8H_0(g_\theta \beta A_\theta)^2} \Big\{ [2A_\perp^2 (A_\parallel^2 + A_\theta^2)]$$

$$\times [I(I+1) - M_I^2] + M_I^2 \left(\frac{g_\parallel g_\perp}{g_\theta^2}\right)^2 (A_\parallel^2 - A_\perp^2) \sin^2 2\theta \Big\} \qquad (5.70)$$

where θ is the angle of the applied field with the symmetry axis and $g_\theta^2 = g_\parallel^2 \cos^2\theta + g_\perp^2 \sin^2\theta$, $g_\theta^2 A_\theta^2 = g_\parallel^2 A_\parallel^2 \cos^2\theta + g_\perp^2 A_\perp^2 \sin^2\theta$. All coupling values are measured in frequency units (hertz); H, in gauss; and $H_0 = h\nu_0/g_\theta\beta$ is the field value where resonance would occur for the angle θ if there were no hyperfine interaction.

Note that the nuclear quantum number appears only as M_I^2 in the second-order term of (5.70). Hence the second-order interaction shifts both the companion components M_I and $-M_I$ for the same $|M_I|$ to higher fields, by the same amount. Because the separations of the M_I and $-M_I$ components are independent of the second-order shifts, one can use these separations to obtain accurate values of the coupling constants without involvement of the second-order theory. For a doublet hyperfine structure $I = 1/2$, the only effect of the second-order term is a shift of the center of the doublet to higher fields relative to H_0. This effect is important for evaluation of the **g** tensor but not for determination of the principal elements of the coupling. For $1/2 < I$, the interaction also destroys the symmetry of the pattern by increasing the separation of successive components on the high-field side of the multiplet and decreasing it on the low-field side. Unfortunately, the sign of the coupling cannot be ascertained from this type of asymmetry. Note the similarity of these corrections to the Breit-Rabi corrections for the isotropic S state (see Chapter III, Section 4.b).

When $1/2 < I$, the nucleus will have a nuclear quadrupole moment and an electric quadrupole interaction that can produce further second-order perturbations of the esr lines, but these perturbations are generally much smaller than the second-order magnetic perturbations described previously and are entirely negligible in the esr of most free radicals. The first-order quadrupole energy (see Chapter III, Section 6) shifts all levels of the same M_I value equally and hence is not detectable in normal esr for which $\Delta M_I = 0$. The second-order quadrupole effects are significant for certain transition elements that have large nuclear quadrupole moments. Bleaney [11] has derived formulas for calculation of second-order quadrupole effects on esr hyperfine structure for axially symmetric couplings. Since these formulas contain terms in M_I and $M_I{}^3$, one can use the shifts, when they are detectable, to find the sign of the magnetic coupling. Because of their complexity and limited applications, these formulas are not reproduced here. They may be found in the treatise by Abragam and Bleaney [12] and in Bleaney's original paper [11].

A prevalent type of second-order perturbation of esr spectra is that caused by an admixture of nuclear spin substates of a particular electron-spin state. Because it does not require large nuclear splittings, this type of perturbation is important for hyperfine structure such as that due to H or F in organic free radicals. This interaction is also field dependent, but the dependence results from the nmr term $g_I\beta_I HM_I$ rather than from the esr term $g\beta HM_S$, as does that of (5.70). The interaction often leads to detectable, and sometimes rather strong, forbidden transitions as well as line shifts. The effects are treated in the following section.

4. SECOND-ORDER, (FORBIDDEN) HYPERFINE TRANSITIONS

In most esr hyperfine structure, the selection rules that are operative are $\Delta M_S = 1$ and $\Delta M_I = 0$. The latter rule indicates that the nucleus does not flip over with the electron-spin transition. When the coupling is anisotropic and the imposed field **H** is not along a principal axis of the coupling tensor, the nuclear spin states can become intermixed, and the nuclear spin transitions are not restricted to $\Delta M_I = 0$. New lines indicated as second-order, or forbidden, transitions may then be observed. When the scrambling of the nuclear-spin states is slight, M_I

may still be treated as a good quantum number and the forbidden components be labeled as $\Delta M_I = \pm 1$ or ± 2 transitions. For small admixtures, however, the forbidden lines will be very weak. Under some conditions the scrambling of the nuclear spin states can be appreciable and the forbidden lines strong. Then M_I can no longer be considered a good quantum number, and the selection rules are determined by the symmetry of the admixed nuclear levels.

Second-order hyperfine transitions due to cross-term interactions between fine structure (when $1/2 < S$) and hyperfine structure, first observed by Bleaney and Ingram [13] and theoretically explained by Bleaney and Rubens [14], has been observed in a number of iron-group salts [12]. Forbidden transitions involving fine structure are not observed in free radicals having doublet spin states, and they are not elaborated here. Their treatment is given by Abragam and Bleaney [12].

4.a. Admixed Nuclear-spin States of Free Radicals

First-order, forbidden hyperfine transitions for free radicals with $S = 1/2$ and $I = 1/2$ were apparently first identified by Miyagawa and Gordy [15,16], who showed that the transitions result from admixture of the nuclear spin states, $M_I = \pm 1/2$, of each electron spin level by interaction of off-diagonal elements of the anisotropic nuclear tensor with the nmr term $g_I \beta_I \mathbf{H} \cdot \mathbf{I}$. Shortly thereafter, McConnell et al. [7] observed weak lines in the esr of irradiated single crystals of malonic acid that they interpreted as due to second-order transitions in the proton hyperfine structure. At about the same time, Atherton and Whiffen [17] observed a nonintegral variation in the relative intensities of hyperfine components of HCHCOOH radicals in irradiated single crystals of glycolic acid that they attributed to a breakdown in the $\Delta M_I = 0$ selection rule in the solid state.

The theory developed by Miyagawa and Gordy [16] for treatment of second-order transitions of spin 1/2 nuclei is described in the following paragraphs. It was found to account completely for the observed effects in the alanine and has since been applied successfully to many other cases.

The spin Hamiltonian for $S = 1/2$ radicals with no nuclear quadrupole interaction is

$$\mathcal{H}_{s,i} = \beta \mathbf{H} \cdot \mathbf{g} \cdot \mathbf{S} + \mathbf{S} \cdot \mathbf{A} \cdot \mathbf{I} - g_I \beta \mathbf{H} \cdot \mathbf{I} \tag{5.71}$$

Effects of the small $\mathbf{g}$ anisotropy of organic free radicals are inconsequential for the second-order effects treated here. The term $\mathcal{H}_0 = \beta \mathbf{H} \cdot \mathbf{g} \cdot \mathbf{S}$ can be solved separately, as described in Section 1. Without loss of generality, we can choose the magnetic field $\mathbf{H}$ to lie along the z axis of an arbitrarily chosen system xyz. With the effects of $\mathbf{g}$ anisotropy negligible, the terms involving the nuclear interaction can be expressed as

$$\mathcal{H}' = (A_{zz}I_z + A_{zx}I_x + A_{zy}I_y)S_z - g_I\beta_I H I_z \tag{5.72}$$

A different coordinate system chosen with reference to the crystal surfaces can be used for finding the principal axis of **A** from the first-order theory, as described in Section 2, and the elements A_{zz}, A_{zx}, and A_{zy} can then be expressed in the principal axes system, regardless of the direction of z.

To find the positions and intensities of the forbidden transitions, one must find both the energies and the wave functions of the perturbed $\mathcal{H}_S$. Because the electron- and nuclear-spin operators commute, the wave function can be expressed by the product:

$$\psi_{e,N} = \psi_{S,M_S}\,\psi_N \tag{5.73}$$

where ψ_{S,M_S} is the electron-spin function and ψ_N is the nuclear-spin function. With the conditions assumed here, the $\mathcal{H}_S$ is diagonal in the electron-spin function, and we express the functions in the ket notation:

$$\psi_{S,M_S} = |M_S), \quad \psi_N = |N(\pm)) \tag{5.74}$$

We carry out the electron-spin operation first and express $\mathcal{H}'$ in terms of the nuclear-spin operator only. Thus for $S = 1/2$, $S_z = \pm 1/2$ and the reduced $\mathcal{H}'$ is

$$\mathcal{H}'(M_S = \pm \tfrac{1}{2}) = \pm \frac{1}{2}\,[(A_{zz} \mp 2g_I\beta_I H)I_z + A_{zx}I_x + A_{zy}I_y] \tag{5.75}$$

Obviously, $\mathcal{H}'$ is not diagonal in the $|I,M_I)$ representation, but its matrix elements can easily be found in this representation, regardless of the value of I. Thus it is appropriate to express ψ_N as a linear combination of the function ψ_{I,M_I}. The general technique for treating this type of problem is described in Chapter III, Section 1. For $I = 1/2$, $M_I = \pm 1/2$, there are two independent, orthogonal nuclear functions corresponding to each electronic state, $M_S = \pm 1/2$. The two nuclear functions of each electronic state have opposite symmetry, and the intermixed, orthogonal functions are expressed as linear combinations of the nuclear spin functions, $|M_I) = |\pm 1/2)$, by

$$|N_l(+)) = a_l|\tfrac{1}{2}) + b_l|-\tfrac{1}{2}) \tag{5.76}$$

$$|N_l(-)) = b_l^*|\tfrac{1}{2}) - a_l^*|-\tfrac{1}{2}) \tag{5.77}$$

$$|N_u(+)\rangle = a_u|\tfrac{1}{2}\rangle + b_u|-\tfrac{1}{2}\rangle \tag{5.78}$$

$$|N_u(-)\rangle = b_u^*|\tfrac{1}{2}\rangle - a_u^*|-\tfrac{1}{2}\rangle \tag{5.79}$$

where $|N_u(\pm)\rangle$ and $|N_l(\pm)\rangle$ signify the nuclear-spin functions ψ_N for the upper, $M_S = 1/2$, and lower, $M_S = -1/2$, electron-spin states, respectively. The a and b values are weighting coefficients corresponding to C_n in the expression of (3.2). They may be evaluated as explained in Chapter III, Section 1 after the secular equation is solved for the eigenvalues of $\mathcal{H}'$. These coefficients are required for calculation of the intensities of the forbidden transitions.

For $I = 1/2$, the matrix elements of $\mathcal{H}'$ of (5.75) in the representation $|I,M_I\rangle$ are easily obtained by use of those for the electron-spin operators (Chapter III, Section 2), by the replacements $S \rightarrow I$, $M_S \rightarrow M_I$. The resulting energy matrix leads to the secular equation:

$$\begin{vmatrix} \frac{1}{4}A_{zz} - \frac{1}{2}(g_I\beta_I H) - E & \frac{1}{4}(A_{zx} - iA_{zy}) \\ \frac{1}{4}(A_{zx} + iA_{zy}) & -\frac{1}{4}A_{zz} + \frac{1}{2}(g_I\beta_I H) - E \end{vmatrix} = 0 \tag{5.80}$$

Solution of this equation yields for the upper level, $M_S = 1/2$:

$$E'_{1/2}(\pm) = \pm\frac{1}{4}\,[(A_{zz} - 2\,g_I\beta_I H)^2 + (A_{zx}^2 + A_{zy}^2)]^{1/2} \tag{5.81}$$

A similar solution of $\mathcal{H}'_{-1/2}$ yields for the $M_S = -1/2$ level:

$$E'_{-1/2}(\pm) = \pm\frac{1}{4}\,[(A_{zz} + 2\,g_I\beta_I H)^2 + (A_{zx}^2 + A_{zy}^2)]^{1/2} \tag{5.82}$$

For simplicity, we designate

$$A_\pm = [(A_{zz} \pm 2\,g_I\beta_I H)^2 + c^2]^{1/2} \tag{5.83}$$

where

$$c^2 = A_{zx}^2 + A_{zy}^2 \tag{5.84}$$

The eigenvalues of $\mathcal{H}_0$ are $\pm\frac{1}{2}g\beta H$. Therefore, the energy levels are

$$E_{M_S} = E_0 + E' \tag{5.85}$$

and

$$E_{1/2}(\pm) = \frac{1}{2}(g\beta H) \pm \frac{1}{4}A_- \tag{5.86}$$

for $|N_u(\pm)\rangle$, and

$$E_{-1/2}(\pm) = -\frac{1}{2}(g\beta H) \pm \frac{1}{4}A_+ \tag{5.87}$$

for $|N_l(\pm)\rangle$.

Figure 5.16 shows the energy levels and possible transitions. The frequencies indicated by the solid lines are given by

$$h\nu_1 = g\beta H + \frac{1}{4}(A_+ + A_-) \tag{5.88}$$

$$h\nu_2 = g\beta H - \frac{1}{4}(A_+ + A_-) \tag{5.89}$$

The doublet spacing between these components, in magnetic field units $(H_1 - H_2)$, is

$$d_+ = \frac{A_+ + A_-}{2g\beta} \tag{5.90}$$

Similarly, the frequencies of the transitions indicated by the broken arrows are given by

$$h\nu_3 = g\beta H - \frac{1}{4}(A_+ - A_-) \tag{5.91}$$

$$h\nu_4 = g\beta H + \frac{1}{4}(A_+ - A_-) \tag{5.92}$$

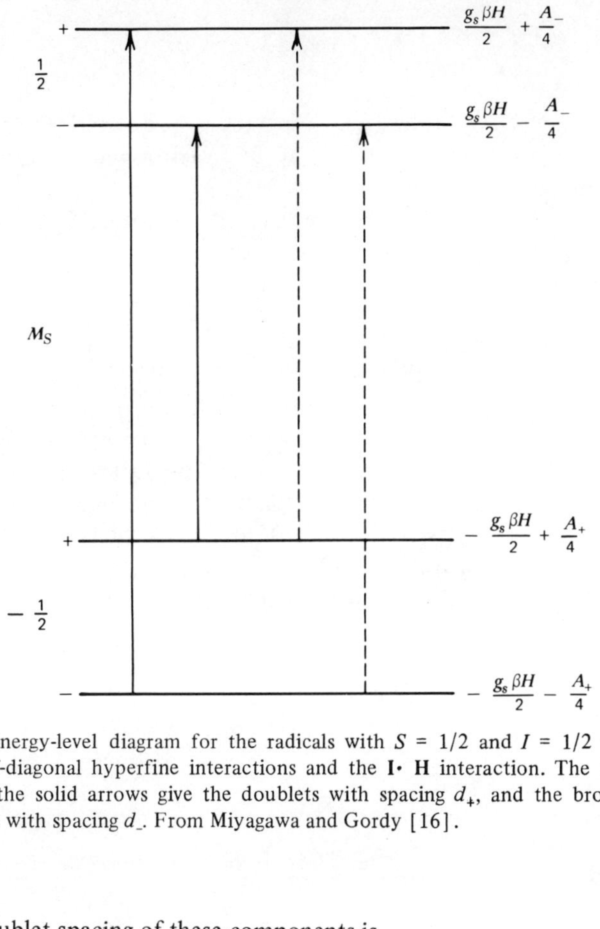

Fig. 5.16 Energy-level diagram for the radicals with $S = 1/2$ and $I = 1/2$ when the $\mathcal{H}_S$ includes off-diagonal hyperfine interactions and the **I**· **H** interaction. The transitions indicated by the solid arrows give the doublets with spacing d_+, and the broken ones give the doublets with spacing d_-. From Miyagawa and Gordy [16].

and the doublet spacing of these components is

$$d_- = \frac{A_+ - A_-}{2g\beta} \tag{5.93}$$

Let us now calculate the relative intensities of the two doublets, d_- and d_+. Because the interaction that induces transitions is that between the alternating magnetic component of the microwave radiation and the rotating perpendicular component of the spin moment, $\mu_x + i\mu_y = g\beta(S_x + iS_y)$, the transition probabilities are proportional to the squared matrix elements:

$$[\int \psi_{1/2}(\pm)\,(\mu_x + i\mu_y)\psi_{-1/2}(\pm)\,d\tau]^2 =$$

$$g^2\beta^2|(1/2|S_x + iS_y|-1/2)|^2\, |(N_u(\pm)|N_l(\pm))|^2 \tag{5.94}$$

The latter form is possible because the nuclear-spin functions indicated by $N_u(\pm)$ and $N_l(\pm)$ are not operated on by the electron-spin operators and are separable from the electron-spin functions.

The intensity ratio of the components of the two doublets d_+ and d_- is, therefore,

$$\frac{I_{d_-}}{I_{d_+}} = \frac{|(N_u(+)\,|\,N_l(+))|^2}{|(N_u(+)\,|\,N_l(-))|^2} = \frac{|(N_u(-)|N_l(-))|^2}{|(N_u(-)|N_l(+))|^2} \tag{5.95}$$

in which the term $g^2\beta^2|(1/2|S_x + iS_y|-1/2)|^2$ has been cancelled from the numerator and denominator of each expression. The nuclear-spin functions are admixtures of the normal nuclear-spin functions corresponding to $M_I = +1/2$ and $M_I = -1/2$ as expressed by (5.76) to (5.79). Since the nuclear-spin functions are normalized orthogonal functions, $(1/2|1/2) = (-1/2|-1/2) = 1$, $(1/2|-1/2) = (-1/2|1/2) = 0$, from the admixed functions we obtain

$$|(N_u\,(+)\,N_l\,(+))|^2 = |a_u^*a_l + b_u^*b_l|^2 \tag{5.96}$$

$$|(N_u\,(+)\,N_l\,(-))|^2 = |a_ub_l - b_ua_l|^2 \tag{5.97}$$

and (5.95) can be expressed as

$$\frac{I_{d_-}}{I_{d_+}} = \frac{|a_u^*a_l + b_u^*b_l|^2}{|a_ub_l - b_ua_l|^2} \tag{5.98}$$

To find the intensity ratio, we need to evaluate the mixing coefficients a and b of the admixed nuclear-spin states. They are the same as the C_n coefficients in (3.6) and can be found by the process described in Chapter III, Section 1. The problem is like that of finding the eigenvectors of the **g** or **A** tensors. For example, from the root $E'_{1/2}(+) = (1/4)A_-$ for $|N_u(+))$ and the secular determinant (second row) for $\mathcal{H}_{1/2}$, one obtains

$$a_u(A_{zx} + iA_{zy}) - b_u(A_{zz} - 2g_I\beta_I H + A_-) = 0 \tag{5.99}$$

From the root $E'_{-1/2}\,(+) = (1/4)A_+$ for $|N_l\,(+))$ and the first row of the secular determinant, one obtains

$$a_l(A_{zz} + 2g_I\beta_I H + A_+) + b_l(A_{zx} - iA_{zy}) = 0 \tag{5.100}$$

Therefore,

$$\frac{a_u}{b_u} = \frac{(A_{zz} - 2g_I\beta_I H + A_-)}{A_{zx} + iA_{zy}} \tag{5.101}$$

and

$$\frac{a_l}{b_l} = \frac{-(A_{zx} - iA_{zy})}{A_{zz} + 2g_I\beta_I H + A_+} \tag{5.102}$$

The normalization requires

$$a_u a_u^* + b_u b_u^* = 1 \tag{5.103}$$

$$a_l a_l^* + b_l b_l^* = 1 \tag{5.104}$$

With the a and b derived from (5.101) to (5.104) and substituted into (5.98), the intensity ratio is found to be

$$r = \frac{I_{d_-}}{I_{d_+}} = \frac{c^2(A_+ - A_- + 4g_I\beta_I H)^2}{[(A_{zz} + A_- - 2g_I\beta_I H)(A_{zz} + A_+ + 2g_I\beta_I H) + c^2]^2} \tag{5.105}$$

where

$$c^2 = (A_{zx}^2 + A_{zy}^2) \tag{5.106}$$

Normalized intensities for the doublet are defined by

$$I_{d_+} = \frac{1}{1 + r} = \cos^2 \psi \tag{5.107}$$

and

$$I_{d_-} = I - I_{d_+} = \frac{r}{1 + r} = \sin^2 \psi \tag{5.108}$$

where $\tan \psi = \sqrt{r}$.

The graphs of Figs. 5.17 and 5.18 give a general impression of the complexities to be encountered if one tries to analyze the hyperfine structure when the

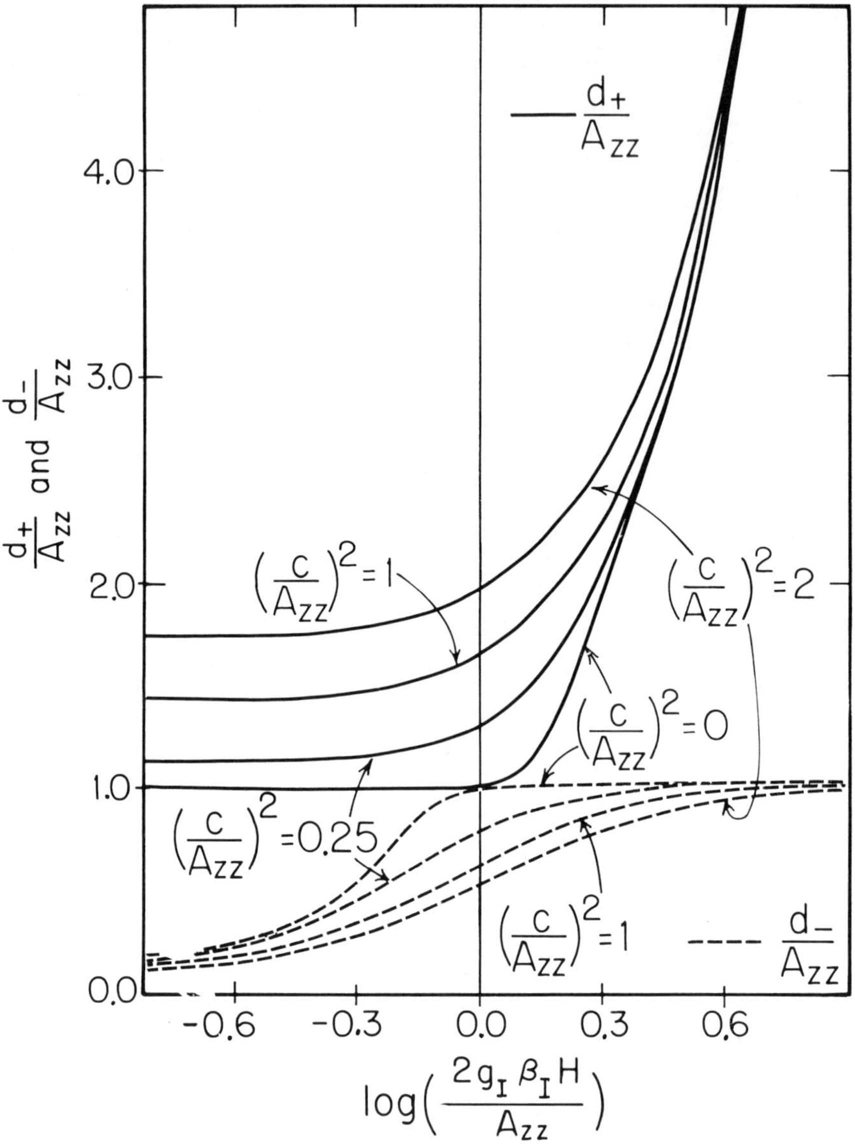

Fig. 5.17 Graphs showing dependence of d_+/A_{zz} and d_-/A_{zz} on $\log(2g_I\beta_I H/A_{zz})$ and $(c/A_{zz})^2$. From Miyagawa and Gordy [16].

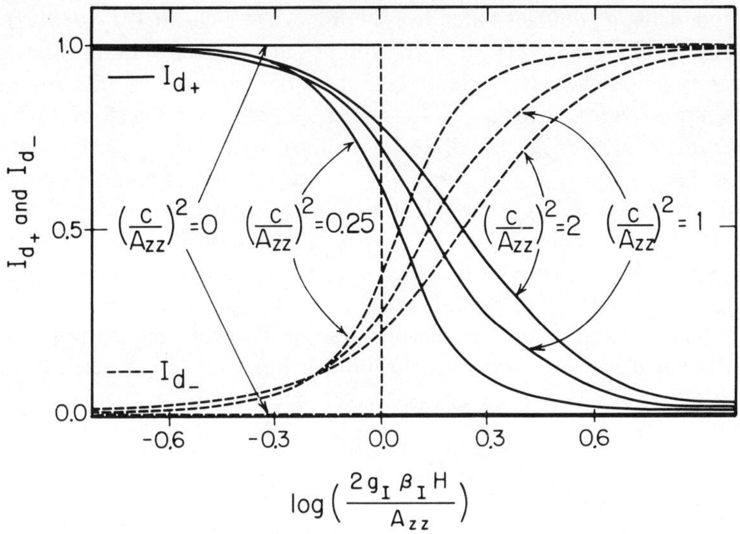

Fig. 5.18 Relative intensities of the d_+ and d_- doublets as a function of $\log(2g_I\beta_I H/A_{zz})$ for various values of $(c/A_{zz})^2$. From Miyagawa and Gordy [16].

observation frequency is such as to make $2g_I\beta_I H \sim A_{\text{eff}}$ for some orientations of the crystal in the observed planes. They should also be useful in the analysis of the esr of other single crystals that have second-order effects of this kind. Such effects are by no means uncommon. When the anisotropy in the coupling is large and all the principal elements are not of the same sign or when all have the same sign and one is small, this critical condition, where $A_{zz} \sim g_I\beta_I H$, will be encountered for some orientations with any convenient observation frequency.

The sign of A_{zz} is not specified but is tacitly assumed to be positive in the preceding derivations. The theory holds for either sign of the coupling. If, for example, A_{zz} is taken as negative in the spin Hamiltonian of (5.72), the A_- and A_+ will be exchanged in (5.86) and (5.87) so that A_+ will apply for $M_S = 1/2$ and A_- for $M_S = -1/2$, but this reversal of levels will have no effect on the doublet spacings d_- and d_+ as calculated from (5.90) and (5.93).

Note from Fig. 5.18 that the doublets d_+ and d_- have approximately the same intensity when the applied field **H** is such as to make $2g_I\beta_I H \sim A_{zz}$. Although the same symmetry selection rules still apply for d_+ and d_-, it is misleading to call either doublet a forbidden transition. For field values of this order, the mixing coefficients a and b for the nuclear-spin functions become approximately equal; that is, the $M_I = 1/2$, $M_I = -1/2$ functions are almost uniformly mixed. Beyond the crossover where $2g_I\beta_I H$ becomes significantly greater than A_{zz}, the

d_- prevails, and the d_+ doublet fades away. For $g_I\beta_I H = 2A_{zz}$, $\log(2g_I\beta_I H/A_{zz}) = 0.6$, and the forbidden $N(\pm) \leftrightarrow N(\pm)$, transitions have normalized intensities of approximately one when $(c/A_{zz})^2 = 0.25$ and the intensity of the normal doublet is almost zero. Then, $d_- \approx A_{zz}$ and $d_+ >> A_{zz}$ (see Fig. 5.17). When $g_I\beta_I H \leqslant A_{zz}/8$, $\log(2g\beta H/A_{zz}) = -0.6$ and the normal doublet d_+ predominates; then $I_{d_+} \approx 1$, $I_{d_-} \approx 0$, $d_+ \approx (A_{zz}^2 + c^2)^{1/2}$ and $d_- << A_{zz}$. When $c^2 = (A_{zx}^2 + A_{zy}^2) = 0$ (isotropic coupling, or **H** along a principal axis of **A**), this crossover occurs exactly and completely at $\log(2g\beta H/A_{zz}) = 0$. For the alanine crystal with **H** along the [001] crystal axis, the crossover frequency was experimentally observed to be 24.7 GHz, as explained below.

One can easily express the off-diagonal tensor elements required for the second-order calculations in terms of the principal elements of **A** when this is desirable. If u, v, and w are the principal axes, one finds by a rotation of the tensor that

$$A_{zz} = A_u l_{uz}^2 + A_v l_{vz}^2 + A_w l_{wz}^2 \tag{5.109}$$

$$A_{zx} = A_u l_{uz} l_{ux} + A_v l_{vz} l_{vx} + A_w l_{wz} l_{wx} \tag{5.110}$$

$$A_{zy} = A_u l_{uz} l_{uy} + A_v l_{vz} l_{vy} + A_w l_{wz} l_{wy} \tag{5.111}$$

where A_u, A_v, and A_w are the principal elements and the l symbols are the direction cosines between the principal axes and the x, y, and z axes, as indicated by the subscripts. When the measurements are made in the principal planes, these expressions can be simplified. Suppose, for example, that the principal axis u is maintained perpendicular to **H**, and that measurements are made with **H** at various orientations in the vw plane at angles θ from the principal axis w. With y and z chosen in the vw plane and hence with x along u, (5.109) to (5.111) reduce to

$$(A_{zz})_{vw} = A_v \sin^2\theta + A_w \cos^2\theta \tag{5.112}$$

$$(A_{zx})_{vw} = 0 \tag{5.113}$$

$$(A_{zy})_{vw} = (A_v - A_w)\sin\theta\cos\theta \tag{5.114}$$

From these expressions, all quantities needed for calculation of d_- and I_{d_-} may be readily obtained. Expressions for the other principal planes are similar.

When there is a known axis of symmetry in the coupling, we are free to choose w as the symmetry axis and y and z in the vw plane. Thus in this case the second-order parameters may be calculated for any orientation of **H** from the similarly reduced transformation formulas

$$A_{zz} = A_{\parallel} \cos^2 \theta + A_{\perp} \sin^2 \theta \tag{5.115}$$

$$A_{zx} = 0 \tag{5.116}$$

$$A_{zy} = (A_{\perp} - A_{\parallel}) \sin \theta \cos \theta \tag{5.117}$$

and hence

$$c^2 = (A_{\perp} - A_{\parallel})^2 \sin^2 \theta \cos^2 \theta \tag{5.118}$$

where θ is the angle of **H** relative to the symmetry axes and where $A_{\parallel}$ and $a_{\perp}$ are the principal elements of A parallel and perpendicular, respectively, to the symmetry axis. Likewise, for this case,

$$A_{\pm} = [(A_{\parallel} \pm 2g_I\beta_I H)^2 \cos^2 \theta + (A_{\perp} \pm 2g_I\beta_I H)^2 \sin^2 \theta]^{1/2} \tag{5.119}$$

where the upper signs are applied together and the lower applied together. These reduced expressions may be substituted directly into (5.90), (5.93), and (5.105) for calculation of the doublet spacings and intensities. These formulas were applied successfully by Lontz and Gordy [10] to account for forbidden transitions observed in the axially symmetric ^{19}F hyperfine structure for irradiated single crystals for trifluoroacetamide.

This second-order theory was rather thoroughly tested on the $C_\alpha H$ proton of the radical $R - \dot{C}_\alpha H - C_\beta H_3$ in a single crystal of irradiated alanine. As in most single crystals, the hyperfine structure of the hydrogen of the $C_\alpha H$ fragment is complicated by the superposition of hyperfine splitting by other hydrogens of the observed radical. In this radical the other couplings arise from the almost isotropic β coupling of the three hydrogens of the methyl group. At room temperature, where the observations were made, the hydrogen couplings of CH_3 are equalized by rotation of this group. Thus the $C_\alpha H$ coupling is superimposed on four, equally spaced components with intensity 1:3:3:1.

The single crystal of alanine, d or l, is orthorhombic, and the unit cell has four molecules. These four molecules give rise to $RCHCH_3$ radicals of four magnetically distinguishable orientations that complicate the spectrum for most directions of the applied field. When, however, the field is applied along one of the crystallographic axes, the four radicals have magnetically equivalent orientations. Furthermore, the principal axes of the $\dot{C}_\alpha H$ coupling proved not to be along the crystal axes, so that the second-order effects could be observed when **H** was imposed along the axes for which the orientations of all the radicals were magnetically equivalent. When the field is applied along the [001] axis, the effective $C_\alpha H$ coupling $(A_{zz} + A_{zx}^2 + A_{zy}^2)^{1/2}$ is accidentally equal to the isotropic $C_\beta H_3$ couplings; and, in the absence of second-order effects, a five-line pattern of four equivalent protons is observed. Such a quintet with sharp components was

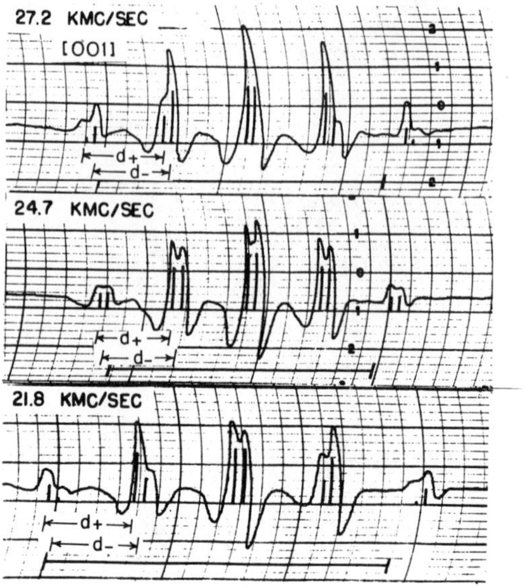

Fig. 5.19 Plots of the esr spectrum of $CH_3\dot{C}HR$ in the critical frequency range where the d_+ and d_- doublets reverse their intensities (see graphs of intensities I_{d_+} and I_{d_-} in Fig. 5.18). From Miyagawa and Gordy [16].

observed at 9000 MHz with only extremely weak satellite lines due to forbidden transitions. The $C_\alpha H$ contribution to the quintet at 9000 MHz is due to the normal d_+ doublet, which at this frequency is predictable by first-order theory. Between these frequencies, the appearance of both d_+ and d_- doublets in significant strength causes a 10-line hyperfine pattern not even qualitatively predictable with first-order theory. Figure 5.19 shows the observed patterns in the critical K-band region where the two types of transitions undergo the reversal of intensities predicted by the theory (see Fig. 5.18). Over the frequency range covered, 21.8 to 27.2 GHz, the doublet spacings do not change significantly, but there is enough difference in spacing of d_+ and d_- to produce the observed quintet of doublets. Because of the coincidence of lines of the inner components, the change is most evident in the outermost components. The bars represent the theoretically predicted spectrum, with the outermost ones in each case due to the d_+ doublet and the next ones due to the d_- doublet, as indicated on the 24.7-GHz graph where the d_+ and d_- doublets become equal in strength. At frequencies lower than 24.7 GHz, the d_+ doublet is the stronger, whereas at the higher frequencies the d_- doublet is the stronger.

Because the strength of the second-order transitions depends on A_{zz} and c^2 as well as on $g_I\beta_I H$, they are likely to be observable at almost any microwave frequency for some orientation of the crystal when the anisotropy in the coupling tensor is large, as for the CH fragment. For alanine, the forbidden transitions appear in significant strength in the X-band region when **H** is along the [100] direction of the crystal. Extra lines due to second-order transitions are now commonly observed in the esr of irradiated molecular crystals. Other examples are given in the sections that follow.

The d_- transitions were called "second-order" transitions by Miyagawa and Gordy [16] because they are induced by cross interactions of the off-diagonal elements of A with the nmr term and are not predicted by first-order theory. It is evident, however, that they do not arise from a second-order perturbation of the nuclear terms on the electron-spin terms $\beta \mathbf{H} \cdot \mathbf{g} \cdot \mathbf{S}$, as do the second-order line shifts treated in Chapter III, Sections 4.k and Chapter V, Section 3. Moreover, when $g_I \beta_I H \sim A_{zz}$, the effects cannot be adequately treated as second-order perturbation of the $\mathbf{I} \cdot \mathbf{H}$ term on the $\mathbf{S} \cdot \mathbf{A} \cdot \mathbf{I}$ term. Generally, the secular equation derived from the energy matrix of the nuclear $\mathcal{H}'$ of (5.75) must be solved for prediction of the effects. For $I = 1/2$, this is not difficult, and the solutions can be expressed in closed formulas, as described earlier by (5.80) to (5.82).

4.b. Forbidden Hyperfine Transitions When $1/2 < I$

Although the theory described in the preceding section applies only for nuclei with spin 1/2, the same phenomena of intermixing will be observed for other spin values when the nmr term is of the order of A_{eff}. For $1/2 < I$, there is an additional complication caused by the nuclear quadrupole interaction. Whenever resultant displacements of the nuclear-spin levels by $\mathbf{S} \cdot \mathbf{A} \cdot \mathbf{I}$, $g_I \beta_I \mathbf{H} \cdot \mathbf{I}$, and the quadrupole interaction $\mathbf{I} \cdot \mathbf{P} \cdot \mathbf{I}$ are such as to cause degeneracy or near-degeneracy of some of the nuclear-spin states M_I, the states will be intermixed by the off-diagonal elements, and similar complications of the hyperfine structure will result when **H** is not along a common principal coupling axis of **A** and eQq.

Derivation of the appropriate secular equation for $1/2 < I$ is straightforward, but its solution is difficult and can best be accomplished with a computer program. When the $\mathbf{S} \cdot \mathbf{A} \cdot \mathbf{I}$ term is much greater than $\mathbf{I} \cdot \mathbf{P} \cdot \mathbf{I}$, and $g_I \beta_I \mathbf{H} \cdot \mathbf{I}$, the latter can be treated as a perturbation on $\mathbf{S} \cdot \mathbf{A} \cdot \mathbf{I}$. The intensities of the forbidden transitions may then be calculated with the perturbed wave functions found from second-order theory without solution of the secular equation. From second-order perturbation theory, Bleaney [11] has derived formulas for the frequencies and intensities of forbidden transitions that apply for axially symmetric coupling cases for which $1 \leqslant I$. These formulas were applied to forbidden transitions of ^{63}Cu and ^{65}Cu in potassium copper sulfate [18]. The formulas are applicable to other similar cases but are not reproduced here because of their complexity.

Rexroad, Hahn, and Temple [19] give approximate formulas for the frequencies and intensities of second-order transitions for doublet-state free radicals $S = 1/2$, for which $I = 1$ and $3/2$. These formulas, however, do not include effects of the nuclear quadrupole interaction that are likely to be important when $1/2 < I$. Close and Rexroad [20] extended the theory to include quadrupole interaction for spin $I = 1$ and applied it to obtain the ^{14}N nuclear quadrupole coupling from the second-order transition of the $N(SO_3)_2^{2-}$ radical observed in an irradiated single crystal of potassium aminedisulfonate. Because of their complexity, these formulas are omitted here. Anyone who observes second-order transitions in nuclei with $I = 1$ should consult the original paper [20].

4.c. Flipping of Environmental Nuclear Spins

In 1954 Zeldes and Livingston [21] discovered weak statellite lines on either side of the normal hyperfine components of H atoms trapped in various acid matrices. On measurement, the spacing of the satellites from the normal lines was found to be that of the nmr proton resonance frequency, $g_I\beta_I H/h$. This suggested that the satellite lines were due to the simultaneous flipping of neighboring proton spins of the acid matrix with the electron spins of the trapped H atoms. A later theoretical treatment by Trammell, Zeldes, and Livingston [22] proved that the observed satellites result from $\Delta M_I = \pm 1$ transitions of environmental protons that are induced by off-diagonal elements in the weak dipole-dipole interactions between the electron spin on the trapped atoms and the spin moments of the neighboring nuclei.

Although we are not concerned here with the spectra of trapped atoms per se, a subject treated in a later chapter, similar satellite lines due to flipping of environmental protons have been observed in irradiated single crystals [23]. Furthermore, the theoretical problem can be treated as an extreme case, for which $A_{zz} << 2g_I\beta_I H$, in the theory described in Section 4.a. Whereas the theory of Trammell, Zeldes, and Livingston [22] is instructive, it is simpler for us at this point to adapt the theory already described here. Because the coupling to the environmental nuclei is generally very small, as compared with $g_I\beta_I \mathbf{H}\cdot\mathbf{I}$, we can expand (5.83) and obtain for this extreme case:

$$A_+ \simeq 2g_I\beta_I H + A_{zz} + \frac{c^2}{4g_I\beta_I H}, \quad A_- \simeq -2g_I\beta_I H - A_{zz} + \frac{c^2}{4g_I\beta_I H} \tag{5.120}$$

where A_+ and A_- are measured in gauss. These approximations are also justified by the fact that the observed satellite doublets have separations:

$$d_- = \frac{A_+ - A_-}{2g\beta} \approx \frac{2A_{zz}}{g\beta} \approx 0 \tag{5.121}$$

$$d_+ = \frac{A_+ + A_-}{2g\beta} \approx \frac{2g_I\beta_I H}{g\beta} \tag{5.122}$$

By substitution of A_+ and A_- from (5.120) into (5.105) with omission of the relatively small A_{zz} and c^2 terms in the denominator, one obtains for the relative intensities of the two doublets:

$$\frac{1}{r} = \frac{I_{d_+}}{I_{d_-}} \simeq \frac{c^2}{(2g_I\beta_I H)^2} = \frac{A_{zx}^2 + A_{zy}^2}{(2g_I\beta_I H)^2} \tag{5.123}$$

Because the d_+ doublet splitting will be well below the line width and hence not resolved, the ratio of the intensity of each satellite to that of the parent line will be half this amount,

$$\frac{1}{r} = \frac{c^2}{8(g_I\beta_I H)^2} \tag{5.124}$$

Nuclei far away from the unpaired electron will not be flipped because c^2 is vanishingly small for them. Such nuclei do not contribute to the intensity of either the satellites or the principal components and are thus excluded from classification as environmental nuclei.

When the electron-spin density is concentrated in an s orbital, as for the trapped H atoms, its dipole-dipole interaction with an environmental nucleus i will be axially symmetric about the radius vector $\mathbf{R}_i$ between the center of spin density and the nucleus i. Even when the unpaired electron is in a p orbital, this interaction has approximately axial symmetry when the R_i values are large, as compared with the spread of the p-orbital density. For the axially symmetric case, the c^2 values are easy to calculate. With $A_\parallel = -2A_\perp = (2g\beta g_I\beta_I/R^3)$ from (6.11) substituted into (5.118), one obtains for the ith nucleus:

$$c_i^2 = \frac{9(g\beta g_I\beta_I)^2 \sin^2\theta_i \cos^2\theta_i}{R_i^6} \tag{5.125}$$

where R_i is the separation between the effective spin density center and the ith nucleus and where θ_i is the angle between R_i and the applied field $\mathbf{H}$. By substitution of this value in (5.124), one can find the contribution of the ith nucleus to the satellite intensity. If, however, there is more than one environmental nucleus of the same kind, that is, nuclei having the same spin moments, their satellites will have approximately the same separations from the parent line, $\sim g_I\beta_I H$, regardless of their values of R_i and θ_i. Thus to obtain the intensity ratio of the individual satellite component to that of the parent line, one must substitute the c_i^2 from (5.125) into (5.124) and sum over all the contributing environmental nuclei of the same kind. The resulting formula is

$$\frac{1}{r} = \frac{9}{8}\frac{(g\beta)^2}{H^2}\sum_i \frac{\sin^2\theta_i \cos^2\theta_i}{R_i^6} \tag{5.126}$$

This formula is the same as that derived from a different approach by Trammell, Zeldes, and Livingston [22]. To apply it, one must know the positions of the various environmental nuclei, that is, their individual θ_i and R_i values. In many cases, however, one may know from the nature of the radical or the crystal structure that only one such nucleus can be sufficiently near the effective spin density center to be able to make significant contribution. Then one can locate the nucleus approximately by measurement of the intensity ratio r as a function of θ_i. The satellite vanishes when **H** is along R_i or perpendicular to it.

If an environmental nucleus is so close to an unpaired p electron that the axially symmetric formula, (5.126), is inadequate, one can use (6.13), (6.15), and (6.16) for calculation of the principal elements of the dipole-dipole coupling and can transform these elements with (5.109) to (5.111) or (5.112) to (5.114) to obtain A_{zx} and A_{zy} elements required in (5.123). In most applications the more approximate formula, (5.126), will be adequate.

It is interesting to note from (5.126) that the satellite intensities are independent of the environmental nuclear moments. This surprising result derives from the fact that the admixing of the nuclear spin states that determines the intensity of the satellites is directly proportional to the coupling coefficients c^2 that depend on $g_I{}^2$ but are inversely proportional to the square of the level separations, which also depend on $g_I{}^2$. Although the intensities of the satellites are independent of g_I, their separations, $d_- = g_I\beta_I H$, depend directly on this quantity and are thus the distinguishing characteristic for satellites of different kinds of nuclei.

The energy-level diagram of Fig. 5.20 indicates the transitions that give rise to environmental proton satellites for an esr in which the first-order spectrum has a proton doublet hyperfine structure, such as that of the hydrogen atom or the $C_\alpha H$ fragment. Note that the nuclear magnetic quantum numbers for the upper sublevels of the environmental nucleus are opposite to those of the normal doublet because of the relative signs of the $A\mathbf{S} \cdot \mathbf{I}$ and $g_I\beta_I\mathbf{H} \cdot \mathbf{I}$ terms. The level-splitting by a second environmental proton will be the same, and its addition to the diagram will be obvious. The additional transitions will have the same frequencies as those indicated by the dotted lines, but the intensities of the satellites will be increased by the added transitions. Of course, the degeneracy of all nuclear spin levels in the sample is lifted by the magnetic field, but only those spin levels sufficiently coupled to the electron spin to undergo simultaneous spin flips with the electron can contribute to the satellite intensities. They constitute the environmental nuclei.

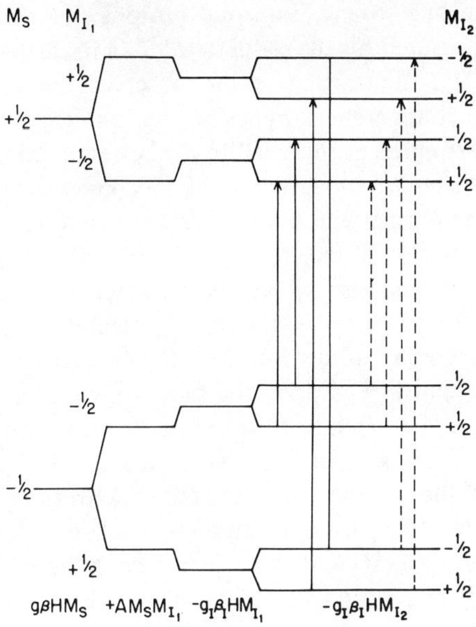

Fig. 5.20 Energy-level diagram indicating the transitions (dotted arrows) that give rise to environmental proton satellites of the lines of the normal transitions (solid arrows).

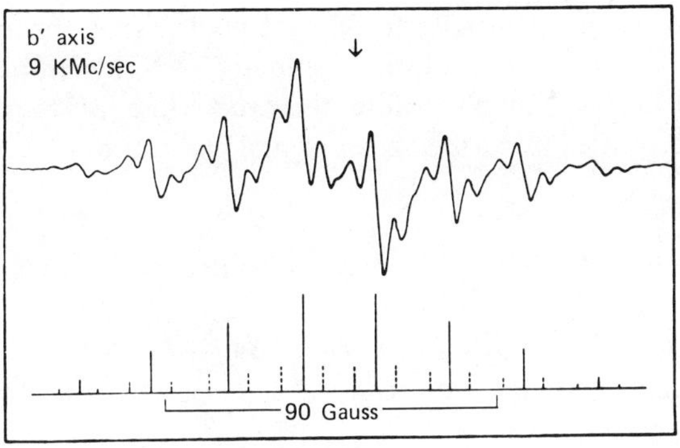

Fig. 5.21 ESR curve of a γ-irradiated single crystal of thymidine showing second-order transitions observed in the first derivative with the field **H** parallel to the b' axis. Solid lines indicate calculated intensity distribution for principal components; dotted lines indicate second-order transitions. From Pruden, Snipes, and Gordy [23].

Satellite structure due to environmental protons of oriented free radicals in single crystals is illustrated by the esr pattern of an irradiated thymidine crystal, shown in Fig. 5.21. The tracing is the observed spectrum, the solid bars represent the normal theoretical hyperfine pattern expected for the radical, and the dotted lines represent the satellite structure spaced according to (5.121) and (5.122) with heights normalized to fit the observed pattern. The satellite separations from the normal line were found to remain approximately constant with change of orientation of the crystal, in agreement with (5.121) and (5.122) and to increase with increase of **H**, as predicted by (5.126). The satellite intensities were observed to vary with crystal orientation and to decrease with increase of **H**, in accordance with (5.126). In observations at 24 kMHz at the same orientation as that in Fig. 5.21, the satellites were barely detectable. With the assumption that the satellites arise primarily from one particular proton located near the electron-spin center, the maximum observed intensity of the satellites (~10% of the normal components), (5.126) indicates that $R = 2.1$Å. The normal octet hyperfine pattern is caused by two protons with approximately isotropic β coupling of 40.5G and three with β coupling of 20.5G. The radical, which is formed by H addition of $C_{(6)}$ of the thymidine crystal, is described in the original paper [23]. An environmental proton satellite structure on a ^{14}N triplet pattern is exhibited in Fig. 5.5 (see Section 2.d).

It is evident from Fig. 5.21 that the environmental satellite pattern, when not correctly identified or analyzed, can confuse and frustrate the observer of esr in single crystals and possibly lead to a misinterpretation of the entire pattern. I well remember making futile attempts to assign the apparent triplet substructure of Fig. 5.21 to overly saturated ^{14}N hyperfine structure of incorrectly postulated radicals before the true cause of the structure was understood and verified by further experimental observations.

References

1. C. Kittel, *Introduction to Solid State Physics,* 5th ed., Wiley, New York, 1976.
2. Y. Kurita and W. Gordy, *J. Chem. Phys.,* **34**, 282 (1961).
3. L. K. Steinrauf, J. Peterson, and L. H. Jensen, *J. Am. Chem. Soc.,* **80**, 3835 (1968).
4. G. Saxebøl and Ø. Herskedal, *Radiat. Res.,* **62**, 395 (1975).
5. K. Akasaka, *J. Chem. Phys.,* **43**, 1182 (1965).
6. P. J. Hamrick, H. Shields, and T. Gangwer, *J. Chem. Phys.,* **57**, 5029 (1972).
7. H. M. McConnell, C. Heller, T. Cole, and R. W. Fessenden, *J. Am. Chem. Soc.,* 82, 766 (1960).
8. J. A. Goedkoop and C. H. MacGillavry, *Acta Crystallogr.,* **10**, 125 (1957).
9. T. Cole and C. Heller, *J. Chem. Phys.,* **34**, 1085 (1961).

10. R. J. Lontz and W. Gordy, *J. Chem. Phys.*, **37**, 1357 (1962).
11. B. Bleaney, *Phil. Mag.*, **42**, 441 (1951).
12. A. Abragam and B. Bleaney, *Electron Paramagnetic Resonance of Transition Ions*, Clarendon, Oxford, 1970.
13. B. Bleaney and D. J. E. Ingram, *Proc. Roy. Soc.*, **A205**, 336 (1951).
14. B. Bleaney and R. S. Rubens, *Proc. Phys. Soc. (Lond.)*, **77**, 103 (1961).
15. I. Miyagawa and W. Gordy, *Bull. Am. Phys. Soc.*, **4**, 260 (1959).
16. I. Miyagawa and W. Gordy, *J. Chem. Phys.*, **32**, 255 (1960).
17. N. M. Atherton and D. H. Whiffen, *Molec. Phys.*, **3**, 1 (1960).
18. B. Bleaney, K. D. Bowers, and D. J. E. Ingram, *Proc. Phys. Soc. (Lond.)*, **A64**, 758 (1951).
19. H. N. Rexroad, Yu. H. Hahn, and W. J. Temple, *J. Chem. Phys.*, **42**, 324 (1965).
20. D. M. Close and H. N. Rexroad, *J. Chem. Phys.*, **50**, 3717 (1969).
21. H. Zeldes and R. Livingston, *Phys. Rev.*, **96**, 1702 (1954).
22. G. T. Trammell, H. Zeldes, and R. Livingston, *Phys. Rev.*, **110**, 630 (1958).
23. B. Pruden, W. Snipes, and W. Gordy, *Proc. Nat. Acad. Sci. USA*, **53**, 917 (1965).

Chapter **VI**

INTERPRETATION OF NUCLEAR COUPLING IN ORIENTED FREE RADICALS

1. COMMON COUPLING MECHANISMS

1.a. Coupling of Atomic Orbitals: Spin Density

The nuclear magnetic coupling constants measured with esr provide a means of obtaining experimental values of electron-spin densities of component atomic orbitals on bonded atoms of molecular free radicals. The spin density of an atomic orbital ψ_i may be expressed as

$$\rho_i = P_i^{\alpha} - P_i^{\beta} \tag{6.1}$$

where P_i^{α} and P_i^{β} are the probabilities that ψ_i is populated by α spins (up) and β spins (down), respectively (see Chapter II). Except for the proton of the $C_{\alpha}H$ fragment (see Chapter II, Section 3.b), the observed nuclear magnetic couplings arise predominately from spin density in atomic orbitals centered on the coupling nucleus.

When an observed nuclear coupling constant is known to arise from a particular atomic orbital of the coupling atom in the free radical, the electron spin density of this atomic orbital is obtained from the ratio of the measured coupling constant to that for the corresponding orbital of the free atom. For example, the spin densities of *s*- and *p*-orbital components are given by

$$\rho_{ns} = \frac{A_{ns}\,(\text{mol. rad.})}{A_{ns}\,(\text{atomic})} \tag{6.2}$$

$$p_{np} = \frac{B_{np}\,(\text{mol. rad.})}{B_{np}(\text{atomic})} \tag{6.3}$$

The relationship between the coupling constants of the molecular orbital and those of the bonded atom is most simply described by expression of the molecular orbital as a linear combination of atomic orbitals (see Chapter 11, Section 3.6).

The required coupling constants of the free atom can be calculated theoretically (see Chapter II), but only approximately for most atoms. The best values are those measured with high-resolution atomic spectroscopy. The atomic coupling constants needed for analysis of the hyperfine structure of the most commonly observed molecular free radicals have already been theoretically calculated or measured by atomic spectroscopy and are tabulated in a number of reference volumes. For convenience, we reproduce these constants in Table A.1. The constants of this table are for *s* or *p* electrons for which the esr or ENDOR hyperfine structure is most often measured in molecular free radicals. Fortunately, the *s* and *p* components can be separated because the *s* coupling is isotropic and the *p* coupling is anisotropic. Coupling constants for each type of orbital can be obtained from measurement of their combined hyperfine splittings.

1.b. Spin Polarization of Bonds

The phenomenon known as *spin polarization of covalent chemical bonds* was first recognized for the $C_\alpha H$ fragment, but it is by no means restricted to this fragment. The unanticipated part of the proton coupling in the $C_\alpha H$ fragment is its large isotropic component. Isotropic coupling can arise only from *s*-orbital spin density on the coupling atom. The 2*s* orbital of the H is too high in energy to have significant spin density, and the predicted coupling of the 2*s* orbital is far too small to account for the observed isotropic component of the $C_\alpha H$ fragment; the 1*s* orbital forms a covalent σ bond to the C_α, the two electrons of which must have antiparallel spins, in accordance with the Pauli exclusion principle. The isotropic coupling, first observed in the naphthalene negative ion by Weissman, et al. [1], was explained by McConnell [2], and other authors [3-5], on the basis of π-σ configuration interaction. The interaction has been treated with self-consistent field theory by McLachlan [6]. It would be too great a diversion at this point to attempt a complete description of this complex interaction. Its principal effects on the $C_\alpha H$ coupling can be attributed to a spin polarization of the σ-electron pair bond so that the spin up (α spin) has a greater probability of being on the H. The result is a negative spin density in the 1*s* orbital of the H relative to the $2p_\pi$ spin density on the C_α. The amount of induced negative spin density on the H is counterbalanced by an equivalent amount of positive spin density in other parts of the orbital so that the net spin density in the CH σ bond is zero, in agreement with the exclusion principle.

Qualitatively, one can account for spin polarization on the basis of Hund's rules for determination of the ground configuration of atoms [7]. The exchange interaction between unpaired electrons in different atomic orbitals is lowest when the electrons have parallel spins, hence the tendency for parallel alignment of the spins of the different orbitals indicated by Hund's rule. A similar exchange interaction between the $p_\pi{}^\alpha$ density on the C_α and the paired electrons of the σ

bond would cause the spin-up alignment (the same as that of the $p_\pi{}^\alpha$ density) to have the greater probability at the C_α end of the σ bond (in the C-bonding orbital), and hence the spin-down alignment to have the greater probability at the H end (in the 1*s* **H** orbital). It is obvious that the spin density thus induced on the H would be negative relative to the $p_\pi{}^\alpha$ density on the C, as was originally shown by McConnell [2].

It is likewise evident that the degree of the spin polarization, and hence the magnitude of the isotropic H coupling, should depend on $\rho_\pi{}^\alpha$, the amount of the inducing spin density on the C_α. A direct proportionality between the isotropic H coupling and the spin density on C_α has been theoretically predicted by McConnell and Chesnut [8] from a molecular-orbital treatment of configuration interaction in aromatic free radicals. The relationship is expressed by the well-known McConnell rule [2]

$$A_f{}^\alpha = \rho_\pi{}^a Q_a \tag{6.4}$$

where $A_f{}^\alpha$ is the isotropic component in the $C_\alpha H$ proton coupling, $\rho_\pi{}^a$ is the π spin density on the C_α, and Q_α is the proportionality constant. This simple rule has been found to hold remarkably well for a wide range of neutral, π-type radicals. It is very useful for obtaining approximate values for $2p_\pi$ densities on the C_α. With some adjustment in the parameter Q_α, it is adaptable to various types of free radicals. Some modifications of the rule, or charge corrections, have been proposed for applications of charged radicals [9,10].

It is evident from the preceding description that the spin polarization of the $C_\alpha H$ bond puts a positive spin density (same alignment as the $p_\pi{}^\alpha$ density) in the σ-bonding orbital of the C_α. This spin density is not detectable with the abundant ^{12}C species, which has spin I=0, but it contributes to the doublet hyperfine splitting of ^{13}C, which has I=½. Since the σ-bonding orbital of the C_α normally has significant *s* character, the spin polarization gives an isotropic component to the ^{13}C coupling.

Because of the prevalence of CH bonds in organic free radicals, together with the large isotropic coupling of the 1*s* orbital of H, the effects of σ-bond spin polarization are most commonly observed on proton splittings in $C_\alpha H$ bonds. The effects are important, however, for other types of bonds and for other coupling nuclei. Configuration interaction might similarly spin polarize any type of covalent σ bond of a free radical. Because this type of polarization depends on exchange interaction of the two electrons of the bond, it does not, however, occur for ionic bonds, and it is reduced by the ionic character of a covalent bond. In bonds such as CF, the effects on the ^{19}F coupling are reduced by the large ionic character of the bond and partly masked by the significant $2p_\pi$ spin density on the F. Nevertheless, these effects must be included for complete and accurate interpretation of the coupling data for the ^{19}F. In a sequence of bonded carbons, positive spin density on C_α induces negative spin on C_β that in turn, induces positive spin density on C_γ. This alternation in sign of the spin density on the carbons is predicted by such treatments as the self-consistent field theory [6], but the negative spin density is not predicted by the popular Hückel theory [11], which neglects this interaction.

Configuration interaction can also occur on free atoms or ions and can involve nonbonding electrons of bonded atoms of free radicals. The large, well-known isotropic hyperfine structure of Mn^{++} ions is due entirely to configuration interaction on the individual ions, which gives some *ns* orbital coupling density, even though the unpaired electron is in the *d* shell. This effect was discovered [12] before spin polarization of the covalent bonds was recognized.

1.c. Hyperconjugation

Isotropic hyperfine structure arising from methyl protons, observed by Venkataraman and Fraenkel in 1955 [13], in methyl-substituted semiquinone ions in alkaline alcoholic solution was later attributed by Bersohn [4] to hyperconjugation. During the same period, a large number of irradiated aliphatic substances in the powdered form were investigated with esr [14-16] and shown to be rich in proton hyperfine structure. The fact that this hyperfine structure was resolvable in the powders indicated that the proton coupling was predominantly isotropic. The total spread of the hyperfine structure in these aliphatic free radicals was much greater than that for aromatic free radicals in solution. Relatively large isotropic splittings by hydrogens evidently bonded to β carbons were observed for many aliphatic free radicals produced by irradiation of the powders. For example, the esr of the ethyl radical, first detected in irradiated $Hg(C_2H_5)_2$ [16], was observed to have, surprisingly, a gross sextet hyperfine pattern with a total spread of about 130 G, which indicated the approximate equality of the coupling of all five hydrogens. Hyperconjugation indicated by admixture of valence-bond structures like **1** and **2**

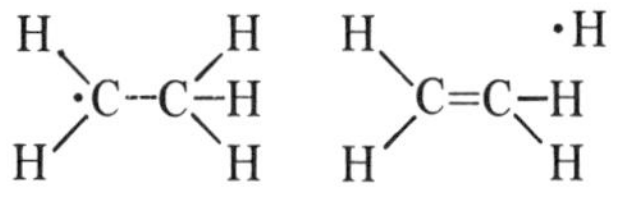

1 2

was proposed [16] to account for the significant coupling detected for β hydrogens in the aliphatic radicals. The CH_3 group must rotate to equalize the coupling of its three protons, and the amount of the hyperconjugation must accidently be such as to approximately equalize the CH_3 coupling with that of the $\cdot CH_2$ to account for the CH_2CH_3 pattern observed for irradiated $Hg(CH_2CH_3)_2$.

These initial results on aliphatic free radicals in powders led Chesnut [17] to treat hyperconjugative proton coupling with molecular-orbital theory [18]. At about the same time, McLachlan [19] treated this type of coupling with valence-bond theory. Both theories predict that the spin density on the β hydrogens is positive (the same as that of the $p_\pi{}^\alpha$ orbital), in contrast to that on the $C_\alpha H$ hydrogen, which is negative. Since these early experimental evidences and the first theoretical treatments of hyperconjugative proton coupling in organic free radicals, numerous measurements of such couplings have been made with esr and ENDOR, and elaborations and refinements have been made in the theoretical treatment of the couplings. See, for example, the discussion by Levy [20].

Despite the complexities of the various theoretical approaches, only moderate success has been achieved in prediction of the spin densities observed on β hydrogens, especially in aliphatic free radicals. The measured parameters for a variety of types of free radicals provide stimulation as well as guidance to quantum theorists in their efforts to improve their calculations. The principal purpose of the treatment given here is to help the experimentalist to understand the basic coupling mechanisms involved and to describe simple theory for derivation of spin densities and structural information from the measured hyperfine structure.

Let us consider a free radical of the form $RC_\alpha HC_\beta H_3$ like that observed in irradiated single crystals of alanine [22] in the coordinate system shown in Fig. 6.1, with x along the axis of the $p_\pi{}^\alpha$ orbital of the unpaired electron on C_α, z along the $C_\alpha C_\beta$ bond, and hence with y perpendicular to the $p_\pi{}^\alpha$ axis and to the $C_\alpha C_\beta$ bond. To simplify the treatment, we assume first that the methyl group does not rotate and that one $C_\beta H$ bond, labeled $C_\beta H^{(1)}$, is in the xz plane, that is, in the plane of the $p_\pi{}^\alpha$ orbital. The three bonding orbitals of C_α will be sp^2 hybrids and will all be in the yz plane perpendicular to $p_\pi{}^\alpha$, but the four bonding orbitals of C_β will be sp^3 hybrids. In designating the C_β orbitals we can, without loss of generality, choose the three p orbitals along the coordinate axis. With this choice of axes, the normalized bonding orbitals of C_β are:

$$\phi_0{}^\beta\ (\text{to } C_\alpha) = \left(\frac{1}{4}\right)^{1/2} s + \left(\frac{3}{4}\right)^{1/2} p_z \tag{6.5}$$

$$\phi_1{}^\beta\ (\text{to } H^{(1)}) = \left(\frac{1}{4}\right)^{1/2} s + \left(\frac{2}{3}\right)^{1/2} p_x + \left(\frac{1}{12}\right)^{1/2} p_z \tag{6.6}$$

$$\phi_2{}^\beta\ (\text{to } H^{(2)}) = \left(\frac{1}{4}\right)^{1/2} s - \left(\frac{1}{6}\right)^{1/2} p_x + \left(\frac{1}{2}\right)^{1/2} p_y - \left(\frac{1}{12}\right)^{1/2} p_z \tag{6.7}$$

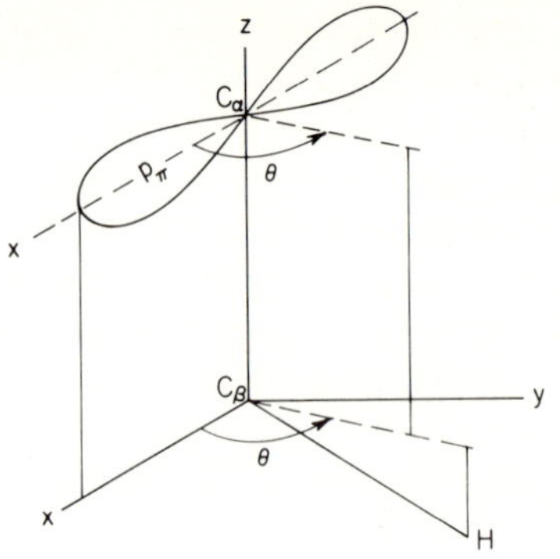

Fig. 6.1 Coordinate system for β-coupling hydrogens. The angle θ is the dihedral angle used in (6.24).

$$\phi_3{}^\beta\ (\text{to } H^{(3)}) = \left(\frac{1}{4}\right)^{1/2} s - \left(\frac{1}{6}\right)^{1/2} p_x - \left(\frac{1}{2}\right)^{1/2} p_y - \left(\frac{1}{12}\right)^{1/2} p_z \quad (6.8)$$

The hyperconjugative π bonding between C_α and C_β will occur entirely between the $p_x{}^\alpha$ and the $p_x{}^\beta$ components of the hybridized C_β orbitals. Since the $p_y{}^\beta$ and $p_z{}^\beta$orbitals are orthogonal to $p_x{}^\alpha$, there will be no overlap and no interaction between these orbitals. Because significant overlap of the $p_x{}^\alpha$ and $p_x{}^\beta$ orbitals occurs, there will be competition between the π_x bond and the $C_\beta H$ σ bonds for use of the $p_x{}^\beta$ orbital. A compromise will be reached, with the degree of π-σ interaction such as to produce the lowest overall energy for the radical. This compromise will depend directly on the amount of overlap of the $p_x{}^\alpha$ and $p_x{}^\beta$ orbitals as well as the relative strengths of the π and σ bonds. For the chosen orientation, one can obtain the relative amounts of the π-σ conjugation with the three $C_\beta H$ bonds and the relative amounts of spin density expected for the three hydrogens simply by comparing the weights – 2/3, 1/6, and 1/6–of the p_x^β orbitals (the squares of their respective coefficients) in ϕ_1^β, ϕ_2^β, ϕ_3^β. Thus for this orientation, the spin densities on $H^{(1)}$, $H^{(2)}$, and $H^{(3)}$ should have the respective ratios of 1:1/4:1/4. For any other orientation of the $C_\beta H_3$ group about the $C_\alpha C_\beta$ bond, the relative spin densities on the three hydrogens may be found by calculation of the component of the $p_x{}^\beta$ orbital in the rotated orbital. Since y and z in the chosen coordinate system (Fig. 6.1) are orthogonal to the $p_\pi{}^\alpha$ orbital, the $p_y{}^\beta$ and $p_z{}^\beta$ components make no contribution to the hyperconjugation for any rotation of $C_\beta H_3$ about the $C_\alpha C_\beta$ bond. The weight of the $p_x{}^\alpha$ component in the rotated bond orbital is $(2/3)\cos^2\theta$ where θ is the dihedral angle between the xz plane and the $C_\alpha C_\beta H$ plane of the particular $C_\beta H$ considered. The factor 2/3 represents the $p_x{}^\beta$ character for the unrotated orbital ϕ_1^β, (6.6).

Thus it is evident that the coupling resulting from hyperconjugation should be proportional to $\cos^2\theta$. Although dependent on the structural parameter θ, the coupling for hydrogen is independent of the orientation of the magnetic field; that is, it is an isotropic Fermi coupling that can be expressed as

$$A_\theta^\beta = B_2 \cos^2\theta \qquad (6.9)$$

where θ is the dihedral angle indicated in Fig. 6.1 and B_2 is the conventionally designated proportionality constant. When the hyperconjugation occurs entirely with π_x spin density on C_α, B_2 can be used to gain knowledge of this spin density ρ_π^α on C_α, as is described in Section 2b.

Note that components of the p_x^β orbital appear in all three of the C_β orbitals, ϕ_1^β, ϕ_2^β, and ϕ_3^β of (6.6) to (6.8); therefore, the hyperconjugation occurs simultaneously with all three of the $C_\beta H$ bonds for the specified orientation. Equation (6.9) shows that for all orientations of the methyl group the interaction occurs simultaneously with all three bonds, except that it is zero for a bond in the yz plane ($\theta = 90°$). This is the reason that valence bond structures such as **1** and **2** shown earlier are incorrect and somewhat misleading unless it is understood that the actual structure is a weighted admixture of valence-bond structures involving all three $C_\beta H$ bonds. Despite their limitations, the valence-bond structures **1** and **2** do provide a pictorial model that is helpful for an understanding of the nature of the interaction and the reason it puts spin density on the hydrogens.

When the methyl group rotates about the $C_\alpha C_\beta$ bond at a frequency appreciably greater than the resonant frequency, the couplings of the three protons are equalized. If $\cos^2\theta$ is averaged over a complete cycle, the equalized coupling values are seen to be $(1/2)B_2$. When the orientation is such that one $C_\beta H$ bond is in the yz plane ($\theta = 90°$ for this bond), the hydrogen coupling for this bond is zero, and that for each of the other two is $(3/4)B_2$. For any fixed orientation, the sum of the hyperconjugative coupling values of the three protons is $(3/2)B_2$.

One should be aware that there can be, and usually are, measurable isotropic contributions to the β-proton coupling from effects of bond spin polarization (see Section 6.1.b). There are also some contributions from anisotropic dipole-dipole interactions like those described in the next section. Hyperconjugation is, however, the dominant factor in determination of the isotropic component of the β coupling in most radicals. For illustrations and further discussion, see Section 2.

Although the hydrocarbon group $C_\alpha C_\beta H$ has been used for this treatment, hyperconjugative coupling is not limited to such groups. This type of coupling might occur for atom Z in the group $X_\alpha Y_\beta Z$ whenever there is π spin density on X_α that interacts with the σ bond of $Y_\beta Z$. When Z is an atom such as F or Cl in which the σ-bonding orbital of Z has p character, the resulting coupling will not be isotropic, as it is for H.

1.d. Interatomic Dipole-dipole Coupling

Calculation of interatomic nuclear coupling requires evaluation of the dipole-dipole interaction of the nucleus of one atom with the electron-spin density of an orbital centered on the nucleus of a second atom. The straightforward approach is to average the Hamiltonian for the interaction over the orbital of the second atom. This type of calculation has been carried out by McConnell and Strathdee [23] for the $C_\alpha H$ fragment where the nucleus of H interacts directly with spin density of the $2p_\pi$ orbital of the C_α. Although this method is formally correct, it is too complex for a precise solution, even for the $C_\alpha H$ fragment. For this reason, we give simpler, semiempirical formulas for approximate evaluation of this type of coupling.

Coupling by *s*-Orbital Spin Density

When a spherically symmetric electron-spin density (*s* electron) ρ_s on atom X couples with the nucleus of atom Y, one can simply, and to a very good approximation, calculate the dipole-dipole coupling with the assumption that the coupling electron-spin density is concentrated at the center of atom X. The coupling will be axially symmetric about the XY direction $\mathbf{R}_{XY}$. When **H** is applied along, or perpendicular to, this axis, both the electron spin and the nuclear spin will be aligned with the field. Two interacting dipoles, both of which are parallel to their line of separation **R**, have an interaction energy of $2\mu_1\mu_2/R_{XY}^3$; when they are perpendicular to **R**, their energy is $-\mu_1\mu_2/R^3$. The aligned spin density moment is $\rho_s g\beta M_s$, and that of the aligned nuclear spin is $g_I\beta_I M_I = (\mu_I/I)\beta_I M_I$, where μ_I is the nuclear magnetic moment in nuclear magnetons (nm). Therefore, the interaction energies for the perpendicular orientation are

$$(A_\mu)_\perp M_I M_s = \frac{-\rho_s g\beta\beta_I(\mu_I/I)M_s M_I}{R_{XY}^3} \tag{6.10}$$

and thus

$$(A_\mu)_\perp = \frac{-\rho_s g\beta g_I\beta_I}{R_{XY}^3} \tag{6.11}$$

where $g_I = (\mu_I/I)$ and R_{XY} is the internuclear distance of atoms X and Y. In frequency units, $A_\mu(\text{Hz}) = A_\mu(\text{erg})/\text{h}$, the coupling constant from (6.11) is

$$(A_\mu)_\perp\ (\text{MHz}) = \frac{-14.1\ \rho_s[\mu_I(\text{nm})/I]}{R_{XY}^3\ (\text{Å})} \tag{6.12}$$

For **H** parallel to the XY direction, the coupling is $(A_\mu)_\parallel = -2(A_\mu)_\perp$ [see (3.108)].

Coupling by *p*-Orbital Spin Density

When the unpaired electron on atom X is in a p_π orbital, the assumption that the spin moment is concentrated at the center of X is not a good approximation, and the interaction is not axially symmetric. However, one can then calculate the coupling with the assumption that the spin density ρ_p is concentrated equally at effective centers of the *p*-orbital lobes on either side of X, as indicated in Fig. 6.2. A precise location of the effective spin centers would be very difficult;

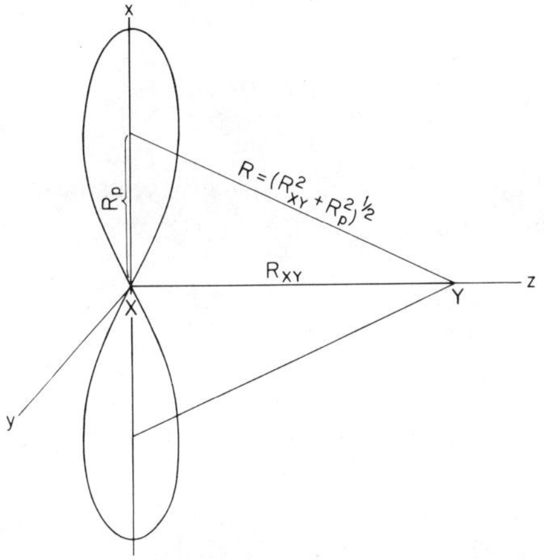

Fig. 6.2 Model used for calculation of the dipole–dipole coupling or unpaired p_π electron-spin density on atom X with the nucleus of atom Y; R_p is the distance of the effective coupling centers of the *p* orbital from the nucleus of X, and R_{XY} is the internuclear distance of X and Y.

but useful, approximate locations can be made with the assumption that they are on the axis at empirically tested, effective distances R_p from the nucleus of X. The value of R_p chosen from a comparison of predicted and observed proton couplings of $C_\alpha H$ fragments (see Section 2.a) is the average of the single-bond and double-bond covalent radii of X. Because of the symmetry, the principal axes in the coupling will be the *x,y,z* system as indicated in Fig. 6.2, with *x* along the p_π orbital, *y* perpendicular to the XY direction and to p_π, and *z* along the XY direction. When **H** is along *y*, we can apply (6.11) to either lobe with $R = (R_{XY}^2 + R_p^2)^{1/2}$, where **R** is the separation of the coupling nucleus Y from the effective spin centers and R_{XY} is the internuclear distance between atoms X and Y. The principal coupling value $(A_\mu)_y$, derived in a manner similar to that for (6.12), is

$$(A_\mu)_y \text{ (MHz)} = \frac{-14.1\, \rho_\pi^X\, [\mu_I(\text{nm})/I]}{(R_{XY}^2 + R_p^2)^{3/2}} \qquad (6.13)$$

where R is in angstrom units and y is perpendicular to the p_π^X orbital and the XY bond.

The prediction of the principal value $(A_\mu)_x$ along the axis of the p_π orbital is not as simple, because the line of separation of the coupling moments is not parallel or perpendicular to the principal axis x. Nevertheless one can calculate the coupling by each lobe from the known angle, $\theta = \arccos (R_p/R)$ of $\mathbf{R}$ with x. Because of symmetry, the coupling by each of the two lobes is equal and additive. Two aligned moments μ_1 and μ_2 with the direction of alignment $\theta°$ from their separation vector $\mathbf{R}$ have the interaction energy

$$E = \frac{\mu_1 \mu_2}{R^3}(3 \cos^2\theta - 1) \tag{6.14}$$

With each p lobe having a moment $(1/2)\, \rho_\pi g\beta M_S$, as before, and with $\cos^2\theta = [R_p^2/(R_{XY}^2 + R_p^2)]$, the coupling $(A_\mu)_x$ is found to be

$$(A_\mu)_x \text{ (MHz)} = 14.1\, \rho_\pi^X \frac{\mu_I(\text{nm})}{I} \frac{2R_p^2 - R_{XY}^2}{(R_{XY}^2 + R_p^2)^{5/2}} \tag{6.15}$$

where x is along the p_π^X orbital. The value of $(A_\mu)_z$, found in a similar way, is

$$(A_\mu)_z \text{ (MHz)} = 14.1\, \rho_\pi^X \frac{\mu_I(\text{nm})}{I} \frac{2R_{XY}^2 - R_p^2}{(R_{XY}^2 + R_p^2)^{5/2}} \tag{6.16}$$

where z is along the XY bond. The consistency of these values can be checked with the relation $(A_\mu)_x + (A_\mu)_y + (A_\mu)_z = 0$.

From the comparisons of calculated and observed anisotropic components of the CH fragments (See Section 2a), it is concluded that usefully approximate values for the effective R_p are

$$R_p = \frac{R_1 + R_2}{2} \tag{6.17}$$

where R_1 and R_2 are the single-bond and double-bond covalent radii of the atom X.

Interatomic dipole-dipole coupling constants predicted by (6.13), (6.15), and (6.16) for the CF fragment with $\rho_\pi = 1, \rho_\sigma = 0, R_{CF} = 1.38$ Å, and $R_C = 0.77$ Å are: $(A_\mu)_x = -5.5$ MHz, $(A_\mu)_y = -19.1$ MHz, and $(A_\mu)_z = 24.6$ MHz, with the coordinate system the same as that in Fig. 6.2. These are maximum values not likely to be observed because the p_π on C in the CF fragment does not closely approach unity. However, unlike H in the CH fragment, the F nucleus has a significant anisotropic coupling component due to π-orbital density on the F.

2. PROTON COUPLING IN HYDROCARBON GROUPS

2.a. α-Coupling Protons: The >CH Fragment

The measurements and analysis of the proton hyperfine structure in oriented $\dot{C}H(COOH)_2$ radicals in irradiated single crystals of malonic acid by McConnell et al. [24,25], described in Chapter V, Section 2.d, provided the first reliable values for the principal elements of the proton coupling of the important $>C_\alpha H$ fragment. In the observation reported by Ghosh and Whiffen [26] for irradiated single crystals of glycine, the hyperfine structure was not completely resolved. Since these first observations, numerous measurements of proton couplings in CH fragments of oriented free radicals of irradiated single crystals have been made. Because of their abundance, only selected examples of the coupling values can be given here. Procedures for obtaining the principal coupling values from the esr spectra of single crystals are described in Chapter V. Here we attempt to show that reasonably consistent interpretation of proton couplings in a variety of organic free radicals can be made with the theory outlined in Section 1. We first treat α coupling in CH fragments and follow with examination of β-proton couplings.

The proton coupling for $C_\alpha H$ fragments has isotropic and anisotropic components of comparable magnitude. The isotropic component is negative; it is due to spin polarization (Section 1.b). The anisotropic component is due to dipole-dipole interaction of the proton moment with ρ_π^α spin density on C_α (Section 1.d). From the measured principal elements of the coupling A_x, A_y, and A_z, both the isotropic component A_f and the principal elements $(A_\mu)_x$, $(A_\mu)_y$, and $(A_\mu)_z$ of the dipole-dipole components can be obtained. In terms of the components, the measured principal elements are:

$$A_x = A_f + (A_\mu)_x \qquad \text{(along the } p_\pi \text{ orbital)} \tag{6.18}$$

$$A_y = A_f + (A_\mu)_y \qquad (\perp \text{ to bond and } p_\pi \text{ orbital)} \tag{6.19}$$

$$A_z = A_f + (A_\mu)_z \qquad \text{(along the bond)} \tag{6.20}$$

where the reference system is chosen to conform to that in Fig. 6.2. Since $(A_\mu)_x + (A_\mu)_y + (A_\mu)_z = 0$, it is evident that

$$A_f = \frac{A_x + A_y + A_z}{3} \tag{6.21}$$

With the value of A_f thus determined, the values of $(A_\mu)_x$, $(A_\mu)_y$, and $(A_\mu)_z$ are found by substitution of the A_f value into (6.18) to (6.20).

Values of the isotropic and anisotropic components for α-proton couplings in a variety of oriented free radicals are given in Table 6.1 for $\rangle C_\alpha H$ fragments and in Table 6.2 for $-C_\alpha H_2$ fragments. These were derived from the principal

Table 6.1 α-Proton Coupling of $\rangle C_{\alpha}H$ Fragments and Derived Constants for Selected Radicals in Irradiated Single Crystals

Radical / *Host Crystal*	*Proton Coupling (MHz)* *Isotropic* A_f	*Proton Coupling (MHz)* *Anisotropic* A_{μ}	C_{α} *Spin Density* $\rho_{\pi}^{\alpha} = (A_f/69.8)$	*Projected* $A_{\mu}(\rho_{\pi}^{\alpha}=1)$ $A_{\mu}/\rho_{\pi}^{\alpha}$	*Source for Coupling Data*
$H\dot{C}(COOH)_2$ / Malonic acid	-60	31 -1 -31	0.860	36.1 -1.2 -36.1	*a, b*
$H\dot{C}(CONH_2)_2$ / Malonamide	-59.0	30.5 -0.1 -30.3	0.845	36.1 -0.1 -35.9	*c*
$F\dot{C}H(CONH_2)$ / Monofluoroacetamide	-63.3	32 0 -33	0.907	35.3 0 -36.4	*d*
$HOOC\dot{C}HCOOK$ / Potassium hydrogen malonate	-57.0	29.0 1.6 -30.6	0.817	35.5 2.0 -37.5	*e,f*
$(HOOC)\dot{C}HCH_2(COOH)$ / Succinic acid	-60	30 1 -32	0.860	34.9 0.2 -37.2	*g,h*
$(HOOC)\dot{C}H(CH_2)_4NH_2$ / *L*-Lysine·HCl·$2H_2O$ at 77°K	-59.6	29.1 0.7 -28.9	0:854	34.1 0:8 -33.9	*i*

$(HOOC)\dot{C}HSCH_2(COOH)$ Thiodiglycollic acid	−43	23 −2 −21	0.616	37.3 −3.2 −34.1	*j*
$HOOC\dot{C}HNHCH_3(HCl)$ Sarcosine (HCl)	−65.2	34.4 0.8 −35.2	0.934	36.8 0.8 −37.7	*k*
$(CO_2H)\dot{C}HCH_2CH(NH_3{}^+Cl)(CO_2H)$ Glutamic acid·HCl	−56	28 0 −28	0.802	34.9 0 −34.9	*l*
$(CO_2)^-\dot{C}HNHCOCH_2(NH_3)^+$ Glycylglycine	−52	27 −1 −26	0.745	36.2 −1.3 −34.9	*m*

[a] McConnell, Heller, Cole, and Fessenden [25].
[b] A. Horsfield, J. R. Morton, and D. H. Whiffen, *Molec. Phys.*, **4**, 327 (1961).
[c] H. N. Rexroad, Y. H. Hahn, and W. J. Temple, *J. Chem. Phys.*, **42**, 324 (1965).
[d] Cook, Rowlands, and Whiffen [70].
[e] R. C. Nicklin, H. A. Farach, and C. P. Poole, *J. Chem. Phys.*, **56**, 1279 (1972).
[f] W. C. Lin and C. A. McDowell, *Molec. Phys.*, **4**, 343 (1961).
[g] Heller and McConnell [28].
[h] D. Pooley and D. H. Whiffen, *Molec. Phys.*, **4**, 81 (1961).
[i] M. Fugimoto, W. A. Seddon, and D. R. Smith, *J. Chem. Phys.*, **48**, 3345 (1968).
[j] Y. Kurita and W. Gordy, *J. Chem. Phys.*, **34**, 1285 (1961).
[k] G. Lassmann and W. Damerau, *Molec. Phys.*, **21**, 555 (1971).
[l] W. C. Lin, C. A. McDowell, and J. R. Rowlands, *J. Chem. Phys.*, **35**, 757 (1961).
[m] M. Katayama and W. Gordy, *J. Chem. Phys.*, **35**, 117 (1961).

Table 6.2 α-Proton Coupling of $-C_{\alpha}H_2$ Fragments and Derived Constants for Selected Radicals in Irradiated Single Crystals

Radical / Host Crystal	Proton Coupling (MHz) Isotropic A_f	Anisotropic A_{μ}	C_{α} Spin Density $\rho_{\pi}^{\alpha} = (A_f/68.2)$	Projected $A_{\mu}(\rho_{\pi}^{\alpha}=1)$ $A_{\mu}/\rho_{\pi}^{\alpha}$	Source for Coupling Data
		32.8		37.8	*a*
$\dot{C}H_2COOH$	−59.1	1.1	0.867	1.3	ENDOR*
Glycine		−33.9		−39.1	
		31.8		36.2	*b*
$\dot{C}H_2COOD$	−59.9	0.6	0.878	0.7	ENDOR*
Glycine HCl		−32.5		−37.0	
		34.7		39.3	*c*
$\dot{C}H_2PO(OH)_2$	−60.2	−1.8	0.883	−2.0	esr
Methylene diphosphoric acid		−32.9		−37.3	
		32		36.4	*d*
$\dot{C}H_2NHCONH_2$	−60	−2	0.880	−2.3	esr
Methylurea		−30		−34.2	

$\dot{C}H_2CONHCH_2CO_2^-$ or $\dot{C}H_2NHCOCH_2NH_3^+$ Glycylglycine	−49	24 1 −27	0.718	33.4 1.4 −37.6	*e* esr
$\dot{C}H_2C(NOH)C(NOH)CH_3$ Dimethylglyoxime	−43.0	22.1 0.5 −22.6	0.630	35.1 0.8 −35.9	*f* ENDOR*
$RXH\text{-}\dot{C}H_2$ Cytidine	−43.86	23.32 0.40 −23.86	0.643	36.3 0.6 −37.1	*g* ENDOR*

*The slightly different values observed for the two $C_\alpha H_2$ protons with the ENDOR technique are averaged.

[a] V. V. Teslenko, Yu. S. Gromovoi, and V. G. Krivenko, *Molec. Phys.*, **30**, 425 (1975).

[b] H. C. Box, E. E. Budzinski, and H. G. Freund, *J. Chem. Phys.*, **50**, 2880 (1969).

[c] M. Geoffroy, L. Ginet, and E. A. C. Lucken, *Molec. Phys.*, **28**, 1289 (1974).

[d] T. S. Jaseja and R. S. Anderson, *J. Chem. Phys.*, **35**, 2192 (1961).

[e] H. Morishima, *Radiat. Res.*, **44**, 605 (1970).

[f] Nelson, Atwater, and Gordy [60].

[g] D. A. Hampton and C. Alexander, Jr., *J. Chem. Phys.*, **58**, 4891 (1973).

elements A_x, A_y, and A_z of the original sources cited in the tables. Although these principal elements are not reproduced in the tables, they may be obtained by addition of the isotropic and anisotropic components.

McConnell's rule, (6.4), is used for calculation of the spin densities on the α-carbon from the isotropic component of the H coupling. These spin densities are listed in column 4 of Tables 6.1 and 6.2. Note that the Q_α value used in the calculation of the ρ_π^α values in Table 6.1 is different from that used for calculation of those in Table 6.2. In their accurate measurements of the isotropic components for numerous rapidly tumbling alkyl radicals in liquids, Fessenden and Schuler [27] noted that the effective Q_α increases with the number of carbon substitutions for hydrogens, from that for the methyl radical for which Q_α = 64.5 MHz. The Q_α = 69.8 MHZ used for $\rangle\dot{C}_\alpha H$ fragments in Table 6.1 is equivalent to the mean value 24.9 for X$\dot{C}$HY radicals in Table 9.10; the Q_α = 68.2 MHZ used for $-C_\alpha H_2$ radicals in Table 6.2 is equivalent to the value of 24.35 G for $CH_3\dot{C}H_2$ (Table 9.9). The radicals included in the tables are neutral, and thus no charge correction is necessary.

The values of the spin densities determined by the isotropic component, as described, are used for calculation of the anisotropic components that would be expected if the unpaired electron were entirely in the $2p_\pi$ orbital of the C_α, that is, for $\rho_\pi^\alpha = 1$. These normalized values are listed in column 5 of the Tables 6.1 and 6.2. If the only interactions were those with $2p_\pi$ spin density on the C_α and if the McConnell rule with constant Q_α were to hold rigorously then one would expect the normalized sets, that is, the projected values of $(A_\mu)_x$, $(A_\mu)_y$, and $(A_\mu)_z$ for $\rho_\pi^\alpha = 1$, to be the same for the different types of radicals. Therefore, the degree of their consistency provides some test of the applicability of McConnell's rule to such diverse radicals in the solid state. Considering the possibility of interactions of the H nucleus with the spin density on atoms other than C_α, especially when ρ_π^α is not close to unity, the degree of correlation of the normalized values is remarkable. Furthermore, the agreement between the normalized values in Table 6.2 with those in Table 6.1 provides support for the two Q_α values used for the $\rangle C_\alpha H$ and $-C_\alpha H_2$ types of radical. We have not included in these tables any radicals for which the coupling group is conjugated with an aromatic ring.

The normalized values, A_μ/ρ_π^α, in column 5 of Tables 6.1 and 6.2 provide experimental sets of dipole-dipole coupling elements for comparison with theoretically predicted ones. They correspond to, but do not agree well with, the dipole-dipole coupling elements calculated by McConnell and Strathdee [23]. They agree much more closely with those calculated by the simple, semi-empirical formulas in (6.13), (6.15), and (6.16). With the effective R_p = 0.720 Å chosen as the mean of the single-bond and double-bond covalent radii for C, (6.17), and with R_{XY} = 1.08 Å as the CH bond length expected for the sp^2-hybridized bonding orbitals of C_α, (6.13), (6.15), and (6.16) predict the principal elements of dipole-dipole coupling for the $C_\alpha H$ fragment when $\rho_\pi^\alpha = 1$ to be

$$(A_\mu)_z = 38.7 \text{ MHz} \quad \text{(along the CH bond)}$$
$$(A_\mu)_x = -2.8 \text{ MHz} \quad \text{(along the } p_\pi \text{ orbital)}$$
$$(A_\mu)_y = -36.0 \text{ MHz} \quad (\perp \text{ to CH and to } p_\pi)$$

These values and those calculated by McConnell and Strathdee [23] are compared in Table 6.3 with the average of the projected experimental values, A_μ/ρ_π^α, from Tables 6.1 and 6.2. Although the projected experimental quantities depend

Table 6.3 Comparison of Experimental and Theoretical Values of A_μ/ρ_π^α for Anisotropic α-Proton Coupling of the Radicals Listed in Tables 6.1 and 6.2

Averaged Experimental Value A_μ/ρ_π^α (MHz)		*Calculated Value A_μ/ρ_π^α (MHz)*	
$\rangle C_\alpha H$	$-C_\alpha H_2$	*Semiempirical (6.13), (6.15), and (6.16)*	*McConnell and Strathdee [23]*
35.7	36.4	38.7	43.5
0	0.0	-2.8	-4.5
-35.9	-36.9	-36.0	-39.0

on ρ_π^α values determined with McConnell's rule, (6.4), it seems unlikely that the errors in ρ_π^α could possibly cause the rather large disagreement with the elements predicted by McConnell and Strathdee. Of the individual radicals, perhaps the simple $H\dot{C}(COOH)_2$ one measured by McConnell, Heller, et al. [25] provides the most reliable test, and its normalized values agree closely with the averaged ones for the group. The values calculated with the semiempirical theory of Section 1.d are in much better agreement with the experimental ones, but we are free, of course, to choose R_p to give the best agreement with one of the three components. If, however, the model on which these formulas are based were not approximately correct, one could not choose R_p to give the degree of agreement achieved for all three components. In both sets of calculated values, the magnitude of the positive component $(A_\mu)_z$ is larger than that of the negative component $(A_\mu)_y$, whereas for the experimental values the magnitude of the negative component is slightly the larger. This disagreement is likewise greater for the McConnell and Strathdee values. It is interesting and convenient that the effective R_p values that gave the best agreement with experiment proved to be the mean of the covalent radii R_1 and R_2. This agreement suggests that (6.17) should be useful for choosing R_p values for fragments of other types.

2.b. β-Coupling Protons

As explained in Section 1.c, the coupling of β hydrogens is due predominantly to hyperconjugation. In addition, there is a small isotropic component due to secondary spin polarization effects and a comparably small anisotropic dipole-dipole component arising from interaction of the β protons with the C_α spin density. The latter interaction may be calculated from the formulas (6.10) to (6.17). Heller and McConnell [28] proposed that the isotropic β coupling is proportional to the π spin density on the α carbon. Their relation may be expressed as

$$A_\beta = Q_\beta(\theta)\rho_\pi^\alpha \tag{6.22}$$

where

$$Q_\beta(\theta) = Q_0^\beta + Q_2^\beta \cos^2\theta \tag{6.23}$$

In experimentally measured β couplings, the spin density ρ_π^α is generally an unknown factor, and the expression applied in the spectrum analysis is

$$A_\beta = B_0 + B_2 \cos^2\theta \tag{6.24}$$

where $B_0 = Q_0^\beta \rho_\pi^\alpha$ and $B_2 = Q_2^\beta \rho_\pi^\alpha$ and θ is the dihedral angle indicated in Fig. 6.1. There is uncertainty about the theoretical nature, origin, magnitude, and even the sign of the small B_0 term [21]; and there is also some question as to whether it should be proportional to ρ_π^α. However, any contribution to B_0 from secondary spin polarization effects should be proportional to the π-spin density on C_α or other atoms. In applications of (6.24), B_0 and B_2 may be regarded simply as nuclear coupling constants to be evaluated from the observed data. Where measurable, B_0 has generally been found to be very small in comparison with the B_2 term. Stone and Maki [29] have attributed the apparent need for B_0 to lack of consideration of the effects of tortional oscillation on the θ-dependent term B_2. Adam and King [30] derived purely theoretical values for the β couplings of methyl groups of alkyl radicals from molecular-orbital treatment of configuration interactions. For these radicals, they predict that $B_0 = 7.6$ MHz and $B_2 = 75.9$ MHz, and they show that higher-order terms such as $B_4 \sin^4\theta$ in the power series expansion of A_β make negligible contribution to the coupling.

Where the applicable Q_2^β is available and where the B_2 term is measurable, the spin density ρ_π^α on the α carbon can be obtained from $\rho_\pi^\alpha = (B_2/Q_2^\beta)$. However, it is difficult to obtain experimental values of Q_2^β because of uncertainties in the value and origin of B_0. The experimental Q_β values are usually derived from the coupling of the methyl group in solution spectra, in which the $C_\beta H_3$ undergoes rapid internal rotation while the molecule as a whole is tumbling rapidly (see Chapter IX, Section 1). Since $(\cos^2\theta)_{av} = 1/2$, the coupling measured under these conditions is

$$A_{rot}^\beta = B_0 + \frac{1}{2}B_2 = Q_\beta \rho_\pi^\alpha \tag{6.25}$$

where

$$Q_\beta = Q_{rot}^\beta = \frac{A_{rot}^\beta}{\rho_\pi^\alpha} = Q_0^\beta + \frac{1}{2}Q_2^\beta \tag{6.26}$$

Probably the "best" experimental values for Q_β^{CH} is that of 29.25 G or 82 MHz derived from the alkyl radicals in solution (see Table 9.9).

Isotropic Coupling: The Methyl Group

As an illustration of β coupling of the methyl group, let us consider in some detail the $CH_3\dot{C}HCOOH$ radical in an irradiated single crystal of l-alanine. This crystal was first studied at room temperature by Miyagawa and Gordy [22], who observed the coupling by the three methyl hydrogens to be equal, approximately isotropic, and with an averaged value of $A^\beta = 70.8$ MHz. The equality of the couplings of the three β hydrogens indicated that the $C_\beta H_3$ was rotating about the $C_\alpha C_\beta$ bond, and the approximate isotropy indicated that the rather large β couplings were due mainly to hyperconjugation. Later, Morton and Horsfield [31] obtained principal coupling values of 76.5 MHz, 67.0, and 67.5 MHz for the rotating $C_\beta H_3$ group at 300°K. The largest value is that for the symmetry axis, which is evidently along the $C_\alpha C_\beta$ direction. Horsfield, Morton, and Wiffen [32] and Miyagawa and Itoh [33] extended these studies to lower temperatures and observed a complete change in the hyperfine pattern that resulted from the cessation of rotation of the methyl group. Note in Fig. 6.3 the comparison of the patterns observed at 300°K with those at 77°K for the same orientation of the crystal in the applied field. The spectral diagram of Fig. 6.4 indicates the origins of the two types of hyperfine patterns shown in Fig. 6.3. Table 6.4 gives the principal values for the coupling elements of the three methyl protons at the two temperatures.

At the lower temperature, 77°K, markedly different isotropic coupling constants are obtained, one for each of the three β hydrogens. Superimposed on the isotropic interactions is a much smaller anisotropic component arising mostly from the direct dipole-dipole interaction of the methyl protons with the π spin

Table 6.4 Hydrogen Coupling Constants of the Methyl Group of $CH_3\dot{C}HCOOH$ in Irradiated Single Crystals of l-alanine at 77°K and 300°K

Coupling Hydrogen	*Principal Value*[a] *(MHz)*		*Isotropic Component*[a] $(A_1+A_2+A_3)/3$	*Isotropic Component*[b] $(A_1+A_2+A_3)/3$	*Calculated Isotropic Component*[c]
			at 77°K		
	A_1	129			
$H_\beta^{(1)}$	A_2	118	121	120	121.7
	A_3	116			
	A_1	84.0			
$H_\beta^{(2)}$	A_2	75.0	77.6	76	76.0
	A_3	73.6			
	A_1	19.3			
$H_\beta^{(3)}$	A_2	15.1	14.8	14	13.9
	A_3	10.1			
Averaged			71.1	70	70.5
			At 300°K		
	A_1	77			
$H_\beta^{(1)}=H_\beta^{(2)}=H_\beta^{(3)}$	A_2	70	72.3	70.3[d]	70.5
	A_3	70			

[a]From Miyagawa and Itoh [33].
[b]From Horsfield, Morton, and Whiffen [32].
[c]Derived from (6.24) with B_0, B_2, and θ values listed in Table 6.5.
[d]From Morton and Horsfield [31].

density on the α carbons. This type of interaction is treated in Section 1.d. The three principal elements of the coupling for each βH as obtained by Miyagawa and Itoh [33] are listed in column 2 of Table 6.4. Since the trace of the anisotropic components for each H is zero, one can obtain the isotropic couplings by simply averaging the three principal elements, as was done in obtaining the values listed in columns 3 and 4 of Table 6.4.

There is good agreement between the isotropic components obtained from the two independent sources. However, Miyagawa and Itoh [33] interpret their results with the assumption that B_0 of (6.24) is zero, and they predict considerable distortion of the methyl group at 77°K. From their analysis of the small anisotropic components in the coupling, they predict dihedral angle values of $\theta_1 = 17°$, $\theta_2 = \theta_1 + 119°$, and $\theta_3 = \theta_1 - 106°$, which are in rough agreement with those predicted from the isotropic components. Since they measured the

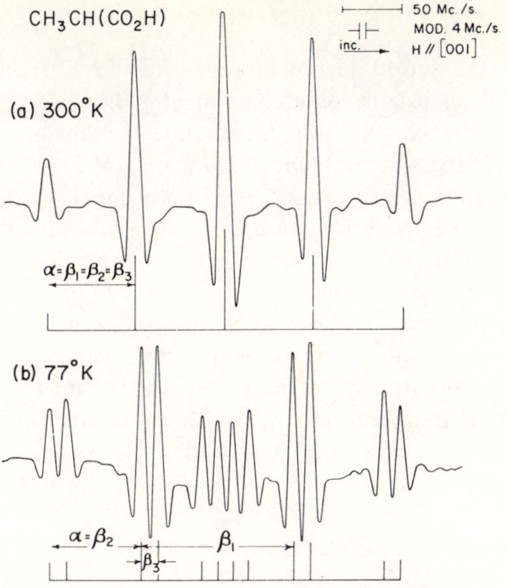

Fig. 6.3 Illustration of temperature effects on the esr hyperfine structure of the $CH_3\dot{C}H$-(CO_2H) radical in an irradiated crystal of D-*l*-alanine. The change results primarily from the cessation of internal rotation by the CH_3 group at the lower temperature. From Horsfield, Morton, and Whiffen [32].

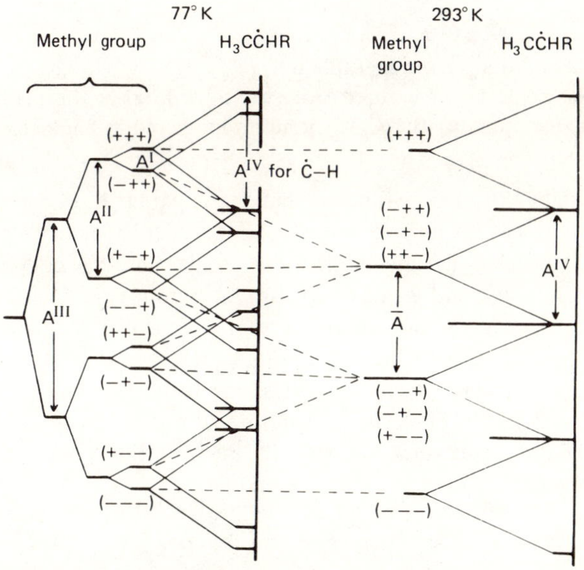

Fig. 6.4 Diagram indicating the transformations that give rise to the temperature-induced changes in the hyperfine patterns shown in Fig. 6.3. From Miyagawa and Itoh [33].

directions of the anisotropic components only to $\pm 10^\circ$, their derivation of θ values from them can be only approximate. Horsfield, Morton, and Whiffen [32] assumed that at the lower temperature the dihedral angles remain equal, 120°, and from (6.24) they found that $B_0 = +9$ MHz, $B_2 = +122$ MHz, and $\theta_1 = 18^\circ$. Since there are three independent, accurately measured isotropic constants these are adequate for solution of the problem with the assumption of equal angles, but this solution provides no test for possible $C_\beta H_3$ distortion.

Because the large distortion of the methyl group predicted by Miyagawa and Itoh seems improbable from energy considerations, we favor the interpretation of Horsfield and co-workers. Consequently, with the assumption of no distortion, we have reanalyzed the problem to obtain a best fitting of the isotropic constants from both observers, with the results shown in Table 6.5. The B_0 value of 8 MHz seems a bit high to arise from spin polarization of the σ bonds of $C_\beta H$, but it is smaller than the earlier value of 9 MHz.

Table 6.5 Molecular Constants of $CH_3CHCOOH$ Derived from the $C_\beta H_3$ Couplings at 77° K[a]

β Coupling Constants (MHz)	*Orientation of $C_\beta H_3$ θ^b*
$B_0 = 8$	$\theta_1 = 17.5^\circ$
$B_2 = 125$	$\theta_2 = 137.5^\circ$
	$\theta_3 = 257.5^\circ$

[a] Data used are given in Table 6.4.
[b] The angle θ is measured from the p_π^α orbital in the plane perpendicular to the $C_\alpha C_\beta$ bond, as indicated in Fig. 6.1.

Stone and Maki [29] have proposed that the apparent need for B_0 in (6.24) arises from failure to take into account the effects of torsional oscillations of the methyl group. Because θ varies with these oscillations, the couplings for the different values θ_1, θ_2, and θ_3 of the nonrotating groups are, in effect, averaged quantities that depend on the amplitude of the torsional oscillations and the equilibrium value of θ. With the assumption of a quadratic potential function, they obtained both a classical and a quantum-mechanical average of the coupling over the torsional states. The classical solution, although less accurate especially for large oscillation amplitude, is simpler; the resulting coupling formula is expressed in the form

$$A^\beta = -(1/2)B_2\phi^2\cos 2\theta_0 + B_2\cos^2\theta_0 \tag{6.27}$$

where θ_0 is the equilibrium value of θ for the particular C_β-H of the nonrotating group and ϕ is the effective amplitude of the torsional oscillation. From comparison of this formula with (6.24), it is evident that $-(1/2)B_2\phi^2 \cos 2\theta_0$ corresponds to B_0. Unlike B_0, however, the first term of (6.27) depends on the orientation θ_0 and averages to zero for the rotating group. The last terms of (6.24) and (6.27) are identical in form, but the resulting values of B_2 obtained by application of the two formulas may be noticeably different because of the first terms. For the rotating group, (6.27) gives the coupling as

$$A^{\beta}_{\text{rot}} = \frac{1}{2} B_2 \tag{6.28}$$

compared with the value $B_0 + (1/2)B_2$ from (6.25).

With (6.27) and the parameter values $B_2 = 140$ MHz, $\phi = 30°$, $(\theta_0)_1 = 18°$, and with the assumption of an undistorted methyl group, that is $(\theta_0)_2 = (\theta_0)_1 + 120°$ and $(\theta_0)_3 = (\theta_0)_1 - 120$, Stone and Maki [29] found that they could satisfactorily predict the isotropic coupling for the three βH's in the nonrotating $CH_3\dot{C}HCOOH$ radical observed in the *l*-alanine crystal at 77°K. They also found that the same B_2 value gave, with (6.28), good agreement with the observed coupling for the rotating methyl group at room temperature. To show the relative weights of the two terms of (6.27), we have repeated their calculations with the result:

$$A^{\beta}_1 = -6.9 + 126.6 = 119.7 \text{ MHz}$$
$$A^{\beta}_2 = -0.9 + 77.3 = 76.4 \text{ MHz}$$
$$A^{\beta}_3 = +7.8 + 6.0 = 13.8 \text{ MHz}$$

with the values of $\theta_0 = 18°$, 138°, and 258°, respectively, and $A^{\beta}_{\text{rot}} = 70$ MHz from (6.28). It is seen that the resulting values agree well with the observed ones given in columns 3 and 4 of Table 6.4, but the agreement is no better than that obtained with (6.24) when $B_0 = +8$ MHz and $B_2 = 125$ MHz. Thus these comparisons provide no basis for choosing one theory over the other and hence no way of knowing whether the first term arises primarily from torsional oscillations or from spin polarization.

Certainly torsional oscillations should influence the coupling to some extent, as the theory of Stone and Maki indicates, but the amplitude of the oscillations, 30°, required to fit the observations for $CH_3\dot{C}HCOOH$ in the *l*-alanine crystal seems abnormally large. If, however, the amplitude is reduced to a more reasonable value of 9° or 10°, the first term is reduced by an order of magnitude and becomes insignificant in this application. In some radicals, effects of both torsional oscillations and spin polarization may make significant contributions. The correct solution would then require that both effects be taken into account. This might be done by simple addition of a small constant B_0 term to (6.27), but there would then be more parameters than observables, with the result that the relative contributions of these terms must be decided by other means. In some applications both these effects may be neglected, and the simpler (6.9) may be used for calculation of the isotropic βH couplings.

The Anisotropic Components

Listed in Table 6.6 are the anisotropic components of the βH coupling of the $CH_3\dot{C}HCOOH$ radical obtained by subtraction of the isotropic components from the principal elements in column 2 of that table. The corresponding theoretical values given in Table 6.6 are calculated from (6.12) with the assumption that the anisotropic term arises entirely from the dipole-dipole interaction of the β-hydrogens with the spin density of $\rho_\pi^\alpha = 0.85$ on the C_α. To simplify the calculation, we have further assumed this spin density to be concentrated at the center of C_α. For correctness, the distribution of the p_π^α orbital should have been taken into account. However, the variation of the orientation of this orbital relative to

Table 6.6 Anisotropic Coupling Constants of the Methyl Group of $CH_3\dot{C}HCOOH$ in Irradiated Single Crystals of *l*-Alanine at 77°K and 300°K

	Observed Components[a] (MHz)				*Calculated Value [From (6.12)] for $\rho_\pi^\alpha = 0.85$*
	$H_\beta^{(1)}$	$H_\beta^{(2)}$	$H_\beta^{(3)}$	*Av.*	
				At 77°K	
$(A_\mu)_1$	+7.8	+6.4	+4.5	+6.2	+7.5
$(A_\mu)_2$	-3.4	-2.6	+0.3	-1.9	-3.8
$(A_\mu)_3$	-4.7	-3.9	-4.7	-4.4	-3.8
		At 300°K ($C_\beta H_3$ rotating)			
$(A_\mu)_\parallel$	5[a]	6.2[b]			7.0
$(A_\mu)_\perp$	-2[a]	-3[b]			-3.5

[a]Derived from data of Miyagawa and Itoh [33].
[b]Derived from data of Morton and Horsfield [31].

the $C_\alpha \cdots H_\beta$ direction as the angle θ is changed, greatly increases the difficulty of the calculation, a difficulty not justified by the accuracy of the measured values. For $\theta = 90°$, the p_π^α orbital is perpendicular to the $C_\alpha \cdots H_\beta$ direction; thus a more accurate calculation can easily be made from (6.13), (6.15), and (6.16). The resulting principal values for $\rho_\pi^\alpha = 0.85$ are +6.1 MHz, -2.7 MHz, and -3.8 MHz, as compared with +7.5, MHz, -3.8 MHz, and -3.8 MHz obtained from the simpler point-dipole assumption. For the rotating CH_3 at 300°K, the values are resolved parallel and perpendicular to the $C_\alpha C_\beta$ bond. Although the accuracies of the values compared in Table 6.6 are not high, the degree of correlation between the predicted and the observed values provides good evidence that the primary source of the small anisotropies observed in β couplings of hydrogens is the dipole-dipole interaction with spin density on C_α. The still smaller dipole-dipole interaction with spin density on more distant atoms is usually negligible. In the $CH_3\dot{C}HCOOH$ radical this distant spin density is mostly on the oxygens.

Other Examples of β Coupling

In Table 6.7 are listed coupling constants of β hydrogens for other selected free radicals observed in X- or γ-irradiated single crystals. In the first two, there is only one $C_\beta H$ group, and the single isotropic coupling constant is insufficient to solve (6.24) for the three unknown parameters θ, B_0, and B_2, as was done for the *l*-alanine radical. Theoretically, it is possible to find the orientation of the $C_\beta H$ from the direction cosines of the three principal elements of the anisotropic dipole-dipole coupling. In practice, however, neither the calculations nor the measurements of these anisotropic components can be made with sufficient accuracy to give reliable results. About all one can conclude from the anisotropic components is that they are of the right signs and relative magnitudes to originate from the dipole-dipole interaction of the β protons with the probable spin density on the C_α. Thus the β-H coupling of $-C_\beta HR_1R_2$ groups is useful mainly in combination with other coupling data or with structural information from other sources. One can, of course, obtain the spin density on the β H from (6.2) with a_{ns} (mol. rad.) = A^β_{iso} and a_{ns} (atomic) = 1420 MHz for H.

When there are two β-coupling hydrogens, $-C_\beta H_2R$ groups, the information is still insufficient for the solution for all three parameters in (3.24), but one can obtain useful, approximate values of θ and B_2 by omission of the small B_0 constant. This is the procedure commonly used in the treatment of such radicals. Although a value of 8 MHz is obtained in the preceding analysis for the *l*-alanine crystal, this value may result partly from large torsional oscillations of the methyl group not present in $-CH_2R$ groups. If, in addition to setting $B_0 = 0$, we assume that the C_β forms four tetrahedral bonds, we can express the two isotropic couplings of $-C_\beta H_2R$ groups by

$$A_\beta^{(1)} = B_2 \cos^2 \theta_1 \tag{6.29}$$

$$A_\beta^{(2)} = B_2 \cos^2(\theta_1 + 120) = B_2 \cos^2(60 - \theta_1) \tag{6.30}$$

Also, it is possible to set $\theta_2 = \theta_1 + 2(120^\circ)$, but this does not give reasonable values of B_2 for any of the radicals in Table 6.8. Thus we conclude that in these radicals the $C_\alpha C_\beta R$ plane has the orientation $\theta_3 = \theta_1 + 240^\circ$ relative to the $p_\pi^\alpha C_\alpha C_\beta$ plane. From (6.29) and (6.30) one obtains

Table 6.7 Examples of β Couplings of Hydrogens Observed in Free Radicals in Irradiated Single Crystals

Radical / Host Crystal	Coupling Atom	Isotropic Component (MHz)	Anisotropic Components (MHz)			Source of Data
		$-C_\beta HR_1R_2$ groups				
$(CO_2H)\dot{C}HC_\beta HR(CO_2H)$ / Fumaric acid	H_β	56	8	-4	-5	*a* esr
$(DO)_2\dot{C}C_\beta H(ND_2)CH_2OD$ / Deuterated DL-serine	H_β	118	5	-3	-3	*b* esr
		$-C_\beta H_2R$ groups				
$(CO_2H)_2\dot{C}C_\beta H_2CH_3$ / Ethyl malonic acid	$H_\beta^{(1)}$ $H_\beta^{(2)}$	70 57	5 5	+1 1	-6 -4	*c* esr
$(CO_2H)\dot{C}HC_\beta H_2(CO_2H)$ / Succinic acid	$H_\beta^{(1)}$ $H_\beta^{(2)}$	100 80	8 9	-2 -1	-7 -8	*d* esr
$HOO\bar{\dot{C}}C_\beta H_2(CO_2H)$ / β-Succinic acid	$H_\beta^{(1)}$ $H_\beta^{(2)}$	78 24	6 11	-3 -2	-3 -8	*e* esr
$(CO_2H)\dot{C}HC_\beta H_2CH_2(CO_2H)$ / Glutaric acid	$H_\beta^{(1)}$ $H_\beta^{(2)}$	134.2 42.3	8.0 7.7	-2.8 -3.4	-5.2 -4.3	*f* ENDOR

$(HO)_2\dot{C}C_\beta H_2CH_2(CO_2H)$ Succinic acid	$H_\beta^{(1)}$ $H_\beta^{(2)}$	76.4 23.4	7.2 10.8	−2.6 −4.7	−4.5 −6.0	*g* ENDOR
		Rotating $-C_\beta H_3$ groups				
$(CO_2H)_2\dot{C}C_\beta H_3$ methyl malonic acid	$C_\beta H_3$	71	5	−2	−2	*h* esr
$(CO_2H)\dot{C}HC_\beta H_3$ *l*-α-Alanine	$C_\beta H_3$	70	6	−3	−3	*i* esr
$(CO_2H)\dot{C}(C_\beta H_3)_2$ α-Amino isobutyric acid	$C_\beta H_3$	66	4	−1	−3	*j* esr

[a] R. J. Cook, J. R. Rolands, and D. H. Whiffen, *J. Chem. Soc.*, 3520 (1963).
[b] D. V. G. L. N. Rao and W. Gordy, *J. Chem. Phys.*, **35**, 764 (1961).
[c] J. R. Rowlands and D. H. Whiffen, *Molec. Phys.*, **4**, 349 (1961).
[d] Heller and McConnell [28]; D. Pooley and D. H. Whiffen, *Molec. Phys.*, **4**, 81 (1961).
[e] H. C. Box, H. G. Freund, and K. T. Lilga, *J. Chem. Phys.*, **42**, 1471 (1965).
[f] A. L. Kwiram, *J. Chem. Phys.*, **55**, 2484 (1971).
[g] H. Muto, K. Nunome, and M. Iwasaki, *J. Chem. Phys.*, **61**, 1075 (1974).
[h] Heller [35].
[i] Morton and Horsfield [31].
[j] A. Horsfield, J. R. Morton, and D. H. Whiffen, *Molec. Phys.*, **4**, 169 (1961).

Table 6.8 Properties of Radicals Derived from the Isotropic $C_\beta H_2$ Coupling Constants[a]

Radical / *Host Crystal*	*Dihedral Angle (°)*			B_2 *(MHz)*	*π Spin Density*[b] *on* C_α ρ_π^α
	$\theta_1(C_\beta H^{(1)})$	$\theta_2(C_\beta H^{(2)})$	$\theta_3(C_\beta C_\gamma)$		
$(CO_2H)_2\dot{C}C_\beta H_2CH_3$ / Ethyl malonic acid	25	145	265	85	0.52
$(CO_2H)\dot{C}HC_\beta H_2(CO_2H)$ / Succinic acid	24.5	144.5	264.5	121	0.74
$HOO^-\dot{C}C_\beta H_2CH_2(CO_2H)$ / Succinic acid	3.5	123.5	243.5	78	0.48
$(HO)_2\dot{C}C_\beta H_2CH_2(CO_2H)$ / Succinic acid	3.5	123.5	243.5	77	0.47
$(CO_2H)\dot{C}HC_\beta H_2CH_2(CO_2H)$ / Glutaric acid	4	124	244	135	0.82

[a] Coupling constants are listed in Table 6.7.
[b] The ρ_π^α is obtained from B_2 with $Q_\beta^2 = 164$ MHz.

$$\frac{A_\beta^{(1)}}{A_\beta^{(2)}} = \frac{\cos^2 \theta_1}{\cos^2 (60 - \theta_1)} \tag{6.31}$$

With (6.31) and the isotropic coupling values listed in Table 6.7, we calculated the values of θ_1 listed in Table 6.8. The values of θ_2 and θ_3 were obtained by simple addition of 120° and 240° to θ_1. The listed values of B_2 were obtained by substitution of the respective θ_1 and $A_\beta^{(1)}$ values in (6.29).

The θ_1 values of Table 6.8 fall into two groups, 3.5° to 4° and 24.5° to 25°. For the larger θ_1 values, the $C_\alpha C_\beta C_\gamma$ plane is almost normal to the p_π^α orbital, and the C_γ is the carbon of the carboxyl group. In the other cases, θ_1 values of 3.5 to 4°, the $C_\alpha C_\beta C_\gamma$ plane is approximately 26° from the plane normal to the p_π^α orbital. In these latter radicals, the C_γ is between C_β and a carboxyl group.

Two of the radicals listed in Table 6.8 have α-coupling protons from which values of the ρ_π^α spin density can be obtained with (6.4). The ρ_π^α values thus obtained from the αH coupling are 0.82 and 0.78, compared with the respective values of 0.74 and 0.82 in Table 6.8. This difference may come about from the neglect of the small constant B_0, inclusion of which would lower the B_2 values and hence the ρ_π^α values obtained from the β couplings.

2.c. Potential Barrier to Internal Rotation of the Methyl Group

Two cases of spectra for $CH_3\dot{C}HCOOH$ have been considered: one for which the $C_\beta H_3$ group undergoes restricted internal rotation about the $C_\alpha C_\beta$ bond (the 300°K spectra) and the other for which the $C_\beta H_3$ group may be considered as frozen in a fixed orientation (the 77°K spectra). At the higher temperature, the rotation of the methyl group must be fast enough to equalize the coupling of its three protons. This condition requires that the frequency of rotation be high in comparison with the hyperfine precessional frequency of the protons (hyperfine splitting). At 77°K the methyl group may still execute restricted rotation through the barrier-tunneling process, but it must do so at a frequency much lower than the hyperfine splitting frequencies observed at 77°K and not greater than $\Delta\nu$, the line widths observed at that temperature. The latter condition requires that any such rotation at 77°K be less than 9 MHz in frequency.

Miyagawa and Itoh [33] and Horsfield, Morton, and Whiffen [32] have used their esr data to derive values for the barrier heights restricting the rotation of the methyl group in $CH_3\dot{C}HCOOH$ in the *l*-alanine crystal. The former applied modified Block equations (Chapter IV, Section 8) with $T_2 = 2.3 \times 10^{-8}$ sec, estimated from exchange rates of D and H, and obtained values for τ, the mean lifetime between reorientations of the $C_\beta H_3$. They then used the classical formula [34]

$$\frac{1}{\tau} = \frac{kT}{h} \exp \frac{-E}{kT} \tag{6.32}$$

and obtained for E, the classical activation energy, a value of 3.6 ± 0.2 kcal/mole = 1258 cm^{-1}.

The method of Horsfield, Morton, et al. [32] is similar in principle but simpler in application. At temperatures between 100°K and 200°K they directly measured the incremental line widths due to interruption of the lifetime in the esr states by the reorientations. The increased line width $\Delta\nu$ is inversely related to the reorientation time τ by $2\pi\Delta\nu = 1/\tau$. They were able to obtain usefully accurate measurements of the small increase in width by comparison of the widths of the inner components with those of the outside components of the hyperfine multiplets. The outside components, which correspond to the parallel alignments of the three methyl proton spins $\alpha\alpha\alpha$ and $\beta\beta\beta$, are not intermixed by the reorientations. Thus the lifetimes in these states are not interrupted by the reorientations as are those of the intermediate states. This may be seen from examination of the hyperfine states of the methyl group at 77°K and 293°K, which are connected by dotted lines in Fig. 6.4. As a consequence, the outermost components remain sharp at the intermediate temperatures. Figure 6.5 shows a comparison in width of an outside component (at +126 MHz) with a close inside component (at +115 MHz). Note how the latter increases in width and decreases in height relative to that at +126 MHz as the temperature is raised

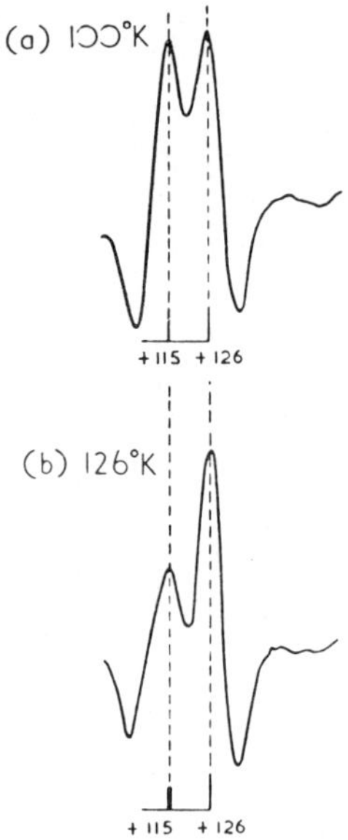

Fig. 6.5 Parts of the second derivative spectra of $CH_3\dot{C}HCO_2H$ demonstrating the different effects of temperature on the line widths of two components. From A. Horsfield, J. R. Morton, and D. H. Whiffen, *Molec. Phys.*, **5**, 115 (1962).

from 100°K to 126°K. In this temperature range the reorientation rate increases rapidly with increase in T. The increase in line width due to the reorientations at 126°K was thus measured as approximately 1.5 MHz. This frequency corresponds to a mean reorientation time $\tau = 1.1 \times 10^{-7}$ sec. If it is assumed that the relaxation mechanism consists of a transition to an excited torsional oscillation state followed by a return with equal probability to any one of the three ground-state orientations, and that such excitation and return to the group state does not produce line broadening, this value of τ would be reduced by 2/3, or to 7×10^{-8} sec. The excitation energy may be obtained by substitution of the τ value into the rate equation, (6.32). Thus Horsfield and colleagues obtained $E = 3$ kcal/mole = 1050 cm^{-1} at 126°K.

It is interesting that Heller [35], in observations of the $CH_3\dot{C}HCOOH$ radical in an irradiated single crystal of methyl malonic acid, has found evidence for rotation of the methyl group even at 4.2°K, although the rotation was highly restricted. This is evidence that the large barrier to internal rotation of the radical in the *l*-alanine crystal is due partly (perhaps mostly) to the intramolecular interactions in the *l*-alanine matrix, as was suggested by Horsfield, Morton, and Wheffen [32]. Heller found, however, that the methyl group in the $CH_3\dot{C}(COOH)_2$ radical, also observed in the methyl malonic acid crystal, undergoes almost free internal rotation at 4.2°K. This indicates that the much more highly restricted internal rotation for $CH_3\dot{C}HCOOH$ in the same crystal results from intermolecular interactions, probably between the $C_\alpha H$ and $C_\beta H_3$ groups.

The barriers to internal rotation of the CH_3 group in stable molecules found by microwave spectroscopy of gases are usually in the range of 0.5 to 3.5 kcal/mole [36]. The barrier for CH_3COOH is only 0.48 kcal/mole, whereas that for CH_3CHO is 1.17 kcal/mole.

2.d. $C_\beta H$ Interaction with Noncarbon X_α Atoms

Many observations of β-proton coupling in hydrocarbon $C_\beta H$ groups have been made in which the isotropic H coupling results from hyperconjugative interaction of the $C_\beta H$ σ bond with π spin density on α atoms other than carbon. Especially large couplings have been observed for N_α atoms, as, for example, $Q_2^\beta = 319$ MHz for the $N_\alpha C_\beta H_2$ radical described in Section 5. The $C_\beta H$ hyperconjugative interaction with π spin density on an S_α is comparable to that with π spin density on a C_α. For example, Q_β is approximately 83 MHz for the monosulfide radical $RC_\beta H_2 S_\alpha$ in γ-irradiated cystine•$(HCl)_2$ (Table 6.31). These and other cases are described in later sections after the spin densities of the X_α atoms are derived from the X-nuclear coupling.

3. COUPLING OF PROTONS BONDED TO NITROGEN

Since the NH or NH_2 group occurs in many organic and most biological chemicals, the nuclear coupling of protons bonded to nitrogen is probably next in importance for free-radical esr to that of protons bonded to carbon. Despite this wide occurrence, the α and β coupling constants of hydrogens bonded to nitrogens are not as accurately known or as well understood as those of hydrocarbon groups. The ^{14}N coupling, which is of the same order of magnitude as the proton coupling, increases the difficulty of resolution and assignment of the hyperfine components. Basically, the NH and CH fragments are similar, and the NH fragment is expected to have an isotropic component due to spin polarization as well as an anisotropic component due to interatomic dipole-dipole coupling, both comparable in magnitude to those of the CH fragment when their ρ_π^α spin densities are the same.

3.a. The NH Fragment

Because N is trivalent, the neutral $N_\alpha H$ fragment must be a terminal group. It could be formed, for example, by H-abstraction from R-NH_2 to produce R_2-$\dot{N}_\alpha$H. The charged species R-$\overset{+}{N}_\alpha H_2$ or X-$\overset{+}{N}_\alpha$H-Y are very unstable and are not often observed in irradiated single crystals. Much more prevalent is the observation of the NH hyperfine structure in X-NH-Y radicals in which the spin density on the N is produced, not by ionization of the molecule or by the breaking of the bond to the NH, but rather by migration of spin density from the X or Y group. This migragtion requires reduction of the electron density of the unshared pair on the N and thus puts positive charge on the N although the resultant charge on the radical is zero. When the NH group in such a radical is perpendicular to the π spin density on the N, the coupling is like that of a $C_\alpha H$ or $N_\alpha H$ fragment, that is, isotropic coupling due to spin polarization rather than to hyperconjugation, as in $C_\beta H$ groups. However, the N is often designated as N_β when the unpaired electron is produced by the breaking of a bond to an adjacent atom on the X or Y group.

A representative example of the R$\dot{N}_\alpha$H species is the radical 3 observed by Hamrick, Shields, and Whisnant [37] in an X-irradiated single crystal of

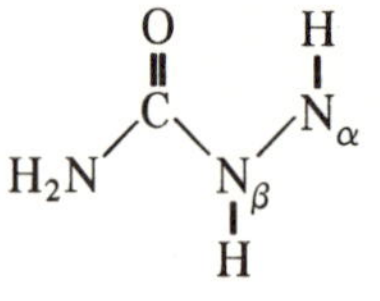

3

semicarbazide. Since both the $^{14}N_\alpha$ and the H_α couplings are measured with good accuracy, this work provides a useful value of Q_α^N for the NH fragment. Also, it provides a further test of the simple formulas for calculation of the interatomic dipole components derived in Section 1.d. In Table 6.9 are listed

Table 6.9 Examples of $^{14}N_\alpha$ and H_α Coupling in X-irradiated Crystals

Radical / *Host Crystal*	*Coupling Atom*	*Principal Value A (MHz)*	*Isotropic Component A_f (MHz)*	*Anisotropic Component A_μ or **B** (MHz)*	*Source for Coupling Data*
$NH_2CONH\dot{N}_\alpha H_\alpha$ / Semicarbazide	$^{14}N_\alpha$	106			
		0	35.3	$B =$ 35.3	
		0			
	H_α	−23.0		32.4	*a*
		−57.0	−55.4	−1.7	
		−86.0		−30.6	
$NCCH_2CONH\dot{N}_\alpha H_\alpha$ / Cyanoacetohydrazide	$^{14}N_\alpha$	103			
		12	39.3	$B =$ 33[c]	
		3			
	H_α	−10.1		38.3	*b*
		−54.9	−48.4	−6.5	
		−80.1		−31.7	

[a] From the results of Hamrick, Shields, and Whisnant [37].
[b] From the results of Lin and Nickel [38].
[c] Derived from $A_{\parallel} = 103$ MHz and $A_{\perp} = 3$ MHz.

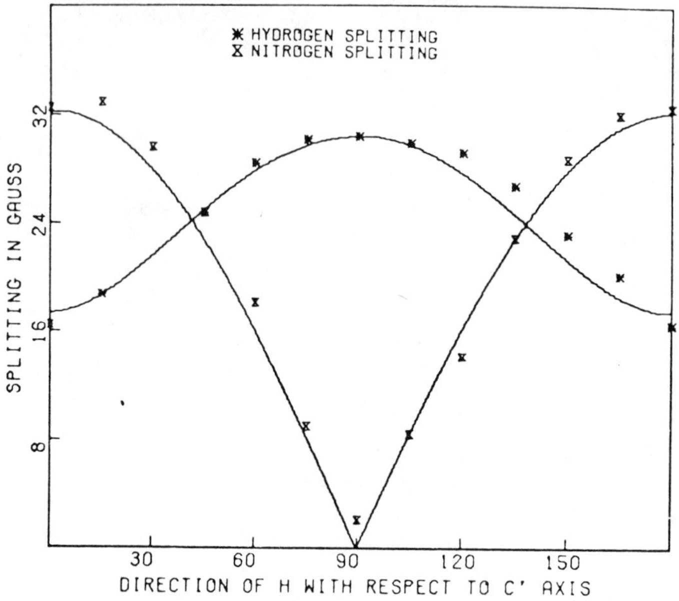

Fig. 6.6 Variations in the splitting of ^{14}N and of H in $NH_2CONH\dot{N}_\alpha H$ as a function of orientation of the host crystal (semicarbazide) in the static magnetic field. Solid curves indicate the theoretical fitting of the observed data points with the coupling constants given in Table 6.9. From Hamrick, Shields, and Whisnant [37].

the principal values of the H_α and the $^{14}N_\alpha$ couplings with their isotropic and anisotropic components. Figure 6.6 shows the agreement of the observed $^{14}N_\alpha$ and H_α splittings in one crystal plane with the splittings calculated from the principal coupling values of Table 6.9.

The isotropic and anisotropic components of the ^{14}N coupling are easily obtainable from the principal values, as described in Section 5.b. The π spin density on N_α obtained from the anisotropic component as corrected for effects of spin polarization of the σ bonds is also described in Section 5.b. The resulting ρ_π^α allows an evaluation of the Q_α^{NH} from the H_α coupling with the relation, $Q_\alpha^{NH} = (A_f^H/\rho_\pi^\alpha)$. These derived constants are summarized in Table 6.10.

The anisotropic components of the H_α coupling listed in Table 6.9 arise predominantly from the dipole-dipole interaction of the proton moment with the large p_π spin density on N_α. Although the ^{14}N coupling indicates that the remaining density of the unpaired electron, 0.22, is in the p_π orbital of N_β, this spin density on N_β gives a maximum contribution to the H_α coupling of less than 2 MHz because of the large $N_\beta \cdots H_\alpha$ distance of 2.2 Å. With the R_p = 0.68 Å chosen as the average of the single-bond and double-bond covalent radii of N [as was done for C in (6.17)], the dipole-dipole coupling for the $N_\alpha H$ frag-

Table 6.10 Spin Densities ρ_π^N and Effective $Q_\alpha^{NH} = Q_{NH}^H$ Values for $N_\alpha H$ Fragments in Selected Radicals[a]

Radical	$A_f^{H_\alpha}$ *(MHz)*	π *Spin Density on* N_α *from Table 6.19*	*Effective* $Q_\alpha^{NH} = (A_f^{H_\alpha}/\rho_\pi^N)$ *(MHz)*
$NH_2CONHN_\alpha H$	55.4	0.78	71
$NCCH_2CONHN_\alpha H$	48.4	0.72	67
$NH_2CON_\alpha HO$	33.8	0.45	75
$NH_2CSNHN_\alpha H$ or $NH_2NHCSN_\alpha H$	49.2	0.66	75
		Mean $Q_{NH}^H =$	72

[a]The signs of A_f and Q_α^{NH}, not measured, are predicted from theory to be negative.

ment was calculated from (6.13), (6.15), and (6.16) with R_{XY} = 1.01 Å, the commonly observed NH distance in gaseous molecules. The resulting anisotropic components obtained for the unpaired electron entirely in a p_π orbital on N_α (i.e., $\rho_\pi^\alpha = 1$) are listed in Table 6.11. For comparison with them, the observed values of Table 6.11 were divided by the spin density of 0.78 to normalize them to $\rho_\pi^\alpha = 1$. The agreement obtained, comparable to that for the CH fragments shown in Table 6.3, provides good evidence that the terminal group of this radical is a true π-type, $N_\alpha H$ fragment. It is interesting that the Q_α^{NH} value obtained, 71 MHz, is close to that for the $\rangle C_\alpha H$ fragment, 69.8 MHz (see Table 9.10).

Table 6.11 Comparison of Theoretical Values of the Anisotropic Proton Couplings of the $N_\alpha H$ Fragment for $\rho_\pi^\alpha = 1$ with Those Projected from the Observed Coupling

Observed[a] $(A_\mu^H/\rho_\pi^\alpha)$ *(MHz)* $NH_2CONHN_\alpha H$	$NCCH_2CONHN_\alpha H$	*Av.*	*Theoretical*[b] A_μ^H *(for* $\rho_\pi^\alpha = 1$*)* *(MHz)*	*Direction*
41.7	53.2	47.5	46.4	‖ to $N_\alpha H$ bond
−2.2	−9.0	−5.6	−2.8	‖ to p_π (N_α)
−39.3	−44.0	−41.7	−43.6	⊥ to p_π and to $N_\alpha H$

[a]Derived from the A_μ^H components of Table 6.9 with ρ_π^α values determined by the ^{14}N couplings, as listed in Table 6.10.
[b]Calculated from (6.13), (6.15), and (6.16) with R_p = 0.68 Å, and $R_{XY}(=R_{NH})$ = 1.01 Å.

Hamrick, Shields, and Whisnant [37] did not obtain the principal values of the H_β coupling but did measure the H_β splitting along the three crystalline axes that are apparently near the direction of the principal axes. Their measurements of the $^{14}N_\beta$ coupling show that the p_π^α spin density on this atom is parallel to that on N_α, and thus it appears that the $N_\beta H$ group is in the same plane as the $N_\alpha H$ group. Its $^{14}N_\beta$ coupling indicates that $\rho_\pi^N = 0.23$. These results suggest that the coupling of $N_\beta H$ is of the same nature as that of $N_\alpha H$ and that the unpaired electron is essentially in an antibonding π orbital between N_α and N_β, perpendicular to the planar —NHNH group.

Lin and Nickel [38] made observations on the radical $NCCH_2CONH\dot{N}H$, produced by X-irradiation of single crystals of cyanoacetohydrazide. They likewise found the —NHNH group to be planar with the unpaired electron in a π orbital between the nitrogens and perpendicular to the —NHNH plane. Thus the radical species is similar in form to $NH_2CONH\dot{N}H$, described already, For comparison, we have treated the data of Lin and Nickel in the same way as those of Hamrick and co-workers described earlier and have summarized the results in the same tables. The Q_α^{NH} values for the two radicals (Table 6.10) are reasonably close. The theoretical values of the dipole-dipole components (Table 6.11) fall between the two sets of observed values but are closer to those for the $NH_2CONH\dot{N}H$ radical. On the whole, the two sets of results are consistent and are probably characteristic of radicals of π-type $N_\alpha H$ fragments.

An example of the XNHY species is the NH_2CONHO radical observed in an X-irradiated single crystal of hydroxyurea ($NH_2CONHOH$) at room temperature. Shields, Hamrick, and Redwine [39], who originally measured the esr of this crystal, reported that they were able to fit the observed data to the radical $H_2NCO\dot{N}H$. However, Fox and Smith [40] showed that the radical observed by Shields, Hamerick, and Redwine [39] was actually NH_2CONHO, which by virtue of the admixture of the valence-bond structure 4 with 5, can have significant π spin density on the NH group, but not nearly so much as that expected for the $RN_\alpha H$ radical. In the hydroxyurea

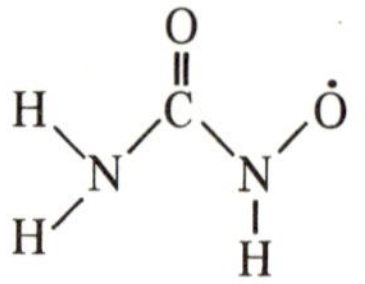

4

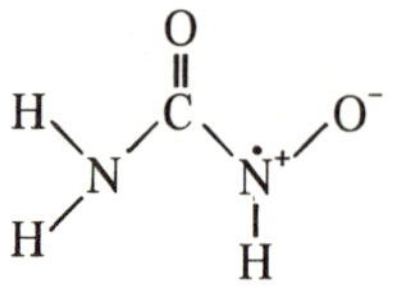

5

the NH group of radical 4 should have an orientation that conforms to the direction cosines observed by Shields and co-workers. Its distinguishing difference from $H_2NCO\dot{N}H$ is the expected smaller magnitude of the coupling due to the lower probable spin density on NH in 4. By comparison of the isotropic components of the NH coupling of the urea radical with the much larger isotropic coupling observed by Smith and Wood [41] for the radical HCONH, which is structurally similar to NH_2CONH, Fox and Smith [40] concluded that the radical observed in the hydroxyurea must be NH_2CONHO. This assignment has been accepted by other investigators [37, 42].

Table 6.12 gives a summary of the NH coupling constants of the NH_2CONHO radical. From the direction cosines of the p_π-orbital spin density on the N it is concluded that the radical is of the π type with the p_π-orbital spin density perpendicular to the NHO plane. A radical having the orbital of the unpaired electron in the NHO plane would be expected to have a much larger, nearly isotropic, H_β coupling due to hyperconjugation.

Table 6.12 Coupling Constants for the NH Group of the Radical $NH_2CON_\alpha HO$ in X-Irradiated Single Crystals of Hydroxyurea[a]

Coupling Atom	*Principal Value*	*A_f(MHz)*	*A_μ or B (MHz)*	*Direction*
	+62.1			‖ to p_π orbital
$^{14}N_\alpha$	+1.5	23.2	20.3	⊥ to p_π orbital
	+1.2			⊥ to p_π orbital
	-4.2		+29.6	‖ to $N_\alpha H$ bond
H_α	-37.8	-33.8	-4.2	‖ to p_π orbital
	-59.4		-25.6	⊥ to p_π and to $N_\alpha H$

[a] From the results of Shields, Hamrick, and Redwine [39].

The Q_α^{NH} value observed for this XNHO radical is comparable to those for the $RNH\dot{N}H$ radicals discussed earlier. A calculation of the anisotropic dipole-dipole component with (6.13), (6.15), and (6.16), and with the probable spin density on the O as well as the N taken into account, predicts the values 28, -6, and -22 MHz as compared with the respective observed values 29.6, -4.2, and -25.6 MHz.

It is of interest that the NH_2CONHO radical has been prepared and observed in liquid solution [43] where the isotropic coupling components can be accurately measured, but no information about the p_π-orbital spin density can be gained. The isotropic components thus observed for the NH group are $A_f(^{14}N) = 22.4$ MHz and $A_f(H) = -32.8$ MHz as compared with the slightly higher values of 23.2 MHz and -33.8 MHz, respectively, for the solid-state values in Table 6.12.

Bower, McRae, and Symons [44] have studied irradiated single crystals of urea and formamide. They reported that at 77°K the urea crystal gave the spectrum of the π-type $NH_2CO\dot{N}_\alpha H$ radical having a high spin density, $\rho_\pi^\alpha = 0.76$, on the N_α. At room temperature both crystals gave an esr characteristic of the $\dot{C}ONH_2$ radical. Claridge and Greenaway [45] have studied the esr of irradiated single crystals of thiosemicarbazide and observed a radical that they report to be either $NH_2NHCS\dot{N}_\alpha H$ or $\dot{N}_\alpha HNHCSNH_2$. The anisotropic component of the ^{14}N coupling is 29.5 MHz when $A_\parallel$ is taken as their maximum coupling value, 94.1 MHz, and $A_\perp$ as their minimum coupling value, 5.6 MHz. This component, when corrected for σ 2p density (Section 5.b), gives $\rho_\pi(N_\alpha) = 0.66$; with their isotropic H_α coupling value of -49.2 MHz, it indicates that $Q_\alpha^{NH} = -75$ MHz, in satisfactory agreement with other values in Table 6.10 for similar radicals.

In calculations of the Q_α^{NH} values for the NH fragment listed in Table 6.10 we have used π spin densities determined by the anisotropic ^{14}N coupling of the particular fragment. These values have been corrected for 2p-orbital density in the σ-bonding orbitals as described in Section 5 and partially for motional perturbations [46] of the $2p_\pi$ orbitals that destroy the cylindrical symmetry of the anisotropic component of the coupling. The corrections applied for these effects are described in Section 5.

3.b. Isotropic β Coupling: The $N_\beta H_3^+$ Group

Large isotropic coupling of β hydrogens through hyperconjugation, very prevalent in $C_\beta H$ groups, is not evident in $RN_\beta H_2$ or $XN_\beta HY$ groups because of the preferred tendency of the unpaired π spin density of the α atom to interact with the unshared pair on the N_β. As explained previously, the isotropic proton coupling of the $N_\beta H$ groups in such radicals is like that of $N_\alpha H$ groups in that it results mainly from spin polarization of the $N_\beta H$ bond by π spin density on N_β. In $-NH_3^+$ groups commonly found in amino acids, however, there is no unshared pair on the N. All its N valence orbitals form σ bonds similar to those formed by carbon. Thus the coupling of the $-N_\beta H_3^+$ group is similar to that of the methyl group; that is, it results mainly from hyperconjugation.

A typical example of $-N_\beta H_3^+$ coupling is that in the $CO_2^-C_\alpha HN_\beta H_3^+$ radical observed in X-irradiated glycine, originally measured by Ghosh and Whiffen [26] and more completely by Collins and Whiffen [47]. Table 6.13 gives the coupling constants that the latter observers obtained at 77°K. The analysis of the results is like that described for the $C_\beta H_3$ group in Section 2.b. At 77°K the internal rotation is frozen out so that each of the three β-hydrogens has a different coupling. The isotropic components can be fitted to (6.24) with the B_0 and B_2 values given in Table 6.14 and with the assumption that the $-NH_3^+$ group is symmetric; that is, its HNH bond angles are equal. To conform to the notation used in Section 2.b, we have labeled the H_β with the largest coupling as $H^{(1)}$ and that with the smallest as $H^{(3)}$ rather than the opposite notation used by Collins and Whiffen.

Table 6.13 β-Hydrogen Couplings in the $-NH_3$ Group of the $CO_2^-CHNH_3^+$ Single Crystal of Glycine at 77°K[a]

Coupling $-NH_3$ *Atom*	*Principal Value (MHz)*	*Isotropic Component* A_f*(MHz)*	*Anisotropic Component* A_μ*(MHz)*	*Calculated Isotropic Component*[b] *(MHz)*
	92.1		+10.1	
$H_\beta^{(1)}$	78.0	82.0	−4.0	82.0
	75.9		−6.1	
	71.3		+6.5	
$H_\beta^{(2)}$	58.8	62.0	−3.2	62.1
	56.0		−6.0	
	+12.5		9.2	
$H_\beta^{(3)}$	+1.5	+3.3	−1.8	+2.9
	−4.0		−7.3	

[a]Data used are taken from Collins and Whiffen [47].
[b]Calculated with (6.24) by use of the B_0, B_2, and θ values listed in Table 6.14.

Table 6.14 Derived Constants of $CO_2^-C_\alpha HN_\beta H_3^+$ in X-Irradiated Glycine at 77°K[a]

Coupling Constants (MHz)	*Orientation of* $N_\beta H_3^+$ *Group* θ^b
$B_0 = 1.5$	$\theta_1 = 23°$
$B_2 = 95$	$\theta_2 = 143°$
$Q_\beta^{NH} = 52^c$	$\theta_3 = 263°$

[a]Data used are given in Table 6.13.
[b]The angle θ is measured from the p_π^α orbital in the plane perpendicular to $C_\alpha N_\beta$.
[c]$Q_\beta^{NH} = [B_0 + (1/2)B_2]/\rho_\pi^\alpha$, where $\rho_\pi^\alpha = (A_f/Q_\alpha^{CH}) = 66/69.8 = 0.945$.

The origin of the small B_0, 1.5 MHz, for the $CO_2^-CHNH_3^+$ radical is uncertain. It may arise from torsional oscillations (see Section 2.b). From the $B_0 + (1/2)B_2$ and the spin density on the C_α, 0.945, obtained from $A_{iso}^\alpha = 66$ MHz and $Q_\alpha = 69.8$ MHz, one can calculate Q_β^{NH} from (6.25). The resulting value of 52 MHz, given in Table 6.14, should be fairly typical for $-N_\beta H_3^+$ radicals. It is significantly less than the value of 82 MHz for the $-C_\beta H_3$ group. The orientations for the σ bond of the $-N_\beta H_3^+$ group relative to the π orbital (Table 6.14) are very close to those of the C_β bonds in the first two radicals of Table 6.8. Agreement between the couplings for the rotating $-N_\beta H_3^+$ group at 300°K with the averaged value for the three isotropic components at 77°K was found to be only approximate.

Observation of $\dot{C}_\alpha H_2 N_\beta H_3^+$ radicals in irradiated single crystals of methylammonium alum has been reported by Kohin and Nadeau [48]. At 300°K the isotropic component of the two H_α couplings was found to be 70 MHz and the isotropic β components of the three H_β couplings, 53.8 MHz. The fact that the three H_β components were equal indicated that the $-N_\beta H_3$ group rotates at 300°. With $Q_\alpha = 68.2$ MHz for $-C_\alpha H_2$ groups (Section 2.b), one obtains $\rho_\pi^\alpha(C) \approx 1$. With the assumption that $B_0 \approx 0$, one then finds that $Q_\beta^{NH} = (1/2)B_2/\rho_\pi^\alpha \approx 54$ MHz. This Q_β^{NH} is close to that for $CO_2^- C_\alpha H N_\beta H_3^+$ (52 MHz) (Table 6.14).

Shrivastava and Anderson [49] observed the $N_\alpha H N_\beta H_3^+$ radical in γ-irradiated crystals of semicarbazide hydrochloride. They found the isotropic component of the H_α coupling to be -69.7 MHz and the isotropic components of the three H_β couplings in the rotating $-NH_3^+$ group to have the same value, 42 MHz. With $Q_\alpha^{NH} = 72$ MHz (the average of those in Table 6.10), we find $\rho_\pi^\alpha(N) = (69.7/72) = 0.97$. With this value, $Q_\beta^{NH} = (42/0.97) = 43$ MHz for the βH coupling of an $N_\beta H_3^+$ group with π spin density on N_α. This is somewhat lower than the 52 MHz for the $CO_2^- C_\alpha H N_\beta H_3^+$ radical. Shrivastava and Anderson also obtained values for the ^{14}N coupling from which the $\rho_\pi^\alpha(N_\alpha)$ can be calculated. However, the anisotropic components have limits of experimental error of ±11 MHz and deviate too far from axial symmetry to be normal 2*p*-orbital couplings. As suggested by the authors, this deviation from axial symmetry probably results from large torsional oscillations of the $-N_\alpha H$ group at the temperature of the measurements (300°K).

4. ^{13}C HYPERFINE STRUCTURE

The ^{13}C atom has a nuclear spin $I = 1/2$, and thus, like the H atom, produces a doublet splitting in the esr lines. Its natural abundance is only 1.1%, and partly for this reason the ^{13}C hyperfine structure has not been as widely measured as has that of H. Although the ^{13}C components have intensities only 1/2% of the unsplit ^{12}C line in the natural concentration, they can be detected with modern, sensitive spectrometers in many irradiated single crystals. These weak satellite lines, which are not widely spaced, are often masked by the wings of the stronger ^{12}C lines and are confused with the weak lines of secondary radicals in the sample. Fortunately, it is now possible to obtain many species of organic chemicals with ^{13}C concentrated at specific positions and in sufficient quantity for growth of crystals of observable size for reliable measurement of the ^{13}C hyperfine structure. This hyperfine structure can be of considerable

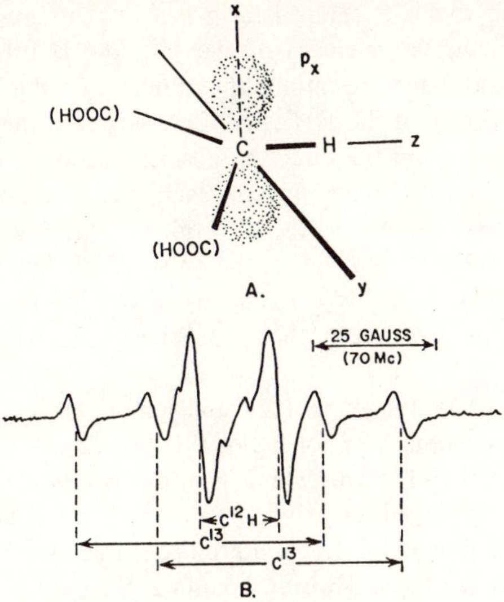

Fig. 6.7 Diagram of free radical (A) and esr spectrum (B) showing the splitting of ^{13}C and of H in the >CH fragment in $\dot{C}H(COOH)_2$. The x, y, z coordinates are along the principal coupling axes of both nucleui. From Cole and Heller [52].

help in measurement of the spin densities on carbon atoms in aliphatic as well as aromatic ringed radicals.

Among the early successful applications of the ^{13}C hyperfine structure was experimental verification that the methyl radical is planar [50] and the proof that the σ-orbital spin density at the C_α nucleus of the $C_\alpha H$ fragment is positive [51, 52] and hence that the induced spin density on the H is negative, in agreement with other evidence (see Chapter V, Section 2.d).

Figure 6.7 illustrates the ^{13}C hyperfine structure obtained by Cole and Heller [52] for the $\dot{C}H(COOH)_2$ radical in an X-irradiated single crystal of malonic acid in which the ^{13}C isotope is concentrated to 39%. For this particular orientation, the ^{13}C splitting is approximately three times that of the superimposed H splitting, which is, of course the same as that for the ^{12}CH central doublet. The H splitting in this radical is described in detail in Chapter V, Section 2.d. The weaker lines not assigned in this figure result from second-order transitions described in Chapter V, Section 4. These second-order transitions can be confused with the ^{13}C hyperfine components for some orientations when one tries to measure the ^{13}C hyperfine structure with the normal isotopic concentrations. The coordinate system used for the analysis of the ^{13}C measurements (Fig. 6.7) is the same as that used in the earlier work on the ^{12}C species, and it is worth noting that the x,y, and z as used in this figure are the principal axes of both the H and the ^{13}C coupling tensors. Since this radical is typical of a π-type radical formed when a bond to the C_α is broken, it is also of interest to note that the maximum principal coupling element of the ^{13}C, the H principal coupling of intermediate magnitude, and the minimum principal g value all occur along x, which is also the direction of the axis of the $2p_x$ orbital of the unpaired electron density on the C. Thus from the direction cosines of any of these principal elements relative to the crystal axes, one can learn the orientation of the radical plane within the crystal. However, the directions of the principal coupling elements of the H_α in the radical plane are the only reliable indicators of the direction of the C–H bond. In Table 6.15 are listed the ^{13}C coupling elements of $^{13}\dot{C}H(COOH)_2$ and those of a few other hydrocarbon radicals observed in irradiated single crystals.

As in previous chapters, the symbol A_f, and sometimes A_{iso}, is used to signify the Fermi-contact, isotropic, component of the coupling in place of the a used for liquids in Chapter IX. The anisotropic p-orbital coupling constant B is used instead of the A_μ employed in previous sections for the interatomic dipole-dipole interactions. This symbol should not be confused with the B_0 and B_2 constants used to signify isotropic β-proton coupling. We continue to use **A** for the general nuclear coupling tensor and A_x, A_y, and A_z or A_1, A_2 and A_3 to signify its principal elements. Many treatments use **T** to signify the nuclear coupling tensor.

Methods of obtaining the principal elements are described in Chapter V. For a central atom such as ^{13}C of a π radical in which the anisotropy in the coupling is caused by spin density of a p_π orbital essentially fixed in space (no rotational or vibrational motion of large amplitude), the coupling tensor will be, at least to a good approximation, axially symmetric about the p orbital, that is, about the x direction of Fig. 6.7. Then, $A_x = A_\parallel, A_y = A_z = A_\perp$, and

$$A_\parallel = A_f + 2B \tag{6.33}$$

$$A_\perp = A_f - B \tag{6.34}$$

From these equations the coupling components A_f and B can be found. For this type of coupling, both A_f and B are positive, and hence $A_\perp$ may be either positive or negative, depending on the relative magnitudes of these constants. One can usually establish the sign of $A_\perp$ by noting whether the splitting drops to zero at some orientation between $A_\parallel$ and $A_\perp$. If $A_\perp$ is zero, obviously $A_f = B$. Because the anisotropic component of the coupling is a traceless tensor, one can always find the isotropic component from the principal elements by

$$A_f = \frac{A_x + A_y + A_z}{3} \tag{6.35}$$

as was done in Section 2, even when the coupling is not axially symmetric. To

Table 6.15 Examples of ^{13}C Coupling in Irradiated Single Crystals

Radical / *Host Crystal*	^{13}C *Coupling* *Principal Value (MHz)*	*Coupling Constant* *Isotropic* A_f *(MHz)*	*Anisotropic*[a] *B (MHz)*	*Source of Data*
$H^{13}C(COOH)_2$ / Malonic acid	213, 43, 23	93	63	*b*
$H^{13}C(SO_3^-)_2$ / $K_2CH_2(SO_3)_2$	260, 62, 55	126	68	*c*
$(CO_2^-)^{13}CH(NH_3)^+$ / Glycine	254, 67, 60	127	65	*d*
$^{13}CH_2CO_2^-$ / Glycine	221, 133, 18	124	(68)	*d*
$(CO_2^-)O^{13}CHO^-$ / Sodium formate	125, 0, 0	42	42	*e*
$(CO_2H)(CH_2)_2{}^{13}CO^-OH$ / Succinic acid	227, 32, 20	93	69	*f*

[a]The *B* coupling constants are derived from maximum and minimum principal values. The *B* value for $^{13}CH_2CO_2^-$ is highly uncertain because of the large deviation from axial symmetry in the coupling.
[b]Cole and Heller [52].
[c]A. Horsfield, J. R. Morton, J. R. Rowlands, and D. H. Whiffen, *Molec. Phys.*, **5**, 241 (1962).
[d]J. R. Morton, *J. Am. Chem. Soc.*, **86**, 2325 (1964).
[e]R. E. Bellis and S. Clough, *Molec. Phys.*, **10**, 31 (1965).
[f]H. C. Box, H. G. Freund, and K. T. Lilga, *J. Chem. Phys.*, **42**, 1471 (1965).

do this, one must, of course, know the signs as well as the magnitudes of the principal elements.

4.a. Isotropic ^{13}C Coupling

From the previous discussion of the spin polarization of the σ bonds, it is evident that each of the three σ bonds to the C_α would be spin polarized by the

$2p_\pi$ unpaired electron density on the C_α. Thus spin polarization would cause a negative spin density to occur in the σ-bonding orbitals of the three atoms bonded to the carbon. This negative spin density must be counterbalanced by an equal, but oppositely polarized, spin density in the three σ-bonding orbitals of the C. A positive spin density will thus be induced in the three σ orbitals of the C. Usually the three σ orbitals can be taken as sp_2 hybrids. If we assume that each of the σ bonds is equally spin polarized, we can, from the proton coupling in the >C-H fragment, estimate the ^{13}C coupling to be expected. Although the spin polarization of the CH bonding orbital of the carbon must have the same magnitude as that of the hydrogen, it contains only 1/3 s character. Thus only one-third as much spin density would be induced by the spin polarization of this particular bond in the s orbital of the carbon as in that of the hydrogen. If, however, the other two σ bonds to the C are likewise spin polarized, as in the methyl radical, for example, the total spin polarization of the s orbital of the C should be equal to that on each H. From the isotropic H coupling of the methyl radical, the positive spin density in the 2s orbital on the C can thus be estimated to be $\rho_{2s}^{C} = 0.0457$. The 2s-orbital coupling of the ^{13}C, $A_{2s}(C)$, is 3110 MHz. Thus one can estimate the 2s contribution to the ^{13}C coupling in the methyl radical to be

$$\rho_{2s}A_{2s} = (0.0454)(3110) = 141 \text{ MHz} \tag{6.36}$$

The isotropic ^{13}C coupling in the methyl radical has been measured to be 41 ± 3 G or 115 ± 8 MHz [50].

The disagreement between the isotropic ^{13}C coupling observed to be 115 MHz and calculated from H coupling as 141 MHz indicates the existence of other interactions on the C that reduce the possible ^{13}C coupling arising from the σ-bond polarization. Karplus and Fraenkel [53] have shown that a negative contribution of the order of −36 MHz might be expected, mainly from exchange interaction between the 1s and 2s orbitals on the C. In effect, positive spin density would be promoted from the 1s orbital to antibonding σ^* orbitals on the C, leaving a negative spin density of the same magnitude in the 1s orbital. Although the promotion adds to the positive spin density in the 2s orbital, resulting effect on the coupling is negative because the magnitude of the coupling of the 1s electron is an order of magnitude greater than that of the 2s electron.

If there is significant unpaired π-electron density on other atoms, X_1, X_2, X_3, bonded to the C_α, the ^{13}C coupling is affected by this, primarily through induced spin polarization of the σ-CX bonds. If all these effects are summed, the isotropic coupling of the ^{13}C can be expressed [53] by

$$A_f^C = \left(S^C + \sum_{i=1}^{3} Q_{CX_i}^{C}\right)\rho_\pi^C + \sum_{i=1}^{3} Q_{X_iC}^{C}\,\rho_\pi^{X_i} \tag{6.37}$$

Karplus and Fraenkel give the theoretically predicted values as $S^C = -35.6$ MHz, $Q^C_{CH} = 54.6$ MHz, $Q^C_{CC'} = 40.3$ MHz, and $Q^C_{C'C} = 38.9$ MHz. These values can be used in the estimation of spin densities ρ_π from measurements of the isotropic components of the ^{13}C coupling or for calculation of the expected ^{13}C splittings when the spin densities are known. However, the empirical value $S^C = -26$ MHz, determined from the CH_3 radical, may be more suitable for such calculations. Note that the first subscript refers to the atom on which the π spin density occurs and that the second subscript refers to the atom to which this atom is bonded. The superscript indicates the coupling atom.

For the methyl radical, for which $\rho^C_\pi = 1$, it is possible to obtain experimental values of the ^{13}C coupling constants in (6.37). Since there is no π spin density on the H's, $Q^C_{X_iC} = 0$. One can obtain S^C by subtraction of the 141 MHz, determined as in (6.36), from the observed isotropic coupling $A_f(^{13}C) = 115$ MHz. Because all these bonds are equivalent, one can obtain $Q^C_{CX_i}$ by dividing by 3 the total 2*s* coupling, 141 MHz. Thus for the methyl radical the experimental values are

$$S^C = -26 \text{ MHz}$$

$$Q^C_{CX_i} = Q^C_{CH_i} = 47 \text{ MHz}$$

As an illustration, let us use the isotropic component of the ^{13}C coupling, $A^C_f = 92.6$ MHz, to calculate the ρ^α_π density on the carbon of the $\dot{C}H(COOH)_2$ radical. With the constants of Karplus and Fraenkel,

$$A^C_f = [-35.6 + 54.6 + 2(40.3)]\rho^C_\pi = 99.6\rho^C_\pi \tag{6.38}$$

which with A^C_f, 92.6 MHz, gives $\rho^C_\pi = 0.93$. This value agrees approximately with the value of 0.86, obtained from the H_α coupling (see Table 6.1). If one uses the experimental values $S^C = -26$ MHz and $Q^C_{CX_i} = 47$ MHz, derived from methyl radical, the predicted spin density is 0.805, also in approximate agreement with that from the H_α coupling. This use of Q^C_{CH} for $Q^C_{CC'}$ is more difficult to justify than the use of the empirical S^C from the methyl radical. The latter involves configuration interaction of ρ^C_π with the subvalence shell on the C and is less sensitive to bond type than is Q^C_{CX}. Furthermore, S^C is more difficult to predict from theory than is Q^C_{CX}. If one uses $S^C = -26$ MHz with the theoretical values $Q^C_{CH} = 54.6$ MHz and $Q^C_{CC'} = 40.3$ MHz for the $\dot{C}H(COOH)_2$ radical, one obtains $\rho^C_\pi = 0.85$, in good agreement with the value from the H_α coupling.

The isotropic ^{13}C coupling in the succinic acid radical $HOOC(CH_2)_2CO\bar{O}H$ in Table 6.1 is the same, 93 MHz, as that for $CH(CO_2H)_2$ described in the preceding paragraph. Thus the ρ^C_π spin density of 0.81 to 0.85 predicted from A^C_f for the two radicals, is the same if one assumes that $Q^C_{CO} = Q^C_{CC'}$. Note, however, that the anisotropic components of the 2*p* orbital differ for the two. There is no α-coupling H in the succinic acid radical from which the ρ_π^C can be derived for comparison, but that derived (see next section) from the anisotropic ^{13}C coupling, 0.85, is in good agreement with the value indicated by A_f.

Radicals listed in Table 6.15 having ^{13}C isotropic coupling of the order of 126 MHz with any combination of the constants in (6.37) would give predicted spin densities ρ_π^C appreciably above unity. One must conclude that these radicals are not strictly of the π type and that the orbital of the unpaired electron on C_α has some small 2*s* character. This possibility, always present in such radicals, makes the spin densities calculated from the isotropic ^{13}C coupling unreliable.

With any combination of the preceding parameters, the $A_f(^{13}C)$ for $(CO_2)^-O^{13}CHO^-$ indicates a ρ_π^C spin density significantly lower than the 0.55 derived from the H_α coupling and also lower than the 0.54 derived from the $2p_\pi$-coupling B. Since the ^{13}C coupling is axially symmetric, the latter should be a good value. Very probably this disagreement results from our neglect of spin density on the β oxygens, which would have the effect of raising the predicted spin density ρ_π^C through the last term of (6.37). Any π spin density on an X_i atom induces negative spin density in the σ-bonding orbitals of the C_α, and thus the $Q_{X_iC}^C$ terms of (6.37) are negative.

It is evident from the above examples that prediction of the ρ_π^C spin densities from the $A_f(^{13}C)$ values is difficult and can be done with useful accuracy only for relatively simple radicals in which the nature of the bonding to C_α is known. Nevertheless, this coupling, when considered with other coupling parameters, can give information about the nature of the bonding.

4.b. Anisotropic 2*p*-Orbital Coupling

The coupling by spin density in a single *p* orbital fixed in space is axially symmetric, and its coupling constant B may be found from the principal coupling values $A_\parallel$ and $A_\perp$ with (6.33) and (6.34) when there is no significant spin density in other *p* orbitals or no other important anisotropic contributions to $A_\parallel$ and $A_\perp$. However, it is evident from Table 6.15 that the ^{13}C coupling in organic free radicals is not always axially symmetric, although it is approximately so in many cases. Obviously, *s* spin density on the ^{13}C that gives rise only to isotropic coupling cannot cause this deviation from axial symmetry. If the orbitals $2p_y$ and $2p_z$ as well as $2p_x$ have spin density, this could cause deviation from axial symmetry in the coupling unless the spin densities of two of the orbitals are equal. In π-type, carbon-centered radicals we follow the conventional designation of the x axis as perpendicular to the radical plane, that is, the sp^2 plane of the C_α orbitals, as indicated in Fig. 6.7. Because of spin polarization of the σ bonds, the $2p_y$ and $2p_z$ orbitals of the ^{13}C in these π-type radicals always have a small positive spin density; unless the bonds are equivalent, this density may differ for the two orbitals. However, difference in spin polarization of the σ bonds is not expected to be large, and calculations by McConnell and Chesnut

[8] indicate that they cause less than a 0.3-MHz difference in the principal values A_y and A_z. Therefore, the marked deviations from axial symmetry in the ^{13}C principal coupling values of Table 6.15 must be attributed to other factors. One possibility is that the radical group to which the $2p_x$ orbital is fixed undergoes restricted internal rotation or torsional oscillations of large amplitude. This effect is like the one explained for the ^{14}N coupling in Section 5. It can be prevented if the measurements can be made at temperatures sufficiently low that the motions are frozen out. A second factor is possible unpaired electron density in an antibonding σ orbital that puts unequal spin density in the $2p_y$ and $2p_z$ orbitals.

For reasons given in Section 5, one can correct, at least partially, for these deviations from axial symmetry by use of the principal element of smallest magnitude as $A_\perp$ rather than the average of the two smaller values when calculating the p-orbital coupling constant B from (6.33) and (6.34). We have used this procedure in deriving the B values listed in Table 6.15.

Although the induced spin polarization of the σ bonds in π-type radicals does not lead to significant deviation from axial symmetry in the coupling of the central atom, it does cause measurable effects on the p-orbital coupling, on the resultant B values, and hence on the unpaired spin densities ρ_π^{C} derived from B. Fortunately, it is not too difficult to make approximate corrections for these effects. For $C_\alpha H$ bonds, the positive spin density in the sp^2-hybrid bonding orbital of C_α is the same in magnitude as the negative spin density on the H and is thus numerically equal to $A_f(H_\alpha)/1420$. We assume that σ bonds formed to other atoms X_i are similarly polarized. When there is no coupling H_α, one can assume the spin polarization to be $Q_\alpha^{\mathrm{H}} \rho_\pi^{\mathrm{C}}/1420$. As an example, $A_f(H_\alpha) = 60$ MHz for the $\dot{C}H(COOH)_2$ radical; consequently, the positive spin density in the C_α bonding orbital is $\rho_\sigma^{\mathrm{C}} = (60/1420) = 0.0425$. Since this sp^2 orbital has 2/3 $2p$ character, it will have a $2p$ spin density of (2/3)(0.0425) = 0.0283. If one assumes that the two $C_\alpha C_\beta$ bonds are similarly polarized, the total $2p_y + 2p_z$ spin density will be 2(0.0425) = 0.085. This will produce a negative contribution to $A_\parallel$ of

$$\Delta A_\parallel = -B_{\mathrm{atomic}}^{\mathrm{C}}(\rho_{2p_y} + \rho_{2p_z}) = -90.8(0.085) = -7.72 \text{ MHz} \quad (6.39)$$

The in-plane contribution will be equivalent to

$$\Delta A_\perp = \frac{B_{\mathrm{atomic}}^{\mathrm{C}}\,[(2\rho_{2p_y} - \rho_{2p_z}) + (2\rho_{2p_z} - \rho_{2p_y})]}{2}$$

$$= \frac{1}{2} B_{\mathrm{atomic}}^{\mathrm{C}}\,(\rho_{2p_y} + \rho_{2p_z}) = +3.86 \text{ MHz} \quad (6.40)$$

To find the coupling by the unpaired electron density of the individual $2p_x$

orbital, one must increase the observed $A_\parallel$ by 7.7 MHz and must decrease the observed $A_\perp$ by 3.8 MHz. If $(A_\parallel)_{obs}$ = 213 MHz and $(A_\perp)_{obs}$ = 23 MHz, the corrected coupling values for the $2p_x$ spin density alone are 221 MHz and 19 MHz. With (6.33) and (6.34) these corrected values give B_{2p_x} = 67.3 MHz and a predicted $2p_x$ orbital spin density of $\rho_\pi^C = (67.3/90.8) = 0.741$. This value is in better agreement with the 0.86 derived from the H_α coupling than is the uncorrected value of 0.70.

In Table 6.16 are given spin densities similarly derived from the ^{13}C couplings listed there. To show the effects of σ-bond spin polarization, the uncorrected

Table 6.16 Spin Densities from Anisotropic $2p$-Orbital Coupling of ^{13}C in π-type Radicals

Radical	ρ_π^C *from* H_α *Coupling*	ρ_π^C *from Anisotropic Coupling*: *Uncorrected* $B(^{13}C)$ / 90.8	*Corrected for Spin-polarization Effects*
$H\dot{C}(COOH)_2$	0.86	0.70	0.74
$H\dot{C}(SO_3^-)_2$	0.87	0.75	0.79
$(CO_2^-)\dot{C}H(NH_3)^+$	0.88	0.72	0.76
$\dot{C}H_2CO_2^-$	0.88	(0.75)	(0.80)
$(CO_2^-)O\dot{C}HO^-$	0.58	0.46	0.49
$(CO_2H)(CH_2)_2\dot{C}O^-OH$		0.76	0.80

values are listed along with those corrected for the induced $2p_y$ and $2p_z$ spin densities described earlier. In making such calculations it is helpful to note that they can be reduced to

$$\rho_\pi^C(\text{cor}) = \rho_\pi^C(\text{uncor}) + \frac{1}{2}(\rho_{2p_y}^C + \rho_{2p_z}^C) \tag{6.41}$$

and when there is H_α coupling and ρ^H and A_f^H are taken as negative,

$$\rho_{2p_y} + \rho_{2p_z} = -2\rho^H = \frac{-2A_f^H}{A_{\text{atomic}}^H} \tag{6.42}$$

Where there is no H_α, as in the last radical given in Table 6.16, one can approximate the in-plane, $2p$-orbital spin densities by

$$\rho_\pi^C(\text{cor}) = \rho_\pi^C(\text{uncor}) + \frac{Q_\alpha^{CH}}{A_{\text{atomic}}^H}\,\rho_\pi^C(\text{cor}) \tag{6.43}$$

With $Q_\alpha^{CH} \approx 73.4$ MHz and $A_{atomic}^{H} = 1420$ MHz, (6.41) then reduces to the very convenient form

$$\rho_\pi^C(\text{cor}) = 1.054\, \rho_\pi^C(\text{uncor}) \tag{6.44}$$

This simpler formula can also be used when there is a coupling H_α.

The degree of agreement between the corrected values of ρ_π^C derived from the anisotropic ^{13}C coupling (last column of Table 6.16) and those derived from the H_α coupling (column 2) is gratifying. It not only tends to verify the method used in correction for the spin polarization effects on the anisotropic ^{13}C coupling, but it also indicates that the Q_α^{CH} values and theory used for calculation of these spin densities from the H_α coupling are essentially correct. The fact that the corrected values are still somewhat lower than those derived from H_α arises at least partly from our failure to correct the A_{max} for reduction caused by torsional oscillations.

5. INTERPRETATION OF ^{14}N HYPERFINE STRUCTURE

The ^{14}N nucleus has a spin $I = 1/2$ and hence produces a triplet splitting of the esr components. Observed under normal conditions without saturation, the three components are of equal intensity. The experimentalist often finds it difficult to separate and measure the splitting because of the superposition of the hyperfind structure of other nuclei such as H or D which have couplings of considerable magnitude. A theoretical description of these coupling constants and the hyperfine structure is given in Chapter III, and methods for obtaining the principal elements of the coupling from experimental measurements are presented in Chapter V. Figure 4.2 shows the energy-level diagram and resulting composite hyperfine pattern expected for two coupling ^{14}N nuclei with various coupling ratios. Here we are concerned not with assignment and measurement of the hyperfine structure, but rather with interpretation of the observed coupling constants. In Section 3.a we have used spin densities derived from the ^{14}N hyperfine structure as an aid in interpretation of the H coupling in NH fragments.

Selected examples of ^{14}N couplings are given in Tables 6.9 and 6.12, with additional examples in Tables 6.17 and 6.18. The isotropic component A_f and the anisotropic component B are derived from the principal values in the same manner as described for ^{13}C in (6.33) to (6.35). When the anisotropic component does not have entirely axial symmetry, as a single p orbital coupling is expected to have, and it is suspected that the asymmetry arises from torsional motions of the orbital, the lowest value rather than the average of the two lower values is used as $A_\perp$ in the calculation of B. Because the most information gained comes from the anisotropic, p-orbital coupling component, measurements on single crystals are very valuable. Whenever possible, these measurements

Table 6.17 ^{14}N Coupling Constants of Selected Radicals

Radical / *Host Crystal*	*Temperature*	*Principal Value* $A(^{14}N)$ *(MHz)*	$A_f(^{14}N)$ *(MHz)*	$B(^{14}N)$ *(MHz)*	*Source of Data*
$\dot{N}(SO_3^-)_2$ / $K_2NH(SO_3)_2$	300°K	106 6 0	37	35	*a*
	77°K	–	37	35	
$\dot{N}HSO_3^-$ / KH_2NSO_3	300°K	97.5 10.3 5.5	38	31	*b*
	77°K	–	38	35	
$\dot{N}^+(CH_3)_3$ / Betaine-HCl	300°K	133.3 37.5 18.5	63	38	*c*
$\dot{N}^+H_2CH_2CO_2^-$ / α-Glycine	77°K	105.8 9.8 8.4	41	33	*d*
$NH_2CSNH\dot{N}H$ or $NH_2NHCS\dot{N}H$ / Thiosemibarbazide	300°K	94.1 7.6 5.6	35.8	29.5	*e*

[a] A. Horsfield, J. R. Morton, J. R. Rowlands, and D. H. Whiffen, *Molec. Phys.*, **5**, 241 (1962).
[b] Rowlands [46].
[c] D. J. Whelan, *Aust. J. Chem.*, **26**, 681 (1973).
[d] K. Nunome, H. Muto, K. Torigama, and M. Iwasaki, *J. Chem. Phys.*, **65**, 3805 (1976).
[e] Claridge and Greenaway [45].

should be made at temperatures sufficiently low to freeze out the motional perturbations of the *p*-orbital coupling.

5.a. Isotropic ^{14}N Coupling

Configuration interactions similar to those described for ^{13}C give rise to isotropic ^{14}N coupling frequently observed in organic free radicals containing nitrogen [54-57]. Stone and Maki [56] have assumed for ^{14}N a formula similar to that of (6.37) for ^{13}C and have attempted to evaluate empirically the S^N and Q^N. However, the ^{14}N coupling is greatly complicated by the unshared electron

Table 6.18 Nuclear Coupling Parameters for Double-bonded, Nitrogen-centered Radicals

Radical / Host Crystal	Coupling Atom	Principal Value A (MHz)	Isotropic Component A_f (MHz)	Anisotropic Component A_μ or B (MHz)	Source of Data
$O\dot{N}{=}C(CH_3)R$ / Dimethylglyoxime (77°K)	^{14}N	130.6 73.9 68.6	91.0	20.7	a
$\dot{N}_\alpha{=}C_\beta HCH_2CONH_2$ / Malonamide	^{14}N	100 19 1	40	33	b,c
	H_β	236 224 223	228	8 −4 −5	
$\dot{N}_\alpha{=}C_\beta H_2$ / KCl	^{14}N		24.1	33.1	d
$\dot{N}_\alpha{=}C_\beta H_2$ / HCN (77°K) (powder)	^{14}N	$(A_\parallel)$ 93 $(A_\perp)$ 0	31	31	e,f,g
	H_β	255	255		

[a] Nelson, Atwater, and Gordy [60].
[b] Cyr and Lin [61].
[c] Lin, Cyr, and Toriyama [64].
[d] N. V. Vugman, M. F. Elia, and R. P. A. Muniz, *Molec. Phys.*, **30**, 1813 (1975).
[e] D. Banks and W. Gordy, *Molec. Phys.*, **26**, 1555 (1973).
[f] J. A. Brivati, K. D. J. Root, M. C. R. Symons, and D. J. A. Tinling, *J. Chem. Soc. A*, 1942 (1969).
[g] Cochran, Adrian, and Bowers [66].

pair in the valence shell and by the varying amounts of hybridization of the σ-bonding orbitals in different radicals. The values of S and Q obtained from the various radicals are not in good agreement, and it seems impossible to be sure what part of S^N arises from the unshared pair in the valence shell and what part arises from exchange interaction involving the 1s electrons. Thus it seems more feasible to use the specific formula:

$$A_f^N = Q_1^N \rho_\pi^N + Q_2^N \sum_i \rho_i \tag{6.45}$$

where $Q_1^N \rho_\pi^N$ is the total coupling resulting from ρ_π^N, the π spin density on the coupling N, and where $Q_2^N \sum_i \rho_i$ is the induced N coupling resulting from the spin density on the other atoms bonded to the N. The probable Q values are: $Q_1^N = 59$ MHz and $Q_2^N = -6$ MHz. Experimentally, Q_1^N values range from 50 to 54 MHz [57]. Often, especially when $\rho_\pi^N > \sum_i \rho_i$, one can approximate the expected isotropic N coupling by the simple relation [54]

$$A_f^N = Q_1^N \rho_\pi^N \tag{6.46}$$

where Q_1^N is of the order of 56 MHz. However, this relation evidently does not hold well for the NH_2 radical, for which the isotropic ^{14}N coupling has been observed to be only 28 MHz [58]. We do not attempt to derive spin densities ρ_π^N from the isotropic coupling. This can be done much more reliably from the anisotropic component.

5.b. Spin Densities from the Anisotropic Component

Most reported p_π spin densities obtained from the hyperfine structure of the central atom represent simply the ratio of the observed B to the corresponding B_{atomic}. Such spin densities are accurate only when the B constant is due entirely to the spin density in the particular p_π orbital with its axis fixed in the crystal. Spin density in the p_y or p_z orbital of the coupling atom or motions of the coupling orbital relative to the crystal axis influence the value of B and hence the apparent spin densities. Also, positive or negative charges on the coupling atom influence the value of B. Approximate corrections for these perturbing factors are described as follows.

Effects of Spin Polarization of σ Bonds

In the planar π-type radical, the unpaired electron spin density on the N is in the $2p_x$ orbital perpendicular to the plane of three sp^2-hybrid orbitals like the π-type carbon-centered radical already treated. Since the neutral N is trivalent, one of the sp^2 hybrids will normally have an unshared pair, and the other two will form σ bonds. In this case the s character of the orbital with the unshared pair is likely to be greater than that of the two σ-bonding orbitals. When the N has a formal positive charge, three equivalent sp^2-bonding orbitals may be formed, and the N^+-centered, π-type radical has a structure like that of the planar π-type carbon-centered radical. For either of these types the corrected spin density is determined by a formula like (6.41) for ^{13}C. Therefore,

$$\rho_\pi^N(\text{cor}) = \rho_\pi^N(\text{uncor}) + \frac{1}{2}(\rho_{2p_y} + \rho_{2p_z}) \tag{6.47}$$

The problem is to find $(\rho_{2p_y} + \rho_{2p_z})$. For the N^+-centered radical with three σ bonds in a common plane, formulas like those for ^{13}C, (6.42) to (6.44), may be applied. For the neutral radicals in which the N forms only 2 σ bonds, we assume that each of the σ-bonding orbitals has 78% $2p$ character, as do the bonding orbitals of NH_3, and we neglect the configuration interaction involving the unshared pair. With these assumptions and with both σ bonds assumed to have the same spin polarization, one finds that

$$\rho_{2p_y}^{N} + \rho_{2p_z}^{N} = 1.56|\rho_s^{H_\alpha}| \qquad (6.48)$$

and hence that

$$\rho_\pi^N(\text{cor}) = \rho_\pi^N(\text{uncor}) + 0.780|\rho_s^{H_\alpha}| \qquad (6.49)$$

where $|\rho^{H_\alpha}| = |A_f^{H_\alpha}|/A_{\text{atomic}}^{H}$. With $Q_\alpha^{NH} = 72$ MHz, $\rho_s(H_\alpha) = 0.051\, \rho_\pi^N(\text{cor})$ and the more convenient form,

$$\rho_\pi^N(\text{cor}) = 1.042\, \rho_\pi^N(\text{uncor}) \qquad (6.50)$$

Table 6.19 Spin Densities from Anisotropic $2p_\pi$ Orbital Coupling of ^{14}N in π-Type Radicals

Radical	B^N *(MHz)*	*Spin Density from Uncorrected* $B^N/47.8$	^{14}*N Coupling Corrected*[a] ρ_π^N
$\dot{N}(SO_3^-)_2$	35	0.73	0.76
$\dot{N}HSO_3^-$	35	0.73	0.76
$\dot{N}^+(CH_3)_3$	38	0.79	0.74[b]
$\dot{N}^+H_2CH_2CO_2^-$	33	0.69	0.72[c]
$NH_2CON\dot{H}NH$	35.3	0.74	0.77
$NH_2CO\dot{N}HO$	20.3	0.42	0.44
$NCCH_2CONH\dot{N}H$	33	0.69	0.72
$NH_2CSNH\dot{N}H$ or $NH_2NHCS\dot{N}H$	29.5	0.62	0.65
$O\dot{N}{=}C(CH_3)R$	20.7	0.43	0.45
$\dot{N}{=}CHCH_2CONH_2$	33	0.69	0.73
$\dot{N}{=}CH_2$	31	0.65	0.69

[a] Corrected for $2p$ spin density in σ orbitals.
[b] Also corrected for formal positive charge, $c^+ = 0.40$ (see Section 6).
[c] The B^N = 41 MHz derived from the measurements of J. Sinclair, *J. Chem. Phys.*, **55**, 245 (1971), yields $(B^N/47.8) = 0.86$ and $\rho_\pi^N(\text{cor}) = 0.91$.

analogous to (6.44) for ^{13}C, is obtained.

In calculation of the ρ_π^N(cor) for the structures given in Table 6.19 for which there is a double bond to the coupling ^{14}N, we use (6.50). The effects of interaction of the unpaired electron with the π component of the double bond are assumed to be similar to that with the σ component, an assumption of questionable validity. This approximation should, nevertheless, be better than no correction.

When the induced $2p_x$ and $2p_y$ spin densities are equal, as is assumed in the preceding calculations, no deviation from axial symmetry in the coupling will result from the spin polarization of the bonds, only a lowering of the observed B value. If the $2p_y$ and $2p_z$ spin densities were to become equal to that of the $(2p_x)_\pi$ orbital, the B value would be reduced to zero, but, of course, this cannot occur when the p_y and p_z spin densities are induced entirely by the π spin density of $2p_x$. Asymmetries in the coupling caused by unequal p_y and p_z spin densities are most likely to occur in the double-bonded structures described in Section 5.c.

Motional Effects

One should be cautious in using the ^{14}N coupling to determine the $2p_\pi$ spin density on the N_α when the observed principal values of this coupling are not axially symmetric, as expected for a p orbital fixed in space. Torsional oscillations or restricted internal rotation of the $-N_\alpha H$ group reduce the maximum component from that of the fixed p orbital and raise the value of the perpendicular component that is in the plane of the oscillation. Since the latter component is negative, the increase in value lowers its magnitude; when the resultant perpendicular component ($2s$ plus $2p$ contribution) is positive, the effect is to increase the magnitude of this resultant.

In calculation of the anisotropic coupling constant B for ^{14}N, many observers take the maximum principal values as $A_\parallel$ and the average of the two lower values as $A_\perp$. Nevertheless, it seems probable that the minimum principal value would be much closer to the correct $A_\perp$ when all the principal values are positive. Consequently, in our calculations we use the minimum observed value as $A_\perp$, with the maximum value as $A_\parallel$. When the torsional oscillations are large, the A_{max} may be significantly less than $A_\parallel$ for the fixed p orbital. Consequently, the effective anisotropic coupling constants for these excited states are generally less than that for the corresponding p orbital fixed in space. Thus when uncorrected, these effective values lead to predicted $\rho_\pi^\alpha(N_\alpha)$ spin densities that are too low and Q_α values that are too high—unless the H_α coupling is likewise reduced. Generally, it can be assumed that the isotropic H_α coupling will not be influenced significantly by the torsional oscillations or even by rotation of the $N_\alpha H$ group. If, however, the p_π^α orbital engages in π bonding with X_β, this orbital may not rotate with $N_\alpha H$. In this case, the H_α coupling will be affected, possibly reduced.

Marked temperature effects on the anisotropic ^{14}N coupling and the derived $p_\pi(N_\alpha)$ spin densities have been demonstrated by Rowlands [46] for the $\dot{N}H_2^+SO_3^-$ and $\dot{N}HSO_3^-$ radicals. Comparison of the coupling constants for $\dot{N}HSO_3^-$ at 300°K and at 77°K is shown in Table 6.17. Note that the B value increases from 31 to 35 MHz at the lower temperature, whereas no change occurs in the isotropic component. Even at 77°K, the B^N value may not have reached its maximum value. The Q_α^{NH} = 72 MHz (Table 6.10) with the observed $A_{iso}(H_\alpha)$ = 63.6 MHz for the $\dot{N}H(SO_3^-)$ radical indicates that $\rho_\pi(N_\alpha)$ = (63.6/75) = 0.88, whereas the value obtained from B^N = 35 MHz is only 0.77 (Table 6.19), In contrast, there is no difference in the coupling constants for $\dot{N}(SO_3^-)_2$ at 300°K and 77°K.

Effects of Charge on the Coupling Atom

A net charge, positive or negative, on the coupling atom will cause errors in the spin densities derived from the observed anisotropic p-orbital coupling if the atomic coupling constant of the neutral atom is employed in the calculation of the spin densities. Methods for approximate calculation of the charge corrections are given in Section 6.

Examples

The spin densities listed in the final column of Table 6.19 have the corrections described previously for $2p_y$ and $2p_z$ spin densities due to σ-bond spin polarization. Only one, $\dot{N}^+(CH_3)_3$, has a charge correction applied. The B^N values used were derived from the principal couplings given in Tables 6.9, 6.12, 6.17, and 6.18. Where the coupling is not axially symmetric, the maximum value was taken as $A_\parallel$ and the minimum value as $A_\perp$. As explained earlier, this was done in an effort to make some correction for the effects of torsional oscillations that cause asymmetry in the coupling. This asymmetry is not, of course, always caused by motions of the p orbitals. It may result from unequal spin denisties in the $2p_y$ and $2p_z$ orbitals.

5.c. Coupling of Double-bonded ^{14}N

In dimethylglyoxime, one of the first irradiated single crystals to be studied with esr [59], it was found that the orbital of the unpaired electron on the ^{14}N of the observed radical

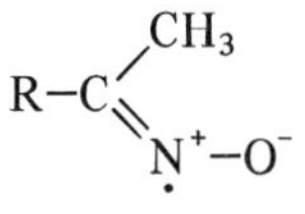

6

was in the plane of the molecule and hence perpendicular to the CN π bond to the N. The ^{14}N coupling has since that time been observed with ENDOR, and this direction of the orbital has been confirmed [60]. Although the unpaired electron orbital on the ^{14}N has some small 2*s* hybridization, it appears that the unpaired electron can correctly be considered as in an antibonding π orbital in the molecular plane, mostly between the N and O, with the unpaired electron shared about equally between the N and O. The principal values of the ^{14}N magnetic coupling obtained from the ENDOR measurements, which are reproduced in Table 6.18, are in good agreement with those from esr.

The radical $N_\alpha{=}C_\beta HC_\gamma H_2CO$, listed in Table 6.18, also has the unparied electron density in a π_x-type orbital perpendicular to the π_y component of the double bond to the coupling ^{14}N. However, the structure is otherwise different from that of the dimethylglyoxime. There is no O with which the ^{14}N can form an antibonding π orbital; as a result, the spin density on the ^{14}N is appreciably higher than that for the $ON{=}CCH_3R$ radical (compare values in Table 6.19). The unpaired electron on the $\dot{N}_\alpha{=}C_\beta HCH_2CONH_2$ radical does engage in π bonding through hyperconjugation with the $C_\beta H$ group. A surprisingly large 1*s* spin density of 0.16 on the H_β is indicated by the isotropic H_β coupling of 228 MHz. If this should originate entirely from hyperconjugation, the Q_β would be exceptionally large (312 MHz). Since the $N_\alpha C_\beta$ bond has a π component, it is reasonable to assume that the $N_\alpha C_\beta HC_\gamma$ group is planar and that the orbital of the unpaired electron is in this plane. For those who may consult the original source for the $\dot{N}{=}CHCH_2CONH_2$ coupling values, it should be mentioned that the observed couplings were first incorrectly assigned by Cyr and Lin [61] to the radical $H_2NOCCH_2CO\dot{N}H$. After this assignment was questioned by Neta and Fessenden [62] and by Symons [63] and a radical of form $N_\alpha{=}C_\beta HR$ proposed, Lin, Cyr, and Toriyama [64] proved by ^{15}N that the identity of the radical originally observed in the irradiated malonamide crystal is that listed in Table 6.18.

The ^{14}N hyperfine structure of the double-bonded radical $\dot{N}_\alpha{=}C_\beta HCH_2CONHCONH_2$ has been measured in an X-irradiated crystal of cyanoacetylurea by Lau and Lin [65], who likewise incorrectly assigned the observed spectrum to a radical of form $RCO\dot{N}H$. This error in assignment was also corrected by Lin, Cyr, and Toriyama [64]. The principal proton coupling values observed by Lau and Lin for the cyanoacetylurea radical, 244, 239, and 237 MHz, are obviously those of a β-coupling H and cannot be those of an $N_\alpha H$ proton. They are comparable in value to those listed in Table 6.18 for the $\dot{N}_\alpha{=}C_\beta HCH_2CONH_2$ radical observed in irradiated malonamide. Their isotropic component, $A_f(H_\beta) = 240$ MHz with the values for Q_0^β and Q_2^β from Table 6.20 for the similar radical in malonamide, yields a spin density $\rho_\pi^{N_\alpha} = 0.77$. This value is reasonable and close to that of 0.73 obtained from the ^{14}N coupling for the malonamide radical. However, it does not agree well with that indicated by principal values of the ^{14}N coupling, 84.6, 55.2, and 24.6 MHz, that were obtained by Lau and

Lin [65]. The closeness of the two larger values suggests that the radical plane and hence the $2p_\pi$ orbital of the unpaired electron on the ^{14}N may be undergoing large torsional oscillations or restricted internal rotations under the conditions of the measurements.

The classic example of a double-bonded radical ($N{=}CR_1R_2$) is CH_2N, originally observed by Cochran, Adrian, and Bowers [66] in photolized mixtures of HI and HCN in a rare-gas matrix at 4.3°K. The radical can be produced directly by exposure of frozen HCN to gaseous H atoms. It is planar in structure, with the p_π orbital of the unpaired electron in the molecular plane and the π_y component of the double bond perpendicular to this plane. Hyperconjugative π bonding occurs equally with the two protons, giving each proton isotropic coupling of $A_f^H = 255$ MHz. The ^{14}N coupling constants listed in Table 6.18 are the more probable of two sets derived from measurements on the radicals in impurity sites in single crystals of KCl. The difference between these values and those listed for the HCN powder probably comes from different perturbations of the matrix. However, the coupling parameters from the three sources cited (Table 6.18*e*, *f*, *g*) are all in good agreement, even though they were measured in different polycrystalline matrix materials. Table 6.19 gives the ρ_π^N spin densities calculated from the B^N for the HCN powder matrix with corrections applied for the small $2p_y$ and $2p_z$ spin densities arising from the polarization of the bonds.

5.d. Derivation of βH Coupling Parameters with Use of ^{14}N Spin Densities

The isotropic β-proton coupling $A_f^H = 255$ MHz for the $N_\alpha C_\beta H_2$ radical reveals a positive $1s$ spin density of $255/1420 = 0.18$ on each H or a combined spin density of 0.36 on the two hydrogens. Since the total unpaired electron-spin density on the radical is unity, only the amount of $1 - \rho_\pi^N = 1 - 0.69 = 0.31$, or 0.155 for each H, can be unpaired electron-spin density resulting from hyperconjugation, if the value of 0.69 derived for ρ_π^N is assumed to be correct. The remaining $1s$ spin density of the hydrogens, $0.36 - 0.31 = 0.05$, or 0.025 on each H, must result from spin polarization of the $C_\beta H$ bonds. With these spin densities, the parameters of the βH coupling formula

$$\begin{aligned} A_\beta^H &= B_0 + B_2 \cos^2\theta \\ &= Q_0^\beta \rho_\pi^N + Q_2^\beta \rho_\pi^N \cos^2\theta \end{aligned} \tag{6.51}$$

can be calculated. Since the spin polarization is independent of orientation, we find that $B_0 = 0.025\,(1420) = 35.5$ MHz. Because the p_π orbital of the unpaired electron of this radical is in the molecular plane, it is evident that $\theta = 0$ and hence $B_2 = 0.155(1420) = 220$ MHz. The correcponding Q values are $Q_0^\beta = 35.5/0.69 = 51$ MHz and $Q_2^\beta = 220/0.69 = 319$ MHz.

The βH coupling parameters for the $N_\alpha{=}C_\beta HC_\gamma H_2CONH_2$ radical derived in a similar way are reasonably consistent with those for $N_\alpha{=}CH_2$. The isotropic βH coupling of 228 MHz reveals a spin density of 0.16 on the $C_\beta H$ hydrogen. Since the unpaired electron-spin density on the ^{14}N is 0.73, there is a remainder of 0.27 on other atoms. Half of this, 0.135, may be assumed to be on the β hydrogen. Thus the spin density on the βH that results from spin polarization of the bond is 0.16 − 0.135 = 0.025, the same as that for the $N_\alpha CH_2$ hydrogens. The parameters of (6.51), calculated as described for $N_\alpha C_\beta H_2$, are listed in Table 6.20.

It is interesting to compare βH coupling parameters for these double-bonded structures with those for $N^+(CH_3)_3$. Except for the hydrogens, this radical is planar, with the unpaired electron mostly in a $2p_\pi(N)$ orbital perpendicular to the molecular plane. For the rotating methyl groups, each of the nine protons has the coupling value A_f^β = 78 MHz [67], and thus the total positive spin density on the hydrogens is ρ_{2p}(total) = 9(78/1420) = 0.49. From Table 6.19, ρ_π^N (cor) = 0.74, and hence the total unpaired electron density on the hydrogens is only 1 − 0.74 = 0.26. Presumably, this unpaired electron-spin density is due to hyperconjugation, and the remaining 1s spin density, 0.49 − 0.23 = 0.26, is due to σ-bond spin polarization. It is surprising that the spin-polarization density for each hydrogen, 0.23/9 = 0.0255, is essentially the same as that for the double-bonded radicals. The part of the isotropic βH coupling which results from spin polarization is B_0 = 36 MHz, and that from hyperconjugative coupling is only 42 MHz. Since this coupling is for the rotating CH_3 group, it is clear that B_2 = 2(42) = 84 MHz. These values with the Q_0^β and Q_2^β derived from them are listed in Table 6.20 along with those obtained by the similar method for the double-bonded structures.

Table 6.20 Some βH Coupling Parameters Derived with the Use of ^{14}N Spin Densities from Table 6.19

	Coupling Parameter[a] (MHz)			
Radical	B_0	B_2	Q_0^{NCH}	Q_2^{NCH}
$N_\alpha{=}C_\beta H_2$	35.5	220	51	319
$N_\alpha{=}C_\beta HCH_2CONH_2$	35.5	192	49	263
$N_\alpha^+(CH_3)_3$	36	84	49	114

[a]The parameters are those in (6.51).

6. CORRECTIONS FOR CHARGES ON THE COUPLING ATOM

Through reduced nuclear screening, a resultant positive charge on a coupling atom in a free radical will increase the anisotropic coupling over that expected for the neutral atom. Conversely, a negative charge will decrease it.

Semiempirical correction constants for formal charge derived by Townes and Schawlow [68] for nuclear quadrupole couplings in stable molecules may be used for correction of anisotropic magnetic coupling constants in free radicals since both couplings depend on the nuclear screening in a similar way, that is, by alteration of the effective $\langle 1/r^3 \rangle$ for the p orbitals. These charge-correction constants for different atoms are tabulated in another volume of this series [36]. Those for atoms commonly observed in free radicals are given in Table 6.21. If there is a net positive charge of c^+ in electron units on the atom, the coupling is increased by a factor of

$$B^+ = B^0 (1 + c^+ \epsilon) \tag{6.52}$$

where ϵ is the charge-correction constant, B is the observed p-orbital coupling constant for the charged atom, and B^0 is the p-orbital coupling constant for the neutral atom. In the derivation of spin densities from the observed B of free radicals, the simplest way to apply the charge correction is with the relation

$$\rho_\pi^{\mathrm{X}} = \frac{B_{\mathrm{obs}}^{\mathrm{X}^+}(\mathrm{rad})}{B_{\mathrm{atomic}}^{\mathrm{X}}(1 + c^+ \epsilon)} \tag{6.53}$$

where c^+ is the net positive charge on the coupling atom X and ϵ is the charge-correction constant for X. When there is a net negative charge of c^- on the coupling atom X, the corresponding formula for the spin density is

$$\rho_\pi^{\mathrm{X}^-} = \frac{B_{\mathrm{obs}}^{\mathrm{X}^-}(1 + c^- \epsilon)}{B_{\mathrm{atomic}}^{\mathrm{X}}} \tag{6.54}$$

Table 6.21 Screening Constants ϵ for Correction of p-Orbital Couplings for Effects of Charges on the Coupling Atom[a]

Atom	ϵ	*Atom*	ϵ	*Atom*	ϵ	*Atom*	ϵ
Be	0.90	Mg	0.70	Ca	0.60	Sr	0.60
B	0.50	Al	0.35	Sc	0.30	Ga	0.20
C	0.45	Si	0.30	Ge	0.25		
N	0.30	P	0.20	As	0.15	Sb	0.15
O	0.25	S	0.20	Se	0.20	Te	0.20
F	0.20	Cl	0.15	Br	0.15	I	0.15

[a]Values are from *Microwave Spectroscopy* by C. H. Townes and A. L. Shawlow, Copyright 1955, McGraw Hill, Inc. Used with permission of McGraw Hill Book Company, New York.

where c^- is the magnitude of the charge in electron units.

For N, $\epsilon = 0.30$ and $B^{N}_{\text{atomic}} = 47.8$ MHz. Thus for a positive N coupling atom,

$$\rho_\pi^{N^+} = \frac{B^{N^+}_{\text{obs}}\ (\text{MHz})}{47.8\,(1 + 0.3c^+)} \tag{6.55}$$

where $B^{N^+}_{\text{obs}}$ is the bona fide ^{14}N coupling constant for the p_π orbital of the radical, that is, after correction of other factors such as spin polarization of σ bonds.

Apart from the approximate nature of the screening constant, the difficulty in applying (6.53) or (6.54) for calculation of spin density is the uncertainty in the net charge, c^+ or c^-, on the coupling atom. In simpler cases, one can approximate the net charges from estimates of the ionic character of the various bonds to the coupling atom by use of the relation [36]

$$i_c = \frac{x - x_i}{2} \tag{6.56}$$

where x is the effective electronegativity of the coupling atom and x_i is that of the atom or group of atoms to which the coupling atom is bonded. If there is a formal charge on the atom, as in $\dot{N}^+(CH_3)_3$, this must also be taken into account. The electronegativity of N, 2.95, is greater than that of H, 2.15, or C, 2.5, the atoms to which it is most frequently bonded in organic free radicals. Therefore, a net negative charge would generally be expected on the coupling ^{14}N of neutral radicals. For example, in such radicals as R(H)NN_αH, the ionic character of the N_αH bond is expected to be of the order of 40%, with the negative charge on the N. Thus if the N-N bond has no ionic character and there is no sharing of the "unshared" electron pair on the N, the charge would be $c^- \approx 0.4$ and (6.54) would indicate an increase in spin density by a factor of 1.12 over that of the uncorrected value, $B_{\text{obs}}/B_{\text{atomic}}$. Because of the large uncertainty in the resultand charge on the coupling ^{14}N, which is influenced by a number of factors, we have not applied charge corrections in the calculation of spin densities on coupling nitrogens of electrically neutral radicals discussed in the preceding sections. Nevertheless, it seems probable that the coupling ^{14}N of most of these radicals has some negative charge and hence that the ρ_π^{N} values are too small by 5 to 10%, and also that the Q_a^{N} values derived from them are too high by comparable amounts.

Without a correction for the positive charge on the radical $N^+(CH_3)_3$ (Table 6.19), the observed ^{14}N coupling indicates an unpaired spin density of 0.83 on the N after corrections have been made for spin-polarization effects. Since the H_β proton coupling for the rotating methyl groups, $A_f\,(C_\beta H_3) = 78$ MHz, shows considerable unpaired electron density on the hydrogens, this value is clearly too high, and a charge correction is warranted. In making this correction we assumed that the indicated unit of positive charge on the N is reduced to 0.40 by 20% ionic character of each of the three equivalent bonds. The remaining charge gives $c^+ = 0.40$ which, in (6.55), yields the more reasonable value of 0.74 for ρ_π^N(cor). Although the radical $\dot{N}^+H_2CH_2CO_2^-$ is electrically neutral, there is a formal positive charge on the $-N^+H_2$ group. We have applied no charge correction in deriving the ρ_π^N(cor) for this radical listed in Table 6.19 because the expected ionic character of each N^+H bond, $i_c^{NH} = 0.4$, combined with the expected $i_c^{NC} = 0.2$ for the N^+C bond approximately neutralizes the formal positive charge.

In a >CH fragment bonded to two carbons, the negative charge on the C_α is similarly estimated to be approximately 20% and with $\epsilon = 0.45$ for C, the uncorrected ρ_π^C values from ^{13}C are estimated to be too low by about 9%. No such charge corrections are made for the ρ_π^C values in Table 6.16. This is probably one reason why the values obtained for the ^{13}C coupling are generally lower than those obtained from the H_α coupling with an independently derived Q_α^C. In the simplest case, $H\dot{C}(COOH)_2$, an increase by a factor of 1.09 in the ρ_π^C value 0.74 derived from the ^{13}C coupling brings it to $\rho_\pi^C(\text{cor}) = 0.81$, in agreement with the value 0.86 derived from the H_α coupling (Table 6.16).

7. FLUORINE HYPERFINE STRUCTURE

The halogen hyperfine structure most commonly observed in oriented free radicals is that of ^{19}F. Like that of H, the ^{19}F nucleus has spin $I = 1/2$ and hence produces a doublet hyperfine structure. However, the coupling tensors of H and F differ markedly because of their unlike electronic configurations. In free radicals, both may have significant interatomic dipole-dipole coupling with spin density on neighboring atoms, but this kind of interaction is more important in H than in ^{19}F because it has the smaller atomic radius.

The first-order ^{19}F hyperfine structure for a representative free radical $\dot{C}F_2COONH_2$ is described in Chapter V, Section 2, and the method for derivation of the principal values in the coupling is given there. Here we deal with higher-order effects and give interpretations of the coupling constants.

7.a. Second-order Effects

Because the ^{19}F hyperfine splittings are often large, about 400 to 500 MHz, second-order displacements of several megahertz occur in the components when the measurements are made at X band, about 9000 MHz. These shifts may be calculated for the $S = 1/2$ state with (5.70). Since they are proportional to M_I^2, they cause no change in the resultant $M_I = \pm 1/2$ doublet splitting and are then important only for the derivation of the principal g values. Because the shifts vary inversely with H_0, they are less, proportionally, at K-band frequencies (24,000 MHz).

Most observers of ^{19}F hyperfine structure in molecular free radicals have detected second-order hyperfine transitions arising from admixed nuclear spin states like those described in Chapter V, section 4.a. In addition to the extra doublet lines in the hyperfine pattern, the separation of the normal doublets can be altered significantly when the admixing is strong, that is, when the observational frequency and angle of the crystal in the field are such that the hyperfine splitting, A_{eff}, approaches $2g_I\beta_I H$. This is shown graphically in Fig. 5.17.

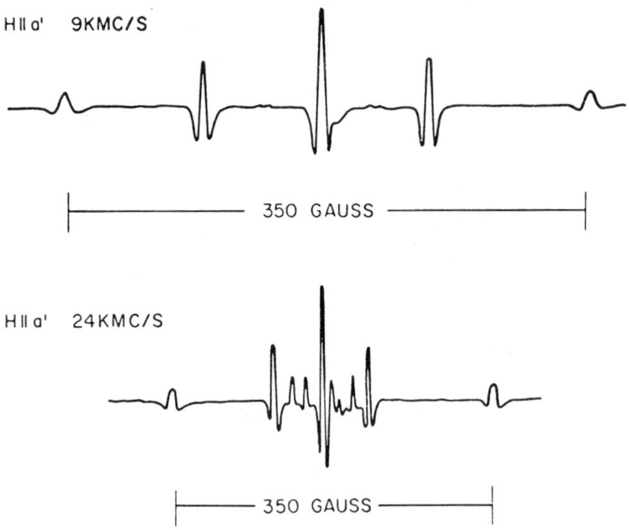

Fig. 6.8 Comparison of the ^{19}F hyperfine patterns of $\dot{C}F_2CONH_2$ for the same crystal orientation observed at two microwave frequencies (*X* and *K* bands). Note the increased intensity of the second-order transitions (indicated in Fig. 6.9) at the higher frequency. From Lontz and Gordy [69].

Figure 6.8 shows a comparison of the ^{19}F hyperfine patterns of the $\dot{C}F_2CONH_2$ radical in an irradiated single crystal of trifluoroacetamide at the two commonly used frequencies, *X* band and *K* band, but at the same orientation of the crystal. As explained in Chapter V, Section 2.d, the apparent quartet is actually a superposition of two triplets caused by the two radicals in unit cell, each having two equivalent ^{19}F nuclei. Only the center components of the triplet coincide. For this orientation of the crystal second-order transitions occur only in the inner triplet, which has component spacing slightly less than half that of the outer triplet. The second-order doublets are closely spaced and barely visible in the upper curve (*X* band). At the higher observation frequency (lower curve), they are significantly stronger and more widely spaced.

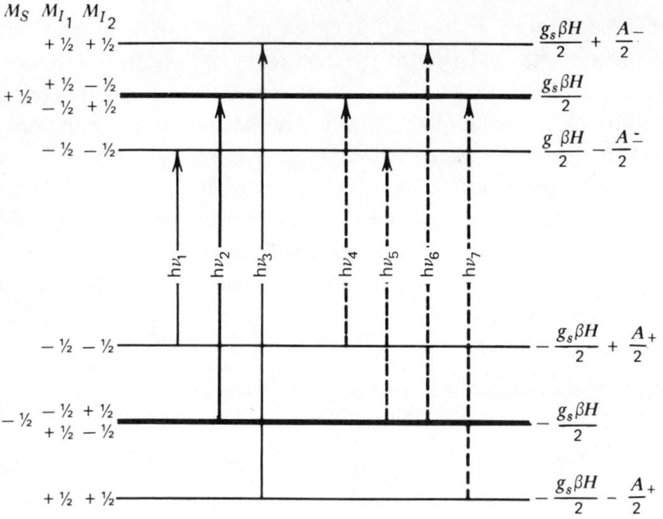

Fig. 6.9 Energy-level diagram with indicated first-order (solid arrows) and second-order (broken arrows) transitions in the $\dot{C}F_2CONH_2$ free radical. Observed lines corresponding to these transitions are shown in Fig. 6.8. From Lontz and Gordy [69].

Figure 6.9 gives the energy level diagram for the $\dot{C}F_2CONH_2$ radical with the primary transitions, which give rise to the triplet (indicated by solid arrows), and the second-order transitions, which give rise to the doublets (indicated by the broken arrows). The separation of the primary triplet components, in frequency units, is

$$(\Delta\nu)_t = \frac{A_+ + A_-}{2} \tag{6.57}$$

the separation of the components of the second-order doublets is

$$(\Delta\nu)_d = \frac{A_+ - A_-}{2} \tag{6.58}$$

For the axially symmetric coupling case, $A_\pm$ may be expressed as

$$A_- = \left[\left(\frac{A_\parallel - 2g_I\beta_I H}{h}\right)^2 \cos^2\theta + \left(\frac{A_\perp - 2g_I\beta_I H}{h}\right)^2 \sin^2\theta\right]^{1/2} \tag{6.59}$$

$$A_+ = \left[\left(\frac{A_\parallel + 2g_I\beta_I H}{h}\right)^2 \cos^2\theta + \left(\frac{A_\perp - 2g_I\beta_I H}{h}\right)^2 \sin^2\theta\right]^{1/2} \tag{6.60}$$

where A_+, A_-, $A_{\parallel}$, and $A_{\perp}$ are in frequency units Hz, H is in gauss, and θ is the angle of the applied field **H** with the symmetry axes of the coupling. These formulas are equivalent to (5.119) with the coupling constants changed to frequency units (see Chapter V, Section 2.a for unit transformation formulas). The transformation of the coefficients A_{zz}, A_{zx}, A_{zy}, and c^2 to the axially symmetric coordinates is given by (5.115) to (5.118). For calculation of the intensity ratio of the primary: second-order transitions, these transformed elements are substituted into (5.105). Table 6.22 shows the remarkably close agreement between the predicted and observed second-order components obtained in the original work [69] on the $\dot{C}F_2CONH_2$ radical.

Table 6.22 Comparison of Observed and Calculated Second-order Transitions of the $\dot{C}F_2CONH_2$ Radical at 24,300 MHz[a]

Angle of Crystal Axis with Fields	*Component Separation of Satellite ν_4 from Primary ν_2 (ΔH in G)*		*Relative Intensity Satellite to Primary*	
	Obs.	*Calc.*	*Obs.*	*Calc.*
65°	48 ± 3	49	0.05 ± 0.02	0.05
70°	40 ± 3	42	0.1 ± 0.03	0.10
75°	33 ± 3	36	0.2 ± 0.04	0.16
80°	29 ± 3	30	0.3 ± 0.05	0.30
85°	25 ± 3	26	0.6 ± 0.06	0.60

[a]From Lontz and Gordy [69].

7.b. α^{19}F Coupling: The $C_\alpha F$ Fragment

It is tempting, but misleading, to compare the >CF fragment with the >CH fragment. As we see later, there are similarities, but also basic differences, in the coupling mechanisms of the H and F in these two species. The obvious difference is the large anisotropic omponent due to the $2p$-orbital coupling of the F, which is not possible for H. The surprising difference is that the predominant α-F coupling is due to positive spin density on the αF rather than to negative spin density, as for the αH.

Listed in Table 6.23 are selected examples of ^{19}F couplings in >C_αF fragments of free radicals in irradiated single crystals. The fact that the spin denisty on the αF is positive became evident when the first two such fluoroorganic free radicals were measured (the first two listed in Table 6.23); a large $2p_\pi$-orbital coupling was observed with the axis of this $2p_\pi$ coupling perpendicular to the $CC_\alpha F$

plane. This showed that the unpaired $2p_\pi$-electron density on the C_α was spilling over onto the F. In molecular-orbital parlance, the unpaired electron is in a 2π molecular orbital comprised of atomic orbital constituents of $2p_x$(F) as well as $2p_x(C_\alpha)$, and perhaps of other atoms. In valance-bond parlance, it means that the $C_\alpha F$ bond has a contributing structure with a $>C^-{=}F^+$ double-bond component.

Certain common features are notable in the coupling values in Table 6.23. The maximum principal value is rather large for an α-coupling nucleus and is approximately the same for all the radicals. There is axial symmetry, or nearly axial symmetry, in all the principal values except those for $CO_2^-CF_2\dot{C}FCO_2^-$. The direction of the largest principal element is perpendicular to the radical plane, which includes the coupling ^{19}F in all except $CO_2^-CF_2\dot{C}FCO_2^-$, for which the directions of the principal elements relative to the molecular structure could not be checked because the crystal structure is unknown.

These common features reveal that the spin density that causes the primary coupling is not induced through π-σ configuration interaction as it is for the CH or NH fragments; instead, the predominant coupling is with $2p_\pi$(F) unpaired electron density. Although there is induced σ-bond spin polarization and interatomic coupling with the π spin density on the C_α, these minor couplings tend to cancel each other. These cancellations cause the near-symmetry in the coupling and make it possible to derive directly from the observed coupling meaningful values for the axially symmetric coupling constant B. Except for $CO_2^-CF_2\dot{C}FCO_2^-$, the anisotropic B_{obs} coupling constants in Table 6.24 were derived from $A_\parallel$, taken to be the maximum principal element, and the isotropic component A_f by use of the relation,

$$B = \frac{A_\parallel - A_f}{2} \tag{6.61}$$

The radical $CO_2^-CF_2\dot{C}FCO_2^-$ has the lowest maximum and the highest intermediate principal coupling value of those listed in Table 6.23. In contrast, the minimum coupling value. 11 MHz, is comparable to those of the other radicals. These features suggest that the $\dot{C}_\alpha F$ group probably was undergoing torsional motions of large amplitude at the temperature of the measurement, presumably about 300°K. If this is true, the minimum principal element is normal to the plane of the oscillations and probably close to the $A_\perp$, which would be observed for the fixed $2p$ orbital. Thus the value $B = A_f - 11 = 188$ MHz should be a rather good one for the static orbital.

Reasonably good values of ρ_π^F, the $2p_\pi$ unpaired electron-spin density on the ^{19}F can be had if one simply takes the ratio of B_{obs}/B_{atomic} with B_{obs} obtained as described previously. However, more refined values can be acquired by the analysis given in the following paragraphs.

Table 6.23 α ^{19}F Couplings in Selected Molecular Free Radicals

Radical / *Host Crystal*	*Principal Value* A_x, A_y, A_z *(MHz)*	*Isotropic Component* A_f *(MHz)*	*Anisotropic Component* A_μ (MHz)	*Direction*	*Ref.*
$\dot{C}F_2CONH_2$	499		288	⊥ to CCF_2 plane	
Trifluoroacetamide	67	211	-144	In CCF_2 plane	*a*
	67		-144		
$\dot{C}HFCONH_2$	530		372	⊥ to CCHF plane	
Monofluoroacetamide	-11	158	-169	In CCHF plane, ⊥ to CF	*b*
	-45		-203	‖ to CF	
$\dot{C}F_2COO(NH_4)$	527		325	⊥ to CCF_2 plane	
Ammonium trifluoroacetate	39	202	-163	In CCF_2 plane	*c*
	39		-163		
$CF_3\dot{C}FCONH_2$	563		357	⊥ to CCFC plane	
Pentafluoroacetate	34	206	-172	In CCFC plane	*d*
	22		-184		

$\dot{C}F(CONH_2)_2$	560		384	⊥ to CFC_2 plane	
Difluoromalonamide	−3	176	−179	In CFC_2 plane	*e*
	−28		−204		
$CO_2^-CF_2\dot{C}FCO_2^-$	421		222		
Sodium perfluorosuccinate	165	199	−34		*f*
	11		−188		
$\dot{C}F_3$	739		337		
Rare-gas matrix (4.2°K)	244	402	−158	Nonplanar	*g*
	222		−180	Pyramidal	
$\dot{S}iF_3$	604		228		
SiF_4	229	376	−77	Nonplanar	*h*
	226		−150	Pyramidal	

[a] Lontz and Gordy [69].
[b] Cook, Rowlands, and Whiffen [70].
[c] Srygley and Gordy [75].
[d] Lontz [84].
[e] M. Iwasaki, S. Noda, and K. Toriyama, *Molec. Phys.*, **18**, 201 (1970).
[f] M. T. Rogers and D. H. Whiffen, *J. Chem. Phys.*, **40**, 2662 (1964).
[g] Maruani, McDowell, Hakajima, and Raghunathan [73]; see also Maruani, Coope, and McDowell [74].
[h] These are averaged values of the nearly equal coupling elements for $F^{(1)}$, $F^{(2)}$, and $F^{(3)}$ obtained by Hasegawa, Sogabe, and Miura [77].

Table 6.24 Anisotropic ^{19}F Coupling Constants and Spin Densities for Selected $C_\alpha F$ Fragments

	Coupling Constants (MHz)[a]			*Spin Densities*[b]	
Radical	B_{obs}	B_π	B_σ	ρ_π^F	$\rho_{2p\sigma}^F$
$\dot{C}F_2CONH_2$	144	141	-10	0.093	-0.007
$\dot{C}HFCONH_2$	186	176	-24	0.116	-0.016
$\dot{C}F_2COO(NH_4)$	163	160	-10	0.106	-0.007
$CF_3\dot{C}FCONH_2$	178	173	-14	0.114	-0.009
$\dot{C}F(CONH_2)_2$	192	185	-18	0.122	-0.012
$CO_2^-CF_2\dot{C}FCO_2$	188			0.124[c]	
CF_3	169	163	-16	0.107	-0.011
SiF_3	114	100	-33	0.066	-0.022

[a] $B_{obs} = (A_{max} - A_f)/2$, except for $CO_2^-CF_2\dot{C}FCO_2$, for which $B_{obs} = A_f - A_{min}$. The values for B_π and B_σ are derived from (6.62) to (6.64), as explained in the text with $\rho_\pi^C = 0.70$ for all except $\dot{C}HFCONH_2$, $\dot{C}F(CONH_2)_2$, and CF_3, for which $\rho_\pi^C = 0.86, 0.80$, and 0.66, respectively.
[b] $\rho_\pi^F = B_\pi/B_{atomic}$, $\rho_{2p\sigma}^F = B_\sigma/B_{atomic}$, with no charge corrections.
[c] Derived from B_{obs}.

Analysis of the Anisotropic Components

For this analysis the conventional choice of the x, y, and z principal axes is made; that is, x is along the axis of the coupling ρ_π^F spin density, y is in the radical plane and perpendicular to the CF direction, and z is along the CF σ bond. The primary term in the anisotropic ^{19}F coupling is due to the ρ_π^F spin density, which contributes the coupling terms $2B_\pi$, $-B_\pi$, and $-B_\pi$ along the respective x, y, z axes. Induced $2p_\sigma$ spin density contributes the terms $-B_\sigma$, $-B_\sigma$, and $2B_\sigma$ along x, y, z, respectively. The only other significant anisotropic terms are those contributed by interatomic dipole-dipole coupling between the ^{19}F nucleus and the ρ_π^C spin density on C_α, which we designate as $(A_\mu^{FC})_x$, $(A_\mu^{FC})_y$, and $(A_\mu^{FC})_z$. Thus one can express the principal anisotropic coupling elements A_μ by

$$(A\mu)_x = A_x - A_f = 2B_\pi - B_\sigma + (A_\mu^{FC})_x \tag{6.62}$$

$$(A\mu)_y = A_y - A_f = -B_\pi - B_\sigma + (A_\mu^{FC})_y \tag{6.63}$$

$$(A\mu)_z = A_z - A_f = -B_\pi + 2B_\sigma + (A_\mu^{FC})_z \tag{6.64}$$

with the assumption that x is along the p_π orbital and z is along CF. The A_μ values are directly measurable quantities given in Table 6.23 for the selected radicals being considered. There are, however, more unknown parameters than equations, and auxiliary relations must be used in the solution. For this purpose, we have used the calculated values of A_μ^{FC} expressed by (6.67) to (6.69). Although these parameters still depend on unmeasured ρ_π^C values, the latter can be estimated with sufficient accuracy for determination of good values for B_π and B_σ, especially for B_π, which is not sensitive to ρ_π^C.

As an illustration, the solution is briefly described for $\dot{C}HFCONH_2$, for which $(A_\mu)_x = 372$ MHz, $(A_\mu)_y = -169$ MHz, and $(A_\mu)_z = -203$ MHz [70]. When these values with $(A_\mu^{FC})_x$, $(A_\mu^{FC})_y$, and $(A_\mu^{FC})_z$ from (6.67) to (6.69) are substituted into (6.62) to (6.64), the solution yields

$$B_\pi = 180.4 - 4.6\ \rho_\pi^C \tag{6.65}$$

$$B_\sigma = -11.3 - 14.6\ \rho_\pi^C \tag{6.66}$$

With the $\rho_\pi^C = 0.86$, as given by the αH coupling (Table 6.1), one obtains $B_\pi = 176$ MHz and $B_\sigma = -24$ MHz, as listed in Table 6.24. Note that the B_π value is not sensitive to ρ_π^C and that a change of 0.1 in the value of ρ_π^C would change B_σ by only 2 MHz. Table 6.24 also gives the B_π and B_σ values derived in a similar way for other radicals. Also recorded in this table are the $2p_\pi$ and $2p_\sigma$ spin densities derived from these couplings. No charge corrections are applied in calculation of the spin densities. Because of counterbalancing effects, such corrections are unknown and probably negligible.

Because of the wide deviation in the A_y and A_z principal values, a complete analysis could not be made for the $CO_2^-CF_2\dot{C}FCO_2^-$, but it is interesting that the $\rho_\pi^F = 0.124$ value determined from B_{obs}/B_{atomic} is close to that for the other radicals listed in Table 6.24.

The Interatomic Dipole-Dipole Interaction

In the preceding analysis, contributions to the anisotropic coupling of the direct dipole-dipole interaction of the ^{19}F nuclear moment with the unpaired electron density ρ_π^C on the C_α were included. These contributions can be predicted approximately with the theory of McConnell and Strathdee [23] or from the semiempirical formulas developed in Section 1.d, which we use here. For the >CF fragment, (6.13), (6.15), and (6.16), and R_p from (6.17) give the principal values of this coupling as

$$(A_\mu^{FC})_x = -5.5\ \rho_\pi^C \text{ MHz} \quad (\parallel \text{ to } 2p_\pi) \tag{6.67}$$

$$(A_\mu^{FC})_y = -19.1\ \rho_\pi^C \text{ MHz} \quad (\perp \text{ to } 2p_\pi \text{ and to CF}) \tag{6.68}$$

$$(A_\mu^{FC})_z = 24.6\ \rho_\pi^C \text{ MHz} \quad (\parallel \text{ to CF}) \tag{6.69}$$

The Induced Components

One can expect configuration interaction on the C_α to produce a spin polarization of the covalent component of the CF σ bond like that in the CH bond. Because of its higher ionic character, this polarization will be less for the CF than for the CH bond. With the effective electronegativity of C in the fluorinated group as 2.7 [69] and that of F as 3.9_5 [36], an ionic character of $i_c = 0.62$ is predicted for the CF σ bond by use of (6.56) compared with an i_c of about 0.20 similarly obtained for the >CH fragment with the electronegativity for C as 2.5 [36]. In the methyl radical with $\rho_\pi^{\mathrm{C}} = 1$, the H coupling indicates a negative spin density of -0.0457 on the H caused by the σ-bond polarization. Since the relative spin polarization of the CF and CH σ bonds is expected to be proportional to the covalent component of the two bonds, 0.38 and 0.80, respectively, the negative spin density in the σ-bonding orbital of the F should be approximately

$$\rho_\sigma^{\mathrm{F}}(-) = \frac{0.38}{0.80}(-0.0457)\ \rho_\pi^{\mathrm{C}} = -0.022\ \rho_\pi^{\mathrm{C}} \tag{6.70}$$

The proper use of the estimated spin density $\rho_\sigma^{\mathrm{F}}(-)$ requires a knowledge of the hybridization of the σ orbital of the F. If this orbital is a normalized *sp* hybrid, $\psi_\sigma^{\mathrm{F}} = a_s\psi_{2s} + b_p\ \psi_{2p}$, then

$$\rho_{2p_\sigma}^{\mathrm{F}}(-) = (1 - a_s^2)\ \rho_\sigma^{\mathrm{F}}(-) \tag{6.71}$$

where a_s^2 is the 2*s* character and $1 - a_s^2$ is the 2*p* character of the orbital. If the F-bonding orbital is pure 2*p*, as we assume here,

$$\rho_{2p_\sigma}^{\mathrm{F}}(-) = \rho_\sigma^{\mathrm{F}}(-) = -0.022\ \rho_\pi^{\mathrm{C}} \tag{6.72}$$

One can also express $\rho_{2p_\sigma}^{\mathrm{F}}(-)$ by

$$\rho_{2p_\sigma}^{\mathrm{F}}(-) = \frac{Q_{2p_\sigma}^{\mathrm{CF}}}{B_{\mathrm{atomic}}^{\mathrm{F}}}\ \rho_\pi^{\mathrm{C}} \tag{6.73}$$

where $Q_{2p_\sigma}^{\mathrm{CF}}$ is a coupling parameter that is directly proportional to the 2*p* character of the F-bonding orbitals and also to the covalent character of the bond. With these assumptions, it is seen from comparisons of (6.72) and (6.73) that

$$Q_{2p_\sigma}^{\mathrm{CF}} = -33\ \mathrm{MHz}$$

The assumption of a 2s character of 0.20 in the σ orbital would give $Q_{2p_\sigma}^{CF} = -28$ MHz. However, we think that the pure $2p_\sigma$ orbital is the more probable.

The negative spin density and associated coupling induced in the σ-bonding orbital of the F by π spin density on C_α is partly canceled by the positive spin density $\rho_{2p_\sigma}^{F}(+)$ induced in this orbital by the unpaired $2p_\pi$ spin density ρ_π^F on the F. The amount of ρ_π^F is known rather accurately from the observed B values; it is given in Table 6.24 for the radicals being considered. However, the corresponding $Q_{2p_\sigma}^{FC}$ values are not well known. At this point we express $\rho_{2p_\sigma}^{F}(+)$ as

$$\rho_{2p_\sigma}^{F}(+) = \frac{Q_{2p_\sigma}^{FC}}{B_{\text{atomic}}^{F}} \rho_\pi^F \tag{6.74}$$

where the $Q_{2p_\sigma}^{FC}$ is yet to be determined. The net 2p spin density of the σ-bonding orbital of the F is

$$\rho_{2p_\sigma}^{F} = \rho_{2p_\sigma}^{F}(+) + \rho_{2p_\sigma}^{F}(-) \tag{6.75}$$

Experimental values of the net spin density can be derived from the B_σ obtained by solution of (6.62) to (6.64). Values for selected radicals are given in the last column of Table 6.24. The fact that the values are all negative shows that the spin density induced by $\rho_\pi^C \approx 0.70$ is greater than that induced by $\rho_\pi^F \approx 0.11$.

Approximate values of $\rho_{2p_\sigma}^{F}(+)$ can be obtained from those for $\rho_{2\pi_\sigma}^{F}$ in Table 6.24 by the use of

$$\rho_{2p_\sigma}^{F}(+) = \rho_{2p_\sigma}^{F} + 0.022\, \rho_\pi^C \tag{6.76}$$

which follows from (6.72) and (6.75). Unlike ρ_π^F, the $\rho_{2p_\sigma}^{F}$ is sensitive to the value used for ρ_π^C. Until accurate ^{13}C couplings have been measured, one must rely on estimated values of ρ_π^C. Also, the factor 0.022 deduced earlier is only approximate. With the values of $\rho_{2p_\sigma}^{F}(+)$ thus obtained, one can derive rough values of $Q_{2p_\sigma}^{FC}$ from (6.74). For the first five radicals and CF_3 of Table 6.24, $Q_{2p_\sigma}^{FC} = 120, 23, 107, 66, 50$, and 45 MHz, respectively. Such large variations are certainly not real; they result mostly from the critical balancing of two small quantities of opposite sign, neither of which is accurately known. One source of error is the assumption that the principal axis z is along the CF σ bond. This will be strictly true only for the two radicals, $\dot{C}F_2CONH_2$ and $\dot{C}F_2COO(NH_4)$, for which the coupling is observed to be axially symmetric about the normal to the radical plane. Note that for these two radicals the largest $Q_{2p_\sigma}^{FC}$ values are obtained, whereas the lowest value is obtained for $\dot{C}HFCONH_2$, which has the largest deviation from axial symmetry. Thus it seems justifiable to omit from the average the lowest value as well as the next lowest, that for CF_3, which is nonplanar [71]. The mean value for the remaining four,

$$Q_{2p_\sigma}^{FC} = 86 \text{ MHz}$$

with (6.74) should be a useful indicator of the $\rho_{2p_\sigma}^{F}(+)$ induced by ρ_π^F spin density in the F of other free radicals.

The CF_3 Radical

The trifluoromethyl radical has been observed by Rogers and Kispert [72] in irradiated single crystals of trifluoroacetamide at 77°K. Rapid reorientations of the radical about the threefold symmetry axis equalized the coupling tensors for the three F nuclei. The ^{13}C splitting was also observed but could not be completely measured because of obscurity of the components for one orientation. Measurement of the isotropic ^{13}C coupling in liquid solution in C_2F_6 by Fessenden and Schuler [71] indicates that the radical is nonplanar. The esr spectra of CF_3 in a rare-gas matrix at 4.2°K [73,74] confirm its pyramidal structure and provide rather accurate principal values for the ^{19}F coupling. These values are given in Table 6.23 instead of the single-crystal values, which apparently are averaged values for radicals rotating about the molecular axis. For further discussion of esr parameters and structure, see Chapter VIII, Section 4.f.

In irradiated single crystals of ammonium trifluoroacetate, Srygley and Gordy [75] observed a secondary radical having three strongly coupling, almost equivalent, ^{19}F nuclei that they tentatively assigned as $(CF_3\dot{C}OONH_4)^-$ or $CF_3\dot{C}OOH$. However, it is not possible to explain the near-equivalence of the three β fluorines in these radicals without rather large torsional oscillations of the CF_3 group. A comparison of the principal elements of the ^{19}F couplings with those later observed for the CF_3 radical in the solid state [74] indicates that the radical they observed is probably the CF_3 radical. The principal coupling elements (in MHz) obtained for the three fluorines are $F^{(1)}$ = 711, 316, and 249; $F^{(2)}$ = 728, 288, and 184; and $F^{(3)}$ = 719, 238, and 186. The averaged values are 719, 281, 206 MHz, compared with those for the CF_3 radical, 739, 244, and 222, listed in Table 6.23. The measured isotropic component A_f = 402 MHz is exactly the same as that for the CF_3 radical. The small doublet splitting of the ^{19}F components that was observed indicates that there is also a weakly coupled proton, but this may be an environmental proton rather than one that is strongly bonded to the radical.

The SiF_3 Radical

Following observations of the esr of SiF_3 in polycrystalline matrices [76], Hasegawa, Sogabe, and Miura [77] succeeded in measuring the spectrum of this radical in an irradiated single crystal of SiF_4. The principal values and their relative directions were obtained separately for each of the three fluorines as well as for ^{29}Si. Despite the symmetry of the radical, the principal coupling values, although close, are not equal for the three fluorines because of unsymmetrical perturbation by the host crystal. In Table 6.23 are listed the averaged values for the corresponding coupling elements of the three fluorines.

An analysis of the ^{19}F coupling elements of SiF can be made by the same procedure as was used for analysis of the fluorocarbon radicals. The interatomic dipole-dipole terms differ somewhat from those of CF because of the differing interatomic distance and the large 3*s* character of the unpaired spin density on the Si: $\rho_{3s}^{Si} \approx 0.30$ with $\rho_{3p}^{Si} \approx 0.40$ indicated by the ^{29}Si coupling. To estimate the A_μ^{FSi} values, we have separately calculated the interaction with $\rho_{3s}^{Si} = 0.30$ spin density by use of (6.12) and with $\rho_{3p}^{Si} = 0.40$ by use of (6.13), (6.15), and (6.16). With $R_{SiF} = 1.56$ Å and $R_p^{Si} = (1.17 + 1.08)/2 = 1.125$ Å [from (6.17)], the interatomic dipole-dipole values $(A_\mu^{FSi})_x = -6$ MHz, $(A_\mu^{FSi})_y = -10$ MHz, and $(A_\mu^{FSi})_z = 16$ MHz were obtained. Substitution of these values with the observed anisotropic components (Table 6.23) into (6.62) to (6.64) gives

$$228 = 2B_\pi - B_\sigma - 6$$

$$-77 = -B_\pi - B_\sigma - 10$$

$$-150 = -B_\pi + 2B_\sigma + 16$$

from which $B_\pi = 100$ MHz and $B_\sigma = -33$ MHz. These are listed in Table 6.24 along with the spin densities derived from them. The uncertainties in these values are due mostly to the uncertainty in the calculated dipole-dipole terms. Since these quantities are much smaller than the observed principal values, they cannot cause large errors in the derived B_π and B_σ. Like the CF_3 radical, SiF_3 is pyramidal in structure [77].

Isotropic Coupling

The isotropic coupling of ^{19}F is so uncertain in origin that it does not provide a reliable source of information about the chemical bonding. It is positive in sign and rather large in radicals such as those listed in Table 6.23. The positive sign indicates that the spin-polarization mechanism that puts negative spin density in the σ-bonding orbital of the F is certainly not the dominant term. If there is no 2*s* character in this orbital, as assumed previously, no isotropic contribution from this source is expected. Indeed, the isotropic coupling may result almost entirely from configuration interaction on the F itself. Radford, Hughes, and Beltran-Lopez [78] observed an isotropic component having about 7.4% of the total coupling in atomic F, which they attributed to an admixture of $2s2p^5ns$ with the ground configuration of $2s^22p^5$. This type of interaction would contribute a positive isotropic coupling term. The configuration interaction in the F atom has been treated theoretically by Goodings [79], who predicted 1*s* and 2*s* spin densities that indicate a resultant $Q_{\text{iso}}^{\text{F}} = +560$ MHz [80]. With $\rho_\pi^{\text{F}} \approx 0.13$, this gives $A_f^{\text{F}} = Q_{\text{iso}}^{\text{F}} \rho_\pi^{\text{F}} \approx 73$ MHz, which is appreciably lower than the approximately 200 MHz reported for the planar radicals in Table 6.23. However, configuration interaction on the bonded atom is expected to differ measurably from that of the free atom.

Note that CF_3 has $A_f^{\text{F}} = 402$ MHz, approximately twice that of the π-type radicals in Table 6.23. The unpaired spin density on the C in this nonplanar radical has an sp^3-hybridized orbital having about 21% 2*s* character [71]. This large 2*s* unpaired electron density on the C may undergo direct exchange interaction with the 2*s* pair on the F to generate positive spin density on the F and thus contribute to the positive isotropic coupling. A similar explanation can be given for the large $A_f^{\text{F}} = 376$ MHz observed for SiF_3, in which the orbital of the unpaired spin density on the Si has appreciable 3*s* character [77].

Although one cannot be sure of the nature of the various interactions that cause the rather large isotropic ^{19}F coupling, it must come primarily from positive spin density of the 2*s* orbital of the F. The σ-bond spin-polarization mechanism as well as atomic configuration interaction with the 1*s* orbital would contribute a negative isotropic term. Such contributions must be small unless there is a very large positive 2*s* spin density. For these reasons, it appears that the $A_f/A_{\text{atomic}} = A_f\,(\text{MHz})/47{,}910$ should represent a good approximation to the positive 2*s* spin density on the coupling F. These values are in the range of $\rho_{2s}^{\text{F}} \approx 0.004$ for all radicals in Table 6.23 except CF_3, for which $(A_f/A_{\text{atomic}}) \approx 0.008$.

Several attempts to calculate isotropic coupling parameters of ^{19}F in organic radicals have disagreed widely in prediction of Q^{F} or Q^{CF} values, also in the models employed [80-83]. Considering the complexity of the problem, this is not surprising.

7.c. β^{19}F Coupling

Perhaps the simplest illustration of βF coupling is that for $CF_3\dot{C}FCONH_2$, in which the internal rotation of the CF_3 group at room temperature equalizes the coupling of the three fluorines. Figure 6.10 shows the hyperfine pattern obtained for one orientation of the host crystal. The αF coupling produces a wide doublet, each component of which is split into a quartet by the three equally coupling $C_\beta F_3$ fluorines. One would expect β coupling caused entirely by hyperconjugation to be symmetrical about the $C_\alpha C_\beta$ axis of rotation. Although the observed coupling does not have complete axial symmetry, the primary terms in the coupling evidently have axial symmetry.

Hyperconjugation gives rise to positive spin density in the σ-bonding orbital of the βF. The mechanism is that already described in Section 1.c. The gross difference in the βF and βH coupling is due to the nature of the σ-bonding orbital,

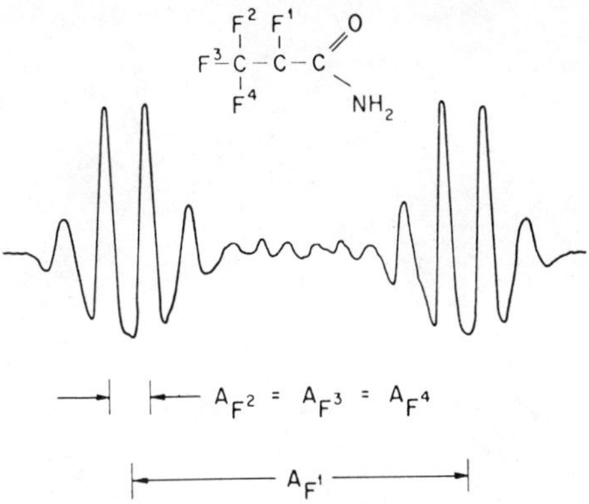

Fig. 6.10 A representative spectrum of the $CF_3\dot{C}FCONH_2$ free radical showing two quartet patterns of the β coupling by the rotating CF_3 group separated from the αF coupling of the CF group. From Lontz [84].

which is essentially $2p_\sigma$ in F and is $1s_\sigma$ in H. For this reason, the βF coupling is strongly anisotropic, in contrast to the βH coupling which is nearly isotropic. Because it is due to $2p_\sigma$ spin density, the βF coupling caused by hyperconjugation should be axially symmetric about the direction of the $C_\beta F$ bond and should have its maximum component in this direction when the bond has a fixed direction. When the $C_\beta F$ is rotating rapidly about $C_\alpha C_\beta$, this coupling will be resolved along the axis of rotation, which will become an axis of symmetry for the observed coupling tensor.

The $2p_\sigma$ β-coupling with reference to a fixed $C_\alpha F$ [see (6.24)] may be expressed as

$$(A_\mu^\sigma)_z = 2B_2^\sigma \cos^2\theta \tag{6.77}$$

and

$$(A_\mu^\sigma)_x = (A_\mu^\sigma)_y = -B_2^\sigma \cos^2\theta \tag{6.78}$$

where z is the $C_\beta F$ direction, θ, the dihedral angle of $C_\beta F$ with the $C_\alpha C_\beta P_\pi$ plane as shown for the $C_\beta H$ group in Fig. 6.1, and B_2^σ is the p-orbital coupling constant for the $\theta = 0$ orientation. The component of $(A_\mu^\sigma)_z$ resolved on the axis of rotation, designated c, is

$$(A_\mu^\sigma)_c = (A_\mu^\sigma)_z \, \frac{1}{2}\,(3\cos^2\alpha - 1)$$

$$= B_2^\sigma \cos^2\theta\,(3\cos^2\alpha - 1) \tag{6.79}$$

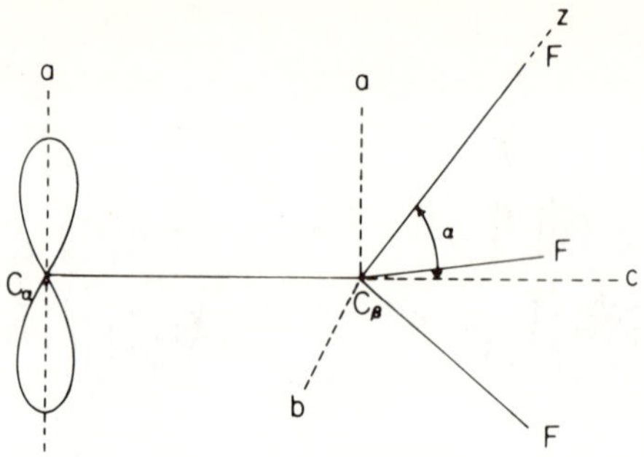

Fig. 6.11 Coordinate system for calculation of β couplings.

where α is the angle between z and c, that is, between $C_\beta F$ and the direction of $C_\alpha C_\beta$, as indicated in Fig. 6.11. In the rotating $C_\beta F$ groups, the $(\cos^2\theta)_{av} = 1/2$; therefore, the coupling along the axis of rotation should be

$$(A_\mu^\sigma)_c^{rot} = \frac{1}{2} B_2^\sigma (3\cos^2\alpha - 1) \tag{6.80}$$

Furthermore, the rotation will symmetrize this coupling about the axis of rotation so that

$$(A_\mu^\sigma)_a^{rot} = (A_\mu^\sigma)_b^{rot} = -\frac{1}{2}(A_\mu^\sigma)_c^{rot} \tag{6.81}$$

With tetrahedral bond angles assumed for the C_β, the angle α equals 70°, and for the rotating radical we have

$$(A_\mu^\sigma)_c^{rot} = -\frac{1}{3} B_2^\sigma \tag{6.82}$$

which with the expected positive B_2^σ would give a negative coupling along the axis of rotation.

If this hyperconjugative coupling were the only anisotropic $\beta^{19}F$ coupling, one could equate the observed $(A_\mu)_\parallel = -68$ MHz on the right side of (6.82) and find B_2^σ and hence $\rho_{2\pi_\sigma}^F$. However, the observed βF coupling for $CF_3\dot{C}FONH_2$ does not have complete axial symmetry, and evidently hyperconjugation is not the only coupling mechanism involved. The interatomic dipole-dipole component is negligibly small and cannot cause the observed asymmetry. It may be caused by a π-π interaction of the unpaired electron density ρ_π^α on C_α with one of the "unshared" pairs on the F. This interaction, similar to the one that puts positive π spin density on the αF, would produce a coupling $2p_\pi(F)$ spin density in the plane of the $2p_\pi$ orbital on C_α, even for the rotating $C_\beta F_3$ group. Therefore, this interaction could account for the observed asymmetry.

Although the $2p_\pi$ spin density on the βF is also expected to depend on the dihedral angle θ, it should retain a direction in the plane of the $2p_\pi$ spin density on C_α. It will give rise to an averaged positive component $2\langle B_\pi \rangle_{av}$ in this plane, which is approximately normal to the axis of rotation. Thus we make the approximation

$$(A_\mu^\pi)_a = 2\langle B_\pi \rangle \tag{6.83}$$

and

$$(A_\mu^\pi)_b = (A_\mu^\pi)_c = -\frac{1}{2}(A_\mu^\pi)_c = -\langle B_\pi \rangle \tag{6.84}$$

where the reference axes a, b, and c are as indicated in Fig. 6.11.

Since both B_2^σ and $\langle B_\pi \rangle$ are expected to be positive, the preceding theoretical considerations indicate that the observed anisotropic couplings should be negative for H parallel to the axis of rotation and positive for H perpendicular to this direction. Although the sign of these couplings could not be ascertained from the symmetrical β couplings observed, here we assume it to be in agreement with this theory and opposite to that assumed in the original investigation [84]. The observed principal values would then all be negative and would give $A_f = -63$ MHz, with the principal anisotropic components $(A_\mu^\beta)_i = -38, +15$, and $+24$ MHz for the $C_\beta F_3\dot{C}_\alpha FCONH_2$ radical at room temperature. By use of these values with the σ and π contributions of (6.81) to (6.84), we obtain for the rotating $C_\beta F$:

$$(A_\mu^\beta)_c^{rot} = -\frac{1}{3}B_2^\sigma - \langle B_\pi \rangle = -38 \tag{6.85}$$

$$(A_\mu^\beta)_b^{rot} = \frac{1}{6}B_2^\sigma - \langle B_\pi \rangle = 24 \tag{6.86}$$

$$(A_\mu^\beta)_a^{rot} = \frac{1}{6}B_2^\sigma + 2\langle B_\pi \rangle = 15 \tag{6.87}$$

The solution of these equations yields

$$B_2^\sigma = +105 \text{ MHz and } \langle B_\pi \rangle = +3 \text{ MHz}$$

and the corresponding spin densities $\rho_{2p_\sigma}^F\ (\theta = 0) = +0.069$ and $\langle \rho_{2p_\pi}^F \rangle = +0.0020$.

It is surprising that the isotropic term A_f is negative and so small in magnitude. Since the $2p_\sigma$ spin density is positive, the $A_f = -63$ MHz may come from a very slight admixture of the 1*s* orbital with the higher *ns* orbitals (see Section 4.a). Because $(A_{1s})_{\text{atomic}} = 9 \times 10^5$ MHz, a 1*s* spin density of only $\rho_{1s} \approx -63/9 \times 10^5 = 7 \times 10^{-5}$ would be required to give the observed A_f. Even if the A_f were entirely from 2*s* spin density, the amount would be only 13×10^{-4}. This upper limit to the magnitude of the 2*s* spin density is less than 2% of the 2*p* spin density found for the σ-bonding orbital and is in agreement with the assumption made in Section 7.b that the σ-bonding orbitals of F are essentially pure 2*p* orbitals.

8. MAGNETIC HYPERFINE STRUCTURE OF Cl, Br, AND I

These halogens have nuclear spins greater than 1/2 and hence have nuclear quadrupole coupling in addition to magnetic coupling. Although the quadrupole interaction is not normally detected in the first-order spectrum, it can cause small, second-order shifts of the first-order transitions and give rise to second-

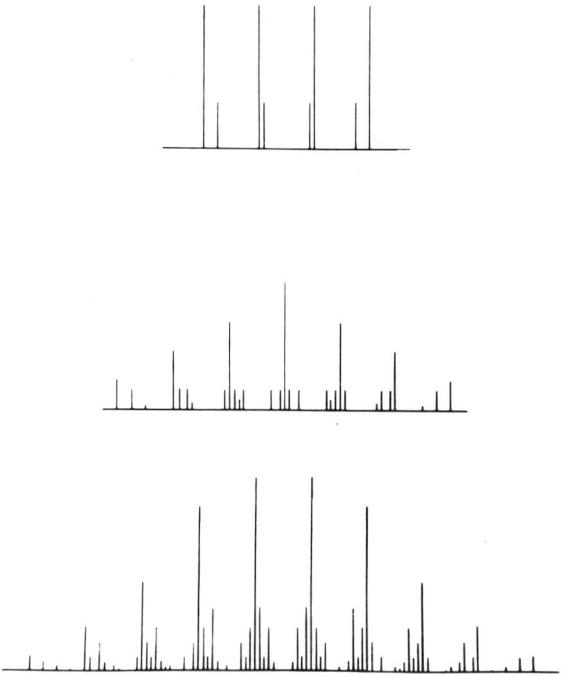

Fig. 6.12 Theoretical hyperfine patterns for one, two, and three equivalent Cl nuclei with ^{35}Cl and ^{37}Cl isotopes in their natural abundance. From Hudson and Root [86].

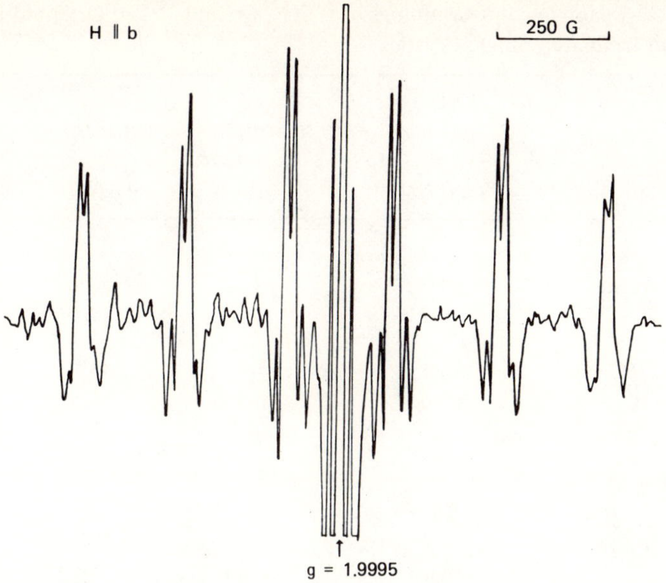

Fig. 6.13 ESR spectrum showing ^{127}I hyperfine structure of the $\dot{C}IHCONH_2$ radical in a γ-irradiated single crystal of pentafluoropionamide at 77°K. The magnetic field is parallel to the *b* axis of the monoclinic crystal. The sextet pattern of ^{127}I, spin 5/2, is split into doublets by the CH proton coupling. The central absorption is an impurity spectrum from the glue of the crystal mount. From Picone and Rogers [85].

order transitions (Chapter V, Section 4), which often complicate the assignment of components. From these second-order transitions or from the small second-order shifts of the first-order lines, the halogen quadrupole coupling constants have been obtained for some of the oriented free radicals to be described here.

The interpretation of the magnetic coupling for chlorine, bromine, and iodine is similar to that for fluorine and is not particularly difficult. However, the identification and assignment of the hyperfine components is complicated for Cl and Br by their two isotopes with unequal concentrations and differing magnetic moments. This complication is indicated by the theoretical patterns for the first-order spectrum of Cl radicals in Fig. 6.12. The relative intensities of the ^{35}Cl and ^{37}Cl components are determined by their natural abundance of 75% and 25%, respectively. Because there is only one stable iodine isotope, the hyperfine patterns of iodide free radicals are a bit simpler than those of chlorine or bromine. Figure 6.13 shows the six components of the hyperfine structure of ^{127}I (spin 5/2) as observed by Picone and Rogers [85] for the radical $\dot{C}IHCONH_2$ in an irradiated single crystal of iodoacetamide. The doublet splitting of the components is from the αH. Theoretically, the six components should be of equal intensity, but the observed intensities are influenced by experimental

Table 6.25 Examples of α Couplings of ^{35}Cl, ^{81}Br, and ^{127}I of Oriented Free Radicals in Irradiated Single Crystals

Radical / *Host Crystal*	*Principal Value* A_x, A_y, A_z *(MHz)*	*Isotropic Component* A_f *(MHz)*	*Anisotropic Component* A_μ *(MHz)*	*Ref.*
$\dot{C}^{35}ClH_2$	57.54		49.69	
Methyl chloride	−14.78	7.85	−22.63	*a*
	−19.21		−27.06	
$\dot{C}^{35}ClHCO_2H$	56.1		45.7	
$ClCH_2CO_2H$	−7.0	10.4	−17.4	*b*
	−17.9		−28.3	
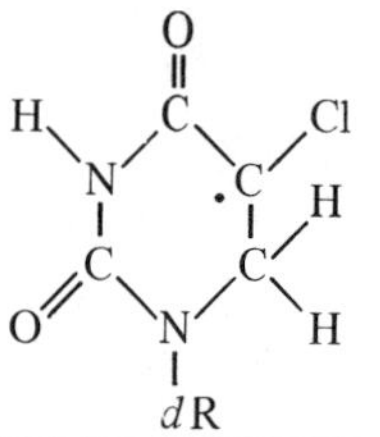	49.99		40.66	
	−10.99	6.33	−17.32	*c*
5-Chlorodeoxyuridine	−17.01		−23.34	
$C_5{}^{35}Cl_5$	18.45		10.06	
Cyclododecane	3.64	8.39	−4.75	*d*
	3.08		−5.33	
$\dot{C}^{81}BrHCONH_2$	810		384	
Bromomalonamide	235	427	−192	*e*
	235		−192	
$\dot{C}IHCONH_2$	678		273	
Iodoacetamide	269	405	−136	*f*
	269		−136	
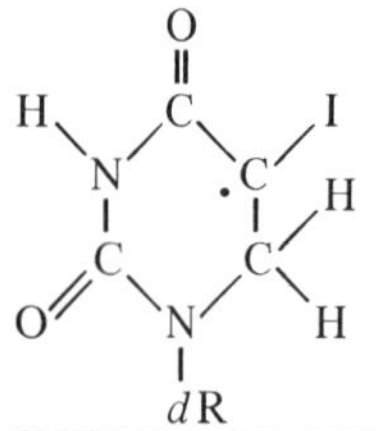	252		84	
	140	168	−28	*g*
5-Iodeoxyuridine	112		−56	

[a]J. P. Michant and J. Roncin, *Chem. Phys. Lett.*, **12**, 95 (1971).
[b]R. P. Kohn, *J. Chem. Phys.*, **50**, *5356 (1969)*.
[c]J. Huttermann, W. A. Bernhard, E. Haindl, and G. Schmidt, *Molec. Phys.*, **32**, 1111 (1976).
[d]P. Bachmann, F. Graf, and H. H. Gunthard, *Chem. Phys.*, **9**, 41 (1975).
[e]Picone and Rogers [88].
[f]Picone and Rogers [85].
[g]J. Huttermann, G. W. Neilson, and M. C. R. Symons, *Molec. Phys.*, **32**, 269 (1976). Values with (+++) rather than (+−−).

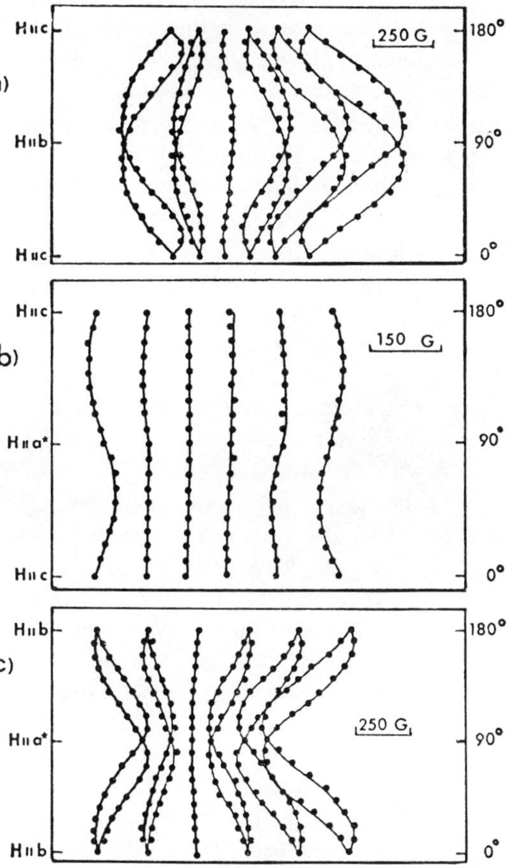

Fig. 6.14 Magnetic-field positions of the ^{127}I lines of the $\dot{C}IHCONH_2$ radical plotted as a function of the field orientation in the three orthogonal reference planes of the host crystal, iodoacetamide: (a) in *bc* plane, (b) in *a**c* plane, (c) in *a**b* plane. From Picone and Rogers [85].

conditions and the superimposed absorption of other radicals in the central region. Picone and Rogers were able to measure all the components in three orthogonal reference planes of the crystal, as shown in Fig. 6.14. An accurate evaluation of the coupling tensor of ^{127}I was made from a fitting of these curves. The resulting principal values are listed in Table 6.25. Relatively few halogen hyperfine patterns have been observed and analyzed in irradiated organic single crystals. Many observations of halogen hyperfine structures have, however, been made in liquids, in salts of the transition elements, and in irradiated inorganic salts. Comprehensive reviews of these studies are available [86,87].

In Table 6.25 are listed the principal elements of the nuclear magnetic coupling tensors of ^{35}Cl, ^{81}Br, and ^{127}I of selected molecular free radicals in irradiated single crystals. Because of the complexity of the composite hyperfine structure in such crystals, the assignments and analysis are often tentative. Nevertheless, the selected data sets given in Table 6.25 appear to be reasonably accurate and internally consistent. Where the signs of the principal values could not be experimentally verified, we have listed the signs that allowed consistent solutions of (6.62) to (6.64). Generally, these signs also correspond with the more reasonable spin-density values.

Because of the large interatomic distances of CCl, CBr, and CI, the dipole-dipole interaction of the nucleus with the spin density on the C_α is almost negligible. From (6.13), (6.15), and (6.16) the terms contributing to $(A_\mu)_x$, $(A_\mu)_y$, and $(A_\mu)_z$, respectively, are (in MHz units): $-0.33\ \rho_\pi^C$, $-0.97\ \rho_\pi^C$, $+1.30\ \rho_\pi^C$ for $C_\alpha{}^{35}$Cl; $-0.34\ \rho_\pi^C$, $-1.94\ \rho_\pi^C$, $+2.48\ \rho_\pi^C$ for $C_\alpha{}^{81}$Br; and $-0.21\ \rho_\pi^C$, $-1.00\ \rho_\pi^C$, $+1.23\ \rho_\pi^C$ for $C_\alpha{}^{127}$I. To obtain numerical values for these terms, we have used ρ_π^C as given by the original observer or as derived from the αH coupling and have substituted these values with the principal coupling values of Table 6.25 into (6.62) to (6.64) and have thus obtained the anisotropic coupling constants B_π and B_σ given in Table 6.26. Obviously, this analysis is not sensitive to possible inaccuracies in the small dipole-dipole terms that depend on ρ_π^C. Furthermore, one can hardly err in choosing the larger (positive) principal element as perpendicular to the C-Hal (carbon-halogen) bond, even if this could not be determined from analysis of the original data. The B_π value obtained is much too large to be attributed to spin polarization and hence cannot be confused with B_σ.

The ρ_π^{Hal} spin density derived from B_π gives a direct measure of the contribution of the double-bonded structures $>C^-{=}Hal^+$ to the ground state of the radical. It is interesting to compare the weights of these structures in the similar radicals H$\dot{C}$(Hal)R. As determined by the $\rho_{np\pi}^{\text{Hal}}$ values in Tables 6.24 and 6.26, they are: $>C^-{=}F^+$ ~11%, $>C^-{=}Cl^+$ ~15%, $>C^-{=}Br^+$ ~25%, and $>C^-{=}I^+$ ~20%. One may attribute variations in this "ionic," double-bond character partly to the decreasing electronegativity of the halogen from fluorine to iodine and partly to the changes in orbital overlap. The reversal of the trend at CBr cor-

Table 6.26 Halogen Coupling Constants and Spin Densities Derived from the Principal Coupling Elements in Table 6.25

Radical	Coupling Halogen	Coupling Constant (MHz) B_π	B_σ	Spin Density[a] ρ_{np_π}	ρ_{np_σ}
$\dot{C}ClH_2$	^{35}Cl	24.0	-2.13	0.172	-0.015
$\dot{C}ClHCO_2H$	^{35}Cl	20.9	-4.17	0.150	-0.030
(dR)$CONHCO\dot{C}ClCH_2$ (ring)	^{35}Cl	19.2	-2.54	0.137	-0.018
C_5Cl_5	^{35}Cl	4.85	-0.53	0.036	-0.004
$\dot{C}BrHCONH_2$	^{81}Br	192	-0.8	0.25	-0.001
$\dot{C}IHCONH_2$	^{127}I	136	-0.6	0.21	-0.0009
(dR)$CONHCO\dot{C}ICH_2$ (ring)	^{127}I	38	-8.8	0.057	-0.013

[a] Atomic coupling constants employed in derivation of the spin densities are those in Table A.1, with no charge corrections applied.

responds to a decrease in the overlap integral from CBr to CI. Picone and Rogers [88] give estimated values of the overlap integral $S(2p_\pi, np_\pi) = 0.122$, 0.147, 0.150, and 0.115 for $n = 2, 3, 4$, and 5, respectively.

Like those for the $C_\alpha F$ fragments (Table 6.24), all the σ-orbital spin densities on Cl, Br, and I are negative and small in magnitude (Table 6.26). Because of the decreasing ionic character of the σ bonds from CF to CI, the negative spin density in the σ orbital of the halogen induced by spin polarization of the covalent component by the ρ_π^C should increase from CF to CI. This effect is noticeable in the comparison of the similar radicals $\dot{C}HFCONH_2$ ($\rho_{2p_\sigma}^F = -0.017$ and $\dot{C}ClHCOOH$ ($\rho_{3p_\sigma}^{Cl} = -0.030$), where the covalent character of the σ bonds increases from about 20% to about 75%. In the corresponding bromide and iodide radicals, the spin polarization induced by ρ_π^C is almost canceled by the opposite spin polarization induced by the rather large ρ_π^{Hal}. This cancellation causes the anisotropic coupling tensors for $\dot{C}BrHCONH_2$ [88] and $\dot{C}IHCONH_3$ [85] to be axially symmetric within the accuracy of the observation (Table 6.25). Unlike that in $\dot{C}F_2CONH_2$, the interatomic coupling in these radicals is too small to cancel the significant anisotropy caused by B_σ. Thus, assuming pure p orbitals on the halogen, we can set $Q_\sigma^{CBr}\, \rho_\pi^C \approx Q_\sigma^{BrC}\, \rho_\pi^{Br}$ and $Q_\sigma^{CI}\, \rho_\pi^C \approx Q_\sigma^{IC}\, \rho_\pi^I$. With ρ_π^C equal to 0.76 for the bromide and to 0.70 for the iodide, as indicated by the $C_\alpha H$ coupling, one finds that $Q_\sigma^{BrC} \approx 3.0\, Q_\sigma^{CBr}$ and $Q_\sigma^{IC} \approx 3.5\, Q_\sigma^{CI}$. Because the σ bonds of CH, CBr, and CI have approximately the same covalent character, we assume that $Q^{CBr} \approx Q^{CH} = 73.4$ MHz and find that $Q_\sigma^{BrC} \approx 220$ MHz. Similarly, $Q_\sigma^{IC} \approx 257$ MHz. In Section 7.b we obtained the averaged value $Q_\sigma^{FC} \approx 86$ MHz. If one multiplies this value by 0.85/0.35 to normalize the covalent character of the CBr and CF, the result is 209 MHz, close to the estimated Q^{BrC}. Of course, one does not expect the configuration interaction to be the same in the different halogens. These comparisons show only that they are of the same order of magnitude. The rather large, negative, $\rho_\sigma^{\mathrm{I}} = -0.013$ for the iodouracil radical is not entirely consistent with the value of -0.0009 for $\dot{\mathrm{C}}\mathrm{IHCONH_2}$, but this is not too surprising when one considers the differences in the structure of the two species.

The isotropic component of the coupling A_f^{Hal}, in all the radicals under consideration is positive and is comparable to the isotropic couplings for αF, in Table 6.23. Because of the possibility of a negative term S^{Hal} from subvalence s electrons (Section 4.a), one cannot calculate accurately the ns spin density of the valence shell from the observed A_f^{Hal}. If, as is commonly done, one attributes the entire isotropic coupling to ns spin density of the valence shell, the required value for all the chlororadicals in Table 6.25 is $\rho_{3s}^{\mathrm{Cl}} \approx 2 \times 10^{-3}$, that for $\dot{\mathrm{C}}\mathrm{BrHCONH_2}$ is $\rho_{4s}^{\mathrm{Br}} = 15 \times 10^{-3}$, and that for $\dot{\mathrm{C}}\mathrm{IHCONH_2}$ is $\rho_{5s}^{\mathrm{I}} \approx 16 \times 10^{-3}$. The much larger value for the Br and I radicals may result from a slight deviation from planarity like that described for CF_3. In consideration of the isotropic halogen coupling, one can neglect any negative ns contribution from spin polarization of the σ bonds because the bonding orbitals of Cl, Br, and I have no significant ns hybridization [36].

9. PHOSPHORUS-CENTERED RADICALS

Three classes of phosphorus-centered radicals have been identified. The three classes may be distinguished by the number of atoms bonding to the phosphorus—two, three, or four. The simplest case, exemplified by PF_2, may be formed by abstraction of an F from PF_3. As far as I am aware, no measurements of oriented PF_2 in single crystals have been made, but good values of the principal coupling elements of both ^{31}P and ^{19}F have been obtained by measurements of the radical trapped in inert matrices at 4.2°K [89]. These show that the unpaired electron is in a π orbital perpendicular to the molecular plane. This structure is like that for NF_2 similarly found from the ^{14}N and ^{19}F in inert matrices [90] (see discussion in Chapter VIII, Section 4.e).

The second type of phosphorus-centered radical is illustrated by $PO_3^{=}$, first observed in irradiated single crystals of disodium orthophosphite pentahydrate by Horsfield, Morton, and Whiffen [91]. These measurements reveal that the unpaired electron is predominantly in an sp^3-hybrid orbital on the P. The radical is more closely represented as $P^+O_3^{\equiv}$, which has axial C_{3v} symmetry with an equal negative charge of nearly unity on each oxygen. It is isoelectronic with SiF_3 (Section 7.b) and has a similar structure. The orbital of the unpaired electron on the P is directed along the symmetry axis of the pyramidal structure. Horsfield and colleagues obtained the following principal values of the ^{31}P coupling:

967 MHz, 1514 MHz, and 1513 MHz. With these values they calculated spin densities on the P: ρ_{3s}^{P} = 0.16 and ρ_{3p}^{P} = 0.527; by deduction they obtained an unpaired spin density of 0.103 on each oxygen. From the ratio of the ρ_{3s}^{P} and ρ_{3p}^{P}, they estimated the OPO bond angle to be 110°, the same as that in the $HPO_3^=$ ion in its magnesium salt [92]. However, in making their estimate of the unpaired 2*s* and 3*p* densities they apparently neglected the effects of spin polarization of the σ bonds on the isotropic coupling and the rather large positive charge effect of the P on the coupling parameters. Without measurements of the ^{17}O hyperfine structure, an attempt to give a more quantitative description of the electronic structure is hardly justified.

Other examples of phosphorus-centered radicals with three atoms or groups bonded to the P are $O\dot{P}(OH)_2$ [93], $C_6H_5\dot{P}(O)OH$ [94], and $(HO)_2(O)PCH_2\dot{P}$ (O)OH [93]. The bond structure and unpaired-electron orbital of the P are evidently similar in all three radicals to that of $PO_3^=$. The ^{31}P coupling constants in $OP(OH_2)$ are close to those of $PO_3^=$.

The electronic structure of the third type of phosphorus-centered radical, that in which four bonds are formed to the central phosphorus, is quite complex and very interesting. Our knowledge of the radical has evolved from the isotropic esr of rotating PF_4, first observed by Morton [95] in γ-irradiated NH_4PF_6 at room temperature and later by Fessenden and Schuler [96] in the SF_6 matrix at −135°K. The latter work revealed that the radical has two pairs of symmetrical fluorine nuclei with equivalent isotropic coupling, one pair with coupling much larger than that of the other pair. These observations provided the impetus for a theoretical treatment of the structure by Higuchi [97]. His treatment, which included both the valence-bond and the molecular-orbital descriptions, began with the assumption that the two PF_2 groups are in orthogonal planes having C_{2v} symmetry. To achieve the pentavalent state, the *d*-shell orbitals of the phosphorus are included in the wave functions of the radicals. Using the isotropic couplings obtained by Fessenden and Schuler, 790 MHz for the two equivalent $F^{(1)}$s and 165 MHz for the $F^{(2)}$s, Higuchi calculated $\angle F^{(1)}PF^{(1)\prime}$ to be 177° and $\angle F^{(2)}PF^{(2)\prime}$ to be 102° with valence-bond theory and 172° and 113°, respectively, with molecular-orbital theory.

Later, Nelson, Jackel, and Gordy [89] observed the PF_4 in the PF_3 polycrystalline matrix at 4.2°K. They found the coupling of both the ^{31}P and the $F^{(1)}$ to be anisotropic and that of the $F^{(2)}$s to be too small to measure. The $F^{(1)}$ isotropic components, $A_{iso}(F^{(1)})$ = 1065 MHz, was noticeably larger than that found earlier in the SF_6 matrix [96] and the $A_{iso}(^{31}P)$ = 3898 MHz only slightly larger than the earlier value, 3727 MHz. The anisotropic components obtained were axially symmetric, with $A_{\parallel}$ = 79 MHz, $A_{\perp}$ = 39 MHz for $F^{(1)}$, and with $A_{\parallel}$ = 336 MHz, $A_{\perp}$ = 168 MHz for ^{31}P. By fitting their observed isotropic components to the molecular-orbital theory of Higuchi, these observers obtained $\angle F^{(1)}PF^{(1)\prime}$ = 180° and $\angle F^{(2)}PF^{(2)\prime}$ = 102°. Thus the $F^{(1)}PF^{(1)\prime}$ is linear in the

PF_3 matrix, or approximately so, and the $PF_2^{(2)}$ is in the plane orthogonal to the $F^{(1)}PF^{(1)\prime}$ direction. Comparison of the anisotropic components for PF_4 in the polycrystalline PF_3 matrix with later values obtained from single-crystal measurements indicates that the anisotropic components observed in the polycrystalline PF_3 matrix were symmetrized and reduced by rotation of the radicals about preferred axes even at 4.2°K.

Hasegawa, Ohnishi, et al. [98] have succeeded in growing single crystals of PF_3 at 77°K in which PF_4 radicals were produced by γ-irradiation. Measurement of the esr hyperfine structure of both ^{31}P and ^{19}F revealed that the radicals maintained fixed orientations without rotating. Thus these observers were able to measure the anisotropic couplings of both ^{31}P and ^{19}F, including the equitorial as well as the axial fluorines. The resulting principal coupling elements are listed in Table 6.27. These measurements provided considerable new information about the bonding of pentavalent phosphorus but at the same time revealed that this bonding is exceedingly complex. Earlier, Gillbro and Williams [99] had obtained information of a similar nature from measurements on oriented $POCl_3^-$ in irradiated single crystals of phosphorus oxychloride. They found a rather large anisotropic component in the coupling of the axial chlorines which indicated a positive $3p_\sigma$ spin density $\rho_{3p_\sigma}^{Cl} \approx 0.29$, much too large to be caused by spin polarization of the σ bonds. They concluded that the unpaired electron is largely in a three-centered molecular orbital directed along the axial $Cl^{(1)}PCl^{(1)\prime}$ direction. It is surprising that no anisotropic ^{31}P coupling was reported, although a large isotropic component indicating $\rho_{3s}^{P} \approx 0.376$ was observed. Except for this difference, these measurements seem to be in accord with the more complete observations of Hasegawa and co-workers on PF_4. The coupling values from both studies are listed in Table 6.27.

With the spin densities obtained from the single-crystal measurements, Hasegawa, Ohnishi, et al. [98] constructed the diagrams showing the distribution of spin density in the PF_4 and $POCl_3^-$ radicals, as shown in Fig. 6.15. The total spin density of the PF_4 is greater than unity, and hence the numbers do not represent true values of the spin densities in the different orbitals, but only approximate values. The inaccuracies result from the neglect of effects of spin polarization, interatomic dipole-dipole coupling, uncertainties in atomic-orbital coupling, incomplete charge corrections, and so on. Nevertheless, these diagrams have relative significance and provide useful qualitative models of the radical structures. The principal difference between the two diagrams is the absence of anisotropic ^{31}P spin density in the $POCl_3^-$. The results on PF_4 suggest that an anisotropic component should be directed in the equitorial plane along the x axis. Note that the phosphoranyl radicals listed in Table 6.27 have anisotropic ^{31}P coupling components similar to those for PF_4. An extensive discussion of the nature of the bonding in the $POCl_3^-$ and related phosphoranyl radicals is given by Gillbro and Williams [99], who emphasized the applicability of Rundle's molecular-orbital theory [100].

Table 6.27 Hyperfine Coupling Constants of Some Phosphorus-centered Free Radicals

Radical / Host Crystal	Coupling Atom	Coupling Constants (MHz) Isotropic A_f	Anisotropic $(A_\mu)_x$	$(A_\mu)_y$	$(A_\mu)_z$	Ref.
PF_4 / Phosphine	^{31}P	3670	+360	-180	-180	*a*
	^{19}F (axial)	857.8	+93.0	-295	+386	
	^{19}F (equi-torial)	170.7	+81.8	-40.9	-40.9	
$POCl_3^-$ / Phosphorus oxychloride	^{31}P	3842				*b*
	^{35}Cl (axial)	112	-84	-84	+168	
	^{35}Cl (equi-torial)	50				
$(HO)_3\dot{P}CH_2PO(OH)_2$ / Methylene diphosphonic acid	^{31}P	2556	+367	-168	-196	*c*
$(HO)_3\dot{P}CH_2CH_2PO(OH)_2$ / Ethylene diphosphonic acid	^{31}P	2287	+370	-171	-199	*d*

[a] Hasegawa, Ohnishi, Sogabe, and Miura [98].
[b] Gillbro and Williams [99].
[c] Geoffroy, Ginet, and Lucken [93].
[d] Whelan [101].

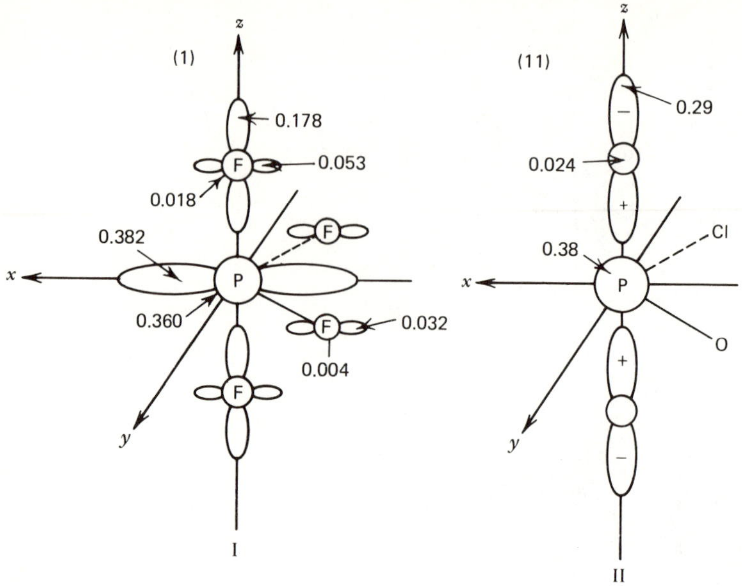

Fig. 6.15 Electronic structure of PF_4 radical I and the $POCl_3^-$ radical II as indicated by their epr spectra. From Hasegawa, Ohnishi, Sogabe, and Miura [98].

The phosphoranyl radicals $R\dot{P}(OH)_3$ are neutral radicals formed simply by H addition to an oxygen of the phosphoric acid. Rather accurate principal values for the ^{31}P coupling are available for the two listed in Table 6.27. If possible *d* contribution to the anisotropic coupling is neglected, the 3*p* spin densities in the phosphorus can be readily derived from the anisotropic components by use of the relations:

$$(A_\mu)_x = 2B_\pi - B_\sigma \tag{6.88}$$

$$(A_\mu)_y = -B_\pi - B_\sigma \tag{6.89}$$

$$(A_\mu)_z = -B_\pi + 2B_\sigma \tag{6.90}$$

The B_π and B_σ values listed in Table 6.28 were obtained from solutions of these equations. The corresponding spin densities are $\rho_{3p}^{P} = B/1.2B_{\text{atomic}}$, where $B_{\text{atomic}} = 287$ MHz and the charge factor 1.2 corrects for an estimated charge on the P of $c^+ = 1.0$ that arises from ionic character of the three P-O(H) bonds (see Section 6). These corrections cause the primary difference between the ρ_{3p_π} spin densities obtained here from those given by the original observers [93,101]. They derived no $\rho_{3p_\sigma}^{P}$ values. Because of the uncertainty in the amount, no

Table 6.28 Comparison of Phosphorus Spin Densities in Some Phosphorus-centered Radicals[a]

Radical	B_π	B_σ	ρ^{P}_{3s}	$\rho^{\mathrm{P}}_{3p_\pi}$	$\rho^{\mathrm{P}}_{3p_\sigma}$
PF_4	180	0	0.36	0.48	0
$(HO)_3\dot{P}CH_2PO(OH)_2$	178	−9.3	0.25	0.52	−0.027
$(HO)_3\dot{P}CH_2CH_2PO(OH)_2$	180	−9.3	0.22	0.53	−0.027

[a]Derived from coupling constants in Table 6.27. The 3*p* spin densities have charge correction factors as explained in the text.

charge correction is applied to the 3*s* spin densities that are derived directly from the A_{iso} with A_{atomic} = 10178 MHz. Such corrections for *s* orbitals are expected to be much less than those for *p* orbitals. In calculation of the ρ^{P}_{3p} spin density for PF_4, an estimated charge of $c^+ = 1.5$ is assumed, and hence the charge correction factor is (1/1.3).

Although the $\rho^{\mathrm{P}}_{3p_\pi}$ for PF_4 is close to that for the phosphoranyl radicals, there is no evidence for spin polarization, that is, $\rho^{\mathrm{P}}_{3p_\sigma} = 0$ for this radical, in contrast to the phosphoranyl radicals for which there is a negative ρ_{3p_σ}, indicating rather strong spin polarization of the σ bonds. This comparison hints of a basic difference in the P-OH and PF bond orbitals in these related radicals.

Despite the considerable new information about the phosphorus bonding that is gained from recent esr measurements on single crystals, there are still uncertainties to be cleared up. Nevertheless, the qualitative features and geometrical structures described for the various types of bonding are rather well established. Discussions of inorganic, phosphorus-centered free radicals are given by Atkins and Symons [102].

10. ^{33}S COUPLING IN SULFUR-CENTERED RADICALS

10.a. Sulfur Oxides

Table 6.29 gives the coupling constants and derived spin densities of three sulfur oxide radicals observed in irradiated single crystals. The ^{33}S coupling tensors for SO_3^- and SO_2^- are axially symmetric, and *B* could be derived in the usual way. However, the relative signs of the principal values, which were not determined by the experiment, appear to be listed incorrectly for SO_2^- by the original investigators [103], who gave the $A_\parallel$ and $A_\perp$ values of 58 ± 0.5 G and 4 ± 4 G for ^{33}S and 30.0 ± 0.5 G and 3 ± 3 G for ^{17}O, respectively. If the resulting isotropic value, $A_{\mathrm{iso}} = (A_\parallel + 2A_\perp)/3$, is to be consistent with that derived from radicals tumbling in liquids, the $A_\parallel$ and $A_\perp$ must be of opposite sign. Furthermore, the nuclear-magnetic moments of ^{17}O and ^{33}S are opposite in sign, and their coupl-

Table 6.29 Coupling Constants and Spin Densities of Some Sulfur Oxide Radicals

Radical / Host Crystal	Coupling Atom	Principal Value A	Coupling Constants (MHz) A_{iso}	B	Spin Densities[a] ρ_{ns}	ρ_{np}	Source of Couplings
SO_3^- / Potassium methane disulfonate	^{33}S	428 314 316	353	38	0.130	0.39	b
SO_2^- / Potassium metabisulfite	^{33}S	162 -11 -11	47	58	0.017	0.59	c
	^{17}O	-84 8 8	-22.7	-31	0.005	0.23	
CH_3SO / Dimethyl sulfoxide	^{33}S	165 -59 -39	22.3	68	0.008	0.73	d

[a]Derived from A_{iso} or B with atomic coupling given in Table A.1 and with charge corrections as explained in the text.
[b]Chantry, Horsfield, Morton, Rowlands, and Whiffen [106].
[c]Reuveni, Luz, and Silver [103].
[d]Nishikida and Williams [105].

ings must also be opposite in sign for positive spin density on each. In Table 6.29 we have made these changes in sign and have converted the values from gauss to megahertz.

On each of the radicals of Table 6.29 there is a positive charge of the order of unity on the sulfur. On the SO_2^- and SO_3^-, we estimate the positive charge to be 1.25 units and that on CH_3SO to be 1.0 units. The positive charge will cause the 3*p*-orbital coupling to be greater than that for the neutral atom. Therefore, a smaller spin density is required to give the observed coupling for the positively charged atom. Procedures for making charge corrections are given in Section 6. For SO_2^- and SO_3^- with $c^+ = 1.25$ the spin density is $\rho_{3p}^{S} = (0.80\, B^S/B^S_{\text{atomic}})$; for CH_3SO with $c^+ = 1$, $\rho_{3p}^{S} = (0.83\, B^S/B^S_{\text{atomic}})$. We have neglected the unknown, but much smaller, charge corrections for the 3*s* orbitals. In the contributing valence-bond structure OSO^- of SO_2^-, no spin density will be on the O^-, and thus we need to correct only for the negative charge on $\dot{O}$ due to the ionic character of the σ bond. Because the electronegativity of the S is increased to about 3 by its 1.25 charge, $i_\sigma \approx 0.25$ and $\rho_{2p}^{O} = (1.06\ B^{O}/B^{O}_{atomic})$. These corrections are applied to the ρ_{np} values given in Table 6.29, but they were not applied by the original investigators. The ρ_{ns} spin densities listed in Table 6.29 apply to the valence shells. Possible contributions to A_{iso} by inner shells are neglected.

The electronic structure of the SO_2^- radical is perhaps the simplest of those in Table 6.29, and the nuclear coupling data, including the ^{17}O coupling, are most complete for it. The maximum principal elements of the ^{33}S and ^{17}O couplings have the same direction, which is that of the minimum g value [103]. These observations indicate that the unpaired electron has a π orbital perpendicular to the molecular plane. The small isotropic components in the couplings are of the magnitude expected from polarization of the σ bonds. The complete equivalence of the coupling of the two oxygens confirms the C_{2v} symmetry of the radical. The S orbital perpendicular to the molecular plane is assumed to be a pure $3p$ orbital and those in the molecular plane, sp^2 hybrids. The $\angle$OSO is predicted to be between 100° and 120° [104]. We designate x as perpendicular to the molecular plane, y as along the symmetry axis, and z as in the OO direction. The unpaired electron density on S is thus $3p_x$ and that on the oxygens, $2p_x$. The antibonding π_x^* orbital of the unpaired electron may be expressed by the following linear combination:

$$\pi_x^* = c_1\,(3p_x^{S}) - c_2\,(2p_x^{O^{(1)}} + 2p_x^{O^{(2)}}) \tag{6.91}$$

With the neglect of the overlap integral, one can set $c_1^2 = \rho_{3p_x}^{S} \approx 0.59$, $c_2^2 = \rho_{2p_x}^{O} \approx 0.23$. The sum, $\rho_{3p_x}^{S} + 2\rho_{2p_x}^{O} = 1.05$, is, within error limits, in agreement with the normalization requirement, $c_1^2 + 2c_2^2 = 1$

The CH_3SO radical is structurally similar to SO_2^-. Its unpaired electron-spin density is mostly in an antibonding π_x^* orbital perpendicular to the CSO plane [105]. The S orbitals in the CSO plane are sp^2 hybrids, two of which form the σ bonds and one of which has an unshared pair. The three sp^2 hybrids probably do not have the same degree of hybridization. Although the electronic structures of the two radicals are similar, the ^{33}S coupling for CH_3SO, unlike that for SO_2^-, is not axially symmetric. We attribute the asymmetry in the coupling primarily to differences in spin polarization of the CS and SO bonds. Because C and S have the same electronegativity, the σ bond of the SC will be essentially covalent, whereas that of the SO will be about 50% ionic. For these reasons alone, the spin polarization of the CS bond is expected to be of the order of twice that of the SO. In addition, the spin polarization of SO induced by the π spin density on S will be partly canceled by opposite spin polarization induced by π spin density on the O. We assume the $3p_y$ orbital to be approximately along the CS direction and neglect the spin polarization along z. Thus we can approximately set

$$A_x - A_{iso} = 143 = 2B_\pi - B_\sigma$$

$$A_y - A_{iso} = -81 = -B_\pi + 2B_\sigma$$

$$A_z - A_{iso} = -61 = -B_\pi - B_\sigma$$

From solution of these equations, we obtain $B_\pi = 68$ MHz, as given in Table 6.29. The β-proton couplings, $A_f(H_1) = A_f(H_2) = 48.5$ MHz, $A_f(H_3) = 0$, reveal a total unpaired electron spin density of 0.07 on the three hydrogens. From normalization of the unpaired spin density to unity, one obtains $\rho^{O}_{2p_x} = 1 - 0.73 - 0.07 = 0.20$. It is interesting that this value is close to the $2p_x$ spin density observed on each O of the SO_2^- radical.

The coupling constants and associated spin densities for SO_3^- and SO_2^- reflect their basic structural differences. For example, the isotropic ^{33}S coupling of SO_3^- is larger by an order of magnitude than that of SO_2^-, whereas its anisotropic component is smaller than that for SO_2^-. The ρ^{S}_{3s} for SO_3^- is much too large to be attributed to spin polarization of the σ bonds. It reveals that the orbital of the unpaired electron density on the S has appreciable 3*s* character. In fact, the 3*s* spin density is one-third that of the 3*p* spin density; if possible 3*d* contributions are negligible, the orbital of the unpaired electron density on the S is $(1/4)^{1/2}\ 3s + (3/4)^{1/2} 3p$. From symmetry, one would then expect the three σ-bonding orbitals of the S to be equivalent sp^3 hybirds and hence the bond angles to be approximately 109° 28′. From their original analysis of the data, Chantry, Horsfield, et al. [106] obtained $\rho^{S}_{3s} = 0.13$ and $\rho^{S}_{3p} = 0.49$ (uncorrected for charge effects), from which they estimated that $\angle OSO \approx 111°$. They concluded that the remaining unpaired spin density is mostly in 2*p* orbitals on the oxygens. This requires that $\rho^{O}_{2p} \approx 0.13$ on each oxygen, or $\rho^{O}_{2p} = 0.16$ if charge corrections are applied. Thus the unpaired electron appears to have an antibonding orbital consisting of an sp^3 hybrid of the S and 2*p* orbitals of each of the three oxygens. The electron pairs of these 2*p* orbitals are correctly in bonding or antibonding molecular orbitals made up from linear combinations of the same set of atomic S and O orbitals so that the structure of the radical is a symmetrical pyramid similar to that of the SO_3 group in the undamaged crystal [106].

The interesting sulfur oxide radical $SO_2Cl_2^-$ has been investigated in an irradiated single crystal of sulfuryl chloride by Gillbro and Williams [107]. They measured a rather large isotropic component for the ^{33}S coupling, about 566 MHz, but they observed no anisotropic component. The isotropic coupling indicates a 3*s* spin density of 0.21 on the S. Evidently, most of the unpaired electron-spin density is in antibonding σ orbitals of the two chlorines. The observed couplings, $A_{iso}(^{35}Cl) = 86.6$ MHz and $B(^{35}Cl) = 46$ MHz, indicate spin populations on each Cl of 0.02 and 0.31 for 3*s* and 3*p*, respectively. The observers conclude that the probable structure is a C_{2v} trigonal-bipyramid, similar to that described for $POCl_3^-$ in Section 9 (see Fig. 6.15). This structure is elaborated in a later paper by Gillbro and Williams [99].

10.b. Sulfur-centered Free Radicals Produced from Amino Acids and Peptides

The free radicals formed by irradiation of the sulfur-containing amino acids are of much interest because of their biochemical and biological significance. Spin centers on the sulfur of these amino acid constituents of proteins tend to develop when the protein is subjected to ionizing radiation [108,109]. In an early attempt to learn the specific form of these radicals, Kurita and Gordy [110] observed and analyzed the esr of a γ-irradiated single crystal of cystine dihydrochloride at room temperature. Since they detected only the ^{32}S species, the information gained came from the **g** tensor alone. They found that the observed **g** tensor could be attributed to the monosulfide radical RCH_2S formed by the breaking of the SS bond. Later, by measurement of ^{33}S hyperfine structure, Hadley and Gordy [111] proved that the radical observed at room temperature in irradiated cystine is actually the disulfide radical RCH_2SS; they were able to reinterpret the **g** tensor, showing that it could also be explained in terms of the disulfide radical. Thus the **g** tensor, which has proved to be widely applicable for identification of sulfur-centered radicals in biochemical powders and polymers, does not provide a critical distinction between the monosulfide and disulfide forms of these radicals. The ^{33}S coupling tensor provides supplementary information of considerable value for clarification of the specific forms and structure of the sulfur-centered radicals. Although it is difficult to obtain the concentrated ^{33}S isotopic species in sufficient quantity to grow single crystals of measurable size, it is possible to obtain the ^{33}S coupling elements from measurements of the hyperfine structure with ^{33}S in its natural concentration of 0.76%. All the ^{33}S couplings treated in this chapter were measured with the naturally occurring concentration of ^{33}S.

After the original work at room temperature on cystine dihydrochloride, a number of investigators observed the esr of ^{32}S-centered radicals in irradiated single crystals of other amino acids or peptides at room temperature and at lower temperatures [112-126]. For the most part, the radicals produced at 77°K or lower temperatures and observed without annealing were interpreted as charged, sulfur-centered radicals. The more recent measurements of the ^{33}S hyperfine structure at lower temperatures have helped to clarify the chemical form and structure of certain of these species (Table 6.30).

Table 6.30 lists the ^{33}S couplings and spin densities of some of the free radicals in irradiated crystals of amino acids and peptides, the chemical form of which has been established by measurements of ^{33}S hyperfine structure. The radicals observed at room temperature are usually found to be of the neutral disulfide form. Four of these, observed in different amino acid crystals, are included in Table 6.30. All of them were found to be stable in the host crystal at room tem-

Table 6.30 ^{33}S Couplings and Spin Densities of Some Sulfur-centered Radicals in Irradiated Amino Acids and Peptides

Radical / *Host Crystal*	*Coupling Atom*	*Coupling Constants (MHz)*		*Spin densities*		*Ref.*
		A_{iso}	B	ρ_{3s}	ρ_{3p}	
RCH_2S / L-Cystine $(HCl)_2$, 77°K, annealed at 200°K	S	68	67.8	0.025	0.87	*a*
$RCH_2S^{(1)}S^{(2)}$ / L-Cystine $(HCl)_2$, 77°K, annealed at 300°K	$S^{(1)}$	35.6	28.3	0.013	0.36	*a*
	$S^{(2)}$	59.7	43.4	0.022	0.56	
$RCH_2S^{(1)}S^{(2)}$ / L-Cysteine (HCl), 77°K, annealed at 300°K	$S^{(1)}$	36.4	28	0.013	0.36	*b*
	$S^{(2)}$	58.8	43	0.022	0.55	
$R'(CH_2)_2S^{(1)}S^{(2)}$ / *N*-Acetyl DL-methionine, 77°K, annealed at 300°K	$S^{(1)}$	36.4	28	0.013	0.36	*b*
	$S^{(2)}$	61.1	40	0.023	0.51	

$R'CH_2S^{(1)}S^{(2)}$ / *N*-Acetyl L-cysteine, 300°K	$S^{(1)}$	29.7	23.0	0.011	0.29	*c*
	$S^{(2)}$	59.6	44.8	0.022	0.57	
$(RCH_2S^{(1)}S^{(2)}CH_2R)^-$ / L-Cystine $(HCl)_2$, 77°K	$S^{(1)}$	89.4	44.3	0.033	0.56	*d,e*
	$S^{(2)}$	89.4	44.3	0.033	0.56	
$(R'CH_2SH)^-$ / *N*-Acetyl L-cysteine, 77°K	S	90	55	0.033	0.85[f]	*c*

$R = (NH_3Cl)\underset{|}{C}HCOOH$ $R' = CH_3CONH\underset{|}{C}HCOOH$

[a] Hadley and Gordy [111].
[b] J. H. Hadley, Jr., and W. Gordy, *Proc. Nat. Acad. Sci. (USA)*, **72**, 3486 (1975).
[c] J. H. Hadley, Jr. and W. Gordy, *Proc. Nat. Acad. Sci. (USA)*, **74**, 216 (1977).
[d] Hadley and Gordy [127].
[e] A. Naito, K. Akasaka, and H. Hatano, *Chem. Phys. Lett.*, **32**, 247 (1975).
[f] Correction applied for unit negative charge on the S by the method described in Section 6.

perature for months or years. The constants for these species were derived, however, from measurements made at 77°K on crystals either irradiated at room temperature or annealed at room temperature after irradiation at lower temperatures. The neutral monosulfide radical, RCH_2S, listed at the top, is not stable at room temperature, nor is it found in the cysteine $(HCl)_2$ crystal irradiated at 77°K and measured without warming. It appears to be formed from a charged primary radical as the temperature of the host crystals is raised to about 150° to 200°K. As the temperature is raised further, the monosulfide radicals react with the molecules of the host crystal to produce the disulfide form. Akasaka and his associates [114] first noted that there is a sequence of three related radicals formed in cysteine dihydrochloride irradiated at 77°K and warmed to 300°K. They labeled the radicals α_1, α_2, and α_3. From correlation of signal strength with rise in temperature, they concluded that one species generates the other in the

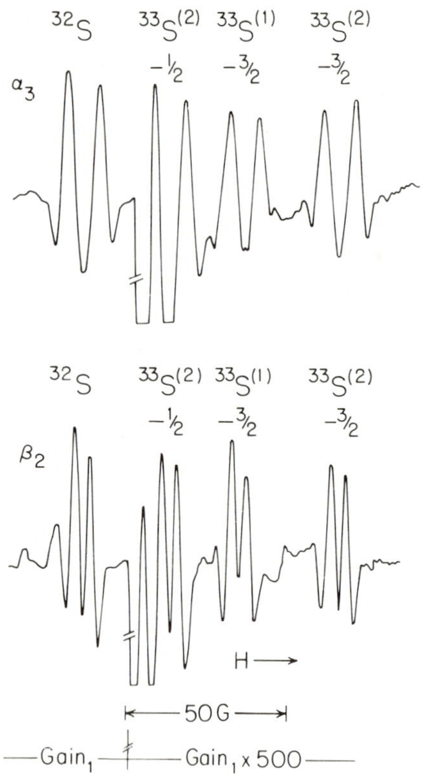

Fig. 6.16 Parts of the esr spectra of the disulfide radicals α_3 and β_2 in a γ-irradiated single crystal of cystine dihydrochloride at 300°K, showing ^{33}S hyperfine components on the high-field side of the ^{33}S spectrum. From Hadley and Gordy [111].

sequence, $\alpha_1 \rightarrow \alpha_2 \rightarrow \alpha_3$. The ^{33}S measurements confirm that α_1 is $(RCH_2SSCH_2)^-$, α_2 is RCH_2S, and α_3 is RCH_2SS. The ^{33}S constants for each of these species are listed in Table 6.30.

Akasaka, Ohnishi, et al. [114] also detected another radical produced at 77°K in irradiated cystine $(HCl)_2$, which they labeled as β_1. Although it appears from its **g** tensor to be the positively charged crystine molecule, its identity has not yet been established from ^{33}S measurements. When the crystal is warmed to room temperature, β_1 generates a second radical, β_2, which has been proved from ^{33}S hyperfine measurements to have the disulfide form [111].

Figure 6.16 shows the ^{33}S hyperfine components on the high-field side of the ^{32}S pattern observed for the two $RCH_2S^{(1)}S^{(2)}$ radicals in irradiated cysteine dihydrochloride, namely, the α_3 and β_2. The ^{33}S components are only 1/500 the strength of those for ^{32}S, and thus the spectrometer gain was increased by a factor of 500 at the break in the curves. The doubling of each component is produced by an isotropic coupling of a proton on the CH_2 group, of about 10 G on α_3 and 5 G on β_2. The ^{33}S hyperfine tensor was found to have axial symmetry about the direction of the minimum g value (g_u), which is perpendicular to the CSS plane (see Chapter VII). Figure 6.17 gives plots of the separation of these ^{33}S components from those of the ^{32}S species as the crystal orientation is varied from the direction of g_u.

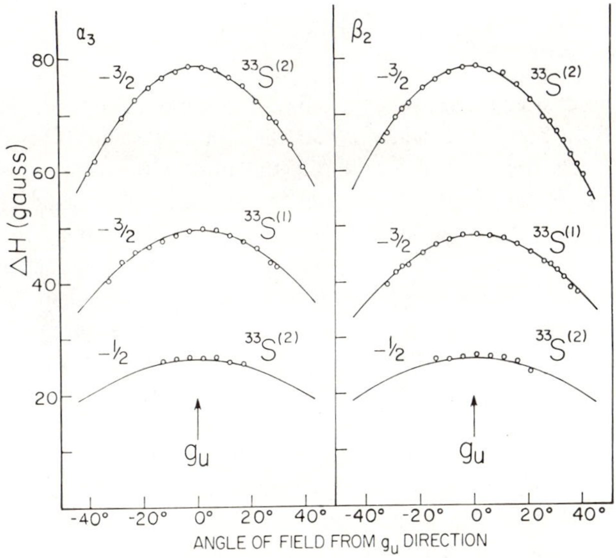

Fig. 6.17 Plots of the ^{33}S component displacements as a function of the angular deviation of the applied field from the principal axis of minimum g, the direction of the p_π-orbital spin density on the coupling S. From Hadley and Gordy [111].

For the four netural disulfide radicals listed in Table 6.30, the sum of the $3p$ unpaired electron density on the $S^{(1)}$ and $S^{(2)}$ is in the range from 0.86 to 0.92. Although these figures are only approximate because of uncertaintity in the B_{atomic} employed, neglect of small charge corrections and spin-polarization effects, they show that the predominant unpaired electron density is on the two sulfurs, more than half of which is on the terminal $S^{(2)}$. The unpaired electron has an anti bonding π_x^* molecular orbital essentially confined to the two sulfurs. This orbital can be expressed approximately by the linear combinations

$$\pi_x^* = c_2(3p_x{}^{(2)}) - c_1(3p_x{}^{(1)}) \tag{6.92}$$

where $c_1^2 = \rho_{3p_x}^{(1)}$ and $c_2^2 = \rho_{3p_x}^{(2)}$. The corresponding normalized, orthogonal, bonding orbital may be expressed approximately as

$$\pi_x = c_1(3p_x^{(1)}) + c_2(3p_x^{(2)}) \tag{6.93}$$

This bonding orbital has two electrons, and the resultant π-bond order between the two sulfurs is

$$N_x = 2(c_1c_2) - c_2c_1 = (\rho_{3p_x}{}^{(1)})^{1/2}(\rho_{3p_x}{}^{(2)})^{1/2} \tag{6.94}$$

For the $RCH_2S^{(1)}S^{(2)}$ radical in L-cystine $(HCl)_2$ (listed in Table 6.30)

$$N_x = (0.36)^{1/2}(0.56)^{1/2} = 0.45$$

Thus the π_x bond strength between $S^{(1)}$ and $S^{(2)}$ is approximately 0.45 times that of a normal SS π bond. This added π bonding probably accounts for the greater stability of the disulfide radical over that of the monosulfide form.

The couplings of the two $C_\beta H_2$ protons in the monosulfide radical listed first in Table 6.30, 47.6 MHz and 58.9 MHz [114], indicate a combined unpaired electron-spin density of (106.5/1420) = 0.075. The combined unpaired electron density of 0.95 on these hydrogens and on the S shows that there is very little spreading of the orbital of the unpaired electron onto the R group. In fact, the possible inaccuracies in evaluation of the ρ_{3p}^{S} could account for the deviation of this sum from unity. Correction for configuration interaction indicated by the isotropic coupling would increase the predicted $\rho_{3p_x}^{S}$, as explained for ^{13}C in Section 4.b. Likewise, there is evidently a negligible spreading of the unpaired electron orbital onto the R, X, or Y groups of the disulfide radicals listed in Table 6.30. In all these species the isotropic ^{33}S coupling is of the right magnitude to result from configuration interaction on the S.

The Negatively Charged Radicals

An esr pattern having a **g** tensor axially symmetric about the SS direction in the undamaged molecule was observed in cystine dihydrochloride crystals irradiated and observed at low temperature by Box and Freund [113] and independently by Akasaka, et al. [114]. This observation was also confirmed by Krivenko, Kayushin, and Pulatova [117]. The fact that these researchers observed only one magnetic site, that is, only one orientation of the radical relative to the crystal axis, revealed that the spin density is symmetrically distributed in the radical. The radicals formed by the breaking of SS or CS bonds in this crystal have two distinguishable orientations. This fact and the observation that the axially symmetric minimum g value has the direction of the SS σ bond of the undamaged molecule strongly indicate that the unpaired electron giving rise to the esr pattern is in an antibonding σ^*_{3p} orbital, mostly localized between the two sulfurs. Akasaka and co-workers showed that the observed **g** tensor corresponds to that theoretically predicted for the negatively charged molecule $(RCH_2SSCH_2R)^-$ in which the unpaired electron is in an antibonding σ^*_{3p} orbital between the sulfurs. This is the species labeled α_1. The later measurements of the ^{33}S hyperfine structure [127] agreed completely with this interpretation. The spin densities on the two sulfurs were found to be identical (Table 6.30) and the coupling $3p$ density, to be directed along the SS σ bond, as is the minimum g value. Note that the total $3p$ spin density, 0.56 + 0.56, is greater than unity. This anomaly is believed to result from the failure to take account of the dipole-dipole coupling of the spin density of the ^{32}S that is directed toward the coupling ^{33}S in the σ bond. Because of the complete symmetry of the $(RCH_2SSCH_2R)^-$ radical one cannot resolve, or distinguish, the coupling of the two forms, $-^{33}S^{(1)}-^{32}S^{(2)}-$ and $-^{32}S^{(1)}-^{33}S^{(2)}-$, but the fact that together they give a signal having twice the strength expected for a radical having distinguishable coupling ^{33}S atoms proves that there are two equivalent positions for the coupling ^{33}S.

Neither the **g** tensor nor the ^{33}S coupling can distinguish between the negatively charged molecule and the positively charged one if the odd electron in the latter should form a one-electron bond between the sulfurs. In both $-(S\cdot\ \cdot S)^-{-}$ and $-(S\cdot S)^+{-}$ forms, the bond strength would be approximately one-half that of the normal SS σ bond. It would appear, however, that a positively charged radical could be formed more easily by removal of one of the nonbonding pairs on the S than by removal of one from the σ-bonding orbital. Thus a radical of the form $-(S\cdot S)^+{-}$ seems improbable. In contrast, an electron captured by the $-SS-$ must go either into a $3d$ orbital or into the antibonding σ^*_{3p} orbital. The esr data indicate that it does the latter [114,127]. The positively charged π^*_{3p}-type of radical is described in Chapter VII, Section 5. Diagrams of the two structures are given in Figs. 6.18 and 6.19.

The specific form of the last radical in Table 6.30, listed as $(R'CH_2SH)^-$, is uncertain, but the ^{33}S hyperfine structure proves that it has the monosulfide form. Its lower B value and abnormal **g** indicate that its electronic structure differs significantly from that of the neutral monosulfide radical listed first in

Table 6.30. The abnormal **g** and probable structure are discussed in Chapter VII.

10.c. β-Hydrogen Coupling in Sulfur-centered Radicals

The measurement of S spin densities with ^{33}S hyperfine structure makes possible an evaluation of the βH coupling constant Q_{β}^{SCH} between the $C_{\beta}H_2$ hydrogen

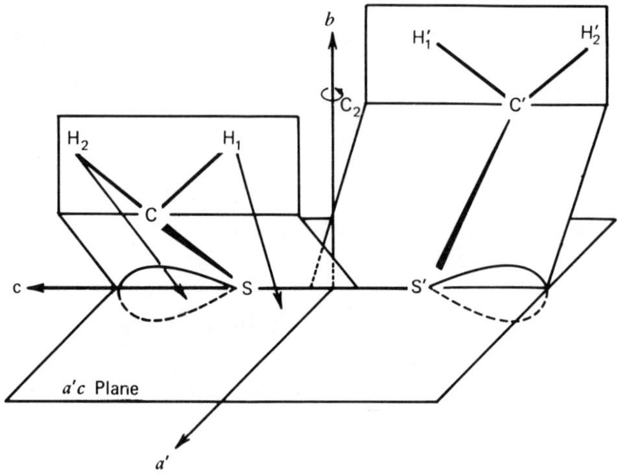

Fig. 6.18 Diagram of the structure of the anion radical $(RCH_2SSCH_2R)^-$. From A. Naito, K. Akasaka, and H. Hatano, *J. Magn. Reson.*, **24**, 53 (1976).

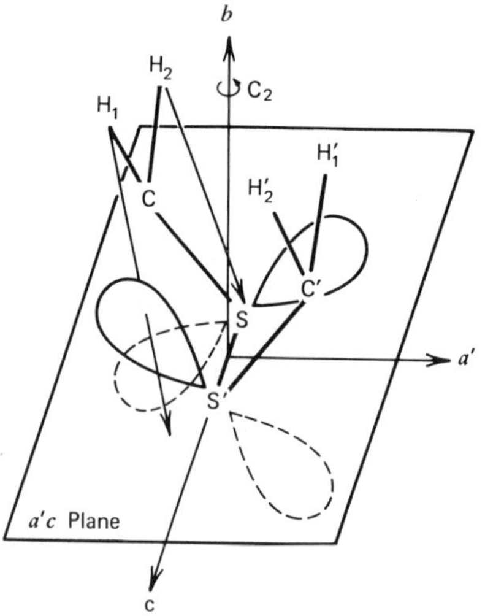

Fig. 6.19 Diagram of the structure of the cation radical $(RCH_2SSCH_2R)^+$. From A. Naito, K. Akasaka, and H. Hatano, *J. Magn. Reson.*, **24**, 53 (1976).

Table 6.31 β-Proton Couplings and Q_β Values for Some Sulfur-centered Radicals

Radical	βH Coupling A_{iso} (MHz)	Source	Dihedral Angle θ^{0} [e]	S_α Spin Density ρ_{3p}	Effective Q_β^{SCH} (MHz)	Calculated A_{iso} (MHz)
CH_3SO	48.5 48.5 0	a	30 150 270	0.73	88	48 48 0
RCH_2S	59 48	b	25 145	0.87	83	59 48
RCH_2SS	28 0	c	20 260	0.36	88	28 1
$(RCH_2SSCH_2R)^-$	28.9 20.1	d	21 141	0.465[f]	72	29 20
$(RCH_2SSCH_2R)^+$	18.4 8.0	d	53.2 293.2	0.481[f]	107	18.5 8.0

[a] Nishikida and Williams [105].
[b] Akasaka, Ohnishi, Suita, and Nitta [114].
[c] Hadley and Gordy [111].
[d] A. Naito, K. Akasaka, and H. Hatano, *J. Magn. Reson.*, **24**, 53 (1976).
[e] $\theta_2 = \theta_1 + 120^\circ, \theta_3 = \theta_1 + 240^\circ$ (assumed).
[f] For spin density derivation, see text.

and the S to which it is bonded (Sections 1.c and 2.b). With the omission of the small constant Q_0^{β}, which cannot be evaluated from these data, (6.22) and (6.23) specialized to the sulfur radicals may be expressed as

$$A_{\beta}^{H} = \rho_{3p}^{S} Q_{\beta}^{SCH} \cos^2 \theta \qquad (6.95)$$

where θ is the dihedral angle shown in Fig. 6.1. Examples of Q_{β}^{SCH} values derived from this relation are given in Table 6.31. For the neutral radicals, the ρ_{3p}^{S} values are those from Table 6.29 and 6.30. For the $(RCH_2S)_2^-$ radicals, the sum of the $\rho_{3p_{\pi}}^{S}$ values derived from the ^{33}S coupling is greater than unity because of the failure to correct for the rather large σ_{3p}^{*} interatomic dipole-dipole interaction. Consequently, the unpaired electron density of each S of this anion radical is assumed to be 0.500 less that on the two β hydrogens, 0.035, as derived from the isotropic βH couplings. In a similar manner, the ρ_{3p}^{S} values for the cation $(RCH_2S)_2^+$ were obtained.

Although the dihedral angle θ can be deduced approximately from the known structure of the undamaged crystal for some of these radicals, we have made the assumption that the β carbon forms its bonds with equivalent sp^3-hybrid orbitals, a condition which requires the dihedral angles to differ by 120°. With this assumption, there is only one independent θ value; and, since there are two observable A_{iso} values for the $C_{\beta}H_2$, or three for $C_{\beta}H_3$, the orientation θ as well as Q_{β}^{SCH} can then be obtained from (6.95). The values thus derived are given in Table 6.31.

Despite the structural differences in the three neutral radicals listed in Table 6.31, their Q_{β}^{SCH} values are remarkably close. The average value, $Q_{\beta}^{SCH} = 86$ MHz, should prove useful in the estimation of π_{3p} spin densities of S_{α} atoms from the isotropic βH coupling in other neutral sulfur-centered radicals, including those measured in liquids. It is interesting that the Q_{β}^{SCH} for the anion is lower, and that for the cation is higher, than those for the neutral radicals; both values are rather close to those of the neutral radicals. Figures 6.18 and 6.19 show the orientations of the $C_{\beta}H_2$ groups relative to the orbital of the unpaired electron density on the S_{α} for the charged cystine radicals.

References

1. S. I. Weissman, J. Townsend, D. E. Paul, and G. E. Pake, *J. Chem. Phys.*, 21, 2227 (1953).
2. H. N. McConnell, *J. Chem. Phys.*, **24**, 632, 764 (1956).
3. S. I. Weissman, *J. Chem. Phys.*, **24**, 890 (1956).
4. R. J. Bersohn, *J. Chem. Phys.*, **24**, 1066 (1956).
5. H. S. Jarrett, *J. Chem. Phys.*, **25**, 1289 (1956).
6. A. D. McLachlan, *Molec. Phys.*, **3**, 233 (1960).

7. F. Hund, *Linienspectren und periodisches System der Elemente,* Springer, Berlin, 1927, p. 124.

8. H. M. McConnell and D. B. Chesnut, *J. Chem. Phys.,* **28**, 107 (1958).

9. J. P. Colpa and J. R. Bolton, *Molec. Phys.,* **6**, 273 (1963); J. R. Bolton, *J. Chem. Phys.,* **43**, 309 (1965).

10. G. Giacometti, P. L. Nordio, and M. V. Pavan, *Theoret. Chim. Acta (Berlin),* **1**, 404 (1963).

11. E. Hückel, *Z. Physik,* **70**, 204 (1931).

12. R. P. Penrose, *Nature,* **163**, 992 (1949).

13. B. Venkataraman and G. K. Fraenkel, *J. Chem. Phys.,* **23**, 588 (1955).

14. W. Gordy, W. B. Ard, and H. Shields, *Proc. Nat. Acad. Sci. (USA),* **41**, 983, 996 (1955).

15. W. Gordy and C. G. McCormick, *J. Am. Chem. Soc.,* **78**, 3243 (1956).

16. C. G. McCormick and W. Gordy, *J. Phys. Chem.,* **62**, 783 (1958).

17. D. B. Chesnut, *J. Chem. Phys.,* **29**, 43 (1958).

18. R. S. Mulliken, C. A. Rieke, and W. G. Brown, *J. Am. Chem. Soc.,* **63**, 41 (1941).

19. A. D. McLachlan, *Molec. Phys.,* **1**, 233 (1958).

20. D. H. Levy, *Molec. Phys.,* **10**, 233 (1966).

21. N. M. Atherton, *Electron Spin Resonance: Theory and Applications,* Wiley, New York, 1973.

22. I. Miyagawa and W. Gordy, *J. Chem. Phys.,* **32**, 255 (1960).

23. H. M. McConnell and J. Strathdee, *Molec. Phys.,* **2**, 129 (1959).

24. T. Cole, C. Heller, and H. M. McConnell, *Proc. Nat. Acad. Sci. (USA),* **45**, 525 (1959).

25. H. M. McConnell, C. Heller, T. Cole, and R. W. Fessenden, *J. Am. Chem. Soc.,* **82**, 766 (1960).

26. D. K. Ghosh and D. H. Whiffen, *Molec. Phys.,* **2**, 285 (1959).

27. R. W. Fessenden and R. H. Schuler, *J. Chem. Phys.,* **39**, 2147 (1963).

28. C. Heller and H. M. McConnell, *J. Chem. Phys.,* **32**, 1535 (1960).

29. E. W. Stone and A. H. Maki, *J. Chem. Phys.,* **37**, 1326 (1962).

30. F. C. Adam and F. W. King, *J. Chem. Phys.,* **58**, 2446 (1973).

31. J. R. Morton and A. Horsfield, *J. Chem. Phys.,* **35**, 1142 (1961).

32. A. Horsfield, J. R. Morton, and D. H. Whiffen, *Molec. Phys.,* **4**, 425 (1961).

33. I. Miyagawa and K. Itoh, *J. Chem. Phys.,* **36**, 2157 (1962).

34. S. Glasstone, K. J. Laidler, and H. Eyring, *The Theory of Rate Processes,* McGraw-Hill, New York, 1946.

35. C. Heller, *J. Chem. Phys.,* **36**, 175 (1962).

36. W. Gordy and R. L. Cook, *Microwave Molecular Spectra*, Interscience New York, 1970, Chapter 12.

37. P. J. Hamrick, Jr., H. Shields, and C. C. Whisnant, *Radiat. Res.*, **48**, 234 (1971).
38. W. C. Lin and J. M. Nickel, *J. Chem. Phys.*, **57**, 3581 (1972); W. C. Lin, *J. Chem. Phys.*, **58**, 2664 (1973).
39. H. W. Shields, P. J. Hamrick, Jr., and W. Redwine, *J. Chem. Phys.*, **46**, 2510 (1967).
40. W. M. Fox and P. Smith, *J. Chem. Phys.*, **48**, 1868 (1968).
41. P. Smith and P. B. Wood, *Can. J. Chem.*, **44**, 3085 (1966).
42. K. Reiss and H. Shields, *J. Chem. Phys.*, **50**, 4368 (1969).
43. J. V. Ramsbottom and W. A. Waters, *J. Chem. Soc. B*, 132 (1966).
44. H. Bower, J. McRae, and M. C. R. Symons, *J. Chem. Soc. A*, 2400 (1971).
45. R. F. C. Claridge and F. T. Greenaway, *J. Magn. Reson.*, **8**, 316 (1972).
46. J. R. Rowlands, *Molec. Phys.*, **5**, 565 (1962).
47. M. A. Collins and D. H. Whiffen, *Molec. Phys.*, **10**, 317 (1966).
48. R. P. Kohin and P. G. Nadeau, *J. Chem. Phys.*, **44**, 691 (1966).
49. K. N. Shrivastava and R. S. Anderson, *J. Chem. Phys.*, **48**, 4599 (1968).
50. T. Cole, H. O. Pritchard, N. R. Davidson, and H. M. McConnell, *Molec. Phys.*, **1**, 406 (1958).
51. H. M. McConnell and R. W. Fessenden, *J. Chem. Phys.*, **31**, 1688 (1959).
52. T. Cole and C. Heller, *J. Chem. Phys.*, **34**, 1085 (1961).
53. M. Karplus and G. K. Fraenkel, *J. Chem. Phys.*, **35**, 1312 (1961).
54. A. Carrington and J. dos Santos-Veiga, *Molec.Phys.*, **5**, 21 (1962).
55. R. L. Ward, *J. Am. Chem. Soc.*, **83**, 3623 (1961); **84**, 332 (1962).
56. E. W. Stone and A. H. Maki, *J. Chem. Phys.*, **39**, 1635 (1963).
57. E. T. Strom, G. A. Russell, and R. Konaka, *J. Chem. Phys.*, **42**, 2033 (1965).
58. S. N. Foner, E. L. Cochran, V. A. Bowers, and C. K. Jen, *Phys. Rev. Lett.*, **1**, 91 (1958).
59. I. Miyagawa and W. Gordy, *J. Chem. Phys.*, **30**, 1590 (1959).
60. W. H. Nelson, F. M. Atwater, and W. Gordy, *J. Chem. Phys.*, **61**, 4726 (1974).
61. N. Cyr and W. C. Lin, *J. Chem. Phys.*, **50**, 3701 (1969).
62. P. Neta and R. W. Fessenden, *J. Phys. Chem.*, **74**, 3362 (1970).
63. M. C. R. Symons, *J. Chem. Phys.*, **55**, 1493 (1971).
64. W. C. Lin, N. Cyr, and K. Toriyama, *J. Chem. Phys.*, **56**, 6272 (1972).
65. P. W. Lau and W. C. Lin, *J. Chem. Phys.*, **51**, 5139 (1969).
66. E. L. Cochran, F. J. Adrian, and V. A. Bowers, *J. Chem. Phys.*, **36**, 1938 (1962).
67. D. J. Whelan, *Aust. J. Chem.*, **26**, 681 (1973).
68. C. H. Townes and A. L. Schawlow, *Microwave Spectroscopy*, McGraw-Hill, New York, 1955.

69. R. J. Lontz and W. Gordy, *J. Chem. Phys.*, **37**, 1357 (1962).
70. R. J. Cook, J. R. Rowlands, and D. H. Whiffen, *Molec. Phys.*, **7**, 31 (1963).
71. R. W. Fessenden and R. H. Schuler, *J. Chem. Phys.*, **43**, 2704 (1965).
72. M. T. Rogers and L. D. Kispert, *J. Chem. Phys.*, **46**, 3193 (1967).
73. J. Maruani, C. A. McDowell, H. Hakajima, and R. Raghunathan, *Molec. Phys.*, **14**, 349 (1968).
74. J. Maruani, J. A. R. Coope, and C. A. McDowell, *Molec. Phys.*, **18** 165 (1970).
75. F. D. Srygley and W. Gordy, *J. Chem. Phys.*, **46**, 2245 (1967).
76. M. F. Merritt and R. W. Fessenden, *J. Chem. Phys.*, **56**, 2353 (1972).
77. A. Hasegawa, K. Sogabe, and M. Miura, *Molec. Phys.*, **30**, 1889 (1975).
78. H. E. Radford, V. W. Hughes, and V. Beltran-Lopez, *Phys. Rev.*, **123**, 153 (1961).
79. D. A. Goodings, *Phys. Rev.*, **123**, 1706 (1961).
80. A. Hinchliffe and J. N. Murrell, *Molec. Phys.*, **14**, 153 (1968).
81. P. H. Anderson, P. J. Frank, and H. S. Gutowsky, *J. Chem. Phys.*, **32**, 196 (1960).
82. A. H. Maki and D. H. Geske, *J. Am. Chem. Soc.*, **83**, 1852 (1961).
83. P. H. H. Fischer and J. P. Colpa, *Z. Naturforsch. A*, **24**, 1980 (1969).
84. R. J. Lontz, *J. Chem. Phys.*, **45**, 1339 (1966).
85. R. F. Picone and M. T. Rogers, *J. Magn. Reson.*, **14**, 279 (1974).
86. A. Hudson and K. D. J. Root, "Halogen Hyperfine Interactions," in *Advances in Magnetic Resonance*, Vol. 5, J. S. Waugh, Ed., Academic, New York, 1971, pp. 1-79.
87. E. T. Kaiser and L. Kevan, Eds., *Radical Ions*, Interscience, New York, 1968.
88. R. F. Picone and M. T. Rogers, *J. Chem. Phys.*, **61**, 4814 (1974).
89. W. Nelson, G. Jackel, and W. Gordy, *J. Chem. Phys.*, **52**, 4572 (1970).
90. J. B. Farmer, M. C. L. Gerry, and C. A. McDowell, *Molec. Phys.*, **8**, 253 (1964).
91. A. Horsfield, J. R. Morton, and D. H. Whiffen, *Molec. Phys.*, **4**, 475 (1961).
92. D. E. C. Corbridge, *Acta Crystallogr.*, **9**, 991 (1956).
93. M. Geoffroy, L. Ginet, and E. A. Lucken, *Molec. Phys.*, **31**, 745 (1976).
94. M. Geoffroy and E. A. Lucken, *Molec. Phys.*, **24**, 335 (1972).
95. J. R. Morton, *Can. J. Phys.*, **41**, 706 (1963).
96. R. W. Fessenden and R. H. Schuler, *J. Chem. Phys.*, **45**, 1845 (1966).
97. J. Higuchi, *J. Chem. Phys.*, **50**, 1001 (1969).
98. A. Hasegawa, K. Ohnishi, K. Sogabe, and M. Miura, *Molec. Phys.*, **30**, 1367 (1975).
99. T. Gillbro and F. Williams, *J. Am. Chem. Soc.*, **96**, 5032 (1974).

100. R. E. Rundle, *Surv. Progr. Chem.*, **1**, 81 (1963); R. J. Hach and R. E. Rundle, *J. Am. Chem. Soc.*, **73**, 4321 (1951).

101. D. J. Whelan, *Aust. J. Chem.*, **26**, 1356 (1973).

102. P. W. Atkins and M. C. R. Symons, *The Structure of Inorganic Free Radicals*, Elsevier, Amsterdam, 1967.

103. A. Reuveni, Z. Luz, and B. L. Silver, *J. Chem. Phys.*, **53**, 4619 (1970).

104. K. P. Dinse and K. Möbius, *Z. Naturforsch. A*, **23**, 695 (1968).

105. K. Nishikida and F. WIlliams, *J. Am. Chem. Soc.*, **96**, 4781 (1974).

106. G. W. Chantry, H. Horsfield, J. R. Morton, J. R. Rowlands, and D. H. Whiffen, *Molec. Phys.*, **5**, 233 (1962).

107. T. Gillbro and F. Williams, *Chem. Phys. Lett.*, **20**, 436 (1973).

108. W. Gordy and H. Shields, *Radiat. Res.*, **9**, 611 (1958).

109. W. Gordy and I. Miyagawa, *Radiat. Res.*, **12**, 211 (1960).

110. Y. Kurita and W. Gordy, *J. Chem. Phys.*, **34**, 282 (1961).

111. J. H. Hadley, Jr. and W. Gordy, *Proc. Nat. Acad. Sci. (USA)*, **71**, 3106 (1974).

112. E. Cipollini and W. Gordy, *J. Chem. Phys.*, **37**, 13 (1962).

113. H. C. Box and H. G. Freund, *J. Chem. Phys.*, **40**, 817 (1964).

114. K. Akasaka, S. I. Ohnishi, T. Suita, and I. Nitta, *J. Chem. Phys.*, **40**, 3110 (1964).

115. K. Akasaka, *J. Chem. Phys.*, **43**, 1182 (1965).

116. H. C. Box, H. G. Freund, and E. E. Budzinski, *J. Chem. Phys.*, **45**, 809 (1966).

117. V. G. Krivenko, L. P. Kayushin, and M. K. Pulatova, *Biofizika*, **14**, (4), 615 (1969).

118. S. Kominami, K. Akasaka, H. Umegaki, and H. Hatano, *Chem. Phys. Lett.*, **9**, 510 (1971).

119. L. P. Kayushin, V. G. Krivenko, and M. K. Pulatova, *Stud. Biophys.*, **33**, (1), 59 (1972).

120. E. L. Thomsen and S. O. Nielsen, *J. Chem. Phys.*, **57**, 1095 (1972).

121. S. N. Dobrayakov, V. G. Krivenko, M. K. Pulatova, and E. Sud'bina, *Biofizika*, **18**, (2), 223 (1973).

122. K. Kawatsura, K. Ozawa, S. Kominami, K. Akasaka, and H. Hatano, *Radiat. Eff.*, **22**, 267 (1974).

123. G. Saxebøl and Ø. Herskedal, *Radiat. Res.*, **62**, 395 (1975).

124. D. G. Cadena, Jr., and J. R. Rowlands, *J. Chem. Soc. B*, 488 (1968).

125. S. Kominami, *J. Phys. Chem.*, **76**, 1729 (1972).

126. L. P. Kayushin, V. G. Krivenko, M. K. Pulatova, and E. Sud'bina, *Studia Biophysica*, **39**, 193 (1973).

127. J. H. Hadley, Jr., and W. Gordy, *Proc. Nat. Acad. Sci. (USA)*, **71**, 4409 (1974).

Chapter VII

RELATION OF THE g TENSOR TO MOLECULAR STRUCTURE

This chapter is concerned with the relationship of the principal elements of the **g** tensor to the electronic structure of molecular free radicals. Basic theory of the **g** tensor has been given in Chapters II and III, and methods for derivation of its principal elements from experimental measurements have been described in Chapter V.

Although the **g** tensor is not useful for measurement of spin densities on the different atoms of a free radical as is the nuclear coupling, it can give valuable information about the structure and orientation of free radicals in single crystals. This information supplements that gained from the nuclear coupling tensors and is sometimes crucial in the decision on the chemical form of the free radical produced by irradiation of molecular crystals. Also, approximate separations of the ground state from higher orbital states of free radicals can be obtained, information not given by the **A** tensors. Most of this information is gained from the anisotropies in **g**, which in molecular free radicals are exceptionally small and difficult to measure accurately.

1. THEORETICAL FORMULATION

1.a. Unpaired Electron Localized on Ions or Bonded Atoms

The second-order formula, (2.15), originally derived by Pryce [1] for interpretation of the **g** tensor of ions of the transition elements, is adaptable, with some modifications [2], to molecular free radicals in condensed matter. Derivation of the formula is given in Chapter II, Section 3. The **g** tensor may be expressed as

$$g_{ij} = g_e + \Delta g_{ij} \tag{7.1}$$

where $i, j, = x, y, z$ and where $g_e = 2.0023$ is the g factor for the free electron spin. For ions of the transition elements, Δg_{ij} is expressed (Chapter II, Section 2) by

$$\Delta g_{ij} = -2\lambda \sum_{\substack{i,j \\ x,y,z}} \sum_{n \neq 0} \frac{(0|L_i|n)(n|L_j|0)}{E_n - E_0} \tag{7.2}$$

where λ is the spin-orbit coupling between **S** and **L**, and L_i and L_j are the angular momentum operators for the x, y, z components of **L**. The notation $|0)$ signifies the wave functions of the orbital ground state and $|n)$, those of higher orbital states. Thus $E_n - E_0$ is the energy difference between the higher $|n)$ states and the orbital ground state.

When there is only one unpaired electron in the paramagnetic element (doublet spin states), **L** and **S** become equivalent to $\boldsymbol{l}$ and **s**, used to designate orbital angular momentum and spin of the individual electrons. The spin-orbit coupling constant λ of **L** and **S**, although numerically the same when $S = 1/2$ as the coupling constant ζ for the single electron ($j = \zeta \boldsymbol{l} \cdot \mathbf{s}$), may be of opposite sign. When there is only one electron in an L shell, $S = 1/2$ and $\lambda = +\zeta$. When an L shell lacks only one electron of being full, $S = 1/2$ and $\lambda = -\zeta$. One can justify this change in sign by treating the missing electron as a hole, equivalent to a positive electron in its rotation about the nucleus or its orbital precession about a magnetic field. Generally,

$$\lambda = \frac{\pm\zeta}{2S} \tag{7.3}$$

where the plus sign applies when the **L** shell is less than half full and the minus sign applies when it is more than half full. From (7.2) it is evident that the sign of λ determines whether the g value is shifted up or down by the residual spin-orbit coupling.

1.b Unpaired Electron in Delocalized Molecular Orbital

A generalized, theoretical treatment of the **g** tensor in molecular free radicals has been made by Stone [2]. An earlier, qualitative explanation of the anisotropies of the **g** tensor to be expected in π-type radicals had been given by McConnell and Robertson [3], and a semiquantitative calculation of the principal g values of the CO_2 radical had been made by Ovenall and Whiffen [4]. Stone's theoretical treatment is too complex for description here; instead, we give a simplified derivation of the molecular **g** based on the development of (7.2) in Chapter II.

The complexity of the molecular **g** tensor arises from the multicentered nature of the molecular orbitals. The **g**-tensor formula of (7.2), cannot be used without modification for molecular orbitals because it is not gauge-invariant [2]. In the derivation of this formula (Chapter II, Section 1), the essential perturbation term is

$$\mathcal{H}' = \lambda \mathbf{L} \cdot \mathbf{S} + \beta \mathbf{L} \cdot \mathbf{H} \tag{7.4}$$

where $\mathbf{L}$ is the angular momentum relative to a coordinate system centered at a fixed point in space. Although this point is chosen as the nuclear center of the particular atom, for convenience, the $\mathcal{H}'$ is independent of the choice of the coordinate center; that is, it is gauge-invariant. For a poly-atomic molecular free radical, the $\lambda\mathbf{L}\cdot\mathbf{S}$ term (where λ is a single spin-molecular coupling constant and $\mathbf{L}$ is the angular momentum of the molecular orbital) is no longer gauge invariant because the spin-orbit coupling is dependent on the separation of the unpaired electron from each atomic nucleus of the radical on which there is significant unpaired electron density. If λ_k is the spin-orbit coupling and $\mathbf{L}_k$ is the orbital angular momentum of the unpaired electron in the component orbital centered on atom k, the gauge-invariant molecular $\mathcal{H}_{\mathbf{L}\cdot\mathbf{S}}$ may be expressed as

$$\mathcal{H}_{\mathbf{L}\cdot\mathbf{S}}^{\mathrm{mol}} = \sum_k \lambda_k \mathbf{L}_k \cdot \mathbf{S} \tag{7.5}$$

where the summation is taken over all atoms of the molecular free radical. The λ_k is a constant closely approximating the atomic spin-orbit coupling constant of atom k. Similarly, we may write

$$\mathcal{H}_{\mathbf{L}\cdot\mathbf{H}}^{\mathrm{mol}} = \beta \sum_{k'} \mathbf{L}_{k'} \cdot \mathbf{H} \tag{7.6}$$

where $\mathbf{L}_{k'}$ is the angular momentum on the k'th atom and where the summation is again taken over all atoms of the free radical. Thus for a molecular free radical with spin density on more than one atom, (7.4) must be altered to the form

$$\mathcal{H}'\text{mol} = \sum_{kk'} (\lambda_k \mathbf{L}_k \cdot \mathbf{S} + \beta \mathbf{L}_{k'} \cdot \mathbf{H}) \tag{7.7}$$

where k and k' are atoms on which the wave function of the unpaired electron density occurs. We now express the ground-state molecular orbital of the unpaired electron by $\psi_0 = |0)$ and the excited states of the molecular orbitals by $|n)$. From comparisons with (2.11), it is seen that the second-order contribution to $\mathcal{H}_S$ may then be expressed as

$$\mathcal{H}_S^{(2)} = -\sum_{\substack{n \\ n \neq 0}} \sum_{k,k'} \frac{(0|\lambda_k \mathbf{L}_k \cdot \mathbf{S} + \beta \mathbf{L}_{k'} \cdot \mathbf{H}|n)(n|\lambda_k \mathbf{L}_k \cdot \mathbf{S} + \beta \mathbf{L}_{k'} \cdot \mathbf{H}|0)}{E_n - E_0} \tag{7.8}$$

By multiplication of the vector products and by expression of all components in the same Cartesian coordinates, this formula may be written as

$$\mathcal{H}_S^{(2)} = -\sum_{\substack{n \\ n \neq 0}} \sum_{i,j=x,y,z} \sum_{k,k'} \frac{(0|\lambda_k L_{ki} S_i + \beta L_{k'i} H_i|n)(n|\lambda_k L_{kj} \cdot S_j + \beta L_{k'j} H_j|0)}{E_n - E_0} \tag{7.9}$$

In the expansion of the numerator, we neglect the spin-spin terms S_iS_j that vanish for $S = 1/2$ and the small diamagnetic terms H_iH_j. The remaining terms that give rise to the **g** anisotropy may be expressed as

$$\mathcal{H}_S^{(2)} = -2\beta \sum_{\substack{n \\ n \neq 0}} \sum_{i,j=x,y,z} \sum_{k,k'} \frac{\lambda_k (0|L_{ki}|n)(n|L_{k'j}|0)}{E_n - E_0} H_i S_j \tag{7.10}$$

By inclusion of the first-order isotropic term [cf. with (2.14)], one can express the molecular-orbital spin Hamiltonian as

$$\mathcal{H}_S = \beta \sum_{i,j=x,y,z} g_{ij} H_i S_i \tag{7.11}$$

where

$$g_{ij} = g_e \delta_{ij} + \Delta g_{ij} \tag{7.12}$$

in which $g_e = 2.0023$ is the free-spin value of the isotropic component and the anisotropic component is expressed by

$$\Delta g_{ij} = -2 \sum_{\substack{n \\ n \neq 0}} \sum_{k,k'} \frac{\lambda_k (0|L_{ki}|n)(n|L_{k'j}|0)}{E_n - E_0} \tag{7.13}$$

The diagonal elements of Δg are

$$\Delta g_{ii} = -2 \sum_{\substack{n \\ n \neq 0}} \sum_{k,k'} \frac{\lambda_k (0|L_{ki}|n)(n|L_{k'i}|0)}{E_n - E_0} \tag{7.14}$$

Because the **L** operators are Hermitian, the matrix product in the numerator of (7.14) is a real, positive quantity. Since E_0 designates the energy of the orbital ground state, the quantity $E_n - E_0$ is positive and hence the spin-orbit couplings, λ_k, determine the sign of Δg_{ii}. In Section 1.C, we describe methods for determination of the signs of λ_k in molecules.

To facilitate the application of (7.14), let us express the molecular orbitals $|0)$ and $|n)$ by the linear combination of normalized atomic orbitals:

$$|0) = \phi_0 = \sum_k c_{0k} \psi_{0k} \tag{7.15}$$

$$|n) = \phi_n = \sum_k c_{nk} \psi_{nk} \tag{7.16}$$

This expression can be simplified by neglect of the effects of orbital overlap and by normalization of the molecular orbitals:

$$(\phi_0|\phi_0) = \sum_k c_{0k}^2 = 1 \tag{7.17}$$

$$(\phi_n|\phi_n) = \sum_k c_{nk}^2 = 1 \tag{7.18}$$

where the coefficients are assumed to be real. When they are not real, c_k^2 is replaced by its complex conjugate $c_k c_k^*$. This normalization is similar to that used for the calculation of nuclear couplings (Chapters II and VI). For nuclear hyperfine structure, it is justified by the inverse cube variation of the coupling with distance of the unpaired electron from the coupling nucleus. Here, it may be similarly justified by the inverse cube variation of the spin-orbit coupling, λ_k, with the distance from the nuclear center of k.

With the wave functions of (7.15) and (7.16) normalized by (7.17) and (7.18), the formula expressed by (7.14) may be written as

$$\Delta g_{ii} = -2 \sum_{\substack{n \\ n \neq 0}} \frac{\sum_k \lambda_k (c_{0k} \Psi_{0k} | L_{ki} | c_{nk} \Psi_{nk}) \sum_{k'} (c_{nk'} \Psi_{nk'} | L_{k'i} | c_{0k'} \Psi_{0k'})}{E_n - E_0}. \qquad (7.19)$$

The coefficient c_{0k}^2 measures the probability that the odd electron will occur in the atomic orbital Ψ_{0k} of atom k, and it is thus equivalent to the spin density of the unpaired electron on atom k, namely, $c_{0k}^2 = \rho_{0k}$. Because of this relationship, c_{0k}^2 can often be obtained from the nuclear coupling of atom k. The coefficients c_{nk}^2 represent the weights of the atomic orbitals Ψ_{nk} in the normalized molecular orbital ϕ_n. The **g** tensor can sometimes be used to give information about the excited molecular orbitals ϕ_n, but usually there are more c_{nk} coefficients than the three measurable principal elements of **g**. Nevertheless, the **g** tensor may contribute significantly to the pool of information required for determination of molecular orbitals in excited states.

Although the derivation of (7.19) is relatively simple, its application to polyatomic molecular radicals is of limited usefulness because of the large number of unknown molecular-orbital parameters, with only one observable **g** tensor. Thus most of the unknown parameters must be assumed or evaluated from other data by calculations which usually can be made only approximately. Furthermore, the application of the formula, even with assumed molecular-orbital parameters, can be quite tedious, primarily because of the several excited-state orbitals usually involved. Another serious difficulty is that the **g** anisotropy in typical organic free radicals is very small and difficult to measure with significant accuracy.

For free radicals having unpaired electron density in p orbitals of more than one component atom, it is sometimes convenient to choose a reference system centered on each of these atoms; the orientations are then chosen with reference to the symmetry of the atomic orbitals, bonding or nonbonding, to that atom. For oriented radicals in single crystals, each of these calculated component tensors must be rotated to common principal axes of the resultant **g** tensor for comparison with the experimental observations. However, for rotating radicals in liquids where only the average $(g_{xx} + g_{yy} + g_{zz})/3$ of the diagonal elements (1/3 trace **g**) is observable, one need not follow this procedure. Stone has shown how to take advantage of the symmetry of certain orbital groups and local structures for calculation of the averaged anisotropy, $\langle \Delta g \rangle = 1/3$ trace Δg, for hydrocarbon and semiquinone ions [5, 6]. His results are useful for comparison with averaged g values observed in liquids (see Chapter IX, Section 4).

1.c. Sign of the Spin-Orbit Coupling Constant and of Δg

Many treatments use instead of λ_k in (7.19) the spin-orbit coupling ζ_k for a single electron which is always a positive quantity. In this case, $\Delta E_n = E_{\phi_n} - E_{\phi_0}$

conforms to the difference in orbital levels rather than to a difference in the energies of the ground state and the excited states of the radical. A change is made in the sign of the transition energies ΔE_n depending on the movement of the electron, that is, whether into or out of the ground-state orbital ϕ_0. As is customary for paramagnetic ions or atoms (Chapter II, Section 2), we have chosen to use an effective spin-orbit coupling λ_k for the configuration centered on atom k with the sign of λ_k positive for the electron transferred out of ϕ_0 and negative for the "hole" that is filled by transfer of an electron into ϕ_0. With this usage, E_0 and E_n signify configuration energies, and $\Delta E_n = E_n - E_0$ is always positive; and, for $S = ½$,

$$\lambda_k = \pm \zeta_k \tag{7.20}$$

with the sign dependent on the nature of the transition that intermixes the ground molecular-orbital state $|0)$ with the excited molecular-orbital state $|n)$, as described in the following paragraphs. Values for ζ are given in Table A.2.

Any transition that removes the single unpaired electron of a molecular free radical (with $S = ½$) from its ground-state orbital ϕ_0 must transfer it to an unoccupied orbital because all other occupied orbitals have electron pairs. In this case, the λ_k constants for the associated atomic orbitals Ψ_{nk} are taken as $+\zeta_k$. The most common transition of this type is to an antibonding molecular orbital. If the induced mixing of the ground state $|0)$ with an excited state $|n)$ involves a transfer of one of the paired electrons into ϕ_0, the associated λ_k constants of (7.19) are taken as $-\zeta_k$. For both types of admixing, E_0 is assumed to be the ground-state configuration energy of the radical, and hence $E_n - E_0$ is positive.

Classically, a particle with charge e and velocity $\mathbf{v}$ in a static magnetic field $\mathbf{H}$ will experience a torque, $e\mathbf{v} \times \mathbf{H}$, which tends to rotate the particle about the direction of the field. Since $\mathbf{v} \times \mathbf{H}$ has no component in the direction of $\mathbf{H}$, a particle moving along the direction of $\mathbf{H}$ is unaffected. If the orbital of the unpaired electron is along x, say, a p_x orbital, a dc magnetic field imposed along y will tend to rotate this orbital in the xz plane. One imposed along z will tend to rotate it in the xy plane. An electron "hole" will be rotated in the sense opposite to that of the electron. However, a field imposed along x will put no torque on an orbital oriented along x. These classical effects of $\mathbf{H}$ are similar to those of the quantum-mechanical, angular-momentum operators described in Section 3. For example, the L_y operator converts the orbital function p_x to $-ip_z$; the L_z operator converts p_x to ip_y, whereas the operator of L_x on p_x gives zero.

2. MATRIX ELEMENTS OF ORBITAL ANGULAR MOMENTUM OPERATORS INVOLVING *p* AND *d* ELECTRONS

Whether calculating the $\mathbf{g}$ tensor for salts of the transition elements or for molecular free radicals, one makes use of angular-momentum matrix elements of atomic orbitals like those in the numerator of (7.2) or (7.19). As is described later, the **g** tensor of a molecular free radical may be expressed in terms of the constituents of its atomic orbitals. In most molecular free radicals observed in irradiated single crystals, the **g** anisotropy results from *p*-orbital constituents of the molecular orbital. The unpaired electrons of the iron-group elements are in the *d* shell and those of the rare earths, in the *f* shell. Because they are needed most often, we give here the matrix elements for the angular momentum as expressed in *p* and *d* orbitals.

In rectangular coordinates, the angular momentum operators for a single electron, in units of ħ, are

$$\begin{aligned} L_x &= l_x = i\left(z\frac{\partial}{\partial y} - y\frac{\partial}{\partial y}\right) \\ L_y &= l_y = i\left(x\frac{\partial}{\partial z} - z\frac{\partial}{\partial x}\right) \\ L_z &= l_z = i\left(y\frac{\partial}{\partial x} - x\frac{\partial}{\partial y}\right) \end{aligned} \tag{7.21}$$

The orientation-dependent factors of the three *p* orbitals are simply x, y, and z, and the orbitals may be expressed as

$$p_x = x\,f(r), \qquad p_y = y\,f(r), \qquad p_z = z\,f(r) \tag{7.22}$$

where the $f(r)$ term is a constant for a particular *np* shell. If the L_x, L_y, and L_z operators of (7.21) are applied to these functions, *p*-orbital functions are generated as given in Table 7.1. For example,

$$L_x|p_x) = i\left(z\frac{\partial}{\partial y} - y\frac{\partial}{\partial z}\right) x\ f(r) = 0 \tag{7.23}$$

$$L_x|p_y) = i\left(z\frac{\partial}{\partial y} - y\frac{\partial}{\partial z}\right) y\ f(r) = iz\,f(r) = i\,p_z \tag{7.24}$$

$$L_x|p_z) = i\left(z\frac{\partial}{\partial y} - y\frac{\partial}{\partial z}\right) z\ f(r) = iy\,f(r) = -ip_y \tag{7.25}$$

With normalized, orthogonal *p* orbitals, $(p_x|p_x) = 1$, $(p_x|p_y) = 0$, and so on. Therefore, the needed matrix elements may also be found from functions given in Table 7.1. For example,

$$(p_x|L_x|p_x) = 0 \tag{7.26}$$

$$(p_x|L_y|p_x) = -i(p_x|p_z) = 0 \tag{7.27}$$

$$(p_x|L_z|p_z) = i(p_x|p_x) = i \tag{7.28}$$

Table 7.1 Orbital Functions Generated by Application of Angular Momentum Operators L_x, L_y, and L_z to p and d Orbitals

ψ_n	$L_x\|\psi_n)$	$L_y\|\psi_n)$	$L_z\|\psi_n)$
p_x	0	$-ip_z$	ip_y
p_y	ip_z	0	$-ip_x$
p_z	$-ip_y$	ip_x	0
d_{xy}	id_{zx}	$-id_{yz}$	$-2id_{x^2-y^2}$
d_{yz}	$(id_{x^2-y^2} + i\sqrt{3}d_{2z^2-x^2-y^2})$	id_{xy}	$-id_{zx}$
d_{zx}	$-id_{xy}$	$(id_{x^2-y^2} - i\sqrt{3}d_{2z^2-x^2-y^2})$	id_{yz}
$d_{x^2-y^2}$	$-id_{yz}$	$-id_{zx}$	$2id_{xy}$
$d_{2z^2-x^2-y^2}$	$-i\sqrt{3}d_{yz}$	$i\sqrt{3}d_{zx}$	0

$$(p_y|L_x|p_z) = -i(p_y|p_y) = -i \tag{7.29}$$

Similarly, the functions generated by operation of L_x, L_y, and L_z on the d orbitals given in (1.27) and (1.28) may be found. For example,

$$L_x|d_{xy}) = i\left(z\frac{\partial}{\partial y} - y\frac{\partial}{\partial z}\right) xy\, f(r) = izx\, f(r) = id_{zx} \tag{7.30}$$

$$L_y|d_{x^2-y^2}) = i\left(x\frac{\partial}{\partial z} - z\frac{\partial}{\partial x}\right)\left(\frac{x^2 - y^2}{2}\right) f(r) = -izx\, f(r)$$

$$= -id_{zx} \tag{7.31}$$

These and other orbital functions thus generated by operation on d orbitals with L_x, L_y, and L_z are listed in Table 7.1. Required matrix elements of d orbitals may be found easily from them. Some examples are

$$(d_{xy}|L_x|d_{x^2-y^2}) = -i(d_{xy}|d_{yz}) = 0 \tag{7.32}$$

$$(d_{yz}|L_x|d_{x^2-y^2}) = -i(d_{yz}|d_{yz}) = -i \tag{7.33}$$

Since the p- and d-shell orbitals are orthogonal, all matrix elements of their mixed components $(p|L_i|d)$ are zero.

3. SYMMETRIC AB_2 RADICALS

In Table 7.2 are listed observed principal g values for a number of related triatomic free radicals having C_{2v} symmetry. Note that CO_2^- and NO_2 have g values that are less than the free-spin value of 2.0023, Δg negative, whereas the other radicals have $\Delta g \sim 0$ or positive. We give a semiquantitative treatment of CO_2^- and SO_2^- that accounts for these basic differences in relative g values.

Table 7.2 Principal g Values for Some AO_2 Radicals

Host Crystal	*Radical*	*Valence-shell Electrons*	*Radical Type*	g_x	g_y	g_z	*Source*
HCO_2Na	CO_2^-	17	σ	2.0032	1.9975	2.0014	*a*
$CaCO_3$	CO_2^-	17	σ	2.00320	1.99725	2.00161	*b*
$NaNO_3$	NO_2	17	σ	2.0057	1.9910	2.0015	*c*
KCl	NO_2^{--}	19	π	2.0038	2.0099	2.0070	*d*
$K_2S_2O_5$	SO_2^-	19	π	2.0019	2.0120	2.0057	*e*
$KClO_4$	ClO_2	19	π	2.0036	2.0183	2.0088	*d*

[a] Ovenall and Whiffen [4].
[b] Marshall, Reinberg, Serway, and Hodges [9].
[c] Zeldes and Livingston [13].
[d] J. R. Morton, *Chem. Rev.*, **64**, 453 (1964).
[e] Reuveni, Luz, and Silver [16].

The molecular orbitals of a bent triatomic molecule of C_{2v} symmetry, like the radicals in Table 7.2, are classified according to their symmetry or antisymmetry with respect to reflection in the two planes of symmetry, the molecular plane and the perpendicular plane that bisects the AB_2 angle [7, 8]. With reference to the coordinate system shown in Fig. 7.1, where x is perpendicular to the molecular plane, y is the BB direction, and z is along the bisector of angle AB_2, the two planes of symmetry are xz and yz. Each molecular orbital is either symmetric or antisymmetric with respect to each of these planes. Those that are symmetric to reflection in both planes (++ symmetry) are conventionally desig-

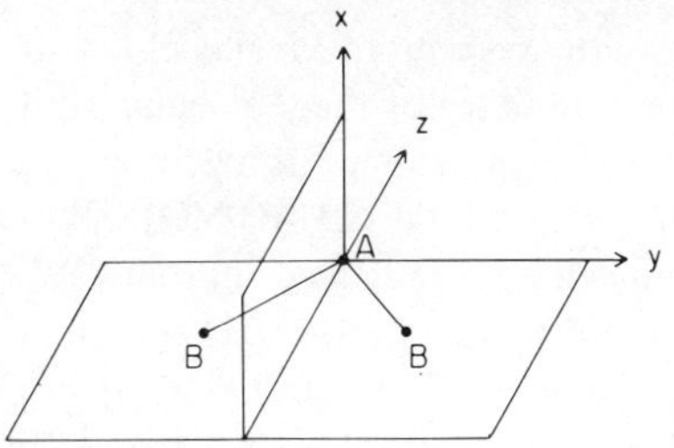

Fig. 7.1 Coordinate system used for **g** calculation of a bent AB_2 molecule.

nated a_1 orbitals; those that are antisymmetric to reflection in both planes (−− symmetry) are designated a_2 orbitals; those that are plus with respect to the xz plane and minus with respect to the yz plane are designated b_1 orbitals; and

Table 7.3 Orbital Classification of Molecules with C_{2v} Symmetry

Orbital Class	*Symmetry Reflection Plane*[a] *xy*	*yz*	*Atomic-Orbital Combinations for* AB_2
a_1	+	+	$(s)_A, (p_z)_A, (s+s')_B, (p_y-p'_y)_B\ (p_z+p'_z)_B$
a_2	−	−	$(p_x - p'_x)_B$
b_1	+	−	$(p_x)_A, (p_x + p'_x)_B$
b_2	−	+	$(p_y)_A, (s - s')_B, (p_y + p'_y)_B, (p_z - p'_z)_B$

[a]Designation of coordinates shown in Fig. 7.2.

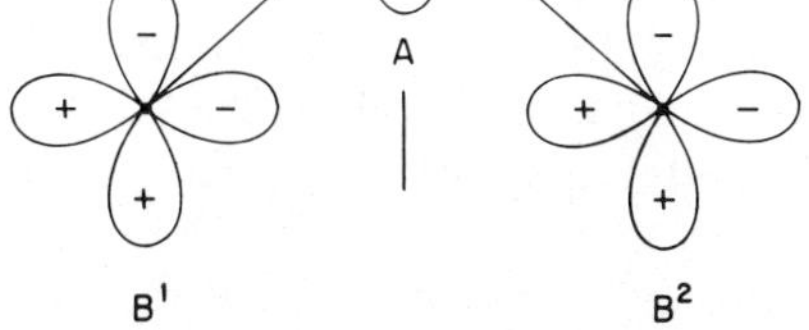

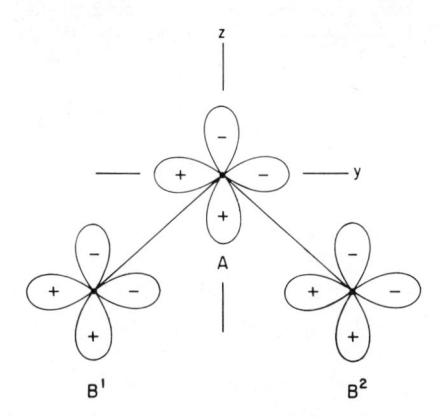

Fig. 7.2 Diagram showing symmetries of the orbitals used in g analysis of the CO_2^- and SO_2^- radicals.

those that are minus with respect to xz and plus with respect to yz are designated b_2 orbitals. The symmetry of the molecular orbital may be determined by the symmetry of the group of atomic orbitals used in the linear combination of atomic orbitals - molecular orbital (LCAO-MO) combinations. Possible combinations for the AB_2 species are indicated in Table 7.3. Figure 7.2 indicates the symmetries of the atomic p_y and p_z orbitals chosen in the LCAO-MO combinations for analysis of CO_2^- and SO_2^- radicals. The p_x orbitals (not shown) are perpendicular to this plane. As indicated approximately by the Walsh diagram [8] of Fig. 7.3, the relative energies of the levels of AB_2 molecules depend on the

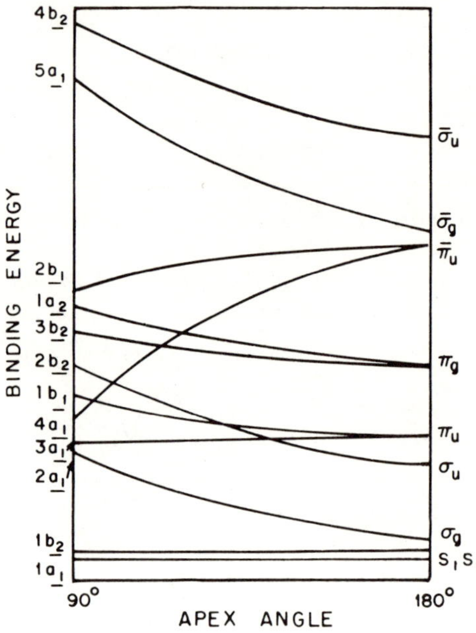

Fig. 7.3 Orbital energy levels of AB_2 molecules as a function of the apex angle. From Walsh [8].

bond angle. Figure 7.4 shows the relative sequence of the occupied molecular orbitals of CO_2^-. The unpaired electron has the orbital of highest energy among those occupied in the ground state. In calculating the g values we require only the specific form of this orbital designated ϕ_0 and of those orbitals sufficiently near to become significantly intermixed with it.

The orbital of the unpaired electron in CO_2^- is the $4a_1$ orbital along the external bisector of the bond angle (along the z axis) [4]. Expressed as a linear combination of the atomic orbitals having the a_1 symmetry, it is

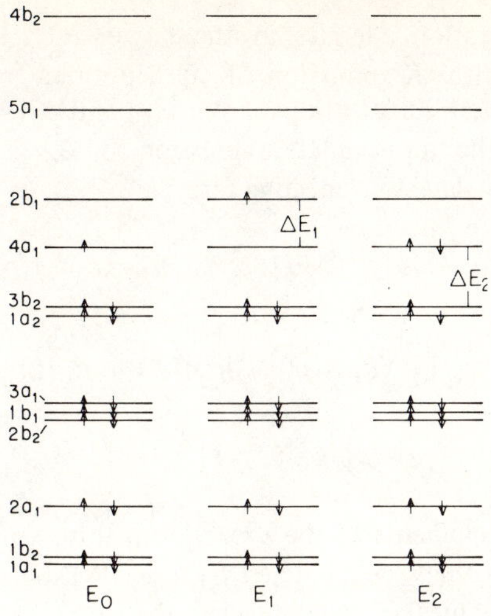

Fig. 7.4 Energy-level diagram of CO_2^- showing excited states important in determination of the **g** tensor.

$$\phi_0(4a_1) = c_1 s(\mathrm{C}) + c_2 p_z(\mathrm{C}) + c_3 [p_z(\mathrm{O}^1) + p_z(\mathrm{O}^2)]$$

$$+ c_4 [(p_y(\mathrm{O}^1) - p_y(\mathrm{O}^2)] + c_5 [s(\mathrm{O}^1) + s(\mathrm{O}^2)] \tag{7.34}$$

Although this orbital satisfies the requirements of symmetry, the s components are isotropic and are included only for normalization of the coefficients. For calculation of the g values, only the second and third terms on the right are significant. The orbital level nearest to the ϕ_0 $(4a_1)$ is that of the $2b_1$ orbital, which is unoccupied in the ground state (see Fig. 7.4). This orbital, which is intermixed with $4a_1$ when the field is applied along y, is an antibonding π orbital perpendicular to the molecular plane; it may be expressed as

$$\phi_1(2b_1) = d_1 p_x(\mathrm{C}) + d_2 [p_x(\mathrm{O}^1) + p_x(\mathrm{O}^2)] \tag{7.35}$$

In these calculations we use the CO_2^- g values measured in $CaCO_3$ by Marshall, Reinberg, et al. [9]. The values measured earlier by Ovenall and Whiffen [4] are in good agreement with them (see Table 7.2). Because of symmetry, the x, y, and z axes as chosen will be principal axes of the **g** tensor. In the observed principal values, the largest deviation from g_e is g_y, for which $\Delta g_y = 1.99727 - 2.0023 = -0.00506$. This deviation arises primarily from the induced admixture of the ground state, containing the unpaired electron in $\phi_0(4a_1)$, with the first excited state, which has the unpaired electron in $\phi_1(2b_1)$, as indicated by Fig. 7.4. The difference in configuration energies of these two states is $\Delta E_1 = E_1 - E_0 = 29{,}400\ cm^{-1}$ [10]. With the omission of contributions from intermixing with other excited states and substitution of $\phi_0(4a_1)$ and $\phi_1(2b_1)$ into (7.14), one obtains, when $i = y$, the approximate expression for Δg_{yy}. All matrix elements of the p_y terms vanish. The nonvanishing terms give

$$\Delta g_{yy} = -\frac{2}{\Delta E_1} \{\lambda_C c_2 d_1 (p_z|L_y|p_x)_C + \lambda_O c_3 d_2 [(p_z|L_y|p_x)_{O^1}$$

$$+ (p_z|L_y|p_x)_{O^2}]\} \{c_2 d_1 (p_x|L_y|p_z)_C + c_3 d_2 [(p_x|L_y|p_z)_{O^1}$$

$$+ (p_x|L_y|p_z)_{O^2}]\} \tag{7.36}$$

Values of the matrix elements of the p orbital in this expression, $(p_z|L_y|p_x) = -i$ and $(p_x|L_y|p_z) = i$, may be obtained by use of Table 7.1, as explained in Section 2. With these values, (7.36) reduces to the form

$$\Delta g_{yy} = -\frac{2}{\Delta E_1} (\lambda_C c_2 d_1 + 2\lambda_O c_3 d_2)(c_2 d_1 + 2 c_3 d_2) \tag{7.37}$$

This final expression for Δg_{yy} is equivalent to that originally obtained by Ovenall and Whiffen [4]. Because the unpaired electron goes into the previously unpopulated orbital $2b_1$ (see Fig. 7.4), the effective λ_C and λ_O are positive, and hence Δg_{yy} is negative.

With $\lambda_C = 28\ cm^{-1}$ and $\lambda_O = 151\ cm^{-1}$ (Table A.2) and $\Delta E_1 = 29{,}400\ cm^{-1}$ [10] substituted along with the observed Δg_{yy} into (7.37), one obtains the following relationship between the orbital coefficients:

$$c_2^2 d_1^2 + 21.7\, c_3^2 d_2^2 + 12.9\, c_2 d_1 c_3 d_2 = 2.66 \tag{7.38}$$

This application illustrates the use as well as the limitations of the **g**-tensor elements. There are four orbital constants in this equation. If three of them can be evaluated by other means, the fourth can be obtained from solution of the equation. As explained in the following paragraphs, c_2 can be obtained from the ^{13}C nuclear coupling, and d_1 and d_2 can be estimated from related coefficients of CO_2.

Ovenall and Whiffen [4] also measured the ^{13}C nuclear coupling in the CO_2^- radical and found the anisotropic component to be 78 MHz along the z, -32 MHz along the x, and -46 MHz along y. We use the later, more accurate values: 82.4 MHz, -37.1 MHz, and -45.2 MHz, respectively, as measured by Marshall et al. [9] in the $CaCO_3$ crystal (see Table 7.2). If one neglects effects of spin polarization and charge correction and attributes the departure from axial symmetry to a small spin density in the p_x orbital, one then can set

$$82.4 = 2B_z - B_x$$
$$-37.1 = -B_z + 2B_x$$
$$-45.2 = -B_z - B_x$$

The solution gives $B_x = 2.7$ MHz and $B_z = 42.6$ MHz. With $B_{\text{atomic}} = 90.8$ (Table A.1), these give for the $2p_z$ spin density $\rho^{C}_{2p_z} = (42.6/90.8) = 0.469$ and for $2p_x$ spin density $\rho^{C}_{2p_x} = 0.03$. Since $\rho^{C}_{2p_z}$ is a measure of the weight of the $2p_z$ (C) orbital in the wave function of the unpaired electron, $c_2^2 = \rho^{C}_{2p_z} = 0.469$ or $c_2 = 0.685$. These values differ appreciably from the 0.66 and 0.81 obtained by Ovenall and Whiffen [4], primarily because they used an estimated B_{atomic} that differs appreciably from that used here. The coefficient c_1 is easily obtainable from the isotropic component in the ^{13}C coupling of 414.4 MHz [9]. Thus $c_1^2 = \rho^{C}_{2s} = (414.4/A_{\text{atomic}}) = (414.4/3110) = 0.133$; hence, $c_1 = 0.365$. For d_1 and d_2, we use 0.75 and -0.47, respectively, which are the values for the corresponding $2\pi_\mu$ orbital of CO_2 [11], renormalized with $d_1^2 + 2d_2^2 = 1$ and used in the earlier analyses [4]. It is interesting that these values of the coefficients are the same as those derived from nuclear coupling for the corresponding orbital in SO_2^- [see Eq. (7.51)]. These values of c_2, d_1, and d_2 substituted into (7.38) yield $c_3 = -0.455$ or 1.11. The latter value is rejected because it violates the normalization requirements. By substitution of the other value, $c_3^2 = (-0.455)^2 = 0.207$ with the preceding $c_1^2 = 0.133$ and $c_2^2 = 0.469$ into the normalizing relation

$$c_1^2 + c_2^2 + 2c_3^2 + 2c_4^2 + 2c_5^2 = 1$$

one obtains

$$c_4^2 + c_5^2 = -0.008 = \sim 0$$

It thus seems probable that the $2s$(O) and the $2p_y$(O) contributions to the odd-electron orbital $\phi_0(4a_1)$ are entirely negligible and that the orbital of the unpaired electron can be expressed approximately as

$$\phi_0(4a_1) = 0.365s(\text{C}) + 0.685p_z(\text{C}) - 0.455\ [p_z(\text{O}^1) + p_z(\text{O}^2)] \qquad (7.39)$$

Now let us consider the principal element perpendicular to the molecular plane, g_x. Since $\Delta g_{xx} = (2.0032 - 2.0023) = 0.0009$ is positive, it must result primarily from an admixing of $\phi_0(4a_1)$ with the next lower, filled orbital, $3b_2$, which from the C_{2v} symmetry may be written as

$$\phi_2(3b_2) = e_1 p_y(\text{C}) + e_2\,[p_y(\text{O}^1) + p_y(\text{O}^2)] + e_3\,[p_z(\text{O}^1) - p_z(\text{O}^2)] + e_4\,[s(\text{O}^1) - s(\text{O}^2)] \qquad (7.40)$$

To calculate Δg_{xx}, we neglect possible small contributions from admixture with other higher excited states and substitute $\phi_2(3b_2)$ with the ground-state orbital $\phi_0(4a_1)$ into (7.14). All matrix elements with the $p_z(\mathrm{O})$ orbitals vanish when $i = x$, and the s orbitals make no contribution because of their spherical symmetry. The resulting expression for Δg_{xx} is

$$\Delta g_{xx} = -\frac{2}{\Delta E_2}(\lambda_\mathrm{C} c_2 e_1 + 2\lambda_\mathrm{O} c_3 e_2)(c_2 e_1 + 2c_3 e_2) \tag{7.41}$$

This equation is similar in form to that for Δg_{yy}, in (7.37). However, it should be noted that the effective spin-orbit couplings λ_C and λ_O are negative here since the excitation may be considered as transfer of an electron "hole" from ϕ_0 to ϕ_2 (see Fig. 7.4). Substitution into (7.41) of the values $\Delta g_{xx} = 0.0009$, $\lambda_\mathrm{C} = -28\ \mathrm{cm}^{-1}$, $\lambda_\mathrm{O} = -152\ \mathrm{cm}^{-1}$, $c_2 = 0.685$, and $c_3 = -0.455$, along with $\Delta E_2 = 35{,}700\ \mathrm{cm}^{-1}$ from optical spectroscopy [10] gives

$$e_1^2 + 9.58e_2^2 - 8.54e_1e_2 = 1.22$$

An extra parameter prevents solution of this equation. We can, however, reduce the number of parameters by the reasonable assumption that the e_4 coefficient in the expansion (7.40) is negligibly small and that $e_3 = e_2$. Normalization of ϕ_2 then requires that $e_1^2 + 4e_2^2 = 1$. Substitution of $e_1^2 = 1 - 4e_2^2$ into the preceding equation and solution yields $e_2^2 = 0.234$ or zero, and the normalization requires that $e_1^2 = 0.06$ or 1.00, respectively. We choose the first set because it is evident that the electron pair cannot be localized in $p_y(\mathrm{C})$. If $e_3 = e_2$, then $\phi_2(3\mathrm{b}_2) \approx (1/2)\,[p_y(\mathrm{O}^1) + p_y(\mathrm{O}^2)] + (1/2)\,[p_z(\mathrm{O}^1) - p_z(\mathrm{O}^2)]$, and thus the electron pair would be essentially localized on the oxygens. Ovenall and Whiffen [4] describe the orbital as a pseudo-π, nonbonding orbital. Unfortunately, we have no sure way of testing this assumption. If we assume that e_3 and e_4 are both zero, the solution yields $e_1^2 = 0.26$ and $e_2^2 = 0.37$.

The most puzzling principal element of the **g** tensor for CO_2^- is that for the z direction, which is along the symmetrv axis of the odd-electron orbital $\phi_0(4a_1)$. Since $L_z|p_z) = 0$, any intermixing of $\phi_0(4a_1)$ with excited-state levels caused by a field imposed along z must be induced through the $c_4\,[p_y(\mathrm{O}^1) - p_y(\mathrm{O}^2)]$; but, as we have seen earlier, the c_4 coefficient, if not zero, is very small. The observed $g_z = 2.00161$ indicates that $\Delta g_{zz} = -0.00069$. The negative sign of Δg would require the induced interaction to occur with one of the unoccupied orbitals, $\phi_0(4a_1)$. Because of the uncertainties involved, we do not attempt a calculation of this small Δg_{zz}.

The **g** tensor of NO_2 is very similar to that of CO_2^- and may be analyzed in a similar way. Each has g_x above, g_y below, and g_z below, but near, g_e (see Table 7.2). In each, the odd electron is evidently in a $\phi_0(4a_1)$ orbital directed along the z axis, as for CO_2^-.

Presumably, CS_2^- also has an electronic structure similar to CO_2^-. Bennett et al. [12] measured the CS_2^- radical, not in oriented crystals but in samples prepared by deposit of Na or K atoms on solid CS_2 at 77°K in a rotating cryostat. They inferred the directions of the observed principal elements relative to the molecular coordinates by comparison with those of CO_2^-. They were able to measure the ^{13}C coupling, from which $\phi_0(4a_1)$ orbital weights $c_1^2 = 0.078$ and $c_2^2 = 0.561$ were derived. From an analysis of Δg_{yy} similar to that described for CO_2^-, they obtained $c_3 = -0.44$ with $\Delta E_1 = 10{,}000 \text{ cm}^{-1}$ and $c_3 = -0.50$ if ΔE_1 is taken as a second possible value of $12{,}000 \text{ cm}^{-1}$. With the first value of $c_3 = 0.19$, the normalization gives $0.078 + 0.561 + 2(0.19) + 2c_4^2 + 2c_5^2 = 1$, from which it is seen that $c_4^2 + c_5^2 = -0.001 = \sim 0$. Thus the odd-electron orbital for CS_2^- appears to be

$$\phi_0(4a_1) = 0.28s(C) + 0.75p_z(C) - 0.44\,[p_z(S^1) + p_z(S^2)] \qquad (7.42)$$

which compares rather closely with that for CO_2, (7.39).

Zeldes and Livingston [13], who measured the g values listed in Table 7.2 for NO_2, also measured the ^{14}N nuclear couplings from which the values $B_z = 19.8$ MHz, $B_x = 2.7$ MHz, and $A_{iso}^N = 153.2$ MHz may be derived. With $A_{atomic}^N = 1540$ MHz and $B_{atomic}^N = 47.8$, these values indicate that $\rho_{2s}^N = 0.10 = c_1^2$ and $\rho_{2p}^N = 0.41 = c_2^2$. With these constants, an analysis of the **g** tensor can be made, like that for CO_2^-. If the c_4^2 and c_5^2 weights are approximately zero, as for the $\phi_0(4a_1)$ orbital for CO_2^-, one can determine $c_3^2 = 0.25$ from the normalization of the orbital by $c_1^2 + c_2^2 + 2c_3^2 = 1$. This value of c_3^2 is near that of 0.21 for CO_2^-.

As the number of the valence-shell electrons in the oxide radicals of Table 7.2 increases by two, the orientation of the orbital of the unpaired electron shifts from the z to the x direction. The first electron added goes into the lowest unfilled orbital, $4a_1$, to fill this orbital and to render the molecule nonmagnetic. The second goes into the next lowest orbital, which is $2b_1$. Thus when there are 19 electrons in the valence shell, the ground-state orbital of the unpaired electron becomes the $2b_1$ orbital, which has its orientation perpendicular to the molecular plane, that is, along the x direction. This shift in the orbital of the odd electron has the effect of changing Δg_{yy} and Δg_{xx} from negative to positive because the lowest excited transition, which, as before, corresponds to the movement of an electron from $4a_1$ to $2b_1$ now shifts the unpaired spin in the opposite sense, from $2b_1$ to $4a_1$. This transfer of spin, in the sense opposite to that for the charge transfer, may be treated most conveniently as the transfer in the same sense of a positive electron (or electron "hole") with normal spin but with negative spin-orbit coupling. The effects on the principal g values may be seen readily from a comparison of the g values for NO_2 and NO_2^{--} in Table 7.2.

The SO_2^- and ClO_2 radicals have two more valence-shell electrons than has CO_2^- or NO_2 and are like NO_2^{--} in having the odd electron in the antibonding, π-type orbital $2b_1$, which is perpendicular to the molecular plane. Such species are usually designated as π-type radicals; those like CO_2^- with the odd electron having an *sp*-hybridized orbital in the molecular plane are often designated as σ-type radicals. This distinction, however, is not always clear. For example, the $4a_1$ orbital may be regarded as a pseudo-π type of molecular orbital because it originates from the π_y orbital of the linear CO_2 molecule and because the unpaired electron density is not localized but occurs on all three atoms. Nevertheless, the two types of radicals can usually be distinguished by the direction of their principal *g* values. Theoretically, the π-type radical has its principal *g* value, which is perpendicular to the molecular plane, equal to, or near, the free-spin value, whereas the σ type has this value near to g_e in the molecular plane. These qualitative features are readily evident in the *g* values for the two classes of radicals in Table 7.2.

The epr spectrum of the SO_2^- radical has been measured by Schneider and coworkers [14, 15] as an impurity in single crystals of KCl and by Reuveni et al. [16] in a γ-irradiated single crystal of $K_2S_2O_5$. Calculations of the Δg values for SO_2^- have been made by Dinse and Möbius [17]. Analysis of the principal *g* values measured by Reuveni and colleagues are given here to illustrate the application of (7.19) to π-type radicals.

The forms of the molecular orbitals for SO_2^- are the same as those already described for CO_2^-, except that the ground-state and first excited-state orbitals are exchanged and the atomic orbitals of S replace those of C. The primary effects on **g** occur through the intermixing of the ground state with the first and second excited states. The molecular orbitals involved are

$$\phi_0(2b_1) = d_1 p_x(\mathrm{S}) + d_2\,[p_x(\mathrm{O}^1) + p_x(\mathrm{O}^2)] \tag{7.43}$$

$$\phi_1(3a_1) = c_1 s(\mathrm{S}) + c_2 p_z(\mathrm{S}) + c_3\,[p_z(\mathrm{O}^1) + p_z(\mathrm{O}^2)] + c_4\,[p_y(\mathrm{O}^1) - p_y(\mathrm{O}^2)] + c_5\,[s(\mathrm{O}^1) + s(\mathrm{O}^2)] \tag{7.44}$$

$$\phi_2(2b_2) = e_1 p_y(\mathrm{S}) + e_2\,[p_y(\mathrm{O}^1) + p_y(\mathrm{O}^2)] + e_3\,[p_z(\mathrm{O}^1) - p_z(\mathrm{O}^2)] + e_4\,[s(\mathrm{O}^1) - s(\mathrm{O}^2)] \tag{7.45}$$

Calculation of the Δg_{ii} for the SO_2^- orbitals is like that already described for CO_2^-.

Preliminary examination of the orbitals and those generated by the L_x operators can greatly simplify such calculations. Since $L_x|p_x) = 0$, all matrix elements involving $L_x|\phi_0)$ vanish, and it is immediately evident that (7.14) predicts that

$$\Delta g_{xx} = 0 \qquad\qquad g_x = g_e \tag{7.46}$$

The operation $L_y|p_x) = -ip_z$ and, hence the function generated by $L_y|\phi_0)$ gives p_z orbitals only when ϕ_0 is $2b_1$. Because of orthogonality, terms involving

$(p_z|p_x)$ or $(p_z|p_y)$ vanish. Therefore, only the z components of the orbitals ϕ_1 and ϕ_2 can lead to intermixing of the ground and excited states by a field imposed along the y axis. Hence in the calculations of Δg_{yy} with (7.14) one needs to include only terms involving p_z orbitals of ϕ_1 and ϕ_2. Effects of the $p_z(O^1)$ and $p_z(O^2)$ components on $\phi_2(2b_2)$ cancel because of their equivalent coefficients, and the s orbitals make no contribution to the anisotropy because of their spherical symmetry. With these simplifications, it is quickly seen that the expression for Δg_{yy} is like that for CO_2^-, with the C notations changed to those for S. The resulting formula for calculations of Δg_{yy} from the orbital coefficients is

$$\Delta g_{yy} = -\frac{2}{\Delta E_1}(\lambda_S d_1 c_2 + 2\lambda_O d_2 c_3)(d_1 c_2 + 2d_2 c_3) \tag{7.47}$$

Although this formula has the same form as (7.37), a negative sign must now be used for λ, and the numerical values of the parameters may differ.

In calculation of Δg_{zz}, note that the operator $L_z|\phi_0)$ generates only p_y orbitals. The only p_y orbitals in $\phi_1(3a_1)$ are those in the c_4 term, which cancels because of its equal and opposite components. Thus one needs to include in the calculation of Δg_{zz} only the e_1 and e_2 terms of $\phi_2(2b_2)$, those involving p_y orbitals. With these exclusions, it is readily seen that

$$\begin{aligned}\Delta g_{zz} = -\frac{2}{\Delta E_2}\{\lambda_S d_1 e_1 (p_x|L_z|p_y)_S + \lambda_O d_2 e_2 [(p_x|L_z|p_y)_{O^1} \\ + (p_x|L_z|p_y)_{O^2}]\}\{d_1 e_1 (p_y|L_z|p_x)_S + d_2 e_2 [(p_y|L_z|p_x)_{O^1} \\ + (p_y|L_z|p_x)_{O^2}]\}\end{aligned} \tag{7.48}$$

which reduces to

$$\Delta g_{zz} = -\frac{2}{\Delta E_2}(\lambda_S d_1 e_1 + 2\lambda_O d_2 e_2)(d_1 e_1 + 2d_2 e_2) \tag{7.49}$$

Like those in (7.47) for Δg_{yy}, the effective spin-orbit couplings are negative, and hence Δg_{zz} is positive.

In the numerical calculation of the Δg values for SO_2^-, it is necessary to estimate or assume the coefficients for the excited-state orbitals. Those for the ground-state orbital ϕ_0 are known from the ^{33}S and ^{17}O nuclear coupling, as described in Chapter VI Section 10.a. Values derived there for $\rho_{3p_x}^S$ and $\rho_{2p_x}^O$ indicate that $d_1^2 \approx 0.59$ and $d_2^2 \approx 0.23$. However, we reduce these values to 0.56 and 0.22, respectively, to conform to the normalization $d_1^2 + 2d_2^2 = 1$. Because $\phi_0(2b_2)$ is an antibonding orbital, opposite signs are chosen for the coefficients, which are $d_1 = +\sqrt{0.56} = 0.75$ and $d_2 = -\sqrt{0.22} = -0.47$. With these d values, with $\lambda_S = -382\ \text{cm}^{-1}$ and $\lambda_O = -152\ \text{cm}^{-1}$, with $\Delta g_{yy} = 0.00097$ (Table 7.2), and with $\Delta E_1 = 34{,}500\ \text{cm}^{-1}$ and $\Delta E_2 = 38{,}500\ \text{cm}^{-1}$ [17], (7.49) reduces to one having only two orbital parameters, c_2 and c_3. One of these must be eliminated before the equation can be solved. For this purpose, we use the normalization relation between orbital coefficients. Although this relation introduces new orbital parameters, these can be eliminated by reasonable assumptions and approximations. For the $4a_1$ orbital, we assume that $c_4^2 + c_5^2 = 0$, as found for the same orbital in CO_2^-, and normalize the other coefficients by $c_1^2 + c_2^2 + 2c_3^2 = 1$. Further reduction of the c coefficients can be obtained from the approximately known bond angle by the Coulson formula [18] relating the angle θ between equivalent sp^2-hybrid orbitals with the degree of hybridization. This relation is usually expressed as $\cos\theta = (-1/\lambda^2)$, where $(1/\lambda^2) = (a_s^2/b_p^2)$ is the ratio of $s{:}p$ character of the orbitals. For a normalized sp hybrid, $a_s^2 = 1 - b^2$, and the relation may be expressed in terms of the s character only by

$$s\ \text{Character} = a_s^2 = \frac{\cos\theta}{\cos\theta - 1} \qquad (7.50)$$

The quantity $[c_1^2/(c_1^2 + c_2^2)]$ is equivalent to the s character in the apex orbital of S in $\phi_1(4a_1)$. Thus each of the equivalent σ-bonding orbitals has s character, $a_s^2 = (1/2)\ [1 - c_1^2/(c_1^2 + c_2^2)]$. This expression for a_s^2 and the bond angle $\theta = 115° \pm 5°$ [17] substituted into (7.50) gives $c_1^2 = 0.683\ c_2^2$, which with the normalization $c_1^2 + c_2^2 + 2c_3^2 = 1$ yields $c_2^2 = 0.594\ (1 - 2c_3^2)$ or $c_2 = \pm 0.77\sqrt{1 - 2c_3^2}$. Since the other parameters are known, use of this relation to eliminate c_2^2 allows a solution of (7.49) for c_3^2. Possible values obtained are $c_3^2 = 0.011$ or 0.453. The first value, the more probable one, requires the associated constants $c_1^2 = 0.396$ and $c_2^2 = 0.581$, or $c_1^2 + c_2^2 = 0.98 \approx 1.00$. This result indicates that the two electrons of the ϕ_1 orbital, that most easily excited, are essentially a nonbonding pair in the sp_z hybrid of the S. The second value of c_3^2, which gives $c_1^2 = 038$, $c_2^2 = 0.056$, and $2c_3^2 = 0.904$, would require the two electrons of the first excited state to be localized mostly on the oxygens. This solution is not acceptable because the corresponding value of ΔE_1 would be appreciably greater than that expected for the electrons localized on the S.

When the observed values $\Delta g_{zz} = 0.0034$ (see Table 7.2) and $\Delta E_2 = 38{,}500\ \text{cm}^{-1}$ [17] with the preceding d_1 and d_2 values and spin-orbit coupling are substituted into (7.49), it reduces to one having only the two coefficients e_1 and e_2 of orbital $\phi_2(2b_2)$. These are reduced by the normalization $e_1^2 + 2e_2^2 + 2e_3^2 + 2e_4^2 = 1$ with the assumptions that $e_3 = e_2$ and $e_4 = 0$. The normalization then gives $e_1^2 = 1 - 4e_2^2$. With the use of this relation and the indicated values for other parameters, (7.49) may then be solved to give $e_2^2 = 0.246$ or 0.062. The first, more probable, value gives $4e_2^2 = 0.984$ and hence $e_1^2 = 0.016 \approx 0$. This solution indicates the pair of electrons of the ϕ_2 orbital to be on the oxygens. The second solution requires with the normalization that $e_2^2 = 0.75$ and hence that 75% of this electron pair be in the $p_y(\mathrm{S})$ orbital. This solution is rejected because the $p_y(\mathrm{S})$ orbitals are taken mostly by the two σ-bond orbitals.

To summarize, this analysis indicates that the ground-state and first two excited-state orbitals of the SO_2^- radical are approximately

$$\phi_0(2b_1) = 0.75p_x(\mathrm{S}) - 0.47[p_x(\mathrm{O}^1) + p_x(\mathrm{O}^2)] \tag{7.51}$$

$$\phi_1(3a_1) \approx 0.64s(\mathrm{S}) + 0.77p_z(\mathrm{S}) \tag{7.52}$$

$$\phi_2(2b_2) \approx \frac{1}{2}[p_y(\mathrm{O}^1) + p_y(\mathrm{O}^2)] + \frac{1}{2}[p_z(\mathrm{O}^1) - p_z(\mathrm{O}^2)]. \tag{7.53}$$

4. ORIENTED ORGANIC FREE RADICALS

4.a. The π and σ Radicals with C_{2v} Symmetry

From its esr spectra, Heller and Cole [19] identified the symmetrical radical T_π in a single crystal of potassium hydrogen maleate, X-irradiated and observed at room temperature. The radical is formed by the removal of the H of the symmetrical hydrogen bridge of the parent molecule **7** and has structure **8**.

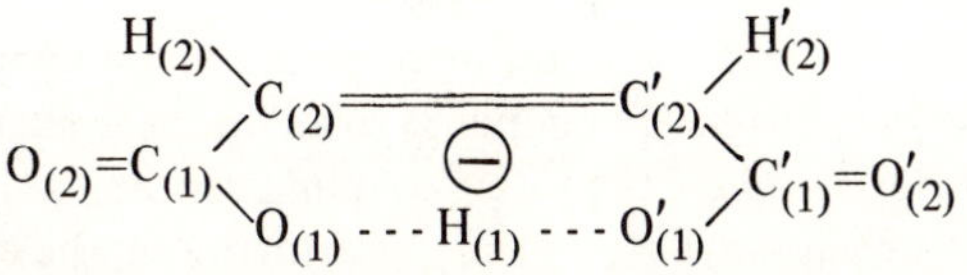

Parent molecule

7

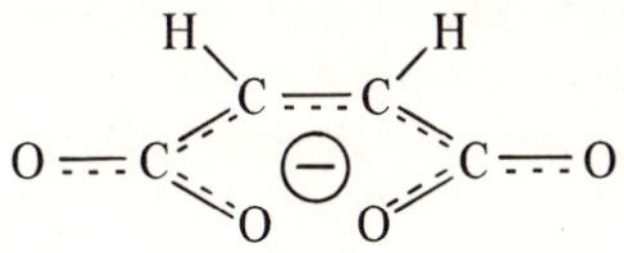

T_π Radical

8

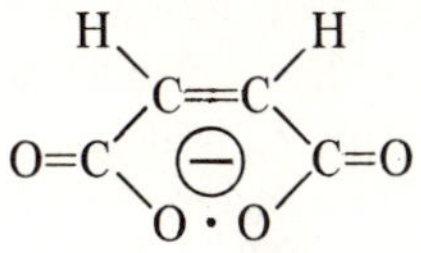

T_σ Radical

9

Like the parent molecule, the T_π radical is planar with C_{2v} symmetry. The observed **g** tensor, as well as the anisotropic component of the CH proton coupling, conforms to that of a π-type radical in which the unpaired electron is in a π orbital that extends over the carbons and oxygens of the two symmetrical halves. Later, Toriyama and Iwasaki [20] interpreted the esr of a radical that they observed in a single crystal of potassium hydrogen maleate, X-irradiated and observed in darkness at 77°K, to be a σ-type radical, that designated as T_σ in structure **9**, with the unpaired electron localized in an antibonding σ^*-orbital between $O_{(1)}$ and $O'_{(1)}$. Like the T_π radical observed earlier, this σ radical has C_{2v} symmetry and, in fact, the same chemical form. Table 7.4 gives the principal g values and their directions for both the π and σ types. Those for the π species are from Iwasaki and Itoh [21].

Because of their C_{2v} symmetry, the **g** tensors of these organic radicals can be treated in the same way as were those of the π and σ radicals of nonlinear AB_2 type (Section 3). The x,y,z reference axes chosen as before with x perpendicular to the moelcular plane and with z along the intersection of the two symmetry planes will, because of symmetry, be principal axes of the **g** tensor. The form of the atomic-orbital combinations in the molecular orbital may be obtained from the symmetry requirements described in Section 3. For example, there are four symmetric (a_2) π-type and four antisymmetric (b_1) π-type atomic combinations. These have the form

$$O_{(1)}(p_x \pm p'_x), \quad C_{(1)}(p_x \pm p'_x), \quad C_{(2)}(p_x \pm p'_x), \quad O_{(2)}(p_x \pm p'_x)$$

where the plus signs apply to the a_2-type orbitals and the minus signs apply to the b_1-type orbitals. They may be used to construct eight linearly independent molecular orbitals, $\phi_1, \phi_2, \ldots, \phi_8$, of which four are of the a_1 (π) type and four are of the $b_1(\pi)$ type. Unfortunately, there are no measurements, as far as I am aware, of the ^{13}C or ^{17}O nuclear couplings to aid in the evaluation of the coefficients, although the isotopic CH proton coupling can be used for estimation of the unpaired spin density on $C_{(2)}$ and $C'_{(2)}$. Furthermore, the energy separations, $\Delta E_n = E_n - E_0$, of the states intermixed by the applied field **H** have not been measured. The application of (7.19), although straightforward, is a bit more complicated here than for the AB_2 molecules because of the increased number of terms in the expansion. Nevertheless, to illustrate the application of **g**-tensor theory to problems of this type, we give a description of the calculation of Δg_{yy} and Δg_{zz} for the T_σ radical made by Toriyama and Iwasaki [20].

Since the odd electron in the σ radical is essentially localized between $O_{(1)}$ and $O'_{(1)}$, its orbital can be expressed as

$$\phi_0[b_2(\sigma)] = \frac{c_{01}}{\sqrt{2}}(p_y + p'_y)_{O_{(1)}} + \frac{c_{02}}{\sqrt{2}}(p_z - p'_z)_{O'_{(1)}} \tag{7.54}$$

where $c_{01} = 3/2$ and $c_{02} = 1/2$ if bond angles of 120° are assumed. A magnetic field applied in the molecular plane, the yz plane, can intermix ϕ_0 only with orbitals perpendicular to this plane, that is, with π orbitals. Thus for calculation of Δg_{yy} and Δg_{zz}, one requires with ϕ_0 only the π orbitals. There are two types of atomic-orbital combinations that meet the C_{2v} symmetry requirements

Table 7.4 Hyperfine and g Tensors of π and σ Symmetrical Radicals in an Irradiated Single Crystal of Potassium Hydrogen Maleate[a]

Radicals		Principal Values	Direction Cosines for (abc)[b] l	m	n	Comparison with X-ray Data: Directions	Diff.
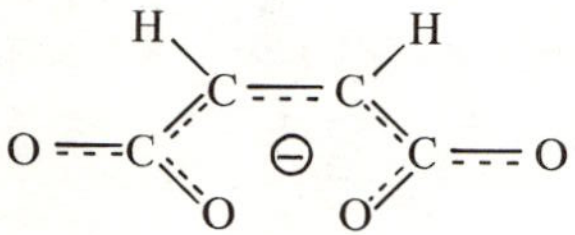	A_1	-2.1_1 G	-0.649	+0.560	-0.515	∥ C-H	6°
	A_2	-7.0_1 G	+0.669	+0.743	0.000	⊥ Radical plane	3°
	A_3	-10.1_3 G	+0.360	-0.370	-0.857	⊥ C-H in plane	6°
	A_{iso}	-6.4_2 G					
	g_{xx}	2.0024	+0.629	+0.778	0.000	⊥ Radical plane	0°
π type	g_{yy}	2.0045	0.000	0.000	+1.000	⊥ C=C	0°
	g_{zz}	2.0040	-0.778	+0.629	0.000	⊥ C=C in plane	0°
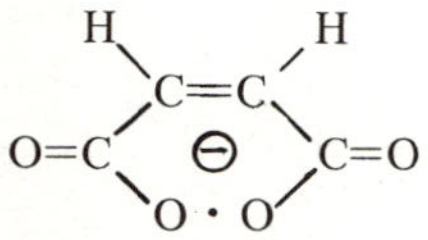	α_H	5 ± 1 G (nearly isotropic)					
	g_{xx}	2.0043	+0.681	+0.732	+0.032	⊥ Radical plane	4°
	g_{yy}	2.0039	+0.022	-0.065	+0.998	∥ $O_1 \cdots O_{1'}$[c]	4°
σ type	g_{zz}	2.0079	+0.732	-0.678	-0.060	⊥ $O_1 \cdots O_{1'}$[c] in plane	5°
x(⊥ Radical plane)			+0.631	+0.776	0.000		
y(∥$O_1 \cdots O_{1'}$)			0.000	0.000	+1.000		
z(⊥$O_1 \cdots O_{1'}$ in plane)			+0.776	-0.631	0.000		

[a] For comparison, the orientations of the parent molecule in the crystal are also given. From Toriyama and Iwasaki [20].
[b] Direction cosines are given for one of the four sites and others have orthohombic symmetry.
[c] $O_1 \cdots O_{1'}$ indicates the ring oxygen atoms.

for these orbitals that must be either $++a_2(\pi)$ or $--b_1(\pi)$. If the hydrogens are omitted, there are eight combinations, four for each symmetry class, which meet these requirements. They may be expressed as

$b_1\ (\pi)$	$a_2\ (\pi)$
$\chi_1^+ = (1/\sqrt{2})\ (p_x + p_x')_{O(1)}$	$\chi_1^- = (1/\sqrt{2})\ (p_x - p_x')_{O(1)}$
$\chi_2^+ = (1/\sqrt{2})\ (p_x + p_x')_{C(1)}$	$\chi_2^- = (1/\sqrt{2})\ (p_x - p_x')_{C(1)}$
$\chi_3^+ = (1/\sqrt{2})\ (p_x + p_x')_{C(2)}$	$\chi_3^- = (1/\sqrt{2})\ (p_x - p_x')_{C(2)}$
$\chi_4^+ = (1/\sqrt{2})\ (p_x + p_x')_{O(2)}$	$\chi_4^- = (1/\sqrt{2})\ (p_x - p_x')_{O(2)}$

To obtain the $b_1(\pi)$ molecular orbitals, one takes linear combinations of the χ^+ orbitals on the left and for the $a_2(\pi)$, the χ^- orbitals on the right. Toriyama and Iwasaki [20] used the Hückel molecular-orbital theory to calculate approximate values of the coefficients in these combinations and the relative energy separations of the levels. The calculated orbitals arranged in sequence of increasing energy are:

$$\phi_1 = 0.675\chi_1^+ + 0.548\chi_2^+ + 0.262\chi_3^+ + 0.418\chi_4^+ \tag{7.55}$$

$$\phi_2 = 0.714\chi_1^- + 0.541\chi_2^- + 0.114\chi_3^- + 0.430\chi_4^- \tag{7.56}$$

$$\phi_3 = -0.664\chi_1^+ + 0.165\chi_2^+ + 0.327\chi_3^+ + 0.652\chi_4^+ \tag{7.57}$$

$$\phi_4 = -0.605\chi_1^- + 0.162\chi_2^- + 0.053\chi_3^- + 0.776\chi_4^- \tag{7.58}$$

$$\phi_5 = -0.053\chi_1^+ + 0.030\chi_2^+ + 0.866\chi_3^+ - 0.496\chi_4^+ \tag{7.59}$$

$$\phi_6 = -0.285\chi_1^- + 0.541\chi_2^- + 0.690\chi_3^- - 0.388\chi_4^- \tag{7.60}$$

$$\phi_7 = -0.317\chi_1^+ + 0.819\chi_2^+ - 0.273\chi_3^+ - 0.393\chi_4^+ \tag{7.61}$$

$$\phi_8 = -0.207\chi_1^- + 0.622\chi_2^- - 0.713\chi_3^- - 0.249\chi_4^- \tag{7.62}$$

The relative energies and symmetries of the orbitals are indicated in Fig. 7.5 (left).

A magnetic field imposed in the molecular plane will intermix the $\phi_0\,[b_2(\sigma)]$ orbital of the odd electron with the π-type orbitals closest to it in energy. Since $L_y|p_y) = 0$ and $L_y|p_z) = ip_x$, a field imposed along y will intermix the π orbitals through the second term on the right of (7.54). The intermixing must occur through interaction on the $O_{(1)}$ oxygens since an L operator centered on one atom gives zero when applied to atomic functions centered on other atoms. Fur-

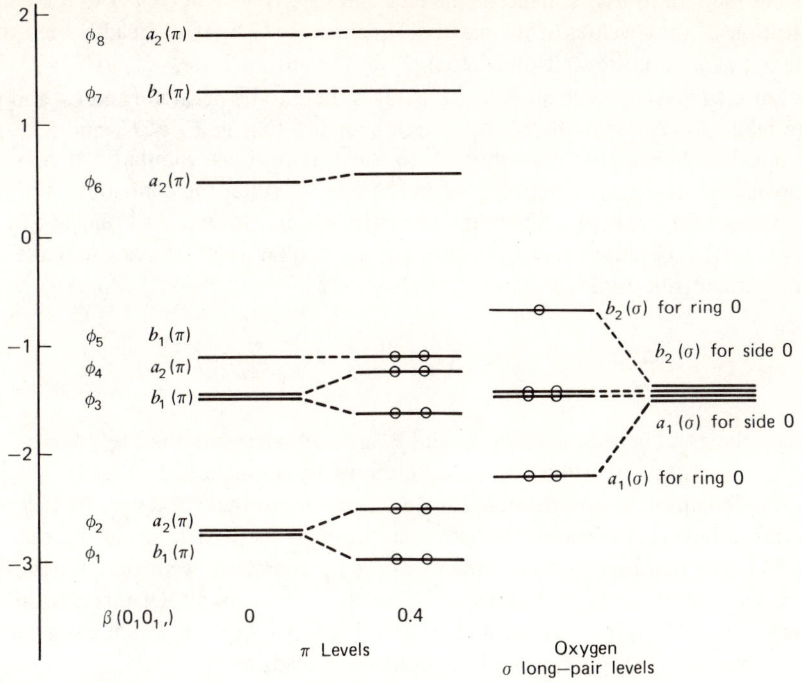

Fig. 7.5 Molecular-orbital energy-level diagram for the symmetrical σ radical formed from potassium hydrogen maleate. From Toriyama and Iwasaki [20].

thermore, the interaction must occur with components of π orbitals of the same symmetry as that of $(c_{02}/\sqrt{2})\, i(p_x - p_x')_{O(1)}$ generated by the L_y and L_z operators on ϕ_0. This means that only $a_2(\pi)$ orbitals need be considered in calculation of Δg_{yy}, and only $(p_x - p_x')_{O(1)}$ components of these. Matrix elements of all other combinations vanish in the calculations. The two $a_2(\pi)$ orbitals closest to ϕ_0 are ϕ_4 and ϕ_6 (see Fig. 7.5). With these combinations and use of Table 7.1, it is easily seen that

$$\begin{aligned}[(\phi_0|L_y|\phi_4) + (\phi_0|L_y'|\phi_4)] &= -i\frac{1}{2}c_{02}c_{41}\;[(p_z - p_z')|(p_z - p_z')]_{O(1)} \\ &= -ic_{02}c_{41}\end{aligned} \tag{7.63}$$

and

$$\begin{aligned}[(\phi_4|L_y|\phi_0) + (\phi_4|L_y'|\phi_0)] &= i\frac{1}{2}c_{02}c_{41}\;[(p_x - p_x')|(p_x - p_x')]_{O(1)} \\ &= ic_{02}c_{41}\end{aligned} \tag{7.64}$$

In the reduction, use is made of the fact that $(p_i|p_i) = 1$ and $(p_i|p_i') \neq 0$. By substitution of these values of the matrix elements into (7.14), it is readily seen that the ϕ_0, ϕ_4 admixture contributes to Δg_{yy} an amount $[-2\lambda_O(c_{02}^2 c_{41}^2)/(E_4 - E_0)]$. Because ϕ_4 has an electron pair, λ_0 is taken here to be negative, and E_4 and E_0 are taken to represent the configuration energies, that is, $E_0 < E_4$ and $E_4 - E_0$ is positive; hence this contribution to Δg_{yy} is positive. Similarly, it may be shown that, since ϕ_6 is unoccupied in the ground state, the contribution of the admixing of ϕ_0 with ϕ_6 contributes a negative amount to Δg_{yy} of $[-2\lambda_O(c_{02}^2 c_{61}^2)/(E_6 - E_0)]$. Because other $a_2(\pi)$ orbitals are too far away from ϕ_0 to make significant contributions, Δg_{yy} may be expressed as

$$\Delta g_{yy} = 2|\lambda_O| c_{02}^2 \left(\frac{c_{41}^2}{|E_4 - E_0|} - \frac{c_{61}^2}{|E_6 - E_0|} \right) \tag{7.65}$$

Since the combination orbitals χ^+ and χ^- are normalized by the $1/\sqrt{2}$ factor, coefficients of these χ orbitals in the ϕ functions are normalized by (7.18).

The derivation of the formula for Δg_{zz} is similar to that for Δy_{yy}. As $L_z|p_z) = 0$, the interaction obviously must occur through the first term on the right in (7.54). Nonvanishing matrix elements of the L_z operation occur only for the $(p_x + p_x')_{O(1)}$ terms in ϕ_n and hence only for the $b_1(\pi)$ orbitals. Of this class, those nearest to $\phi_0\,[2b_2(\sigma)]$ are ϕ_5 and ϕ_3 (see Fig. 7.5), both of which give positive contributions to Δg_{zz}. The resulting interaction leads to

$$\Delta g_{zz} = 2|\lambda_O| c_{01}^2 \left(\frac{c_{31}^2}{|E_3 - E_0|} + \frac{c_{51}^2}{|E_5 - E_0|} \right) \tag{7.66}$$

Equations (7.65) and (7.66), with the measured values $\Delta g_{yy} = 0.0016$ and $\Delta g_{zz} = 0.0056$, provide an approximate means of testing the molecular model and the Hückel calculations against experimental measurement of the **g** tensors. For this purpose, we use $c_{01}^2 = 3/4$ and $c_{02}^2 = 1/4$, which are squares of the coefficients in (7.54), obtained with the assumption of bond angles of 120°. The constants for the π orbitals are obtained from the theoretical equations (7.57) to (7.60). For example, c_{31} is the coefficient of χ_1 in the expansion of ϕ_3, c_{41} is that of χ_1 in ϕ_4, and so on. By substitution of these values with $|\lambda_O| = 152\ \text{cm}^{-1}$ into (7.65) and (7.66) one obtains

$$\Delta g_{yy} = 76 \left(\frac{0.37}{\Delta E_4} - \frac{0.08}{\Delta E_6} \right) \tag{7.67}$$

and

$$\Delta g_{zz} = 288 \left(\frac{0.44}{\Delta E_3} + \frac{0.03}{\Delta E_5} \right) \tag{7.68}$$

where $\Delta E_n = E_n - E_0$. Although the energy differences are unknown, the approximate, relative values obtained from the Hückel calculations are: $\Delta E_4 = 0.95\Delta E_3$, $\Delta E_5 = 0.55\Delta E_3$, and $\Delta E_6 = 1.37\Delta E_3$. By use of (7.66) with the observed $\Delta g_{zz} = 0.0056$, one obtains $\Delta E_3 = 20{,}000$ cm^{-1} and hence $\Delta E_4 = 19{,}000$ cm^{-1}, $\Delta E_5 = 11{,}000$ cm^{-1}, and $\Delta E_6 = 27{,}000$ cm^{-1}. By substitution of these values of ΔE_4 and ΔE_6 into (7.67), one obtains $\Delta g_{yy} = 0.0013$, which is in satisfactory agreement with the measured value of 0.0016. Although we use the relative theoretical values for deriving these results, it is evident that from the observed Δg values one could obtain good approximate values of ΔE_4 and ΔE_3 by omitting the very small, last terms of both equations. This approximation yields $\Delta E_3 = 18{,}000$ cm^{-1} and $\Delta E_4 = 17{,}500$ cm^{-1}.

A magnetic field imposed along x, perpendicular to the molecular plane, would intermix ϕ_0 $[2b(\sigma)]$ with the σ-bonding and σ-nonbonding pair orbitals of $O_{(1)}$ and $O'_{(1)}$. Since these orbitals have electron pairs, the admixture would cause Δg_{xx} to be positive, as is observed. No attempt is made here to calculate the value of this component.

The stable π-type radical, T_π, observed at room temperature is not formed from the T_σ simply by an excitation of the unpaired electron from the $b_2(\sigma)$ to the $a_2(\pi)$ orbital (Fig. 7.5) because T_σ decays before the crystal is brought to room temperature. Incidentally, the formation of T_π from T_σ would require a change in the $O_{(1)}$-$O'_{(1)}$ distance as well as a shift in the odd-electron orbitals. Toriyama and Iwasaki [20] conclude that T_π is formed directly by the loss of the H at room temperature rather than from a conversion of T_σ. They offer the suggestion that T_π may have the form

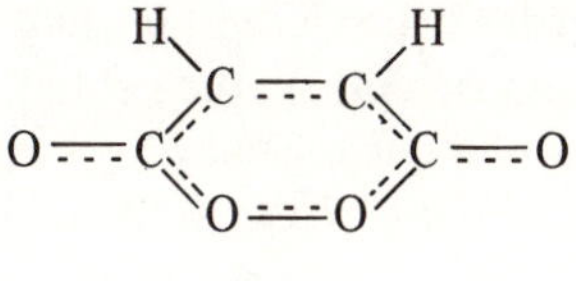

10

in which a σ bond is formed between $O_{(1)}$ and $O'_{(1)}$ and the π bonds are delocalized over the completed ring. This structure would require $O_{(1)}$ and $O'_{(1)}$ to move closer together after the loss of $H_{(1)}$. In this form of T_π, the $b_2(\sigma)$ level would be raised considerably above its T_σ value, and the $a_1(\sigma)$ level would be correspondingly lowered by the formation of the σ bond by the $O_{(1)}O'_{(1)}$. In this case the odd electron would go into the $a_2(\pi)$ orbital. In the open-ring model of the T_π radical in which there is no bonding between $O_{(1)}$ and $O'_{(1)}$, the $b_2(\sigma)$ orbital would be lowered below the level of the ϕ_5 $[b_1(\pi)]$ bonding orbital; consequently, one electron of the $b_1(\pi)$ pair would fall into the $b_2(\sigma)$ orbital to leave the unpaired electron in ϕ_5 $[b_1(\pi)]$.

In either of these forms of the T_π radical, in open or closed rings, the spin density of the unpaired electron on the different atoms would be in p_x or p'_x orbitals, and the L_x or L'_x operations on the component atomic orbitals of either the $\phi_6\,[a_2(\pi)]$ or $\phi_5\,[b_1(\pi)]$ would give zero. Therefore, for either form, the theoretical $\Delta g_{xx} = 0$ and $g_{xx} = g_e$, in agreement with the observation of $g_{xx} = 2.0024$ (see Table 7.4). A magnetic field imposed either along the symmetry axis z or along the y axis would tend to intermix the in-plane, σ-bonding or non-bonding pair orbitals with the orbital of the unpaired electron in the ground state, whether $\phi_6\,[a_2(\pi)]$ or $\phi_5\,[b_1(\pi)]$. These σ orbitals are expected to have much lower energy than either ϕ_6 or ϕ_5; hence the ΔE values in the denominator of (7.19) would be large, and the predicted Δg_{yy} or Δg_{zz} would be small. Since the interactions would occur with an electron-pair orbital, the signs of both Δg_{yy} and Δg_{zz} are expected to be positive. These qualitative predictions are in agreement with the observations of $\Delta g_{yy} = 0.0022$ and $\Delta g_{zz} = 0.0017$. We do not attempt a quantitative prediction of these small Δg values.

4.b. Comparison of *g* Values of Selected Free Radicals in Irradiated Single Crystals

The esr spectra of oriented free radicals in many organic single crystals have now been measured. The observed species are usually produced by ionizing X- or γ-irradiation and are measured at room temperature. Some observations are made at 77°K and a few at liquid helium temperature. Tables 7.5 and 7.6 show principal values of selected examples of these free radicals. Most of these radicals are too complex in structure for a quantitative analysis of their g values to be made. Nevertheless, useful information can be gained from qualitative deductions.

Most, but not all, free radicals produced by irradiation of organic molecules are found to be π-type radicals; that is, the orbital of the electron is proved, either by the **g** tensor or by the nuclear coupling tensors, to be in a delocalized π molecular orbital. The π orbital can usually be represented by a linear combi-

Table 7.5 Principal g Values of Selected Aliphatic Free Radicals in Irradiated Single Crystals

Radical / *Host Crystal*	⊥ *to Plane of* π *Orbital of Unpaired electron*	*In Plane of* π *Orbital of Unpaired Electron*		*Source*
$H\dot{C}(CO_2H)_2$ / Malonic acid	2.0026	2.0033	2.0035	a
$HO_2CCH_2\dot{C}HCO_2H$ / Succinic acid	2.0019	2.0026	2.0045	b

Table 7.5 Continued

Radical / *Host Crystal*	⊥ *to Plane of π Orbital of Unpaired electron*	*In Plane of π Orbital of Unpaired Electron*		*Source*
$HO_2C\dot{C}HCOH$ / Urea oxalate	2.0024	2.0047	2.0048	*c*
$HO_2CCH_2\dot{C}HCO_2H$ / Maleic acid	2.0026	2.0036	2.0039	*d*
$H_3CCONH\dot{C}HCO_2H$ / *N*-Acetylglycine	2.0027	2.0032	2.0042	*e*
$H_3N^+CH_2CONH\dot{C}HCO_2^-$ / Glycylglycine	2.0028	2.0033	2.0035	*f*
$CH_3\dot{C}HNHCOC_6H_5$ / *N*-Benzoylalanine	2.0020	2.0025	2.0026	*g*
$HCO_2CH_2O\dot{C}HCO_2H$ / Diglycolic monohydrate	2.0022	2.0044	2.0061	*h*
$HCO_2CH_2S\dot{C}HCO_2H$ / Thiodiglycolic acid	2.0020	2.005	2.011	*i*
$\dot{C}F(CONH_2)_2$ / Difluoromalonamide	2.0029	2.0043	2.0047	*j*
$\dot{C}F_2CONH_2$ / Trifluoroacetamide	2.0026	2.0036	2.0039	*k*
$\dot{C}HBrCONH_2$ / Bromomalonamide	1.9993	2.0428	2.0428	*l*
$\dot{C}HICONH_2$ / Iodoacetamide	1.9902	2.0423	2.0423	*m*

[a]H. M. McConnell, H. C. Heller, T. Cole, and R. W. Fessenden, *J. Am. Chem. Soc.*, **82**, 766 (1960).
[b]D. Pooley and D. H. Whiffen, *Molec. Phys.*, **4**, 81 (1961).
[c]D. V. G. L. N. Rao and W. Gordy, *J. Chem. Phys.*, **35**, 362 (1961).
[d]J. B. Cook, J. P. Elliott, and S. J. Wyard, *Molec. Phys.*, **12**, 185 (1967).
[e]I. Miyagawa, Y. Kurita, and W. Gordy, *J. Chem. Phys.*, **33**, 1599 (1960).
[f]M. Katayama and W. Gordy, *J. Chem. Phys.*, **35**, 117 (1961).
[g]G. Lassmann and W. Damerau, *Molec. Phys.*, **21**, 551 (1971).
[h]Y. Kurita, *J. Chem. Phys.*, **36**, 560 (1962).
[i]Y. Kurita and W. Gordy, *J. Chem. Phys.*, **34**, 1285 (1961).
[j]M. Iwasaki, S. Noda, and K. Toriyama, *Molec. Phys.*, **18**, 201 (1970).
[k]R. J. Lontz and W. Gordy, *J. Chem. Phys.*, **37**, 1357 (1962).
[l]R. F. Picone and M. T. Rogers, *J. Chem. Phys.*, **61**, 4814 (1974).
[m]R. F. Picone and M. T. Rogers, *J. Magn. Reson.*, **14**, 279 (1974).

nation of p atomic orbitals having parallel axes. The plane perpendicular to these p-orbital components is referred to as the *radical plane*, although it is obvious that all the atoms of the radical need not be in this plane. This plane over which the π orbital spreads is designated in this treatment as the yz plane. Hence, the p orbitals that constitute the π orbital of the unpaired electron are p_x orbitals. Because of the axial symmetry of the p-orbital components, the x direction will be a principal axis of the **g** tensor of the π-type radicals. Because $L_x|p_x) = 0$, the theoretically predicted g_x value is $g_e = 2.0023$. It is evident that this prediction applies only to molecular orbitals comprised of pure p_x orbitals fixed in space. For example, d-orbital contributions can cause deviations. A more common cause of the deviation of g_x from g_e is that the plane of the π orbital is not well defined because of torsional oscillations or chain vibrations, especially when observations are made at room temperature.

It is seen that the g_x values, those perpendicular to the radical plane, of all the chain radicals in Table 7.5 are close (probably within error limits) to the free-spin value of 2.0023—except those of the last two, in which d-orbital components on the Br or I are likely responsible for the deviation. The average value of g_x for the first 11 radicals in this table is 2.0024. Likewise, the g_x values for the ringed radicals in Table 7.6, except the iodide, are near the free-spin value and have the average value of 2.0024. The observed direction of this free-spin, or near free-spin, value in the π-type radicals can be useful for indicating the orientation of the radical plane relative to the crystal faces or to the orientations of the undamaged parent molecule when these orientations are known from X-ray diffraction. These directions are also useful for confirmation of the directions of the principal axes of the p-orbital nuclear coupling.

The two in-plane, principal g values of the π radicals in Tables 7.5 and 7.6 are all greater than the free-spin value of 2.0023. With the omission of the four radicals that have significant spin density on O, S, Br, or I, the in-plane g values of the remaining nine in Table 7.5 are in the range from 2.0025 to 2.0048. Further-

Table 7.6 Principal g Values of Selected Organic, Ringed Radicals

Radical / *Host Crystal*	*⊥ to Plane of Ring*	*In Plane of Ring*		*Source*
$OC\dot{C}HNHCOCH_2NH$ / Diketopiperazine	2.0025	2.0040	2.0045	*a*
$HN\dot{C}H(CH_2)_4CO$ / ε-Caprolactam	2.0025	2.0032	2.0038	*b*
$XCH_2\dot{C}CHCHCOCHCH$ / L-Tyrosine-HCL	2.0023	2.0045	2.0067	*c*

Table 7.6 Continued

Radical / Host Crystal	⊥ to Plane of Ring	In Plane of Ring		Source
dR-$NCH_2\dot{C}(CH_3)CONHCO$ / Thymidine	2.0024	2.0030	2.0042	d
H_3C-$\dot{N}CHCH_2CONHCO$ / 1-Methyl uracil	2.0020	2.0030	2.0030	e
R-$NCH_2\dot{C}ClCONHCO$ / 5-Chlorodeoxyuridine	2.0021	2.0069	2.0086	f
R-$NCH_2\dot{C}ICONHCO$ / 5-Iododeoxyuridine	1.987	2.039	2.050	g
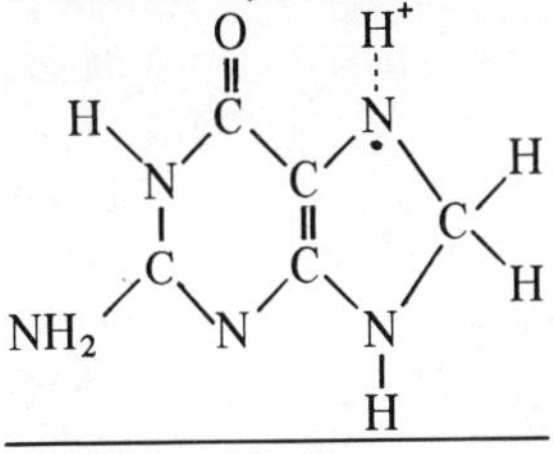 Guanine HCl dihydrate	2.0023	2.0048	2.0048	h
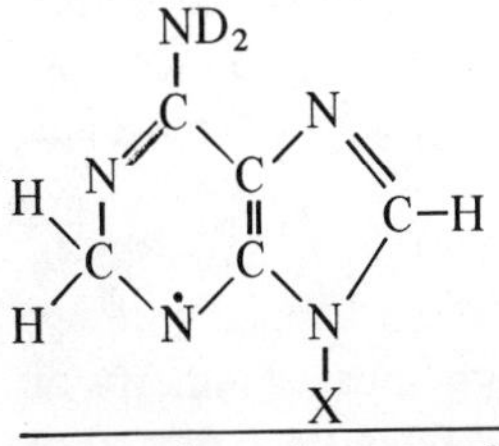 Deoxyadenosine monohydrate	2.0030	2.0050	2.0050	i

[a]I. Miyagawa, *Tech. Rep. Inst. Solid State Physics, Univ. Tokyo*, Series A, No. 27 (1961).
[b]M. Kashiwagi and Y. Kurita, *J. Chem. Phys.*, **40** 1780 (1964).
[c]E. L. Fasanella and W. Gordy, *Proc. Nat. Acad. Sci. (USA)*, **62**, 299 (1969).
[d]B. Pruden, W. Snipes, and W. Gordy, *Proc. Nat. Acad. Sci. (USA)*, **53**, 917 (1965).
[e]W. Flossmann, J. Hüttermann, A. Müller, and E. Westhof, *Z. Natürforsch.*, **28c**, 523 (1973).
[f]J. Hüttermann, W. A. Bernhard, E. Haindi, and G. Schmidt, *Molec. Phys.*, **32**, 1111 (1976).
[g]J. Hüttermann, G. W. Nelson, and M. C. R. Symons, *Molec. Phys.*, **32**, 269 (1976).
[h]C. Alexander and W. Gordy, *Proc. Nat. Acad. Sci. (USA)*, **58**, 1279 (1967).
[i]J. J. Lichter and W. Gordy, *Proc. Nat. Acad. Sci. (USA)*, **60**, 450 (1968).

more, seven of them are axially symmetric or nearly so. The fact that the in-plane values are greater than 2.0023 indicates that the admixing, which causes these Δg shifts, involves the transfer of an electron from a σ-bond pair ($\sigma \rightarrow \pi$ transition) or from an in-plane, nonbonding pair ($n \rightarrow \pi$ transition) to the π orbital of the unpaired electron. This means, in effect, that these electron-pair orbitals are the nearest in-plane orbitals to the π orbital of the unpaired electron. An admixture involving a transfer of the odd electron to an in-plane, antibonding orbital would contribute a negative Δg. For the same atoms, the σ-bonding pairs usually have lower energy than do the nonbonding pairs; that is, $\Delta E_{\sigma \rightarrow \pi} > \Delta E_{n \rightarrow \pi}$, where n signifies a nonbonding pair.

The magnitudes of the in-plane Δg values depend directly on the spin-orbit coupling constants of the atoms on which significant unpaired electron-spin density occurs and inversely on the energy differences ΔE of the admixed orbitals. The magnitudes of the spin-orbit coupling for valence p electrons of the atoms considered here are: C (29 cm^{-1}), N (76 cm^{-1}), O (151 cm^{-1}), F (272 cm^{-1}), S (382 cm^{-1}), Cl (587 cm^{-1}), Br (2460 cm^{-1}), and I (5060 cm^{-1}) [22] (see Table 7.5). The organic free radicals that have the unpaired electron concentrated in p_π orbitals of carbon generally have very small Δg values because the spin-orbit coupling of C is small and because there are normally no nonbonding pairs on the C's so that the perturbing interactions must occur with the σ bonds. These considerations account for the larger magnitudes of the in-plane Δg values for radicals in Tables 7.5 and 7.6 that have p_x spin density on noncarbon atoms. By comparison of the relative Δg values and spin-orbit couplings, one can gain a rough estimate of the spin density on the heavier atoms.

In π-type organic radicals where the unpaired electron density is primarily in p_x orbitals of C's, the three σ bonds to each C are most often formed with approximately equivalent sp^2 hybrids of the C. Because of the near symmetry of these in-plane bonds to the C, the in-plane Δg_{yy} and Δg_{zz} values are expected to be approximately equal and the **g** tensor to be nearly axially symmetric. This approximate symmetry is evident in several of the radicals in Tables 7.5 and 7.6 and is widely observed in hydrocarbon free radicals. With the assumption that the σ bonds are equivalent, the mean in-plane Δg is expressed approximately by

$$\frac{\Delta g_{yy} + \Delta g_{zz}}{2} \approx \frac{2|\lambda_C|}{\Delta E_{\sigma \rightarrow \pi}} \tag{7.69}$$

for the carbon-centered, π-type radical. The unpaired electron is mostly, but not exclusively, concentrated on carbon in the first seven radicals of Table 7.5, for which the average is $(\Delta g_{yy} + \Delta g_{zz})/2 = 0.0013$. With this value, (7.69) indicates that $(\Delta E_{\sigma \rightarrow \pi})_{av} \approx 44{,}600$ cm^{-1}, or 127 kcal/mol.

5. PEROXIDE FREE RADICALS

Peroxide free radicals are commonly detected in organic or biological compounds that are irradiated in air or exposed to air or oxygen after irradiation under vacuum. Gaseous oxygen molecules diffuse into the sample and become attached to free-radical centers in the sample, converting them into peroxide radicals in which the unpaired electron is concentrated on the O_2. This attachment is probably the most common initial step in the well-known oxygen effect in the enhancement of radiation damage. The oxygen effect in esr is most frequently observed in irradiated powdered samples into which the O_2 molecules can most easily diffuse. These studies are discussed in Chapter VIII. The **g** tensors of oriented peroxide radicals have been measured in a few single crystals. These are useful in interpretation of the results on irradiated powders and polymers.

The principal g values for the simplest peroxide free radical, HOO, as measured in a γ-irradiated single crystal of $H_2O_2 \cdot CO(NH_2)_2$, are listed in the top row of Table 7.7. At the temperature of the observation, 77°K, the radicals may be considered as fixed in space without internal rotation. These g values are reasonably

***Table** 7.7* Principal g Values of Some Oriented Peroxide Free Radicals

Host Crystal / *Radical*	T *(K°)*	g_u	g_v	g_w	*Source*
H_2O_2-Urea					
HOO	77°	2.001_8	2.008_1	2.049_5	*a*
Trifluoroacetamide					
NH_2COCF_2OO	77°	2.002_2	2.007_4	2.038_4	*b*
	300°	2.008_2	2.018_2	2.021_0	
Tetralinperoxide					
$C_{10}H_{11}OO$ Site I	123°	2.005	2.015	2.039	*c*
$C_{10}H_{11}OO$ Site II		2.005	2.011	2.045	

[a]T. Ichikawa and M. Iwasaki, *J. Chem. Phys.*, **44**, 2979 (1966).
[b]Toriyama and Iwasaki [24].
[c]Melamud, Schlick, and Silver [23].

close to those observed at 77°K for the NH_2COCF_2OO peroxide radical produced by admission of air to a γ-irradiated single crystal of trifluoroacetamide (second row of Table 7.7). Note, however, that there is a marked change in the g values of the latter radical when the observations are made at room temperature. These changes are attributed to an averaging of the "frozen-radical" values by a restricted internal rotation of the peroxide group at 300°K. From a comparison of the values, one may conclude that there are, particularly for site I, torsional oscillations of the peroxide group in $C_{10}H_{11}OO$ that are of rather large amplitude at the temperature of the measurements, 123°K.

The ^{17}O hyperfine structure of the two oxygens, as well as the **g** tensors, of the $C_{10}H_{11}OO$ radical was measured by Melamud et al. [23], who used a $^{17}O_2$ concentrated sample in the preparation of their single crystals. These measurements showed the unpaired π-electron density on the $C_{10}H_{11}OO'$ to be 0.41 on O and 0.62 on O′. The maximum ^{17}O coupling values of the two oxygens, 168 MHz for O and 247 MHz for O′ are parallel and colinear with g_u. Thus the unpaired electron appears to be entirely on the oxygens and is evidently in an antibonding π^* orbital perpendicular to the COO plane. Because the necessary structural information on the crystal was not available, the authors could not find the direction of the principal values relative to the coordinates of the tetralin group (1,2,3,4-tetrahydronaphthalene). From theoretical calculations of the **g** tensor like those described for the disulfide radical RCH_2SS in Section 6.a, one can expect g_u to be perpendicular to the COO plane, with the maximum in-plane values g_w approximately along the OO direction.

Although the structure of the trifluoroacetamide crystal is still unknown, Toriyama and Iwasaki [24] were able to determine the approximate structure and the orientation of the COO peroxide group relative to the $CCONH_2$ plane, as indicated by the ^{19}F coupling of the preoxygenated radicals (see description in Chapter V, Section 52.d). The structure deduced from the **g** tensor at 77°K is indicated by the diagram of Fig. 7.6. Note that the CO bond of the peroxide

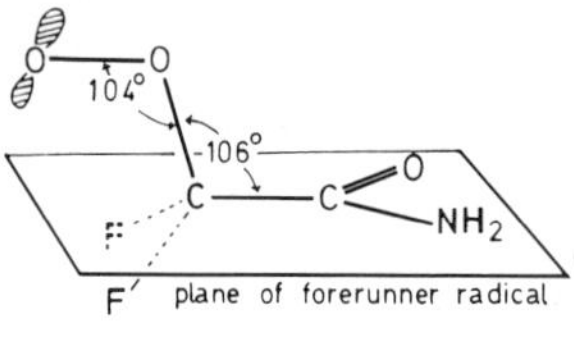

Fig. 7.6 Structure of the peroxide radical formed by addition of O_2 to the CF_2CCONH_2 radical in an irradiated crystal of trifluoroacetamide. From Toriyama and Iwasaki [24].

group replaced, in effect, the CF bond broken by the irradiation. The OO direction is approximately coplanar with the $CCONH_2$. The surprising finding is that the p_π orbital of the unpaired electron, as indicated by the direction of g_u, is not perpendicular to the COO plane, as would be expected for the most stable π bonding. This feature of the structure would seem to require confirmation.

Toriyama and Iwasaki [24] were able to explain the **g** tensor that they observed for the NH_2COCF_2OO radical at 300° in terms of that for 77°K, with the assumption that the OO rotates about the CO bond at 300°K, producing a partial averaging of the **g**-tensor elements for the rigid radical observed at 77°K.

6. THE g TENSOR OF SULFUR-CENTERED FREE RADICALS IN IRRADIATED AMINO ACIDS AND PEPTIDES

The spin-orbit coupling of *p* orbitals of S is appreciably greater than that for *p* orbitals in other atoms comprising the amino acids. Thus for radicals produced by irradiation of amino acids, peptides, and proteins, the spread of the esr pattern due to **g** anisotropy is noticeably greater when the spin density of the unpaired electron is concentrated on sulfur rather than on other common constituent atoms of such biochemicals. For this reason, the **g** anisotropy of sulfur-centered radicals provides a useful, quick means for detection of electron-spin density localized on S atoms in irradiated biochemicals. Spin centers on such metallic trace elements as Cu and Fe in the natural proteins can be distinguished from those of sulfur by their still larger **g** anisotropy. Although the limited information that electron spin density occurs on the S of irradiated biochemicals is useful in studies of radiation damage, further knowledge about the structure and mechanisms of these sulfur-centered radicals is obviously desirable. This need for more detailed information led Kurita and Gordy [25] to study the esr of oriented sulfur-centered radicals in irradiated single crystals of L-cystine dihydrochloride. Since this study, the **g** tensors of oriented sulfur-centered radicals have been observed in irradiated single crystals of the more common sulfur-containing amino acids and simpler peptides. Despite careful measurement and usually good agreement among different observers of these **g** tensors, there remained considerable uncertainty about the structure or chemical form of the radicals (viz., whether monosulfide or disulfide) until the ^{33}S hyperfine structure of their esr was measured (see Chapter VI, Section 10.b). The principal *g* values of those believed now to be correctly identified are listed in Table 7.8.

The cause for the ambiguities in the earlier assignments of the radical structures from the **g** tensor alone is evident from comparison of the principal *g* values of the different species observed in single crystals of irradiated L-cystine dihydrochloride. The ^{33}S hyperfine structure has confirmed the identification of all of them except the cation, which is believed to be assigned with reasonable certainty from other considerations. Both the radicals, α_3 and β_2, observed at room temperature, have the disulfide form. Their distinguishable principal values and proton splittings (10-G doublet for α_3 and 5-G doublet for β_2) evidently result from two stable conformations of the disulfide radicals. These two sets of characteristic *g* values are also found for the disulfide radicals observed at room temperature in irradiated L-cysteine (HCl) and *N*-acetyl L-cysteine. All sufur-centered radicals stable at room temperature and definitely identified in any of the host crystals of Table 7.8 have the disulfide form. The neutral monosulfide radical α_2 observed in the L-cystine $(HCl)_2$ matrix is formed from the anion α_1 in the temperature range from 100° to 200°K and apparently reacts further to form α_3 before the crystal reaches room temperature. Now that its monosulfide

Table 7.8 Principal *g* Values of Some Suifur-centered Free Radicals in Irradiated Single Crystals of Amino Acids and Peptides

Host Crystal / *Radical*	*Species Label*	*Stabilization Temperature, °K*	g_u	g_v	g_w	*Source*
L-Cystine dihydrochloride						
$(RCH_2SSCH_2R)^-$	α_1	~ 77°	2.002*	2.018**	2.018**	*a,b*
$(RCH_2SSCH_2R)^+$	β_1	4.2°	2.005	2.028	2.033	*c*
	β_1	~ 77° (in dark)	2.001	2.029	2.033	*d*
RCH_2S	α_2	~180°	2.000	2.010	2.066†	*a*
RCH_2SS	α_3	Room	2.003	2.025	2.053	*e*
	α_3	Room	2.000††	2.026	2.053	*f*
RCH_2SS	β_2	Room	2.002**	2.027	2.067	*f*
L-Cysteine hydrochloride						
		77°	1.99	1.99	2.29	*g*
$(RCH_2SH)^-$?		77°	1.985	2.004	2.251†	*h*
		77°	1.985	1.998	2.278	*i*
RCH_2SS	α_3	Room	1.998	2.023	2.055	*j*
	α_3 ?	Room	2.002	2.023	2.054	*i*
RCH_2SS	β_2	Room	1.999	2.023	2.062	*j*
	β_2 ?	Room	2.001	2.022	2.061	*i*
N-Acetyl-L-cysteine						
$(R'CH_2SH)^-$?	–	77°	1.990	2.006	2.214	*k*
$R'CH_2SS$	T_1	Room	2.003	2.024	2.053	*l*
$R'CH_2SS$	S_1	Room	2.002	2.025	2.064	*l*
N-Acetyl methionine						
$R'(CH_2)_2S^+CH_3$	–	77°	2.002	2.0134	2.0226	*m*
$R'(CH_2)_2SS$	–	Room	2.004	2.026	2.063	*j*
DL-Methionine						
$R''(CH_2)_2S^+CH_3$	–	77°	2.002	2.013	2.022	*n*
$R''(CH_2)_2SS$?	–	Room	2.0055	2.0195	2.0593	*o*

$R = (NH_3Cl)CHCOOH$ $R' = CH_3CONHCHCOOH$, $R'' = NH_2CHCOOH$

Table 7.8 Continued

*Parallel to SS.
**Perpendicular to SS.
†Parallel to CS.
††Perpendicular to CSS plane.
[a]Akasaka, Ohnishi, Suita, and Nitta [26].
[b]H. C. Box and H. G. Freund, *J. Chem. Phys.*, **40**, 817 (1964).
[c]H. C. Box and H. G. Freund, *J. Chem. Phys.*, **41**, 2571 (1964).
[d]K. Akasaka, S. Kominami, and H. Hatano, *J. Phys. Chem.*, **75**, 3746 (1971).
[e]Kurita and Gordy [25].
[f]Hadley and Gordy [27].
[g]Akasaka [29].
[h]Box, Freund, and Budzinski [30].
[i]Kayushin, Krivenko, and Pulatova [31].
[j]J. H. Hadley and W. Gordy, *Proc. Nat. Acad. Sci. (USA)*, **72**, 3486 (1975).
[k]Saxebøl and Herskedal [32].
[l]J. H Hadley and W. Gordy, *Proc. Nat. Acad. Sci. (USA)*, **74**, 216 (1977).
[m]K. Kawatsura, K. Ozawa, S. Kominami, K. Akasaka, and H. Hatano, *Radiat. Eff.*, **22**, 267 (1974).
[n]S. Kominami, *J. Phys. Chem.*, **76**, 1729 (1972).
[o]D. G. Cadena, Jr., and J. R. Rowlands, *J. Chem. Soc. (B)*, 488 (1968).

form has been proven, it can be distinguished from the disulfide form by its differing g_y value and by its proton hyperfine splitting (wide triplet).

6.a. Theoretical Derivation of the Principal *g* Values

Because of their lower symmetry and general complexity, it is not feasible to derive theoretical expressions for the *g* values of the sulfur-centered radicals in the same way as for the radicals with C_{2v} symmetry (section 4.a). As examples, we describe semiquantitative treatments for the neutral monosulfide and disulfide species RCH_2S and RCH_2SS and for the disulfide anion $[RCH_2SSCH_2R]^-$. The approaches used are applicable to the other related structures.

The Monosulfide Radical RCH_2S

In this species the unpaired spin density and orbital momentum may be considered as centered on the single S. The small delocalization of the unpaired electron results from hyperconjugation with the CH_2 group and puts spin density only on the H's. This *s*-orbital density on the H's makes no contribution to the orbital momentum or to the **g** anisotropy but reduces the spin density on the S by 0.08, as indicated by the proton coupling. Because any significant *s* hybridization of the σ-bonding orbital requires an unreasonbly high excitation energy for prediction of the observed g_y (see discussion to follow), we shall assume a pure p-bonding orbital, as was done by Akasaka et al. [26]. With the x axis chosen along the $(p_\pi)_{\dot{S}}$ orbital of the unpaired electron and with the z axis along the CS bond, the normalized orbitals involved in calculations of the tensors are

$$\phi_0 = (0.92)^{1/2} (p_x)_S + (0.08)^{1/2} (s_1 + s_2)_H \text{ (unpaired electron)} \tag{7.70}$$

$$\phi_1 = (p_y)_S \text{ (nonbonding pair)} \tag{7.71}$$

$$\phi_2 = \left(\frac{1}{2}\right)^{1/2} [(p_z)_S + \psi_C] \text{ (σ bond)} \tag{7.72}$$

The valence 3s orbital of the S also has a nonbonding pair, but this orbital has no influence on the calculations. The ground-state configuration is $\phi_2^2\phi_1^2\phi_0^1$ with energy E_0. Excited-state configurations that are intermixed with the ground state by the applied field are $\phi_2^2\phi_1^1\phi_0^2$ and $\phi_2^1\phi_1^2\phi_0^2$ with energies E_1 and E_2, respectively. In the first excited configuration, one of the electrons of the nonbonding pair is promoted to ϕ_0; in the second, one of the σ-bonding electrons is promoted to ϕ_0. Since the bonding electrons are expected to have lower energy than the nonbonding pair, one can expect $E_0 < E_1$. From (7.14), the Δg_{ii} values may be expressed as

$$\Delta g_{ii} = -2\lambda_S \sum \frac{|(\phi_n|L_i|\phi_0)|^2}{E_n - E_0} \tag{7.73}$$

Since $L_x|\phi_0) = L_x|p_x = 0$, $\Delta g_{xx} \doteq 0$. Because $L_y|\phi_0) = -(0.92)^{1/2} ip_z$, one may write $|(\phi_n|L_y|\phi_0)|^2 = 0.92(\phi_n|p_z)^2$. Thus only the p_z component of ϕ_2 has nonvanishing matrix elements that contribute to Δg_{yy}. Similarly, $L_z|\phi_0) = (0.92)^{1/2} ip_y$, and only the p_y orbital, or ϕ_1, contributes to Δg_{zz}. It is readily seen that

$$\Delta g_{xx} = 0 \tag{7.74}$$

$$\Delta g_{yy} = \frac{-0.92\lambda_S}{E_2 - E_0} \tag{7.75}$$

$$\Delta g_{zz} = \frac{-1.84\lambda_S}{E_1 - E_0} \tag{7.76}$$

Because the configuration interaction occurs with filled orbitals, the effective spin-orbit coupling is negative, $\lambda_S = -382 \text{ cm}^{-1}$. The predicted principal elements may be expressed as

$$g_u = g_x = g_e \tag{7.77}$$

$$g_v = g_y = g_e + \frac{351}{E_2 - E_0} \tag{7.78}$$

$$g_w = g_z = g_e + \frac{703}{E_1 - E_0} \tag{7.79}$$

With $g_e = 2.0023$ and the observed $g_v = 2.010$ and $g_w = 2.066$ from Table 7.8, (7.78) and (7.79) yield

$$E_1 - E_0 = 11{,}000 \text{ cm}^{-1}, \qquad E_2 - E_0 = 46{,}000 \text{ cm}^{-1}$$

The $E_1 - E_0$ is the energy required to lift an electron of the nonbonding pair to ϕ_0 ($n \to \pi$ transition), and $E_2 - E_0$ is that required to lift an electron from the CS σ bond to ϕ_0 ($\sigma \to \pi$ transition). Although the values obtained appear reasonable, no spectral values are available for comparison.

It is of interest that the theoretical g_z, and hence the derived $E_1 - E_0$ value, is independent of the amount of s hybridization, whereas the theoretical g_y and hence $E_2 - E_0$ is sensitive to s character in the σ (S)-bonding orbital. One can completely rule out the one-third s character previously assumed for this orbital [25] because the observed $g_v = 2.010$ would then indicate an infinite $E_2 - E_0$. An s character of only 10% yields an unacceptably high value of 240,000 cm^{-1} for $E_2 - E_0$. Kurita and Gordy [25] were able to obtain a reasonable value of $E_2 - E_0$ with the assumption of one-third s character only because the observed $g_v = 2.029$ that they used for RCH_3S was actually that for RCH_2SS, as the later ^{33}S hyperfine measurements proved [27]. The g_v value for the correctly assigned RCH_2S radical, if accurate, effectively proves that the hybridization on the S is negligibly small, as assumed for ϕ_1 and ϕ_2 in (7.71) and (7.72).

One may ask why an energy of 11,000 cm^{-1}, or 31 kcal/mole, would be required to shift an electron from the $(p_y)_S$ to the $(p_z)_S$ orbital. Perhaps the principal reason is that the $n \to \pi$ transition breaks up the rather strong hyperconjugation of the unpaired electron with the CH_2. The large proton couplings (Chapter VI, Section 10.c) indicate that the $(p_x)_S$ orbital has the most favorable orientation for this conjugation.

The Disulfide Radical RCH_2SS

Whereas the observed $g_v = 2.010$ of the RCH_2S indicates little or no hybridization of the σ-bonding orbital of the S, a reason for the larger $g_v = 2.025$ is that the σ-bonding orbitals of the central $S^{(1)}$ in the $RCH_2S^{(1)}S^{(2)}$ radical have appreciable s character. If there were no hybridization on either S, the predicted g_v would be approximately the same as it is for the RCH_3S radical. Furthermore, the measured CSS bond angle of 103.8° in the undamaged molecule of the L-cystine $(HCl)_2$ crystal [28] indicates an s character of 0.19 in each of these bonds [see (7.50)]. In their studies of the irradiated crystal of L-cystine $(HCl)_2$, Hadley and Gordy [27] found that a stable configuration for the α_3 of the RCH_2SS radical could be achieved in the crystal by a rotation of the SS bond about the CS bond by 52° and an opening of the CSS bond angle to about 120°. The predicted directions for this model were found to agree satisfactorily with those observed for the α_3 species. This bond angle is similar to that for the S_2O molecule, for which ∠OSS = 118°. The bond angles formed by sulfur vary over a wide range from that of 92° for H_2S to 119°19′ for SO_2. A correspondingly wide variation in orbital hybridization is expected, conforming to these various structures. However, the terminal S would be expected to have essentially the same type of bonding with little or no hybridization of its orbitals since it need not change to accommodate differing bond angles.

The procedure used by Hadley and Gordy [27] for the α_3 radical in irradiated L-cystine $(HCl)_2$ will serve as an illustration of an approximate calculation of the **g** tensor of a disulfide radical. The same method may be applied to the peroxide radicals described in Section 5. For the reasons given, we assume a $CS^{(1)}S^{(2)}$ angle of 120° and sp^2 hybridization of the three $S^{(1)}$ orbitals in the CSS plane. However, we assume pure p orbitals for the terminal $S^{(2)}$, like those found in RCH_2S. The ^{33}S coupling shows that the spin density of the unpaired electron is in an antibonding π^* orbital perpendicular to the plane, with $\rho_\pi^{(1)} = 0.36$ and $\rho_\pi^{(2)} = 0.56$. The remaining spin density is mostly on the hydrogens, and its effects on Δg are negligible. Although the $p_x^{(1)}$ and $p_x^{(2)}$ components of the π^* orbital are parallel, the other p orbitals of $S^{(1)}$ and $S^{(2)}$ are not parallel. Since the principal element, g_x, is expected to be perpendicular to the molecular plane, the directions of the two in-plane principal elements cannot be chosen from symmetry considerations, as was done for the radicals having C_{2v} symmetry (Section 3). One can express the in-plane, p-orbital components of $S^{(1)}$ and $S^{(2)}$ in common yz coordinates arbitrarily chosen, but the **g** tensor would not be diagonal in such a system, and derivation of the elements and diagonalization of the tensor could be tedious.

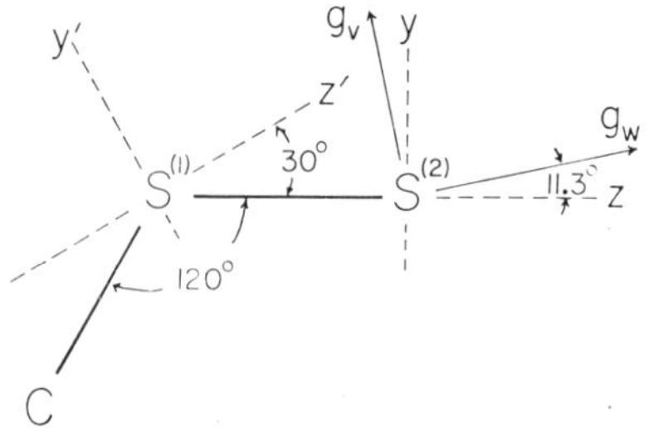

Fig. 7.7 Coordinate system for calculation of the **g** tensor of the disulfide radical. From Hadley and Gordy [27].

To simplify the problem, one can express the in-plane orbitals in separate axes chosen according to the symmetry of the orbitals of the particular S. Figure 7.7 shows the systems chosen for $S^{(1)}$ and $S^{(2)}$. The CS and SS σ bonds are assumed to be equivalent and formed of sp^2 hybrids of $S^{(1)}$. If the unpaired electron were entirely on $S^{(1)}$, the reference coordinates y' and z' would be principal axes of **g**; if it were entirely on $S^{(2)}$, y and z would be principal axes for the radical. For these two idealized cases we can easily calculate the principal elements of **g** in the separate axis systems. In this approximate treatment, we assume that the resultant tensor may be expressed as the tensor sum

$$(\mathbf{g}) = g_e + \rho_\pi^{(1)}\,(\Delta\mathbf{g})^{(1)} + \rho_\pi^{(2)}\,(\Delta\mathbf{g})^{(2)} \tag{7.80}$$

where $(\Delta\mathbf{g})^{(1)}$ is the theoretical **g** tensor expected if the unpaired electron were entirely on $S^{(1)}$ and $(\Delta\mathbf{g})^{(2)}$ is that expected if it were entirely on $S^{(2)}$, and where $\rho_\pi^{(1)}$ and $\rho_\pi^{(2)}$ are the unpaired ($3p_x$) spin densities on $S^{(1)}$ and $S^{(2)}$. We first calculate $(\Delta\mathbf{g})^{(1)}$ in its principal axes system x, y', z', as indicated in Fig. 7.7. The relevant orbitals are

$$\phi_0 = (p_x)_{S^{(1)}} \tag{7.81}$$

$$\phi_1' = \left[\left(\frac{1}{3}\right)^{1/2} s + \left(\frac{2}{3}\right)^{1/2} (p_{y'})\right]_{S^{(1)}} \tag{7.82}$$

$$\phi_2' = \left(\frac{1}{2}\right)^{1/2} \left\{\left[\left(\frac{1}{3}\right)^{1/2} s - \left(\frac{1}{6}\right)^{1/2} p_{y'} + \left(\frac{1}{2}\right)^{1/2} p_{z'}\right]_{S^{(1)}} + (\psi_\sigma)_{S^{(2)}}\right\} \tag{7.83}$$

$$\phi_3' = \left(\frac{1}{2}\right)^{1/2} \left\{\left[\left(\frac{1}{3}\right)^{1/2} s - \left(\frac{1}{6}\right)^{1/2} p_{y'} + \left(\frac{1}{2}\right)^{1/2} p_{z'}\right]_{S^{(2)}} + (\psi_\sigma)_{C}\right\} \tag{7.84}$$

The nonvanishing matrix elements

$$(\phi_n'|L_x|\phi_0), \qquad (\phi_n'|L_{y'}|\phi_0), \qquad (\phi_n'|L_{z'}|\phi_0)$$

in the x, y', z' reference system are obtained in the usual manner, as described for the monosulfide radical. By substitution of these into (7.14) with $\lambda_S = -382$ cm^{-1} and with $E_3' = E_2'$, one finds that

$$\Delta g_{xx}^{(1)} = 0 \tag{7.85}$$

$$\Delta g^{(1)}_{y'y'} = \frac{382}{E'_2 - E_0} \tag{7.86}$$

$$\Delta g_{z'z'} = \frac{510}{E'_1 - E_0} + \frac{128}{E'_2 - E_0} \tag{7.87}$$

with all off-diagonal elements zero. The assumption that $E_3 = E'_2$ is justified by the approximately equal SS and CS bond energies. The off-diagonal terms $\Delta g^{(1)}_{ij}$ vanish because x, y', z' are principal axes of $(\Delta \mathbf{g})^{(1)}$.

The tensor $(\Delta \mathbf{g})^{(2)}$ is diagonal in the x, y, z system, and its principal diagonal elements are found in the same way as those described for the monosulfide radical except that the spin density is factored out here. The resulting values are

$$\Delta g^{(2)}_{xx} = 0 \tag{7.88}$$

$$\Delta g^{(2)}_{yy} = \frac{382}{E_2 - E_0} \tag{7.89}$$

$$\Delta g^{(2)}_{zz} = \frac{764}{E_1 - E_0} \tag{7.90}$$

with all off-diagonal elements zero.

For the addition, the elements of $(\Delta \mathbf{g})^{(1)}$ and $(\Delta \mathbf{g})^{(2)}$ must be expressed in a common coordinate system. To do this, we rotate $(\Delta \mathbf{g})^{(1)}$ to the x, y, z system of $(\Delta \mathbf{g})^{(2)}$. Since the x direction is common, this is achieved by rotation of the y' and z' components of $(\Delta \mathbf{g})^{(1)}$ about x to the y and z directions. As y and z are not principal axes of $(\Delta \mathbf{g})^{(1)}$, off-diagonal elements are generated by this rotation. If θ is the angle between the primed and unprimed axes, the elements of $(\Delta \mathbf{g})^{(1)}$ in the unprimed system are

$$\Delta g^{(1)}_{yy} = \Delta g^{(1)}_{y'y'} \cos^2\theta + \Delta g^{(1)}_{z'z'} \sin^2\theta \tag{7.91}$$

$$\Delta g^{(1)}_{zz} = \Delta g^{(1)}_{y'y'} \sin^2\theta + \Delta g^{(1)}_{z'z'} \cos^2\theta \tag{7.92}$$

$$\Delta g^{(1)}_{yz} = \Delta g^{(1)}_{zy} = (\Delta g^{(1)}_{y'y'} - \Delta g^{(1)}_{z'z'}) \sin\theta\cos\theta \tag{7.93}$$

The resultant $\Delta \mathbf{g}$ tensor in the x, y, z system is

$$(\Delta \mathbf{g}) = \rho_1 \begin{pmatrix} \Delta g^{(1)}_{yy} & \Delta g^{(1)}_{yz} \\ \Delta g^{(1)}_{zy} & \Delta g^{(1)}_{zz} \end{pmatrix} + \rho_2 \begin{pmatrix} \Delta g^{(2)}_{yy} & 0 \\ 0 & \Delta g^{(2)}_{zz} \end{pmatrix}$$

$$= \begin{pmatrix} \Delta g_{yy} & \Delta g_{yz} \\ \Delta g_{zy} & \Delta g_{zz} \end{pmatrix} \tag{7.94}$$

where

$$\Delta g_{yy} = \rho_1 \; \Delta g_{yy}^{(1)} + \rho_2 \; \Delta g_{yy}^{(2)} \tag{7.95}$$

$$\Delta g_{yz} = \Delta g_{zy} = \rho_1 \; \Delta g_{yz}^{(1)} = \rho_1 \; \Delta g_{zy}^{(1)} \tag{7.96}$$

$$\Delta g_{zz} = \rho_1 \; \Delta g_{zz}^{(1)} + \rho_2 \; \Delta g_{zz}^{(2)} \tag{7.97}$$

To find the theoretical principal elements, one must diagonalize the resultant tensor ($\Delta\mathbf{g}$) by solution of the secular equation

$$\begin{vmatrix} \Delta g_{yy} - \Delta g_i & \Delta g_{yz} \\ \Delta g_{zy} & \Delta g_{zz} - \Delta g_i \end{vmatrix} = 0 \tag{7.98}$$

which gives

$$\Delta g_v = \frac{1}{2}\{\Delta g_{yy} + \Delta g_{zz} - [(\Delta g_{yy} - \Delta g_{zz})^2 + 4(\Delta g_{yz})^2]^{1/2}\} \tag{7.99}$$

$$\Delta g_w = \frac{1}{2}\{\Delta g_{yy} + \Delta g_{zz} + [(\Delta g_{yy} - \Delta g_{zz})^2 + 4(\Delta g_{yz})^2]^{1/2}\} \tag{7.100}$$

For the model chosen (Fig. 7.7), $\theta = 30°$, the specific values of $\Delta g_{yy}^{(1)}$, $\Delta g_{zz}^{(1)}$, and $\Delta g_{yz}^{(1)}$ are obtained from (7.91) to (7.93) by substitution of $\Delta g_{y'y'}^{(1)}$ and $\Delta g_{z'z'}^{(1)}$ from (7.86) and (7.87) with $\cos\theta = \sqrt{3}/2$ and $\sin\theta = 1/2$. The values of $\Delta g_{yy}^{(2)}$ and $\Delta g_{zz}^{(2)}$ are given by (7.89) and (7.90). By substitution of these various elements with the unpaired-electron densities $\rho_\pi^{(1)} = 0.36$ and $\rho_\pi^{(2)} = 0.56$ from Table 6.30 into (7.95) to (7.97), the parameters Δg_{yy}, Δg_{yz}, and Δg_{zz} may be expressed in terms of the excitation energies $E_n - E_0$. There are then only two equations, (7.99) and (7.100), linking the $E_n - E_0$ values to the measured Δg values. However, one can set $E_2' = E_2$ since ϕ_2' and ϕ_2 are orbitals of the same σ SS bond. There are still three unknown parameters: $E_2 - E_0$, $E_1' - E_0$, and $E_1 - E_0$. The orbitals ϕ_1' and ϕ_1 have nonbonding pairs on $S^{(1)}$ and $S^{(2)}$, respectively. Since a different hybridization is assumed for each orbital, E_1' is only approximately equal to E_1. To make the solution possible, however, we set $E_1' = E_1$. This leaves only two unknown parameters, $E_2 - E_0$ and $E_1 - E_0$; with the measured $\Delta g_v = 0.023$ and $\Delta g_w = 0.051$, (7.99) and (7.100) can be solved for these parameters. The result is:

$$E_1 - E_0 = 12{,}000 \text{ cm}^{-1}, \qquad E_2 - E_0 = 18{,}000 \text{ cm}^{-1}$$

The $E_1 - E_0$ thus obtained is evidently a weighted average of the two values $E_1' - E_0$ and $E_1 - E_0$. As a test of the consistency of the solution, these energy values were substituted back into (7.99) and (7.100); they led to the predicted values

$$g_v = 2.026, \qquad g_w = 2.054$$

which agree satisfactorily with those observed for the α_3 radicals (Table 7.7). The orientations of the resultant principal elements g_v and g_w may be found by the method described in Chapter V, Section 1 [see (5.19) to (5.21)]. The angle ϕ_{wz} between the maximum principal element g_w and the z axis (SS direction) may be found from the first of these equations:

$$(\Delta g_{yy} - g_w)l_{wy} + \Delta g_{yz} l_{wz} = 0 \tag{7.101}$$

Since $l_{wz} = \cos\phi_{wz}$, $l_{wy} = \cos\phi_{wy} = \sin\phi_{wz}$, it is seen that

$$\tan\phi_{wz} = \frac{\Delta g_{yz}}{\Delta g_w - \Delta g_{yy}} \tag{7.102}$$

The calculated values $\Delta g_{yy} = 0.022$, $\Delta g_{yz} = -0.0044$, with $\Delta g_w = 0.051$ gives $\phi_{wz} = -8.6°$, as measured from w to z (see Fig. 7.7).

In the earlier calculations with sp^2 hybrids assumed on both $S^{(1)}$ and $S^{(2)}$, the values $E_1 - E_0 = 11{,}000$ cm^{-1}, $E_2 - E_0 = 25{,}000$ cm^{-1}, and $\phi_{wz} = -11.2°$ were obtained for the α_3 radical in the L-cystine dihydrochloride crystal [27]. This ϕ_{wz} agrees somewhat better with the observed value $-15°$, and the energies appear more reasonable than do those obtained with the assumption of no hybridization of the orbitals of $S^{(2)}$. It may be that the measured g_v value of RCH_2S that indicates no hybridization on the terminal S is a bit too low and that some hybridization, perhaps as much as 10%, occurs on $S^{(2)}$ of α_3. Neither the values of the Δg elements nor their directions are measured with high accuracy, and the analysis is only approximate because of the necessary assumptions. The problem is worked out here in some detail as an example. Solutions for other degrees of hybridization or the application of this method to other related radicals will be obvious.

The Cystine Anion $(RCH_2SSCH_2R)^-$

Both the **g** tensor and the ^{33}S coupling of the cation radical observed at low temperature in irradiated L-cystine dihydrochloride are axially symmetric about the SS direction in the undamaged molecule (see Chapter VI, Section 10.b). Figure 7.8 shows the structural arrangement of the relevant part of the molecule relative to the crystalline axes. The SS bond is along the *c* axis, which proved to be the axis of minimum *g* and maximum ^{33}S coupling for the anion radical. The unpaired electron is unquestionably in an antibonding σ^* SS orbital with the geometry and orientation of the radical essentially that of the undamaged molecule. The SS bond length, however, is expected to be greater for the radical because of the weakening of the σ bond to about one-half strength by the antibonding electron. The CSS bond angles of 103.8° [28] indicate about 19% *s* character in the S-bonding orbitals in the undamaged molecule, whereas the ^{33}S nuclear coupling (Chapter VI, Section 10.b) indicates essentially pure p_σ orbitals in the radical. The greater SS bond length would decrease the CSS angle, causing it to approach 90°. The dihedral angle between the $C^{(1)}S^{(1)}S^{(2)}$ and $S^{(1)}S^{(2)}C^{(2)}$ planes in the undamaged molecule is 79.2°, but the complete axial symmetry observed in the **g** tensor indicates that these planes are orthogonal in the anion. Therefore, in calculation of the **g** tensor, the assumption of pure *p* orbitals on each S seems justifed.

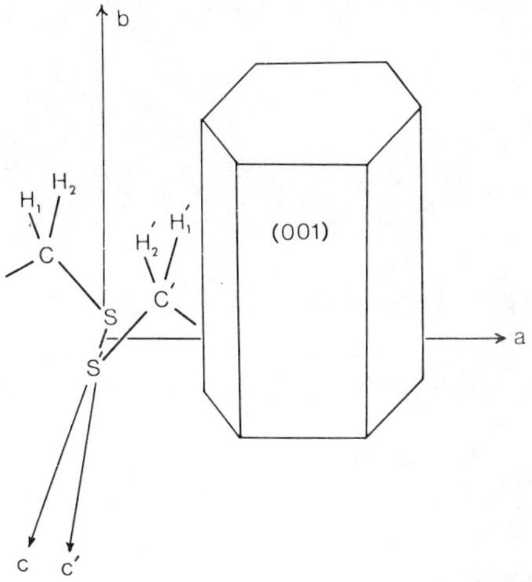

Fig. 7.8 Crystal form of cystine dihydrocholoride and the orientation of the sulfur bonds of the cystine molecule relative to the crystal axes. From Hadley and Gordy [27].

To simplify the problem of calculation of the *g* values, we treat the unpaired electron density on the two sulfurs in separate coordinate systems, as was done for the RCH_2SS radical. For $S^{(1)}$, we designate the reference system x, y, z with x perpendicular to SS and to the $C^{(1)}S^{(1)}S^{(2)}$ plane and y perpendicular to SS and in the $C^{(1)}S^{(1)}S^{(2)}$ plane. For $S^{(2)}$, we designate the axes as x', y', z with x' perpendicular to the $S^{(1)}S^{(2)}C^{(2)}$ plane, with y' in this plane and with z along the SS bond. With the assumptions made, it is evident that x' has the direction of y, that y' has the direction of x, and that elements of the two component **g** tensors are directly additive. For $S^{(1)}$, the pertinent orbitals are

$$\phi_0 = \left(\frac{1}{2}\right)^{1/2} [p_z{}^{S(1)} - p_z{}^{S(2)}] \tag{7.103}$$

$$\phi_1^{(1)} = p_x{}^{S(1)} \tag{7.104}$$

$$\phi_2^{(2)} = \left(\frac{1}{2}\right)^{1/2} [p_y{}^{S(1)} + \psi^{C(1)}] \tag{7.105}$$

The $\Delta g^{(1)}$ values derived from (7.14) with these functions are

$$\Delta g_{zz}^{(1)} = 0 \tag{7.106}$$

$$\Delta g_{xx}^{(1)} = \frac{-(\lambda_S/2)}{E_2^{(1)} - E_0} \tag{7.107}$$

$$\Delta g_{yy}^{(1)} = \frac{-\lambda_S}{E_1^{(1)} - E_0} \tag{7.108}$$

From symmetry, it is evident that $\Delta g_{zz}^{(2)} = \Delta g_{zz}^{(1)}$, $\Delta g_{x'x'}^{(2)} = \Delta g_{yy}^{(1)}$, $\Delta g_{y'y'}^{(2)} = \Delta g_{xx}^{(1)}$, and $E_n^{(1)} = E_n^{(2)} = E_n$. The resultant principal values are

$$\Delta g_{\parallel} = \Delta g_{zz} = 0 \tag{7.109}$$

$$\begin{aligned} \Delta g_{\perp} = \Delta g_{xx}^{(1)} + \Delta g_{y'y'}^{(2)} &= \Delta g_{yy}^{(1)} + \Delta g_{x'x'}^{(2)} \\ &= \frac{-(\lambda_S/2)}{E_2 - E_0} - \frac{\lambda_S}{E_1 - E_0} \\ &= \frac{191}{\Delta E_2} + \frac{382}{\Delta E_1} \end{aligned} \tag{7.110}$$

Because there is only one $\Delta g_{\perp}$, (7.110) cannot be solved for ΔE_1 and ΔE_2. For an estimate of ΔE_1, Akasaka, Ohnishi, et al. [26] assumed $\Delta E_2 \gg \Delta E_1$ and neglected entirely the term in ΔE_2. With this assumption, they obtained $\Delta E_1 \approx$ 24,000 cm^{-1}. The ΔE_1 energy corresponds to that for an $n \rightarrow \sigma_{SS}^*$ transition; the

ΔE_2 energy, to a $\sigma_{SC} \rightarrow \sigma^*_{SS}$ transition. A reasonable estimate is $E_2 \approx 2E_1$. With this assumption and the observed $\Delta g_{\perp} = 0.016$, (7.110) gives $\Delta E_1 = (E_1 - E_0) \approx$ 30,000 cm^{-1}.

The Cysteine Anion

In a single crystal of L-cysteine (HCl) irradiated with 1.5-MeV electrons at 77°K and observed at that temperature, Akasaka [29] found the principal *g* values of 1.99, 1.99, and 2.29, unusual for S-centered radicals. Note that most of the *g* values in Table 7.8 are in the range from 2.00 to 2.065. Following this experiment, Box, Freund, and Budzinski [30] measured principal *g* values of 1.985, 2.004, and 2.251 of radicals produced in an L-cysteine (HCl) crystal at 77°K with ultraviolet (uv) irradiation. They found the largest value to be approximately along the CS bond of the undamaged molecule. Kayushin, Krivenko, and Pulatova [31] observed principal *g* values of 1.985, 1.998, and 2.278 after irradiation of the L-cysteine (HCl) crystal with γ-rays. Because of the similarity of these rather atypical *g* values, we have assigned them in Table 7.8 to the common anion radical $(RCH_2SH)^-$, originally proposed by Kayushin, Krivenko, and Pulatova et al. [31]. Although Akasaka [29] proposed the neutral monosulfide radical as the source of the signals which he observed, this form of the radical does not seem to be the probable source of these *g* values that differ markedly from those of the now well established monosulfide radical in L-cystine $(HCl)_2$. Compare the values in Table 7.7.

In single crystals of *N*-acetyl-L-cysteine irradiated with 4-MeV electrons at 77°K, Saxebφl and Herskedal [32] observed the set of *g* values as 1.990, 2.006, and 2.214. Because of the similarity of these values to those for the anion radical observed in irradiated L-cysteine (HCl), we have tentatively assigned them in Table 7.8 to the acetylcysteine anion $(R'CH_2SH)^-$. The radical giving these *g* values has been proved by ^{33}S hyperfine measurements (Chapter VI, Section 10.b) to have the monosulfide form.

These **g** tensors ascribed to the cysteine anion are exceptional, not only because g_w is unusually large, but also because g_u is significantly less than the free-spin value. Although the ^{33}S hyperfine structure revealed spin density on only one S in the acetylcysteine radical, it indicated a 3*p* spin density of only 0.70 and a 3*s* spin density of 0.03 on the S. The rather large CH_2 proton couplings of 62 MHz and 82 MHz [32] show that 0.10 of the unpaired electron density is on the CH_2. The 0.03 3*s*(S) spin density is expected because of effects of σ-bond spin polarization. The normalization $\Sigma \rho_n = 1$, therefore, indicates a residual unpaired electron density of $1 - 0.80 = 0.20$, which is undetected by measurement of hyperfine structure. Although one might assume this density of 0.20 to be on other atoms of the radical, the mechanism for such transfer is not apparent, nor would the transfer provide an explanation for the anomalously large g_w and the negative Δg_u. Consequently, Hadley and Gordy [33] proposed that in the formation of the cysteine anion the valency of the S is expanded by utilization of one of the 3*d* orbitals. With the orbital of the unpaired electron of the $(R'CH_2SH)^-$ assumed to be

$$\phi_0 = \left(\frac{7}{10}\right)^{1/2} p_x(S) + \left(\frac{1}{5}\right)^{1/2} d_{xy}(S) + \left(\frac{1}{10}\right)^{1/2} \psi(CH_2) \qquad (7.111)$$

as indicated by the hyperfine measurements, they were able to show that Δg_u should be negative, as observed, and to predict the $g_v = 2.006$ value with the reasonable $\sigma(CS) \rightarrow \phi_0$ transition energy of 34,000 cm^{-1}. Information about the orbital energies is insufficient to allow a meaningful prediction of the large g_w value. Structural details of the cysteine anion radical are still uncertain.

References

1. M. H. L. Pryce, *Proc. Phys. Soc.*, **A63**, 25 (1950).
2. A. J. Stone, *Proc. Roy. Soc. (Lond.)*, **A271**, 424 (1963).
3. H. M. McConnell and R. E. Robertson, *J. Phys. Chem.*, **61**, 1018 (1957).
4. D. W. Ovenall and D. H. Whiffen, *Molec. Phys.*, **4**, 135 (1961).
5. A. J. Stone, *Molec. Phys.*, **6**, 509 (1963).
6. A. J. Stone, *Molec. Phys.*, **7**, 311 (1964).
7. R. S. Mulliken, *Rev. Mod. Phys.*, **14**, 204 (1942).
8. A. D. Walsh, *J. Chem. Soc.*, 2266 (1953).
9. S. A. Marshall, A. R. Reinberg, R. A. Serway, and J. A. Hodges, *Molec. Phys.*, **8**, 225 (1964).
10. G. W. Chantry and D. H. Whiffen, *Molec. Phys.*, **5**, 189 (1962).
11. J. F. Mulligan, *J. Chem. Phys.*, **19**, 347 (1951).
12. J. E. Bennett, B. Mile, and A. Thomas, *Trans. Faraday Soc.*, **63**, 262 (1967).
13. H. Zeldes and R. Livingston, *J. Chem. Phys.*, **35**, 563 (1961).
14. F. Schneider, W. Heinze, and W. Sudars, *AEG-Mitt.*, **55**, 232 (1965).
15. J. Schneider, B. Dischler, and A. Rauber, *Phys. Status Solidi*, **13**, 141 (1966).
16. A. Reuveni, Z. Luz, and B. L. Silver, *J. Chem. Phys.*, **53**, 4619 (1970).
17. K. P. Dinse and K. Möbius, *Z. Natürforsch.*, **23a**, 695 (1968).
18. C. A. Coulson, *Valence*, Oxford Un. Pr., London, 1952, p. 195.
19. H. C. Heller and T. Cole, *J. Am. Chem. Soc.*, **84**, 4448 (1962).
20. K. Toriyama and M. Iwasaki, *J. Chem. Phys.*, **55**, 2181 (1971).
21. M. Iwasaki and K. Itoh, *Bull. Chem. Soc. Jap.*, **37**, 44 (1964).
22. D. S. McClure, *J. Chem. Phys.*, **17**, 905 (1949); **20**, 682 (1952).
23. E. Melamud, S. Schlick, and B. L. Silver, *J. Magn. Reson.*, **14**, 104 (1974).

24. K. Toriyama and M. Iwasaki, *J. Phys. Chem.*, **73** 2663 (1969).

25. Y. Kurita and W. Gordy, *J. Chem. Phys.*, **34**, 282 (1961).

26. K. Akasaka, S. -I. Ohnishi, T. Suita, and I. Nitta, *J. Chem. Phys.*, **40**, 3110 (1964).

27. J. H. Hadley, and W. Gordy, *Proc. Nat. Acad. Sci. (USA)*, **71**, 3106 (1974).

28. L. K. Steinrauf, J. Peterson, and L. H. Jensen, *J. Am. Chem. Soc.*, **80**, 3835 (1958).

29. K. Akasaka, *J. Chem. Phys.*, **43**, 1182 (1965).

30. H. C. Box, H. G. Freund, and E. E. Budzinski, *J. Chem. Phys.*, **45**, 809 (1966).

31. L. P Kayushin, V. G. Krivenko, and M. K. Pulatova, *Studia Biophysica (Berlin)*, **33**, No. 1, 59 (1972).

32. G. Saxebøl and Ø. Herskedal, *Radiat. Res.*, **62**, 395 (1975).

33. J. H. Hadley, and W. Gordy, *Proc. Nat. Acad. Sci. (USA)*, **74**, 216 (1977).

Chapter VIII

RANDOMLY ORIENTED RADICALS IN SOLIDS

Although the most precise esr information about radicals in solids is obtained from measurements on single crystals, such measurements are possible for only a limited number of substances. When observations on single crystals are not feasible, much can be learned from esr measurements on polyoriented free radicals in polycrystalline solids or powders.

The amount and kind of information which can be gained from esr measurements on randomly oriented radicals in solids depends on the nature of the **g** tensor and the nuclear coupling tensors. If both **g** and **A** are isotropic, the same information can be gained as is obtained from observations on single crystals. When both **g** and **A** are decidely anisotropic, the esr of the powder is sometimes hopelessly scrambled, and one can obtain very little useful information. However, for many types of randomly oriented free radicals in solids, it has proved possible to obtain rather accurate values for the principal elements of the **g** and **A** tensors from analysis of the shape of the esr patterns. Description of theoretical methods for such analysis will follow. Numerous esr spectroscopists have contributed to the development of these methods. Among the earliest of them are Bleaney [1], Sands [2], Singer [3], Searl, Smith, and Wygard [4], Blinder [5], Kneubühl [6], and Ibers and Swalen [7].

1. EFFECTS OF g ANISOTROPY ON LINE SHAPES

1.a. Shapes due to Axially Symmetric g Tensors

When the spread of the esr pattern from all other causes is negligibly small as compared with that caused by **g** anisotropy, the usual line shape produced by factors other than **g** anisotropy is expressed by the delta function $\delta(H - H_0)$, which is unity when $H = H_0$ and zero when $H \neq H_0$. The esr absorption at a fixed frequency ν_0 then occurs entirely within the limiting field values $H_{\|} = h\nu_0/g_{\|}\beta$ and $H_{\perp} = h\nu_0/g_{\perp}\beta$, where $g_{\|}$ and $g_{\perp}$ are the principal values of the axially symmetric g being considered. With the additional assumption that the absorption coefficient is constant within the field range from $H_{\|}$ to $H_{\perp}$, the total absorption or integrated intensity may be expressed as

$$\mathscr{I} = \int_{H_{\|}}^{H_{\perp}} I(H)dH \tag{8.1}$$

The fractional power absorption $I(H)dH/\mathscr{I}$ in the field range from H to $H + \Delta H$ results entirely from the fraction dN/N_0 of the randomly oriented radicals that have their magnetic symmetry axes oriented between θ and $\theta + \Delta\theta$, or between $(\pi - \theta)$ and $(\pi - \theta + \Delta\theta)$, relative to the direction of H where $\Delta\theta$ corresponds to the spread in the effective $g = h\nu_0/\beta H$ to $g = h\nu_0/(H + \Delta H)$. With the assumed line shape of the δ function, only radicals having orientations within this range can absorb within the field range from H to ΔH. The spherical surface area included within the two surface rings corresponding to $\Delta\theta$ is $2(2\pi r)\sin\theta(r\,d\theta)$, and the total surface area of the sphere is $4\pi r^2$. Thus it is evident that

$$\frac{I(H)dH}{\mathscr{I}} = \frac{dN}{N_0} = \frac{4\pi r^2 \sin\theta \, d\theta}{4\pi r^2} = \sin\theta \, d\theta \tag{8.2}$$

For convenience, we normalize the total absorption expressed by (8.1) to unity. With this simplification, we find from (8.2) that

$$I(H) = \sin\theta \frac{d\theta}{dH} \tag{8.3}$$

It is possible to express $\sin\theta$ and $d\theta$ in terms of the magnetic field by use of the g-factor expression

$$g = (g_{\|}^2 \cos^2\theta + g_{\perp}^2 \sin^2\theta)^{1/2} \tag{8.4}$$

and the use of $g = h\nu_0/\beta H$, $g_{\|} = h\nu_0/\beta H_{\|}$, $g_{\perp} = h\nu_0/\beta H_{\perp}$, and $\cos^2\theta = \sin^2\theta - 1$. With these transformations of the variables, $I(H)$ is found from (8.3) to be

$$I(H) = \frac{H_{\|}H_{\perp}^2}{H^2[(H_{\perp}^2 - H^2)(H_{\perp}^2 - H_{\|}^2)]^{1/2}} \tag{8.5}$$

for H between $H_{\perp}$ and $H_{\|}$. For H outside this region, $I(H) = 0$. The upper solid curve of Fig. 8.1 is a plot of the function represented by (8.5) when $H_{\|} < H_{\perp}$; the broken curve is an approximate representation of the real absorption curve as modified by other line-broadening factors neglected in this derivation. The first derivatives of the absorption curves are detected and displayed by the typical esr spectrometer. The first derivatives of the absorption curves (upper) are displayed in the lower graph of Fig. 8.1.

Although the approximate derivation given in the preceding paragraph takes into account only the line spread due to **g** anisotropy, it is easy, at least in principle, to include other line-broadening factors. For the axially symmetric case, closed formulas can still be derived for the overall line shape [6,8]. These formulas, however, are rather cumbersome, and the problem is most easily solved numerically by a computer program that includes the various broadening factors [9].

At this point we designate the line-shape function exclusive of g anisotropy by the function $f(H - H_r)$, where H_r is the resonance field for the peak intensity of the line. In the polycrystalline material the broadening function is superimposed on that due to **g** anisotropy as described by (8.5), or to **A** anisotropy to be described in later sections. Although the resonance field H_r depends on θ, the shape of the function $f(H - H_r)$ remains approximately the same for all orientations of the radical in the field. Thus the shape function for a group of radicals in the polycrystalline material that have resonance field values in the range from H_r to $H_r + \Delta H_r$ would still be $f(H - H_r)$ when ΔH_r is small as

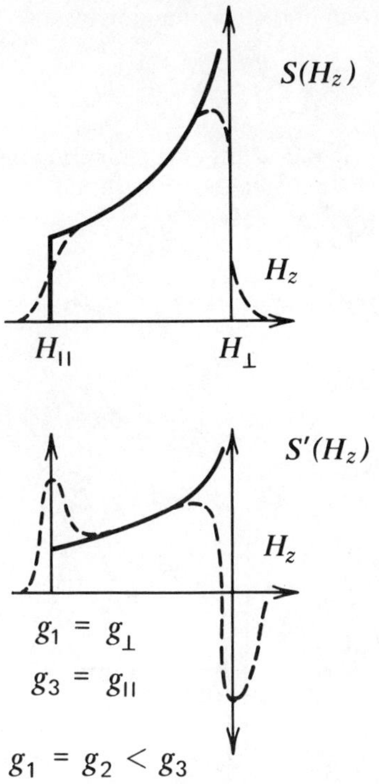

Fig. 8.1 Theoretical shape functions due to axially symmetric **g** tensors in radomly oriented free radicals. Solid curves represent idealized functions due to **g** anisotropy alone. Broken curves include illustrative corrections for other broadening factors. From Kneubühl [6].

compared with the width of the shape function in the single crystal. Because of the presence of H in $f(H - H_r)$, it is evident that radicals with orientations outside the $\Delta\theta_r$ range corresponding to ΔH_r will now also absorb in the field range from $H = H_r$ to $H_r + \Delta H_r$. Therefore, to get the power absorption $I(H)\Delta H$ in the narrow band ΔH, one must integrate the shape function over all orientations of the radicals, that is,

$$I(H)\Delta H \propto \int_0^{\pi/2} f(H - H_r)\sin\theta\, d\theta \tag{8.6}$$

Now, $\sin\theta\, d\theta$ may be expressed in terms of H_r and dH_r by comparison of (8.3) and (8.5). From (8.3) and (8.4), one sees that

$$\sin\theta\, d\theta \propto H_r^{-2}\,[(H_\perp^2 - H_r^2)(H_\perp^2 - H_\parallel^2)]^{-1/2}\, dH_r \tag{8.7}$$

The prevalent Lorentzian line-shape function is

$$f(H - H_r) \propto [(H - H_r)^2 + \Delta H_{1/2}^2]^{-1} \tag{8.8}$$

where $\Delta H_{1/2}$ is one-half the width of the function between half-intensity points [see (4.43)]. With (8.7) and (8.8) substituted into Eq. (8.6) and with ΔH chosen as a constant interval, one finds that

$$I(H) \propto \int_{H_\parallel}^{H_\perp} [(H - H_r)^2 + \Delta H_{1/2}^2]^{-1} H_r^{-2} [(H_\perp^2 - H_r^2)(H_\perp^2 - H_\parallel^2)]^{-1/2} dH_r \tag{8.9}$$

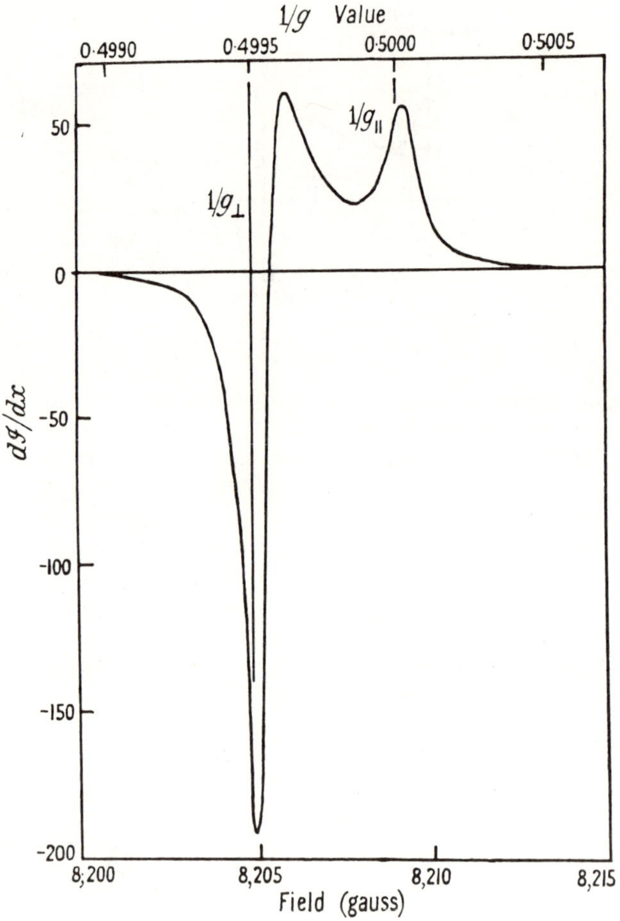

Fig. 8.2 First derivative of esr absorption curve at ν_0 = 23,000 MHz for randomly oriented radicals having axially symmetric **g** tensors with $g_\parallel$ = 2.000, $g_\perp$ = 2.002 and Lorentzian shape with $\Delta H_{1/2}$ = 1.025 G. From Searl, Smith, and Wyard [4].

By integration of an expression similar to that of (8.9), Searl, Smith, and Wyard [4] derived the shape function for the axially symmetric **g** tensor with the assumption that the lines of a particular orientation of the radical in the field would have a Lorentzian shape. The formula that they obtained, although expressed in closed form, is rather complex and is not repeated here. Usually the integrations are made numerically by computer. From the resulting integral, they computed the first-derivative function $dI(H)/dH$ for the frequency $\nu_0 = 23$ kMHz, with the assumed parameters: $g_{\parallel} = 2.000$ G, $g_{\perp} = 2.002$ G, and $\Delta H_{1/2} =$ 1.025 G. The resulting curve, which is rather typical of organic free radicals having axially symmetric **g** with no nuclear coupling, is reproduced in Fig. 8.2. Later, they modified their calculations to include a small factor due to the dependence of the transition probability on **g** [8]. However, this factor is not significant for organic free radicals where $g_{\parallel} \approx g_{\perp}$.

Ibers and Swalen [7] similarly calculated the polycrystalline line shape for axially symmetric **g**, but with a variety of Lorentzian line widths and with the assumptions that $H_{\parallel} = 3600$ G and $H_{\perp} = 3300$ G. The $I(H)$ expression that they employed is like that of (8.9) except that they included under the integral the factor $[1 + (H_r/H_{\parallel})^2]$ to correct for the variation of the transition probability with **g** [1]. The various shape functions they obtained are reproduced in Fig. 8.3. These shape functions are useful in demonstrating the importance of the line widths relative to the spread of **g** in influencing the shapes of the absorption curves. The curves displayed by the usual esr spectrometer are the first or second derivatives of these absorption curves.

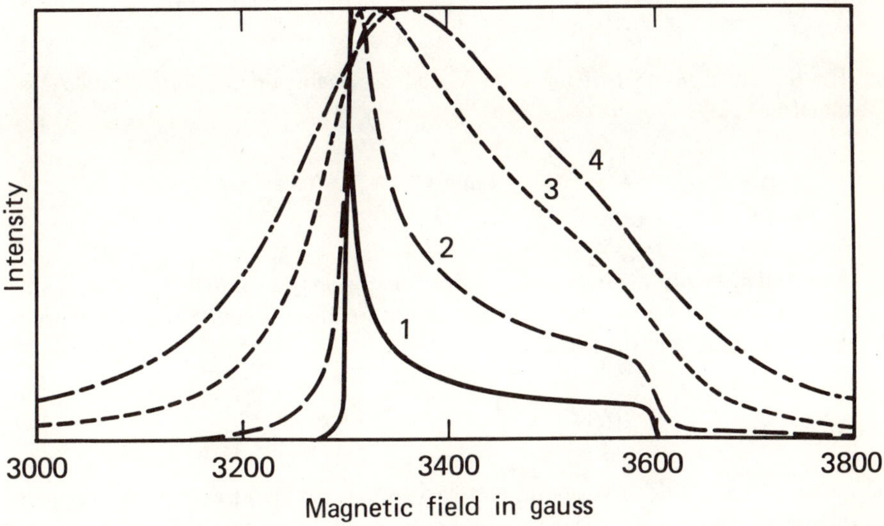

Fig. 8.3 Theoretical esr absorption functions for randomly oriented radicals having axially symmetric **g** tensors with Lorentzian line widths: (1) 1 G, (2) 10 G, (3) 50 G, and (4) 100 G, when $H_{\parallel} = 3600$ G and $H_{\perp} = 3300$ G. From Ibers and Swalin [7].

It is evident from Fig. 8.3 that the accuracy with which evaluation of $g_{\parallel} = h\nu_0/\beta H_{\parallel}$ and $g_{\perp} = h\nu_0/\beta H_{\perp}$ can be made from the powder spectrum depends on the line widths $\Delta H_{1/2}$ relative to $|H_{\parallel} - H_{\perp}|$. Rather accurate values can be obtained for both $g_{\parallel}$ and $g_{\perp}$ if $\Delta H_{1/2} < |H_{\parallel} - H_{\perp}|/100$. Since $\Delta H_{1/2}$, which depends primarily on spin-lattice relaxation time, can be significantly reduced by a lowering of the temperature of the sample and $(H_{\parallel} - H_{\perp})$ is often essentially independent of temperature, one can usually improve the accuracy of the values of $g_{\parallel}$ and $g_{\perp}$ by lowering the temperature of the observations. From Fig. 8.3 it is also obvious that $H_{\perp}$ and hence $g_{\perp}$ can be found with greater accuracy than can $H_{\parallel}$ and hence $g_{\parallel}$. This applies especially when the Lorentzian widths are large.

1.b. Polycrystalline Line Shapes when $g_x \neq g_y \neq g_z$

When no two principal g values of the radical are equal (no axial symmetry), the fraction of randomly oriented radicals in a polycrystalline sample that absorb in the field range from H to $H + dH$ depends on both the polar angles θ and ϕ. In an approximate treatment, we again neglect the broadening due to factors other than g anisotropy. The dN radicals that undergo resonance absorption in the field range of H to $H + dH$ have angular orientations within the range from θ to $\theta + d\theta$ and from ϕ to $\phi + d\phi$ corresponding to the spread in the effective $g = h\nu_0/\beta H$ to $g = h\nu_0/(H + dH)$. They are enclosed within the solid angle

$$d\Omega = \frac{dS}{r^2} = \frac{r \sin\theta (r\, d\theta) d\phi}{r^2} = \sin\theta\, d\theta\, d\phi \tag{8.10}$$

The fractional number of the total N_0 radicals having orientations within $d\Omega$ is, therefore,

$$\frac{dN}{N_0} = \frac{d\Omega}{4\pi} = \frac{\sin\theta\, d\theta\, d\phi}{4\pi} = \frac{d(\cos\theta) d\phi}{4\pi} \tag{8.11}$$

As for the axially symmetric case, this fraction may be equated to $I(H)dH/\mathcal{I}$; again, if the total integrated intensity $\mathcal{I}$ is normalized to unity, the relationship may be expressed in the form

$$I(H)dH = \frac{1}{4\pi}\, d\Omega = \frac{1}{4\pi}\, d(\cos\theta) d\phi \tag{8.12}$$

For this problem, it is convenient also to express the effective g in terms of the polar angles θ and ϕ. This transformation of (3.32) gives

$$g = (g_x^2 \sin^2\theta \cos^2\phi + g_y^2 \sin^2\theta \sin^2\phi + g_z^2 \cos^2\theta)^{1/2} \tag{8.13}$$

The corresponding resonance field values of H at constant frequency may be expressed as

$$H = \frac{h\nu_0}{\beta}(g_x^2 \sin^2\theta \cos^2\phi + g_y^2 \sin^2\theta \sin^2\phi + g_z^2 \cos^2\theta)^{-1/2} \quad (8.14)$$

where x, y, z are the principal axes of **g**.

The desired shape function $I(H)$ may be evaluated most easily with numerical computer methods [9]. Griscom, Taylor, et al. [10] give details of a computer program for this evaluation. Note from the relations of (8.11) that equal areas in $(\cos\theta)$–ϕ space correspond to equal elements of solid angle $\Delta\Omega$. Their computer was programmed to calculate the resonance field for all orientations corresponding to equal elements of a fine grid in $(\cos\theta)$–ϕ space. The number of times this program gave a resonance field in the range of $H_i - H_{i-1} = \Delta H$ was taken as a measure of the intensity function $I(H_i)$ in the $H = H_i$ to H_{i-1} range. Constant intervals ΔH were chosen, and the procedure was applied to all intervals between H_{min} and H_{max}. The envelope of the resulting values of $I(H_i)$ in this region is the shape function $I(H)$.

The intensity function for a particular field value H_z between H_{min} and H_{max} may also be expressed from (8.12) by the integral

$$I(H_z) = \frac{1}{4\pi\,\Delta H_z}\int_{H_z}^{H_z+\Delta H_z} d\Omega \quad (8.15)$$

By transformation of the variables, Kneubühl [6] was able to express this integral in terms of the complete elliptic integral of the first kind and to derive an analytical expression of $I(H_z)$ in the range from H_{min} to H_{max}. The analytical formulas are too complex for reproduction here. A plot of the resulting function is given by the upper solid curve of Fig. 8.4. The lower solid curve gives the first derivative of $I(H)$. The dotted curves in each case are examples of the modification of these idealized shapes by line broadening due to spin-lattice relaxation, and so on. Note that there are characteristic features at field values H_1, H_2, and H_3 that provide a means of measurement of the corresponding g values, g_1, g_2, and g_3. The outside values, H_{min} and H_{max}, are best defined in the derivative curves; generally, g_{min} and g_{max} can be most accurately evaluated from the derivative presentation. As for the axially summetric case (see Fig. 8.3), the g value can be most accurately obtained when the line broadening caused by other factors is small as compared with the spread, $H_{max} - H_{min}$, caused by **g** anisotropy.

The line-shape function $f(H - H_r)$ for each single crystallite may be introduced into the calculation of $I(H)$ in a manner similar to that described for the

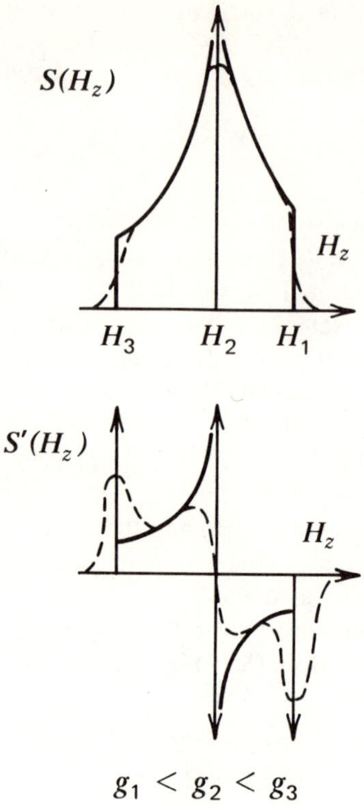

Fig. 8.4 Theoretical shape functions due to anisotropy in **g** for randomly oriented free radicals with no principal elements equal. Broken curves include illustrative corrections for other broadening factors. From Kneubühl [6].

axially symmetric case. Because of its field spread, this function causes absorption by all crystallites (radicals of all orientations) for each field value. One now defines the elemental solid angle $d\Omega = \sin\theta\, d\theta\, d\phi$ as that enclosing orientations of all crystallites that have their peak resonance field H_r within the range H_i to $H_i + \Delta H$. Absorption by these crystallites will, however, occur outside this field range because of the spread in $f(H - H_r)$. Likewise, radicals with orientations outside $d\Omega$ will absorb within this field range. Thus for each interval H_i to $H_i + \Delta H_i$ between $H_{\min}$ and $H_{\max}$, the function $f(H - H_r)$ must be averaged over all orientations of the radicals with the field. If the fields are chosen at equal intervals ΔH_i across the band $H_{\min}$ to $H_{\max}$, the constant factor ΔH_i may be dropped from the proportional relationship, and the shape function may be expressed by

$$I(H) \propto \int_0^{4\pi} f[H - H_r(\theta,\phi)]\, d\Omega \tag{8.16}$$

With the Lorentzian line-shape function and with $d\Omega = \sin\theta\, d\theta\, d\phi$, this relation may also be written as

$$I(H) \propto \int_{\phi=0}^{2\pi} \int_{\theta=0}^{\pi} \frac{\sin\theta\, d\theta\, d\phi}{[H - H_r(\theta,\phi)]^2 + \Delta H_{1/2}^2} \tag{8.17}$$

These integrals, to the best of my knowledge, have been evaluated only with numerical methods by a computer. Swalen and Gladney [9] outline, or reference, computer programs for evaluating them.

Since the line-shape function $f(H - H_r)$ is essentially independent of the orientation of the individual crystallite, or radical, a simple approach to this complex problem is to calculate the weight $I(H_r)$ for each interval $H_r = H_i$ to

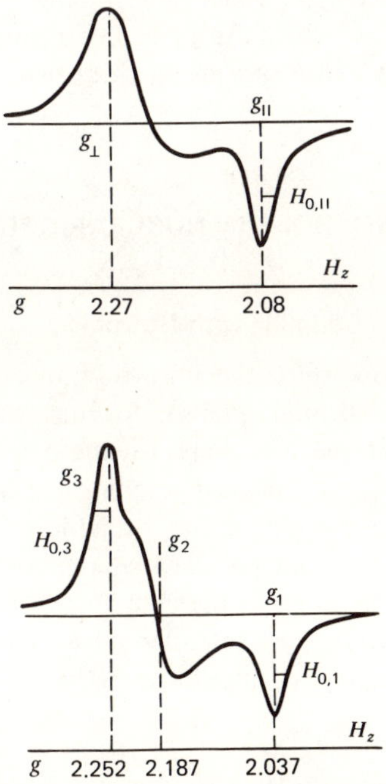

Fig. 8.5 Derivative presentation of epr spectrum of polycrystalline $CuSO_4 \cdot 5H_2O$ with axially symmetric **g** tensor (upper figure) and of polycrystalline $CuCl_2 \cdot 2H_2O$ with the three principal g values different (lower figure). From Kneubühl [6].

$H_i + \Delta H_i$ by the computer program outlined earlier for (8.12). One then multiplies each $I(H_i)$ determined in this way by an equivalent line-shape function $f(H - H_i)$ and sums the result over all H_i values. The envelope of the resulting sum is the desired shape function

$$I(H) \propto \sum_i f(H - H_i) I(H_i) \tag{8.18}$$

where the H_i values range from H_{min} to H_{max} in equal intervals. The values of $I(H_i)$ are equivalent to those of $I(H_z)$ in (8.15) and may also be obtained by numerical integration of

$$I(H_i) \propto \int_{H_i}^{H_i + \Delta H_i} d\Omega$$

Figure 8.5 shows a comparison between observed first-derivative curves for polycrystalline substances for which none of the principal g values are equal (lower curve) and those for which the **g** tensor is axially symmetric (upper curve). The line breadths and **g**-tensor spreads for these two copper salts are comparable in magnitude.

2. EFFECTS OF ANISOTROPY IN NUCLEAR COUPLING ON LINE SHAPES

2.a. Axially Symmetric Coupling with Isotropic g

Here we treat species for which the anisotropy in **g** is negligible in comparison with that in the nuclear coupling tensor **A**. First, we neglect other broadening factors and assume that the line shape for the polycrystalline substance arises entirely from anisotropy in the nuclear coupling and that the line shape exclusive of anisotropy in nuclear coupling is the δ function, $\delta(H - H_r)$. We also consider only one coupling nucleus and assume that the coupling is axially symmetric, as it would be for a p electron on the coupling nucleus. With these conditions, radicals having their symmetry axes oriented θ degrees from the applied field will absorb radiation quanta $h\nu_0$ at field strengths of

$$H = H_0 - \frac{M_I}{g\beta} A \tag{8.19}$$

where $H_0 = h\nu_0/g\beta$ is a constant, independent of θ, and where

$$A = (A_\parallel^2 \cos^2\theta + A_\perp^2 \sin^2\theta)^{1/2}$$
$$= [(A_\parallel^2 - A_\perp^2)\cos^2\theta + A_\perp^2]^{1/2} \tag{8.20}$$

From a procedure similar to that given for anisotropic **g** broadening, the line shape of the hyperfine components for randomly oriented radicals is found to be

$$I(H) = \frac{dN}{dH} = \frac{\sin\theta}{dH/d\theta} \tag{8.21}$$

From differentiation of (8.19) with A from (8.20), one obtains

$$\frac{dH}{d\theta} = \frac{M_I}{g\beta A}(A_\parallel^2 - A_\perp^2)\cos\theta \sin\theta \tag{8.22}$$

and hence

$$dN/dH = A\left[\frac{M_I}{g\beta}(A_\parallel^2 - A_\perp^2)\cos\theta\right]^{-1} \tag{8.23}$$

With $A = g\beta(H_0 - H)/M_I$ from (8.19) and $\cos\theta$ from (8.20) substituted into (8.23), the resonance shape may be expressed as a function of magnetic field strength by

$$I(H) = (H_0 - H)\left\{\left(\frac{M_I}{g\beta}\right)^2 (A_\parallel^2 - A_\perp^2)\left[(H_0 - H)^2 - \left(\frac{M_I}{g\beta}\right)^2 A_\perp^2\right]\right\}^{-1/2} \tag{8.24}$$

Except for a scale factor, each component has the same shape function, but the appearance of the overall multiplet is influenced by the overlapping of these components. However, for $M_I = 0$, the nuclear interaction will be zero for all orientations and will have no influence on the line shape. Figure 8.6 shows the appearance of the spectrum containing two components corresponding to $M_I = \pm 1/2$ and $A_\perp = -(1/2)A_\parallel$. The $H_\parallel$ and $H_\perp$ values are determined by (8.19) with $A = A_\parallel$ and $A_\perp$, respectively. When the broadening caused by other factors is much smaller than that caused by **A** anisotropy, reasonably accurate values of the coupling constants can be obtained by measurement of the separation of the two strong peaks and of the two outer shoulders.

The constants $A_\parallel$ and $A_\perp$ may, of course, have isotropic as well as anisotropic parts. The most commonly observed coupling results from both *s*- and *p*-orbital

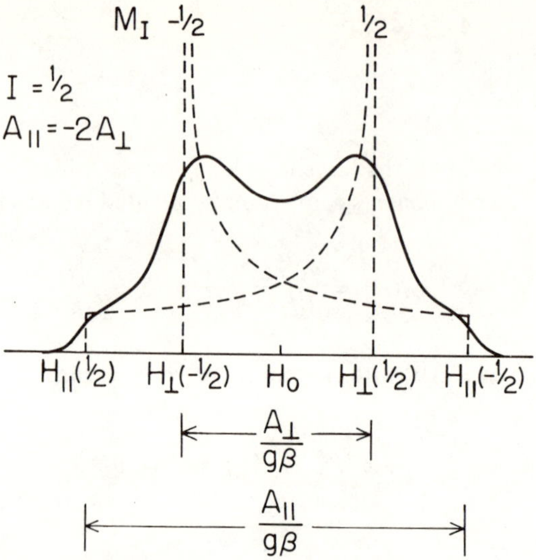

Fig. 8.6 Theoretical line-shape function resulting from anisotropic dipole–dipole nuclear coupling in randomly oriented radicals where $I = 1/2$. For the solid curve, this distortion is superimposed on the normal Gaussian line shape.

spin density. With the isotropic s component indicated by A_f and the p-orbital coupling constant indicated by B, the relationships between the constants are

$$\begin{aligned} A_{\parallel} &= A_f + 2B \\ A_{\perp} &= A_f - B \end{aligned} \tag{8.25}$$

and (8.24) may be expressed as

$$I(H) = (H_0 - H)\left\{\left(\frac{M_I}{g\beta}\right)^2 3B(2A_f + B)\left[(H_0 - \mathrm{H})^2 - \left(\frac{M_I}{g\beta}\right)^2 (A_f - B)^2\right]\right\}^{-1/2} \tag{8.26}$$

Figure 8.7 illustrates the expected shape of a multiplet of a polycrystalline sample caused by a single coupling nucleus with $I = 1$, with $A_f = 3B$ (upper curve) and $A_f = -3\mathrm{B}$ (lower curve). The dotted curves are plots of (8.26) with the $H_{\parallel}$ and $H_{\perp}$ determined by (8.19) with $A = A_{\parallel}$ and $A_{\perp}$, respectively, as given by (8.20) with $\theta = 0$ and $\pi/2$. The solid curves include a Lorentzian line shape in addition to the anisotropic nuclear broadening. The assumed Lorentzian shape is evident for the central component which corresponds to $M_I = 0$, for which the nuclear coupling vanishes. The three states $M_I = 1,0,-1$ have very

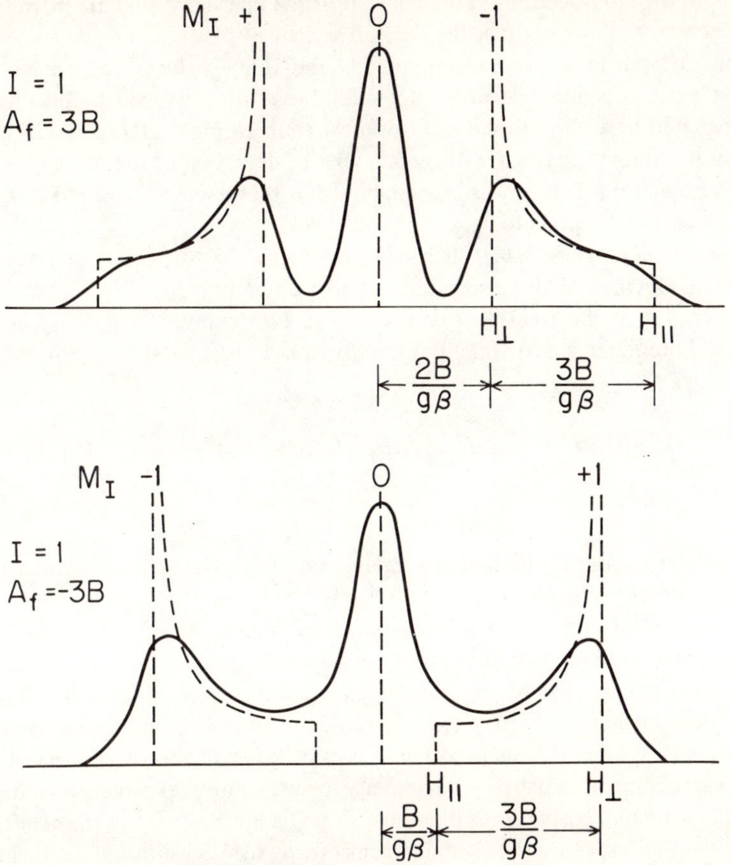

Fig. 8.7 Theoretical line-shape function caused by combined isotropic and anisotropic components in nuclear coupling in randomly oriented radicals where $I = 1$. For the solid curve, this distortion is superimposed on the normal Gaussian line shape.

nearly equal probability; in the absence of line distortion, there would be three components of equal height. This is obviously not true in Fig. 8.7. However, the integrated intensity, or area under the absorption curve, remains equal for the three components. A comparison of the upper and lower curves of Fig. 8.7 makes it evident that the relative signs of A_f, $A_{\parallel}$, and $A_{\perp}$ can be easily found from the appearance of the spectrum of the powder.

The theoretical curves of Figs. 8.6 and 8.7 represent absorption intensity versus field strength at constant frequency. The appearance of esr signals depends on the nature of the detection employed. The usual receivers detect first or second derivatives of the actual absorption curves. In deriving the hyperfine coupling constants from polycrystalline measurements, one must take into account the modifications of the contour of the line shape that are introduced by the detector. For example, the derivative display of the upper curve of Fig. 8.7 might appear to have five components and that of the lower, to have three. Even with straight absorption detection, the top curve might be mistakenly interpreted as a quintet and the lower, as an isotropic triplet. It is advantageous in the interpretation to have different types of detection of the resonances and to use as much information as is available from other sources to predict the types of patterns to be expected.

The correction of (8.24) for broadening factors, as of (8.26), can be made by the introduction of the associated line-shape function $f(H - H_r)$, as was done in Section 1.a in the treatment of resonances with g anisotropy. When $f(H - H_r)$ is the Lorentzian shape function, comparison with (8.9) shows that

$$I(H) \propto \int_{H_{\parallel}}^{H_{\perp}} [(H-H_r)^2 + \Delta H_{1/2}^2]^{-1} (H_r-H_0)[(H_r-H_0) - \left(\frac{M_I}{g\beta}\right)^2 A_{\perp}^2]^{-1/2} dH_r \tag{8.27}$$

Numerical computer methods for evaluation of this function are similar to those already described in Section 1.

Most nuclear coupling tensors encountered in molecular free radicals are either isotropic (s-orbital coupling) or axially symmetric (p-orbital coupling). Interatomic coupling, that is, coupling by spin density on one atom with the nucleus of a neighboring atom, may tend to cause anisotropic coupling, which does not have axial symmetry. Such coupling is usually small, however, except that in the CH fragment. Anisotropic coupling, for which no two principal elements are equal, is treated by methods like those described in Section 1.b for **g** anisotropy. The adaptations of these methods to anisotropy in **A** tensors will be obvious.

2.b. Line Shapes Due to Combined g and A Anisotropy

In the most commonly observed molecular free radicals, the anisotropic nuclear coupling tensors result from p-orbital spin density. In these radicals one principal element of the **g** tensor is generally found from single-crystal measurements and from theoretical calculations to be along the axis of the p_π or p_σ orbital of the unpaired electron density on the coupling atom. This direction is that of the axis of the coupling p orbital and is also the principal axis of symmetry of the nuclear coupling. Because there is no preferred direction for the principal element A in the perpendicular plane, the **g** and **A** tensors of such radicals may be treated as having common principal axes, even when $g_x \neq g_y \neq g_z$. Furthermore, the principal axes of coupling for all nuclei in a π-type free radical generally have the same directions. These features in the most common type of spin Hamiltonian greatly simplify the problem of calculation of resonance shapes for randomly oriented samples.

For an axially symmetric **g** and **A** with common symmetry axes, the resonance field values at a constant frequency ν_0 may be expressed as

$$H = \frac{h\nu_0}{g\beta} - \frac{M_I}{g\beta} A \tag{8.28}$$

where

$$g = [(g_{\parallel}^2 - g_{\perp}^2)\cos^2\theta + g_{\perp}^2]^{1/2} \tag{8.29}$$

and

$$A = \frac{1}{g}[(A_{\parallel}^2 g_{\parallel}^2 - A_{\perp}^2 g_{\perp}^2)\cos^2\theta + g_{\perp}^2 A_{\perp}^2]^{1/2} \tag{8.30}$$

By differentiation of H with respect to θ and substitution of the result into (8.21) one obtains the shape function

$$I(H) = \frac{dN}{dH} = \frac{\beta g^3}{(g_{\parallel}{}^2 - g_{\perp}^2)(h\nu_0 - 2M_I A) + (M_I/A)(A_{\parallel}^2 g_{\parallel}^2 - A_{\perp}^2 g_{\perp}^2)\cos\theta} \tag{8.31}$$

where θ ranges in value from 0° to 90°. An expression of this kind was first obtained by Neiman and Kivelson [11]. There are singularities in (8.31) at the limiting values, $\theta = 0$ and $\theta = (\pi/2)$, similar to those in (8.23). At $\theta = 0$, the absorption begins with a sudden jump from zero to a finite value; at $\theta = 90°$, $I(H)$ becomes infinite. These abrupt end points of the spectrum result from the neglect of line broadening caused by relaxation processes. The magnetic field values corresponding to these limiting values of θ as obtained from (8.28) are

$$H_{\parallel}(\theta = 0) = \frac{h\nu_0}{g_{\parallel}\beta} - \frac{M_I A_{\parallel}}{g_{\parallel}\beta} \tag{8.32}$$

$$H_{\perp}(\theta = 90°) = \frac{h\nu_0}{g_{\perp}\beta} - \frac{M_I A_{\perp}}{g_{\perp}\beta} \tag{8.33}$$

Experimentally, these $H_{\parallel}$ and $H_{\perp}$ values can be most accurately identified from the first derivative $dI(H)/dH$ of the absorption contours.

Theoretical plots of the idealized shapes indicated by (8.31) with the $H_{\parallel}$ and $H_{\perp}$ values determined by (8.32) and (8.33) are shown in Fig. 8.8 for a spin of $I = 3/2$ when $g_{\parallel} > g_{\perp}$ and $A_{\parallel} > A_{\perp}$. The H values depend on g only; $H_{\parallel} =$

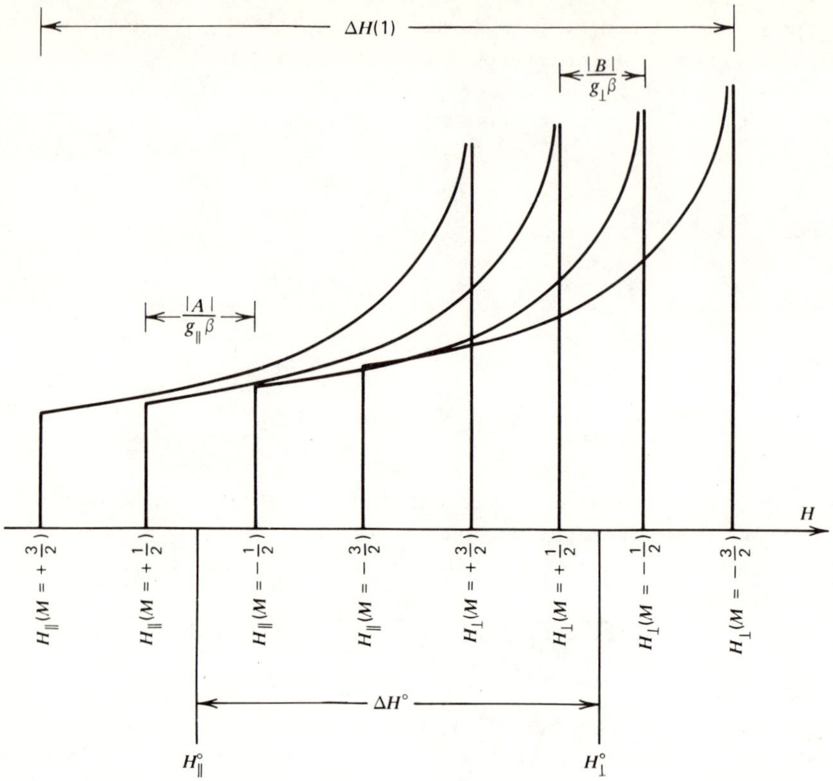

Fig. 8.8 Theoretical hyperfine structure for randomly oriented free radicals having axially symmetric anisotropy in **g** and in nuclear coupling when $I = 3/2$, $g_{\|} > g_{\perp}$, $|A| > |B| > 0$. Distortions due to other broadening factors are weighted. From Lee and Bray [12].

$h\nu_0/g_{\|}\beta$, $H_{\perp}^0 = h\nu_0/g_{\perp}\beta$, and $\Delta H^0 = H_{\perp}^0 - H_{\|}^0$. It should be noted that the spread of the spectrum caused by **g** anisotropy

$$\Delta H^0 = \frac{h\nu_0}{\beta}\left(\frac{1}{g_{\perp}} - \frac{1}{g_{\|}}\right) \tag{8.34}$$

varies directly with operating frequency ν_0, whereas the spread due to anisotropy in the nuclear coupling is independent of this frequency. The spacing between the $H_{\|}$ peaks and between the $H_{\perp}$ peaks of the adjacent hyperfine component

$$\Delta H_{\|} = \frac{A_{\|}}{g_{\|}\beta}, \quad \Delta H_{\perp} = \frac{A_{\perp}}{g_{\perp}\beta}$$

indicated in Fig. 8.8 gives a direct measure of the nuclear coupling constants. It is evident that the appearance of the spectrum, even the number of observable peaks, depends on the relative values of the constants, $g_{\parallel}$, $g_{\perp}$, $A_{\parallel}$, and $A_{\perp}$, as well as upon the operating frequency.

Figure 8.9 illustrates the appearance of the spectrum for $I = 3/2$ with relative values of the parameters, $g_{\parallel}$, $g_{\perp}$, $A_{\parallel}$, and $A_{\perp}$, somewhat different from those for Fig. 8.8. The solid line of the middle graph represents the sum of the idealized theoretical components of the top graph; the dashed curve (middle graph)

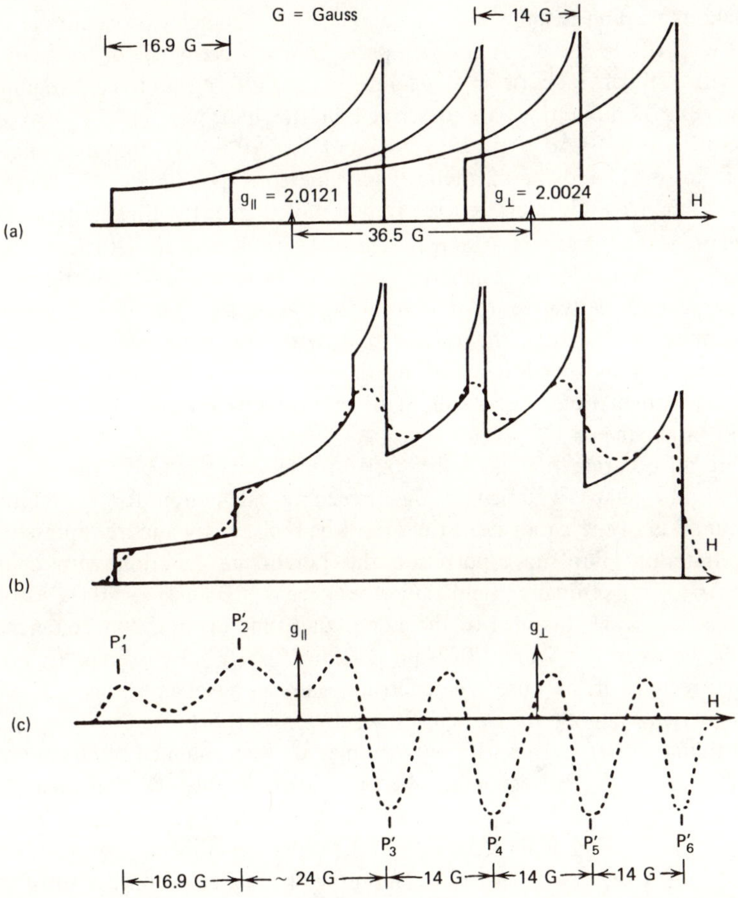

Fig. 8.9 Theoretical shape functions due to axially symmetric anisotropy in **g** and **A** for *K*-band frequency when $I = 3/2$. Top curve is idealized line for individual hyperfine components; middle curve is summation of these components with dashed curve illustrating possible modifications by other line-broadening factors; bottom curve is first derivative of dashed middle curve. From Lee and Bray [12].

illustrates how this idealized pattern is likely to be altered by line broadening due to relaxation processes. The bottom graph is an approximate first-derivative display of the dashed curve of the middle graph. Note the correspondence between the peak separations in this derivative curve and those of the idealized component of the top graph. The curves of Figs. 8.8 and 8.9 are sets constructed by Lee and Bray [12] to assist in their analysis of the esr of irradiated glasses containing boron. Other helpful diagrams may be found in their paper.

By use of (8.28) to (8.30), θ can be eliminated from (8.31), and the shape function can be expressed in terms of H, as was done for (8.24). However, the resulting expression is unnecessarily complicated. A simpler procedure is to use (8.31) with (8.28) to (8.30) in a computer program designed for calculation of $I(H)$ with estimated or assumed parameters, $g_{\parallel}$, $g_{\perp}$, $A_{\parallel}$, and $A_{\perp}$. Programs to achieve this computation are described in the literature [13,14]. At equal intervals θ_i to $\theta_i + \Delta\theta$ between $\theta = 0$ and $\theta = 90°$, the computer calculates $I(H_i)$ from (8.31) and the magnetic field value corresponding to θ_i from (8.28). The resulting $I(H_i)$ values plotted as a function of H gives the first estimate of the line shape. Since most spectra are recorded as derivatives of the actual absorption, it is usually desirable to design the computer program for calculation of the first or second derivative of the resulting shape function thus derived. By comparison of this first theoretical plot with the experimentally observed curve, one can make adjustments of the estimated parameters for a second theoretical calculation, and so on until one obtains the best fitting with the experimental curve.

Before an accurate fitting of the experimental curve can be achived, the δ-function line shape assumed in the preceding calculation for a preliminary determination of the parameter values must be replaced by a more representative shape function. For most purposes, the Lorentzian function with constant width $\Delta H_{1/2}$ for resonance field values between $\theta = 0$ and $\theta = 90°$ is adequate. With this condition, the area of the Lorentzian function at each resonance field value H_r between $\theta = 0$ and $90°$ is proportional to $I(H_r)$. By examination of the terminal regions of the observed spectrum, one can make an estimate of $\Delta H_{1/2}$. A Lorentzian function is chosen for each interval field value H_r with area proportional to $I(H_r)$. These Lorentzian lines are then summed by the computer for all values of H_r corresponding to the chosen intervals $\Delta\theta$ between $0°$ and $90°$. From comparison of the resulting curves with the experimentally observed resonance, a better choice of $\Delta H_{1/2}$ and the other experimental parameters may be made for a second calculation. This process is repeated until a satisfactory fitting with the experimental curves is achieved. Computer programs for these calculations are available [13,14].

Rollmann and Chan [13] include second-order nuclear magnetic interactions, also nuclear quadrupole interactions, in a calculation of the shape of the esr spectrum of randomly oriented radicals for which all interaction tensors have the same axis of symmetry. They also describe a computer program for numerical solution of the rather complicated equation obtained for dN/dH. Details are given in the original paper [13].

When neither the **g** tensor nor the **A** tensor has an axis of symmetry, analytical derivation of the shape function for the powder spectrum is not feasible, and the line-shape problem must be solved by numerical computer methods like those described in Section 1.b for **g** anisotropy alone.

Here we assume that the **g** and **A** tensors have common principal axes, and we express the effective **g** and **A** with reference to these axes, assumed to be x, y, z. From (8.11) it is then evident that the shape function may be expressed as

$$I(H) = \frac{1}{4\pi}\left(\frac{dH}{d\Omega}\right)^{-1} = \frac{1}{4\pi}\,\frac{d(\cos\theta)d\phi}{dH} \tag{8.35}$$

In this case,

$$H = \frac{h\nu_0}{g\beta} - \frac{M_I}{g\beta}A \tag{8.36}$$

where

$$g = (g_x^2 \sin^2\theta \cos^2\phi + g_y^2 \sin^2\theta \sin^2\phi + g_z^2 \cos^2\theta)^{1/2} \tag{8.37}$$

and

$$A = \frac{1}{g}(g_x^2 A_x^2 \sin^2\theta \cos^2\phi + g_y^2 A_y^2 \sin^2\theta \sin^2\phi + g_z^2 A_z^2 \cos^2\theta)^{1/2} \tag{8.38}$$

when θ and ϕ are the polar angles of **H** with the principal axes z and x, respectively, as indicated in Fig. 8.10. A computer program for numerical solution of (8.35) has been designed by Griscom, Taylor, et al. [10]. This program is described briefly in Section 1.b.

The preceding analysis includes only the spread caused by anisotropy. If a Lorentzian or Gaussian line-shape function indicated by $f[H - H_r(\theta\phi)]$ is assumed for the single crystallite, that is, for the shape function exclusive of **g** and **A** anisotropy, this function must be averaged over all values of θ and ϕ for each resonance-field value $H_r(\theta,\phi)$, as described in Section 1.b. The shape function for the polycrystalline substance is then given by (8.16) or (8.17), but with resonance-field values determined by (8.36) rather than by (8.14). Computer programs for numerical evaluation of the resulting integrals are given by Swalen and Gladney [9], also by Lefebvre and Maruani [15]. Taylor and Bray

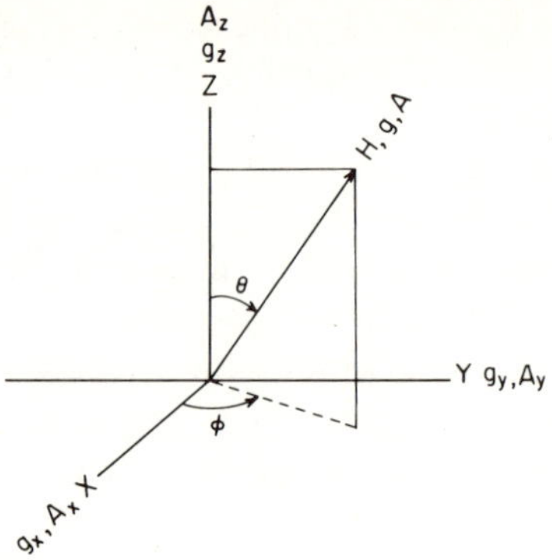

Fig. 8.10 Coordinate system for expression of g and A in polar angles of **H**, θ and ϕ, with the common principal axes x, y, z of **g** and **A**.

[16] describe methods for computer simulations of esr spectra of this kind observed for polycrystalline and glassy samples.

3. THE ESR OF TRAPPED ATOMS

3.a. Atoms Having $^2S_{1/2}$ Ground States: Hydrogen and the Alkali Atoms

The simplest of all free radicals, and one of the most important for the radiation chemist, is the hydrogen atom. Experience has shown that when organic chemicals are subjected to ionizing irradiation, the light hydrogen atom is the component most often removed. Because of its light weight and spherical shape, it is most easily displaced from its position in the lattice. Many of the more complicated organic free radicals identified with esr in irradiated solids appear to be produced simply by the removal of an H atom from the parent molecule.

The strong-field magnetic resonance of gaseous H atoms was first observed by Beringer and Heald [17]. An almost identical esr was observed for H atoms produced and trapped within solids at 88°K by Livingston, Zeldes, and Taylor [18]. The characteristic hyperfine pattern is a doublet separated by approximately 507 G at constant frequency or 1420 MHz at constant magnetic field. This doublet separation, which is due to the nuclear coupling of the H atom (spin $I = 1/2$), has been found to vary by only one percent or so for H trapped in many kinds of solids at low temperature. Table 8.1 gives a comparison of the effects of the inert-gas matrices on the esr parameters. The g factor for the resonance is isotropic and is close to 2.0023, the value for the spin of the free electron. However, the center of the resonance and the "apparent g value" are noticeably shifted by the Breit-Rabi effect described in Chapter III, Section 4.b unless the operating magnetic field is large as compared with the doublet splitting. For H atoms, this condition is achieved to a good approximation when the

Table 8.1 Hyperfine Coupling Constants and *g* Factors for Hydrogen, Nitrogen, Phosphorus, and Arsenic Atoms in Nonpolar Matrices at Low Temperature

Atom	*Matrix*	*g*	*Coupling in MHz*	*Ref.*
H	Free	2.002256	1420.4057	*a*
	Ne	2.00207	1426.56	*b*
	Ar	2.00161	1436.24	*b*
	Kr	1.99967	1427.06	*b*
	Xe	2.00057*	1405.57*	*b*
	H_2	2.00230	1417.11	*c*
	CH_4	2.00207	1411.09	*c*
^{14}N	Free	2.0022	10.45	*d*
	Kr	2.0021	12.3	*e*
	Xe	2.0019	12.4	*e*
^{31}P	Free	2.0019	55.08	*f,g*
	Ar	2.0012	80.61	*e*
	Kr	2.0001	83.48	*e*
	Xe		86.51	*e*
^{75}As	Free	1.9965	-66.59	*g*
	Ar	1.9960	-35.22	*e*
	Kr	1.9951	-31.50	*e*
	Xe	1.9943		*e*

*Calculated from the central component (see Fig. 8.11).
[a] P. Kusch, *Phys. Rev.*, **100**, 1188 (1955).
[b] Foner, Cochran, Bowers, and Jen [19].
[c] C. K. Jen, S. N. Foner, E. L. Cochran, and V. A. Bowers, *Phys. Rev.*, **112**, 1169 (1958).
[d] W. W. Holloway and R. Novick, *Phys. Rev. Lett.*, **1**, 367 (1958).
[e] Jackel, Nelson, and W. Gordy [23].
[f] H. G. Dehmelt, *Phys. Rev.*, **99**, 527 (1955).
[g] J. M. Pendlebury and K. F. Smith, *Proc. Phys. Soc. (Lond.)*, **84**, 849 (1964).

resonance is observed in the centimeter region where the imposed field is of the order of 10 kG. Even at these frequencies (~30,000 MHz), the Breit-Rabi corrections must be applied for precise values of g and A.

In addition to the differences caused by the matrix materials, illustrated in Table 8.1, certain nonpolar matrices have discrete multiple effects on the esr parameters of H atoms. In place of single, sharp components like those observed in many matrices including argon, closely spaced clusters of lines are detected in certain ones, as shown for the xenon matrix in Fig. 8.11. These multiple

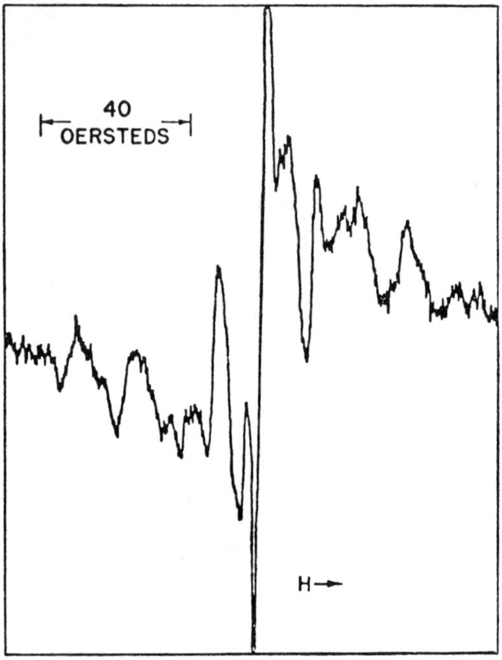

Fig. 8.11 Cluster of H-atom lines for lower-field component that illustrate the effects of different trapping sites in the xenon matrix. From Foner, Cochran, Bowers, and Jen [19].

components, first detected by Foner, Cochran, et al. [19], have been explained by Adrian [20] in terms of varied trapping sites having different perturbations on the esr parameters. The sharp central component of each cluster for the xenon may arise from the H atoms trapped in substitutional sites and its satellites, from the more cramped, interstitial sites. When there are bonded hydrogens near the trapped H atoms, satellite lines appear that are due to second-order flipping of the nuclei of the environmental hydrogens, as explained in Chapter V, Section 4.c. These satellites are easily distinguishable from those due to multiple trapping sites by the variation of the operating field.

Jen, Bowers, et al. [21] have observed the esr of trapped alkali atoms — Li, Na, K, Rb, and Cs — in inert rare-gas matrices at low temperature. Like hydrogen, these atoms have $^2S_{1/2}$ ground states. Except for the increase in hyperfine components due to their different nuclear spin, these alkali atoms have an esr similar to that of the H atom. Component clusters resulting from multiple trapping were also observed for them. Although similar in kind to those for the H atoms, these matrix effects are more pronounced for the larger alkali atoms.

3.b. Atoms Having $^4S_{3/2}$ Ground States: Nitrogen, Phosphorus, and Arsenic

The free N, P, and As atoms have valence-shell configurations, $(ns)^2(np_x)^1(np_y)^1(np_z)^1$. Thus the p valence shells are symmetrically filled with unpaired electrons, and the ground state is $^4S_{3/2}$. In the absence of configuration interaction, the g factor would be that of the free electron, 2.0023, and the hyperfine splitting would be zero. Configuration interaction, however, gives the free atoms an appreciable isotropic hyperfine splitting as well as a slight perturbation of g from the free-spin value. Matrix interactions alter further the configuration interactions in the trapped atoms. These effects on the esr spectrum are demonstrated in Table 8.1. The van der Waals interaction theory devised by Adrian [22] to explain matrix effects on the hyperfine structure of N atoms has been generalized and applied with good success to P and As atoms in argon, krypton, and xenon matrices at 4.2°K by Jackel, Nelson, and Gordy [23]. Description of the calculations and comparison of theoretical and observed parameters may be found in the original paper.

3.c. Atoms Having $^2P_{1/2}$ Ground States: Aluminum and Gallium

Very interesting esr measurements on matrix-isolated atoms of Al and Ga have been made by Ammeter and Schlosnagle [24]. Their work appears to be the first successful identification and measurement of trapped atoms having singularly occupied p orbitals in the ground state. The esr of Al atoms had evidently been measured before by Knight and Weltner [25], but they ascribed the spectra to a molecular Al-X complex. Both atoms have the valence-shell configuration $(ns)^2(np)^1$ or the $^2P_{1/2}$ ground state. The orbital angular momentum in the rare-gas matrices, X = Ne, Ar, Kr, and Xe at 4.2°K, was found to be highly quenched, approaching that of atoms of ionic crystals or free radicals. The **g** tensor and hyperfine tensor of the atoms were observed to be axially symmetric with **g** approaching closely the free-spin value of 2.0023. The g factor expected for free atoms in the $^2P_{1/2}$ state, that given by (1.10), is 2/3, or approximately one-third the observed value. Furthermore, the smallness of the line widths observed also indicates effective quenching of the orbital momentum of the trapped atoms.

The observed esr parameters for the trapped Al atoms are listed in Table 8.2. Note that there are distinguishable trapping sites in a given matrix as well as distinguishable effects of the various matrices on the parameters. To a good approximation, however, the parameters remain the same and are characteristic of those of Al atoms with quenched orbital angular momentum. Similar effects to those of Al were observed for the trapped Ga atoms. A tabulation of the corresponding Ga constants is given in the original publication.

The distinguishable trapping sites result from different van der Waals complexes MX_n with the matrix atoms. Figure 8.12 illustrates the octahedral MX_6 complex in which the np orbital of the unpaired electron is oriented along

Table 8.2 Nuclear Couplings and g Values for Aluminum Atoms in Rare-gas Matrices at 4.2°K[a]

Matrix	Site	Principal g Values	Couplings[b] (MHz)		Line Widths (MHz)
Neon	1	$g_\parallel = 2.000$	$A_\parallel = 139$	$A_f = -24$	$\parallel: 18$
		$g_\perp = 1.925$	$A_\perp = -106$	$B = 82$	$\perp: 18$
	2	$g_\parallel = 2.000$	$A_\parallel = 139$	$A_f = -24$	$\parallel: 18$
		$g_\perp = 1.927$	$A_\perp = -106$	$B = 82$	$\perp: 18$
Argon	1	$g_\parallel = 2.000$	$A_\parallel = 143$	$A_f = -20$	$\parallel: 9.8$
		$g_\perp = 1.951$	$A_\perp = -102$	$B = 82$	$\perp: 9.8$
	2	$g_\parallel = 2.000$	$A_\parallel = 143$	$A_f = -20$	$\parallel: 9.8$
		$g_\perp = 1.956$	$A_\perp = -101$	$B = 81$	$\perp: 9.8$
Krypton	1	$g_\parallel = 2.001$	$A_\parallel = 136$	$A_f = -15$	$\parallel: 8.4$
		$g_\perp = 1.989$	$A_\perp = -90$	$B = 75$	$\perp: 9.8$
	2	$g_\parallel = 2.001$	$A_\parallel = 136$		$\parallel: 8.4$
		$g_\perp = ?$	$A_\perp = ?$		($\perp$ not observed)
	3	$g_\parallel = 1.997$	$A_\parallel = 174$	$A_f = -7.4$	$\parallel: 5.6$
		$g_\perp = 1.962$	$A_\perp = -76$	$B = 83$	$\perp: 20$
Xenon	1	$g_\parallel = 2.001$	$A_\parallel = 140$		$\parallel: 42$
		$g_\perp = 2.02$	$\lvert A_\perp \rvert < 75$		$\perp: \sim 56$

[a] Data taken from Ammeter and Schlosnagle [24], Table VI.
[b] $A_f = (A_\parallel + 2A_\perp)/3,\ B = (A_\parallel - A_\perp)/3.$

the X_5 - X_6 axis. A quantitative discussion of the perturbations of the esr parameters by the varied trapping configurations such as MX_6, MX_{12}, is given in the original paper [24].

4. SMALL MOLECULAR FREE RADICALS TRAPPED IN INERT MATRICES AT LOW TEMPERATURE

Simple free radicals composed of only a few atoms such as CN, NH_2, and CH_3 that are not generally stable in solids at room temperature can be observed most advantageously in inert matrices at low temperature. The inert matrices most commonly employed are those made by freezing the rare gases, neon, argon, krypton, or xenon, although other frozen gases such as H_2, N_2, CH_4, CF_4, are also used.

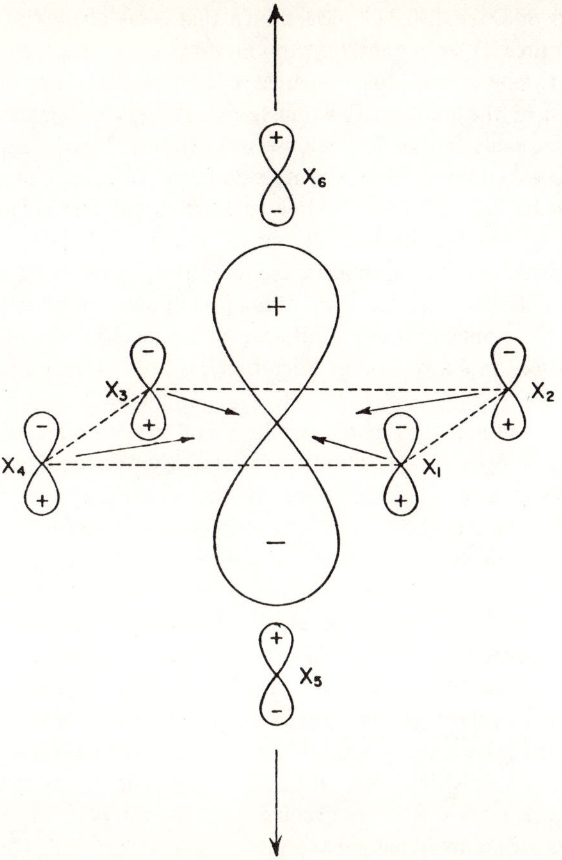

Fig. 8.12 Diagram of a metal atom trapped in an octahedral MX_6 van der Waals complex, with figures indicating relative alignment of *p* orbitals. From Ammeter and Schlosnagle [24].

In one procedure for these observations [26], the free radicals are produced in the gaseous state by some method such as an electric discharge and are then trapped with the inert matrix material on a sapphire rod maintained at the temperature of liquid helium. The sapphire rod with the sample is then inserted directly into the microwave cavity of the spectrometer. In an alternative procedure, neutral molecules from which the free radicals can be produced by irradiation of the frozen solid are trapped with the inert gas. The irradiation can be achieved by uv light through a quartz window in the liquid helium flask [27,28], or it can be achieved with X-rays or γ-rays directly through the metallic walls of the microwave cavity [29].

Experiments at Duke [30,31] have shown that γ-ray energy intercepted from a cobalt 60 source by argon and krypton matrices migrates to, and is effectively transferred to, simple molecules such as CH_4 and PH_3, that are trapped in dilute solution in the matrix. The transferred energy dissociates the molecules into simple free radicals that can be observed with esr. This energy transfer does not occur for all combinations of matrix and molecule, but for certain combinations it can be used to produce free radicals essentially isolated within the matrix. This is possible because extremely dilute concentrations of neutral molecules trapped within the matrix are effectively dissociated with moderate dosage of irradiation. Electron-spin resonance allows study not only of the properties of the trapped radicals, but also of the mechanisms of energy migration and transfer. Similarly, energy migration has been observed to occur in H_2O at 4.2°K [29].

The effects of an inert gas matrix on the esr of trapped radicals, though small, are not negligible even for the simple, spherically symmetric, H atoms described in the previous section. The interactions of the matrix with the trapped radicals are complex and varied; they are usually most pronounced for polar matrices on molecular free radicals in which the electron spin density is not spherically symmetric.

Weltner and his associates (see references for Table 8.3) at the University of Florida have measured esr spectra of a large number of diatomic molecules having $^2\Sigma_{1/2}$ ground states. Their studies were made primarily on dilute concentrations of the molecules in inert gas matrices at 4.2°K. Considerable information about the magnetic properties and electronic structures of the molecules was obtained from these studies, which were either combined or correlated with optical spectral measurements on the molecules similarly trapped. Their more important esr results are summarized here.

4.a. Diatomic Metal Hydrides in $^2\Sigma$ States

In Table 8.3 are given the basic esr spectral constants of a number of diatomic metal hydrides trapped in inert-gas matrices. Descriptions of experimental techniques for producing, trapping, and measuring the esr spectra of these molecules, as well as theoretical discussions of the data, may be found in the papers cited in the table. Each of these hydrides has a single unpaired electron and a $^2\Sigma$ ground state. The **g** tensor and nuclear coupling tensors are found to be axially symmetric for all of them. From the probable electronic structures described in the following paragraphs, it is evident that the magnetic axes of symmetry should be the internuclear axes.

The alkaline-earth atoms M are nonmagnetic and have two electrons in their outermost populated shell with $(ns)^2$ configuration. If the metal were to form a normal electron-pair bond with the H, one of these electrons would have to be promoted to a higher np orbital; or, alternatively, two hybrid orbitals may be

Table 8.3 ESR Parameters for Some Diatomic Metal Hydrides in Argon Matrix at 4.2°K

Species	$g_{\parallel}$	$g_{\perp}$	Coupling Isotope	Coupling (MHz) $A_{\parallel}$	$A_{\perp}$	A_{iso}^{Af}	A_{dipole}^{B}	ρ_s^H	Ref.
^{9}BeH	2.0021	2.0022	H	201	190.8	194	3.3	0.137	*a*
			^{9}Be	-208	-194.8	-199	-4.3		*a*
^{9}BeD	2.0020	2.0021	D	30.5	29.4	29.8	0.4		*a*
*MgH	2.0020	2.0002	H	298	295	296	1.0	0.208	*b*
^{25}MgH	2.0020	2.0002	^{25}Mg	(-)226	(-)218	(-)221	(-)2.7		*b*
*CaH	2.0013	1.9966	H	138	134	135	1.4	0.095	*b*
*SrH	2.0004	1.9865	H	123	121	122	0.6	0.086	*b*
*BaH	1.9984	1.9746	H	47	46	46	0.3	0.032	*b*
*ZnH	2.0003	1.9855	H	477	479	479	-0.3	0.337	*c*
*CdH	1.9970	1.9529	H	515	514	514	0.3	0.362	*c*
^{111}CdH			^{111}Cd	(-)4358	(-)3966	(-)4097	(-)131		*c*
*HgH	1.976	1.8280	H	707	711	710	-1.3	0.500	*c*
^{199}HgH			^{199}Hg	7790	6608	7002	394		*c*
PdH	1.965	2.293	H	103	106	105	~0	0.074	*d*
			^{105}Pd	(-)867	(-)801	(-)788	(-)36		*d*
YbH	1.9953	1.9402	H	226	224	225	0.6	0.159	*e*
			^{171}Yb	5724	5266	5433	145		*e*

*Isotopes with spin $I = 0$.

[a] Knight, Brom, and Weltner [33].

[b] L. B. Knight and W. Weltner, *J. Chem. Phys.*, **54**, 3875 (1971).

[c] L. B. Knight and W. Weltner, *J. Chem. Phys.*, **55**, 2061 (1971).

[d] L. B. Knight and W. Weltner, *J. Molec. Spectrosc.*, **40**, 317 (1971).

[e] R. J. Van Zee, M. L. Seely, and W. Weltner, *J. Chem. Phys.*, **67**, 861 (1977).

constituted, one of which forms the bond while the other is occupied by the unpaired electron. In each case, the unpaired electron density would be entirely on the metal atom. However, the hyperfine coupling constant of the H reveals considerable spin density on the H atom, up to 0.50 in HgH, as may be seen from comparison of ρ_H values in Table 8.3. The observed H spin densities are much too large to arise from spin polarization of the covalent components of these highly ionic bonds. Thus their H-spin densities indicate that the orbital of the unpaired electron in these hydrides is an antibonding σ^* orbital which can be approximately expressed by the linear combination

$$\Phi_{\sigma^*} = a\psi(\mathrm{M}) - b\psi_{1s}(\mathrm{H}) \tag{8.39}$$

With neglect of orbital overlap, the normalization of Φ_{σ^*} requires that $a^2 + b^2 = 1$. Since $b^2 = \rho_{1s}^{\mathrm{H}}$, $a^2 = 1 - \rho_{1s}^{\mathrm{H}}$. Now, a^2 is equivalent to the unpaired electron density of the orbital $\psi(\mathrm{M})$, which is assumed to be an *ns-np*-hybridized orbital because of the significant isotropic as well as anisotropic coupling components observed for the metal atoms for which nuclear couplings were measured. Although the anisotropic component of the hydrogen coupling can be attributed in all cases to dipole-dipole interaction between the spin density on the metal (M) and the nucleus of H, only a small fraction of the anisotropic coupling of M can arise from dipole-dipole interaction of the smaller spin densities on H. For example, the approximate formula (6.12) with spin density equivalent to $a^2 = 1 - 0.21 = 0.79$ on Mg indicates that H, with its large $\mu_{\mathrm{I}} = 2.793$ nm, would have an axially symmetric dipole-dipole coupling of 10 MHz, much larger than the observed value. In contrast, the anisotropic component of the ^{25}Mg coupling that arises from $\rho_s^{\mathrm{H}} = 0.21$ is indicated by this relation to be only 0.17 MHz in magnitude, as compared with the observed value of 2.7 MHz. In both cases this estimate with (6.12) assumes the coupling spin density to be concentrated at the center of the neighboring atom. This assumption should be much more reliable for spin density on the small H atom than for that in the much larger metal atoms.

Because $A_{\mathrm{atomic}}^{\mathrm{H}}$ is accurately known, the value of b^2 and hence of a^2 is rather accurately measured. Therefore, except for the degree of hybridization of $\psi(\mathrm{M})$, the σ^* orbital of the unpaired electron is determined. For BeH and MgH, this hybridization can be derived from the measured molecular couplings of ^{9}Be and ^{25}Mg. For this purpose, the corresponding atomic orbital couplings are required. They can be obtained from theoretical values of $|\Psi(0)|^2$ and $\langle (a_0/r)^3 \rangle_{\mathrm{av}}$ which are available in the literature. For conversion of these quantities to the desired isotropic and anisotropic atomic-orbital couplings, the following relations are convenient to use:

$$(A_f)_{\mathrm{atomic}}(\mathrm{MHz}) = 800.1 \frac{\mu_I}{I} |\Psi(0)|^2 \tag{8.40}$$

$$B_{\text{atomic}}(\text{MHz}) = 38.2 \frac{\mu_I}{I} \langle (a_0/r)^3 \rangle_{\text{av}} \tag{8.41}$$

where $a_0 = 5.29167$ Å is the Bohr radius, I is the nuclear spin, and μ_I is the nuclear magnetic moment [in n.m.). With the theoretical value $\Psi(0)^2 = 0.5687$ in atomic units (au) [32,33] for the $2s$ orbital of Be and with $\mu_I(^9\text{Be}) = -1.1776$ nm and $I = 3/2$, (8.40) yields $A_f(\text{atomic}) = -357$ MHz. Similarly, the theoretical value $\langle (a_0/r)^3 \rangle_{\text{av}} = 0.173$ for the $2p$ orbital given by Barnes and Smith [34] yields with (8.41) $B_{\text{atomic}} = -5.19$ MHz for ^{9}Be. However, from the electronegativity difference, $x_H - x_{Be} = 1.7$, a σ-bond ionic character of 0.35 is estimated from (6.56) and a charge $c^+ = 0.35$ is calculated for the Be. With $\epsilon = 0.90$ (see Table 6.21), the charge-corrected atomic coupling is $B_{\text{atomic}}(\text{Be}^{+0.35}) = 5.19[1 + 0.35\ (0.90)] = -6.8$ MHz. The spin density on H makes no contribution to the isotropic coupling of ^{9}Be in the molecule, but it makes a contribution of -0.46 MHz [as calculated with (6.12)] to the anisotropic coupling. Thus the anisotropic ^{9}Be component due to the spin density of the $2p$ orbital is $-4.3 - (-0.46) = -3.84$ MHz. From these various constants, the Be spin densities are seen to be

$$\rho_{2s}^{\text{Be}}(\text{BeH}) = \frac{-199}{-357} = 0.56$$

$$\rho_{2p}^{\text{Be}}(\text{BeH}) = \frac{-3.84}{-6.8} = 0.56$$

The sum

$$\rho_{2s}^{\text{Be}} + \rho_{2p}^{\text{Be}} + \rho_{1s}^{\text{H}} = 0.56 + 0.56 + 0.14 = 1.26$$

agrees poorly with the normalization, $a^2 + b^2 = 1$. The uncertainty in B_{atomic} (Be) is believed to be responsible for this discrepancy. If $\rho_{2p}^{\text{Be}} = 0.30$ (as determined from the normalization) is assumed to be correct, the orbital $\psi(\text{Be})$ in (8.39) is 65% $2s$ and 35% $2p$.

The σ-bonding orbital of BeH, which contains an electron pair, is similarly made up of a linear combination of a $3s$-$3p$ hybrid on the Be and the $1s$ orbital of H. Thus the esr data provide a semiquantitative description of the nature of the bonding of this small molecule, which has been the subject of many theoretical studies [35-39]. The availability of quantitative esr data on magnetic properties should stimulate and assist in more rigorous theoretical investigations.

A similar treatment can be given for MgH. To obtain the $3p$-orbital coupling of ^{25}Mg, one reduces the B value of -2.7 MHz given in Table 8.3 by -0.2

MHz, the amount due to dipole-dipole coupling with the spin density on H. Thus $B_{3p}^{\mathrm{Mg}}(^{25}\mathrm{MgH}) = -2.5$ MHz. The value of $\langle (a_0/r)^3 \rangle_{\mathrm{av}}$ given by Barnes and Smith [34] for the $3p$ orbital is 0.77. From (8.41) the $3p$ atomic orbital coupling is found to be $B_{\mathrm{atomic}}(^{25}\mathrm{Mg}) = -10$ MHz. However, from the electronegativity difference, $x_{\mathrm{H}} - x_{\mathrm{Mg}} = 1.0$, a charge of $c^+ = 0.50$ is estimated from (6.56) to be on Mg. Since $\epsilon(\mathrm{Mg}) = 0.70$ (see Table 6.21), $B_{\mathrm{atomic}}(\mathrm{Mg}^{+0.5}) = -10(1 + 0.35) = -14$ MHz. The spin density is thus estimated to be

$$\rho_{3p}^{\mathrm{Mg}} = \frac{-2.5}{-14} = 0.18$$

For calculation of the $3s$ spin density, we use the value $A_f(\text{atomic}) = 333$ MHz obtained from the value $|\Psi(0)|^2 = 1.216$ au by Fischer [32] with the conversion relation of (8.40). The observed isotropic molecular coupling for ^{25}Mg then gives

$$\rho_{3s}^{\mathrm{Mg}} = \frac{-221}{-333} = 0.66$$

The sum of the molecular spin densities is

$$\rho_{3s}^{\mathrm{Mg}} + \rho_{3p}^{\mathrm{Mg}} + \rho_{1s}^{\mathrm{H}} = 0.66 + 0.10 + 0.21 = 0.97$$

Considering the approximation of some of the constants and the neglect of spin-polarization effects, this sum is in satisfactory agreement with the normalization value of 1.00. The orbital $\psi(\mathrm{Mg})$ is indicated to have 79% $3s$ and 21% $3p$ character.

Because the np orbital contributing to the σ^* orbital of the unpaired electron is oriented along the molecular axis, the $g_{\parallel}$ value is expected from the theory described in Chapter VII, Section 1 to be the free-spin value $g_e = 2.0023$. The fact that the $g_{\parallel}$ values given in Table 8.3 tend to be slightly lower than g_e is attributed to matrix interactions. Experimentally, the amount of the shift is found to depend on the constitution of the matrix. This matrix effect appears for other trapped radicals to be described. Exclusive of matrix interactions, the $\Delta g_{\perp} = g_{\perp} - g_e$ values are expected from theory to be negative; accordingly, $g_{\perp}$ is found to be lower than $g_{\parallel}$ for the molecules in Table 8.3. A field imposed at right angles to the magnetic axis would tend to intermix the $^2\Sigma$ state with the higher $^2\pi$ states. Since the intermixing would transfer unpaired electron density into a previously empty π orbital, the intermixing would cause a negative $\Delta g_{\perp}$ for reasons explained in Section 1.c. The magnitude of the $\Delta g_{\perp}$ is inversely proportional to the energy difference ΔE of the two intermixed levels. We do not give here the quantitative calculations of Δg, which may be carried out according to the procedures described in Chapter VII.

4.b. Diatomic Metal Fluorides in $^2\Sigma$ States

Several diatomic metal fluorides trapped in rare-gas matrices at 4°K have been investigated by Knight, Easley, et al. [40]. The observed esr parameters are summarized in Table 8.4. Like the diatomic metal hydrides, these fluorides

Table 8.4 ESR Parameters for Some Alkaline-earth Monofluorides in Neon Matrix at 4°K[a]

				Coupling (MHz)			
Species	$g_\parallel$	$g_\perp$	*Isotope*	$A_\parallel$	$A_\perp$	A_f	B
MgF	2.0020	2.0010	^{19}F	331	143	206	62.7
CaF	2.0020	2.000	^{19}F	149	106	120	14.3
SrF	2.0020	1.9970	^{87}Sr	591	570	577	7.0
			^{19}F	126	95	105	10
BaF	2.0010	1.9950	^{137}Ba	2453	2401	2418	17
			^{19}F	67	59	62	2.7

[a] From L. B. Knight, W. C. Easley, and W. Weltner, *J. Chem. Phys.*, **54**, 322 (1971).

have $^2\Sigma$ ground states. Figure 8.13 shows the observed first-derivative esr pattern for BaF in the neon matrix. Since this pattern is for the Ba isotope having $I = 0$, the nuclear splitting indicated by $A_\parallel$ and $A_\perp$ is due entirely to ^{19}F.

Unlike the corresponding hydrides, the metal fluorides have essentially ionic structure. The ionic character is given approximately by $i_c = (x_F - x_M)/2$, and the electronegativity difference is greater than two for all molecules listed in Table 8.4. It is, therefore, reasonable to assume that one of the $(ns)^2$ electrons of the metal (M) is captured by the fluorine on formation of the fluoride MF, and that a single unpaired electron is left in the valence shell. However, this unpaired electron does not remain entirely in the *ns* orbital. Considerable *ns-np*, and possibly some *ns-nd*, hybridization is induced in the orbital of the odd electron by the strong electrostatic forces of the ionic bond. Evidence for this is the anisotropic component in the coupling of the ^{87}Sr and ^{137}Ba. Because the couplings of the corresponding atomic np orbitals are unknown, the np spin densities cannot be derived from the B values listed in Table 8.4. Evidence that they are rather large can be gained from the isotropic couplings. The atomic A_{iso} for $^{87}Sr^+$ is 992 MHz [41]; that for $^{137}Ba^+$ is 4050 MHz [42], as determined from optical measurements. With these values and the corresponding molecular A_f values in Table 8.4, the *ns* spin densities are estimated to be

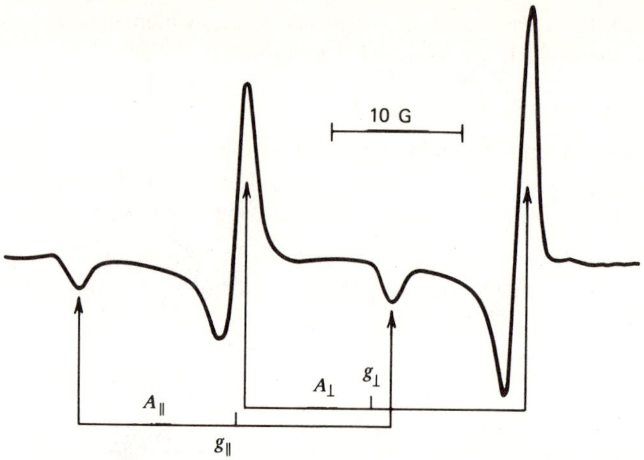

Fig. 8.13 Observed esr spectrum of ^{nat}BaF in neon matrix at 4°K. From Knight, Easley, Weltner, and Wilson [40].

$$\rho_{ns}^{\mathrm{Sr}}\ (\mathrm{Sr^+F^-}) = 0.58, \qquad \rho_{ns}^{\mathrm{Ba}}\ (\mathrm{Ba^+F^-}) = 0.60$$

From the normalization of the total unpaired-electron density to unity, one obtains a combined spin density of the order of 0.40 for the *np* and *nd* orbitals. This density is expected to be mostly *np*.

Despite the relatively large ^{19}F coupling observed, the 2*s* spin density on the F is much less than 1% because the large $A_f{}^{19}$F (atomic) ≈ 48,000 MHz. Nevertheless, the 2*p* spin density ρ_{2p}^{F}(MgF) = 0.04 is surprisingly large. For all the other molecules listed, ρ_{2p}^{F} is only 1% or less. Since the bonding is expected to be ionic in all these molecules, the origin of the F spin density is complex. It must arise, at least in part, from exchange polarization [40]. Various distortions for explaining the F^- and M^+ coupling in the molecules are discussed by Knight, Easley, et al. [40].

4.c. The BO, BS, CN, and AlO Molecules in $^2\Sigma$ States

The diatomic oxides and sulfides of B and Al have a single unpaired electron and a $^2\Sigma$ ground state. Their esr spectra have been studied by Weltner and his group with the molecules isolated in inert matrices at 4°K. The principal *g* values and nuclear couplings they obtained are listed in Table 8.5. These species are isoelectric with the CN radical, for which the corresponding parameters are also listed. The esr spectrum of CN in a rare-gas matrix was first observed by Cochran, Adrian, and Bowers [43]. The parameters listed in the table are later ones by Easley and Weltner [44]. The magnetic parameters for all species listed

Table 8.5 Observed ESR Parameters of BO, BS, and CN

					Coupling (MHz)		
Species	*Matrix*	$g_{\parallel}$	$g_{\perp}$	*Coupling Isotope*	$A_{\parallel}$	$A_{\perp}$	*Source*
BO	Neon 4°K	2.0012	2.0015	^{10}B	340.9*	346.2*	*a*
				^{11}B	1018*	1034*	
BS	Neon 4°K	2.0016	1.9942	^{10}B	285.7	256.8	*b*
				^{11}B	853.3	766.7	
	Argon 4°K	2.0015	1.9944	^{10}B	293.7	263.2	
				^{11}B	878.6	787.5	
AlO	Neon 4°K	2.0015	2.0004	^{27}Al	872	713	*c*
	Argon 4°K	2.0014	1.9997	^{27}Al	1006	845	
	Krypton 4°K	1.9993	1.9973	^{27}Al	1022	869	
CN	Argon 4°K	2.0015	2.0003	^{13}C	678	543.1	*d*
				^{14}N	18.2	-28.0	
	Rotating 7°K	2.0008		^{13}C	A_{iso} =	588.1	*d,e*
				^{14}N	A_{iso} =	-12.6	

*In BO, $A_{\parallel} < A_{\perp}$ implies that the dipole-nuclear coupling is negative, a unique condition in diatomics.

[a] L. B. Knight, W. C. Easley, and W. Weltner, *J. Chem. Phys.*, **54**, 1611 (1971).

[b] J. M. Brom and W. Weltner, *J. Chem. Phys.*, **57**, 3379 (1972).

[c] Knight and Weltner [25].

[d] Easley and Weltner [44].

[e] Cochran, Adrian, Bowers [43].

in Table 8.5 were found to be axially symmetric, with a common axis for **g** and **A**. The molecules were prepared in the gaseous state and were trapped at 4°K on a sapphire rod in the desired proportions with molecules of the rare gas forming the matrix.

Figure 8.14 shows the esr spectrum of BO obtained in the neon matrix. The ^{10}B and ^{11}B isotopes were in their natural concentrations of 20% and 80%, respectively. The four strong components are from the ^{11}BO species, $I(^{11}B) = 3/2$; the weaker septet spectrum is from ^{10}BO, $I(^{10}B) = 3$. Although the spectrum appears to be isotropic on this condensed scale, the anisotropy is clearly evident and measurable on an expanded scale. However, the BO molecules trapped on the sapphire rod in the neon matrix, unlike those in the argon matrix, did not have completely random orientations. The analysis of the observed line shapes indicated that the BO molecules have strong preferential orientations perpendicular to the condensing surface.

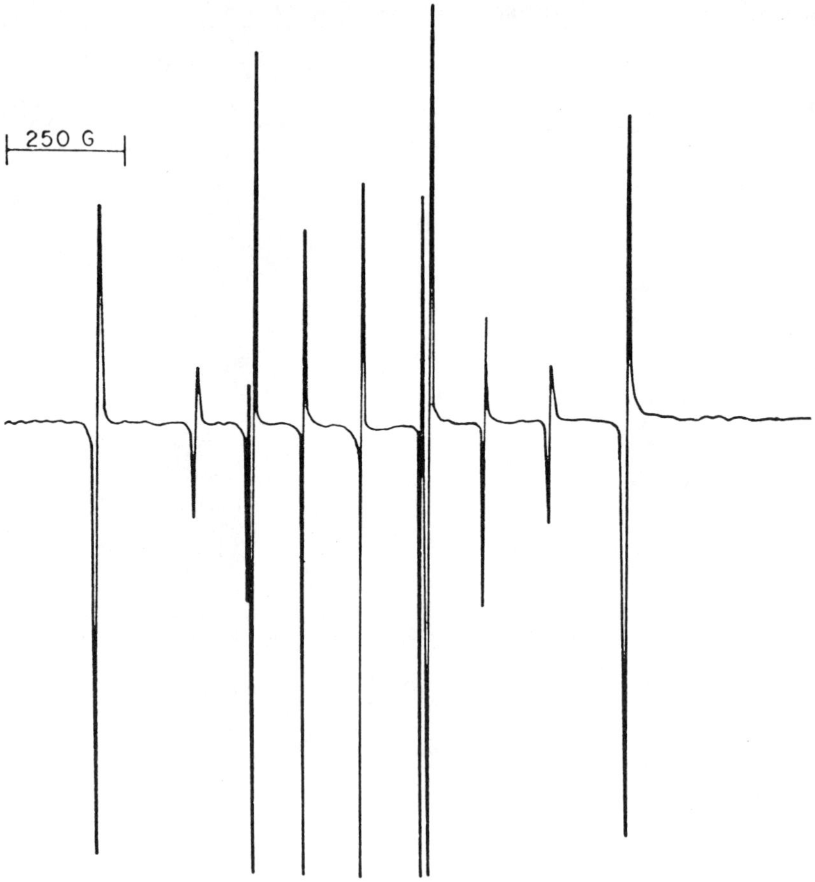

Fig. 8.14 Observed esr spectrum of ^{nat}BO in neon matrix at 4°K. From L. B. Knight, W. C. Easley, and W. Weltner, *J. Chem. Phys.*, **54**, 1610 (1971).

In Table 8.6 are given isotropic and anisotropic coupling constants with spin densities derived from them. Because of the semialignment of the BO molecules and the opposing term caused by an unknown spin density on the O, no evaluation of the anisotropic coupling constant B could be made for BO. Hence, neither the $2p$ spin density on the B nor the spin density on the O could be derived. Nevertheless, it can be concluded that aggregately they amount to $1 - \rho_{2s}^{B} = 0.49$. For BS, the sum of the densities $\rho_{2s}^{B} + \rho_{2p}^{B} = 0.92$ on B indicates an unpaired electron density of only 0.08 on the S. Thus the unpaired electron is mostly in an sp hybrid on the B. In calculation of the ρ_{2p}^{B} value we have used an atomic $2p$-orbital coupling constant of 53 MHz (from Table A.1) for ^{11}B corrected for a formal charge $c^{+} = 0.25$ due to the estimated ionic character of

Table 8.6 Coupling Constants and Spin Densities for BO, BS, CN, and AlO

Molecule	Matrix	Coupling Atom	Coupling Constants[a]		Spin Densities	
			A_f	B	ρ_{ns}	ρ_{np}
BO	Ne	^{11}B	1028	–	0.51	–
BS	Ar	^{11}B	818	30.4	0.41	0.51
CN	Ar	^{13}C	588	45	0.20	0.45
		^{14}N	-12.6	15.4	~0	0.35
AlO	Ne	^{27}Al	766	53	0.28	0.55
	Ar	^{27}Al	899	54	0.33	0.56
	Kr	^{27}Al	920	51	0.34	0.53

[a] Derived from the principal coupling values in Table 8.5 with $A_f = (A_{\parallel} + 2A_{\perp})/3$, and $B = A_f - A_{\perp}$.

the BS bond. The corrected value is $B_{\text{atomic}}(B^{+0.25}) = 53(1 + c^{+}\epsilon) = 60$ MHz. The correction may, however, be too large since the effective electronegativity on B is probably higher than the value used (2.0) because of the large *s* character in the σ-bonding orbital.

The BS bond structure consistent with these spin density values may be explained as follows. Two linear *sp* hybrids are formed on the B in the direction of the bond axis. One of these has the unpaired electron, and the other forms a σ bond with an orbital of the S. The remaining electron on the B has a pure *p* orbital perpendicular to the bond axis and forms a π bond with the S. The orbital of the unpaired electron on the B is 45% 2*s* and 55% 2*p*, while the conjugate orbital used for the σ bond is 55% 2*s* and 45% 2*p*. From a comparison of the 2*s* spin density on BO and BS, one may assume that the structure of BO is similar to that of BS, only more ionic.

For CN, there is a 2*p* spin density of 0.35 on the N with no significant 2*s* spin density on this atom. The total spin density on the C and N is 0.20 + 0.45 + 0.35 = 1.00, as is expected for normalization of the density of the unpaired electron. In calculation of the 2*p* spin densities we have made a charge correction of $c^{+} = 0.25$ on the C and $c^{-} = -0.25$ on the N, in accordance with the estimated ionic character. Uncorrected, the 2*p* spin densities are 0.50 on C and 0.32 on N. The large spin density of the 2*p* orbital on N suggests that the unpaired electron has an antibonding π^* molecular orbital.

The spin densities for the 3*p* orbital of the Al of AlO given in Table 8.6 are corrected for a charge $c^{+} = 1.00$ on the Al, estimated from (6.56). The uncorrected value, $B_{\text{atomic}}(\text{Al}) = 71$ MHz was obtained from $\langle\langle(a_0/r)^3\rangle\rangle_{\text{av}} = 1.28$ given by Barnes and Smith [34]. For the ionic form, the corrected value is $A_{\text{atomic}}(\text{Al}^{+1}) = 96$ MHz. From the sum of the 3*s* and 3*p* densities on the Al, it is concluded that the unpaired electron of AlO is mostly in an *sp* hybrid on the Al. For example, the sum of the spin densities for AlO in the argon matrix is $\rho_{3s} + \rho_{3p} = 0.89$. We can, therefore, conclude that a double bond, with considerable ionic character, is formed between the Al and O, leaving the unpaired electron essentially in a $3s$-$3p$-hybrid orbital on the Al, having about 37% $3s$ and 63% $3p$ character. Probably this hybrid orbital has the direction of the bond axis.

All the $g_{\parallel}$ values in Table 8.5 are slightly less than g_e, the free spin value. Theoretically, these $g_{\parallel}$ should equal g_e (see Section 1). The deviation is attributed to matrix perturbations, as was done for the deviation for the diatomic hydrides discussed in the previous section. Except that for BO, all $g_{\perp}$ values are less than the $g_{\parallel}$ values. If it is assumed that the matrix perturbations are the same for $g_{\parallel}$ and $g_{\perp}$, then $\Delta g_{\perp} = g_{\perp} - g_{\parallel}$ gives a measure of the intermixture of the $^2\Sigma$ ground state with the higher $^2\Pi$ states induced when the applied field is perpendicular to the symmetry axis. The negative values of $\Delta g_{\perp}$ indicate that this intermixing involves transfer of the odd electron to an empty Π level, and that the effective spin-orbit coupling is positive (see Section 1.c). There is no apparent reason for the deviant behavior of the $g_{\perp}$ value of BO, for which $g_{\parallel} < g_{\perp}$. Also for BO, in contrast to the other molecules, $A_{\parallel} < A_{\perp}$. Possibly the partial alignment of the BO molecules on the surface of the substrate sapphire rod caused a misassignment of the axis of symmetry of BO despite the special precautions taken by the observers. A theoretical calculation of $\Delta g_{\perp}$ for CN is made by Easley and Weltner [44].

4.d. Hydride Radicals of the Group IV Elements

The esr of the important methyl radical was first detected in X-irradiated $Zn(CH_3)_2$ at 77°K [45]. It has since been observed in numerous irradiated organic solids. The most precise measurements of the esr constants have been achieved with the radicals isolated in nonpolar matrices at low temperatures. Table 8.7 illustrates the relatively slight differences in the g values and proton couplings observed in a variety of nonpolar matrices at low temperature. Also shown are the rather narrow line widths observed for CH_3 in some nonpolar matrices. The narrowness of the lines observed is due in part to the restricted rotation of the radicals that tends to average out the **g** anisotropy and the anisotropic dipole-dipole interaction of the spin density on the C with the proton moments.

Except for those in the xenon matrix, the CH_3 radicals were made in the gaseous state by the action of an electric discharge on a mixture of methane and the inert gas forming the matrix. The mixture was then trapped on a sapphire rod at 2.4°K and observed. The concentration of the methane gas was maintained at a low level, of the order of 1%, so that the trapped radicals were for the most part isolated at substitutional sites in the inert matrix. In the xenon matrix, the CH_3 radicals were produced by X- or γ-irradiation of the frozen

Table 8.7 ESR Parameters of CH_3 Radicals Observed in Various Nonpolar Matrices at 4.2°K

Matrix	g_{av}	*H Splitting* A_f *(MHz)*	*Line Width* $\Delta H_{1/2}(G)$	*Source*
Ar	2.00203	64.64	3.7	[a]
Xe	2.0020	64.37		[b]
H_2	2.00266	65.07	1.4	[a]
N_2	2.00203	64.64		[a]
CH_4	2.00242	64.39	4.5	[a]

[a]C. K. Jen, S. N. Foner, E. L. Cochran, and V. A. Bowers, *Phys. Rev.*, **112**, 1167 (1958).
[b]Jackel and Gordy [49].

matrix containing methane in dilute solution. With this method, H atoms released by the radiation may be trapped in the vicinity of the CH_3, but they are not likely to cause noticeable perturbations of its esr constants. This method is used to produce trapped radicals of the other Group IV hydrides for measurement of their esr parameters given in Table 8.8.

Table 8.8 Nuclear Coupling Constants and Electron Spin Densities of Group IV Hydride Radicals

		Isotropic Coupling A_f *(MHz)*		*Spin Density*		
XH_3	*Matrix*	*H*	*X*	ρ^{H}_{1s}	ρ^{X}_{ns}	*Ref.*
CH_3	Kr	64.46		0.045		[a]
	Kr	64.46	^{13}C 108	0.045	0.035	[b]
SiH_3	Kr	22.7	^{29}Si (−)745	0.016	0.22	[a]
	Xe		^{29}Si (−)535		0.16	[c]
GeH_3	Kr	42	^{73}Ge (−)210	0.030	0.14	[d]
	Xe		^{73}Ge (−)210		0.14	[c]
SnH_3	Kr	73		0.05		[a]
	Xe		$^{117,119}Sn$ (−)1065		0.12	[c]

[a]Morehouse, Christiansen, and Gordy [47].
[b]Fessenden and Schuler [63].
[c]Jackel and Gordy [49].
[d]Jackel, Christiansen, and Gordy [48].

Structure of the Group IV Hydride Radicals

The structure of the methyl radical was shown to be planar by measurement of the ^{13}C hyperfine coupling in the esr of the $^{13}CH_3$ species produced by X-irradiation of CH_3I at 77°K with the ^{13}C isotope concentrated to 53% of the sample [46]. Only the isotropic component of the ^{13}C coupling was measured, probably because the anisotropic component was largely averaged out by motions of the radicals. This component is adequate, however, to prove that the radical is planar. The proof is simple. The 2*s* spin density on the C as revealed by the isotropic ^{13}C coupling (see Table 8.8) is only 0.035. This entire amount is expected from effects of spin polarization of the unpaired electron density in the *p* orbital on the C; in fact, it is less than the negative spin density, 0.045, induced on the hydrogens by the spin polarization of the σ bond. Thus there can be no significant 2*s* character in the orbital of the unpaired electron, which may, therefore, be assumed to be a pure 2*p* orbital. Because of the expected C_{3v} symmetry, the three bonding orbitals must be sp^2 hybrids in a plane perpendicular to the 2*p* orbital of the unpaired electron.

In contrast to the esr of CH_3, that of the other Group IV hydride radicals indicates that they have nonplanar pyramidal structures [47-49]. Evidence for this is the appreciable *ns* character in the orbital of the unpaired electron revealed by the isotropic nuclear coupling of isotopes of the central atom: ^{29}Si, ^{73}Ge, or $^{117,119}Sn$. These couplings and the *ns* spin densities on the central atoms computed from them are listed in Table 8.8. The bond angles of the pyramidal structure may be calculated from the *s* character s_b of the bonding orbital with (7.50). Obviously, $s_b = (1 - s_u)/3$, where s_u is the *ns* character of the unpaired electron, obtainable from the *ns* spin density. However, the amount of *ns* spin density obtained directly from the coupling of the central atom includes spin density in orbitals other than that of the odd electron. By taking from the observed spin density that due to bond spin polarization, as estimated from the H coupling, one can obtain a good approximation of the *ns* spin density of the unpaired electron and hence of s_u. Table 8.9 shows values of bond angles and

Table 8.9 Structures of XH_3 Radicals in the Xenon Matrix[a]

Radical	*s Character of Bond Orbitals*	*Bond Angle* θ	*Pyramidal Angle*[b] β
CH_3	0.333	120°	90° (planar)
SiH_3	0.285	113.5°	74°
GeH_3	0.295	115°	76.5°
SnH_3	0.31	117°	83°

[a] From Jackel and Gordy [49].
[b] Angle between perpendicular height and an edge of pyramid.

pyramidal angles thus derived for radicals in the Xe matrix. The structures are the same in the Kr matrix, except that of SiH_3, for which the bond angle was found to be 110.6° [47]. Details of the calculations may be found in the papers cited in Table 8.8. The most accurate structure is that for SiH_3 in the Kr matrix. The anisotropy in the ^{29}Si coupling was also measured in this matrix, and the anisotropic coupling constant of the $3p$ orbital was derived [47]. The $B(^{29}Si)$ = 24 G = 67 GHz showed the $3p$ character to be 0.78, which with the observed ρ_{3s} = 0.22 gave a normalized spin density of 1.00, that expected for the unpaired electron density. This agreement indicates that the resultant spin-polarization effects on the ^{29}Si coupling are negligible. Additional evidence for this is the small $\rho_{1s}^{H} = 0.015$.

There is uncertainty as to whether the spin density on the H is negative or positive in the heavier Group IV hydride radicals. In CH_3, it is definitely negative, but the significant hybridization of the odd-electron orbital in SiH_3, GeH_3, and SnH_3 may lead to a reversal of the sign of the H spin density [47].

Principal g Values

The **g** anisotropy is so small for CH_3 that only the averaged values are measured. These averaged values are very close to the free spin, g_e = 2.0023, as is shown in Table 8.7. For SiH_3, GeH_3, and SnH_3, the **g** anisotropy, although not large, has been measured and been found to be axially symmetric, as would be expected from their structures. It was not expected that $g_{\parallel}$ would be greater than $g_{\perp}$, as was observed (see Table 8.10). The spectrum shown in Fig. 8.15 demonstrates this effect for GeH_3.

Table 8.10 Observed and Calculated g Values for XH_3 Radicals[a]

	Observed Values for Rotating Radicals		*Derived Values for Static Radicals*	
Radical	$g_{\parallel}$	$g_{\perp}$	$g_{\parallel}$	$g_{\perp}$
SiH_3	2.007	2.005	2.003	2.007
GeH_3	2.017	2.010	2.003	2.017
SnH_3	2.025	2.014	2.003	2.025

[a] From Morehouse, Christiansen, and Gordy [47].

If matrix effects on **g** are negligible and if the radicals are fixed in space with valence orbitals conforming to the structures in Table 8.9, the g values calculated with (7.12) and (7.14) are [47]

$$g_{\parallel} = g_e, \quad g_{\perp} = g_e - \frac{\rho_{np}^{X}(\lambda_{\text{eff}})}{E_1 - E_0} \tag{8.40}$$

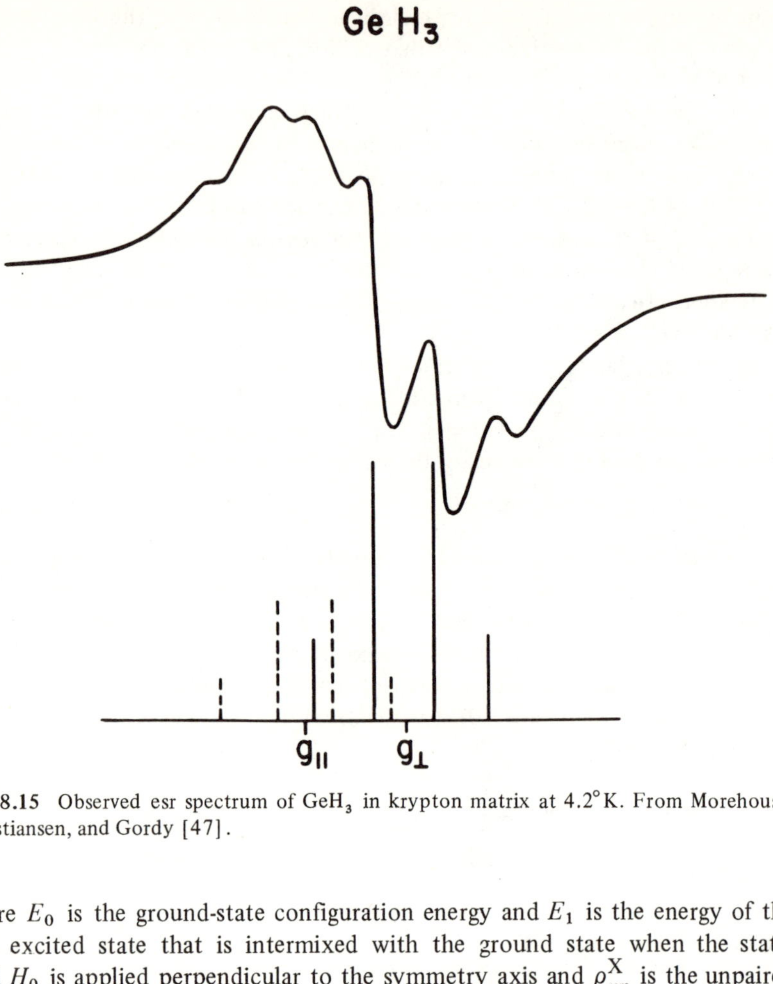

Fig. 8.15 Observed esr spectrum of GeH_3 in krypton matrix at 4.2°K. From Morehouse, Christiansen, and Gordy [47].

where E_0 is the ground-state configuration energy and E_1 is the energy of the first excited state that is intermixed with the ground state when the static field H_0 is applied perpendicular to the symmetry axis and ρ_{np}^{X} is the unpaired electron-spin density in the *np* orbital on the central atom. For SiH_3, the ρ_{3p}^{Si} value is measured to be 0.78. For GeH_3 and SnH_3, the values $\rho_{4p}^{Ge} = 0.89$ and $\rho_{5p}^{Sn} = 0.93$ are deduced from the isotropic coupling of X and H and the normalization relation. It seems probable that the effective spin-orbit coupling λ_{eff} is negative here; that is, $E_0 \to E_1$ conforms to the transfer of an electron from one of the bonding orbitals to the orbital of the unpaired electron (see Section 1.c). If this is true, $g_\perp > g_e$, and also $g_\perp > g_\parallel$. From Table 8.10 it is seen that $g_\perp > g_e$, but that $g_\perp < g_\parallel$. Also, it is seen that $g_\parallel$ is significantly greater than g_e, even though the matrix effects are expected to reduce $g_\parallel$ below g_e. These anomalies were cleared up with the assumption that the radicals are not completely static in the matrix but undergo restricted rotation about the axis of least moment of inertia [47]. This would require that they rotate about an axis perpendicular to the symmetry axis as illustrated in Fig. 8.16. Such rotation would average the $g_{\parallel}$ and $g_{\perp}$ expected for the static radicals. If $g_{\parallel}^{r}$ and $g_{\perp}^{r}$ signify the values for the radicals undergoing the restricted rotation and $g_{\parallel}^{s}$ and $g_{\perp}^{s}$ signify the values for the static radicals, it is evident that

$$g_{\perp}^{r} = \frac{(g_{\parallel}^{s} + g_{\perp}^{s})}{2} \tag{8.41}$$

$$g_{\parallel}^{s} = g_{\perp}^{r} \tag{8.42}$$

and

$$g_{\parallel}^{s} = 2g_{\perp}^{r} - g_{\parallel}^{r} \tag{8.43}$$

The $g_{\parallel}^{s}$ and $g_{\perp}^{s}$ values derived with these relations from the observed (g^{r}) values are listed in Table 8.10. It is seen that the $g_{\parallel}$ values for the static radicals are all very near to the free-spin value g_e, and $g_{\perp} > g_{\parallel}$, as expected from theory. This comparison provides strong evidence that the radicals are undergoing restricted rotation even at the temperature of 4.2°K.

4.e. Triatomic Hydride and Fluoride Radicals of B, N, and P

The Hydrides

In Table 8.11 are listed the g values and isotropic nuclear couplings obtained for NH_2 and PH_2 measured in isolation within inert matrices. Differences in the

Table 8.11 ESR Parameters and Spin Densities of NH_2 and PH_2

Radical	*Matrix (4.2°K)*	g_{av}	*Isotope*	*Coupling A_f(MHz)*	*Spin Density* ρ_{ns}	*Ref.*
NH_2	Ar	2.0038[b]	H	67.0	0.047	*a*
			^{14}N	29.0	0.0188	*a,b*
	Kr	2.0036	H	67.1	0.0473	*b*
			^{14}N	29.9	0.0194	*b*
PH_2	Kr	2.0087	H	50	0.035	*c*
			^{31}P	244	0.022	*c*
	Xe	2.0050	H	48	0.034	*d*
			^{31}P	229	0.0225	*d*

[a] Foner, Cochran, Bowers, and Jen [53].
[b] Fischer, Charles, and McDowell [54].
[c] Morehouse, Christiansen, and Gordy [50].
[d] Jackel and Gordy [49].

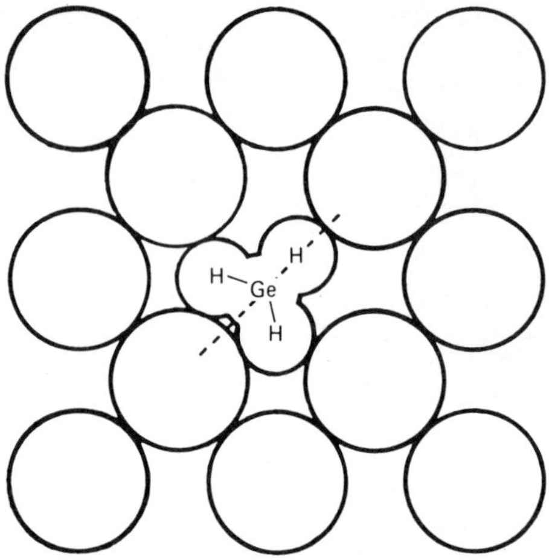

Fig. 8.16 Diagram indicating GeH_3 radicals trapped in a substitutional site in the krypton matrix at 4.2°K. The esr evidence suggests that the radical rotates about an axis perpendicular to the symmetry axis, represented by the dashed line.

matrix effects are small, probably within the experimental error. The anisotropy in **g** and that in **A** are effectively eliminated by restricted rotations of the radicals at 4.2°K in these matrices. Because g_{av} is greater than g_e, it appears that for the static radicals $g_{\parallel} < g_{\perp}$, as expected from theory, and also that the anisotropy in **g** is small since g_{av} is near g_e.

Although the anisotropic *p*-orbital coupling cannot be measured because of the rotations, one can conclude from the smallness of the 2*s* and 3*p* spin densities determined by the isotropic couplings of the ^{14}N and ^{31}P that the orbital of the unpaired electron is essentially pure *p* in both NH_2 and PH_2. The small *ns* spin densities observed are of the right magnitude to result from atomic configuration interaction and bond spin-polarization effects. Thus both are π-type radicals with the electron in a pure *np* orbital perpendicular to the molecular plane.

The isotropic coupling can be used for calculation of the value of the bond angle in NH_2 [50]. The method is simple. Since the spin density averaged over the NH σ bond must be zero to comply with the exclusion principle, the spin polarization of the bond must put equal but opposite (in sign) spin densities on the N and H. Note, however, that the isotropic spin density of 0.019 on N is appreciably lower than that on the two H atoms, 2(0.047) = 0.094. This lower spin density in the 2*s* orbital on the N is due to the large 2*p* character, and hence small 2*s* character, of the N-bonding orbitals. With some minor approximations, the observed 2*s* spin density on N can be used for calculation of the degree of *sp* hydridization, from which the bond angle may be obtained.

In addition to that due to bond spin polarization, there is a small amount of isotropic spin density caused by interaction of the unpaired electron with non-bonding pairs on the N, which can be estimated from the atomic ^{14}N coupling (10.3 MHz), in the krypton matrix (Table 8.1). This isotropic atomic coupling results from spin density induced in the *ns* shells by three unpaired 2*p* electrons. Since there is only one inducing 2*p* electron on the NH_2 as compared with three on the N atom, we take a third of the value, 1/3(10.3) = 3.4 MHz, as an estimate of the coupling induced in the nonbonding N orbitals for NH_2. Since this is expected to be of the same sign as that induced in the bonding N orbitals, the 2*s* spin density in each of the two bonding orbitals is calculated to be

$$\rho^{N}_{\sigma 2s} = \frac{1}{2} \frac{29.9 - 3.4}{A^{N}_{\text{atomic}}} = 0.0086 \tag{8.44}$$

This amount, which can also be obtained from the H coupling, is equivalent to $a_s^2 \rho_s^H$, where a_s^2 is the 2*s* character of the N-bonding orbitals and ρ_s^H is the magnitude of the spin density on the bonded H. Therefore, for the krypton matrix,

$$a_s^2 \rho_s^H = 0.0473\ a_s^2 = 0.0083$$

from which $a_s^2 = 0.175$. Thus the bonding orbitals of the N have 17.5% 2*s* character and 82.5% 2*p* character, b_p^2. From this hybridization, the bond angle θ can be calculated from the Coulson relation (see Chapter VII, Section 3) to be

$$\cos\theta = \frac{-1}{\lambda^2} = \frac{-a_s^2}{b_p^2} = -0.212 \tag{8.45}$$

which gives $\theta = 102°\ 45'$. This value is in close agreement with 103° 25′, that obtained by Dressler and Ramsey [51] from optical spectroscopy of the gas. The good agreement is an indication that the esr calculation for NH_2 is sound. A similar calculation for PH_2 indicates that the bonding orbitals have 27% 3*s* character [50]. However, derivation of the bond angle of PH_2 by use of the Coulson relation, (8.45), is questionable because Jordan's theoretical calculations [52] indicate that a small amount of *d* hybridization influences the bond angle significantly in PH_2. As far as we know, the bond angle for PH_2 has not been measured in the gaseous state.

Marked differences in the dissociation of the Group V hydrides under irradiation have been encountered. The NH_2 radicals observed by Foner, Cochran, et al. [53] were produced by uv irradiation of the argon matrix containing small concentrations of NH_3. Those observed by Fischer, Charles, and McDowell [54] were produced as a secondary product in uv photolytic decomposition of HN_3 in dilute concentration in various rare-gas matrices at 4.2°K. Only very weak signals of NH_2, with strong signals of N atoms, were detected from γ-irradiated NH_3 in dilute solution in the Kr and Xe matrices [49]. The PH_2 radicals listed in Table 8.11 were produced by γ-irradiation of the Kr or Xe matrix containing 2% PH_3 (4.2°K). From PH_3 irradiated in an argon matrix, Adrian, Cochran, and Bowers [55] reported detection of P and H atoms only. Likewise, only P and H atoms were observed for γ-irradiated PH_3 in the argon matrix [49], and strong signals from P atoms along with those of PH_2 radicals were observed for the krypton matrix. Only As and H atoms were observed when AsH_3 was γ-irradiated in an argon matrix at 4.2°K, but an esr spectrum, apparently of AsH_2 radicals, was observed when AsH_3 was γ-irradiated in a xenon matrix [49]. The AsH_2 pattern could not be sufficiently resolved for measurement of the principal g values and coupling constants. Repeated attempts to detect SbH_2 radicals by irradiation of SbH_3 in rare-gas matrices have failed.

The Fluorides

The esr of NF_2 has been detected by a number of observers [55-60]. Farmer, Gerry, and McDowell [58] measured its spectra in the krypton matrix. They obtained only the isotropic components of the nuclear coupling but observed an axially symmetric **g** anisotropy (see Table 8.12). With the NF_2 radicals isolated in the neon metrix at 4.2°K, Kasai and Whipple [60] were able to measure the anisotropy in the ^{14}N and ^{19}F coupling as well as in **g**. Their values are listed in Table 8.12. The first derivative of the esr absorption curve that they obtained is shown in Fig. 8.17. Note the well-defined peaks, which allow accurate measurement of the nuclear coupling. The indicated 212-G splitting is that of ^{19}F and the 49-G splitting, that of ^{14}N. The magnified inset shows the effects of the small anisotropy in **g**.

Also reproduced in Table 8.12 are the isotropic and anisotropic esr parameters for PF_2 radicals isolated in the xenon matrix [61]. Earlier reported measurements of esr of PF_2 are inconsistent with these values, but Colussi, Morton, et al. [62] have shown that in these earlier reports the spectra of other radicals were incorrectly assigned to PF_2. To clear up these inconsistencies, Colussi, Morton, et al. [62] prepared PF_2 radicals by irradiation of 5-mole% of PF_3 in a C_2F_6 matrix at −140°C and observed for the rotating radicals the following esr parameters: $g = 1.9994$, $A_{iso}(^{31}P) = 228$ MHz, and $A_{iso}(^{19}F) = 93$ MHz. They also prepared PF_2 radicals by uv photolysis of PF_2Cl dissolved in Freon 13 at −100°C and observed the following isotropic values: $g = 1.9997$, $A_{iso}(^{31}P) = 230$ MHz, and $A_{iso}(^{19}F) = 91$ MHz. These parameters agree satisfactorily with those listed in Table 8.12 for the rotating radicals in the xenon

Table 8.12 ESR Parameters for BF_2, NF_2 and PF_2

Radical	Matrix T (°K)	Principal g Values	Isotope	Coupling Constants (MHz) $A_x = A_\parallel$	Coupling Constants (MHz) $A_y = A_z = A_\perp$	Source
NF_2	Ne 4.2°	$g_x = 2.0011$ $g_y = 2.0042$ $g_z = 2.0042$	^{14}N ^{19}F	137 594	0 −47.4	*a*
NF_2	Kr 4.2°	$g_\parallel = 2.0022$ $g_\perp = 2.0059$	^{14}N ^{19}F		$A_{iso} = 48$ $A_{iso} = 168$	*b*
PF_2	Xe 4.2°	$g_x = g_\parallel = 2.0027$ $g_y = g_z = g_\perp = 2.0016$	^{31}P ^{19}F	863 352	−76.2 −37.8	*c*
PF_2	Xe 77°	$g_{iso} = 2.0020$	^{31}P ^{19}F		$A_{iso} = 237$ $A_{iso} = 91$	*c*
BF_2	Xe 4.2°	$g_{iso} = 2.0012$	^{10}B ^{11}B ^{19}F		$A_{iso} = 278.0$ $A_{iso} = 826.2$ $A_{iso} = 532$	*d*

[a] Kasai and Whipple [60].
[b] Farmer, Gerry, and McDowell [58].
[c] Nelson, Jackel, and Gordy [61].
[d] W. Nelson and W. Gordy, *J. Chem. Phys.*, **51**, 4710 (1969).

matrix at 77°K. The small deviations are attributable to the significant differences in the trapping media.

Listed in Table 8.13 are isotropic and anisotropic coupling constants A_f and B derived from the $A_\parallel$ and $A_\perp$ values in Table 8.12. Also listed are the *ns* and *np* spin densities derived from the A_f and B values. In derivation of the *np* spin densities, corrections were applied for the positive charges on the N and P that are caused by the rather large ionic character in the bond. (see Chapter VI, Section 6 for a description of the method). No charge corrections were applied for the small *ns* spin density because the charge corrections are expected to be much less significant for the Fermi contact coupling A_f and because the corrections to be applied are not known. Likewise, no corrections are made for the 2*p* spin density on the fluorines because of uncertainty in the charge effects for F, which are believed to be small. The ionic characters, $i_c = 0.45$ for NF_2 and $i_c = 0.90$ for PF_2, were estimated from (6.56) with the electronegativity

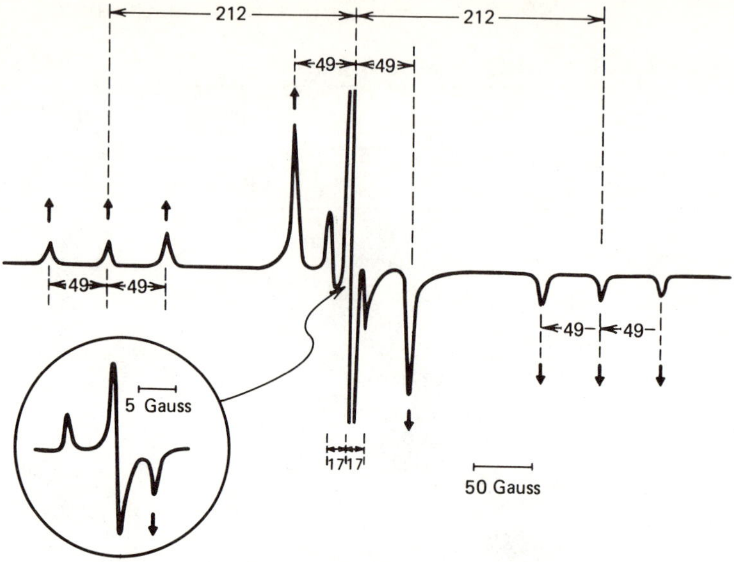

Fig. 8.17 First-derivative esr spectrum of NF_2 radicals isolated in a neon matrix at 4°K. The magnetic field increases from left to right. The inset shows the strong central line on an expanded scale with reduced gain. The observation frequency is 9430 MHz. From Kasai and Whipple [60].

Table 8.13 Nuclear Coupling Constants and Spin Densities of NF_2 and PF_2

		Coupling Constants[a]		*Spin Densities*	
Radical	*Coupling Atom*	A_f *(MHz)*	*B (MHz)*	ρ_{ns}	ρ_{np}[b]
NF_2	^{14}N	45.7	45.7	0.030	0.75
	^{19}F	166	213	0.0035	0.14
PF_2	^{31}P	237	313	0.023	0.80
	^{19}F	92	130	0.002	0.09

[a] Derived from data in Table 8.12.
[b] Corrections are applied for formal charges on N and P atoms as described in the text.

difference of $x_N - x_F = 0.9$ and $x_P - x_F = 1.8$, respectively. This calculation gives the charges as $c^+ = 0.9$ for N in NF_2 and $c^+ = 1.8$ for P in PF_2.

The *ns* spin densities in both NF_2 and PF_2 are of the magnitude to be caused by spin polarization effects. The sum of the ρ_{np} spin densities on the three atoms is 1.03 for NF_2 and 0.98 for PF_2. It is evident that the orbital of the unpaired electron of both radicals is an antibonding π^* orbital that spreads over the molecular plane. With the x coordinate chosen as perpendicular to the molecular plane, the orbital for PF_2 may be expressed by the linear combination

$$\phi_{\pi^*} = c_1 \psi_{3p_x}(\mathrm{P}) + c_2 \psi^{(1)}_{2p_x}(\mathrm{F}^{(1)}) + c_3 \psi^{(2)}_{2p_x}(\mathrm{F}^{(2)}) \qquad (8.46)$$

where

$$c_1^2 + c_2^2 + c_3^2 = 1$$

and

$$c_1^2 = \rho^{\mathrm{P}}_{3p} = 0.80 \qquad \text{and} \qquad c_2^2 = c_3^2 = \rho^{\mathrm{F}}_{2p} = 0.09$$

The orbital for NF_2 is similar. Although the sum of the *np* densities for NF_2 is 1.03, this deviation from unity is certainly within the possible error in the derivation. The charge corrections applied are only approximate; but if no charge corrections were applied, the sum of the *np* spin densities for NF_2 would be 1.22 and for PF_2, 1.24, both unacceptably high.

There are interesting differences in the g values for NF_2 and PF_2. While $g_\parallel$ for both is near g_e, as expected from theory, $g_\perp > g_\parallel$ for NF_2 and $g_\perp < g_\parallel$ for PF_2. The observed positive $\Delta g_\perp$ is expected for NF_2 because its structure is similar to that of NO_2^{--} (see Table 7.2), which has Δg positive. The odd electron is in an antibonding π orbital in both radicals, and the lowest excited state is one in which an electron in a σ bond is transferred to this orbital via a $\sigma \rightarrow \pi^*$ transition. For such a transition, the effective spin-orbit coupling is negative and Δg is positive (see Chapter VII, Section 1.c). Evidently, in PF_2 the transition to the lowest excited state is one in which the odd electron of the π^* orbital is transferred to a previously unoccupied d orbital on the P. This would correspond to a positive λ_{eff} and a negative $\Delta g_\perp$, as observed. The evident reasons for the differences are the availability of the d orbitals on P and the higher ionic character of the σ bonds in PF_2 over those of NF_2.

Only the isotropic components of g and A could be measured for BF_2. These are listed in Table 8.12. Their analysis by the authors cited indicates that 93% of the unpaired electron density is on the B, in an sp^2-hybrid orbital having 44% 2*s* character, and that the remaining 7% is in hybridized bonding orbitals of the two fluorines. The bond angle is predicted to be 112°. Thus BF_2, unlike NF_2 and PF_2, is a σ^*-type radical with the orbital of the unpaired electron in the molecular plane.

The esr spectra and structure of the interesting PF_4 radical are discussed in Chapter VI, Section 9.

4.f. Fluorinated Methyl Radicals

Fessenden and Schuler [63] measured the isotropic components in the nuclear coupling of the fluorinated methyl radicals shown in Table 8.14. These couplings revealed the interesting change in the molecular structure from CH_3 to CF_3 shown in Table 8.15.

Table 8.14 ESR Parameters of Fluorinated Methyl Radicals

	Measured Parameters[a]				*Derived Unpaired Electron Density*[b]	
Radical	g_{av}	*Coupling Isotope*	A_{iso} *(MHz)*	A_{iso}/A_{at}	*2s*	*2p*
CF_3	2.0031	^{13}C	761	0.245	0.245	0.434
		^{19}F	399	0.008	0	0.107
CHF_2	2.0041	H	62.2	0.044		
		^{13}C	417	0.134	0.12	0.66
		^{19}F	236	0.005	0	(0.11)
CH_2F	2.0045	H	59.1	0.042		
		^{13}C	154	0.050	0.03	0.86
		^{19}F	180	0.004	0	(0.11)

[a] Measured in the xenon matrix at −188°C by Fessenden and Schuler [63].
[b] Derived as explained in text.

Table 8.15 Hybridization and Bond Angles Derived from ESR Parameters of the Fluorinated Methyl Radicals in Table 8.14

	2s Character of Orbitals on C			
Radical	*Odd Electron*	*σ-Bonding Orbital*	*Bond Angle*	*Structure*
CF_3	0.360	0.213	106°	Pyramidal
CHF_2	0.154	0.282	113°	Pyramidal
CH_2F	0.034	0.322	118°	Pyramidal
CH_3	0.000	0.333	120°	Planar

As already explained, the isotropic ^{13}C coupling showed the methyl radical to be planar: the coupling is so small as to be explained entirely by the spin-polarization effects, with the orbital of the unpaired electron a p_π orbital. This is not true for CF_3, as is evident from the larger ^{13}C isotropic coupling, 761 MHz as compared with 108 MHz for CH_3. The $(A_{iso}/A_{atomic}) = 0.245$ is much too large to be attributed to configuration interaction; it must result primarily from 2*s* character in the orbital of the odd electron. Because of the high ionic character of the σ CF bond, spin-polarization effects are much less than for CH_3. From analysis of the ^{19}F coupling of CF_3 in Chapter VI, Section 7.b, the $2p_\sigma$ spin density on F is found to be only -0.011 (see Table 6.24). Since the σ-bonding orbitals are approximately pure 2*p*, we assume that the counter-spin polarization of the σ-bonding orbital of C is +0.011. With an estimate of 25% 2*s* character in these bonding orbitals, the total spin density in the three bonding orbitals is +0.008. In derivation of the 2*s* spin density of the odd-electron orbital, this amount must be subtracted from the value 0.245. There is also a small negative contribution to the coupling that is due to spin polarization of the inner electron core, described in Chapter VI, Section 4.a. For an inducing density $\rho_\pi^C = 1.00$, the contribution to the coupling is calculated by Karplus and Fraenkel [64] to be $S^C = -36$ MHz, but the value of $S^C = -26$ MHz is derived in Chapter VI, Section 4.a from the measured coupling of the CH_3 radical. If the latter value of S^C is used, the estimated contribution to the ^{13}C coupling from this source is $-26\ \rho_\pi^C = \sim -14$ MHz. This is equivalent to a 2*s* spin density of -0.005. If the theoretical value $S^C = -36$ MHz is used, the value for the spin density is -0.008, which is equal in magnitude but opposite in sign to that estimated for the σ-bond spin polarization. For this reason, the A_{iso}/A_{atomic} may be taken to be the same as the 2*s* unpaired electron spin density in CF_3. The spin-polarization terms are so small that the approximations used for this evaluation cannot seriously influence the value of this derivation.

Unlike CF_3, CHF_2 and CH_2F have a significant positive 2*s* spin density on C which is due to the large spin polarization of the CH bonds. The amount of the CH contribution may be determined from the H spin densities. When this is done and the CF σ-bond effects are estimated as described for CF_3, the unpaired electron spin density ρ_{2s}^C is 0.12 for CHF_2 and 0.03 for CH_2F, as listed in Table 8.14. These values indicate that CHF_2 and CH_2F are not planar but are more nearly planar than is CF_3. To derive the degree of orbital hybridization on the C and the structural angles, one must take into account the unpaired electron density on the fluorines.

Because of the unpaired electron densities on the fluorines, the 2*s* character of the odd-electron orbital on C is not equivalent to the ρ_{2s}^C value, as it is for CH_3. The amount of the unpaired spin density on the fluorines in CF_3 is known from the measurements of Maruani, McDowell, et al. [65,66]. The resulting value of 0.107 for ρ_{2p}^F is given in Table 8.14. From the normalization of the spin density of the unpaired electron,

$$\rho_{2s}^{C} + \rho_{2p}^{C} + 3\,\rho_{2p}^{F} = 1$$

with $\rho_{2s}^{C} = 0.245$ and $\rho_{2p}^{F} = 0.107$ one obtains $\rho_{2p}^{C} = 0.434$ as given in Table 8.14. Although the anisotropic ^{19}F couplings in CHF_2 and CH_2F, to our knowledge, have not been measured, ρ_{π}^{F} is close to 0.11 for all the fluorocarbons in Table 6.24, and their $\rho_{2p_{\sigma}}^{F}$ values are all approximately −0.01. Thus we assume $\rho_{2p}^{F} = 0.11$ for both CHF_2 and CH_2F and from the normalizations, $\rho_{2s}^{C} + \rho_{2p}^{C} + 2\rho_{2p}^{F} = 1$ for CHF_2 and $\rho_{2s}^{C} + \rho_{2p}^{C} + \rho_{2p}^{F} = 1$ for CH_2F, we obtain the values of ρ_{2p}^{C} listed in Table 8.14. The 2*s* character of the unpaired-electron orbital on the carbon is $\rho_{2s}^{C}/(\rho_{2s}^{C} + \rho_{2p}^{C})$, and the 2*s* character of each bonding C orbital is $a_s^2 = (1/3)[1 - \rho_{2s}^{C}/(\rho_{2s}^{C} + \rho_{2p}^{C})]$. The resulting values of a_s^2 are given in Table 8.15. The bond angles listed in Table 8.15 are derived from (8.44) with $(1/\lambda^2) = a_s^2/(1 - a_s^2)$.

The derived bond-orbital hybridizations listed in Table 8.15 differ from those obtained by Fessenden and Schuler [63] primarily because we have taken into account the appreciable unpaired-electron density on the fluorines in evaluating the orbital hybridization on the C. In these radicals the odd electron is not entirely localized on the carbon but is in an antibonding pseudo-π^* orbital that spreads over the fluorines as well as the C.

Theoretically, one should also be able to derive the bond angles of CF_3 from comparison of the directions of the principal coupling elements of the ^{19}F coupling for the three fluorines. However, these directions cannot be measured accurately; in the single-crystal measurements of Rogers and Kispert [67], the radicals were evidently rotating about the symmetry axis. The structural derivation from the isotropic ^{13}C coupling has the advantage of being independent of the motions of the radical. Edlùnd, Lund, et al. [68] have treated the ^{19}F coupling in CF_3 with intermediate neglect of differential overlap (INDO) and *ab initio* theory and have shown that this coupling has no principal axis along the corresponding CF bond. They conclude that the bond angle is in the range of 109.5° to 112.0°.

4.g. The Ethyl Radical

The ethyl radical, detected in 1956 in X-irradiated $Hg(CH_2CH_3)_2$ at 77°K, was among the first trapped organic free radicals to be observed with esr [69]. As explained in Chapter VI, Section 1.c, its spectrum was used in the early formulation of theories of β coupling based on the mechanism of hyperconjugation. In the intervening years, many observations of this radical have been made in irradiated organic powders. The CH_2 and CH_3 couplings with their overlapping anisotropies are so nearly equal that their group hyperfine structures are not separable for fixed, randomly oriented radicals. As a consequence, the esr pattern in some solids appears as a gross sextet of five roughly equivalent protons. However, partially resolved peaks due to distinguishable CH_2 and CH_3 couplings have been observed in a number of irradiated solids [70-75]. The patterns in these solids are apparently simplified by the restricted rotation of the CH_2 group about the CC σ-bond, which reduces the anisotropy and symmetrizes the coupling of the CH_2 as well as CH_3 protons. The most precise solid-state measurements are those made on the radicals isolated in inert-gas matrices at 4.2°K. Cochran, Adrian, and Bowers [71] observed the esr of CH_3CH_2 radicals in the Ar matrix produced by photolyzing ethyl iodide in the matrix. Subsequently, McDowell, Raghunathan, and Shimokoshi [75] produced the radicals in this matrix by uv photolysis of the matrix containing equal concentrations of very dilute C_2H_4 and HI mixtures (1:1:500). Hydrogen atoms released from the HI migrated through the matrix and became attached to the C_2H_2, forming isolated CH_3CH_2 radicals. The esr pattern they observed for the CH_3CH_2 is shown in Fig. 8.18 with a stick diagram to indicate its origin. The pattern is made axially symmetric by relative, internal rotations of

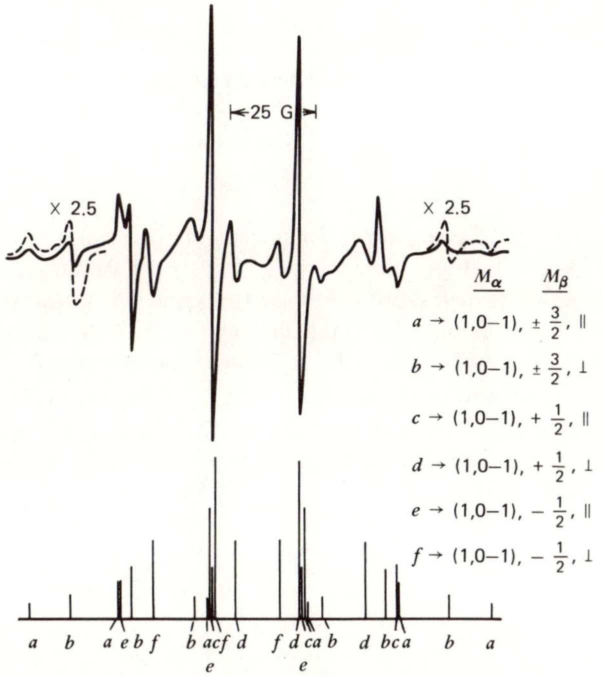

Fig. 8.18 The esr spectrum of trapped ethyl radicals obtained by uv photolysis of C_2H_4 + HI + Ar (ratios 1:1:500) at 4.2°K. Broken curves at the extremities correspond to an increased amplification of 2.5 times. The stick diagram at the base represents the assigned spectrum as indicated. From McDowell, Raghunathan, and Shimokoshi [75].

the CH_2 and CH_3 groups about the direction of the CC bond. Accurate principal values of both the α-coupling (CH_2) and β-coupling (CH_3) protons were obtained by computer simulation of the observed pattern. The computer program used is that described by Le Febvre and Maruani [15] (see Section 2.b). The resulting esr parameters are listed in Table 8.16.

Table 8.16 ESR Parameters for CH_3CH_2 Radicals Isolated in the Argon Matrix at 4.2°K[a,b]

g_{av}	Coupling Group	Proton Couplings (MHz)			
		Principal Value		A_{iso}	A_μ
2.0023 ±0.0003	CH_2	$A_\parallel^\alpha$	(−)83.0 ± 1.4	(−)63.9	(−)19.1
		$A_\perp^\alpha$	(−)54.4 ± 1.4		(+) 9.5
	CH_3	$A_\parallel^\beta$	(+)80.4 ± 1.4	(+)75.1	(+) 5.3
		$A_\perp^\beta$	(+)72.6 ± 1.4		(−) 2.5

[a] From the data by McDowell, Raghunathan, and Shimokoshi [75]. The signs within parentheses were not measured but are expected from theory.
[b] The constants are symmetrized by restricted internal rotations about the CC σ bond.

To compare the dipole-dipole components $(A_\mu)_\parallel$ and $(A_\mu)_\perp$ with those calculated by the methods described in Chapter VI, section 1.d, one must transform the calculated values $(A_\mu)_x$, $(A_\mu)_y$, and $(A_\mu)_z$ for the static CH fragment to one rotating about the CC σ-bond. For the theoretical values described in Chapter VI, Section 2.a, x is along the $2p_\pi^C$ orbital, y is perpendicular to this orbital and to the CH bond, and z is along the CH bond. By transforming the tensor of the CH fragment to the CC axis of rotation and by averaging the transformed values over a complete cycle of rotation, one obtains the relationships [75]

$$(A_\mu)_\parallel^\alpha = (A_\mu)_y^\alpha \sin^2\theta + (A_\mu)_z^\alpha \cos^2\theta \tag{8.47}$$

$$(A_\mu)_\perp^\alpha = \frac{1}{2}\left[(A_\mu)_x^\alpha + (a_\mu)_y^\alpha \cos^2\theta + (A_\mu)_z^\alpha \sin^2\theta\right] \tag{8.48}$$

where θ is the angle between the CH bond and the CC direction. For the planar $-CH_2$ group, $\theta = 60°$ is a reasonable assumption. The A_μ values calculated for the CH fragment with (6.13), (6.15), and (6.16) are: $(A_\mu)_x = -2.8\ \rho^\alpha_{2p_\pi}$, $(A_\mu)_y = -36.0\ \rho^\alpha_{2p_\pi}$, and $(A_\mu)_z = 38.7\ \rho^\alpha_{2p_\pi}$ (see Table 6.3). The spin density $\rho^\alpha_{2p_\pi} = 0.937$ is determined by the $A^\alpha_{\mathrm{iso}} = 63.9$ MHz and the $Q_\alpha = 68.2$ MHz of Fessenden and Schuler [76]. With these parameters, (6.45) - (6.46) yield the predicted values: $(A_\mu)_\parallel = -16.2$ MHz and $(A_\mu)_\perp = 8.1$ MHz. Similar calculations made with the parameters of McConnell and Strathdee from Table 6.3 yield: $(A_\mu)_\parallel = -17.2$ MHz and $(A_\mu)_\perp = 8.7$ MHz. This is in somewhat better agreement with the observed values: $(A_\mu)_\parallel = (-)19.1$ MHz and $(A_\mu)_\perp = (+)9.5$ MHz, as given in the last column of Table 8.16.

Barfield has calculated the anisotropic proton hyperfine coupling tensors for the ethyl radical both with valence-bond theory and with the unrestricted Hartree-Fock (UHF) method, using INDO [77]. McDowell, Raghunathan, and Shimokoshi et al. [75] have transformed the predicted parameters for comparison with their observed $A_\parallel$ and $A_\perp$ values for both the α and β protons. The agreement is satisfactory for the β protons but is rather poor for the α protons.

5. TRAPPED RADICAL IONS

The esr of hundreds of radical ions (+ and −) have been observed in both solids and liquids. A comprehensive volume, *Radical Ions,* edited by Kaiser and Kevan [78] and published in 1968, gives extensive tabulation of esr parameters for both organic and inorganic ion radicals in solid and liquid solutions, together with theoretical discussion of these parameters. Specialists in studies on ion radicals should consult this work. We can give here only selected data on a few of the simpler molecular ion radicals, most of which have been reported since publication of this volume.

5.a. Trihydride and Trifluoride Ions of Boron and Nitrogen

The esr of the BH_3^- radical was first observed by Catton, Symons, and Wardale [79] in γ-irradiated potassium borohydride at low temperature. Although the hyperfine pattern was incompletely resolved, the radical was correctly identified, and isotropic coupling constants of ^{11}B and H were obtained which are in approximate agreement with the more accurate values listed in Table 8.17. These listed values were obtained by Sprague and Williams [80] from observations at 25°C on radicals isolated in the $(CH_3)_4NBH_4$ matrix. The lines were observed to be considerably sharper at this temperature than at −100°C, evidently because of the increased molecular motions at the higher temperature. As a result, the ^{11}B hyperfine structure was completely resolved. By computer simulation of this well-resolved spectrum, accurate values of the isotropic g and of the couplings were obtained, as listed in Table 8.17. Similarly, the rotational motions of the NH_3^+ radicals observed at room temperature in X-irradiated single crystals of

Table 8.17 Isotropic ESR Parameters and Derived Properties of Isoelectric Ion Radicals of Boron and Nitrogen

Radical / *Matrix*	g_{av}	*Isotope*	A_{iso}	ρ_{ns} A_{iso}/A_{at}	*Structure*	*Source*
BH_3^- / $(CH_3)_4NBH_4$	2.0023	^{11}B	58.0	0.029	Planar	*a*
		H	42.3	0.030		
NH_3^+ / NH_4ClO_4 (xtal), 25°C	2.0035	^{14}N	54.6	0.035	Planar	*b*
		H	72.5	0.051		
BF_3^- / $Si(CH_3)_4$ 84°K	2.0021	^{11}B	429	0.212	Pyramidal	*c*
		^{19}F	499	0.001		
NF_3^+ / $NF_4^+AsF_6^-$ 77°K	2.007	^{14}N	275	0.179	Pyramidal	*d*
		^{19}F	468	0.001		

[a] Sprague and Williams [80].
[b] Cole [81].
[c] R. L. Hudson and F. Williams, *J. Chem. Phys.*, **65**, 3381 (1976).
[d] Mishra, Symons, Christe, Wilson, Wagner [83].

ammonium perchlorate by Cole [81] allowed an accurate evaluation of the isotropic esr parameters, as listed in Table 8.17. It is remarkable that these two ionic species are sufficiently stable for observation in the solid matrices at room temperature.

The *ns* spin densities on the atoms of BH_3^- and NH_3^+ as derived from the isotropic coupling in the usual way are also listed in Table 8.17. Those for H and B in BH_3^- are essentially equal and are of the amounts expected from spin polarization of the σ bonds. Although there are three bonds to B, each has only one-third 2*s* character. Therefore, the 2*s* spin density on B that is due to spin polarization of the three bonds should be $\rho_{2s}^{B} = 3(1/3)\,\rho_{1s}^{H} = \rho_{1s}^{H}$, provided that configuration interaction with the inner 1*s* shell on the B produces negligible effects. This appears to be true for BH_3^- but not for NH_3^+, for which the spin density on N (0.035), is significantly less than that on H (0.051). Since the 2*s* spin densities in both radicals are sufficiently small to arise entirely from bond spin polarization, we may assume that the orbital of the unpaired electron is a pure 2*p* orbital on the central atom and hence that both structures are planar.

Spin density induced in the *s* shell of the inner core is expected to be negative (see Chapter VI, Section 4.a) and, therefore, to reduce the A_{iso} and the calculated spin density if this polarization is neglected. The complicating effects of the unshared pair on neutral N radicals as well as of unpaired electron density on atoms bonded to the N, as described in Chapter VI, Section 5.b, are absent in NH_3^+, but the small, though unknown, effects of the plus charge on the isotropic coupling complicates the calculations. Because of the ionic character of the three σ bonds, the resulting charge on the N is expected to be approximately zero. If the charge effects are negligible, one would expect A_{iso}^N to be expressed by (6.37), which applies to ^{13}C coupling in hydrocarbon radicals. For NH_3^+, this equation takes the form

$$A_f^N = (S^N + 3Q_{NH}^N)\,\rho_\pi^N \tag{8.49}$$

For NH_3^+, $\rho_\pi^N = 1$, and with $A_f^N = 54.6$ MHz (from Table 8.17), (8.49) becomes

$$S^N + 3Q_{NH}^N = 54.6$$

The value of Q_{NH}^N may be estimated from the H coupling and A_{atomic}^N, as described for CH_3 in Chapter VI, Section 4.a. Since it is due to spin polarization of the NH bond, $Q_{NH}^{^{14}N} = \rho_{1s}^H A_{atomic}^{^{14}N} = 0.051(1540) = 78$ MHz. With this value, (8.49) gives $S^N = -179$ MHz. A similar analysis for NH_2 with the constants in Table 8.11 gives $Q_{NH}^{^{14}N} = 72$ MHz and $S^N = -115$ MHz. Since $\rho_\pi^N = 1$, $Q_{NH}^H = 72.5$ for NH_3^+ as compared with $Q_{NH}^H = 67$ MHz for NH_2. The averaged value of Q_{NH}^H is 72 MHz for a variety of radicals (see Table 6.10).

Unlike the corresponding hydrides that are planar, BF_3^- and NF_3^+ are pyramidal in structure. This is shown by the 2*s* spin densities on the central atom, which are much too large to be attributed to induced spin polarization of the bonds or inner-core electrons. Without a knowledge of the anisotropic couplings on B or the fluorines in BF_3^-, one can calculate neither the amount of *s* hybridization of the orbitals nor the bond angles. From comparison with CF_3 (Table 6.24), one would expect appreciable 2*p* spin density on the fluorines. Thus the unpaired 2*p* spin densities in Table 8.17 probably represent the lower limit of the 2*s* spin density on the B. By taking into account the probable 2*p* density on the three fluorines (~0.3), we estimate the 2*s* character of the unpaired electron orbital on the boron to be 20 to 25% and the bond angles to be approximately tetrahedral. The esr of the related neutral radical BF_2 measured in the xenon matrix at 4.2°K [82] shows that the unpaired electron has a σ-type orbital in the molecular plane with 93% of the electron-spin density on the boron in an sp^2-hybrid orbital having 44% 2*s* character.

Mishra, Symons, et al. [83] measured the anisotropic components in the NF_3^+ and found $A_\parallel(^{14}N) = 322$ MHz and $A_\perp(^{14}N) = 252$ MHz, from which $B_\pi(^{14}N) =$

23 MHz. We assume the unit positive charge to be on the N and find $\rho_{2p_\pi}^{N}$ = 23/(1 + 0.3) = 0.17. The normalization, $\rho_{2s}^{N} + \rho_{2p}^{N} + 3\rho_{\pi}^{F} = 0.18 + 0.17 + 3\rho_{\pi}^{F} = 1$, indicates that $\rho_{\pi}^{F} = 0.22$. Mishra, Symons, et al. [83] also give $A_{\parallel} = 841$ MHz and $A_{\perp} = 224$ MHz for ^{19}F, but these yield $\rho_{\pi}^{F} = 0.28$, which gives the total unpaired electron density as 1.19, a bit too high. If we accept the values for ^{14}N as correct, the indicated 2*s* character of the odd electron on the N^+ is a_s^2 = (0.179/0.35) = 0.51. This indicates that the N^+-bonding orbitals have only (0.49/3) = 0.16 2*s* character and that the bond angles of NF_3^+ are approximately 101°.

5.b. The Anion Halides: CF_3Cl^-, CF_3Br^-, and CF_3I^-

The esr spectra of trapped radicals by electron attachment to CF_3Cl, CF_3Br, and CF_3I have been measured by Hasegawa and Williams [84,85]. The anion radicals were produced by γ-irradiation of dilute concentrations (⩽5%) of the neutral CF_3X molecules in solid tetramethylsilane and other matrices at 77°K. The observed spectra showed that the esr parameters, both **g** and the nuclear coupling tensors, are axially symmetric. Figure 8.19 shows the first-derivative esr pattern they observed for CF_3Cl^- in the tetramethylsilane matrix at 101°K. Below the observed pattern are bar graphs of the composite hyperfine patterns

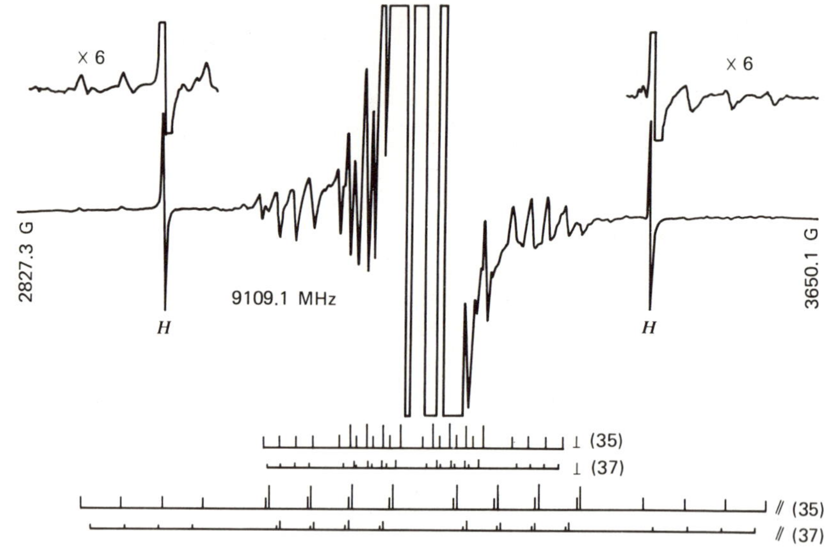

Fig. 8.19 First-derivative esr spectrum of a γ-irradiated solid solution of 5 mole % CF_3Cl in $(CH_3)_4Si$, observed at 101°K. The stick plots represent the calculated positions of perpendicular (upper) and parallel (lower) components for $CF_3{}^{35}Cl^-$ and $CF_3{}^{37}Cl^-$. From Hasegawa and Williams [84].

for the three equivalently coupling ^{19}F nuclei and the weighted ^{35}Cl and ^{37}Cl patterns for the parallel and perpendicular orientations of the magnetic symmetry axis in the applied field. The selected esr parameters that produce the best fitting of the observed spectra are listed in Table 8.18.

Table 8.18 ESR Parameters for CF_3X^- Radicals in a $(CH_3)_4Si$ Matrix[a]

Radical T	$g_{\parallel}$	$g_{\perp}$	Coupling Nucleus	Hyperfine Couplings (MHz) $A_{\parallel}$	$A_{\perp}$
CF_3Cl^- 101°K	2.0021	2.0070	^{35}Cl	121	49.6
			^{37}Cl	101	41.2
			^{19}F	541	245
CF_3Br^- 121°K	2.0036	2.0212	^{79}Br	685	297
			^{81}Br	739	320
			^{19}F	494	222
CF_3I^- 98°K	2.0002	2.0483	^{127}I	1046	501
			^{19}F	405	188

[a] From Hasegawa and Williams [84].

Hasegawa and Williams interpreted the axially symmetric esr parameters that they observed as indicating that the unpaired electron has an $a_1(\sigma^*)$ antibonding orbital composed largely of the axial p orbital of the carbon and the X halogen. However, the ^{13}C coupling was not measured. Although the anisotropic principal coupling values $A_{\parallel}$ and $A_{\perp}$ for ^{19}F were measured, the $2p$-orbital spin densities on the F atoms were not derived, presumably because of uncertainties about orientations of those orbitals. Since less than a third of the unpaired electron density is on the X, the proposed model is incompletely described. Obviously, a measurement of the ^{13}C coupling is highly desirable to help complete the description. In the absence of data for the ^{13}C coupling, we have attempted an interpretation of the ^{19}F couplings to gain more information about the distribution of electron spins and to obtain an indirect evaluation of the spin density on the C.

At the temperature of these observations (~100°K), it seems highly probable that the CF_3 group would undergo rapid reorientation or restricted rotation about the C_{3v} symmetry axis, which would equalize the coupling of the three fluorines and cause them to have the observed axial symmetry. The CF_3 radicals were found to undergo rapid reorientation about the symmetry axis in the trifluoroacetamide matrix at 77°K [67]. If the spin density on the fluorines were in $2p$ orbitals directed along the symmetry axis, they may, of course, have axial symmetry and may be equal without such rotation, but it appears that rotational averaging is a more likely cause of the symmetry. A possible assumption is that the electron density on each F is in a $2p$ orbital directed along the FC bond, but this does not appear likely; an unreasonably high $2p_\sigma$ spin density of 0.20 on each F, or 0.60 on the three fluorines, would be required to give the averaged ^{19}F coupling observed along the symmetry axis for CF_3Cl^-. Because of the high electronegativity of F, this distribution seems unlikely. A more probable model is one in which the spin density on the fluorines is in a p_π-type orbital in the XCF plane and perpendicular to the CF bond. This axially symmetric orbital would be approximately 20° from the molecular symmetry axis, and the averaged value of its anisotropic coupling along the molecular symmetry axis could be obtained by rotation of the axially symmetric tensor component with the relation

$$B_{\text{obs}} = \frac{B_\pi(3\cos^2\theta - 1}{2} \tag{8.50}$$

when $\theta = 20°$. This transformation gives $B_\pi = 1.21\, B^{\text{F}}_{\text{obs}}$. Spin densities of these p_π orbitals for the fluorines as determined by B_π/B_{atomic} are given in Table 8.19.

Table 8.19 Coupling Constants and Spin Densities for CF_3X^- Radicals

Radical Ion		*Coupling Constants (MHz)*[a]				*Spin Densities*[b]	
		A_f	B_{obs}	B_σ	B_π	ρ_s	ρ_p
CF_3Cl^-	^{35}Cl	73.3	24	24		0.016	0.167
	^{19}F	344	98.5		119	0.007	0.079
	C	–	–		–	0.145	0.435[c]
CF_3Br^-	^{81}Br	460	140	140		0.020	0.201
	^{19}F	313	90.5		110	0.007	0.072
	C	–	–		–	0.141	0.422[c]
CF_3I^-	^{127}I	683	182	182		0.033	0.285
	^{19}F	260	72.5		88	0.006	0.058
	C	–	–	–	–	0.127	0.381[c]

[a] Values of A_f and B_{obs} are derived from $A_\parallel$ and $A_\perp$ values in Table 8.18, and B^{F}_π is obtained from B_{obs}, as explained in the text.

[b] For Cl, Br, and I, the spin densities are those given by Hasegawa and Willaims [84]. For *F*, $\rho_s = (A_f/A_{\text{atomic}})$, $\rho_p = (B_\pi/B_{\text{atomic}})$.

[c] These ρ_s and ρ_p values for C are obtained from normalization of the spin densities, as explained in the text.

The unpaired spin density on ρ^C is calculated from the normalization

$$\rho^C = 1 - (\rho_s^X + \rho_p^X) - 3\rho_p^F \tag{8.51}$$

with the assumption that the small ρ_s^F results from configuration interaction. It is assumed that the valence orbitals of C are sp^3 hybrids and that $\rho_s^C = (1/4)\rho^C$ and $\rho_p^C = (3/4)\rho^C$. The resulting values are given in Table 8.19.

If the preceding analysis is correct, the spin density in the antibonding σ^* orbital of the CX bond in the CF_3X^- radicals undergoes a σ-π exchange interaction with an electron pair on the F. The spin density of the odd electron thus spreads over all atoms of the ion molecules, with the amount in reverse order to the electronegativity. Approximately half the unpaired electron density is on the central carbon.

Radicals formed by electron attachment to more complex fluorocarbons have also been studied by Hasegawa, Shiotani, and Williams [85].

6. IRRADIATED MOLECULAR POWDERS

Free radicals observable with esr can be produced by exposure of almost any powder or polycrystalline solid to ionizing radiation. Survey studies of the esr produced by irradiation of many classes of organic and inorganic substances have been made. Although many incorrect assignments were made, particularly in the initial period of esr spectroscopy of irradiated substances, the knowledge gained from these broad surveys of irradiated powders is considerable. The feedback of the accumulating information and the elimination of misinformation continuously improved the ability of the esr spectroscopist to assign the observed esr patterns to the proper radicals and to analyze them correctly. Such studies assisted in the selection of compounds that would be suitable for single-crystal studies. The more definite information obtainable from the study of single crystals was, in turn, an aid to analysis of powder spectra. Likewise, these analyses have been enhanced by the precise measurements of isotropic nuclear coupling components and averaged g values now being made on free radicals in liquid solutions (Chapter IX). The introduction of computer simulation of powder spectra was a great advance in the techniques of their analysis (see Section 2). Because of the many substances that cannot, for practical reasons, be studied as single crystals or in the liquid state, these improvements in spectroscopy of powders are highly desirable.

The literature on the esr of irradiated powders encompasses essentially all classes of organic compounds, including the biologically significant amino acids, purines and pyramidines, hormones, and vitamins. We cannot give a survey of these studies here, nor do we attempt to make a critical analysis of selected spectra. Many of the simple radicals found have been reexamined in single crystals isolated in inert matrices or in liquids and are described in other sections. The remaining sections of this chapter are devoted to irradiated polymers and to solid-state reactions that cannot be investigated in single crystals.

7. SYNTHETIC POLYMERS

Radiation effects on polymers that compose plastics and synthetic fibers are of considerable practical interest. Electron-spin resonance of the radiation-induced free radicals provides one of the most effective means of studying these effects. For most of these polymers it is not possible to obtain samples in which the induced free radicals are preferentially oriented, as are free radicals in single crystals. Some success has been achieved, nevertheless, in observation with partially oriented radicals in certain polymers, including Teflon and polyethylene. Fortunately, it has proved possible in a number of important polymers to find the chemical forms of the stable free-radical groups and to learn much about the nature of the radiation damage from a study of randomly oriented radicals.

In addition to radiation studies, much use has been made of esr for observations of propagating radicals in the polymerization process, for study of oxidation, cross-linking, mechanical molecular fractures, deformations, and thermal degradations of polymers. These studies are described in the comprehensive source book, *ESR Spectroscopy in Polymer Research,* by Rånby and Rabek [86]. It contains more than two thousand references to esr studies of polymers, with summaries of the information gained for the more important ones. Here, we can give only a few examples of these extensive investigations.

7.a. Polyethylene and Polyfluoroethylene

The esr studies of Voevodsky and his associates [87] showed that the principal free radical formed in polyethylene by irradiation at 77°K is

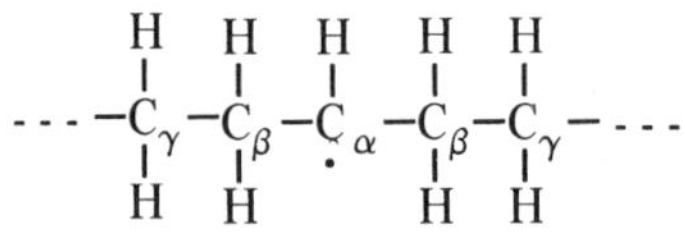

11

which is produced by the breakage of a C-H bond. The **g** anisotropy of this radical is small. The couplings in the α and the β protons are so nearly equal that they cannot be distinguished, and the coupling to the γ protons is too small to produce observable splitting. Consequently, the observed spectrum reproduced in Fig. 8.20 appears to be a sextet pattern of five equally coupling protons. This free radical has been confirmed by Charlesby and Omerod [88], who produced

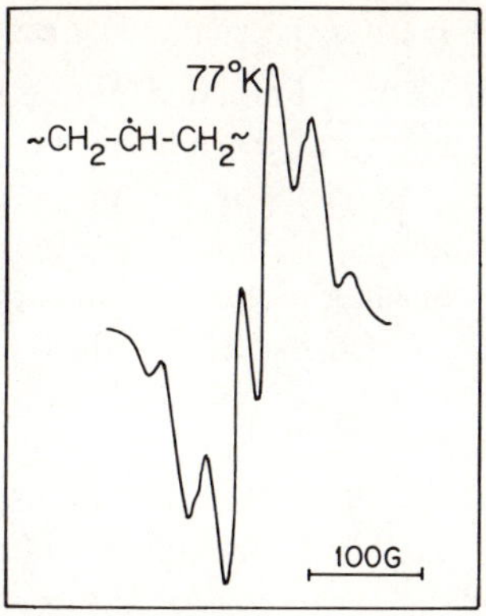

Fig. 8.20 An esr spectrum of irradiated polyethylene powder at 77°K. From Molin, Koritzky, Bouben, and Voevodsky [87].

orientation effects with stretched samples, and by Lawton, Balwit, and Powell [89], who used oriented crystalline samples. The radical is not stable at room temperature. When the sample is warmed, the pattern changes to that of some other radical, or radicals, more difficult to identify. The predominant radical at room temperature may, however, be the allyl type:

$$\cdots -\underset{H}{\overset{H}{\underset{|}{C}}}-\underset{\cdot}{\overset{H}{\underset{|}{C}}}-\overset{H}{\underset{|}{C}}=\overset{H}{\underset{|}{C}}-\underset{H}{\overset{H}{\underset{|}{C}}}- \cdots$$

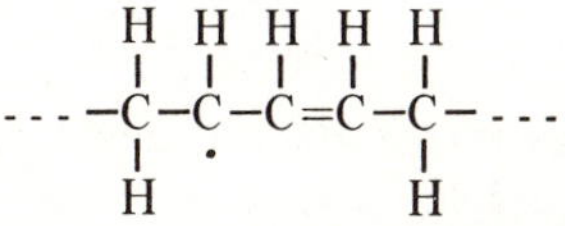

12

Ohnishi, Sugimoto, et al. [90], who studied the temperature dependence of the esr of stretched polyethylene, have obtained evidence for the allyl radical. A thorough study of the physical condition of the sample and the environmental effects has been made by Lawton, Balwit, and Powell [89]. A theoretical treatment of the nuclear couplings in radicals of these types is given by Morokuma and Fukui [91].

It was shown by Rexroad and Gordy [92] that the principal free radical found on irradiation of polytetrafluoroethylene is

$$\cdots -\underset{F}{\overset{F}{\underset{|}{C}}}-\underset{F}{\overset{F}{\underset{|}{C}}}-\underset{\cdot}{\overset{F}{\underset{|}{C}}}-\underset{F}{\overset{F}{\underset{|}{C}}}-\underset{F}{\overset{F}{\underset{|}{C}}}- \cdots$$

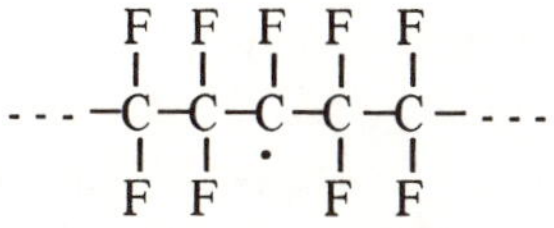

13

formed by the loss of an F from the chain. The esr of the propagating radical RCF$_2$ĊF$_2$ has been observed in γ-irradiated polytetrafluoroethylene by Siegel and Hedgpeth [93]. Comparison of the relative intensities of the esr signals of the two radicals indicates the yield of the propagating radicals to be about 1/10 that of the chain radicals. Despite the large anisotropy in the fluorine coupling, the torsional or wiggling motions of the chain reduce the anisotropies in both of these radicals so much that their ^{19}F hyperfine structure can be resolved. The effects of thermal motions on the esr of the radicals have been studied by Toriyama and Iwasaki [94]. The four β-fluorines of the chain radicals have equivalent couplings of 33 G (92 MHz), and the αF has a coupling of 92 G (258 MHz). Ovenall [95] made experiments on ordered samples of Teflon and observed orientation effects of the resonance of these radicals.

Molecular oxygen combines with the chain radicals to form the relatively stable peroxide radical [92,96]:

$$\begin{array}{c} \text{F F F F F} \\ \text{| | | | |} \\ \cdots \text{–C–C–C–C–C–} \cdots \\ \text{| | | | |} \\ \text{F F O F F} \\ \text{|} \\ \text{O} \end{array}$$

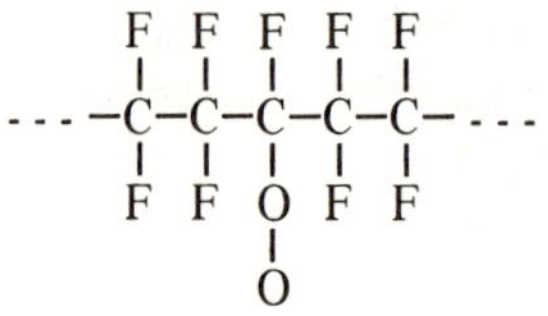

14

in which the spin density, concentrated on the O$_2$, does not interact with the F nuclei sufficiently to produce a hyperfine structure. The corresponding peroxide radical RCF$_2$CF$_2$OO is formed by the addition of O$_2$ to the propagating radicals [93]. The ^{17}O hyperfine structure for both of these peroxide radicals has been measured by Che and Tench [97], who exposed the irradiated Teflon to molecular oxygen with the ^{17}O isotope concentrated. At 77°K, the observed ^{17}O hyperfine structure of both peroxide radicals indicates that 99% of the unpaired electron-spin density is in a $2p_\pi$ orbital on the oxygens with 69% of this density on the terminal oxygen. Similar polymer peroxide radicals have been observed in irradiated polyethylene, polypropylene, and many other irradiated polymers [86].

7.b. Polystyrene: Interaction with Hydrogen Atoms

Wall and Ingalls [98] have shown with esr that at room temperature gaseous H or D atoms diffuse into finely divided polystyrene and add directly to its phenyl rings, forming radicals of cyclohexadienyl type. No evidence was found for an H-abstraction reaction from the polystyrene, but additional H atoms were found to abstract the extra hydrogens from the radicals and thus to restore the phenyl rings. These investigators estimate the rate of reaction of the H atoms with the cyclohexadienyl radicals to be 4×10^3 liters/(mole)(sec) at room temperature.

8. POLYAMINO ACIDS AND PROTEINS

8.a. Radiation Effects

Despite the complexity of proteins, their radiation-induced free radicals have relatively simple esr patterns when observed at room temperature [99-101]. The most common patterns can be assigned either to a >C-H fragment or to a radical in which the spin density is concentrated on sulfur. The sulfur pattern is preferentially formed when proteins having a relatively high cystine or cysteine content are irradiated at room temperature. When the proteins are irradiated and observed at 77°K, several other kinds of esr centers are observed [102-103]. When those containing sulfur are irradiated at 77°K and warmed to 300°K, the characteristic room-temperature, sulfur-centered radicals are formed [102, 103]. It has been possible to show that the >C-H fragment is formed on the polypeptide backbone in silk [100,101], although it is not possible to be sure that the >C-H fragment is not also produced in a side-chain group of some proteins. Investigations of irradiated single crystals of amino acids and simple peptides have helped greatly in the interpretation of the esr of irradiated proteins (see Chapters VI and VII). It appears that the two most frequently observed radicals in irradiated proteins at room temperature have the forms

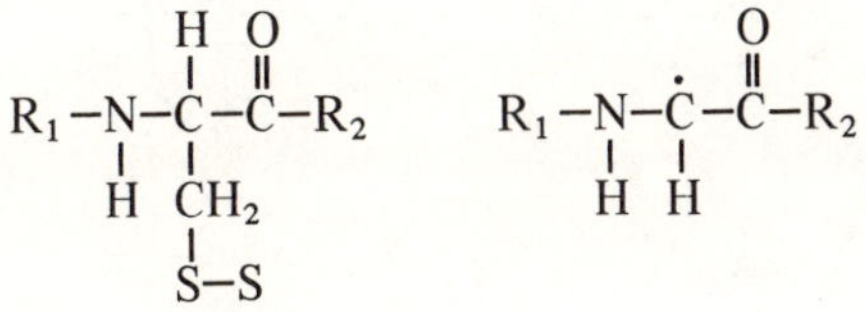

$$
\begin{array}{cc}
\underset{15}{\text{R}_1-\text{N}-\overset{\text{H}}{\underset{\underset{\underset{\text{S}-\text{S}}{|}}{\text{CH}_2}}{\overset{|}{\text{C}}}}-\overset{\text{O}}{\overset{\|}{\text{C}}}-\text{R}_2} &
\underset{16}{\text{R}_1-\text{N}-\overset{\cdot}{\underset{\text{H}}{\overset{|}{\text{C}}}}-\overset{\text{O}}{\overset{\|}{\text{C}}}-\text{R}_2}
\end{array}
$$

The unpaired electron of the disulfide radical 15 is mostly in an antibonding π^* orbital between the two sulfurs (Chapter VI, Section 10.b).

Studies have been made of the esr patterns induced by γ-irradiation in 19 polyamino acids [104,105]. The doublet characteristic of the >C-H fragment was observed for a few, but most gave patterns unlike those observed in similarly irradiated proteins. This evidently results from the fact that they did not contain the glycine constituent nor the sulfur amino acid constituent. Figure 8.21 shows the spectra for irradiated polyglycine and polyalanine. These patterns clearly must come from the polymeric radicals of structures 17 and 18, respectively.

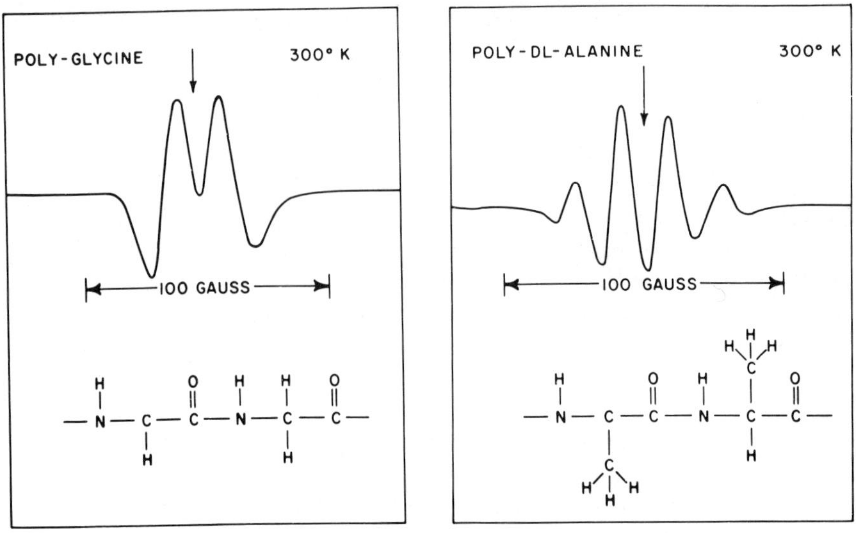

Fig. 8.21 Second-derivative esr spectra of polyglycine and polyalanine, γ-irradiated and observed in a vacuum at room temperature. The observed radicals are formed by loss of an H atom from the polypeptide chain, as indicated by the formulas. From Drew and Gordy [104].

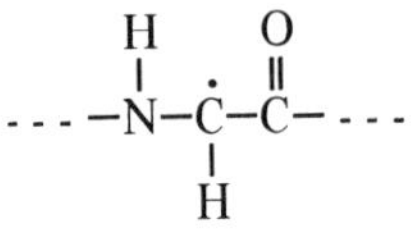

17

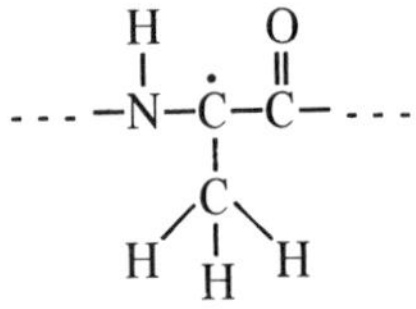

18

No other radicals which could reasonably be expected to be produced by irradiation of these polymers could give the observed patterns. Each is formed by the loss of an H atom from the C_α of the peptide chain. The doublet for **17** arises from the α coupling of the >C-H proton; the quartet for the polyalanine radical comes from the coupling of the three β protons of the CH_3 group that are equalized by rotation about the CC bond. The anisotropic components of the $C_\alpha H$ proton coupling are mostly canceled by the torsional motions of the polyglycine so that the doublet splitting is approximately equivalent to A_f. From the doublet splitting of 48 MHz with Q_α = 73.4 MHz (see Chapter VI, Section 2.a), the π spin density on the C_α of the polyglycine radical, **17**, is estimated to be $\rho_\alpha = 0.65$. The methyl proton coupling of 50 MHz with Q_β^{rot} = 82 MHz (see Chapter VI, Section 2.b) indicates $\rho_\alpha = 0.61$ for the π spin density on the C_α in the polyalanine radical.

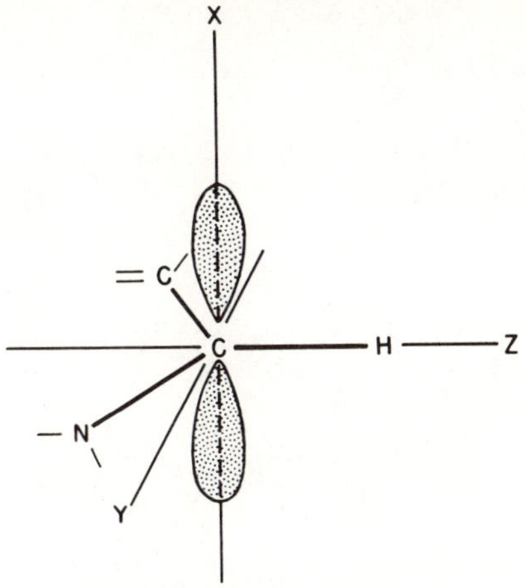

Fig. 8.22 Diagram of the >CH fragment radical in the polypeptide chain. From Gordy and Shields [101].

Gordy and Shields [101] have observed orientation dependence of the C-H resonance in a number of irradiated native proteins. For oriented strands of silk, it was found possible to analyze the data in terms of the structure of silk and to prove that the observed doublet arises from a >C-H fragment with its C-H bond oriented approximately normal to the silk strands. Figure 8.22 shows the directions of the principal axes of the proton coupling expected for the >C-H in the polypeptide chain. The X axis is along the carbon p orbital, Z is along the C-H bond, and Y is in the plane of the radical and perpendicular to the C-H bond. From the known structure of silk [106], it can be concluded that the Y axis is approximately parallel to the silk strands and that X and Z are perpendicular to the oriented strands. Thus measurements with the strands parallel to the magnetic field should give directly the principal elements A_y in the proton coupling; because the $C_\alpha H$ bonds are randomly oriented in the perpendicular x-z plane, measurements with the strands perpendicular to **H** should give an average of A_x and A_z. The principal elements of the >C-H fragment are known, from theory and from measurements on single crystals, to have the same sign (negative). Thus the separation of the components for the perpendicular orientation corresponds to $A_\perp = (1/2)(A_x + A_z)$, and the extra line width for this orientation over that for the parallel orientation gives an approximate measure of $(A_x - A_z)$. The values for the principal elements obtained from a fitting of the theoretical with experimental curves of Fig. 8.23 for oriented silk strands are [101]:

A_x = −50 MHz (⊥ to NCHC plane)
A_y = −73 MHz (⊥ to C-H bond and in NCHC plane)
A_z = −21 MHz (along the C-H bond)

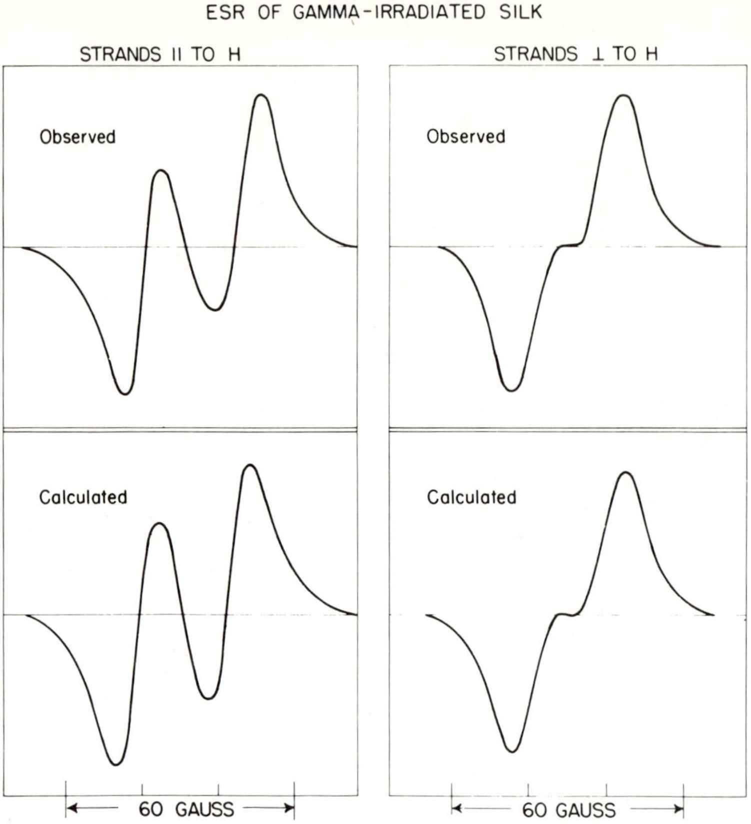

Fig. 8.23 Observed and theoretical first-derivative esr curves for the >CH fragment radical on the polypeptide chain of silk, for the parallel and perpendicular orientations of the strands in the magnetic field. From Gordy and Shields [101].

These values compare favorably with the values of the principal elements for the >C-H fragment in irradiated acetylglycine: $A_x = -48$ MHz, $A_y = -76$ MHz, and $A_z = -28$ MHz, as obtained by measurement of single crystals (see Table 6.1). Since all these principal elements have the same sign, the isotropic component of the protein coupling in the silk is $A_f = -48$ MHz. This gives for the π-spin density on the C_α, $\rho_\alpha = 0.65$, the same as that for polyglycine.

In a coiled protein chain such as the α helix [107] the >C-H fragments formed on the polypeptide backbone would have the Y axis tilted away from the axis of the coil. Although the problem of finding the principal elements of the $C_\alpha H$ proton coupling for such proteins is considerably more complicated than those for silk, the quantity $(A_\parallel - A_\perp)$ gives an indication of the orientation of Y, or the C-H bond, with the structural axis. It is evident that the difference in the doublet splitting for $H_{\parallel}$ and $H_{\perp}$, or $(A_{\parallel} - A_{\perp})$, will be greatest when the Y axis of Fig. 8.22 is along the axis of the fiber, as in silk; when the Y axis is tilted away from the figure axis, as in the α helix, the difference between $A_{\parallel}$ and $A_{\perp}$ becomes smaller. Table 8.20 provides a comparison of the $A_{\parallel}$ and $A_{\perp}$ values

Table 8.20 Orientation Dependence of the ESR Doublet in Some Irradiated Native Proteins[a]

	Observed Doublet Splitting (MHz)	
Specimen	*Specimen Axis* ∥ *to* **H**	*Specimen Axis* ⊥ *to* **H**
Silk	72.9	36.4
Chicken-feather quill	56.0	46.2
Rat-tail tendon	51.0	45.4
Fish-fin bone	50.4	44.3
Bird-leg bone	50.0	43.7

[a]From Gordy and Shields [101].

for silk with those for feather quill and some specimens in which the protein giving the resonance is collagen. If the >C-H fragment is assumed to be on the polypeptide backbone, these values indicate that the Y axis, which gives a measure of the pitch of the helix, is more than 30° off the axis of the helix in feather quill and in these collagen samples.

About 34% of the residues in collagen is glycine; about 10% is alanine. It is interesting that the observed pattern for irradiated collagen [101] appears to be a superposition of the polyglycine pattern and the polyalanine pattern, in apprxoimately the same proportions.

The cystine-like resonance observed in the proteins has a large anisotropy in $\mathbf{g}$ and therefore should be a sensitive indicator of the orientation of sulfur groups relative to the direction of the protein polymer strands. Nevertheless, the sulfur resonance has been examined for a number of oriented protein specimens—including hair, chicken-feather quill, and porcupine quill—and has been found to give the sulfur pattern characteristic of powdered samples [101]. Evidently, the sulfur group on which the spin density is concentrated has multiple orientations in these proteins. In contrast, one of these specimens (feather quill) has a superimposed >C-H doublet resonance that was found to be orientation dependent. The magnitude of the principal elements in $\mathbf{g}$ in the sulfur radicals in the proteins may be found by the methods described in Section 1. These values agree well with those found for the sulfur radicals in irradiated amino acid crystals at room temperature (Chapter VII, Section 6).

Molecular oxygen reacts directly with the >C-H fragment in certain irradiated proteins [108], either killing the resonance or converting it to a singlet characteristic of the peroxide radical R-O-O. Small quantities of certain compounds, especially the sulfur-containing chemical protectors when dissolved into the protein via H_2O solution and then dried, are found to influence drastically the esr patterns formed in the protein by ionizing radiations [109,110].

Yields of free radicals per 100 eV of radiation dosage, or G values, have been measured for a number of proteins and amino acids by Henriksen, Sanner, and Pihl [103] and Müller (111). Descriptions of other radiation studies of proteins with esr are found in a review by Zimmer and Müller [112] and one by Keighley [113].

8.b. Interactions with Hydrogen Atoms

Free radicals produced by exposure of the polyamino acids and certain proteins to gaseous hydrogen atoms have been observed with esr spectroscopy [114, 115]. For the aliphatic polyamino acids [115], the observed free radicals were generally similar to those produced by ionizing radiations. Most of them could be interpreted as resulting from H abstraction from the CH on the polypeptide chain. For example, a doublet was produced in polyglycine like that shown in Fig. 8.20. However, the spectra in polyalanine indicated breakage of the CN bond of the polypeptide chain, probably by H addition to an NH to form -$CHNH_2$ and the radical $\dot{C}HCH_3CO$-.

Free radicals formed by H addition to the ringed side groups of L-polyphenylalanine, poly-L-tyrosine, and poly-DL-tryptophan were observed, in addition to those produced by H abstraction from other groups [114]. Similar H-addition radicals were observed for the constituent monomers. Figure 8.24 shows the esr spectra of powdered samples of the polymers after exposure to gaseous H atoms at 77°K. The stick diagrams indicate the esr multiplets expected for the H-addition radicals indicated in Fig. 8.24. Similar patterns were also obtained by exposure of the respective monomers to H atoms as well as by γ-irradiation of the polymers.

The hyperfine structure for the H-addition phenylalanine radicals consists of a wide triplet, with splitting of 43 G, from the CH_2 of the ring and a superimposed quartet due to three equivalent $C_\alpha H$ hydrogens with splittings of 11 G each. The only way for the three approximately equivalent $C_\alpha H$ groups to occur is through H-addition on the meta carbon, as indicated. It may be shown from either valence-bond theory or Hückel molecular-orbital theory that H addition on the ortho or the para carbons on the phenylalanine ring would give equivalent splitting by two $C_\alpha H$ groups and a triplet substructure, in disagreement with the observed pattern. However, H addition on either the ortho or the meta carbons of the tyrosine ring should lead to two approximately equivalent $C_\alpha H$ groups and a triplet substructure, in agreement with observations. The meta position is chosen as the more likely for tyrosine because it is predicted to have the higher free valency. To give rise to the observed hyperfine structure for the poly-DL-

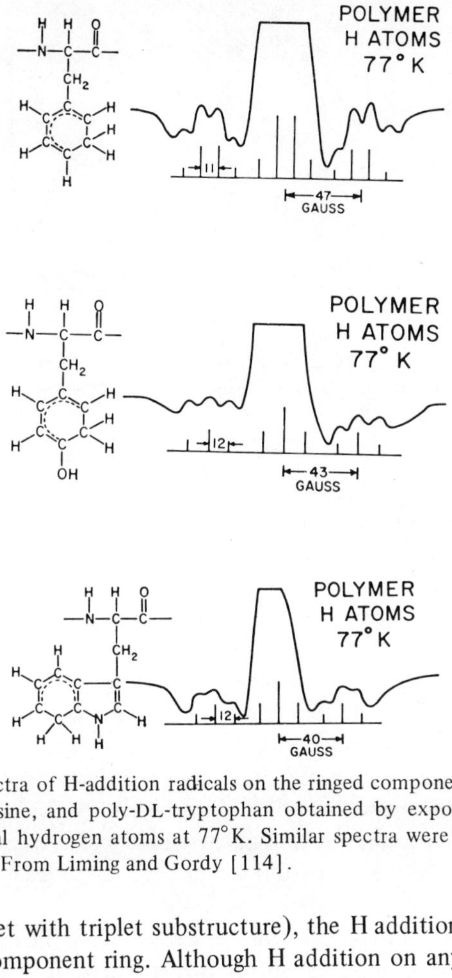

Fig. 8.24 ESR spectra of H-addition radicals on the ringed components of poly-L-phenylalamine, poly-L-tyrosine, and poly-DL-tryptophan obtained by exposure of the respective polymers to thermal hydrogen atoms at 77°K. Similar spectra were obtained by γ-irradiation of the samples. From Liming and Gordy [114].

tryptophan (triplet with triplet substructure), the H addition must occur on the five-membered component ring. Although H addition on any C-H group on this ring might give the observed pattern, the one indicated was thought to be the most probable from consideration of free valencies and steric factors.

Evidence for H-addition radicals like those just described was also observed for certain proteins—carboxypeptidase, insulin, and papain—after exposure to gaseous H atoms. Carboxypeptidase has 5% phenylalanine, 6% tyrosine, and 2% tryptophan in its constituents, but the observed pattern indicates that the H addition occurs primarily on the phenylalanine ring. Papain has 2% phenylalanine, 7% tyrosine, and 2% tryptophan; the esr pattern indicates that the H addition occurs on the tyrosine or tryptophan rings. Compare the observed spectra in Fig. 8.25 with those in Fig. 8.24.

In contrast to these polymers, the γ-irradiated monomers of L-phenylalanine and L-tyrosine gave no evidence for the H-addition radicals in their esr spectra.

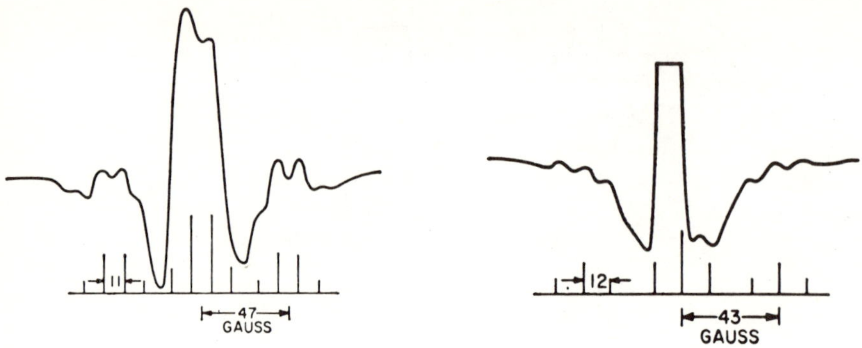

Fig. 8.25 ESR spectra obtained by exposure of powdered proteins to thermal H atoms at 77°K: left, carboxypeptidase; right, papain. From Liming and Gordy [114].

Measurements and detailed analysis of a γ-irradiated single crystal of L-tyrosine (HCl) [116] showed that the principal esr pattern is that of a radical formed by H abstraction from the OH on the ring. The H couplings range from 3 to 14 G, and the principal g values of 2.0023, 2.0045, and 2.0067 are such as to make this hyperfine structure unresolvable in the polymer spectra. This H-abstraction radical could, however, account for much of the strong, unresolved central component in the poly-L-tyrosine or papain spectra.

Thus it appears that the tyrosine ring in the proteins easily gives as well as accepts H atoms, a dual property that could be of significance in biological functions including radiation protection. It is of interest that random copolymers composed of equal numbers of L-tyrosine and L-alanine units, and likewise random copolymers composed of equal numbers of L-tyrosine and glutamic acid units, showed only the esr pattern of the poly-L-tyrosine when exposed to H atoms [114].

In the original paper [114], spin densities derived from the proton coupling of these ringed radicals are compared with those calculated from the Hückel theory and from the McLachlan self-consistent field theory. The latter method gave somewhat better agreement with the experimental values. Because the $C_\alpha H$ couplings have a rather large dipole-dipole component that cannot be evaluated from these patterns, the isotropic coupling components from which the spin densities are derived are not accurately measured. Although these measurements have only qualitative significance, they are of importance in establishing the chemical form of the radicals produced and the nature of radiation effects on the proteins. More reliable tests of the various molecular-orbital theories can be obtained from measurements of irradiated single crystals of the constituent monomers (Chapter VI).

9. THE NUCLEIC ACIDS AND POLYNUCLEOTIDES

The study of radiation effects on DNA and RNA polymers with esr, like that of the proteins, has been accompanied by parallel studies of their constituents [117-137]. Most of the irradiated constituents gave characteristic and distinguishingly different hyperfine patterns that were resolvable, or partly resolvable, even in powdered samples [117]. Most of the relatively pure, dry samples of the nucleic acids themselves, when X- or γ-irradiated at room temperature [118] or at 4.2°K [138], gave a singlet resonance of 30 to 40 G in width, in some instances with evidence of unresolved structure. This signal, which has been observed in a number of laboratories, does not closely resemble the pattern of any of the constituent components. However, in certain types of DNA or with special conditions of chemical or physical treatment, resonance patterns closely resembling those given by basic-ring constituents have been observed. By bombarding DNA samples with heavy dosage (10 Mrad) of high-energy electrons (1 MeV) at

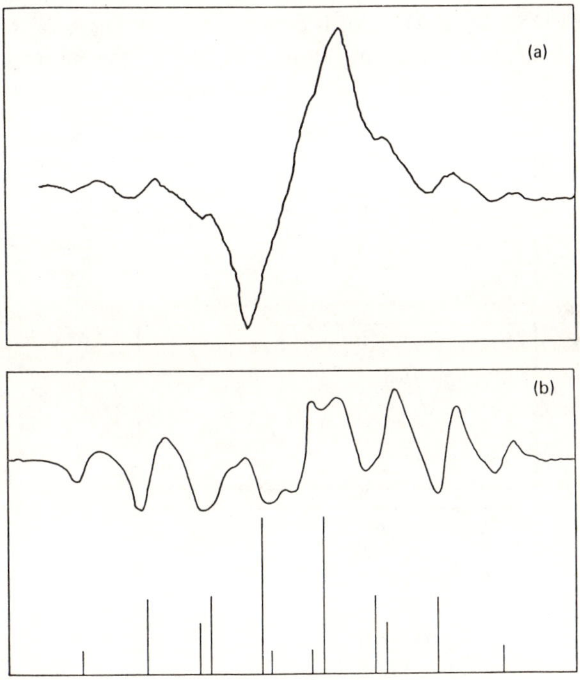

Fig. 8.26 Comparison of the esr signals observed at 77°K in calf-thymus DNA after irradiation with 10 Mrad of 1.0-MeV electrons (a) and with ~10^7 ergs/mm^2 of uv light (b). The stick diagram at the base corresponds to signals of randomly oriented thymine H-addition radicals. From Pershan, Shulman, Wyluda, and Eisinger [123].

200°K and then annealing the samples at temperatures intermediate between 77°K and 300°K, Salovey, Shulman, and Walsh [139] were able to produce an esr signal closely resembling that of irradiated thymine or thymidine. Likewise, Ehrenberg, Ehrenberg, and Löfroth [120] produced a thymidine-like resonance in DNA by special chemical treatment. In an irradiated sample of trout-sperm DNA containing 20% H_2O and 1/10% protein, Dorlet, Van de Vorst, and Bertinchamps [140] detected a broad triplet closely resembling that of the irradiated purine nucleotides. In T2-bacteriophage DNA, Müller observed as one component a doublet that may arise from a >C-H fragment formed in the deoxyribose group [133]. The thymidine-like resonance observed in DNA could be enhanced by irradiation of the sample under hydrogen gas [125], by absorbed H_2O [123], and by irradiation with uv light [123].

The enhancement of the thymidine-type esr of DNA by uv irradiation over ionizing radiation is demonstrated by Fig. 8.26. The bar graph at the base of the lower diagram indicates the esr components for irradiated thymidine powder. The enhancement of the thymidine-type resonance by pretreatment of the sample of calf-thymus DNA by H_2O vapor is demonstrated by Fig. 8.27. Measurements of γ-irradiated single crystals of thymidine [141] and observation of the

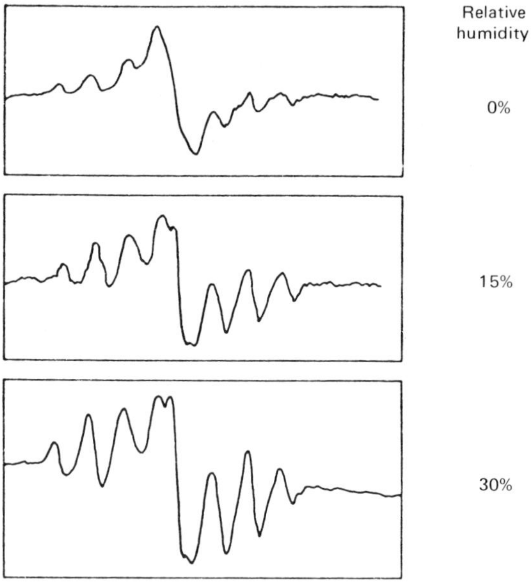

Fig. 8.27 Demonstration of the effects of absorbed H_2O on the production of thymine hydrogen-addition radicals by uv irradiation of calf-thymus DNA. The samples were exposed to moisture at 298°K and were irradiated and observed at 77°K. From Pershan, Shulman, Wylunda, and Eisinger [123].

differences between the esr hyperfine structures of thymine and of thymidine powders produced when they were exposed to gaseous hydrogen and deuterium atoms [127] proved that the esr observed earlier in irradiated powders of thymidine was due to the radical

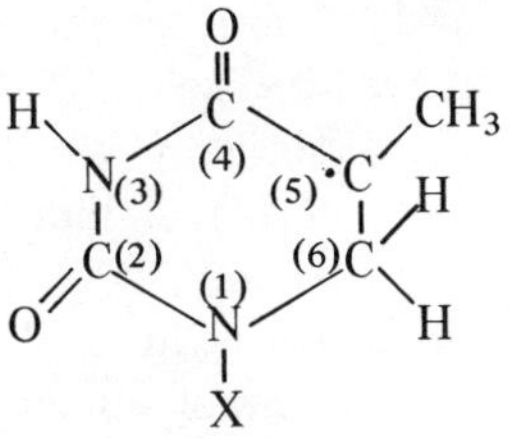

19

formed by direct H addition on C_6 of the thymidine ring. This was also demonstrated by observation of the esr of partially deuterated samples of uv irradiated powders of thymidine and comparison with signals produced from normal

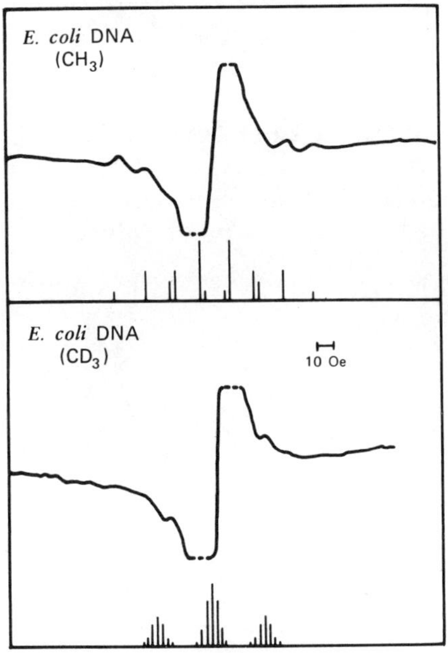

Fig. 8.28 Comparison of esr spectrum of uv irradiated *E. coli* DNA grown with normal thymine (upper) with that grown with CD_3-substituted thymine (lower). The stick diagrams represent the lines expected for the thymine H-addition radicals for the two species. From Pershan, Shulman, Wyluda, and Eisinger [123].

thymidine [123]. Objective proof that the thymidine-like hyperfine pattern observed in irradiated DNA originates from its thymidine constituent was obtained by Pershan, Shulman, et al. [123] from comparison of the esr of uv-irradiated *E coli* DNA that had been grown with normal thymine and that grown with partially deuterated thymine (see Fig. 8.28).

It is now clear that the production of the thymidine-like resonance by irradiation of DNA depends on an available source of H atoms. For DNA, the primary source of the H atoms is the dissociation of adsorbed H_2O by the radiation. Without excessive dosage (~10 Mrad) of ionizing radiation, only the singlet resonance is produced in very dry, pure DNA [139]. It is surprising that the hydrogens are not more readily produced in the *d*-ribose component, but perhaps H atoms released in the *d*-ribose cannot effectively migrate to the $C_{(6)}$ of the thymine ring. Only the singlet esr was detected for samples irradiated and observed at 4.2°K [138]. This is not surprising since H atoms do not easily migrate at that temperature. However, at 4.2°K the singlet was observed to be very strong, which is evidence that it arises from primary radicals induced by the irradiation. Although the nature of these primary radicals is not yet known, their production at 4.2°K suggests that they are ionized species. The fact that this central singlet is not as readily produced by uv as by ionizing radiation is also in accord with the assumption that its production requires ionization of the molecules. However, the enhancement of the thymidine resonance by uv irradiation is not yet completely explained.

Hyperfine components in the esr observed for the purine constituents of the nucleic acids—adenosine, adenylic acid, guanidine, and guanylic acid—can be assigned to radicals formed by the H-addition reactions **20** and **21**.

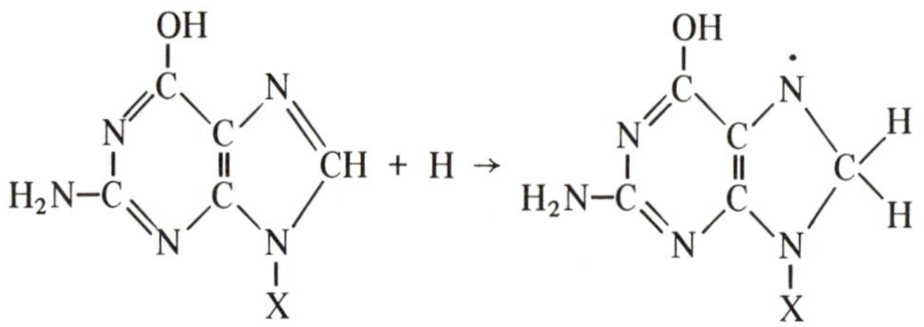

(Guanylytic component)

20

Proof that H-addition reactions occur in both the purine and the pyrimidine bases (cytosine at 77°K) was provided by Herak and Gordy [127], who produced the esr signals expected for such radicals by exposure of powdered samples of the bases to gaseous H and D atoms. That the absorbed H atoms added directly to

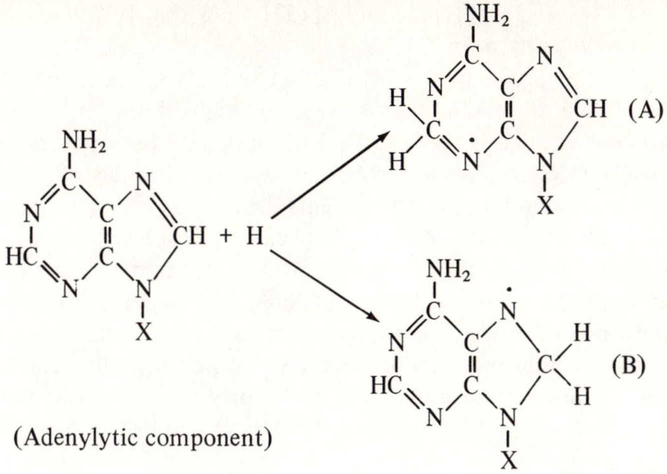

21

the purine and pyrimidine rings was proved by observation of the changes in the hyperfine structure to be expected if D were substituted for H in the bombarding gas. Figure 8.29 shows the characteristic esr for the H- and D-addition radicals on guanine and adenine in the powder form. Although it is not possible from the powder spectra to distinguish between radicals **21**(A) and **21**(B), single-crystal measurements on γ-irradiated deoxyandenosine monohydrate at room temperature indicate that radical **21**(A), H-addition on $C_{(2)}$, is formed in this case [142]. Measurements on an irradiated single crystal of guanine hydrochloride

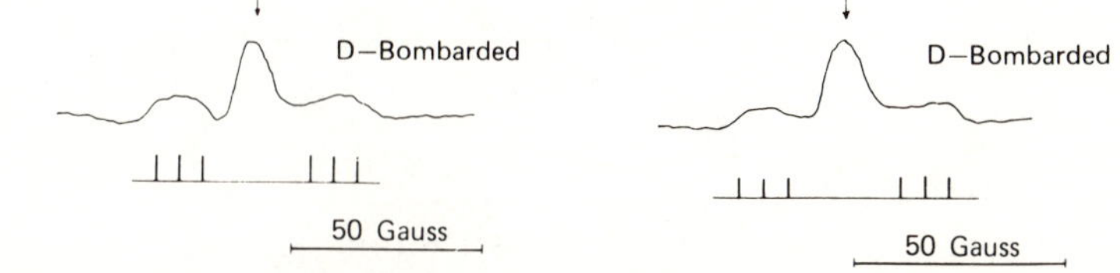

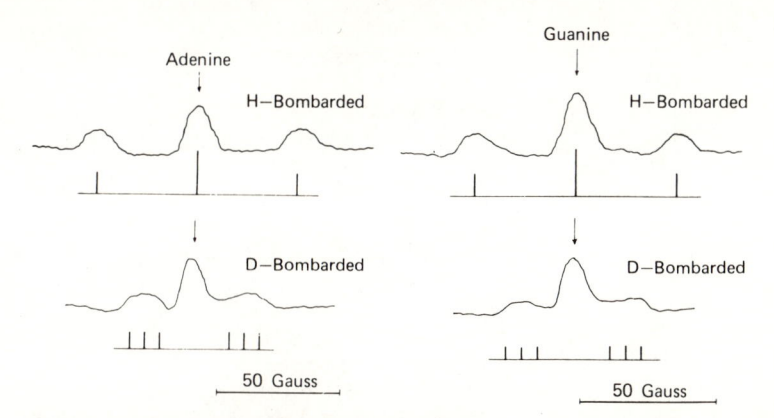

Fig. 8.29 Characteristic esr patterns for H- and D-addition radicals of adenine and guanine in the powdered form. From Herak and Gordy [127].

dihydrate by Alexander and Gordy [143] show that H-addition occurs on $C_{(8)}$ of the guanine ring, as in **2.0**.

Despite the production of H-addition radicals in the purine bases and the nucleosides or nucleotides through H-atom bombardment [144], no signals from corresponding radicals could be similarly produced by exposure of DNA or the polynucleotides to gaseous streams of H atoms. Possibly this is due to steric factors or scavenger action that prevents the absorbed H atoms from reaching the reactive site on the ring. As mentioned earlier, a triplet resonance similar to that expected for a radical formed by H-addition to the purine rings has been observed in an irradiated sample of DNA containing 20% H_2O and 1/10% protein [140], but this triplet may arise from a radical produced in the protein component.

In one experiment, very dry samples of polyadenylic acid γ-irradiated at 300°K or lower temperature did not show the esr signals expected for H-addition radicals [145]; but when the samples that had absorbed small amounts of H_2O were irradiated at 200°K and annealed at 240°K, signals from the H-addition radicals were observed, as demonstrated for polyadenylic acid in Fig. 8.30. Compare this with the adenine spectrum in Fig. 8.29. In Fig. 8.30, there is a superimposed doublet esr believed to arise from an H-abstraction radical in the *d*-ribose component. Evidence that the H atom added to the rings came from H_2O was provided by observation of changes produced when the samples were moistened with D_2O rather than H_2O.

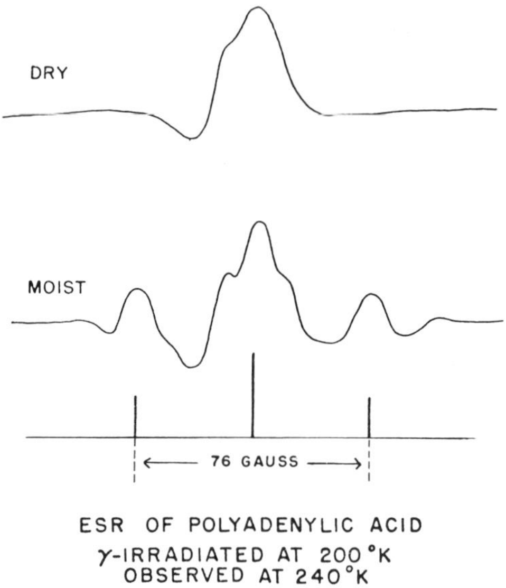

Fig. 8.30 ESR patterns of γ-irradiated samples of dry and of moist polyadenylic acid (cf. with Fig. 8.29). From Herak and Gordy [144].

The chemical composition of RNA differs from that of DNA only in having the base uracil replace thymine and ribose replace deoxyribose. It is interesting that no esr spectrum attributable to an H-addition radical on the uracil ring has, to our knowledge, been reported for irradiated RNA at room temperature, although the H-addition radical 22

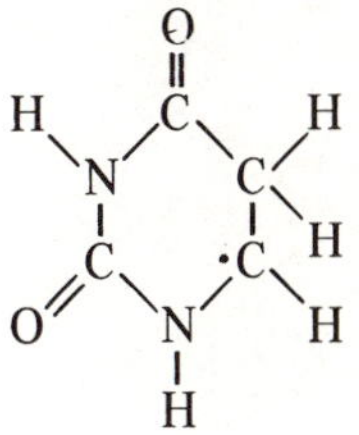

Uracil + H

22

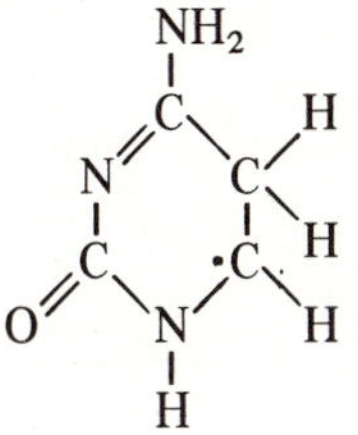

Cytosine + H

23

is rather stable at room temperature. By exposure of powdered RNA to gaseous H atoms at 77°K, Herak and Gordy [146] produced esr signals characteristic of H-addition radicals on both purine and pyrimidine constituents. Figure 8.31 shows the resulting esr observed at 77°K and after the sample was warmed to room temperature. Superimposed on the spectrum of the H-addition radical is a central singlet like that produced by exposure of the RNA to ionizing radiation. The radical source of this singlet is unknown. The triplet component indicated by the solid bar graph is like the triplet produced by exposure of the purine bases, either guanine or adenine, to H atoms (cf. with Fig. 8.29). The dashed bar graphs under the RNA pattern for 77°K in Fig. 8.31 corresponds closely with the hyperfine components of H-addition radicals on cytosine or uracil (Fig. 8.32).

Figure 8.32 shows the esr spectra of the H-addition radicals produced for both uracil and cytosine by subjection of powdered samples to gaseous H atoms at 77°K. The signal for the cytosine radical disappeared before the sample reached room temperature and did not appear on exposure of the sample to H atoms at room temperature. Note that the esr patterns of uracil and cytosine are almost indistinguishable at 77°K. The changes in the uracil pattern when the sample is warmed are completely reversible. They result from a reorientation of the $C_\beta H_2$ group relative to the plane of the ring. Because the pyrimidine component of the RNA spectrum in Fig. 8.31 is unstable at room temperature, it seems probable that this component comes from an H-addition radical on the cytosine ring rather than on the uracil ring. Nevertheless, a strong signal characteristic of the uracil H-addition radical has been produced by γ-irradiation of moist polyuradilic acid at 195°K [145].

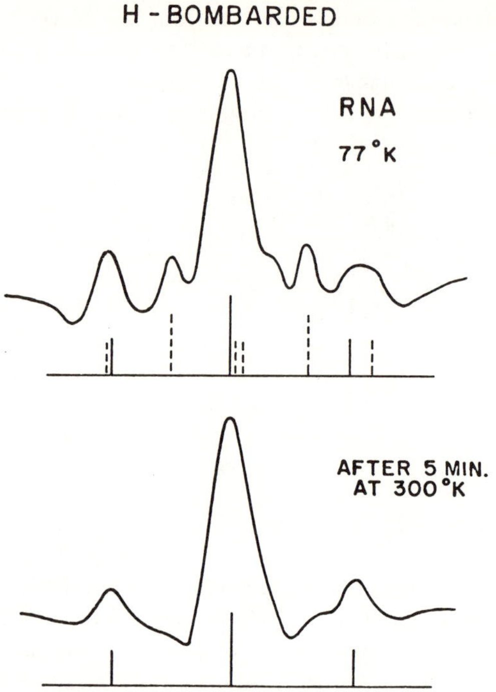

Fig. 8.31 ESR patterns produced by exposure of RNA to thermal H atoms at 77° K (upper) and after warming to 300° K (lower). The upper diagram shows evidence for both purine and pyrimidine H-addition radicals (compared with Figs. 8.29 and 8.32), whereas only the purine component remains at 300° K. From Herak and Gordy [145].

There is evidently more than one mechanism for H addition to purine or pyrimidine rings of the nucleic acids and more than one site for the addition. Certainly the experiments in which the nucleic acid bases were exposed to gaseous deuterium atoms and found to have the characteristic esr hyperfine patterns for D-addition radicals when there was no source of D atoms within the sample has proved that neutral atoms add directly to these bases even at 77°K. However, Omerod [128] found that dry DNA which showed no esr evidence for either H-addition radicals or H atoms when subjected to ionizing radiations at 77°K did give the characteristic esr pattern for the thymidine H-addition radicals when the sample was warmed to 300°K. Therefore, he concluded that in this case the production of the addition radicals consisted of a two-step process in which the thymine ring first captured an electron (at 77°K) and later, as the temperature was raised, captured a proton to form the neutral H-addition radicals. This two-step process may be an important one for formation of H-addition radicals in moist samples or in biological cells where the H^+

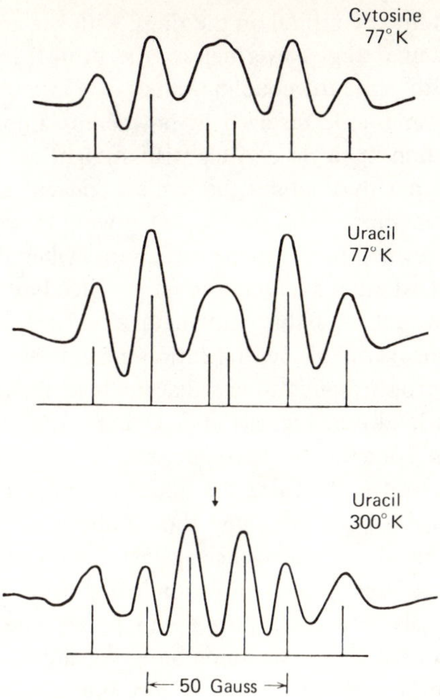

Fig. 8.32 ESR spectra of H-addition radicals of cytosine and uracil produced by exposure of the powdered samples to thermal H atoms at 77°K. Signals of the cytosine radicals could not be detected at 300°K, but those of the uracil radicals remained strong at room temperature. From Herak and Gordy [145].

ions or H_2O molecules are attached to nitrogens on the rings. Evidence for this process has been obtained by other investigators [147-149].

The primary anion radicals, which may be the precursors of the H-addition radicals on the rings of the nucleic acids, have been observed with esr in single crystals irradiated at low temperature. The first such esr study was apparently that of Herak and Galogaza [150] on a single crystal of cytosine monohydrate at 77°K. They were able to make detailed measurements and analysis of the cation radical formed by loss of an electron from the ring. They postulated that the anion was formed by capture of the electron in another ring, presumably in an antibonding π^* orbital, but they could not resolve the hyperfine structure sufficiently to specify the orbital. As the crystal was warmed to room temperature, the esr signals from both anion and cation radicals disappeared, apparently because of mutual neutralization, without formation of the H-addition radicals. Anion radicals in purines as well as pyrimidines were measured later by Box and his associates [151-154] with ENDOR spectroscopy in single crystals irradiated and observed at 4.2°K. Their measurements proved that the captured electron is indeed in an antibonding π^* orbital on the rings, with the greatest concentration on $C_{(6)}$ of the thymine ring. Nevertheless, the proton hyperfine coupling is sufficiently small, with principal coupling values of −55.5 MHz, −29.9 MHz, and −13.0 MHz, that it cannot be resolved in powdered samples of thymidine or DNA, and its resoltuion in single crystals with straight esr spectroscopy would be difficult. In the other pyrimidines the highest concentration of spin density was also on $C_{(6)}$; in adenine it was on $C_{(8)}$. One would certainly expect similar anion radicals to be formed in the nucleic acid rings when they are irradiated at low temperature, but whether such radicals, if produced by ionizing radiation at room temperature, would be stable until neutralized by proton capture is unknown. However, this form of neutralization may well occur when H_2O is attached through hydrogen bonds to an adjacent N of the ring. This would lead to formation of the addition radicals and ^{-}OH. Furthermore, in the Watson-Crick helix of DNA, one might postulate that protons may be transferred across the hydrogen bridges to form the addition radicals on the anion rings. However, the difficulty of observing the addition radicals in dry DNA irradiated at room temperature is evidence against this postulate.

Müller and his associates [155] have concluded from extensive studies of irradiated single crystals containing purine and pyrimidine rings that the two-step process of (1) formation of the anion base ring and (2) capture of a proton to produce the neutral addition radical occurs preferentially on $C_{(6)}$ for the cytosine and uracil rings, whereas the direct addition of H atoms (the one-step process) occurs preferentially on $C_{(5)}$. They designate the two-step process as the "ionization path" and the one-step process as the "excitation path." They conclude that in adenine derivatives the excitation path leads to H addition on $C_{(8)}$ and the ionization path to H addition on $C_{(2)}$. They observed that the $C_{(5)}$ addition radicals of the pyrimidines could be transformed to $C_{(6)}$ addition radicals by optical irradiation and that $C_{(8)}$ addition radicals of the purines could be transformed to $C_{(2)}$ addition radicals by a similar process. In this connection, it is interesting that OH addition on $C_{(5)}$ has been found to occur [156] when uracil powder is exposed to gaseous OH radicals but that the OH replaces the $C_{(5)}$ hydrogen, which, in turn, transfers to $C_{(6)}$ to produce the addition radical **24** (right-hand structure).

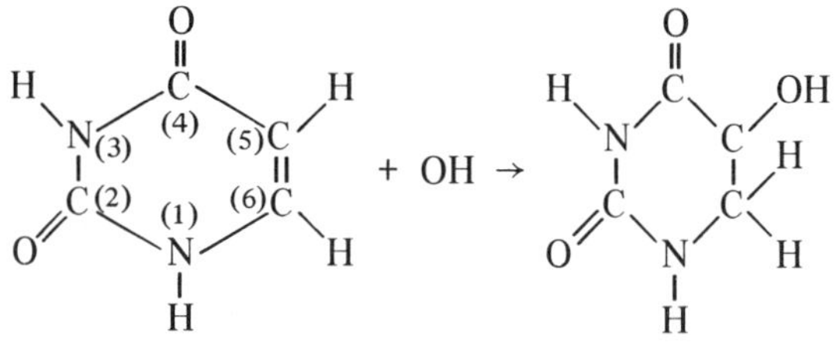

24

In agreement with Müller's results on single crystals, the direct additions to uracil and cytosine from gaseous H atoms were found to occur on $C_{(5)}$ [156].

A review of the esr work on irradiated single crystals of the nucleic acid constituents to 1973 is given by Herak [135]; a 1973 review of the work on short-lived free radicals formed from these constituents in aqueous solutions is provided by Nicolau [136]. Pullman [157,158] has given theoretical discussion of the electronic structure of the constituents and has made predictions of spin densities on various atoms with molecular-orbital theory. Müller [133] gives a comprehensive review of the earlier esr studies (to 1967) of the formation of free radicals in the nucleic acids and their constituents by ionizing radiations. Although considerable information has been gained already, it is clear that much is yet to be learned from esr spectroscopy about the complex genetic molecules, the nucleic acids. As in the past, these investigations will include observations on powders as well as single crystals and liquid solutions and will encompass a wide range of temperatures and other physical conditions.

References

1. B. Bleaney, *Proc. Phys. Soc. (Lond.),* **A63**, 407 (1950); *Phil. Mag.,* **42**, 441 (1951).
2. R. H. Sands, *Phys. Rev.,* **99**, 1222 (1955).
3. L. S. Singer, *J. Chem. Phys.,* **23**, 379 (1955).
4. J. W. Searl, R. C. Smith, and S. J. Wyard, *Proc. Phys. Soc. (Lond.),* **A74**, 491 (1959).
5. S. M. Blinder, *J. Chem. Phys.,* **33**, 748 (1960).
6. F. K. Kneubühl, *J. Chem. Phys.,* **33**, 1074 (1960).
7. J. A. Ibers and J. D. Swalen, *Phys. Rev.,* **127**, 1914 (1962).
8. J. W. Searl, R. C. Smith, and S. J. Wyard, *Proc. Phys. Soc. (Lond.),* **A78**, 1174 (1961).
9. J. D. Swalen and H. M. Gladney, *IBM Journal.,* **8**, 515 (1964).
10. D. L. Griscom, P. C. Taylor, D. A. Ware, and P. J. Bray, *J. Chem. Phys.,* **48**, 5138 (1968).
11. R. Neiman and D. Kivelson, *J. Chem. Phys.,* **35**, 156 (1961).
12. S. Lee and P. J. Bray, *J. Chem. Phys.,* **39**, 2863 (1963).
13. L. D. Rollmann and S. I. Chan, *J. Chem. Phys.,* **50**, 3416 (1969).
14. F.-D. Tsay and H. R. Gray, *J. Chem. Phys.,* **54**, 3760 (1971).
15. R. Lefebvre and J. Maruani, *J. Chem. Phys.,* **42**, 1480 (1965).
16. P. C. Taylor and P. J. Bray, *J. Magn. Reson.,* **2**, 205 (1970).
17. R. Beringer and M. A. Heald, *Phys. Rev.,* **95**, 1474 (1954).
18. R. Livingston, H. Zeldes, and E. H. Taylor, *Phys. Rev.,* **94**, 725 (1954).
19. S. N. Foner, E. L. Cochran, V. A. Bowers, and C. K. Jen, *J. Chem. Phys.,* **32**, 963 (1960).

20. F. J. Adrian, *J. Chem. Phys.,* **32**, 972 (1960).
21. C. K. Jen, V. A. Bowers, E. L. Cochran, and S. N. Foner, *Phys. Rev.,* **126**, 1749 (1962).
22. F. J. Adrian, *Phys. Rev.,* **127**, 837 (1962).
23. G. S. Jackel, W. M. Nelson, and W. Gordy, *Phys. Rev.,* **176**, 453 (1968).
24. J. H. Ammeter and D. C. Schlosnagle, *J. Chem. Phys.,* **59**, 4784 (1973).
25. L. B. Knight and W. Weltner, *J. Chem. Phys.,* **55**, 5066 (1971).
26. A. M. Bass and H. P. Broida, Eds., *Formation and Trapping of Free Radicals,* Academic, New York, 1960.
27. D. Brown, R. Florin, and L. Wall, *J. Phys. Chem.,* **66**, 2606 (1962).
28. R. Florin, D. Brown, and L. Wall, *J. Phys. Chem.,* **66**, 2672 (1962).
29. H. N. Rexroad and W. Gordy, *Phys. Rev.,* **125**, 242 (1962).
30. W. V. Bouldin, R. A. Patten, and W. Gordy, *Phys. Rev. Lett.,* **9**, 98 (1962).
31. W. V. Bouldin and W. Gordy, *Phys. Rev.,* **135**, A806 (1964).
32. C. F. Fischer, "Tabulation of Hartree-Fock Results," thesis, University of British Columbia, January 1968.
33. L. B. Knight, J. M. Brom, and W. Weltner, *J. Chem. Phys.,* **56**, 1152 (1972).
34. R. G. Barnes and W. V. Smith, *Phys. Rev.,* **93**, 95 (1954).
35. S. Aburto, R. Gallardo, R. Munoz, R. Daudel, and R. Lefebvre, *J. Chim. Phys.,* **56**, 563 (1959).
36. P. E. Cade and W. M. Huo, *J. Chem. Phys.,* **45**, 1063 (1966); **47**, 614 (1967).
37. R. F. W. Bader, I. Keaveny, and P. E. Cade, *J. Chem. Phys.,* **47**, 3381 (1967).
38. A. C. Chan and E. K. Davidson, *J. Chem. Phys.,* **49**, 727 (1968).
39. C. F. Bender and E. R. Davidson, *Phys. Rev.,* **183**, 23 (1969).
40. L. B. Knight, W. C. Easley, W. Weltner, and M. Wilson, *J. Chem. Phys.,* **54**, 322 (1971).
41. M. Heyden and H. Kopfermann, *Z. Phys.,* **108**, 232 (1938).
42. N. Comaniciu, V. Drăgănescu, and V. Tatu, *Rev. Roum. Phys.,* **11**, 399 (1966).
43. E. L. Cochran, F. J. Adrian, and V. A. Bowers, *J. Chem. Phys.,* **36**, 1938 (1962).
44. W. C. Easley and W. Weltner, *J. Chem. Phys.,* **52**, 197 (1970).
45. W. Gordy and C. G. McCormick, *J. Am. Chem. Soc.,* **78**, 3243 (1956).
46. T. Cole, H. O. Pritchard, N. R. Davidson, and H. M. McConnell, *Molec. Phys.,* **1**, 406 (1958).
47. R. L. Morehouse, J. J. Christiansen, and W. Gordy, *J. Chem. Phys.,* **45**, 1751 (1966).

48. G. S. Jackel, J. J. Christiansen, and W. Gordy, *J. Chem. Phys.*, **47**, 4274 (1967).

49. G. S. Jackel and W. Gordy, *Phys. Rev.*, **176**, 443 (1968).

50. R. L. Morehouse, J. J. Christiansen, and W. Gordy, *J. Chem. Phys.*, **46**, 1747 (1966).

51. K. Dressler and D. Ramsey, *Phil. Trans. Roy. Soc. (Lond.)*, **A251**, 533 (1959).

52. P. C. Jordan, *J. Chem. Phys.*, **41**, 1442 (1964).

53. S. N. Foner, E. L. Cochran, V. A. Bowers, and C. K. Jen, *Phys. Rev. Lett.*, **1**, 91 (1958).

54. P. H. H. Fischer, S. W. Charles, and C. A. McDowell, *J. Chem. Phys.*, **46**, 2162 (1967).

55. F. J. Adrian, E. L. Cochran, and V. A. Bowers, *Adv. Chem.*, **36**, 50 (1962).

56. L. H. Piette, F. A. Johnson, K. A. Booman, and C. B. Colburn, *J. Chem. Phys.*, **35**, 1481 (1961).

57. H. E. Doorenbos and B. R. Loy, *J. Chem. Phys.*, **39**, 2393 (1963).

58. J. B. Farmer, M. C. L. Gerry, and C. A. McDowell, *Molec. Phys.*, **8**, 257 (1964).

59. C. B. Colburn, R. Ettinger, and F. A. Johnson, *J. Inorg. Chem.*, **3**, 455 (1964).

60. P. H. Kasai and E. B. Whipple, *Molec. Phys.*, **9**, 497 (1965).

61. W. Nelson, G. Jackel, and W. Gordy, *J. Chem. Phys.*, **52**, 4572 (1970).

62. A. J. Colussi, J. R. Morton, K. F. Preston, and R. W. Fessenden, *J. Chem. Phys.*, **61**, 1247 (1974).

63. R. W. Fessenden and R. H. Schuler, *J. Chem. Phys.*, **43**, 2704 (1965).

64. M. Karplus and G. K. Fraenkel, *J. Chem. Phys.*, **35**, 1312 (1961).

65. J. Maruani, C. A. McDowell, H. Hakajima, and R. Raghunathan, *Molec. Phys.*, **14**, 349 (1968).

66. J. Maruani, J. A. R. Coope, and C. A. McDowell, *Molec. Phys.*, **18**, 165 (1970).

67. M. T. Rogers and L. D. Kispert, *J. Chem. Phys.*, **46**, 3193 (1967).

68. O. Edlund, A. Lund, M. Shiotani, J. Sohma, and K.-Å. Thuomas, *Molec. Phys.*, **32**, 49 (1976).

69. W. Gordy and C. G. McCormick, *J. Am. Chem. Soc.*, **78**, 3243 (1956).

70. B. Smaller and M. S. Matheson, *J. Chem. Phys.*, **28**, 1169 (1958).

71. E. L. Cochran, F. J. Adrian, and V. A. Bowers, *J. Chem. Phys.*, **34**, 1161 (1961).

72. P. B. Ayscough and C. Thomson, *Trans. Faraday Soc.*, **58**, 1477 (1962).

73. P. J. Sullivan and W. S. Koski, *J. Am. Chem. Soc.*, **86**, 159 (1964).

74. H. W. Fenrick and J. E. Willard, *J. Am. Chem. Soc.*, **88**, 412 (1966).

75. C. A. McDowell, P. Raghunathan, and K. Shimokoshi, *J. Chem. Phys.*, **58**, 114 (1974).
76. R. W. Fessenden and R. H. Schuler, *J. Chem. Phys.*, **39**, 2147 (1963).
77. M. Barfield, *J. Chem. Phys.*, **53**, 3836 (1970); for corrections, see *J. Chem. Phys.*, **55**, 4682 (1971).
78. E. T. Kaiser and L. Kevan, Eds., *Radical Ions,* Wiley, New York, 1968.
79. R. C. Catton, M. C. R. Symons, and H. W. Wardale, *J. Chem. Soc. A,* 2622 (1969).
80. E. D. Sprague and F. Williams, *Molec. Phys.*, **20**, 375 (1971).
81. T. Cole, *J. Chem. Phys.*, **35**, 1169 (1961).
82. W. Nelson and W. Gordy, *J. Chem. Phys.*, **51**, 4710 (1969).
83. S. P. Mishra, M. C. R. Symons, K. O. Christe, R. D. Wilson, and R. I. Wagner, *Inorg. Chem.*, **14**, 1103 (1975).
84. A. Hasegawa and F. Williams, *Chem. Phys. Lett.*, **45**, 275 (1977).
85. A. Hasegawa, M. Shiotani, and F. Williams, *Faraday Disc. Chem. Soc.*, **63**, 157 (1977).
86. B. Rånby and J. F. Rabek, *ESR Spectroscopy in Polymer Research,* Springer-Verlag, Berlin, 1977.
87. Yu. N. Molin, A. T. Koritzky, N. Ya. Bouben, and V. V. Voevodsky, *Akad. Doklad. Nauk USSR* **124**, 127 (1959).
88. A. Charlesby and M. G. Omerod, "Low Temperature Reactions of Radicals in Irradiated Polymers," *Fifth International Symposium on Free Radicals,* Uppsala, July 6-7, 1961, paper 11-1.
89. E. J. Lawton, J. S. Balwit, and R. S. Powell, *J. Chem. Phys.*, **33**, 395, 405 (1960).
90. S. Ohnishi, S. Sugimoto, and I. Nitta, *J. Chem. Phys.*, **37**, 1283 (1962).
91. K. Morokuma and K. Fukui, *Bull. Chem. Soc. Jap.*, **36**, 534 (1963).
92. H. N. Rexroad and W. Gordy, *J. Chem. Phys.*, **30**, 399 (1959).
93. S. Siegel and H. Hedgpeth, *J. Chem. Phys.*, **46**, 3904 (1967).
94. K. Toriyama and M. Iwasaki, *J. Phys. Chem.*, **73**, 2919 (1969).
95. D. W. Ovenall, *J. Chem. Phys.*, **38**, 2448 (1963).
96. W. B. Ard, H. Shields, and W. Gordy, *J. Chem. Phys.*, **23**, 1727 (1955).
97. M. Che and A. J. Tench, *J. Chem. Phys.*, **64**, 237 (1976).
98. L. A. Wall and R. B. Ingalls, *J. Chem. Phys.*, **41**, 1112 (1964).
99. W. Gordy, W. B. Ard, and H. Shields, *Proc. Nat. Acad. Sci. (USA),* **41**, 983 (1955).
100. W. Gordy and H. Shields, *Proc. Nat. Acad. Sci. (USA),* **46**, 1124 (1960).
101. W. Gordy and H. Shields, *Memoires Acad. Roy. Belgigue,* **33**, 191 (1961).
102. F. Patten and W. Gordy, *Proc. Nat. Acad. Sci. (USA),* **46**, 1137 (1960).
103. T. Henriksen, T. Sanner, and A. Pihl, *Radiat. Res.*, **18**, 147 (1963).

104. R. C. Drew and W. Gordy, *Radiat. Res.,* **18**, 552 (1963).
105. F. G. Liming, Jr. and W. Gordy, *Proc. Nat. Acad. Sci. (USA),* **60**, 794 (1968).
106. L. Pauling and R. B. Corey, *Proc. Nat. Acad. Sci. (USA),* **37**, 729 (1951).
107. L. Pauling, R. B. Corey, and H. R. Branson, *Proc. Nat. Acad. Sci. (USA),* **37**, 205 (1951).
108. R. A. Patten and W. Gordy, *Radiat. Res.,* **22**, 29 (1964).
109. W. Gordy and I. Miyagawa, *Radiat. Res.,* **12**, 211 (1960).
110. T. Henriksen, T. Sanner, and A. Pihl, *Radiat. Res.,* **18**, 163 (1963).
111. A. Müller, Int. *J. Radiat. Biol.,* **5**, 199 (1962).
112. K. G. Zimmer and A. Müller, "New Light on Radiation Biology from Electron Spin Resonance Studies," in *Current Topics in Radiation Research,* M. Ebert and A. Howard, Eds., North Holland, Amsterdam, 1965, pp. 1-47.
113. J. H. Keighley, "Electron Spin Resonance," in *Introduction to the Spectroscopy of Biological Polymers,* D. W. Jones, Ed., Academic, London, 1976, pp. 221-270.
114. F. G. Liming, Jr., and W. Gordy, *Proc. Nat. Acad. Sci. (USA),* **60**, 794 (1968).
115. F. G. Liming, Jr., *Radiat. Res.,* **39**, 252 (1969).
116. E. L. Fasanella and W. Gordy, *Proc. Nat. Acad. Sci. (USA),* **62**, 299 (1969).
117. H. Shields and W. Gordy, *Bull. Am. Phys. Soc.,* **1**, 267 (1956).
118. H. Shields and W. Gordy, *Proc. Nat. Acad. Sci. (USA),* **45**, 269 (1959).
119. P. G. Shen, L. A. Blyumenfeld, A. E. Kalmanson, and A. G. Pasynskii, *Biofizika,* **4**, 263 (1961).
120. A. Ehrenberg, L. Ehrenberg, and G. Löfroth, *Nature,* **200**, 376 (1963).
121. A. Van de Vorst, "Étude par Résonance Paramagnètique Électronique de l'Action des Rayonnements Ionisants sur les Acids Nucléiques et Leur Constituants," thesis, Université de Liège, Belgium, 1963.
122. W. Köhnlein and A. Müller, *Int. J. Radiat. Biol.,* **8**, 141 (1964).
123. P. S. Pershan, R. G. Shulman, B. J. Wyluda, and J. Eisinger, *Physics,* **1**, 163 (1964).
124. A. Van de Vorst and F. Villée, *C. R. Acad. Sci., (Paris),* **259**, 928 (1964).
125. W. Gordy, B. Pruden, and W. Snipes, *Proc. Nat. Acad. Sci. (USA),* **53**, 751 (1965).
126. H. C. Heller and T. Cole, *Proc. Nat. Acad. Sci. (USA),* **54**, 1486 (1965).
127. J. N. Herak and W. Gordy, *Proc. Nat. Acad. Sci. (USA),* **54**, 1287 (1965).
128. M. G. Omerod, *Int. J. Radiat. Biol.,* **9**, 291 (1965).
129. K. V. Rajalakshmi, K. Venkateswarlu, and A. Van de Vorst, *C. R. Acad. Sci. (Paris),* **261**, 4879 (1965).

130. B. B. Singh and A. Charlesby, *Int. J. Radiat. Biol.*, **9**, 157 (1965).
131. D. E. Holmes, L. S. Myers, and R. B. Ingalls, *Nature*, **209**, 1017 (1966).
132. A. Pihl and T. Sanner, *Radiat. Res.*, **28**, 96 (1966).
133. A. Müller, "The Formation of Radicals in Nucleic Acids, Nucleoproteins, and their Constituents by Ionizing Radiations," in *Progress in Biophysics and Molecular Biology*, Vol. 17, J. A. V. Butler and H. D. Huxley, Eds., Pergamon, New York, 1967, pp. 99-147.
134. W. Gordy, *Ann. N. Y. Acad. Sci.*, **158**, 100 (1969).
135. J. N. Herak, "EPR of Irradiated Single Crystals of the Nuclear Acid Constituents," in *Physico-Chemical Properties of Nucleic Acids*, J. Duchesne, Ed., Academic, London, 1973, pp. 197-221.
136. C. Nicolau, "Short-lived Free Radicals in Aqueous Solutions of Nucleic Acid Components," in *Physico-Chemical Properties of Nucleic Acids*, J. Duchesne, Ed., Academic, London, 1973, pp. 143-195.
137. H. W. Shields, "Electron Spin Resonance of Irradiated Biomolecules," in *Experimental Methods in Biophysical Chemistry*, Wiley, London, 1973, pp. 417-482.
138. R. A. Patten and W. Gordy, *Nature*, **201**, 361 (1964).
139. R. Salovey, R. G. Shulman, and W. M. Walsh, Jr., *J. Chem. Phys.*, **39**, 839 (1963).
140. C. Dorlet, A. Van de Vorst, and A. J. Bertinchamps, *Nature*, **194**, 767 (1962).
141. B. Pruden, W. Snipes and W. Gordy, *Proc. Nat. Acad. Sci. (USA)*, **53**, 917 (1965).
142. J. J. Lichter and W. Gordy, *Proc. Nat. Acad. Sci. (USA)*, **60**, 450 (1968).
143. C. Alexander and W. Gordy, *Proc. Nat. Acad. Sci. (USA)*, **58**, 1279 (1967).
144. J. N. Herak and W. Gordy, *Proc. Nat. Acad. Sci. (USA)*, **56**, 7 (1966).
145. J. N. Herak and W. Gordy, *Proc. Nat. Acad. Sci. (USA)*, **55**, 698 (1966).
146. J. N. Herak and W. Gordy, *Proc. Nat. Acad. Sci. (USA)*, **55**, 1373 (1966).
147. R. A. Holroyd and J. W. Glass, *Int. J. Radiat. Biol.*, **14**, 445 (1968).
148. A. Graslund, A. Ehrenberg, A. Rupprecht, and G. Ström, *Biochim. Biophys. Acta*, **254**, 172 (1971).
149. M. D. Savilla, *J. Phys. Chem.*, **75**, 626 (1971).
150. J. N. Herak and V. Galogaza, *J. Chem. Phys.*, **50**, 3101 (1969).
151. H. C. Box and E. E. Budzinski, *J. Chem. Phys.*, **62**, 197 (1975).
152. H. C. Box, W. R. Potter, and E. E. Budzinski, *J. Chem. Phys.*, **62**, 3476 (1975).
153. H. C. Box and E. E. Budzinski, *J. Chem. Phys.*, **64**, 1593 (1976).
154. H. C. Box, *Faraday Disc. Chem. Soc.*, **45**, 264 (1977).
155. E. Westhof, W. Flossmann, H. Zehner, and A. Müller, *Faraday Disc. Chem. Soc.*, **45**, 248 (1977).

156. J. N. Herak and W. Gordy, *Science,* **153**, 3744 (1966).

157. B. Pullman and A. Pullman, *Quantum Biochemistry,* Wiley-Interscience, New York, 1963.

158. B. Pullman, "Aspects of the Electronic Structure of the Nucleic Acids and Their Constituents," in *Molecular Biophysics,* B. Pullman and B. Weissbluth, Eds., Academic, New York, 1965, pp. 117-189.

Chapter **IX**

FREE RADICALS IN LIQUID SOLUTIONS

Many aromatic free radicals, both charged and neutral, are stable enough to be chemically prepared and observed with esr when in solution in certain organic liquids. These free radicals have their unpaired electrons in π orbitals which spread over one or more of the aromatic rings. Aliphatic free radicals are much more unstable, but they are now commonly observed in liquid solution by special techniques. The esr spectra are measured for short-lived free radicals produced by continuous irradiation of liquids while under observation in the microwave cavity. Rapid-flow methods have also been developed that permit observations of short-lived free radicals produced by chemical reactions in flowing liquid systems.

Free radicals in liquid solutions of low viscosity have effective spin Hamiltonians which are essentially isotropic. Their esr lines are sharp because the anisotropies in both the **g** tensor and the nuclear couplings are averaged out by the rapidly tumbling motions. Since the average of a dipole-dipole coupling over all directions in space is zero, measurements of the esr hyperfine structure for radicals in liquid solutions give only the Fermi contact interactions. However, these components can be measured with high accuracy and are very useful. Not only do they provide information about the radicals per se; they also aid greatly in the interpretation of esr in solids.

When the anisotropies are effectively averaged out, the lines are so sharp that closely spaced hyperfine components can be resolved and measured. To achieve this result, it is necessary to have the free radicals in solution sufficiently dilute that interradical exchange interactions do not wipe out the isotropic hyperfine structure. Because they can give readily interpretable and valuable information about short-lived free radicals, many esr measurements are now made on free radicals in liquid solutions. The well-resolved hyperfine patterns provide distinguishing labels for various radical species. Many types of chemical reactions in the liquid state can be studied by esr observations of free-radical intermediaries.

The units and symbols used here are as follows. The isotropic nuclear coupling measured in solution spectroscopy is customarily designated by a rather than by the A_{iso} or A_f that we have used in the chapters on solids, where anisotropic as well as isotropic coupling components are involved. We use the customary designation of a in this chapter on liquids. Because the effective g values for free radicals in solution are also isotropic and are nearly the same for a variety of molecular free radicals, it is possible to compare directly the nuclear coupling values expressed in the convenient field units of gauss. For this reason, it is customary to express the isotropic coupling for solutions in gauss, and we follow this convention also. With accurately measured g values, the a values can be accurately converted to frequency units (Section 2.1) by the relation

$$a(\text{MHz}) = 2.8025 \frac{g}{g_e} a(\text{G})$$

or, when the slight deviations of g from g_e are neglected, simply by a(MHz) = 2.8025 a (G).

1. THEORY OF SOLUTION SPECTRA

1.a. Spin Hamiltonian and esr Frequencies

Let us consider a free radical with $S = 1/2$, having k coupling nuclei all with the same coupling axes p, q, r, which are also the principal axes of the **g** tensor. Expressed in terms of these principal axes, the spin Hamiltonian for the magnetic interactions in the static radical has the form

$$\mathcal{H}_{S,I} = \beta(g_p H_p S_p + g_q H_q S_q + g_r H_r S_r) + \sum_k (A_p^k S_p I_p^k + A_q^k S_q I_q^k + A_r^k S_r I_r^k) \quad (9.1)$$

If the free radical is in solution, the p, q, r axes are not fixed in space as they are in a single crystal but are varying in time because of the tumbling motions. Thus it is necessary to express the Hamiltonian with reference to a space-fixed system, x,y,z. Without loss of generality, z can be chosen along the laboratory-fixed field **H**, with the polar angles $\theta_p, \theta_q, \theta_r$ and the azimuthal angles ϕ_p, ϕ_q, ϕ_r indicating the direction p,q,r relative to the space-fixed system. In a liquid solution, however, θ and ϕ are constantly changing. For simplicity, we express these time-dependent angles for a particular radical at time t by $\theta_p, \theta_q, \theta_r$ and ϕ_p, ϕ_q, ϕ_r. The transformed, time-dependent Hamiltonian for this radical may then be expressed [1] as

$$\mathcal{H}_{S,I} = \mathcal{H}_0 + \mathcal{H}_t \quad (9.2)$$

where

$$\mathcal{H}_0 = \frac{1}{3}\beta H S_z(g_{xx} + g_{yy} + g_{zz}) + \frac{1}{3}\sum_k (A_{xx}^k + A_{yy}^k + A_{zz}^k)\mathbf{S} \cdot \mathbf{I}^k \quad (9.3)$$

and

$$\mathcal{H}_t = \sum_k \Big\{[(\beta H \Delta g_{xy} + \Delta A_{xy}^k I_z^k)(\cos^2\theta_p - \tfrac{1}{3})$$
$$+ (\beta H \Delta g_{zy} + \Delta A_{zy}^k I_z^k)(\cos^2\theta_r - \tfrac{1}{3})]\, S_z$$

$$+ \frac{1}{2} [\Delta A^k_{xy} \sin \theta_p \cos \theta_p (I^k_+ e^{-i\phi_p} + I^k_- e^{+i\phi_p})$$

$$+ \Delta A^k_{xy} \sin \theta_r \cos \theta_r (I^k_+ e^{-i\phi_r} + I^k_- e^{+i\phi_r})] S_z$$

$$+ \frac{1}{2} (\beta H \, \Delta g_{xy} + \Delta A^k_{xy} I^k_z) \sin \theta_p \cos \theta_p (S_+ e^{-i\phi_p} + S_- e^{+i\phi_p})$$

$$+ \frac{1}{2} (\beta H \, \Delta g_{zy} + \Delta A^k_{zy} I^k_z) \sin \theta_r \cos \theta_r (S_+ e^{-i\phi_r} + S_- e^{+i\phi_r})$$

$$+ \frac{1}{4} \Delta A^k_{xy} \sin^2 \theta_p (S_+ I^k_+ e^{-2i\phi_p} + S_- I^k_- e^{+2i\phi_p})$$

$$+ \frac{1}{4} \Delta A^k_{zy} \sin^2 \theta_r (S_+ I^k_+ e^{-2i\phi_r} + S_- I^k_- e^{+2i\phi_r})$$

$$- \frac{1}{4} [\Delta A^k_{xy} (\cos^2 \theta_p - \frac{1}{3}) + \Delta A^k_{zy} (\cos^2 \theta_r - \frac{1}{3})] (S_+ I^k_- + S_- I^k_+) \Big\} \qquad (9.4)$$

where

$$\Delta g_{xy} = g_{xx} - g_{yy} \quad , \qquad \Delta g_{zy} = g_{zz} - g_{yy}$$

$$\Delta A^k_{xy} = A^k_{xx} - A^k_{yy} \quad , \qquad \Delta A^k_{zy} = A^k_{zz} - A^k_{yy}$$

$$I^k_+ = I^k_x + i I^k_y \quad , \qquad I^k_- = I^k_x - i I^k_y$$

$$S_+ = S_x + i S_y \quad , \qquad S_- = S_x - i S_y$$

and z is chosen along **H**. A special case of the above Hamiltonian, that for axially symmetric **g** and **A**, was first obtained by McConnell [2].

A general solution of this complex, time-dependent $\mathcal{H}'_{SI}$ is not feasible. Nevertheless, its solution for two important limiting cases is relatively simple. For very slow tumbling rates, that is, when $\nu_t \ll \nu_0$, where ν_t is the mean tumbling frequency and ν_0 is the Larmor precessional frequency of the spins, the tumbling motions may be neglected and the magnetic axes considered as fixed in space at angles from the reference axes corresponding to fixed values of θ and ϕ, as in a glass matrix. The solution for this case is the same as that already obtained in Chapter VIII for randomly oriented radicals in powders or glasses.

The special solution most important for the applications in this chapter is that in which the molecular tumbling frequency ν_t is very high in comparison with the magnetic resonance frequencies ν_0, $\nu_t >> \nu_0$. When this is true, the time-dependent terms in (9.4) will be averaged out, or nearly so, and the spin Hamiltonian may be expressed to a good approximation as

$$\mathcal{H}_{S,I} = \beta H S_z \langle g \rangle + \sum_k A^k_{\text{iso}} \mathbf{S} \cdot \mathbf{I}^k \tag{9.5}$$

where

$$\langle g \rangle = \frac{1}{3} \text{ trace } \mathbf{g} = \frac{1}{3}(g_{xx} + g_{yy} + g_{zz}) \tag{9.6}$$

and

$$A^k_{\text{iso}} = \frac{1}{3} \text{trace } \mathbf{A}^k = \frac{1}{3}(A^k_{xx} + A^k_{yy} + A^k_{zz}) \tag{9.7}$$

The last expression holds because the anisotropic dipole-dipole coupling $\mathbf{B}^k$ is a traceless tensor; that is, it has no diagonal elements, whereas the isotropic Fermi contact interaction has no off-diagonal elements. Although the observable $\langle g \rangle$ is the average of the diagonal elements, the off-diagonal elements in the $\mathbf{g}$ tensor are averaged out by the tumbling motions so that $\langle \Delta g \rangle = \frac{1}{3}$ (sum of the principal elements), as in (9.57). Solution of the $\mathcal{H}_0$ of (9.3) is simple. For the resonant peak field values at constant frequency, it yields

$$H = H_0 - \sum_k A^k_{\text{iso}} M_{I_k} \tag{9.8}$$

where

$$H_0 = \frac{h\nu_0}{\beta \langle g \rangle} \tag{9.9}$$

and where A^k_{iso} is in magnetic field units, $A_{\text{iso}}(\text{G}) = [h/(\beta\langle g \rangle)] A_{\text{iso}}$ (Hz). These formulas are used for frequency analysis of most of the solution spectra measured to date and most of those to be described in this chapter. The incomplete averaging out of the anisotropies in dilute solutions does not significantly affect their peak frequencies. Hence, the first-order formulas (9.8) and (9.9) are widely applicable to solution spectra except for those of high viscosity.

1.b. Relaxation Times and Line Widths

It is apparent that the time-dependent parts of $\mathcal{H}_t$ of (9.4) will average to zero over a complete cycle of the angular variables because

$$\left\langle \cos^2\theta - \frac{1}{3}\right\rangle_{av} = \frac{1}{4\pi}\int_0^{2\pi}\int_0^{\pi}(\cos^2\theta - \frac{1}{3})\sin\theta d\theta\ d\phi = 0$$

$$\left\langle \sin\theta\cos\theta\, e^{\pm i\phi}\right\rangle_{av} = \frac{1}{4\pi^2}\int_0^{2\pi}(\cos\phi \pm i\sin\phi)d\phi\int_0^{\pi}\sin\theta\cos\theta\ d\theta = 0$$

$$\left\langle \sin^2\theta\, e^{\pm 2i\phi}\right\rangle_{av} = \frac{1}{4\pi^2}\int_0^{2\pi}(\cos 2\phi \pm i\sin 2\phi)d\phi\int_0^{\pi}\sin^2\theta\ d\theta = 0$$

However, it is not probable that all the radicals in the solution will undergo such rotations in a time that is short compared with $1/\Delta\nu_0$. This is not required to ensure that the anisotropies will be sufficiently reduced to produce very sharp lines. We can expect a random distribution of tumbling frequencies. Anisotropies for most, but not all, of the radicals will be averaged to zero when the mean tumbling frequency is high in comparison with the frequency spread of the static radicals that is due to the anisotropy.

Through the anisotropic **g** and **A** tensors, principal axes for which are fixed in the molecular frame, the esr line widths are modulated by the random rotations of the molecular radicals in the liquid solution. Thus there will be a spectrum of frequency modulations of the line widths corresponding to the spectrum of tumbling rotational frequencies. The statistical problem of calculation of the effect of these random modulations on the line breadth is like that encountered in the interpretation of nmr line widths in liquids. The problem is successfully treated in the classic paper by Bloembergen, Purcell, and Pound [3] on relaxation effects in nuclear resonance absorption of liquids. Adaptations to esr line widths of solutions have been made by McConnell [2] and Kivelson [4]. Special effects in esr spectra such as alternating line widths of certain hyperfine components, which are not involved in nmr theories, are treated by Freed and Fraenkel [5,6].

In the Bloembergen-Purcell-Pound theory, the line-shape function is expressed in terms of the correlation function $K(\tau)$ of the random modulation $F(t)$ defined by

$$K(\tau) = \left\langle F(t)\, F^*(t+\tau)\right\rangle_{av} \tag{9.10}$$

where $F(t)$ signifies the time-dependent functions in the spin Hamiltonian, (9.4).

The intensities of the Fourier component frequencies of these random modulations are then expressed by

$$J(\omega) = \int_{-\infty}^{\infty} K(\tau)e^{i\omega\tau}\, d\tau \tag{9.11}$$

where $\omega = 2\pi\nu$ is the angular frequency. The correlation function is an even function of τ and is independent of t. Bloembergen et al. [3] assumed it to be an exponential function of $|\tau|$ and expressed it as

$$K(\tau) = \langle F(t)F^*(t)\rangle_{av} e^{-|\tau|/\tau_c} \tag{9.12}$$

where τ_c is the "correlation time" for the particular modulation considered. Since the average $\langle F(t)F^*(t)\rangle_{av}$ is independent of τ, it can be removed from the above integral of (9.11), which may then be integrated to give

$$J(\omega) = \langle F(t)F^*(t)\rangle_{av}\, 2\tau_c/(1 + \omega^2\tau_c^2) \tag{9.13}$$

These random frequency modulations of the magnetic field acting on the electron spin induce spin transitions when the random frequency happens to be such as to overlap some of the Larmor frequencies of the spin system in the laboratory-applied field **H**. The probability of such induced transitions will be proportional to the intensity of the Fourier components of the random spectrum at the Larmor frequency. With use of the correlation function, this probability can be calculated readily with time-dependent perturbation theory [7].

The probability of an induced transition $m \leftrightarrow n$ at angular frequency ω_{mn} is

$$p_{mn} = \hbar^{-2} \int_{-\infty}^{\infty} K(\tau)e^{-i\omega_{mn}\tau} d\tau \tag{9.14}$$

which with (9.13) may be integrated to give

$$p_{mn} = \frac{\hbar^{-2} \langle F(t)F^*(t)\rangle_{av} 2\tau_c}{1 + \omega_{mn}^2\tau_c^2} \tag{9.15}$$

The probability of relaxation by the induced transitions $m \rightarrow n$ is the inverse of the lifetime in the state m; that of the reverse transition $m \leftarrow n$ is the inverse of the lifetime in the state n. Since these probabilities are equal, it is reasonable to assume the inverse of the relaxation time to be

$$\left(\frac{1}{T_1}\right)_{mn} = p_{m\rightarrow n} + p_{m\leftarrow n} = 2p_{mn} \tag{9.16}$$

Hence from (9.15)

$$\left(\frac{1}{T_1}\right)_{mn} = \frac{16\pi^2}{h^2}\left\langle F(t)F^*(t)\right\rangle_{\text{av}} \frac{\tau_c}{1 + \omega_{mn}^2\tau_c^2} \tag{9.17}$$

In the present problem, the functions $F(t)$ are the terms of the time-dependent $\mathcal{H}_t$ of (9.4). Thus for calculation of T_1, it is convenient to have the averages of the time-dependent angular factors in the $\mathcal{H}_t$ of (9.4). These averages are

$$|\langle\cos^2\theta - 1/3\rangle_{\text{av}}|^2 = \frac{4}{45}$$

$$\langle(\cos\theta\sin\theta\, e^{+i\phi})(\cos\theta\sin\theta\, e^{-i\phi})\rangle_{\text{av}} = \frac{2}{15}$$

$$\langle(\sin^2\theta\, e^{+2i\phi})(\sin^2\theta\, e^{-2i\phi})\rangle_{\text{av}} = \frac{8}{15}$$

where $\theta = \theta_p$ or θ_r and $\phi = \phi_p$ or ϕ_r. The operator averages that we likewise need are $|(M_S|S_z|M_S)|^2 = M_S^2$, $|(M_I|I_z|M_I)|^2 = M_I^2$, $|\langle M_S|S_-|M_S + 1\rangle_{\text{av}}|^2 = |\langle M_S + 1|S_+|M_S\rangle_{\text{av}}|^2 = S(S + 1) - M_S(M_S + 1)$ with similar averages for I_- and I_+. Here we are considering free radicals in the doublet state for which $S = 1/2$, and the squared averages of the nonvanishing matrix elements of S_- and S_+ are

$$|\langle -\tfrac{1}{2}|S_-|\tfrac{1}{2}\rangle_{\text{av}}|^2 = 1, \qquad |\langle \tfrac{1}{2}|S_+|\tfrac{1}{2}\rangle_{\text{av}}|^2 = 1$$

The rotating components of the anisotropic magnetization about z corresponding to $e^{+i\phi}$ and $e^{-i\phi}$ due to the tumbling rotations induce transitions $M_S \rightarrow M_S' = -1/2 \rightarrow 1/2$ and $1/2 \rightarrow -1/2$ via these coupling matrices of S_- and S_+. These transitions are the principal contributors to the spin-lattice or spin-molecular relaxation, and thus to T_1. Nuclear transitions $M_S, M_I \rightarrow M_S, M_I \pm 1$ will be produced similarly by the rotating components through I_- and I_+, but because of the smallness of the nuclear magnetic moments as compared with the electron-spin moment and because of the low-correlation spectrum at the nuclear transition frequencies, these make negligible contributions to T_1 for nonviscous liquids. We neglect these contributions as well as the still less probable transitions $M_S, M_I \rightarrow M_S \pm 2, M_I$ and $M_S, M_I \rightarrow M_S, M_I \pm 2$ that are induced by the last terms of (9.4).

Since only the terms containing the factors $S_+e^{-i\phi}$ and $S_-e^{+i\phi}$ make significant contributions to $1/T_1$, substitution of the averages for these terms into (9.17) yields

$$\frac{1}{T_1} \gtrsim \frac{8\pi^2}{15h^2} \Big[\big(\beta H \, \Delta g_{xy} + \Sigma \, \Delta A^k_{xy} M_{I_k}\big)^2 + \big(\beta H \, \Delta g_{zy} + \Sigma \, \Delta A^k_{zy} M_{I_k}\big)^2\Big] \frac{\tau_c}{1 + \omega_0^2 \tau_c^2} \tag{9.18}$$

When the tensors **g** and $\mathbf{A}^k$ are axially symmetric with a common symmetry axis, $\Delta g_{xy} = g_{xx} - g_{yy} = 0$, $\Delta A^k_{xy} = A^k_{xx} - A^k_{yy} = 0$. With $\Delta g_{zy} = \Delta g$, $\Delta A^k_{zy} = \Delta A^k$, (9.18) reduces to

$$\frac{1}{T_1} \gtrsim \frac{8\pi^2}{15h^2} \big(\beta H \, \Delta g + \Sigma \, \Delta A^k M_{I_k}\big)^2 \frac{\tau_c}{1 + \omega_0^2 \tau_c^2} \tag{9.19}$$

To find the total effects of the tumbling motions on the line widths, we must also obtain an expression for T_2', the spin-spin relaxation time or the transverse relaxation time that applies to liquid solutions. A description of T_2' is given in Chapter IV, Section 9.d. For the nmr of liquids, Bloembergen et al. [3] derived $(1/T_2')^2$ by integration of $J_0(\nu)d\nu = \langle F_0(t)F_0^*(t)\rangle_{\text{av}} \times [2\tau_c/(1 + 4\pi^2 \nu^2 \tau_c^2)]\, d\nu$ from $\nu = [\nu_0 - 1/(\pi T_2')]$ to $\nu = [\nu_0 + 1/(\pi T_2')]$. Justification of their procedure is given in the original paper. In the present problem (esr spectra), the spin-spin relaxation arises from the terms in which there is no change in the resultant component of S_z nor I_z. For the $\mathcal{H}_t$ of (9.4), it results mainly from the first term in the brackets. This term corresponds to $F_0(t)$ in the expression (9.13) for $J_0(\omega)$. Thus approximately,

$$\left(\frac{1}{T_2'}\right)^2 \approx \hbar^{-2} \int_{\omega_0 - \frac{1}{T_2'}}^{\omega_0 + \frac{1}{T_2'}} J_0(\omega)\,d\omega \tag{9.20}$$

On substitution of the relevant $J_0(\omega)$ terms followed by integration, (9.20) becomes

$$\left(\frac{1}{T_2'}\right)^2 \gtrsim \left(\frac{32\pi^2}{45h^2}\right) [(\beta H \,\Delta g_{xy} + \Sigma\, \Delta A_{xy}^k\, M_{I_k})^2$$

$$+ (\beta H \,\Delta g_{zy} + \Sigma\, \Delta A_{zy}^k\, M_{I_k})^2] \tan^{-1} \frac{2\tau_c}{T_2'} \quad (9.21)$$

For nonviscous solutions where $\tau_c << T_2'$, one can set $\tan^{-1}(2\tau_c/T_2') \approx (2\tau_c/T_2')$ and express (9.21) in the form

$$\frac{1}{T_2'} \gtrsim \frac{64\pi^2}{45h^2} [(\beta H \,\Delta g_{xy} + \Sigma\, \Delta A_{xy}^k M_{I_k})^2 + (\beta H \,\Delta g_{zy} + \Sigma\, \Delta A_{zy}^k M_{I_k})^2]\tau_c$$

$$(\text{viscosity low, } \tau_c << T_2') \quad (9.22)$$

For the axially symmetric case,

$$\frac{1}{T_2'} \gtrsim \frac{64\pi^2}{45h^2} (\beta H\, \Delta g + \Sigma\, \Delta\, A^k M_{I_k})^2\, \tau_c \quad (9.23)$$

Also, for rapidly tumbling radicals for which $\omega_0^2\tau_c^2 << 1$, one can set $[\tau_c/(1 + \omega_0^2\tau_c^2)] \approx \tau_c$ and obtain from (9.18):

$$\frac{1}{T_1} \gtrsim \frac{8\pi^2}{15h^2} [(\beta H \,\Delta g_{xy} \Sigma\, \Delta A_{xy}^k M_{I_k})^2 + (\beta H \,\Delta g_{zy} + \Sigma\, \Delta A_{zy}^k M_{I_k})^2]\tau_c \quad (9.24)$$

When the viscosity is sufficiently low that $\tau_c << T_2'$ and $\tau_c < 1/\omega_0$, the anisotropies will be reduced to insignificance, and the line will be homogeneously broadened. For such lines, one can set

$$\Delta\omega_{1/2} = \frac{1}{T_2} = \frac{1}{T_2'} + \frac{1}{2T_1} \quad (9.25)$$

or

$$\Delta\nu = \frac{1}{\pi T_2'} + \frac{1}{2\pi T_1} \quad (9.26)$$

where $\Delta\omega_{1/2} = 2\pi\Delta\nu_{1/2} = \pi\Delta\nu$ and $\Delta\nu$ is the width of the line between half-intensity points, in frequency units. If there is significant residual anisotropic broadening, this expression does not hold; T_2 is not equivalent to the Bloch transverse relaxation time (see discussion in Chapter IV, Section 9.e). In this case, it is customary to define a pseudorelaxation time, $T_2^* = 1/\Delta\omega_{1/2}$. Then, obviously, $T^* < T_2$. Thus it is probable that line widths calculated with the preceding theory represent minimum values and that actual widths corresponding to $1/T_2^*$ will be somewhat greater because some residual anisotropies are not averaged out.

One can combine (9.22), (9.24), and (9.25) into the following line-width formula for free radicals in solutions of low viscosity:

$$\Delta\nu \gtrsim 5.3\,\{[(\Delta\nu_g)_{xy} + (\Delta\nu_h)_{xy}]^2 + [(\Delta\nu_g)_{zy} + (\Delta\nu_h)_{zy}]^2\}\tau_c \quad (9.27)$$

or for radicals with axial magnetic symmetry:

$$\Delta\nu \gtrsim 5.3(\Delta\nu_g + \Delta\nu_h)^2\tau_c \quad (9.28)$$

where $\Delta\nu_g = \beta H \Delta g/h$, $\Delta\nu_h = \Sigma\,\Delta A^k M_{I_k}/h$, and $\Delta\nu = 2\Delta\nu_{1/2} = \omega_{1/2}/\pi$ is the width of the line between half-intensity points. The $\Delta\nu_g$ and $\Delta\nu_h$ correspond to the frequency displacements for the oriented radicals caused by **g** and $\mathbf{A}^k$ anisotropy. It is evident that the hyperfine components of the solution spectra will have different line widths because of the difference in values of $\Sigma\,\Delta A^k M_{I_k}$ associated with the particular component.

The observed line widths should be greater than those given by the expressions on the right in (9.27) or (9.28), not only because some terms in the $\mathcal{H}_t$ of (9.4) are neglected in the derivation, but, even more, because other factors not included in this $\mathcal{H}_t$ are likely to shorten the relaxation times and increase the line widths. Among the more important of these factors are internal vibrational modes, particularly torsional bending modes which spread the effective magnetic axes within the radicals. Such modes are excited by the random collisions of the radicals with the solvent molecules. Also neglected are the effects of intermolecular interactions such as dipole-dipole interaction of the radical spins, electron or nuclear, with those of the solvent molecules or other radicals in the solutions. These interactions are moderated by the diffusion of the radicals, not included in the calculation. The effects of molecular diffusion are much more significant in nmr than in esr relaxation [3]. The significance of the approximate theory based on tumbling rotation is that it provides an explanation for the essential elimination of the large inhomogeneous broadening of esr lines characteristic of randomly oriented radicals in powders or in viscous liquids and the production of the sharp, homogeneous lines characteristic of nonviscous solutions.

To apply the theory for estimation of relaxation times, values of τ_c are required. A commonly used, if questionable, measure of correlation times involved in the esr of liquid solutions is the modified Debye formula [3],

$$\tau_c = \frac{4\pi\eta a^3}{3kT} = 3 \times 10^{16} \frac{\eta a^3}{T} \qquad (9.29)$$

for calculation of the effective reorientation time or persistence of orientation in dielectric dispersion of polar liquids. This formula contains a factor of 1/3, introduced in the original Debye formula by Bloembergen et al. as a correction for the difference in the correlation functions involved in magnetic resonance from the cos θ function involved in dielectric dispersion. In this formula η signifies the viscosity of the liquid, a is the effective radius of the molecules that are treated as spheres, and T is the absolute temperature of the liquid. In nmr experiments on pure liquids the observed nuclei are those of the liquid molecules, and the agreement with the modified Debye formula is close [3]. In most esr experiments the tumbling radicals are in dilute solution in the solvent liquid, and the use of the Debye τ_c is more questionable, or more difficult to apply. However, if the solute radicals are comparable in size to the solvent molecules, (9.29) should give a reasonable estimate of the effective τ_c for free radicals in dilute solution in the liquid, and one generally finds that the esr lines are sharp; that is, the anisotropies are effectively averaged out for such solutions when $\tau_c < (1/\nu_0)$, where ν_0 is the esr frequency and τ_c is calculated from (9.29). For example, sharp-line esr spectra of many types of free radicals in aqueous solution or in nonviscous organic liquids are commonly observed. Some are described later. At $T = 300°$K, water has viscosity $\eta = 0.01$ poise and $a = 1.5$ Å; (9.29) indicates that $\tau_c = 3.4 \times 10^{-12}$ sec, which is three orders of magnitude less than the $(1/\nu_0) = 10^{-9}$ sec for X-band esr frequencies. However, $1/\tau_c$ thus calculated corresponds to the mean tumbling frequency of the solvent water molecules rather than to the solute radicals. For effective averaging out of the anisotropies, the mean tumbling frequencies of the free radicals in the aqueous solution should be appreciably greater than the esr frequency. If these radicals are large in size relative to the H_2O molecules, or if they interact strongly with the solvent, the effective τ_c may be significantly larger than that calculated for the pure water.

Illustrations

Figures 9.1 and 9.2 illustrate the high resoltuion achievable with solution spectra when the correlation time τ_c is sufficiently short that the anisotropies in **g** and **A** are averaged out by the tumbling rotation of the radicals. These figures also illustrate the residual effects of these anisotropies on the line

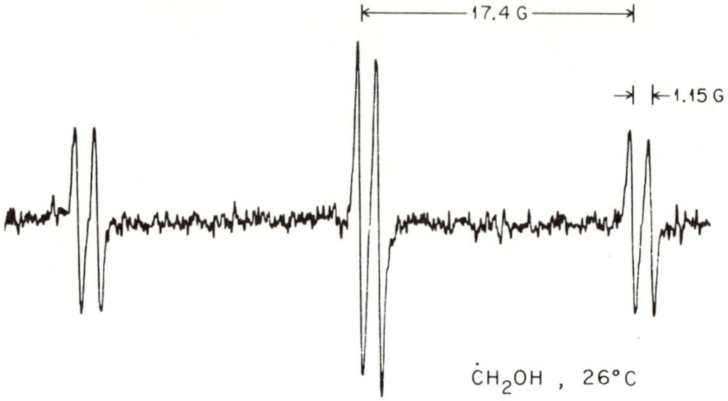

Fig. 9.1 ESR spectrum of $\dot{C}H_2OH$ formed by photolysis of methyl alcohol containing 1% H_2O_2. From Livingston and Zeldes [8].

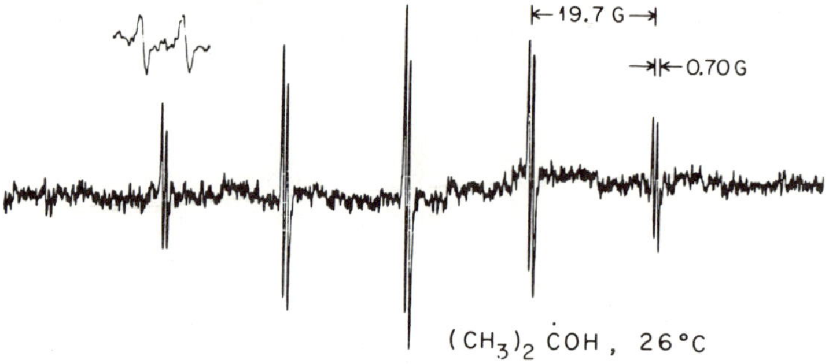

Fig. 9.2 ESR spectrum of $(CH_3)_2\dot{C}OH$ formed by photolysis of isopropyl alcohol containing 0.4% H_2O_2. From Livingston and Zeldes [8].

widths of the solution spectra. The spectra for $\dot{C}H_2OH$ were observed in CH_3OH solution with slight concentration of H_2O_2 used in production of the radicals. Similarly, the spectra of $(CH_3)_2\dot{C}OH$ were observed in $(CH_3)_2CHOH$ with slight concentration of H_2O_2. Both spectra were measured at the same temperature, 26°C. Although the viscosity of methanol is only one-fourth that of isopropyl alcohol at that temperature, the lines of $\dot{C}H_2COH$ are noticeably broader than those of $(CH_3)\dot{C}OH$; $\Delta H \approx 0.29$ G or $\Delta\nu \approx 0.81$ MHz for the former and $\Delta H \approx 0.07$ G or $\Delta\nu \approx 0.20$ MHz for the latter.

The measured line width for $\dot{C}H_2OH$ is listed in Table 9.1 with the isotropic proton coupling and $\langle g \rangle$ values obtained from the spectra. Note that the small OH proton coupling (~1 G) is measured to three figures. Temperature effects on both the coupling and the line width are noticeable. Livingston and Zeldes

Table 9.1 Measured Parameters for $\dot{C}H_2OH$[a]

Temp. (°C)	a_α (G)	a_{OH} (G)	*g*	*Approx. Line Width* (G)
41	17.26	1.00	2.00333	
26	17.38	1.15	2.00334	0.29
-29	17.78	1.66	2.00331	0.20
-50	17.84	1.75	2.00336	0.18

[a]From Livingston and Zeldes [8].

[8] did not give a value of $\Delta\nu$ for the $(CH_3)_2\dot{C}OH$, but the value of 0.07 G can be obtained from measurement of the doublet inset of Fig. 9.2. The greater line width for $\dot{C}H_2OH$ than for $(CH_3)_2\dot{C}OH$ is expected because of the greater anisotropy in the proton coupling for the static radicals of the former (α coupling) over that of the latter (β coupling). The $\langle g \rangle$ values for the two radicals indicate that their **g** anisotropy is small and comparable in magnitude. As an example, we apply the preceding theory to obtain a semiquantitative calculation of the line width for the isopropyl alcohol radical.

Application of (9.27) to the $(CH_3)_2\dot{C}OH$ radical is simpler and more certain than to the $H_2\dot{C}$-OH radical because the anisotropy of the proton coupling is small in the former (<5 G) and can be neglected in an approximate calculation. Although the Δg values have not been measured, they can be assumed from comparison with other similar carbon-centered, π-type, radicals having $\langle g \rangle$ comparable to that for $(CH_3)_2\dot{C}OH$ (2.00317). For example, the $\dot{C}H_2COOH$ radical observed in single crystals of malonic acid has $\langle g \rangle = 2.0032$, $\Delta g_1 = 0.0014$, and $\Delta g_2 = 0.0022$ [9]. At the observation frequency of $\nu_0 = 9500$ MHz, $\Delta g_1 = 0.0014$ and $\Delta g_2 = 0.0022$ correspond to $\Delta\nu_g = (\Delta g/g)(\nu_0)$ of 6.7 MHz and 10.5 MHz. Substitution of these values for $(\Delta\nu_g)_{xy}$ and $(\Delta\nu_g)_{zy}$ into (9.27), with the neglect of the small $\Delta\nu_h$ anisotropies, gives for $(CH_3)_2\dot{C}OH$:

$$\Delta\nu \text{ (MHz)} = 5.3\,[(6.7)^2 + (10.5)^2] \times 10^6\, \tau_c \text{ sec.} = 8.22 \times 10^8\, \tau_c \text{ sec}$$

which, with the measured line width, $\Delta\nu = 0.20$ MHz, gives $\tau_c = 2.4 \times 10^{-10}$ sec. For comparison, we can obtain a value of τ_c from (9.29). At the observation temperature of 26°C, the viscosity of isopropyl alcohol is $\eta = 0.0206$ poise. From the molecular dimensions of $(CH_3)_2\dot{C}OH$, we estimate its effective radius as $a \approx 4.5$ Å, three times that for H_2O. Use of these values in the Debye equation (9.29) gives $\tau_c \approx 1.9 \times 10^{-10}$ sec for the isopropyl alcohol solvent at 26°C. When this τ_c is substituted into (9.27), a theoretical line width, $\Delta\nu \approx 0.16$ MHz, is obtained, somewhat less than the observed value of 0.20 MHz.

Inclusion of the small anisotropies in the proton coupling would improve the agreement. The most significant facts are that the calculated line widths, as well as those observed, are more than an order of magnitude smaller than the frequency spread in $\Delta\nu_g \approx 10$ MHz.

The anisotropies in the two α-coupling protons in $\dot{C}H_2OH$ indicate $\Delta\nu_h \approx 48$ MHz for each. The anisotropies in g are comparable to those for $(CH_3)_2\dot{C}OH$. Although both η and a are smaller for methanol than for isopropyl alcohol, these decreases are counterbalanced by the large $\Delta\nu_h$, with the result that the theoretically estimated $\Delta\nu$ is somewhat greater for the $\dot{C}H_2OH$ than for the $(CH_3)_2\dot{C}OH$. The calculation, however, is complicated by the fact that the principal coupling axes of the two $C_\alpha H$ protons have different orientations in the molecular plane, neither of which coincides with the directions of the principal g axes. Because of these complications, we do not attempt a detailed calculation.

2. SPECIALIZED PERTURBATIONS OF SOLUTION SPECTRA

The production of very sharp spectral lines by the rapid tumbling motions of radicals in dilute solutions makes possible the study of a number of specialized effects such as those of restricted internal rotation, cis-trans isomerization, conformational interconversions, electron-spin exchange, proton exchange, and radical-solvent interactions. Certain of these effects can also be observed in the sharp-line spectra of single crystals. An example of this, the restricted rotation of the CH_3 group in the alanine radical R-$\dot{C}HCH_3$, is described in Chapter VI, Section 2.b. The number of hyperfine components observed, as well as the line widths of certain of these components, depends on the rates of these various transformations relative to the frequency separations of the hyperfine components in the "frozen" states in which the exchange does not occur or is very slow in comparison with the nuclear precession. By measurement of the component widths and separations over a range of temperatures, one can often find the reaction rates or correlation times from which one can derive the activation energy required for transformation from one configuration to the other, as described in Chapter VI, Section 2.c for the restricted rotation of the CH_3 group in the alanine radical.

A free radical that is tumbling or rotating in the solvent so that only its isotropic nuclear couplings and $\langle g \rangle$ are observable may have two or more stable isomeric forms, each of which has isotropic parameters distinct from the others. If the different isomeric species are not interconvertible or if the frequency of such conversion is very low in comparison with the electron or nuclear precession, each form will have its own characteristic esr pattern distinct from the other forms, as if it were a chemically distinct radical. If the interconversion rate is very fast in comparison with the spin precessional rates, the esr patterns will coalesce into one esr pattern that is characteristic of the averaged patterns of the separate isomeric species. While the peak frequencies for these two conditions will be essentially those for the unperturbed and the averaged spectra, the line widths may be measurably influenced by the mean conversion frequency or the correlation time τ_c for the conversion. At conversion rates between the slow and fast extremes, especially when $1/\tau_c$ becomes comparable to the difference between the spin precessional frequencies in the interconverting states, both the line widths and frequencies may be drastically influenced, with the result that parts of the spectra will be unresolvable.

Comprehensive theoretical treatments of the diverse configurational changes that can markedly influence esr spectra are rather complex. They follow two general approaches. One approach is to add the resulting time-dependent modulations of the magnetization to the Bloch rate equations (4.107) to (4.109) and then to solve these equations for the modified frequencies and relaxation times. This method was first used by Gutowsky, McCall, and Slichter [10] for calculation of nuclear-exchange effects on nmr spectra. Further development of this approach was made by McConnell [11], and outlines of the method as applied to esr spectra are given by Johnson [12], Sullivan and Bolton [13], and Atherton [14]. Solution of the modified Bloch equations is cumbersome, but the most significant results for application to isotropic esr of nonviscous solutions can be stated in relatively simple closed formulas [e.g., (9.41)].

The other approach, that used most extensively for treatment of nonviscous solutions, is the density-matrix formulation. It seems to have evolved from the original successes of Bloembergen, Purcell, and Pound [3] in explaining nmr relaxation and line widths of liquids (see Section 1). Rigorous formulation and generalization of this theory have been made by Wangness and Bloch [15], and Kubo and Tomita [16]. Many applications and expansions of the theory have since been made. Excellent reviews of the subject are given by Redfield [17] Johnson [12], and Sullivan and Bolton [13]. Among the books that give lucid treatments of the subject is one by Slichter [18], Poole and Farach [19], and Atherton [14]. The papers by Freed and Fraenkel [5,6] adapt these theories to the calculation of alterations in line widths of esr spectra caused by periodic configurational changes in the structures of free radicals in solution. Descriptions of the essential results of the theory as it applies to dilute, nonviscous solutions of free radicals are given in a review by Fraenkel [20].

2.a. Single Nucleus with Fluctuating Coupling

Consider first a single nucleus having modulated isotropic coupling within a rapidly tumbling free radical. The modulation is assumed to result from periodic changes between two configurations, A and B. The type of configuration change need not be specified. The coupling nucleus may oscillate between two coupling sites or have its coupling spin density modulated by periodic shifts of charges or by reorientations of molecular groups in other parts of the radical. We assume that when the radical has configuration A the nucleus has coupling a_A and that for configuration B it has coupling a_B. In these two configurations, the angular frequencies of the hyperfine components are $\omega_A(M_I) = \gamma_e a_A M_I$ and $\omega_B(M_I) = \gamma_e a_B M_I$, where a_A and a_B are in gauss and where $\gamma_e = (g\beta/\hbar)$. If the time in each configuration A or B is long in comparison with the nuclear precessional times [i.e., $1/\tau_A$, $1/\tau_B << (\omega_A - \omega_B)$] (slow exchange), a separate hyperfine structure will be observed for each configuration. When there are no other coupling nuclei in the radical, each hyperfine multiplet will consist of $2I + 1$ components. The relative strength of the multiplets will be proportional to p_A and p_B, the probabilities for the occurrence of the respective configurations A and B. When the exchange rate is very fast, $1/\tau_A$, $1/\tau_B > (\omega_A - \omega_B)$, and the exchange is adiabatic (secular process) so that the coupling of the precessing nucleus jumps back and forth between a_A and a_B many times during the lifetime in the spin states, the effective field on the nucleus will be the averaged field for the two sites. There will be only one hyperfine structure observed with $2I + 1$

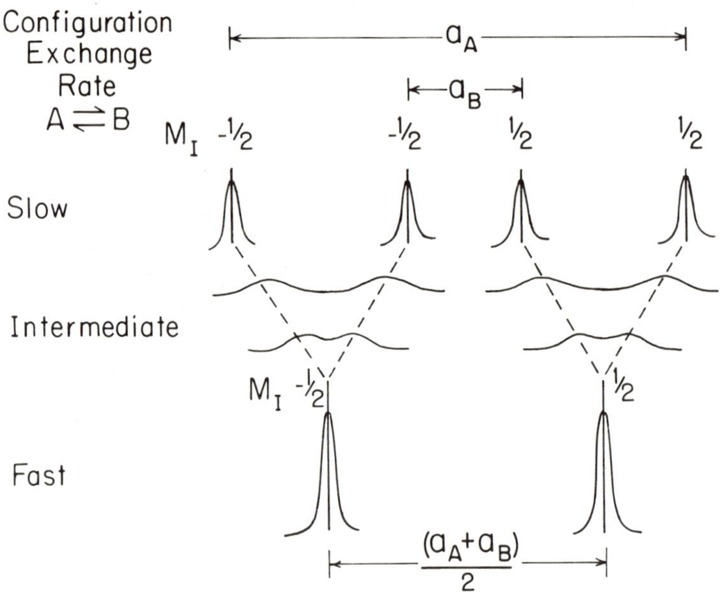

Fig. 9.3 Diagram showing qualitative effects on the esr hyperfine structure of periodic change of coupling values $a_A \rightleftharpoons a_B$ of a single coupling nucleus having spin $I = 1/2$. The top spectrum is for two equally probable configurations A and B in which the interconversion effects are negligible. The bottom spectrum is that for a very fast interconversion $A \rightleftharpoons B$ in which the precessing nucleus responds only to the averaged coupling field $(a_A + a_B)/2$ for the two configurations. At intermediate exchange rates for which $\tau_c \sim (1/a_A)$, $(1/a_B)$, the components for the separate configurations are broad and overlapping.

components and with splitting $\bar{a} = (p_A a_A + p_B a_B)$. Figure 9.3 illustrates the spectra for these cases when $I = 1/2$, $a_A = 4a_B$ and $p_A = p_B$.

When the lifetimes of conformations A and B, τ_A and τ_B, are long in comparison with $T_{2,0}$, the components are essentially those for two different radicals in which the line widths are

$$(\Delta\omega_{1/2})_A = \left(\frac{1}{T_{2,0}}\right)_A \quad \text{and} \quad (\Delta\omega_{1/2})_B = \left(\frac{1}{T_{2,0}}\right)_B \tag{9.30}$$

As the rate of exchange A $\rightleftharpoons$ B increases, the lines begin to broaden from those given by (9.30). The increased broadening may be attributed to the decrease in the lifetime τ_A or τ_B of the two conformations considered as separate radicals in very slow exchange. Even though the transformation may be adiabatic, the phase relations of the precessing spins are destroyed by the considerable difference in the nuclear precessional frequency at the two sites. As a result, the transverse relaxation time is shortened. Therefore, contributions of $1/\tau_A$ or $1/\tau_B$ must be added to the line width as the slow exchange speeds up, that is, for the semi-slow exchange. In angular frequency units, the line half-widths may then be expressed by

$$(\Delta\omega_{1/2})_A = \left(\frac{1}{T_{2,0}}\right)_A + \left(\frac{1}{\tau_A}\right) \tag{9.31}$$

$$(\Delta\omega_{1/2})_B = \left(\frac{1}{T_{2,0}}\right)_B + \left(\frac{1}{\tau_B}\right) \tag{9.32}$$

where τ_A and τ_B are mean lifetimes in configurations A and B and $\Delta\omega_{1/2} = \pi\Delta\nu$ when $\Delta\nu$ is the full width (in Hz) between half-intensity points.

In the intermediate region of exchange rates, the problem of calculation of line width is much more complicated and is best accomplished with a numerical computation program [12,13].

The line widths for rapid alternation of coupling $a_A \leftrightarrow a_B$ can be calculated with either the modified Bloch equations or with the spectral-density or relaxation matrix method similar to that used for the treatment of the tumbling motions. In the Bloch formulation, the varying field at the coupling nucleus causes an increased rate of decay in the transverse magnetization due to the diminished phase coherence of the precessing spins. There is no corresponding decay in the M_z magnetization (adiabatic fast passage). In the spectral-density-matrix method, this adiabatic phase corresponds to a secular process in which only the diagonal matrix elements of the density matrix are included.

For a single coupling nucleus, the spin Hamiltonian for the rapidly tumbling radicals may be expressed as

$$\mathcal{H}_{S,I} = \langle g \rangle \beta H_z S_z + a(t)\mathbf{S} \cdot \mathbf{I} \tag{9.33}$$

In the rapid site exchange, the time-independent part can be taken as the completely averaged interaction, and the time-dependent part can be treated by first-order perturbation theory in the relaxation-matrix method. The Hamiltonian $\mathcal{H}_{S,I}$ is then expressed as

$$\mathcal{H}_{S,I} = \mathcal{H}_0 + \mathcal{H}(t) \tag{9.34}$$

where

$$\mathcal{H}_0 = g\beta H_z S_z + \bar{a}\mathbf{S} \cdot \mathbf{I} \tag{9.35}$$

$$\mathcal{H}(t) = [a(t) - \bar{a}]\,\mathbf{S} \cdot \mathbf{I} \tag{9.36}$$

and

$$\bar{a} = p_A a_A + p_B a_B \tag{9.37}$$

Derivation of the line widths by the relaxation matrix method are given in several reference works [12, 14, 17] and are not repeated here. The resulting line width for rapid configuration exchange is

$$(\Delta\omega_{1/2})_{M_I} = \frac{1}{T_2(M_I)} = \frac{1}{T_{2,0}} + \frac{1}{(T_2)_{\text{exc}}} \tag{9.38}$$

where

$$\frac{1}{T_{2,0}} = p_A\left(\frac{1}{T_{2,0}}\right)_A + p_B\left(\frac{1}{T_{2,0}}\right)_B \tag{9.39}$$

and

$$\left(\frac{1}{T_2}\right)_{\text{exc}} = \hbar^{-2}(g\beta)^2\, p_A p_B (a_A - a_B)^2 M_I^2 \tau \tag{9.40}$$

in which $\tau = \tau_A \tau_B/(\tau_A + \tau_B)$ and the couplings a_A and a_B are in gaussion units (G).

Solution of the modified Bloch equations gives results in complete agreement with the results described above. In this formulation the results are usually expressed in angular frequencies rather than coupling constants. With the transformation,

$$\omega(M_I) = \gamma_e a M_I = \frac{g\beta}{\hbar} a M_I \tag{9.41}$$

(9.40) may be expressed as

$$\left(\frac{1}{T_2}\right)_{\text{exc}} = p_A p_B [\omega_A(M_I) - \omega_B(M_I)]^2 \tau \tag{9.42}$$

In this expression $\omega_A(M_I)$ and $\omega_B(M_I)$ are the angular frequencies ($\omega = 2\pi\nu$) for the hyperfine components corresponding to the particular M_I values when the radical is characterized by the respective configuration A or B (slow exchange where $\tau_c \approx 0$). The illustration of Fig. 9.3 is for $I = 1/2$, and M_I has only the values 1/2 and $-1/2$. The preceding expressions apply to any value of I, but only to a single coupling nucleus having fluctuating coupling. Although nuclei with $I > 1/2$ generally have nuclear quadrupole coupling, this interaction is canceled by the rapidly tumbling motions and, hence it makes no significant contribution to the line broadening.

When the configurations A and B are equally probable, $p_A = p_B = 1/2$, $\tau_A = \tau_B = 2\tau = \tau_c$ and (9.42) becomes

$$\left(\frac{1}{T_2}\right)_{\text{exc}} = \frac{1}{8} [\omega_A(M_I) - \omega_B(M_I)]^2 \tau_c \tag{9.43}$$

2.b. Exchange of Two Coupling Nuclei

The qualitative effects on isotropic esr hyperfine structure caused by peridoic exchange of two inequivalent protons are illustrated by Fig. 9.4. The protons may be exchanged by internal rotation or by other means. The same effects would be achieved by periodic reversal of their coupling spin densities without exchange of the nuclei. For a very slow rate of exchange in which $\tau_c \gg 1/(a_1 + a_2)$, where τ_c is the mean time for the exchange, a four-component hyperfine structure expected for two inequivalent protons with couplings a_1 and a_2 will be observed. This structure is indicated by the top diagram of Fig. 9.4, where it is assumed that $a_1 = 3a_2$ and that both couplings are positive. Note that this spectrum is like that for a single coupling proton that exchanges coupling sites at a low exchange rate (see the top diagram of Fig. 9.3). For fast exchange,in

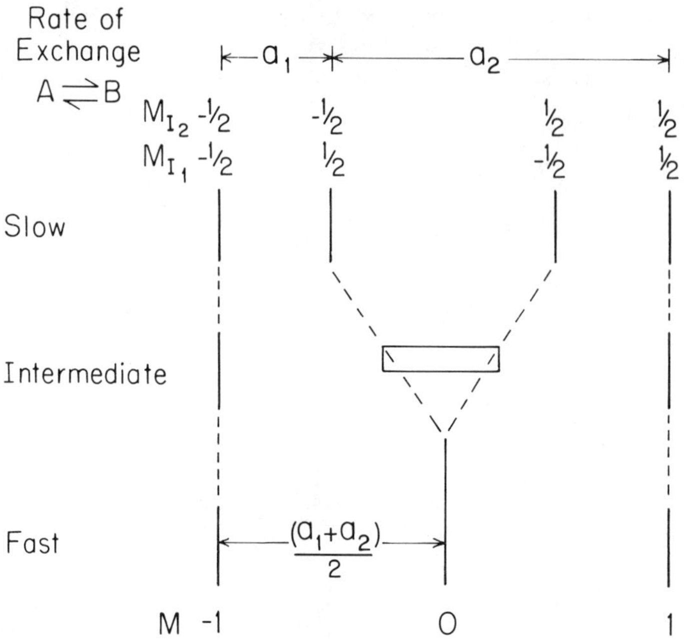

Fig. 9.4 Diagram indicating qualitative effects on the spectrum caused by exchange of coupling values a_1 and a_2 of two nuclei having equal spins of $I = 1/2$. Compare this diagram with that of Fig. 9.3 for a single coupling nucleus.

which $\tau_c \ll 1/(a_1 + a_2)$, the two couplings will be averaged, and a three-component hyperfine structure of two equivalent protons having couplings $\bar{a} \approx (a_1 + a_2)/2$ will be observed, as indicated by the bottom stick diagram of Fig. 9.4. In the intermediate region, where $\tau_c \approx 1/(a_1 + a_2)$, the two inside components of the top diagram approach each other and become very broad whereas the two outside components remain essentially unchanged. Finally, when τ_c continues to decrease (exchange frequency continues to increase), the two inside components converge to form the relatively sharp central line having twice the statistical weight of each outside component.

If the two isomeric forms are again designated as A and B, their angular hyperfine frequencies for no interconversion, or for very low rates of interconversion, are

$$\omega^{A}_{M_{I_1}, M_{I_2}} = \omega_0 + \gamma_e(a_1 M_{I_1} + a_2 M_{I_2}) \tag{9.44}$$

$$\omega^{B}_{M_{I_1}, M_{I_2}} = \omega_0 + \gamma_e(a_2 M_{I_1} + a_1 M_{I_2}) \tag{9.45}$$

from which

$$\omega^{A}_{M_{I_1},M_{I_2}} - \omega^{B}_{M_{I_1},M_{I_2}} = \gamma_e(a_1 - a_2)(M_{I_1} - M_{I_2}) \tag{9.46}$$

Since for the outermost components, $M_{I_1} = M_{I_2}$ when $I_1 = I_2$, these two components are the same for the two configurations. This holds for any spin value I for the exchanged, identical nuclei. Let us assume that A and B are equally probable configurations, that there is no significant flipping of the spins during the exchange (secular or adiabatic process), and that the time for the exchange (jumping time) is short compared with the lifetimes in the two equilibrium configurations A and B. A rapid exchange of I_1 and I_2, one that occurs many times during the nuclear precessional period, will average out and equalize the effective couplings of the two exchanged nuclei. Thus for the rapid exchange, we can replace a_1 and a_2 by their averages in (9.44) and (9.45) and obtain

$$\bar{\omega}_M = \frac{(\omega^{A}_{M_{I_1},M_{I_2}} + \omega^{B}_{M_{I_1},M_{I_2}})}{2} = \gamma_e \bar{a} M \tag{9.47}$$

where $M = M_{I_1} + M_{I_2}$ and $\bar{a} = [(a_1 + a_2)/2]$.

In the exchange process, the magnetic field at each nucleus is modulated as the coupling constant of nucleus (1) is converted to that originally for nucleus (2), and vice versa. When $a_1 > a_2$, the coupling of nucleus (1) is decreased by the exchange and that of (2) is increased by a like amount; that is, the modulations of the magnetic fields at the two nuclei are inversely correlated, or out of phase. As a result, the sum of their coupling is not altered by their exchange, whatever the rate.

The contribution of the exchange process to the component line widths for exchange of two nuclei may be calculated in a manner similar to that for site exchange by a single nucleus. When the rate of site exchange is slow, the line-width formula is like (9.31) with $\tau_A = 2\tau$, where τ is the mean exchange time when $\tau_A = \tau_B$. When the rate of site exchange is fast, the line width can be obtained by substitution of the angular frequency difference from (9.46) into (9.42). The result is

$$\left(\frac{1}{T_2}\right)_{\text{exc}} = \gamma_e^2 p_A p_B (a_2 - a_1)^2 (M_{I_1} - M_{I_2})^2 \tau \tag{9.48}$$

where $\gamma_e = (g\beta/\hbar)$ and where a_1 and a_2 are in G. This term is added to $1/T_{2,0}$, the line width in the absence of exchange. Thus for a component corresponding to $M = M_{I_1} + M_{I_2}$,

$$(\Delta\omega_{1/2})M = \frac{1}{T_2(M)} = \frac{1}{T_{2,0}} + \left(\frac{1}{T_2}\right)_{\text{exc}} \tag{9.49}$$

For equally probable conformations,

$$(\Delta\omega_{1/2})M = \frac{1}{T_{2,0}} + \frac{1}{8}\,\gamma_e^2(a_1 - a_2)^2(M_{I_1} - M_{I_2})^2\tau_c \tag{9.50}$$

When the exchange is very rapid, $\tau_c \approx 0$, it makes no contribution to the widths, which become like those in the absence of exchange. Regardless of the value of τ_c, it is evident that the exchange term in (9.48) vanishes when $M_{I_1} = M_{I_2}$. The outermost components always have $M_{I_1} = M_{I_2}$ when $I_1 = I_2$; hence these components are never broadened by exchange. In the diagram of Fig. 9.4, the center line for the fast exchange is made up of the $M_{I_1} = 1/2$ and $M_{I_2} = -1/2$ components. For this component, (9.48) gives the exchange contribution to the width of $(1/8)\gamma_e^2(a_1 - a_2)^2\tau_c$. From measurement of the extra width of this central line over the widths of the outside components, one can obtain an experimental value for this exchange contribution. If a_1 and a_2 can be obtained from measurements at lowered temperature, an experimental evaluation of τ_c can be made.

The vinyl radical is a relatively simple example of the cis-trans conversion in which two protons are exchanged. The esr of this radical has been measured in the argon matrix at 4°K by Adrian, Cochran, and Bowers [21] and found to have an eight-line hyperfine pattern of three inequivalent protons, as illustrated in the top diagram in Fig. 9.5. This observed pattern conforms to either of the equivalent forms **25A** or **25B**

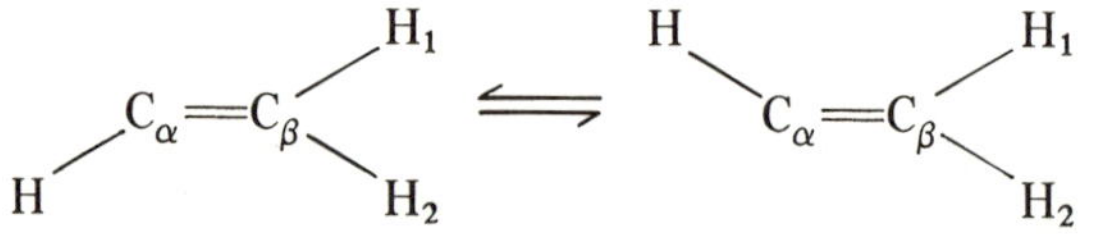

25A **25B**

with the isotropic proton coupling constants $a_\alpha = 15.7$ G, $a_{\beta_1} = 68.5$ G, $a_{\beta_2} = 34.2$ G for form 25A and $a_\alpha = 15.7$ G, $a_{\beta_1} = 34.2$ G, and $a_{\beta_2} = 68.5$ G for form 25B. (These couplings are quoted by Fessenden and Schuler [22] as remeasured values, privately communicated by Adrian, Cochran, and Bowers.) The spectra of these two equivalent tautomers cannot be distinguished. Hence, if there is no interconversion, or if only a very slow interconversion occurs, the spectra characteristic of a single species is observed. The conversion of one form to the

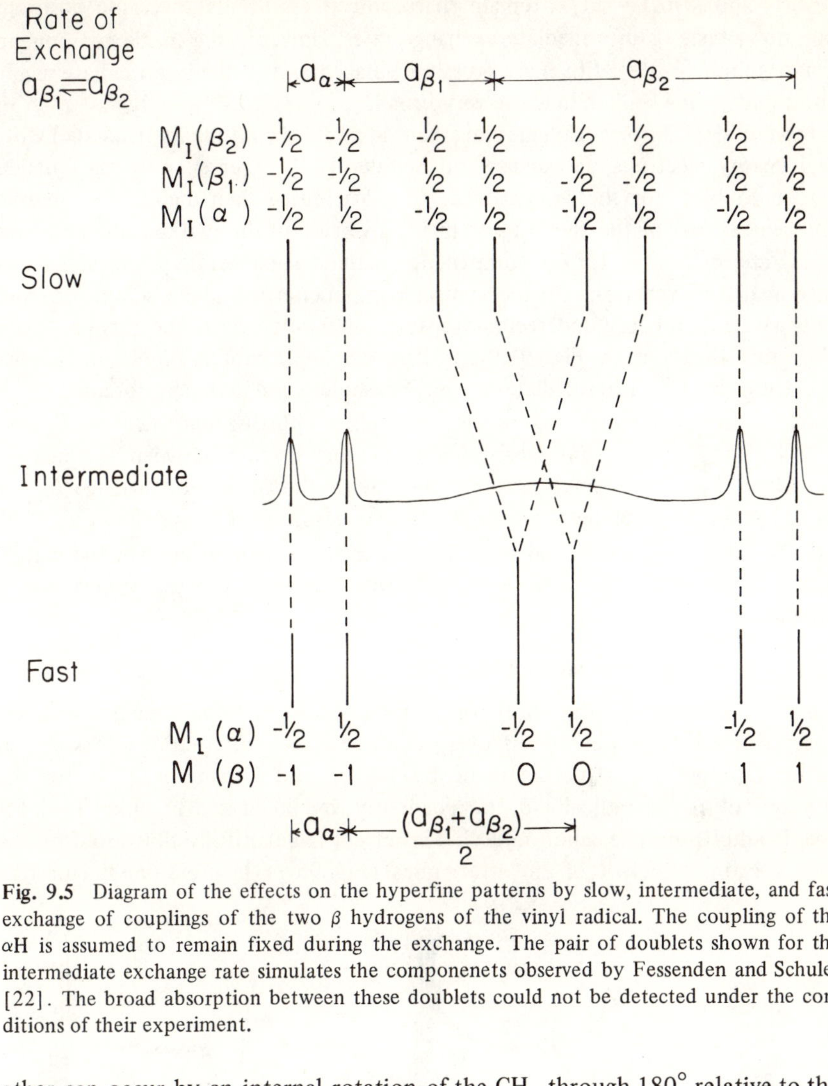

Fig. 9.5 Diagram of the effects on the hyperfine patterns by slow, intermediate, and fast exchange of couplings of the two β hydrogens of the vinyl radical. The coupling of the αH is assumed to remain fixed during the exchange. The pair of doublets shown for the intermediate exchange rate simulates the componenets observed by Fessenden and Schuler [22]. The broad absorption between these doublets could not be detected under the conditions of their experiment.

other can occur by an internal rotation of the CH_α through 180° relative to the CH_2. This would exchange the coupling values of the β protons as indicated without changing the α protons. If such internal rotation were to occur at a sufficiently rapid rate (fast exchange), the β couplings would be equalized, and a six-component hyperfine structure would be observed, as indicated in the bottom diagram of Fig. 9.5. The H_α coupling is equivalent for forms 25A and 25B and gives rise to the same doublet splitting in both the top and bottom patterns. For reasons already given, the outside doublets, which conform to $M_I(\beta_1)$ =

$M_I(\beta_2)$ and $M_I(\alpha) = \pm 1/2$, remain sharp and at essentially the same positions for slow, fast, or intermediate exchange rates. However, the inside components for which $M_I(\beta_1) \neq M_I(\beta_2)$ are grossly displaced and in the intermediate region indicated by the broken lines are very broad.

In a liquid mixture of ethylene and ethane at $-180°C$ continually irradiated with high-energy electrons, Fessenden and Schuler [22] observed only the outside set of doublets for the vinyl radical. They attributed their inability to observe the central part of the spectrum to interconversion of the two tautomeric forms at a rate such that $1/\tau_c$ is comparable to the resonance frequencies. At this intermediate exchange rate the central components would be scrambled, and only a broad, probably undetectable, absorption would occur. The curve between the stick diagrams in Fig. 9.5 represents an approximate simulation of the spectrum in this intermediate range. Fessenden and Schuler obtained a_α = 13.4 G from measurement of the α doublet splitting and $\langle a_{\beta_1} + a_{\beta_2} \rangle$ = 102.4/2 = 51.2 G from measurement of the spacing between the pair of doublets. The latter value agrees closely with the arithmetic average $(a_{\beta_1} + a_{\beta_2})/2$ = 51.3 G obtained from the matrix measurement of Adrian et al. [21] at 4°K. However, the a_α value of 13.4 G is somewhat lower than the matrix value of 15.7 G, an indication that the inversion process lowers the α coupling.

2.c. Alternating Line-width Effect

An early impetus to the study of periodic conformational changes with esr spectra of solutions was the discovery of the effect of alternating line width on isotropic hyperfine structure made by Bolton and Carrington [23] for the duroquinol cation radical and, independently, by Freed and Fraenkel [24] for the 1,4-dinitrodurene anion radical. This effect is beautifully illustrated by the esr hyperfine structure of the latter radical (Fig. 9.6). The gross fine-line pattern

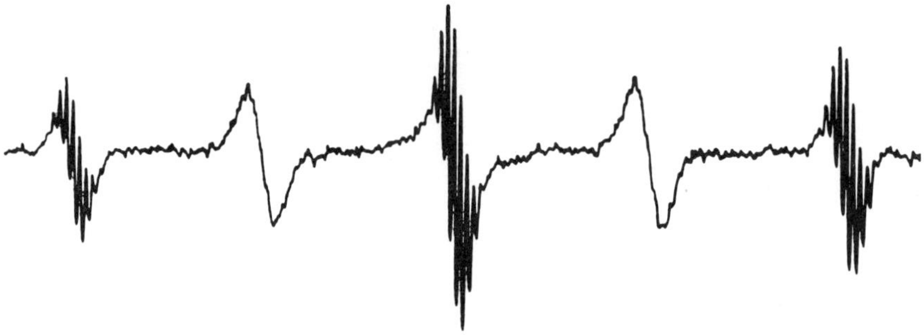

Fig. 9.6 First-derivative esr spectrum of the dinitrodurene anion radical in dimethyl formamide solution at room temperature. The gross quintet pattern arises from rapid exchange of coupling values of two ^{14}N nuclei, as described in the text. This pattern illustrates the alternation in line widths diagrammed in Fig. 9.7. From Freed and Fraenkel [24].

results from coupling by two equivalent ^{14}N nuclei, and the substructure evident in three of the ^{14}N components comes from equivalent coupling by the protons of the four methyl groups. The lack of resolution of this substructure in all five ^{14}N components is caused by anomalous broadening of two of them. As explained in detail by Freed and Fraenkel [5,6], the alternating line width exhibited in this spectrum is due to rapid periodic exchange of coupling values of the two ^{14}N nuclei. The probable mechanism for out-of-phase modulation of the two ^{14}N couplings is the oscillation between forms **26A** and **26B**

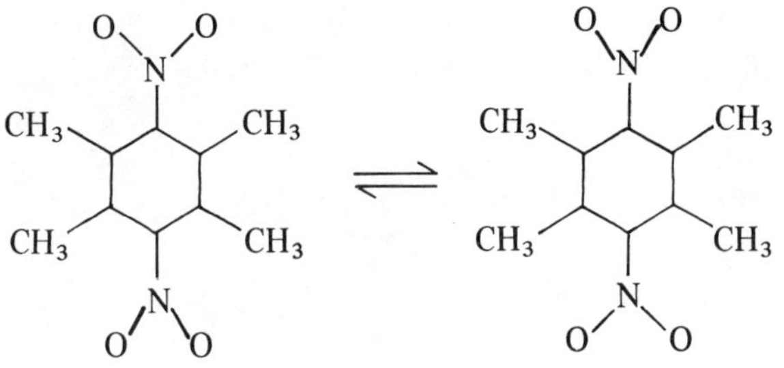

26A **26B**

in which one NO_2 group is in plane and the other is normal to the plane of the ring. To produce the observed spectrum, the tumbling motions must average out all the anisotropy, and the interconversion between forms **26A** and **26B** must be sufficiently rapid to equalize the isotropic ^{14}N coupling of the two orthogonal NO_2 groups. Exclusive of line widths, the ^{14}N hyperfine structure is, under these conditions, the quintet pattern indicated by the bottom stick diagram of Fig. 9.7. The five components of this diagram correspond to $M_N = M_{I_1} + M_{I_2} = -2,-1,0,1,2$. The top stick diagram of Fig. 9.7 indicates the ^{14}N hyperfine pattern expected for the radical if there were no interconversion, or only slow interconversion, between forms **26A** and **26B**. As the exchange rate between forms **26A** and **26B** increases, certain of the lines, those formed by unequal values of M_{I_1} and M_{I_2}, begin to shift in frequency, as indicated by the broken lines; but others, those formed from equal values of M_{I_1} and M_{I_2}, do not change positions and are not broadened by the exchange. That there is no exchange broadening of these latter components when the exchange is rapid is shown by substitution of their M_{I_1} and M_{I_2}, values into (9.48). It is seen that the $(1/T_2)_{exc}$ term vanishes regardless of the relative coupling values. The broadening of the M_N components for which $M_{I_1} \neq M_{I_2}$ does not vanish. Their widths are calculable by (9.49) when the parameters of this equation are known. Note that the two outside lines, those for $M_N = -1$ and 1, have $M_{I_1} = M_{I_2}$, and that both are sufficiently sharp to have resolvable substructure. The central $M_N = 0$ line is

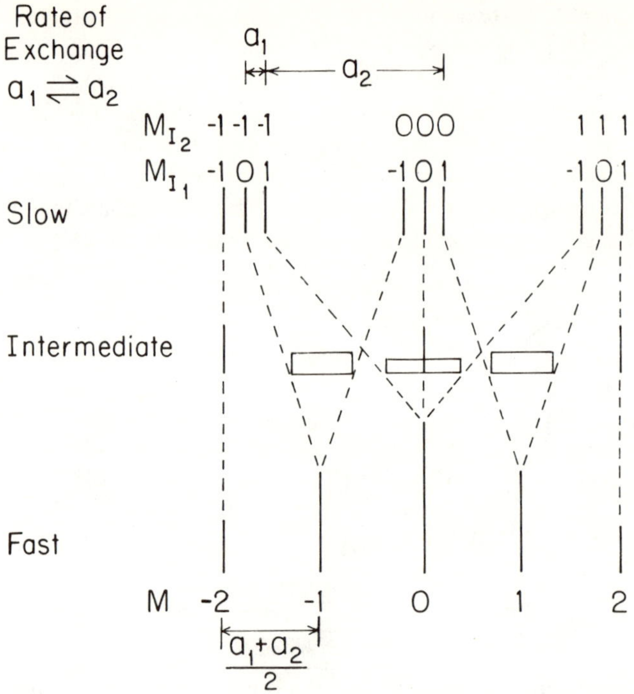

Fig. 9.7 Diagram of effects on the hyperfine patterns for different rates of exchange of the couplings of two nuclei having spins $I = 1$. Note that the three components for $M_{I_1} = M_{I_2}$ remain essentially unchanged.

made up of two components that are broadened by the exchange, those for $M_{I_1} + M_{I_2} = (\pm 1) + (\mp 1)$, and one component that is not, for which $M_{I_1} = M_{I_2} = 0$. It is this latter, very sharp, component superimposed on the much broader absorption of the other two that makes resolvable the substructure in the central line of the methyl group. No constituent of the $M_N = \pm 1$ lines has $M_{I_1} = M_{I_2}$; consequently, they are so broad that the substructure is not resolvable. The widths of the resolvable proton hyperfine lines give a measure of $1/T_{2,0}$. It is obviously difficult to measure the exchange broadening of the $M_N = -1$ and 1 components because of the spreading of these lines by the unresolved substructure, but it is evident that this broadening is greater than the widths of the lines in the unresolvable substructure. These latter widths should be nearly equal to those for the resolved substructure in the other M_N components.

Many other examples of alternating line widths in esr hyperfine structure are described in the review of the subject by Sullivan and Bolton [13].

2.d. Coupling Nuclear Groups

When there are more than two equivalent nuclei of a radical with fluctuating coupling, it is convenient to classify them as *equivalent* or *completely equivalent.* Freed and Fraenkel [5,6] define completely equivalent nuclei as those having equivalent coupling at any instant of time, $a_i(t) = a_j(t)$, and they define equivalent nuclei as those for which the couplings are equivalent only when averaged over a period of time. For example, the three hydrogens of a β-coupling methyl group will have equivalent coupling when the group is undergoing rapid rotation about its symmetry axis, but at any instant of time during the rotation they are inequivalent. The two nuclei illustrated in Fig. 9.4 are equivalent in their coupling for fast exchange but are inequivalent at any instant during the exchange and are inequivalent when the exchange is stopped by cooling. For calculation of modulation effects on line widths and frequencies, two completely equivalent nuclei may be considered as having a common spin, $I_T = I_1 + I_2$, with magnetic quantum numbers $M_T = I_T, I_T - 1, I_T - 2 \cdots -I_T$ and component hyperfine spacing of $a_1(t) = a_2(t)$. Because of degeneracies, however, the relative intensities of the hyperfine components will not be the same as those for a single nucleus with spin I_T. Let us suppose that the couplings of two sets, I and II, of completely equivalent hydrogens are modulated by an alternation between two molecular conformations, A and B, which exchanges the coupling of set I with that of set II. Since the couplings of the nuclei of each set remain equivalent throughout the exchange process, each set can be treated as a single nucleus with total spin I_T. The diagram for exhibiting the effects of slow, fast, and intermediate exchange of the coupling values of the two sets, each having two completely equivalent hydrogens ($I_T = 1/2 + 1/2 = 1$), is like that of Fig. 9.7 for the exchange of two ^{14}N nuclei ($I = 1$), except that in slow exchange the triplet of each set has an intensity ratio of 1:2:1 rather than the 1:1:1 for ^{14}N; and in fast exchange, the quintet intensity ratio is 1:4:6:4:1 rather than the 1:2:3:2:1 for the ^{14}N pair. The same type of alternation of line widths will occur in the fast exchange quintet as that illustrated for the ^{14}N pair.

If the nuclei of set I or II are equivalent in configuration A or B but are not completely equivalent, they will become inequivalent during the transition A $\leftrightarrows$ B. Therefore, in calculation of the modulation effects, their spins cannot be combined and treated as a single spin of I_T, but each must be treated as a separate spin, the coupling of which is modulated individually.

Fraenkel [20] gives a number of examples illustrating the effects of coupling modulations on equivalent and completely equivalent spin groups.

2.e. Electron Exchange

Exchange of a solute free radical with one of the paired electrons of molecules of a diamagnetic solvent converts the solvent molecule to a free radical and takes away the spin of the original free radical. Because of the large difference in energies of the orbitals of the usual free radical solute and solvent molecules, this kind of exchange is not prevalent, except when the solute radicals are molecular ions of the solvent molecules. Many aromatic anion radicals studied in liquids are produced by ionization of alkali atoms placed in the solution for the purpose of supplying electrons to the aromatic solute molecules. The alkali atom most commonly used is sodium. The first such ionic species to be studied was the naphthalene negative ion, produced when sodium and naphthalene were dissolved in dilute solution in tetrahydrofuran [25]. Although the electron-transfer process between naphthalene and its anion is probably the one most thoroughly studied [26,27], the effects of such transfer on esr spectra have been observed in many other liquid systems [12].

The jumping of the electron between two molecules does not cause significant flipping of the electron spin. Hence the problem can usually be treated as a secular process (adiabatic transfer). The jumping of the electron back and forth between two molecules produces an out-of-phase modulation of the hyperfine structure of the interacting molecules. Line broadening and frequency conversion of the nuclear hyperfine components, as already described in Sections 2.a and 2.b, are expected. The exchange time τ is inversely related to the rate constant k for the reactions

$$\text{mol}\,(1) + \text{mol}^-(2) \leftrightharpoons \text{mol}^-(1) + \text{mol}\,(2)$$

For slow exchange, the line width $(\Delta\omega_{1/2})$ is given by (9.31), where

$$\tau_A = \tau_{\text{mol}^-(1)} = \tau_{\text{mol}^-(2)} = 2\tau$$

It is evident that the lifetime for a particular molecular anion, τ_{mol^-}, should depend on the concentration of the similar, neutral species in the solution. Thus one can set $\tau_{\text{mol}^-} = (1/Ck_2)$, where C is the concentration of the interacting neutral species and k_2 is the reaction rate constant in liter/mole-sec. From measurement of the line widths as a function of C, one can find k_2. The rate constants and associated activation energies for electron-transfer reactions between a number of aromatic molecules and their anions are tabulated by Johnson, who gives an informative treatment of this and other chemical rate processes as measured by esr and nmr [12].

Electron exchange rates between unpaired electrons of free radicals in solution can also be studied by their effects on hyperfine patterns and line widths. These effects were first studied by Lloyd and Pake on the ^{14}N hyperfine spectrum of $NO(SO_3)_2^{--}$ radical in aqueous solutions [28]. Because relatively concentrated solutions of radicals are required for study of electron spin-spin exchange, these phenomena are not often observed in the dilute solutions of radicals commonly studied in esr of the liquid state.

2.f. Proton Exchange

Using the rapid-mixing techniques for production of transient free radicals (Section 3.e), Fischer [29] observed that the OH proton splitting of the esr lines of $\dot{C}H_2OH$ radicals in aqueous solutions depended on the pH of the solutions, which was varied by addition of H_2SO_4. He explained the effect in terms of an exchange of the OH proton of the radical with protons in the acidic solution, attributed the observed effects to the exchange reactions,

$$\dot{R}OH + H_3O^+ \rightleftharpoons \dot{R}OH_2^+ + H_2O$$

and related the mean exchange frequency $1/\tau$ to the reaction constant k by

$$\frac{1}{\tau} = \left(\frac{1}{[\dot{R}OH]}\right)\left(\frac{d[\dot{R}OH]}{dt}\right) = k[H_3O^+] \tag{9.51}$$

in which $[\dot{R}OH]$ represents the molar concentration of the free radicals (in this case $R = \dot{C}H_2$) and $[H_3O^+]$ represents the concentration of H_3O^+ ions assumed to be the source of the solution protons. He obtained the $1/\tau$ values from the OH doublet separation at the particular pH or $[H_3O^+]$ values from an approximate formula that may be expressed in the form,

$$\frac{1}{\tau} = \frac{1}{\sqrt{2}}\left[(\Delta\omega_0)^2 - (\Delta\omega)^2\right]^{1/2} = \frac{\gamma_e}{\sqrt{2}}\left[(\Delta H)_0^2 - (\Delta H)^2\right]^{1/2} \tag{9.52}$$

where ΔH is the OH splitting in G for the particular pH or H_3O^+ concentration and $(\Delta H)_0 = (a_H^{OH})_0$ is the OH splitting when the H_3O^+ concentration is zero (no exchange). Gutowsky and Holm [30] originally derived a formula of this kind for treatment of proton-exchange effects in nmr. One can obtain this relation by equating the derivative of $I(\omega)$ to zero where $I(\omega)$ is the more general shape function obtained from solution of the Bloch equations with the exchange modulation included. It holds only in the region where ΔH is small relative to $(\Delta H)_0$, that is, in the regions where the two components of the OH doublet begin to coalesce [12]. For acid concentrations at which complete coalescence occurs, $\Delta H = 0$, (9.52) reduces to

$$\frac{1}{\tau} = \frac{\gamma_e}{\sqrt{2}}(\Delta H)_0 \tag{9.53}$$

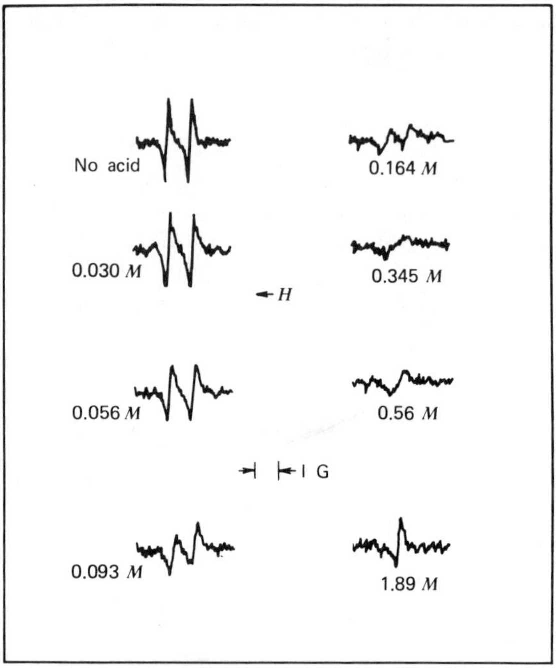

Fig. 9.8 Sequence showing effects of an acidic solution on the OH splitting of the $\dot{C}H_3OH$ radical. The spectral segment shown is the upper-field doublet of Fig. 9.1. It was produced here by photolysis of an aqueous solution containing 10% acetone and 5% methyl alcohol, to which various concentrations of HCl were added as indicated. From Zeldes and Livingston [31].

Zeldes and Livingston [31] applied the critical condition of (9.53) to obtain rather accurate values for $1/\tau$ and hence the reaction rate constant k for both the alcohol radicals $\dot{C}H_2OH$ and $(CH_3)_2\dot{C}OH$ in photolyzed aqueous solutions to which HCl was added. Figure 9.8 shows OH doublets of $\dot{C}H_2OH$ (see complete spectrum in Fig. 9.1) as the HCl concentration is increased from zero to 1.89 M. This sequence provides a beautiful example of the effects of chemical exchange-modulation of a single coupling proton, described in Section 2.a. Zeldes and Livingston chose the concentration of 0.345 M as that for which the doublet splitting vanishes and obtained $k = 3.6 \times 10^7$ liter/mole-sec for the reaction constant [see (9.51)] for $\dot{C}H_2OH$ in the solution at 28°C. By a similar procedure, they found that for $(CH_3)_2\dot{C}OH$ in the same solution, $k = 7.2 \times 10^7$ liter/mole-sec. This value of k for $\dot{C}H_2OH$ does not agree well with Fischer's earlier value of 1.76×10^8 liter/mole-sec for 17°C. From the nature of its derivation, it would appear that (9.52) should be most applicable in the critical region used by Zeldes and Livingston, and hence their values of k may be the more reliable.

However, variation in techniques for generation of the radicals and differences in other constituents of the solutions may account for much of the discrepancy in these independent evaluations of k for $\dot{C}H_2OH$.

Interesting chemical exchange effects were observed by Zeldes and Livingston [32] for the acetoin radical $CH_3\dot{C}OHCOCH_3$ when an acid, HCl, was added to the photolyzed solutions of isopropyl alcohol containing 2% diacetyl. Marked changes were observed to occur in the spectrum (at room temperature) as the HCl concentration was increased. In the photolyzed solution at 32°C that contains no acid, the spectrum is the upper curve in Fig. 9.9. This is the spectrum to be expected for the radical $CH_3\dot{C}OHCOCH_3$ in which the spin density is concentrated mostly on a single carbon. The proton splittings by the two CH_3 groups, 13.4 G and 2.6 G, are decidedly different, and the doublet splitting of the OH hydrogen, 2.1 G, is well resolved. The lower curve of Fig. 9.9 shows the spectrum (not to the same scale) obtained when 27.6% HCl is added to the solution. The latter spectrum can be explained in terms of a rapid, four-way, exchange indicated by the transformations,

$$
\begin{array}{ccc}
\text{*CH}_3\dot{C}\text{(OH}_\alpha\text{)}-\text{CCH}_3\text{(=O)} & \underset{K_2}{\overset{K_2}{\rightleftharpoons}} & \text{*CH}_3\dot{C}\text{(=O)}-\text{CCH}_3\text{(OH}_\alpha\text{)} \\
K_1 \updownarrow K_1 & \times \underset{K_2\ K_2}{\overset{K_2\ K_2}{}} & K_1 \updownarrow K_1 \\
\text{*CH}_3\dot{C}\text{(OH}_\beta\text{)}-\text{CCH}_3\text{(=O)} & \underset{K_2}{\overset{K_2}{\rightleftharpoons}} & \text{*CH}_3\dot{C}\text{(=O)}-\text{CCH}_3\text{(OH}_\beta\text{)}
\end{array}
$$

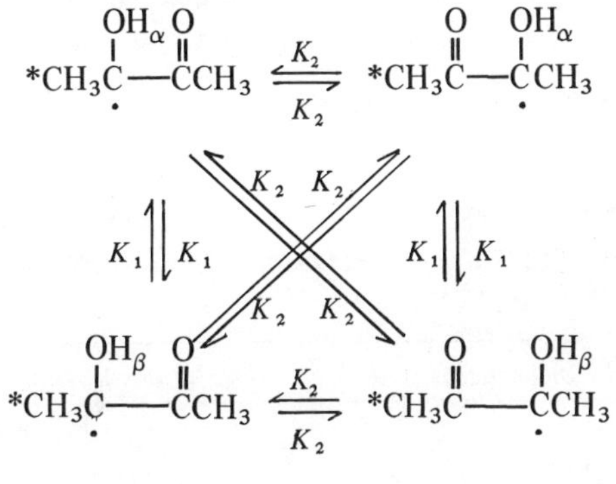

27

which equalizes the proton coupling of the two CH_3 groups by the K_2 exchange and neutralizes the OH splitting by the exchange indicated by K_1. The asterisk makes a distinction between the two CH_3 groups, and the α and β indicate the two possible spin orientations of the OH protons. The transformations are evidently caused by a chemical exchange of the OH proton with the acid protons in the solution. However, this exchange is not always a direct process. When a proton from the solution is added to the carbonyl oxygen, the proton of the OH group is lost, and vice versa, as indicated in the transformations. As the concentration of HCl is reduced, the exchange rate is slowed, and the spectral changes corresponding to the slower configuration-exchange effects are observed.

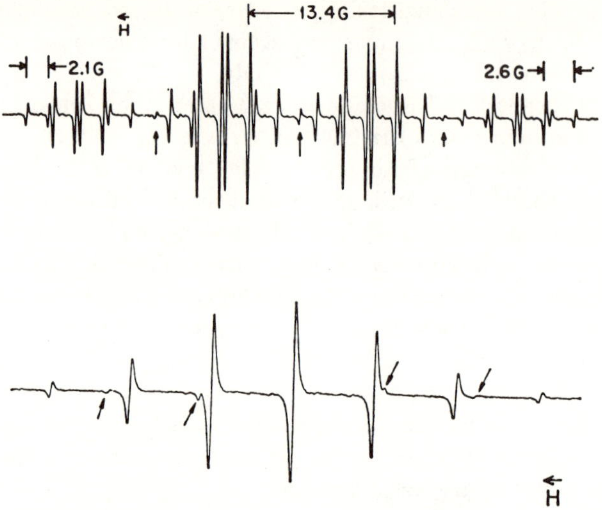

Fig. 9.9 Upper curve is the esr spectrum of the acetoin radical $CH_3\dot{C}OHCOCH_3$ in isopropyl alcohol containing 2% diacetyl at 32°C. Lines designated by arrows are from another radical. Lower curve is the esr spectrum of the acetoin radical after 27.6% concentrated HCl is added to the above solution. From Zeldes and Livingston [32].

Table 9.2 Comparison of the Proton-exchange Rate K_2 for the Acetoin Radical with Percentage Concentration of HCl in the Photolyzed Solution[a]

Percent HCl in Solution	*Exchange Rate K_2 (sec^{-1})*
0.107	2.30×10^6
0.74	1.07×10^7
5.0	7.0×10^7
14.2	5.2×10^8
27.2	3.0×10^9

[a]From data of Zeldes and Livingston [32].

Zeldes and Livingston [32] used the modified Bloch equations to calculate theoretical curves for comparisons with those observed for various concentrations of HCl. By adjustment of the rate constant for a best fit of the predicted to the observed spectra, they obtained values K_2 for the different percentages of HCl concentrations, as given in Table 9.2.

3. PRODUCTION AND OBSERVATION OF FREE RADICALS IN SOLUTION

3.a. Charged Aromatic Radicals Generated by Oxidizing and Reducing Agents

The study of the esr of charged aromatic radicals in dilute solution in organic solvents such as tetrahydrofuran was begun in 1953 by Weissman and his associates [33-36], who showed that closely spaced hyperfine components arising from different protons on the rings could often be resolved, or partially resolved. The Weissman group originally studied negative ions of naphthalene, anthracene, nitrobenzene, and other similarly ringed systems. The hyperfine structure of the naphthalene negative ion consists of 28 or more hyperfine components spread over about 25 G. Since this initial work, numerous ionic free radicals, both positive and negative, of complex organic ringed systems have been studied, some of which are described here. The negative ion radicals are generally formed by the reaction of the aromatic molecules with the alkali metals in solutions such as

$$C_{10}H_8 \rightarrow C_{10}H_8^- + Na^+$$

The same negative radicals have been prepared from Li and from Na.

When the neutral aromatic molecules are in sufficient concentration with their negative ions, an exchange of the electron occurs between ion and molecule, causing a broadening of the resonance (Section 2.e). Eventually, as the concentration of the neutral molecules is increased, the hyperfine components fuse into a single, broad resonance. Ward and Weissman [37] have shown how these effects on the esr can be used for measurement of the rate of transfer of the electron between the ion and the neutral molecule.

Positively charged aromatic ions can also be prepared and their esr observed in solution. One method of production is the oxidation of the molecule by sulfuric acid [38]. Hyperfine structure of the same type and of the same order of magnitude of splitting is observed as was noted for the negative ion. One of the more complicated patterns resolved is that of Wurster's blue ion

$$[(CH_3)_2N-\langle\rangle-N(CH_3)_2]^+$$

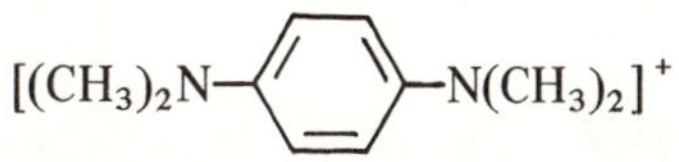

It was found by Weissman [39] to have a hyperfine structure of at least 39 components, consisting of 13 triplets separated by 7.4 G, with the triplet components spaced 2.1 G apart.

The esr of many multiringed cation radicals has been observed by Lewis and Singer [40], who produced the cations by oxidation of the aromatic hydrocarbons with $SbCl_5$ in dilute solution in CH_2Cl_2. Most of the resolved spectra had 50 to 100 closely spaced hyperfine components with line widths in the range from 0.04 to 0.40 G. As examples, the two-ringed naphthalene cation at $-91°C$ had 81 components and line widths of 0.12 G, the three-ringed anthracene cation at $-85°C$ had about 50 observable components and widths of 0.20 G, the four-ringed pyrene cation at $-38°C$ had 75 components and widths of 0.15 G, and the six-ringed cation of 9,9′-bifluorene at $-91°C$ had about 100 observable components and widths of only 0.04 G.

3.b. Alkali Metal-Ion, Aromatic-Ion Complexes

In their observation of the naphthalene negative ions in radicals produced by the reaction of sodium with naphthalene in solution, Atherton and Weissman [26] detected a four-component substructure on each of the hyperfine components of the naphthalene negative ion. They interpreted this substructure as being due to the ^{23}Na nuclear splitting in the (naphthalene)$^{-}$–Na^{+} complex. With nuclear spin of 3/2, ^{23}Na would be expected to split each of the lines of the naphthalene radical into four equally intense components, provided that the associated Na^{+} ions were sufficiently close that some of the electron spin density of the naphthalene negative ions would spread into the 3*s* orbital of the Na^{+} ion. The nuclear coupling constant of an electron of the free ^{23}Na atom is 316 G, whereas Atherton and Weissman observed the splitting of the ^{23}Na in the (naphthalene)$^{-}$-Na^{+} complex in tetrahydropyran solution to be $A_{Na} = 1.26$ G. Thus the 3*s* spin density on the Na in the complex is indicated to be only $\rho_{Na} = (1.26/316) = 0.004$.

Figure 9.10 demonstrates the splitting by K^{+} ions ($I = 3/2$) in the (biphenyl)$^{-}$–K^{+} complex in tetrahydropyran solution as obtained by Nishigushi, Nakai, et al.

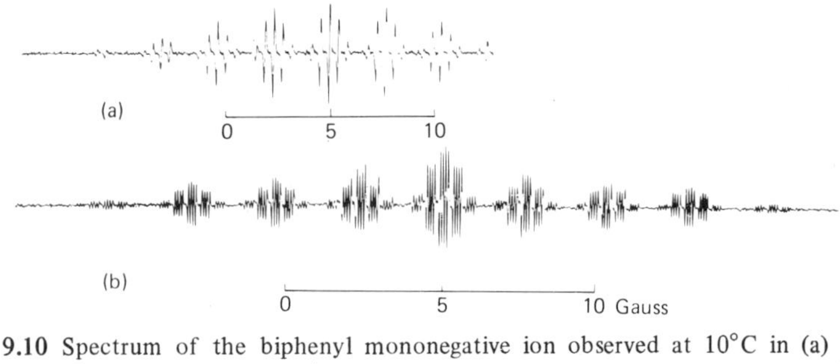

Fig. 9.10 Spectrum of the biphenyl mononegative ion observed at 10°C in (a) dimethoxyethane solution containing K^{+} ions, and (b) tetrahydropyran solution containing K^{+} ions. In (b) each line of (a) is split further, into four lines of equal intensity, by association with the K^{+} ions (nuclear spin, 3/2). From Nishiguchi, Nakai, et al. [41].

[41]. The ^{39}K coupling constant of the complex was observed, under the conditions of the experiment, to be A_K = 0.083 G, whereas the splitting for the atom ρ_K = 1 is 82.83 G. The spin density in the 4*s* orbital of the K^+ in the biphenyl ion complex is thus indicated to be ρ_K = (0.83/82.8) = 0.001. These observers also resolved the hyperfine structures for Li, Na, Rb, and Cs in biphenyl-alkali metal complexes.

The nuclear splitting by the metal in the aromatic-ion, metal-ion complex depends on the nature of the solution as well as on temperature. The solvent dependence is illustrated by the data of Nishigushi, Nakai, et al. [41] in their Tables I and II. No K splitting was observed for solutions in dimethoxyethane. Hyperfine structure of alkali metals bound in organic radicals has also been studied by de Boer and Mackor [42].

3.c. Semiquinones

Measurements of magnetic susceptibility by Michaelis, and his associates [43] led to the detection of semiquinone radicals long before the discovery of esr. Many semiquinone radicals have been studied in liquid solution with esr [44-47]. The proton hyperfine structure, which has been resolved for many of them, has given considerable information about their electronic structure.

The semiquinone radicals are prepared by chemical reactions, and most studies of them have been made on alkaline solutions in alcohols, in which the radicals have a sufficiently long life for esr observations. Venkataraman and Fraenkel [45] have used a continuous-flow process to maintain a constant concentration of radicals during the course of esr measurements.

Of the many esr studies made on the semiquinones, only a few can be mentioned. The simplest and perhaps the most basic of these radicals is the *p*-benzosemiquinone anion, which gives the beautiful quintet pattern shown in Fig. 9.11. Because of the high order of the symmetry of the radical, its pattern can be predicted either with molecular-orbital theory or in terms of the conjugated valence-bond structures.

The captured electron is in an antibonding π^* orbital that extends over the ring and on the oxygens. It is customary to indicate such an anion radical by an encircled negative charge at the center of the ring.

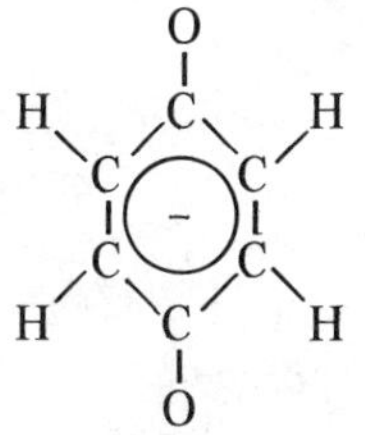

28

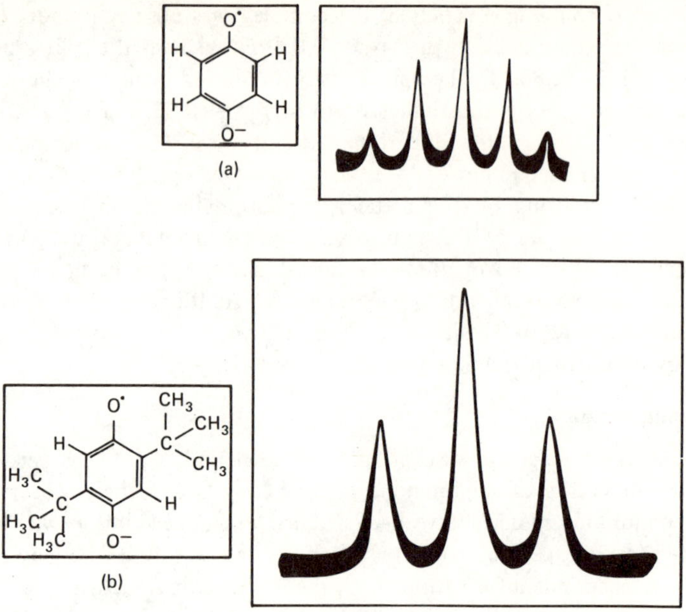

Fig. 9.11 (a) ESR spectra of unsubstituted *p*-benzosemiquinone (from Venkataraman and Fraenkel [45]); (b) esr spectra of 2,5-di-*t*-butyl-*p*-benzosemiquinone from Fraenkel [46].

From symmetry, it is evident that the unpaired π^*-electron density would have equal weight on the two oxygens and likewise on the ortho and meta carbons, which would give equivalent couplings to the four protons. The proton couplings can be considered as arising from the four equivalent spin-polarized >C-H fragments, the anisotropic components of which are averaged out by the tumbling motions. From the observed Fermi contact coupling of 2.37 G, the spin density on each of the C-H carbons can be estimated from (6.4) to be $\rho = 0.10$. Thus the four equivalent structures that put π^* density on the C-H groups constitute approximately 40% of the ground state of the radical. The remaining π^* density is mostly on the two oxygens, about 30% on each.

If a monovalent atom, or group X, is substituted for one of the four hydrogens on the *p*-benzosemiquinone anion, the symmetry is destroyed, and the couplings of the three remaining hydrogens will not be exactly equal, although they will remain approximately the same provided that the group does not drastically disturb the symmetry of the π^* orbital of the unpaired electron. If two equivalent X groups are substituted, there is again symmetry, and the couplings of the two remaining hydrogens on the ring are equivalent. The effects of this kind of substitution are illustrated by 2,5-di-*t*-butyl-*p*-benzosemiquinone

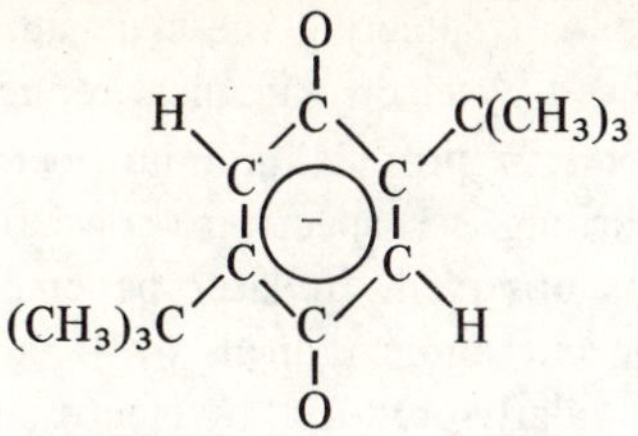

29

which gives the well-resolved triplet, also shown in Fig. 9.11. The (CH_3) groups are too far removed from the ring to give detectable splitting.

3.d. Electrochemically Generated Free Radicals

Geske and Maki [48] initiated a highly effective electrochemical method for generation of organic free radicals for study in dilute liquid solutions. They placed a small mercury pool electrode directly in the center of a microwave cavity and produced charged free radicals by continuous electrolysis of the solute molecules in the selected organic solvents. Details of the design of the system and a description of its operation are given in their original paper. A sketch of the essential parts is given in Fig. 9.12. In this paper they demonstrated the

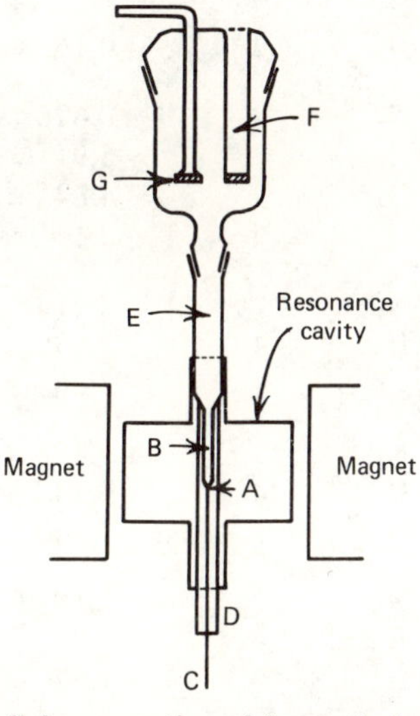

Fig. 9.12 Electrochemical cell for generation of free radicals directly within the resonance cavity of an esr spectrometer. The mercury pool electrode indicated by A is connected to the voltage source by a platinum wire C. The second electrode is outside the resonant cavity. From Geske and Maki [48].

effectiveness of the method by detecting the spectrum of nitrobenzene anion radicals ($C_6H_5NO_2$) produced by electrochemical reduction of nitrobenzene in acetonitrile solution with tetra-*n*-propylammonium perchlorate as a supporting electrolyte. A relatively strong esr spectrum consisting of 54 well-resolved hyperfine components was observed. In later papers, the observers reported studies of three dinitrobenzene anion radicals [49] and of 14 para-substituted nitrobenzene anions [50] similarly produced by electrochemical reduction of the parent molecule in the acetonitrile solvent. The esr spectra of a number of electrochemically prepared polyazine anions have been measured in dimethyl sulfoxide solutions by Stone and Maki [51]. Despite the numerous, closely spaced, hyperfine components observed, they were able to detect the ^{13}C splitting in pyrazine, with ^{13}C in its natural concentration of 1.1%. Table 9.3

Table 9.3 ESR Data for Polyazine Anions[a]

Anion of	*Nucleus*	n^b	*Coupling Constant*[c] *(G)*	*Line Width*[d] *(G)*
s-Tetrazine	N	4	5.275 ± 0.006	0.110
	H	2	0.212 ± 0.002	
	^{13}C	2	<0.8 or 10.6*n* ± 0.8	
Pyridazine	N	2	5.90 ± 0.08	0.087
	H	2	6.47 ± 0.08	
	H	2	0.16 ± 0.01	
Phthalazine	N	2	0.876 ± 0.005	0.150
	H	2	5.91 ± 0.02	
	H	2	4.64 ± 0.02	
	H	2	2.14 ± 0.02	
Pyrazine	N	2	7.213 ± 0.008	0.158
	H	4	2.639 ± 0.008	
	^{13}C	4	2.88 ± 0.02	
Diprotopyrazine (cation)	N	2	7.426 ± 0.008	0.188
	H	4	3.147 ± 0.005	
	H	2	7.883 ± 0.004	
Phenazine	N	2	5.15 ± 0.02	0.110
	H	4	1.80 ± 0.02	
	H	4	1.57 ± 0.02	

[a] From data of Stone and Maki [51].
[b] Number of equivalent nuclei.
[c] Absolute values.
[d] Full width measured between derivative extrema.

gives the observed coupling and line width obtained from a close fitting of the observed esr patterns to theoretically computed curves, as demonstrated in Fig. 9.13 for the phthalazine anion. Glarum and Synder [52] used electrochemical reduction to generate the anion radical of phenanthrene for esr measurements.

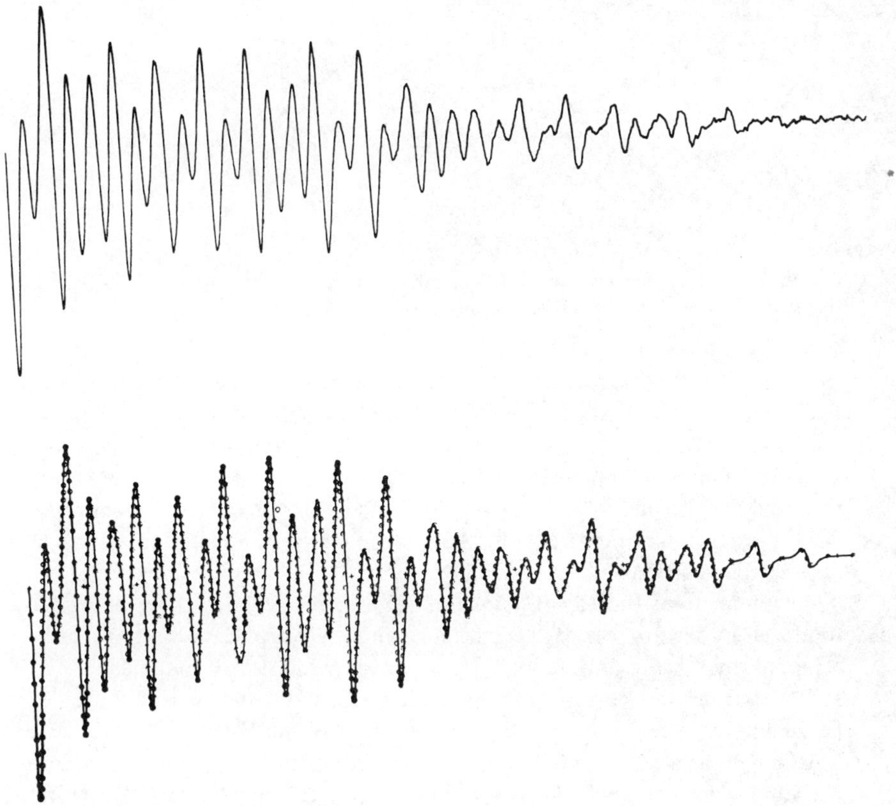

Fig. 9.13 Segment of the observed esr spectrum of the phthalazine anion (upper) and the calculated segment (lower). From Stone and Maki [51].

Fraenkel and co-workers [53-56] have made extensive use of the method and have notably improved the experimental techniques to avoid the introduction or production of impurities that broaden the esr lines. As a result, they have been able to obtain spectra of exceptional resolution from electrochemically generated free radical in organic liquid solutions.

The esr spectrum of the positive ion radical of *N,N'*-tetramethylbenzidine (TMB^+) obtained by Galus and Adams [57] is shown in Fig. 9.14. The lower curve represents the similar spectra of TMB^+ with the ring hydrogens (but not

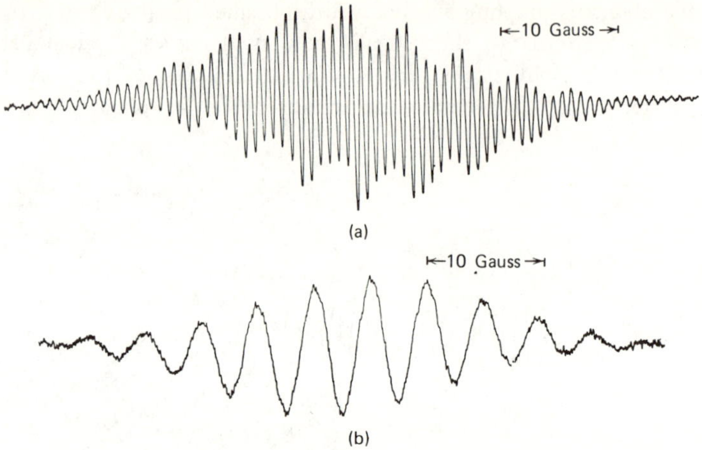

Fig. 9.14 (a) EPR spectrum of TMB^+ in 50% acetone-aqueous buffer, pH 3.8; (b) EPR spectrum of ring-deuterated TMB^+. From Galus and Adams [57].

the methyl hydrogens) replaced by deuteriums. The esr of the latter sample shows that the larger splitting is due to the methyl hydrogens. The spectrum consists of 13 components from the 12 equivalent methyl protons, each having a substructure of nine components caused by the eight-ring protons. The coupling observed for the methyl protons is 4.70 G and that for the ring protons is 0.78 G. The sample was prepared by oxidation of TMB^+ at a platinum electrode in 50% acetone-aqueous buffer at pH 3.8.

Measurement of the esr of anions of 2,5-dimethyl-2,4-hexadiene and glyoxal produced by electrolysis of the molecules in tetrahydrofuran solutions has been achieved by Tolles and Moore [58]. Anion radicals of butadiene and its substituted derivatives produced by electrolytic reduction of the molecules in liquid ammonia have been studied with esr by Levy and Myers [59].

Although a mercury pool electrode was used in the original experiment, platinum electrodes have been used by many later investigators. Most of the charged radicals produced by this method have been anions. For their production, the anode is placed in the sensitive region of the microwave cavity, and the cathode is placed outside the cavity. The polarity can be reversed, however, for the production of cation radicals, as in the work of Galus and Adams [57]. Various electrolytes have been used to increase the conductivity of the solution. Aqueous solvents have been employed in some applications. The selected studies mentioned provide only a few examples of the many free radicals electrochemically generated and observed with esr.

3.e. Transient Free Radicals Generated by Rapid Mixing of Liquids

In 1961 Saito and Bielski [60] observed the esr of $HO\text{-}\dot{O}$ radicals produced in aqueous solutions at room temperature by the rapid mixing of acidified solutions of ceric sulfate with H_2O_2. The reaction producing the radicals is

$$Ce^{4+} + H_2O_2 \rightarrow (Ce^{3+}H^{+}) + HO\dot{O}$$

Following this experiment, Dixon and Norman [61-64] developed a practical, rapid-mixing, liquid-flow method for observation of short-lived free radicals (lifetime ~0.01 sec) that are chemically produced by reacting liquids as they flow into the microwave absorption cavity. Figure 9.15 shows their apparatus for mixing the flowing reactants at the entrance to the microwave cavity. Figure 9.16 shows examples of the spectra obtained from organic liquids.

Most of the free radicals investigated with the Dixon-Norman flow method have been produced by hydrogen abstraction from an organic molecule by $\dot{O}H$ radicals. Dixon and Norman first used the transition metal ions in acidic aqueous solutions containing H_2O_2, producing $\dot{O}H$ from reactions of the type

$$Ti^{3+} + H_2O_2 \rightarrow Ti^{4+} + OH^{-} + \dot{O}H$$

When an additional organic substance such as ethyl alcohol is introduced into the mixture, a secondary reaction of the type

$$CH_3CH_2OH + \dot{O}H \rightarrow CH_3\dot{C}HOH + H_2O$$

produces radicals by H abstraction from the organic substance. The latter can

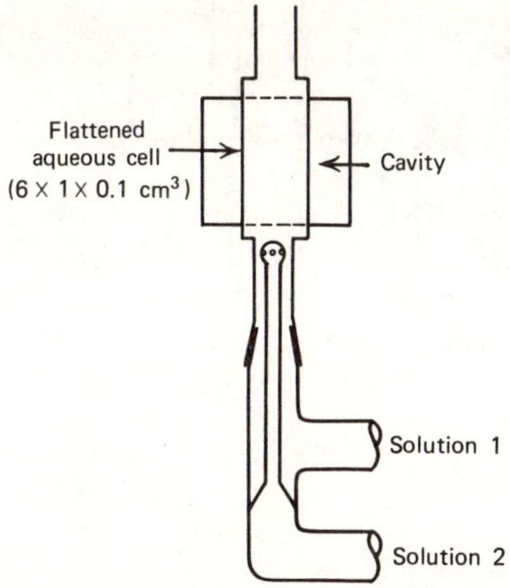

Fig. 9.15 Diagram of glass cell for rapidly mixing reactive solutions at entrance to microwave cavity. From Dixon and Norman [62].

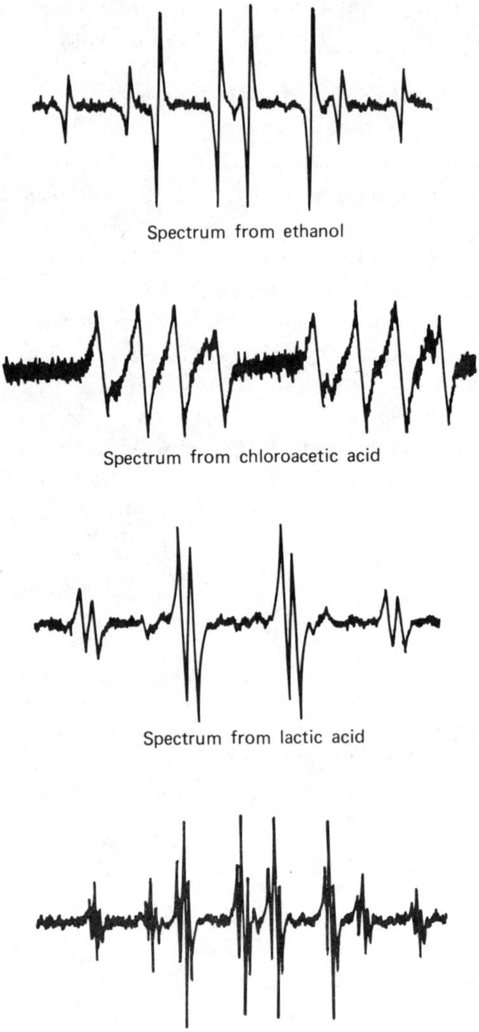

Fig. 9.16 Examples of esr spectra of transient radicals produced in rapidly mixing solutions by H-abstraction from various substances as indicated. From Dixon and Norman [62] and Dixon, Norman, and Buley [63].

often be unambiguously identified by its esr hyperfine structure. For example, the esr observed by Dixon and Norman for the preceding reaction (see Fig. 9.16) is that expected for the CH$_3$ĊHOH and cannot be assigned to other radicals such as CH$_3$CH$_2$CO, which might be postulated.

When two or more radicals are formed in the liquid reactions, their separate patterns can often be resolved and identified, as is illustrated by Fig. 9.17 from Smith, Pearson, et al. [65]. By comparison of the intensity ratios of the separate patterns, the relative concentrations of the different radicals are readily ascertained.

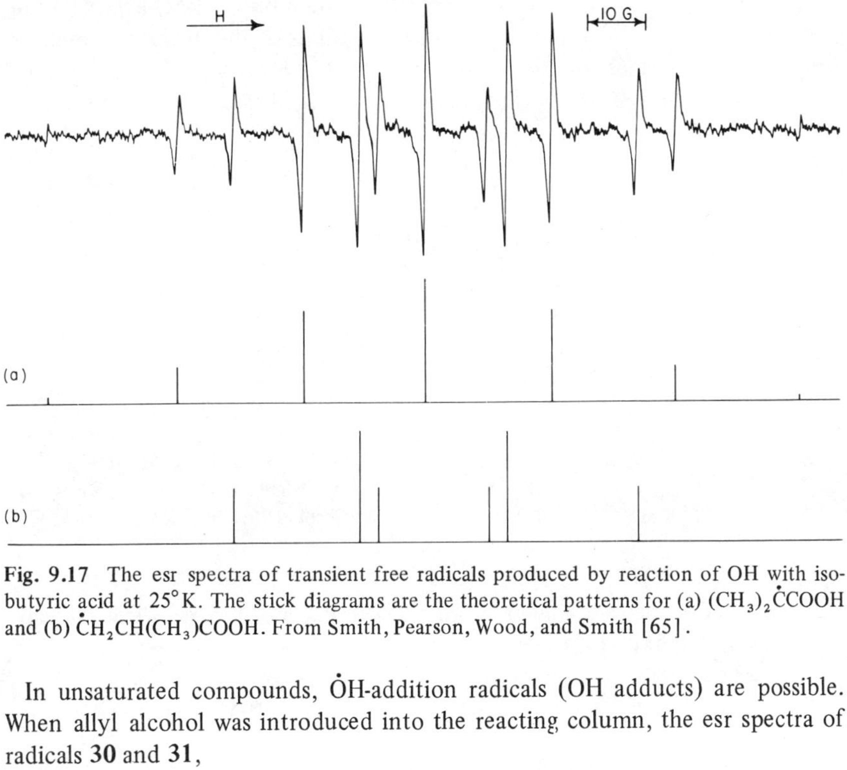

Fig. 9.17 The esr spectra of transient free radicals produced by reaction of OH with isobutyric acid at 25°K. The stick diagrams are the theoretical patterns for (a) (CH$_3$)$_2$ĊCOOH and (b) ĊH$_2$CH(CH$_3$)COOH. From Smith, Pearson, Wood, and Smith [65].

In unsaturated compounds, ȮH-addition radicals (OH adducts) are possible. When allyl alcohol was introduced into the reacting column, the esr spectra of radicals **30** and **31**,

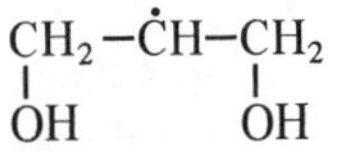

$$\begin{array}{cc} \text{CH}_2-\overset{\cdot}{\text{CH}}-\text{CH}_2 & \overset{\cdot}{\text{CH}}_2-\text{CH}-\text{CH}_2 \\ |\quad\quad\quad\quad | & \quad\quad |\quad\quad | \\ \text{OH}\quad\quad\text{OH} & \quad\text{OH}\;\;\text{OH} \\ \mathbf{30} & \mathbf{31} \end{array}$$

both of which are formed by OH addition to the CH$_2$=CHCH$_2$OH, were observed [63] with radical **30** having the greater concentration.

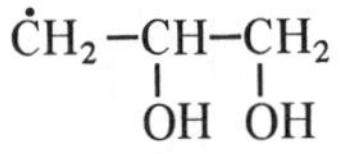

Even the large anisotropic couplings of the halogen nuclei are eliminated by the tumbling motions so that the hyperfine structure is resolved for the reacting liquids as they flow through the absorption cell, as is demonstrated by the spectrum of the radical $\dot{C}HClCCO_2H$ produced by H abstraction from chloroacetic acid (Fig. 9.16). The doublet splitting caused by the CH proton has a quartet substructure arising from the Cl coupling. Both ^{35}Cl and ^{37}Cl have a spin of 3/2 and thus yield 2(3/2) + 1 = 4 equally intense hyperfine components. Because their magnetic moments are nearly equal, their respective quartets are not well resolved, although the slight difference does cause a broadening of the lines. The only stable fluorine isotope, ^{19}F, has a spin of 1/2. Thus, the fluorinated hydrocarbon radicals in the liquid-flow system should have hyperfine structure qualitatively similar to that of hydrocarbon radicals although the spacings of the components are different. The esr of the $CF_3\dot{C}HOH$ (Fig. 9.18) has been obtained by Smith, Pearson, and Tsina [66],

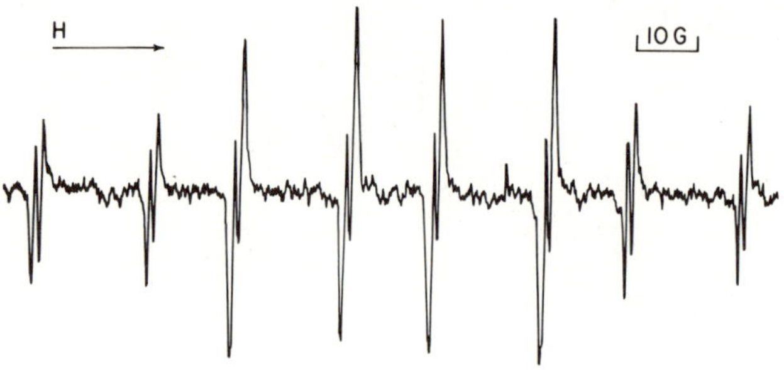

Fig. 9.18 The esr spectrum of $CF_3\dot{C}HOH$ produced by $\dot{O}H$ reaction with CF_3CH_2OH in a rapid, liquid mixing system. From Smith, Pearson, and Tsina [66].

who employed a method similar to that of Dixon and Norman. The radicals were formed by the reaction

$$CF_3CH_2OH + \dot{O}H \rightarrow CF_3\dot{C}HOH + H_2O$$

The gross structure of the esr consists of a quartet caused by the equally coupling ^{19}F nuclei of CF_3, $A_F(CH_3) = 32$ G, with a doublet splitting $A_H(CH)=$ 18.2 G of these components caused by the CH proton. Finally, there is a still smaller doublet splitting, $A_H(OH) = 1.11$ G, caused by the OH proton.

Hyperfine structure of ^{14}N in the esr of a number of free radicals produced through hydrogen abstraction by OH in flowing liquids has been observed [63]. The radical $(CH_3)_2\dot{C}—C{\equiv}N$, studied by Pearson, Smith, and Smith [67], shows a ^{14}N triplet structure with component spacing $A_N = 3.6$ G.

Fischer [68] has observed a ^{14}N splitting, A_N = 3.53 G, for the acrylonitrile monomer radical $(HOCH_2)\dot{C}H(CN)$.

A series of papers by Norman and his associates, who employ the Dixon-Norman system for study of various kinds of chemical reactions in the mixed liquids, has been published in the *Journal of the Chemical Society* [69-72]. In one variant of the system [71] the CO_2^- anion radical was used as a one-electron reducing agent for elimination of the halogen from aliphatic halides to produce various types of observable organic free radicals in the flowing aqueous solutions.

No elaboration is needed to justify the importance to the chemist of esr "fingerprints" of short-lived radical species in a column of reacting chemicals. After being postulated for many years, such reactants can be directly observed, and their chemical form can be learned. Table 9.4 gives a partial list of the free

Table 9.4 Coupling Constants of Transient Radicals Derived from Various Compounds in Liquid Flow System[a]

		Isotropic Coupling (G)				
		H coupling				^{14}N
Parent Compound	*Radical Derived*	C_α-*H*	C_β-*H*	C_γ-*H*	*O-H*	*Coupling*
n-Propylamine	$\dot{C}H_2(CH_2)_2NH_2$	22.5	26.9			
Ethanolamine	$\dot{C}H(OH)CH_2NH_2$	18.1	11.8			10.3
Diethanola-mine	$\dot{C}H(OH)CH_2NHC_2H_4OH$	18.4	10.8			11.1
Acetone	$\dot{C}H_2COCH_3$	20.3		<1		
Diethyl ketone	$\dot{C}H_2CH_2COC_2H_5$	22.1	24.7			
Formaldehyde	$\dot{C}H_2OH$	17.8			1.0	
Acetaldehyde	$\dot{C}H(OH)CH_3$	16.0	22.2			
Diethyl ether	$\dot{C}H(CH_3)OC_2H_5$	13.8	21.9	1.4		
Acetic acid	$\dot{C}H_2CO_2H$	21.8				
Trimethyl acetic acid	$\dot{C}H_2C(CH_3)_2CO_2H$	21.8		0.7		
Propionic acid	$\dot{C}H_2CH_2CO_2H$	22.4	26.6			
Glycolic acid	$\dot{C}H(OH)CO_2H$	17.8			2.6	
Lactic acid	$\dot{C}(OH)CH_3CO_2H$		17.1		2.0	

[a] From Dixon, Norman, and Buley [63].

radicals, with their esr constants, that have been observed with the rapid mixing technique. A review of this development with emphasis on its significance in chemical reactions is given by Waters [73]. Chemical exchange effects on the spectra are described by Sullivan and Bolton [13] (see Section 2).

The Dixon-Norman method has been adapted directly or in variant forms to the study of reactions in the liquid state involving important biochemicals such as the amino acids and the nucleic acid bases. Numerous free radicals formed by $\dot{O}H$ or $\dot{N}H_2$ addition to the purines and pyrimidines in aqueous solutions have been detected from their esr spectra. This work is covered in a review by Nicolai [74]. Results on amino acid solutions are reported by Poupko et al. [75]. A variant of the system involves the oxidation of hydrazine, instead of reduction of H_2O_2, to form the cation radical $(N_2H_4)^+$ as an intermediary [76]. Burlamacchi and Tiezzi [77] have observed hydrazine adducts to vitamin C and other biologically significant compounds.

3.f. Photolysis of Rapidly Flowing Solutions

A notable innovation in the rapid-flow systems for study of short-lived radicals in liquids was made by Livingston and Zeldes [8]. Initially, they mixed small quantities (~1%) of H_2O_2 with organic liquids and photolyzed the mixture with intense uv radiation. The photolysis breaks the peroxide bonds and produces OH radicals that interact with the organic solvent molecules to generate the observed free radicals. The radiation is applied continuously to the liquid as it flows

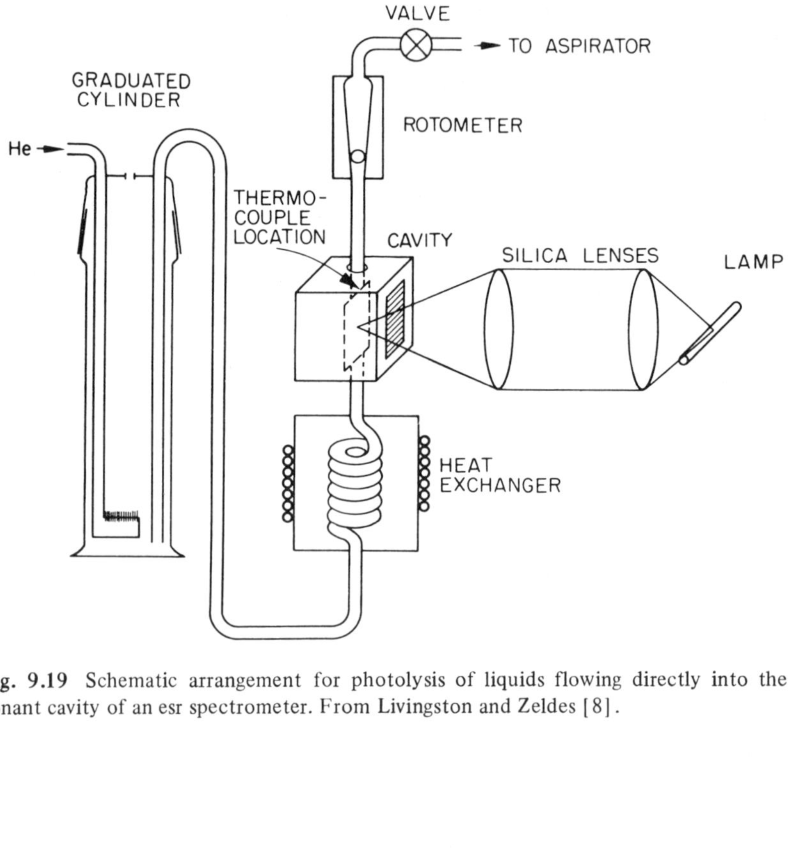

Fig. 9.19 Schematic arrangement for photolysis of liquids flowing directly into the resonant cavity of an esr spectrometer. From Livingston and Zeldes [8].

through the most sensitive region of the microwave absorption cavity. A sketch of the apparatus is shown in Fig. 9.19. This method combines the liquid-flow techniques (Section 3.e) with the photochemical method introduced earlier by Gibson, Ingram, et al. [78] for the production of secondary radicals in the solid state.

In their first applications of the technique, Livingston and Zeldes studied a number of radicals produced by abstraction of H atoms from alcohols through the reactions

$$h\nu + H_2O_2 \rightarrow 2(\dot{O}H)$$

$$\dot{O}H + RCH_2OH \rightarrow R\dot{C}HOH + H_2O$$

where R represents various hydrocarbon groups of the different alcohols. For the same alcohols, their observed results were generally in agreement with those from the Dixon-Norman techniques. There were certain differences in radical formation that may have arisen from the fact that the irradiated structures were not acidic. Fischer [68] has shown that the type of radical formed in the rapid-flow experiments depends somewhat on the acidity of the solvent. In many instances, Livingston and Zeldes observed splitting by the OH hydrogens not observed by Dixon and Norman, who used acidic solutions. In strongly acidic solutions, rapid exchange between the OH protons and the H^+ ions can completely eradicate the OH splitting (see Section 2.f).

A qualitative difference in the results of the two experiments occurred for allyl alcohol $CH_2{=}CHCH_2OH$. Dixon and Norman observed radicals formed only by OH addition reactions, whereas Livingston and Zeldes observed only radicals formed by H abstraction. It is of interest that the latter workers detected distinguishable differences in the esr for the two steric isomers

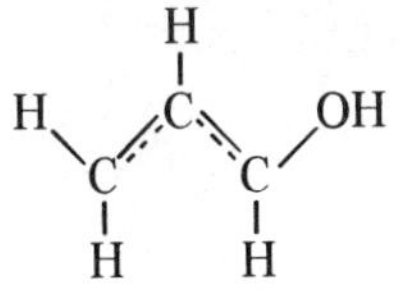

32

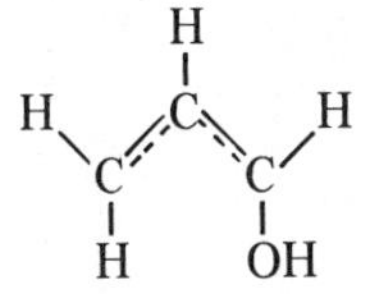

33

of the radicals formed by H abstraction from the CH_2 group adjacent to the OH.

Livingston and Zeldes made their observations on free radicals formed from alcohols at various temperatures ranging from $-70°C$ to $60°C$. Samples of the spectra obtained are given in Figs. 9.1 and 9.2. Generally, the spectra of the observed free radicals were similar at the different temperatures although slight differences in coupling were observed, and in some instances the relative concentrations of different radicals changed. For example, the spectrum from ethyl alcohol at room temperature is that for $CH_3\dot{C}HOH$ with no OH splitting, whereas at $-70°C$ an OH splitting of 1.13 G was observed for this radical, and weak lines from a second radical, $\dot{C}H_2CH_2OH$, were also formed by H abstraction from the CH_3 group. In contrast, OH splitting was observed for $(CH_3)_2\dot{C}OH$ at room temperature (see Fig. 9.2), but no OH splitting could be detected for this radical at $-22°C$ nor at $-40°C$. The OH splitting was resolved for $\dot{C}H_2OH$ at all temperatures at which this species was observed, and the splitting was found to increase at the lower temperatures (see Table 9.1).

In their second paper on this method, Zeldes and Livingston [31] observed the radicals produced from the photolysis of acetone solutions of alcohols or ethers (RH molecules). In the primary reaction, photolysis produced a triplet excited state in the acetone which, in turn, abstracted an H atom from the RH molecule. The resultant transformation is

$$(CH_3)_2CO + RH + h\nu \rightarrow (CH_3)_2\dot{C}OH + \dot{R}$$

Thus two neutral organic free radicals are produced, both of which were observed. Note that the acetone H-adduct radical $(CH_3)_2\dot{C}OH$ is the same as that obtained by H abstraction from isopropyl alcohol, the spectrum of which is shown in Fig. 9.2. Thus when isopropyl alcohol was used as RH donor in acetone solution, the spectrum of the one radical only was observed. When diethyl ether was used as the H donor, spectra were observed for both $CH_3\dot{C}HOCH_2CH_3$ and $(CH_3)_2\dot{C}OH$. When pure acetone was photolyzed, the two radicals produced by the reaction

$$(CH_3)_2CO + (CH_3)_2CO + h\nu \rightarrow (CH_3)_2\dot{C}OH + \dot{C}H_2COCH_3$$

were observed in addition to CH_3 radicals. It was proposed that the methyl radicals originated from the secondary transformation

$$\dot{C}H_2COCH_3 \rightarrow CH_2CO + \dot{C}H_3$$

These examples suggest the possible difficulties in identification of plural transient radicals that may be produced when organic liquid mixtures are photolyzed in a microwave-absorption cell. To aid in identification of the radicals, there is now a rich bank of characteristic esr solution spectra of all kinds of free radicals. When an observed esr pattern cannot be assigned by comparison with known spectra, its source radical can often be ascertained by variation of the components of the photolyzed mixture or by variation of other experimental conditions.

Livingston, Dohrmann, and Zeldes [79] have observed ^{13}C hyperfine structure for several short-lived radicals with the isotope concentration in its natural abundance of 1.1%. From the small values of its ^{13}C coupling, 32.09 G and 13.85 G for the α and β coupling, respectively, it was concluded that $\dot{C}H_2COO$ is a planar π radical. From the larger value $a_{(^{13}C_\alpha)} = 65.05$ G for the central carbon, it was concluded that the skeleton of the radical $(CH_3)_2\dot{C}OH$ is not planar and not a pure π-type radical. For a discussion of the relationship of radical structure and isotropic ^{13}C coupling, see Chapter VIII, Section 4.f. In strongly acid solutions, the ^{13}C couplings of the two central carbons of the acetion radical $CH_3\dot{C}OH\text{-}COCH_3$ are equivalent, in agreement with the earlier finding [32] that the form of the radical is symmetrized by chemical exchange of its OH protons. In the absence of acid, the ^{13}C couplings for the two central carbons were found to be inequivalent. Chemical exchange effects on the alcohol radicals as well as those on the acetoin radical are described in Section 2.f.

Smith et al. [80] compared the esr of free radicals they produced from aliphatic nitriles by use of the $TiCl_3\text{-}H_2O_2$ reacting method of Dixon and Norman with those produced by Livingston and Zeldes [81] by the liquid-flow photolysis method and concluded that the differences in the type of radicals produced generally indicate a more selective abstraction of hydrogen by the photo-flow technique. With the latter method, the observed radicals were produced almost exclusively by H-abstraction from the carbon adjacent to the CN, whereas with the $TiCl_3\text{-}H_2O_2$ method Smith et al. [80] found that, with the possible exception of the cyanoacetates, the H-abstraction can occur from the carbon at any position.

The flowing liquid photolysis method has been adapted to study of transient free radicals produced by H abstraction and by H or OH addition in biologically significant molecules such as the pyrimidines [82-83]. These studies are reviewed by Nicolai [74].

3.g. Electron-beam Radiolysis of Liquids

Fessenden and Schuler [22] have developed a method for observation of the esr of short-lived free radicals produced by radiolysis of various liquids in static systems. They irradiated the pure liquids by a beam of 2.8-MeV electrons focused directly and continuously into the microwave-absorption cell while the induced esr spectra were being recorded. With this technique they measured the liquid-state spectra of a large number of alkyl radicals produced from hydrocarbons at temperatures well below room temperature. Similar experiments have been made on fluorocarbons [84]. Although the *in situ* radiolysis method can be used on liquid mixtures and solutions [22,85,86], its applicability to pure liquids has the advantage that the radicals produced are usually easier to identify than are those produced in a mixture of reacting liquids. The method is readily adaptable to pulse techniques for accurate measurements of the lifetimes of radicals (Section 3.h). Here we describe the characteristic esr spectra of a few of the simpler, more basic, free radicals which Fessenden and Schuler obtained.

The Methyl Radical

The esr pattern of the methyl radical in liquid solution is a quartet with component separations of 23 G and intensity ratios 1:3:3:1. Although the quartet splitting varies slightly in different media, the pattern can usually be identified readily in either solids or liquids. This important free radical thus has a simple characteristic esr, useful for its identification, for measurement of its concentration in solutions, and for observation of its interactions. The quartet splitting is a consequence of the equivalent isotropic couplings of its three hydrogens. In the completely static radicals, the proton couplings have sizeable anisotropic components caused by the dipole-dipole interaction with the spin density on the carbons. In the calculation of its hyperfine structure, the $\dot{C}H_3$ radical may be considered as a composite of three $>\dot{C}$-H fragments oriented 120° apart in a plane, each with a spin density of $\rho_{\pi}^{C} \approx 1$. Since the coupling of the >C-H fragment depends on the orientation of the C-H bond relative to the applied magnetic field, the three protons in the methyl radical have strictly equivalent coupling only when the radicals are tumbling or rotating sufficiently fast to average out the anisotropic components. In liquid solutions, and even in most solids, this small, nonpolar radical undergoes rotational motions at temperatures well below 300°K.

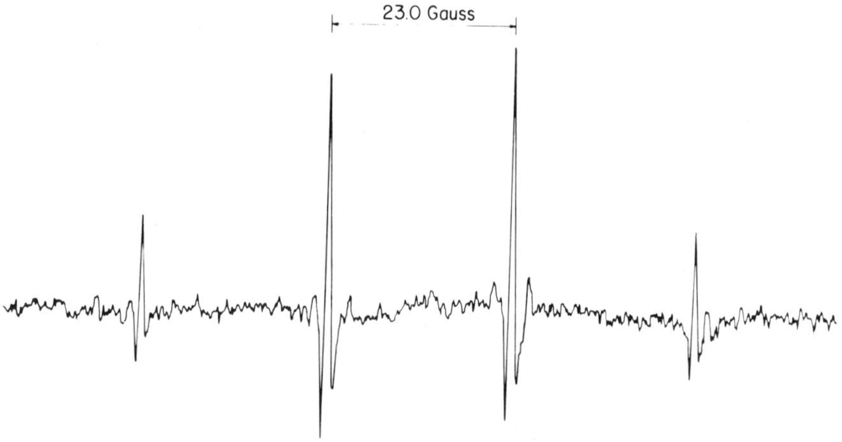

Fig. 9.20 The esr spectrum of the methyl radical in liquid methane at −176°C produced by 2.8-MeV electron radiolysis of the liquid during observation. From Fessenden and Schuler [22].

Figure 9.20 shows the characteristic esr pattern of the $\dot{C}H_3$ radical in the liquid state. To obtain this spectrum, Fessenden and Schuler [22] produced the radicals by bombarding liquid methane with 2.8-MeV electrons at -176°C. At this low temperature, the tumbling motions are still sufficient to average out the anisotropic components in the proton coupling as well as dipole-dipole interactions with the nuclei of the undamaged molecules. The coupling is strictly equivalent for the three protons, and the pattern observed is a pure quartet. The negative spin density in the 1s orbital of each H is measured by the ratio of the coupling a_α^H = 23.0 G to that of atomic hydrogen and is ρ_{1s}^H = (23/504) = -0.046. This spin density provides an accurate evaluation of the spin polarization of the C-H σ bond by an unpaired electron entirely in the $2p_\pi$ orbital of an α carbon and hence an accurate value of Q_α^{CH} in the McConnell relation for calculation of π spin densities in other radicals.

The Ethyl Radical

Figure 9.21 shows the liquid-state esr pattern for the ethyl radical observed by Fessenden and Schuler [22] in liquid ethane at -180°C, continuously

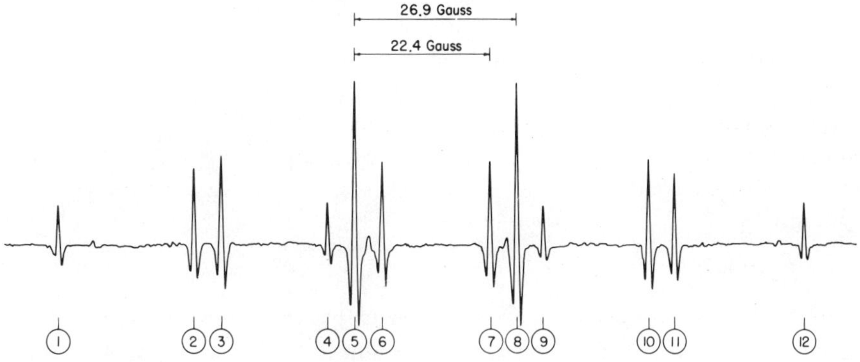

Fig. 9.21 The esr spectrum of the ethyl radical in liquid ethane at −180°C produced by 2.8-MeV electron-beam bombardment of the liquid during observation. Lines are indicated by numbers. From Fessenden and Schuler [22].

bombarded with 2.8-MeV electrons. The indicated splittings provide accurate values of the $C_\alpha H_2$ and $C_\beta H_3$ couplings, a_α^H = 22.4 G and a_β^H = 26.9 G. These may be compared with the corresponding solid-state values of 22.8 G (63.9 MHz) and 26.8 G (75.1 MHz) in the argon matrix at 4.2°K (see Table 8.16). It is interesting that the a_α^H for the ethyl radical is so close to the value, 23.0, for the methyl radical. This near equality indicates that the hyperconjugation does not reduce the H_α coupling in proportion to its reduction of the C_α spin density. The characteristic pattern of Fig. 9.21 provides an identifying "fingerprint" for the ethyl radical in solutions. Compare it with the solid-state curve in Fig. 8.18.

Propyl Radicals

Electron bombardment of liquid propane at 93°K was found to produce both propyl and isopropyl radicals, **34** and **35**, respectively.

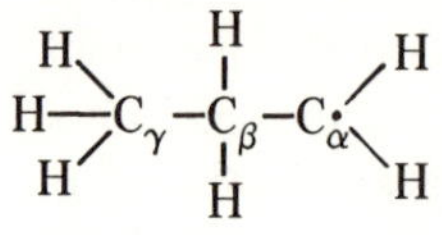

34

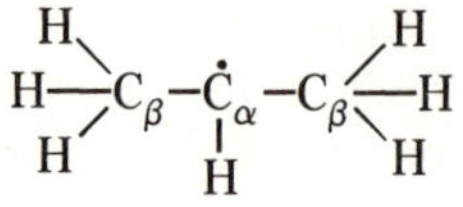

35

The propyl radical **34** gives a component hyperfine structure consisting of a triplet splitting by the equivalent α hydrogens, each component of which is again split into a triplet by the two equivalently coupling β hydrogens. The former leads to a component separation of 22.1 G; the latter, to a component separation of 33.2 G. The isotropic splitting observed for the $C_\alpha H_2$ arises from spin polarization of the C_α-H bonds, whereas the C_β-H_2 splitting arises from hyperconjugation.

The isopropyl radical **35** gives a 14-component hyperfine structure that consists of a pair of septets with intensity ratios 1:6:15:20:15:6:1. This structure is just that expected from the six equivalently coupling β hydrogens of the two CH_3 groups, with the additional doublet splitting by one α hydrogen of radical **35**. The component splitting by the β hydrogens is 24.7 G, and that by the α hydrogen is 22.1 G.

Butyl Radicals

Figure 9.22 shows the esr spectrum obtained by electron bombardment of isobutane at −145°C. The lines marked 3-8 are the six central components of the 10-component hyperfine pattern expected for the *tert*-butyl radical.

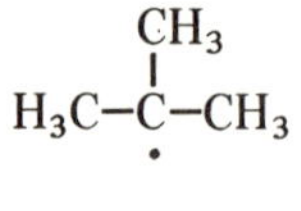

tert-Butyl radical

36

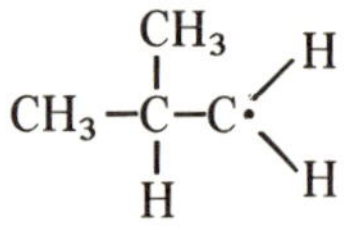

Isobutyl radical

37

The equivalent coupling constant of the nine protons is 22.7 G. The six weaker lines indicated in the lower figure represent two sets of triplets arising from the

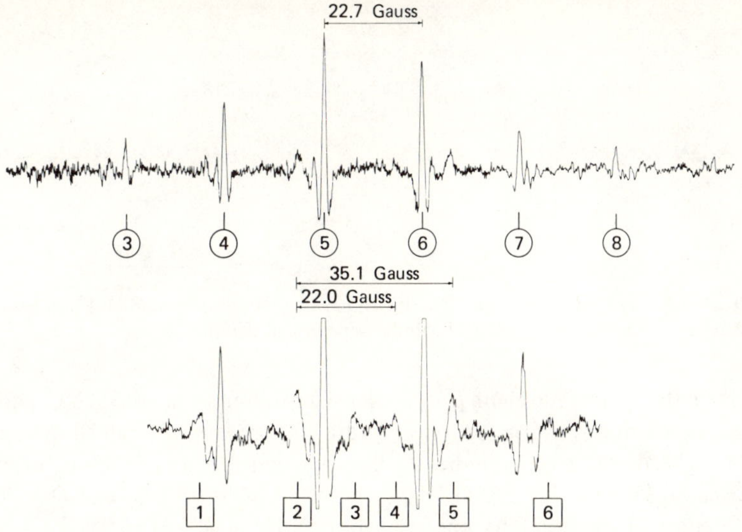

Fig. 9.22 The esr obtained by 2.9-MeV electron-beam radiolysis of liquid isobutane at −145°C. The indicated lines in the top graph are the six central components of the *tert*-butyl radical. From Fessenden and Schuler [22].

isobutyl radical. The triplet splitting of 22.0 G arises from the two equivalent protons on the α carbon; the doublet splitting of 35.1 G arises from the single hydrogen bonded to the β carbon. The latter splitting is greater than the 26-G splitting expected for the rotating C_β-H group. With the assumption that the spin density on the α carbon is like that in the ethyl radical, the dihedral angle θ of the C_β-H bond relative to the π orbital of the unpaired electron is predicted from the coupling to be approximately 36°.

Cycloalkyl Radicals

In addition to the normal-chain or forked-chain hydrocarbons, Fessenden and Schuler [22] irradiated a number of cycloalkyl liquids. Figure 9.23 shows the 10-component esr hyperfine pattern as obtained from cyclopentane at −80°C. It has the relative intensities and the spacing for two quintets. The spectrum can be assigned to the cyclobutyl radical

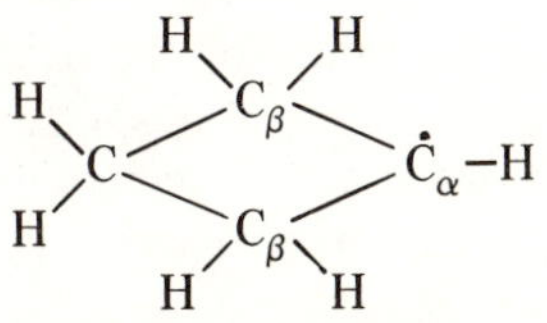

38

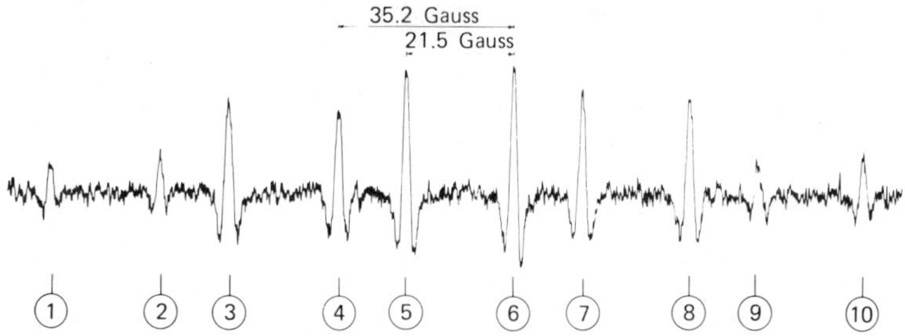

Fig. 9.23 The esr spectrum of the cyclopentyl radical obtained by 2.8-MeV electron-beam radiolysis of cyclopentane at −80 C. From Fessenden and Schuler [22].

in which the four equivalent β hydrogens have couplings of 35.2 G and give rise to the quintet splitting; the one α hydrogen separates the pair of quintets by 21.2 G. The latter coupling indicates that the spin density of the $2p_\pi$ orbital of the α carbon should be 0.844. With $\rho = 0.844$, $a_\beta = 35.2$ G, and $Q = 58.5$ G (see Chapter VI, Section 2.b), one can estimate the angle θ of (6.9):

$$35.2 = (0.844)\, 58.5 \cos^2 \theta$$

The value thus obtained is $\theta = 32^\circ$, in good agreement with the value of 30° expected for tetrahedral bonds to the β carbons. Results on a number of higher cycloalkyl radicals are given by Fessenden and Schuler [22]. Although the esr patterns differ in some detail from that of cyclopentane, the free radicals observed are formed in the same way as the cyclopentane radical, by loss of an H atom from a carbon of the ring. In some, the pattern is made complex by unequal coupling of the β hydrogens.

A listing of the alkyl free radicals observed by Fessenden and Schuler [22] with their measured couplings is given in Table 9.5. In Table 9.9 are listed the Q_α and Q_β values they derived from the couplings. These Q values are widely used in analysis of proton coupling in other organic free radicals. They were used in the analysis of the esr of single crystals described in Chapter VI.

Because of the high precision that is possible with the sharp lines of esr of the liquid state, slight, second-order Breit-Rabi shifts of the hyperfine components (Chapter III, Section 4.b) can be measured, even the very small shifts caused by the proton splittings of the order of 23 G for the alkyl radical at X-band frequencies. The corresponding shifts for the deuterated species are negligible because of the smallness of the deuterium splitting. The Breit-Rabi correction brings the g values for the two isotopic species into agreement. It is of interest that the precise g values for the methyl and ethyl radicals in the liquid state,

Table 9.5 Isotropic Hyperfine Coupling Constants for Some Alkyl and Cycloalkyl Radicals[a]

Radical	a_α *(G)*	a_β *(G)*	a_γ *(G)*
$\dot{C}H_3$	23.04	—	—
$CH_3\dot{C}H_2$	22.38	26.87	—
$C_2H_5\dot{C}H_2$	22.08	33.2	0.38
$(CH_3)_2\dot{C}H$	22.11	24.68	—
$CH_2{=}CHCH_2\dot{C}H_2$	22.23	29.7	0.63 0.35
$C_2H_5\dot{C}HCH_3$	21.8	{ 24.5 (CH_3) 27.9 (CH_2)	*b*
$CH_3)_2CH\dot{C}H_2$	22.0	35.1	*b*
$(CH_3)_3\dot{C}$	—	22.72	—
$C_2H_5\dot{C}HC_2H_5$	21.8	28.8	*c*
$(CH_3)_2\dot{C}CH_2CH_3$	—	{ 22.8 (CH_3) 17.6 (CH_2)	*b*
$(CH_3)_3C\dot{C}H_2$	22.7	—	*b*
$(C_2H_5)_3\dot{C}$	—	17.3	*b*
$(C_2H_5)_3C\dot{C}HCH_3$	21.7	24.9	*b*
Cyclo-$\dot{C}_4H_7$	21.20	36.66	1.12
Cyclo-$\dot{C}_5H_9$	21.48	35.16	0.53
Cyclo-$\dot{C}_6H_{11}$	21.15	41 and 5	0.71
Cyclo-$\dot{C}_7H_{13}$	21.8	24.7	*b*

[a] From Fessenden and Schuler [22].
[b] Additional structure due to γ protons expected but not resolved.
[c] Additional structure due to γ protons observed but not assigned.

2.00255 and 2.00260, respectively, are very close to 2.00232, the value for the free electron spin. The *g* values for the other alkyl radicals are similar.

Fessenden and Schuler [84] measured the esr of $\dot{C}F_3$ radicals in liquid mixtures of C_2F_6 and CF_4 under bombardment with 2.8-MeV electrons at $-163°C$. They observed sharp lines with widths of about 0.3 G. Because of the large isotropic ^{19}F coupling, $a^F = 144.75$ G, Breit-Rabi corrections to fourth order (Chapter III, Section 4.b) were required for a precise theoretical fitting of the spectrum. Solid-state measurements of the ^{13}C coupling show that the molecule is pyramidal in structure, as explained in Chapter VIII, Section 4.f. Also measured by Fessenden and co-workers are the ^{13}C couplings in a number of free radicals produced by electron radiolysis of aqueous solutions of organic acids [85,86].

Neta [87] has used *in situ* electron-beam radiolysis combined with esr spectroscopy to study radiation effects on biological pyrimidines in solutions having various constituents including ·OH donors as well as OH scavengers.

3.h. Pulsed Radiolysis: Radical Reaction Rates

The electron-beam radiolysis method described in the previous section is readily adaptable to short-pulse radiolysis that, combined with esr spectroscopy, provides a powerful method for study of reaction kinetics in the liquid state. This method is similar to that of pulsed photolysis used with optical spectroscopy for detection of short-lived radicals and measurement of their decay times. The esr detection makes identification of the radicals easier and more certain; overlapping lines from different types of radicals are not prevalent, and measurement of the decay rates of a particular species in a solution containing several radicals is possible. Because the lines of transient free radicals in solution are not strong, special techniques are required for measurement of these lifetimes or decay rates. For this purpose, a high-energy, pulsable source such as an electron-beam accelerator or a high-intensity, pulsable laser beam is desirable if not necessary. In addition, special esr receivers that integrate signals over many pulses are required. To achieve a signal:noise ratio of the order of 100 on the alkyl radicals that have lifetimes of the order of 5×10^{-3} sec, Fessenden and Schuler [22] employed an esr receiver with a time constant of 3 sec. Obviously, one cannot watch the radical signal decay on a CRO scope after the generator is cut off.

Electronic computer techniques are now available for storing a sequence of signals from many synchronized pulses in a memory bank and of integrating them into a common signal for amplification and display. Since the noise accumulated in the sequence has random phases, it tends to cancel, with the result that the noise does not increase as does the phase-coherent esr signal in the integrated sequence. Thus the signal:noise ratio can be made to increase, within limits, by an increase in the number of pulsed signals. This technique has the effect of increasing the signal:noise ratio by increasing the effective time constant of the receiver by summation of the signals for many short time intervals.

First, the esr of the radical under study is observed by the continuous irradiation process already described (Section 3.g). After a strong, well-resolved component has been selected, the esr spectrometer is tuned to its peak intensity. The radiation source is then pulsed at regular intervals longer than the anticipated lifetime of the radical with pulse lengths that are short compared with the radical lifetime. The esr signal is observed during intervals Δt, at a time $(t_1 - t_p)$ from each pulse. The sequence of signals is stored and integrated into a common signal for detection. The sampling interval Δt is then moved to a new time $(t_2 - t_p) = (t_1 + \Delta t)$, and the process is repeated until $(t - t_p)$ becomes so long relative to the radical decay time τ_s that the signal is not detectable. The decay

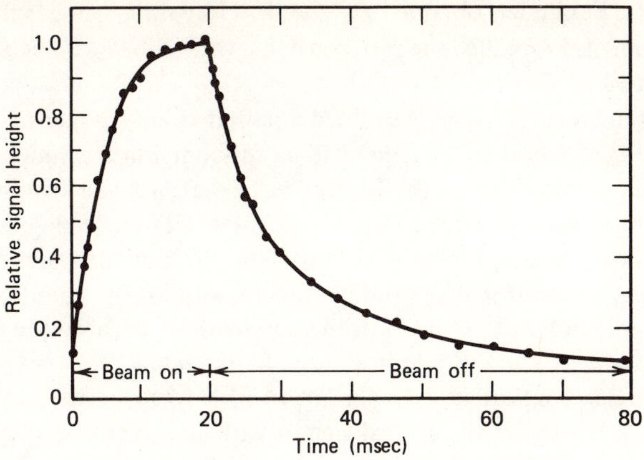

Fig. 9.24 Growth and decay curve of the esr signal of the ethyl radical in liquid ethane at −177°C, as measured by pulsed radiolysis. From Fessenden [88].

curve for the radical is thus obtained. If the sampling process begins at the time the pulsing beam is turned on, the growth of the signal from zero to its peak height can also be observed.

The technique just described was used by Fessenden [88] to trace the growth and decay curve shown in Fig. 9.24 for the ethyl radical produced by pulsed electron bombardment of liquid ethane. He used a 2.8-MeV Van de Graaff generator to provide the pulsed electron beam. The relative signal strength is proportional to $[R]/[R]_s$, where [R] is the radical concentration at the particular time and $[R]_s$ is the steady-state concentration. The curves for the buildup and decay of the radical have the forms

$$[R]/[R]_s = \tanh \frac{t}{\tau_s} \qquad \text{(formation)} \qquad (9.54)$$

$$[R]/[R]_s = \left(1 + \frac{t}{\tau_s}\right)^{-1} \qquad \text{(decay)} \qquad (9.55)$$

where τ_s is the steady-state lifetime defined by $\tau_s = ([R]_s/P_R)$, where P_R is the rate of production of the radicals. It is evident from (9.55) that the reciprocal of the decay curve is a straight line, $([R]_s/[R]) = 1 + (t/\tau_s)$, the slope of which is $1/\tau_s$. This slope, thus determined by the "beam-off" region of Fig. 9.24, gives $\tau_s = 6.93$ msec for the ethyl radical under the conditions of the experiment. The reaction rate constant k for decay of the radicals is equivalent to $k = (1/2)P_R/[R]_s^2$, which with $\tau_s = ([R]_s/P_R)$ can be expressed as $k = (2P_R\tau_s^2)^{-1}$. From the estimated radical yield of 4.7 radicals/100 eV for the ethyl radicals, Fessenden obtained $P_R = 5.4 \times 10^{-5}$ msec^{-1} under the conditions of his experiment. From this, the rate constant was estimated to be $k = 1.7 \times 10^8$ M^{-1} sec^{-1} at $-177°$C.

Smaller, Remco, and Avery [89] used a pulsed radiolysis method for study of radical scavenger action of oxygen- and sulfur-containing compounds in liquid systems. In principle, their method is similar to that of Fessenden, but they used a much stronger pulsed source and more sensitive data processing that employed a multichanneled analog memory system. For the pulsed source, they used a 15-MeV linear accelerator that produced pulses of 0.5-μsec duration with beam intensities of the order of 10 mA. For measurement of the absolute rate constant k, it is necessary to know the radical production rate P_R or peak radical concentration under continuous bombardment. This was measured by comparison of the peak intensity of the radical signals with the intensity of signals from a calibrated sample of DPPH in benzene solution. Decay of the cycloalkyl radicals produced by pulsed radiolysis of cyclopentane and of cyclohexane were studied. The second-order rate constant k_1 for the radical recombination in the pure liquid

$$\dot{R} + \dot{R} \xrightarrow{k_1} \text{Reaction products}$$

was measured as described above for the ethyl radicals in liquid ethane. The scavenger compounds were then added in relatively high concentration so that the radical decay was dominated by the pseudo-first-order reaction of the radicals with the scavenger. For the sulfur scavengers, the measured rate constant is for the reaction

$$\dot{R} + R'S \xrightarrow{k_2} \text{reaction products}$$

and for the molecular oxygen scavenger, the rate constant is for

$$\dot{R} + O_2 \xrightarrow{k_3} R\dot{O}_2$$

With the excess scavenger present, so that the radical-scavenger reaction is a pseudo-first-order process, the decay curve is exponential. The experimental decay rate constant obtained directly from an observed decay curve is K, defined as

$$K = k_i[X] \tag{9.56}$$

where [X] is the molar concentration of the scavenger X and $k_i = 2,3$ is the rate constant for radical-scavenger reactions indicated earlier. Incidentally, the rate constant k_2 was measured by the growth of the signal of the peroxide radical $R\dot{O}_2$ as well as by the decay of the $\dot{R}$ signal, and satisfactory agreement was obtained between the values. Table 9.6 gives some reaction-rate constants measured by Smaller, Remco, and Avery [89] with this procedure. Details of

Table 9.6 Rate Constants k_i (T) for Reactions of Cycloalkyl Radical (R)[a]

Reaction	k_1	*Cyclopentyl* $\dot{R}$ M^{-1} sec^{-1} *(T)*	*Cyclohexyl* $\dot{R}$ M^{-1} sec^{-1} *(T)*
$\dot{R} + \dot{R}$	k_1	4.0×10^8 (−50°C)	2.5×10^9 (+10°C)
$\dot{R}$ + Sulfur	k_2	6×10^7 (−40°C)	
$\dot{R}$ + Naphthalenethiol	k_2	4.1×10^7 (−40°C)	2.1×10^8 (+18°C)
$\dot{R}$ + Benzylmercaptan	k_2	3.3×10^7 (−32°C)	1.5×10^8 (+17°C)
$\dot{R} + O_2$	k_3	3.9×10^6 (−40°C)	4.3×10^7 (+25°C)

[a]From data of Smaller, Remco, and Avery [89].

their experiment and examples of the observed decay curves are given in their paper [89]. This pulsed radiolysis method with the high-current, pulsed, linear accelerator source has been used with much success by Nucifora, Smaller, et al. [90] for study of scavenger "protectors" in solution with biological pyrimidines. Various reactions of hydrated electrons and OH radicals with the pyrimidines were observed.

4. THE g VALUES FROM SOLUTION SPECTRA

Measured g values for tumbling radicals in solution correspond to 1/3 trace **g** [see (9.6)]. For rapidly tumbling radicals usually observed, the off-diagonal elements are completely averaged out, and the measured values are

$$\langle g \rangle = \frac{1}{3}(g_u + g_v + g_w)$$

$$= g_e + \langle \Delta g \rangle = g_e + \frac{1}{3}(\Delta g_u + \Delta g_v + \Delta g_w) \qquad (9.57)$$

where g_u, g_v, and g_w are the principal elements. Because exceedingly careful measurements are required for meaningfully accurate values of the very small $\langle \Delta g \rangle$ for organic free radicals and because quantitative theoretical interpretation of these small, "lumped" parameters is even more difficult, there has been relatively little emphasis on studies of g values of radicals in solution. Most spectroscopists have concentrated their efforts on isotropic nuclear couplings. Nevertheless, the pioneering efforts by Blois, Brown, and Maling [91] to obtain

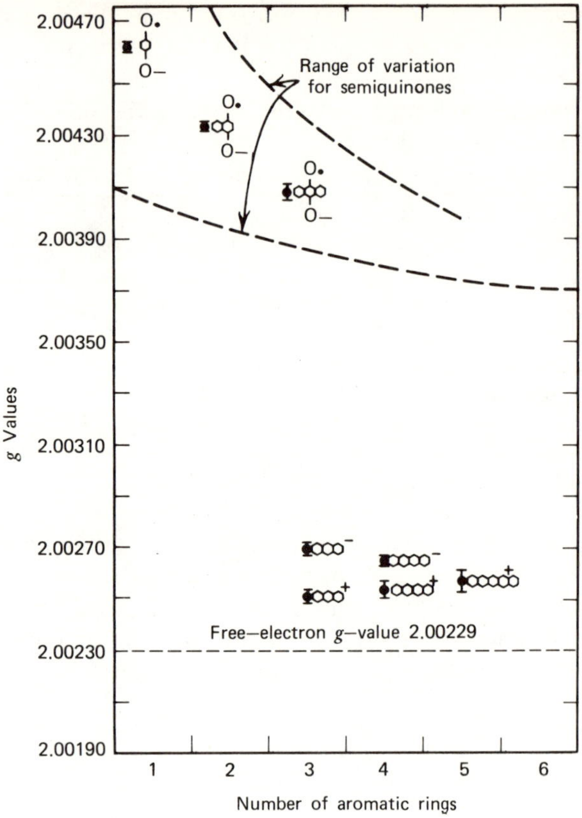

Fig. 9.25 Plot of *g* values of aromatic ions as a function of the number of rings. From Blois, Brown, and Maling [91].

precise *g* values of long-lived aromatic free radicals in solution proved rewarding. Qualitative differences in the $\langle g \rangle$ values of different classes of compounds were clearly evident, and, for at least one radical, *p*-benzosemiquinone, orderly variation of $\langle g \rangle$ with temperature was observed. Most of the class differences can be attributed to probable spin density on atoms having different spin-orbit couplings. These effects are evident in Figs. 9.25 and 9.26. The *g* values obtained for the aromatic hydrocarbon ions (+ or −) by Blois, Brown, and Maling [92] are all near the free spin value, but those for the semiquinones are noticeably higher, as Fig. 9.25 shows. The higher values of the semiquinones may be attributed to significant unpaired electron-spin density on the oxygens, for which the 2*p* spin-orbit coupling $\zeta_O = 151\ \text{cm}^{-1}$ as compared with $\zeta_C = 29\ \text{cm}^{-1}$ for carbon (Table A.2). The effects of spin-orbit coupling of the constituent halogen atoms are especially noticeable in Fig. 9.26, where $\langle g \rangle$ is plotted against λ

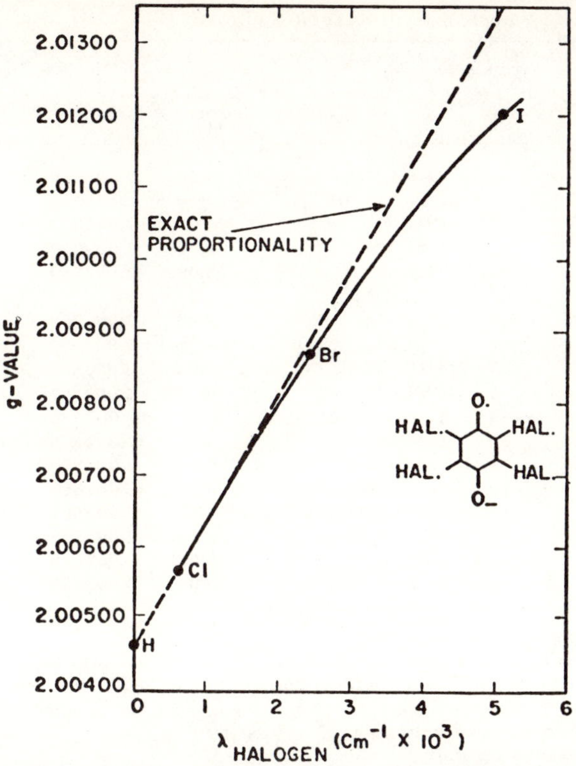

Fig. 9.26 Relation of g values of tetra-halogenated benzosemiquinones as a function of the spin–orbit coupling of the halogen. From Blois, Brown, and Maling [91].

for the substituted halogen. These relations can be helpful in the identification of unknown radicals in a solution or in a qualitative determination of the spin densities on the different atoms.

Following the measurements of Blois, Brown, and Maling [91], Stone developed a generalized, semiquantitative theory of $\langle g \rangle$ for solution radicals [92], and Segal, Kaplan, and Fraenkel [93] extended the precise $\langle g \rangle$ measurements of aromatic radicals in solution and improved the accuracy of the earlier values [91]. Table 9.7 gives their values for ringed hydrocarbon radicals. The $\langle g \rangle$ values for most of these radicals had been measured earlier by Blois, Brown, and Maling [91] and the agreement between the independent measurements is close. There is a slight, uniform difference in the values because the various researchers did not use identical values of the conversion constant, $k = g_e(\mu_p/\mu_e)$ [93].

Since the observable $\langle g \rangle$ for liquids is the sum of the diagonal elements, it is possible to calculate the diagonal elements contributed by the constituent atoms

Table 9.7 The *g* Values of Hydrocarbon Radicals[a]

Radical	Position in Fig. 9.28	Solvent System[b]	Linewidth (G)[c]	*g* Value[d]	Deviation from g_e ($\Delta g \times 10^5$)[e]
Benzene$^-$	1	THF:DME(2:1)/Na-K alloy; −101°	0.30	2.002850	53.1
Naphthalene$^-$	2	DME/Na; −58°	0.05	2.002752	43.3
Coronene$^-$	3	DME/Na	0.70	2.003067	74.8
Pyrene$^-$	4	THF or THF:DME (2:1)/Na	0.08	2.002719	40.0
Anthracene$^-$	5	DME/Na	0.08	2.002709	39.2
Fluoroanthene$^-$	6	DME/Na	0.05	2.002692	37.3
Perylene$^-$	7	THF or DME/Na	0.06	2.002667	34.8
Tetracene$^-$	8	DME/Na	0.07	2.002682	36.3
Pentacene$^-$	9	THF or DME/Na	0.07	2.002686	36.7
Triphenylmethyl	10	See Table 9.8	0.05	(2.00260)[f]	(28)
Cyclooctatetraene$^-$	11	See Table 9.8	0.17	2.002579	26.0
Perinaphthenyl	12	See Table 9.8	0.05	(2.00266)[f]	(34)
Pentacene$^+$	13	Conc. H_2SO_4	0.11	2.002605	28.6
Tetracene$^+$	14	Conc. H_2SO_4	0.18	2.002598	27.9
Perylene$^+$	15	Conc. H_2SO_4	0.06	2.002578	25.7
Anthracene$^+$	16	Conc. H_2SO_4	0.09	2.002565	24.8

[a] From Segal, Kaplan, and Fraenkel [93].
[b] Entries are: solvent/reducing agent; temperature (°C). All measurements are at room temperature unless otherwise specified. Abbreviations: DME, 1,2-dimethoxyethane; THF, tetrahydrofuran.
[c] Full width between derivative extrema δ.
[d] Corrected for second-order shifts.
[e] $\Delta g = g_{exp} - g_e$, where g_{exp} is the *g* value corrected for second-order shifts and $g_e = 2.002319$ is the *g* value of the free electron.
[f] See Table 9.8.

or bonds and from their sum to obtain a predicted $\langle g \rangle$. Stone [92] noticed that many bonds to atoms and nonbonding orbitals of the atoms in aromatic free radicals are of the same kind and that their contribution to $\langle g \rangle$ can be grouped advantageously in calculations of $\langle g \rangle$. For example, the σ-bond energies of all the CC bonds may be assumed to be equal; the CH bond energies may also be taken as equal. Thus the CC bonding σ orbitals and the antibonding σ* orbitals can be grouped together, as can the CH σ and σ* orbitals in the calculation of $\langle g \rangle$ for the molecular radicals. Likewise, the nonbonding oxygen orbitals of the semiquinones may be grouped. Such grouping not only simplifies the calculation of $\langle g \rangle$ for an individual radical, but also offers the advantage that parts of the calculations are transferable to similar radicals having the same groups. In calculations of $\langle g \rangle$ for aromatic radicals Stone assumed the ground state to be an orbital singlet with an unpaired electron occupying the highest populated orbital and with all lower orbitals filled by electron pairs. This theory is illustrated below.

If the contribution to $\langle \Delta g \rangle$ by a group s centered on different atoms is $\langle \Delta g \rangle_s$, the total $\langle g \rangle$ is obtained by a summation over the various s groups:

$$\langle g \rangle = g_e + \frac{1}{3} \sum_s \text{trace}\,(\Delta \mathbf{g}_s) \tag{9.58}$$

The theoretical (Δg_s) values are obtained from a formula like (7.19) with the summation taken over only those atoms on which s groups are centered. For the diagonal elements,

$$(\Delta g_s)_{ii} = 2 \sum_{n_s} \sum_{k_s} \zeta_{k_s} \frac{(c_{0k_s} \psi_{0k_s} | L_{k_s i} | c_{n_s k_s} \psi_{n_s k_s}) \sum_{k'_s} (c_{n_s k'_s} \psi_{n_s k'_s} \psi_{n_s k'_s} | L_{k'_s i} | c_{0k'_s} \psi_{0k'_s})}{\epsilon_0 - \epsilon_{n_s}} \tag{9.59}$$

where $i = x$, y, or z. To conform to the usage by Stone, the orbital energy levels ϵ_0 and ϵ_n are used in place of the configuration energies E_0 and E_n, and ζ_k is used in place of the λ_k of (7.19), as described in Chapter VII, Section 7.1.c. Values for ζ are given in Table A.2.

In calculation of $\langle \Delta g \rangle$ for aromatic radicals, the odd electron is assumed to be in a delocalized π orbital, $\phi_0 = \Sigma_k c_{0k} \psi_k$, with energy ϵ_0. All intermixable excited orbitals, $\phi_n = \Sigma_k c_{nk} \psi_k$, are assumed to be localized σ or σ^* orbitals with orbital energies ϵ_n. This assumption applies to aromatic free radicals where the odd-electron orbital does not have spin-orbit coupling with other π orbitals. Consider the C_1C_2 bond of Fig. 9.27, where the p_π components of the unpaired electron are in the z direction. The part of ϕ_0 that contributes unpaired spin density on C_1 and C_2 may be expressed as

$$c_1 p_{z1} + c_2 p_{z2} = \frac{1}{2}(c_1 + c_2)(p_{z1} + p_{z2}) + \frac{1}{2}(c_1 - c_2)(p_{z1} - p_{z2}) \tag{9.60}$$

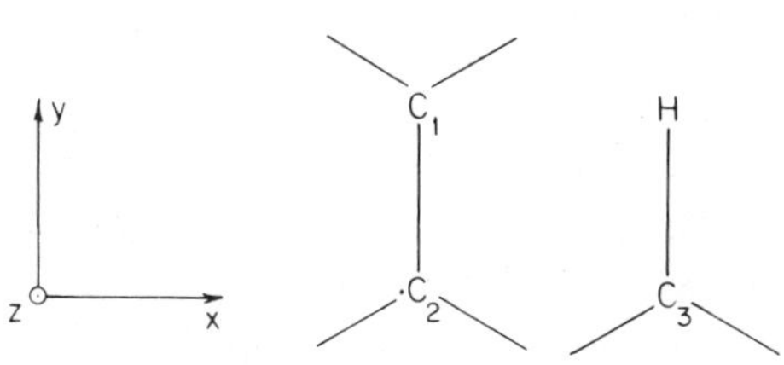

Fig. 9.27 Reference system used for calculation of group contributions to g values in hydrocarbon radicals.

where c_1 and c_2 are weighting coefficients; c_1^2 and c_2^2 give the unpaired electron density on C_1 and C_2 when the coefficients of $\phi_0 = \Sigma_k c_k \psi_k$ are normalized by $\Sigma c_k^2 = 1$. The localized bonding and antibonding orbitals between C_1 and C_2 are

$$\phi_\sigma = \left(\frac{1}{6}\right)^{1/2} s_1 - \left(\frac{1}{3}\right)^{1/2} p_{y1} + \left[\left(\frac{1}{6}\right)^{1/2} s_2 + \left(\frac{1}{3}\right)^{1/2} p_{y2}\right] \quad (9.61)$$

$$\phi_{\sigma*} = \left(\frac{1}{6}\right)^{1/2} s_1 - \left(\frac{1}{3}\right)^{1/2} p_{y1} - \left[\left(\frac{1}{6}\right)^{1/2} s_2 + \left(\frac{1}{3}\right)^{1/2} p_{y2}\right] \quad (9.62)$$

The latter correspond to the orbitals ϕ_{nk} that are intermixed with ϕ_0 by the applied field. With the coordinates indicated in Fig. 9.27, the Δg values contributed by this bond are found by substitution of these functions into (9.59) and performance of the indicated operations. All matrix elements of the y and z operators are found to vanish. Thus $\Delta g_{yy}(\mathrm{CC}) = 0$ and $\Delta g_{zz}(\mathrm{CC}) = 0$. This may be verified with the expressions given in Table 7.1. The nonvanishing elements of L_x with (9.59) give

$$\Delta g_{xx}(\mathrm{CC}) = \frac{1}{6}\frac{(c_1 + c_2)^2 \zeta_\mathrm{C}}{\epsilon_\pi - \epsilon_{\sigma*}} |(p_{z1}|L_{x1}|p_{y1}) + (p_{z2}|L_{x2}|p_{y2})|^2$$

$$+ \frac{1}{6}\frac{(c_1 - c_2)^2 \zeta_\mathrm{C}}{\epsilon_\pi - \epsilon_\sigma} |(p_{z1}|L_{x1}|p_{y1}) + (p_{z2}|L_{x2}|p_{z2})|^2 \quad (9.63)$$

Since $(p_z|L_x|p_y) = i$, the squared magnitude in each term on the right is 4, and the contributions of the C_1C_2 to $\langle \Delta g \rangle$ are

$$\Delta g_{yy}(\mathrm{CC}) = \Delta g_{zz}(\mathrm{CC}) = 0 \quad (9.64)$$

$$\Delta g_{xx}(\mathrm{CC}) = (\tfrac{2}{3}\zeta_\mathrm{C})\left[\frac{(c_1 + c_2)^2}{\epsilon_\pi - \epsilon_{\sigma*}} + \frac{(c_1 - c_2)^2}{\epsilon_\pi - \epsilon_\sigma}\right] \quad (9.65)$$

For the >C-H fragment (see Fig. 9.27), the part of the delocalized π_z orbital of the unpaired electron on the C will similarly interact with the C-H bond to produce contributions to $\langle \Delta g \rangle$. This component of the π_z orbital may be designated $c_3 p_z(C_3)$, where c_3 is the weighting coefficient for the atomic orbital of C_3 in the linear combination of atomic orbitals of ϕ_{π_z}. If the approximation is made that the C-H bond is a pure covalent one, the bonding and antibonding molecular orbitals may be expressed as

$$\phi_\sigma = \left(\frac{1}{2}\right)^{1/2} \left\{ s_H + \left[\left(\frac{1}{3}\right)^{1/2} s + \left(\frac{2}{3}\right)^{1/2} p_y \right]_{C_3} \right\} \tag{9.66}$$

$$\phi_{\sigma *} = \left(\frac{1}{2}\right)^{1/2} \left\{ s_H - \left[\left(\frac{1}{3}\right)^{1/2} s + \left(\frac{2}{3}\right)^{1/2} p_y \right]_{C_3} \right\} \tag{9.67}$$

Again applying (9.59) with $\phi_{0k} = p_z(C_3)$, with ϕ_σ and $\phi_{\sigma *}$, the orbitals ϕ_{nk} intermixed with ϕ_{0k}, one obtains

$$\Delta g_{yy}(\text{CH}) = 0, \qquad \Delta g_{zz}(\text{CH}) = 0 \tag{9.68}$$

$$\Delta g_{xx}(\text{CH}) = \frac{2}{3} c_3^2 \zeta_\text{C} \left(\frac{1}{\epsilon_\pi - \epsilon_\sigma} + \frac{1}{\epsilon_\pi - \epsilon_{\sigma *}} \right) \tag{9.69}$$

It should be noted that $\epsilon_\sigma < \epsilon_\pi < \epsilon_{\sigma *}$; since ζ_C is positive, the contribution of the last term to $\langle g \rangle$ is negative. Thus the two terms tend to cancel, but usually $(\epsilon_\pi - \epsilon_\sigma) < (\epsilon_{\sigma *} - \epsilon_\pi)$, and hence the positive term is the greater for both the CC and CH bonds.

An obvious advantage of calculating the bond contribution to $\langle \Delta g \rangle$ in this way is that the symmetry axis of the bond can be chosen as the reference axis and the calculations can thus be simplified. Only the in-plane component perpendicular to the σ bond is nonvanishing. If the chosen axis happens to converge with the principal axis of **g** for the entire radical, the calculated components are simply added to trace **g**. In the planar aromatic radicals, other bonds of the same variety will be oriented differently in the molecular plane. The theoretical terms, calculated as in the preceding paragraph, must be rotated to the common principal axes to give the contributions of these other bonds to $\langle g \rangle$. This rotation will change the above-calculated terms by a constant factor that can be obtained from the known structure of the ringed radicals. Off-diagonal elements generated by the rotations need not be calculated since they are averaged out by the tumbling motions and thus make no contribution to $\langle g \rangle$.

From these considerations it is evident that the contributions from all theoretical terms to $\langle g \rangle$ for an aromatic radical will have ϵ_π, the energy of the orbital of the unpaired π electron. In the Hückel molecular-orbital approximation, the energy of this orbital may be expressed as

$$\epsilon_\pi = \alpha + \lambda\beta \tag{9.70}$$

where α is the Coulomb integral, β is the resonance integral, and λ is the Hückel energy parameter obtained from solution of the secular equation for the π-orbital energy of the particular radicals. Stone [92] substituted ϵ_π from

(9.70) into the theoretical expressions for the different bond contributions to $\langle \Delta g \rangle$ and expanded the terms in powers of λ. On grouping the terms centered on particular carbon atoms, he found that the terms centered on any given carbon C_i could be expressed by

$$\langle \Delta g \rangle_{C_i} = b_i + c_i \lambda \tag{9.71}$$

and that the sum over all the atoms in an aromatic hydrocarbon radical could be written as

$$\langle \Delta g \rangle = \langle g \rangle - g_e = b + c\lambda \tag{9.72}$$

where $b = \Sigma b_i$, $c = \Sigma c_i$. Thus a simple relationship is found between $\langle \Delta g \rangle$ and the Hückel parameter λ for the orbital.

Although the constants a and b of (9.72) are not easy to calculate, Stone [92] suggested that they be treated as empirical constants to be evaluated from observed Δg values. He plotted the values measured by Blois, Brown, and Maling [91] against theoretical values of λ for the aromatic hydrocarbon radicals and found a linear relationship, in agreement with (9.72), with the constants $b = (24.7 \pm 0.8) \times 10^{-5}$ and $c = -(19.3 \pm 2.4) \times 10^{-5}$. With their remeasured, more precise $\langle g \rangle$ values, Segal, Kaplan, and Fraenkel [93] plotted $\langle \Delta g \rangle$ versus λ for the aromatic hydrocarbon radicals listed in Table 9.7 and found the linear relationship shown in Fig. 9.28. From this plot they obtained $b = (31.9 \pm 0.4) \times 10^{-5}$ and $c = -(16.6 \pm 1.0) \times 10^{-5}$.

The benzene, coronene, and cyclooctatetraene anions were omitted from the plot of Fig. 9.28 because in these radicals the unpaired electron occupies a degenerate orbital that leads to Jahn-Teller distortions of the structure that complicate the calculations of g. The deviations in g caused by this degeneracy are treated in a later paper by Moss and Perry [94].

In a related work, Stone [95] predicted that for even-alternant radical ions of aromatic hydrocarbons made up of six-membered rings, the anions and cations have the same principal axes for **g** and that the differences between their in-plane components $(g_{xx} - g_{yy})$ have the same magnitude but are opposite in sign. These predictions were tested by Fassaert and de Boer [96] through precise measurements on the plus and minus ions of anthracene, tetracene, and perylene in dilute solutions. In agreement with theory, the $(g_{xx} - g_{yy})$ was found in all three to change sign when passing from the cation to the anion form, but the quantitative agreement of the predicted and observed magnitudes was not close. For the neutral, odd-alternant radicals, Stone predicted that the in-plane values would be equal, with $g_{xx} = g_{yy} = 2.0027$ and $g_{zz} = 2.0024$.

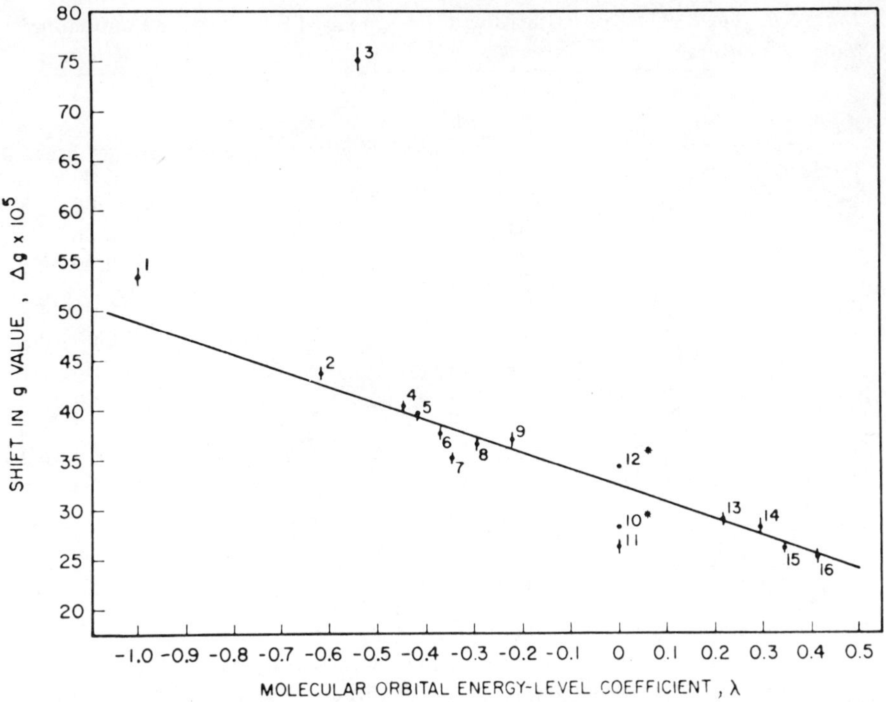

Fig. 9.28 Plot of $\langle \Delta g \rangle$ values of hydrocarbon radicals as a function of the Hückel energy-level coefficient for the π orbital of the unpaired electron. Numbers identify the points with the radicals listed in Table 9.7. From Segal, Kaplan, and Fraenkel [93].

When the unpaired electron is localized in a nonbonding orbital, the resonance integral is zero, and hence the Hückel parameter λ is also zero. For aromatic hydrocarbons of this type, (9.72) becomes simply

$$\langle g \rangle = g_e + b \tag{9.73}$$

and with the preceding experimental value of b [91],

$$\langle g \rangle = 2.002319 + 31.9 \times 10^{-5} = 2.002638$$

for this class of radicals. Table 9.8 gives the g values of three such radicals measured by Segal, Kaplan, and Fraenkel [93] in various solvents at different temperatures. In each of these radicals the unpaired electron is concentrated primarily on a single carbon, and the measured value is close to the predicted one.

Table 9.8 The *g* Values of Radicals with the Unpaired Electron in a Nonbonding Molecular Orbital[a]

Radical	*Solvent*[b]	*Temp.* (°C)	*g Value*[c]
Triphenylmethyl	Toluene	23	2.002588
	CCl_4	23	2.002598
	TGL	-24	2.002597
	CS_2	-101	2.002626
Perinaphthenyl	DME	23	2.002608
	DME	-65	2.002598
	$CBrCl_3$	23	2.002637
	$CHCl_3$	23	2.002660
	CH_2Cl_2	23	2.002660
	CCl_4	23	2.002682
	CCl_4	-18	2.002688
Cyclooctatetraene$^-$	TGL/Li[d]	23	2.002588
	TGL/Li[d]	-23	2.002584
	DME[e]	23	2.002578

[a] From Segal, Kaplan, and Fraenkel [93].
[b] Abbreviations: DME, 1,2-dimethoxyethane; TGL, tetraglyme.
[c] Corrected for second-order shifts. Errors are ±0.000006 or less.
[d] Reduced with lithium.
[e] Electrolytic reduction.

Absolute **g**-tensor calculations for planar hydrocarbon radicals and an extension of Stone's *g*-factor theory to nonplanar radicals have been made by Biehl, Plato, and Möbius [97]. The theoretical calculations for planar radicals included the plus and minus biphenyl ions, plus and minus ions of naphthalene, anthracene, tetracene, and the neutral benzyl and perinaphthenyl radicals. The calculated values were found to fit quite closely the Stone relation $\langle \Delta g \rangle = b + c\lambda$ [see (9.72)], with the constants $b = 35 \times 10^{-5}$ and $c = -7.6 \times 10^{-5}$, which agree reasonably well with the experimental values, 31.9×10^{-5} and -16×10^{-5}, obtained from the plot of Fig. 9.28.

For nonplanar phenyl-substituted hydrocarbon radicals, the calculations of Biehl, Plato, and Möbius et al. [97] revealed evidence for direct π-σ mixing, which puts unpaired electron-spin density in the σ system. This could account, at least in part, for anomolies observed in the *g* factor for nonplanar systems [98,99].

5. NUCLEAR COUPLINGS FROM SOLUTION SPECTRA

Only the isotropic component of nuclear couplings is measurable in solution spectra, whereas measurement of both isotropic and anisotropic components is possible in single crystals (Chapter VI). However, the isotropic components obtained from solution spectra are generally more accurate than those measured with single crystals, and they are certainly much easier to obtain from solution measurements.

5.a. Hydrogen Couplings and Q_{CH}^{H} Values in Aliphatic Radicals

In organic free radicals, the hydrogen hyperfine structure is the most prevalent, and the greatest amount of information gained from esr solution spectra is gained from the isotropic hydrogen nuclear coupling. Since the valence shell of H is composed entirely of the 1*s* orbital, the interatomic H coupling is entirely isotropic. There is no subshell to be polarized, and the populations of the higher excited levels of the bonded H are entirely negligible. Thus the proton coupling measured for solutions with the accurately known H atomic coupling provide very precise values of the electron-spin density on the bonded hydrogens. From the remarkably simple McConnell [100] formula, described in Chapter VI, Section 1.b

$$a_{\alpha}^{H_i} = Q_{XH}^{H} \rho_{\pi}^{X} \tag{9.74}$$

the observed coupling a_{α}^{H} gives the π spin density ρ_{π}^{X} on the atom to which H_i is bonded. The accuracy of ρ_{π}^{X} thus obtained is limited by that of the proportionality constant, which varies with the X atom, and also for the same X in different classes of radicals. Thus Q_{α}^{H} is not, strictly speaking, a constant but rather a coupling parameter having so nearly the same value for certain large groups of radicals that (9.74) has proved to be enormously useful for mapping π spin densities in organic radicals, particularly in aromatic ringed structures having CH groups distributed about the rings. The principal problem is not in obtaining reliable values for the proton couplings, which usually can be measured easily with high accuracy, but in finding the Q_{α}^{H} appropriate to the type of radical being considered. The Q_{α} values can, in principle, be derived theoretically [100-108] from a calculation of the π-σ configuration interaction by use of molecular-orbital theory, but the values most commonly used (and probably the more reliable ones) are experimental or semiempirical. Examples of experimental Q_{α} values are those for which the π spin density on the X atom can be derived from measurement of the anisotropic hyperfine structure of single crystals. This experimental ρ_{π}^{X} is substituted into (9.74) with the measured value of a_{α}^{H} to give an empirical value for $Q_{\mathrm{XH}}^{\mathrm{H}}$. The anisotropic ^{13}C coupling together with a_{α}^{H} has been measured for a number of >CH fragments in single crystals. Theoretical or semiempirical evaluations of Q_{α}^{H} are more often made for symmetric aromatic hydrocarbon radicals for which the π spin densities can be calculated with rather good accuracy from molecular-orbital theory. This theoretical ρ_{π}^{C} is then substituted into (9.74) with the measured a_{α}^{H} for a determination of $Q_{\mathrm{CH}}^{\mathrm{H}}$. The $Q_{\mathrm{CH}}^{\mathrm{H}}$ values thus obtained for aromatic hydrocarbons are in the range of 22 to 32 G, as are most of those for aliphatic radicals. It should be remembered that the induced H_{α} spin densities are negative (see Chapter VI, Section 1.b); and, since ρ_{π}^{X} is positive, the sign of Q_{α} is negative. For convenience, only the magnitudes are usually stated.

Fessenden and Schuler [22] have derived the $Q_{\mathrm{CH}}^{\mathrm{H}}$ values for simple alkyl radicals from their esr measurements described in Section 3.g. Their Q^{H} values (listed in Table 9.9) are widely used for derivation of π spin densities in alkyl

Table 9.9 Empirical Values of ρ_{π}, Q_{α}, and Q_{β} [a]

Radical	Spin Density $\rho_{\pi}(C_{\alpha})$	Coupling Constants (G) a_{α}	$a_{\beta}(CH_3)$	Q_{α}^{H}	$Q_{\beta}(CH_3)$ [b]
$\dot{C}H_3$	1	(−)23.04	–	(−)23.04	–
$CH_3\dot{C}H_2$	0.919	(−)22.38	26.87	(−)24.35	29.25
$(CH_3)_2\dot{C}H$	0.844	(−)22.11	24.68	(−)26.20	29.25
$(CH_3)_3\dot{C}$	0.776	–	22.72	–	29.30

[a] From Fessenden and Schuler [22].
[b] Constant for rotating CH_3 group.

free radicals. To obtain the ρ_{π}^{C} values used for their derivations, Fessenden and Schuler [22] started with the reasonable assumption that $\rho_{\pi}^{\mathrm{C}} = 1$ for the methyl radical and obtained $\rho_{\mathrm{C}}^{\mathrm{H}}$ for the substituted methyl radical from the relation

$$\rho_{\pi}(\mathrm{C}_{\alpha}) = (1 - 0.081)^{m} = (0.919)^{m} \tag{9.75}$$

where m is the number of substituted methyl groups. They obtained (9.75) from the assumption, justified by Chesnut's theory [109], that Q_{β}^{H} for the substituted methyl radicals is constant. From their experimental observations they conclude that the substitution reduces the βH coupling by very nearly 8.1% per substituted CH_3 group. This rule gives $\rho_{\pi}^{\mathrm{C}\alpha} = 0.919$ for the ethyl radical, a value in good agreement with that of 0.911 derived later by Lazdins and Karplus [107] from molecular-orbital theory. A relation similar to that of (9.75) has been proposed by Fischer [110] for radicals of other types.

Fischer derived Q_α^H values for methyl-substituted radicals of type $CH_3\dot{C}HX$ by assuming that Q_β^H for the $CH_3\dot{C}$ group has the constant value of 29.25 G for the series that includes X = COOH, $COCH_2CH_3$, OH, and OCH_2CH_3 [110]. He then derived the π spin densities ρ_π^α from the a_β^H coupling of the CH_3 group. With these ρ_π^α values he obtained Q_α^H values from the αH coupling, using the McConnell equation (9.74). The resulting Q_α^H values for the preceding sequence of X groups were 23.7, 23.9, 19.5, and 18.3, respectively, in G. The last two are abnormally low, and, for reasons that follow, should not be regarded as Q_α^H values for pure π-type radicals.

From comparison of the ratios of the a_α(H) and a_β(H) couplings of an extended group of $CH_3\dot{C}HX$ radicals and from comparisons of the H couplings with the available $a_\alpha(^{13}C)$ couplings and with the αH spin-spin coupling from nmr, Dobbs, Gilbert, and Norman [111] concluded that the unpaired electron densities on the C_α in $CH_3\dot{C}HOH$ and in $CH_3\dot{C}HOC_2H_5$ are not pure p_π spin densities but have some *sp* hybridization. In effect, this means that the three σ-bonding orbitals to C_α are not exactly in a common plane. As pointed out by these observers, a small *s* character in the unpaired electron orbital on C_α will have negligible effects on the a_β(H) but may reduce the a_α(H) and increase the $a_\alpha(^{13}C)$ by significant amounts. For example, the hyperconjugative β coupling is directly proportional to the p_α character of the unpaired electron density on C_α, and a 3% *s* character in this orbital will reduce a_β(H) by only 3%. However, the isotropic $^{13}C_\alpha$ coupling, which results from about 3% spin polarization of the s components of the σ-bonding orbitals and is positive in sign, will be approximately doubled. The *s* component in the unpaired electron orbital on C_α cannot, of course, induce spin polarization in the $s(C_\alpha)$ orbital itself, but it can induce a positive spin polarization of the 1*s* orbital of αH through a direct spin-spin interaction. This interaction tends to put positive spin density in the αH orbital, which tends to cancel the negative spin density indirectly induced through the unpaired *p*-orbital component. The result is a lowering of the magnitude of the negative a_α(H). This is, in part, a mechanistic explanation for the positive sign of the αH coupling of certain σ-type radicals. Experimentally, the lowering of the magnitude of the αH coupling in the Group IV hydrides is found to decrease rapidly with increase of *s* character of the unpaired electron orbital on the central atom, and the sign of the αH coupling appears to reverse as the hybridization increases in the sequence C,Si,Ge,Sn (see Table 8.8).

From these considerations, it is evident that for nonconjugated aliphatic radicals, the βH coupling is a more reliable indicator of spin in the C_α than is the αH coupling or the isotropic $^{13}C_\alpha$ coupling, $a_\alpha(^{13}C)$, for radicals in which there may be small departures from planarity in the radical center. Unfortunately, many such radicals do not have β-coupling groups, and from solution spectra one cannot obtain *p*-orbital couplings, which give a more realiable measure of the unpaired π-electron densities. Thus it is important to find Q_α^H values that remain relatively constant for selected groups of radicals.

The ρ_π^α spin densities for the $CH_3\dot{C}HX$ radicals in Table 9.10 were calculated from the a_β^H coupling of the rotating CH_3 group with $Q_\beta = 29.25$ G. The Q_α

Table 9.10 Proton Couplings and Q_{CH}^H Values (in Gauss) for Radicals of the Type $CH_3\dot{C}HX$

X	$a_\beta(CH_3)$	$\rho_\pi(C_\alpha)^a$	$a_\alpha(H)$	$Q_{CH}^{H\ b}$	Source of a_α and a_β
CH_3	24.68	0.8438	22.11	26.20	c
C_2H_5	24.5	0.8376	21.8	26.03	c
CO_2H	24.98	0.8540	20.18	23.63	d
CO_2CH_3	24.9	0.8513	20.3	23.85	e
COC_2H_5	22.59	0.7723	18.45	23.89	d
CN	22.99	0.7860	20.33	25.87	e
CH_2OH	25.3	0.8650	21.7	25.09	e
$CH(OH)CO_2H$	25.81	0.8824	21.86	24.77	e
CH_2CO_2H	25.1	0.8581	21.2	24.71	e
OH	22.19	0.7590	15.37	(20.29)[g]	f
OC_2H_5	22.28	0.7617	13.96	(18.33)[g]	d
			Average =	24.79 G	
			=	69.8 MHz	

[a] $\rho_\pi(C_\alpha) = [a_\beta(CH_3)/29.25]$.
[b] $Q_{CH}^H = [a_\alpha(H)/\rho_\pi(C_\alpha)] = Q_\alpha^H$.
[c] Fessenden and Schuler [22].
[d] Fischer [110].
[e] Dobbs, Gilbert and Norman [111].
[f] Livingston and Zeldes [8].
[g] Omitted in average.

values were calculated with these densities and the a_α^H coupling. It is seen that the Q_α is approximately the same for all except the two radicals in which an oxygen is bonded to the C_α. The mean value for those in which a carbon of the X group is bonded to the C_α is $Q_\alpha^H = 24.8$ G = 69.8 MHz. In Chapter VI we used this mean Q_α^H value for analysis of the αH coupling of $X\dot{C}HY$ radicals in the single crystals listed in Table 6.1. In analysis of the proton hyperfine structure of the $X\dot{C}H_2$ type of radical in single crystals (Table 6.2), we calculated the ρ_π^α spin densities from the isotropic αH coupling components with the $Q_\alpha^H = 24.35$ G = 68.2 MHz for CH_3CH_2 (Table 9.9).

With the ρ_π^α spin densities obtained in this way for the $X\dot{C}HY$ and XCH_2 radicals listed in Tables 6.1 and 6.2, we normalized their anisotropic CH coupling components to the spin-density value of $\rho_\pi^\alpha = 1$. If the relative ρ_π^α values are correct, these normalized anisotropic components should be approximately the same for all the radicals since the CH distances, and hence the dipole-dipole interactions for $\rho_\pi^\alpha = 1$, are approximately the same. Comparison of the normalized values shows that they are nearly the same. Among these radicals are some in which an F, N, P, or S is bonded to the C_α. This comparison provides good evidence that the Q_α^H values used are approximately correct for these rather diverse radicals and that they should be applicable to others of similar types. Note that the Q_α values used for the $X\dot{C}H_2$ and $X\dot{C}HY$ species, 68.2 and 69.8 MHz, are nearly equal. Had we used their averaged value, Q = 69 MHz, the results would not have been noticeably different.

It is evident that s character in the unpaired electron orbital of C_α leads to a decrease in the effective Q_α^H. When $Q_{X_\alpha H}^H$ can be measured, it can be used, as can $a_\alpha(X)$, for an indicator of the planarity of the radical site. If one finds that the Q_α^H = 24.8 G for $C_\alpha H$ with the observed $a_\alpha(H)$ indicates a significantly lower spin density on C_α than does the observed $a_\beta(H)$ with Q_β^H = 29.8 G, one is justified in suspecting nonplanarity of the radical site. In this case the spin density indicated by $a_\beta(H)$ should be accepted rather than that indicated by $a_\alpha(H)$.

Fischer [110] has suggested that the deviations in Q_α^{CH}, like those shown in Table 9.10, are due to the inductive effects by the different electron-attracting substituent X groups. He attributes the decrease in Q_α to the lowering of the electron density of the CH σ bond (i.e., reduction in its covalent character). Such effects are expected, but the increasing positive charge on C_α by highly electronegative X groups also increases the hybridization of the unpaired electron orbital on the C_α; therefore, the two effects cannot be distinguished by the lowering of Q_α. The increasing *s* character in the unpaired electron orbital and the deviation from planarity of the trigonal bonds to C_α by the electronegative fluorine are demonstrated in Table 8.14. However, the substitution of only one F, as in H_2CF, causes only a small (if any) deviation from planarity. If one assumes the observed a_α^H = 59.1 MHz = 21.1 G to arise from induced negative spin density on the hydrogens and accepts the $\rho_\pi^{C_\alpha}$ = 0.86 deduced in the analysis (Chapter VIII, Section 4.f), one predicts the perfectly normal Q_α^{CH} = 24.5 G for CH_2F. Furthermore, the planar $FCH(CONH_2)$ radical in Table 6.1 appears to have a normal Q_α^{CH} of about 24.9 G, like the other radicals in this table. A derivation for the definitely nonplanar CHF_2 in Table 8.1, like that described for CH_2F, would indicate the exceptionally high value of 33.6 G for Q_α^{CH}. The large αH coupling in this σ-type radical is most likely positive, and (9.74) is not applicable to it.

5.b. Hydrocarbon-ring Systems: Even and Odd Alternant Radicals

The $Q_{\mathrm{CH}}^{\mathrm{H}}$ values for symmetrical, planar, ringed free radicals constructed entirely of linked >CH fragments like those in Table 9.11 can be evaluated

Table 9.11 Isotropic Coupling and Derived $Q_{\mathrm{CH}}^{\mathrm{H}}$ Values for Ringed Hydrocarbon Radicals

	Coupling Parameter[a] *(G)*			
Radical	a_i^{H}	$\sum_i a_i^{\mathrm{H}}$	$Q_{\mathrm{CH}}^{\mathrm{H}}$	*Source of* a_i^{H}
C_5H_5	5.98	29.90	29.90	*d*
$C_6H_6^+$	4.44	26.64	24.79[b]	*e*
$C_6H_6^-$	3.82	22.94		*f*
C_7H_7	3.91	27.37	27.37	*g*
$C_8H_8^-$	3.21	25.68	27.18[c]	*h*
		Average =	27.3 G	

[a] The signs of these coupling parameters are negative.
[b] Average for anion and cation.
[c] $Q_{\mathrm{CH}}^{\mathrm{H}} = \sum_i a_i^{\mathrm{H}} + 12\,(1/8)$, where 12 (1/8) is the charge correction from (9.77).
[d] P. J. Zandstra, *J. Chem. Phys.*, 40, 612 (1964).
[e] M. K. Carter and G. Vincow, *J. Chem. Phys.*, 47, 292 (1967).
[f] R. W. Fessenden and S. Ogawa, *J. Am. Chem. Soc.*, 86, 3591 (1964).
[g] A. Carrington and I. C. P. Smith, *Molec. Phys.*, 7, 99 (1963).
[h] H. L. Strauss, T. J. Katz, and G. K. Fraenkel, *J. Am. Chem. Soc.*, 85, 2360 (1963).

directly from the measured proton coupling with the assumption that the unpaired π-electron spin density is equally divided among the carbons. This assumption may be verified by the equality of splitting by all the hydrogens. If there are n carbons in the radical, the π spin density on each is $1/n$, and this measured splitting by each hydrogen is $Q_{\mathrm{CH}}^{\mathrm{H}}(1/n)$; the summation of the splitting by the n protons is simply $Q_{\mathrm{CH}}^{\mathrm{H}}$. The cyclical radicals in Table 9.11 are of this type, and the effective $Q_{\mathrm{CH}}^{\mathrm{H}}$ for them is $\sum_n a_i^{\mathrm{H}}$. The positive and negative ions of benzene have differing values of $Q_{\mathrm{CH}}^{\mathrm{H}}$. To obtain a $Q_{\mathrm{CH}}^{\mathrm{H}}$ of benzene for comparison with those of the neutral radicals, we have averaged those for the anion and the cation. Also, we have projected the Q_α for $C_8H_8^-$ to the neutral value by a charge correction to be described later [see (9.77)]. The average of the $Q_{\mathrm{CH}}^{\mathrm{H}}$ values in Table 9.11 is −27.3 G, very near the value of −27 G, the one most often obtained or used for aromatic ringed systems.

Conjugated hydrocarbons are classified as even alternant, odd alternant, or nonalternant depending on the symmetry of the structures or the electronic orbitals. The even and odd alternant systems are illustrated in Fig. 9.29 with asterisks to distinguish the alternant carbons. Free radicals from these systems are described in later paragraphs. Nonalternant conjugated hydrocarbons have ringed

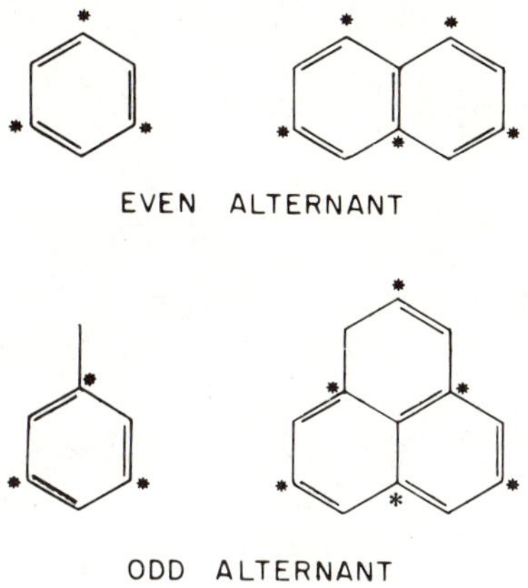

Fig. 9.29 Examples of alternant hydrocarbons.

structures that are inequivalent, with differing bond angles, a familiar example of which is azulene. The most extensive esr studies of aromatic hydrocarbon radicals have been for the even-alternant ones. There are relatively fewer of the odd-alternant radicals, the most thoroughly studied of which are the benzyl, perinaphthenyl, triphenylmethyl, and allyl radicals.

The even-alternant hydrocarbons are characterized by equal numbers of bonding and antibonding π molecular orbitals with Hückel energies $\alpha + \lambda\beta$ and $\alpha - \lambda\beta$, where α and β are the Coulomb and resonance energies and λ is the Hückel parameter obtained from solution of the secular equation for the system. The value of λ is the same for the bonding and antibonding orbitals. The unpaired electron is in the highest bonding π orbital in the cation radicals and in the lowest antibonding orbital in the anion radicals.

Effects of Charge on Couplings in Anions and Cations

A special interest in the esr of alternant hydrocarbon radicals results from the pairing theorem, which predicts that anion and cation forms of the π-orbital electron spin densities on corresponding carbons should be identical [112-114]. However, numerous measurements of the proton hyperfine structure of the monopositive and mononegative ions of many alternant hydrocarbons have shown that the total spread of the hyperfine pattern for the cation is greater than that for the anion. Also, corresponding couplings for the cations are found to be greater in magnitude than are those for the anions, as shown for $C_6H_6^+$ and $C_6H_6^-$ in Table 9.11. Colpa and Bolton [115,116] proposed that these differences in coupling for the positive and negative ions are due not to differences in the π spin densities (in agreement with the pairing theorem), but rather to an alteration in the effective Q_{CH}^H caused by the charges. They suggested modification of the McConnell relation to the form

$$a_i^H = [Q_{CH}^H(0) + K\epsilon_i^\pi]\rho_i^\pi \tag{9.76}$$

where a_i^H is the proton coupling of the C_iH fragment of the radical ion, ϵ_i^π is the excess charge on C_i, K is a constant, $Q_{CH}^H(0)$ is the Q_α value for a carbon in which there is no excess charge, ρ_i^π is the π spin density on C_i. Approximately, $|\epsilon_i| = \rho_i^\pi$ so that (9.76) may be expressed [117] as

$$\left(a_i^H\right)_\pm = Q_{CH}^H(0)\,\rho_i^\pi \pm K_{CH}^H(\rho_i^\pi)^2 \tag{9.77}$$

where $Q_{CH}^H(0)$ and K are both negative in sign and the $\pm$ sign corresponds to the sign of the radical ion.

From measurements of a_i^H for the corresponding positive and negative ion radicals, one can obtain empirical values of the constants in (9.77). It is evident that

$$(a_i^H)_{av} = \frac{(a_i^H)_+ + (a_i^H)_-}{2} = Q_{CH}^H(0)\rho_i^\pi \tag{9.78}$$

$$\Delta a_i^H = \frac{(a_i^H)_+ - (a_i^H)_-}{2} = K_{CH}^H(\rho_i^\pi)^2 \tag{9.79}$$

For example, from the coupling values for the plus and minus ions of benzene in Table 9.11 one obtains $(a_i^H)_{av} = -4.13$ G and $\Delta a_i^H = -0.31$ G. With the proportioned π spin density on each C_i of $\rho_i^\pi = 1/6$, (9.78) and (9.79) give $Q_{CH}^H(0) = -24.78$ G and $K_{CH}^H = -11.2$ G.

From careful measurements of the ^{13}C hyperfine structure of the positive and negative ions of anthracene, Bolton and Fraenkel [114] found the splitting to be almost identical in the two ions and thus concluded that the distribution of the π spin density in the cation and the anion is the same, in agreement with the pairing theorem. They further concluded that the observed difference in the αH splitting results from a change in the effective Q_α^H, as proposed by Colpa and Bolton [115]. From the a_i^{H} coupling of both ions with $(Q_{\mathrm{CH}}^{\mathrm{H}})_{\mathrm{av}} = -27$ G, they calculated π spin densities on the individual carbons of the anthracene ions that they compared to values predicted with Hückel's theory, with McLachlan's treatment [103] (which includes configuration interactions), and with the calculations by Snyder and Amos [118], who used the spin-polarized Hartree-Fock method. The agreement of these two methods was found to be reasonably good. These comparisons provide support for Eq. (9.77) and for the value of -27 G for $Q_{\mathrm{CH}}^{\mathrm{H}}(0)$, in agreement with the averaged values in Table 9.11.

It is easily shown from (9.78) and (9.79) that

$$\Delta a_i^{\mathrm{H}} = \frac{K_{\mathrm{CH}}^{\mathrm{H}}}{[Q_{\mathrm{CH}}^{\mathrm{H}}(0)]^2}(a_i^{\mathrm{H}})_{av}^2 \tag{9.80}$$

Thus, if $K_{\mathrm{CH}}^{\mathrm{H}}$ and $Q_{\mathrm{CH}}^{\mathrm{H}}(0)$ are constant for a group of radicals, the plot of Δa_i^{H} against $(a_i^{\mathrm{H}})_{av}^2$ should be a straight line. Bolton [117] plotted Δa_i^{H} against $(a_i^{\mathrm{H}})_{av}^2$ for the radical ions of anthracene, tetracene, and pentacene and found that the points did fall closely along a straight line. With the assumed value of $Q_{\mathrm{CH}}^{\mathrm{H}}(0) = -27$ G, he found from the slope of the line that $K_{\mathrm{CH}}^{\mathrm{H}} = -12$ G.

Lewis and Singer [119] constructed plots of measured values of a_i^{H} versus Hückel orbital spin densities for a large number of radical ions. They made separate plots for the anion and cation radicals and found the straight-line relationship for both, with all points falling rather close to the line. The slopes of the lines differed because of the charge effects on Q_α^{H} described previously. From the two plots they obtained

$$|a_i^{\mathrm{H}}|_- = 28.6\,\rho_i^\pi \text{ (anions)} \tag{9.81}$$

$$|a_i^{\mathrm{H}}|_+ = 35.7\,\rho_i^\pi \text{ (cations)} \tag{9.82}$$

with standard deviations of ±0.43 G for the anions and ±0.38 G for the cations. These relations show a strong charge-dependence of $Q_{\mathrm{CH}}^{\mathrm{H}}$. Consequently, the authors correlated all the data with the Colpa-Bolton relation, (9.77), and found that the best fit could be achieved with $Q_{\mathrm{CH}}^{\mathrm{H}}(0) = -32.2$ G and $K_{\mathrm{CH}}^{\mathrm{H}} = -16$ G. However, odd-alternant radicals in which strong π-π configuration interactions put negative π spin density on some of the carbons (see discussion to follow) were included in the correlations. The Hückel Molecular Orbital (HMO) value applies best to the even-alternant systems in which the configuration interactions are less pronounced. We have correlated the coupling of even-alternant radical ions listed in Table 9.12 with (9.77) and found that good agreement of the observed coupling is achieved with the HMO values of ρ_i^π and

Table 9.12 Comparison of Proton Couplings of Some Even-alternant Radical Ions with Those Calculated with HMO Spin Densities

			Magnitude of Proton Coupling (G)			
			Cation		*Anion*	
Hydrocarbon	*Carbon Position*	*HMO*[a] ρ_i^π	*Obs.*[b]	*Calc.*[c]	*Obs.*[b]	*Calc.*[c]
Anthracene	9	0.193	6.533	6.62	5.337	5.73
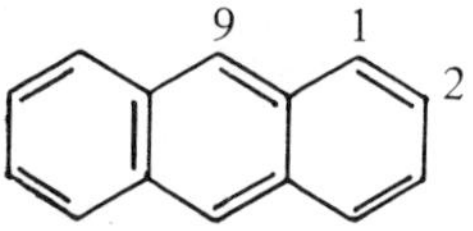	1	0.097	3.061	3.22	2.740	2.99
	2	0.048	1.379	1.56	1.509	1.51
Naphthacene	5	0.147	5.06	4.96	4.25	4.44
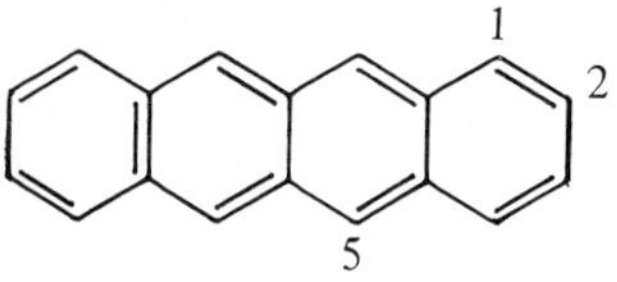	1	0.056	1.689	1.83	1.55	1.75
	2	0.034	1.020	1.10	1.15	1.07
Pentacene	6	0.141	5.083	4.75	4.263	4.27
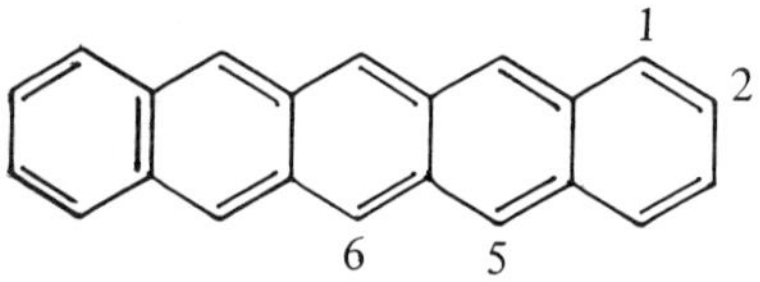	5	0.106	3.554	3.53	3.032	3.26
	1	0.035	0.975	1.13	0.915	1.11
	2	0.025	0.757	0.81	0.870	0.79
Perylene	3	0.108	4.10	3.60	3.53	3.32
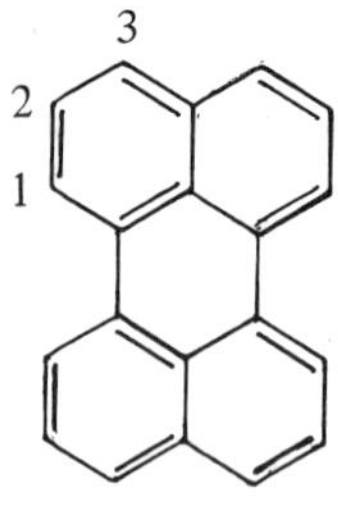	1	0.083	3.10	2.74	3.08	2.57
	2	0.013	0.46	0.42	0.46	0.41

Table 9.12 (Continued)

			Magnitude of Proton Coupling (G)			
			Cation		Anion	
Hydrocarbon	Carbon Position	HMO[a] ρ_i^π	Obs.[b]	Calc.[c]	Obs.[b]	Calc.[c]
Dibenzo(*a*,*c*)-triphenylene	4	0.067	2.28	2.20	2.06	2.09
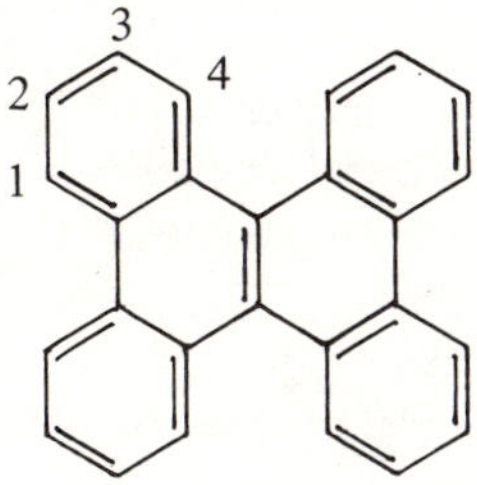	2	0.056	1.99	1.83	1.71	1.75
	1	0.027	0.60	0.87	0.62	0.85
	3	0.002	<0.03	0.06	<0.03	0.06

[a]Hückel molecular-orbital values of ρ_i^π from Lewis and Singer [119].
[b]Observed values from tabulation of Vincow [130].
[c]Calculated from (9.83) with ρ_i^π (HMO) values from column 3.

the $Q_{\text{CH}}^{\text{H}}(0) = -32$ G but with the $K_{\text{CH}}^{\text{H}} = -12$ G obtained by Bolton from even alternant radicals as described previously. Comparison of the observed and calculated coupling magnitudes indicate the following relationship between the proton coupling and the HMO spin density for the even-alternant hydrocarbon radical ions:

$$|a_i^{\text{H}}|_\pm = 32\,\rho_i^\pi \pm 12(\rho_i^\pi)^2 \tag{9.83}$$

where + refers to the cations and − to the anions, ρ_i^π is the HMO density on C_i, and where $|a_i^{\text{H}}|$ is the magnitude of the proton coupling in gauss.

Snyder and Amos [118] calculated the proton coupling for a large number of alternant as well as nonalternant hydrocarbon radicals with π spin densities derived with unrestricted Hartree-Fock (UHF) orbitals For the neutral odd-alternant radicals, they employed the McConnell relation, (9.74), with $Q_{\text{CH}}^{\text{H}} = -27$ G; for even-alternant and nonalternant ions, they used the Colpa-Bolton relation, (9.77), with $Q_{\text{CH}}^{\text{H}} = -27$ G and $K_{\text{CH}}^{\text{H}} = -12.8$ G. Although the agreement between the calculated and observed couplings for the even-alternant ions was not as close as that shown in Table 9.12, these calculations have the advantage that they achieve a semiquantitative correlation of a wider, more diverse group of radicals with a single Q_{CH}^{H} value of −27 G.

The basic mechanism for the change in the effective Q_α by a formal charge on the radical is not entirely clear. Presumably, the mechanism is the same as that by which a formal charge on an atom alters its spin-spin coupling. There is less nuclear screening in a positively charged atom; the electronic orbitals are pulled closer to the nucleus; hence the interacting spins are closer to each other. A negative formal charge has the inverse effect. From such qualitative considerations, one might expect the π-σ interactions to be stronger and the effective Q_α greater in magnitude for the cation radicals. However, this mechanism should also be operative in neutral radicals where charges on the atoms are produced by bond ionic character. Perhaps the effect of bond ionic character on Q_α in neutral radicals is masked by the accompanying change in the covalent character of the spin-polarized σ bond. The problem is obviously complex, and at this stage the justification of (9.77) is largely empirical.

Giacometti, Nordio, and Pavan [120] derived a relation similar to (9.77) by including the cross term in the π spin-density matrix. Their relation may be expressed in the form

$$a_i^{\mathrm{H}} = Q_1 \rho_{ii}^\pi + Q_2 \sum_j \rho_{ij}^\pi \tag{9.84}$$

which, in terms of the HMO coefficients, may be expressed as

$$a_i^{\mathrm{H}} = Q_1 c_i^2 + Q_2 \sum_j c_i c_j \tag{9.85}$$

where the summation is taken over the atoms to which the coupling fragment C_iH is bonded. The ρ_{ij}^π represents the π spin density in the bond-overlap region. Snyder and Amos [118] also correlated this relation with the observed coupling by employing values ρ_{ii}^π and ρ_{ij}^π calculated with UHF functions. With $Q_1 = -27$ G and $Q_2 = -6.3$ G, they obtained agreement with the observed coupling values comparable to that achieved with the Colpa-Bolton relation. The parameter values originally proposed by Giacometti, Nordio, and Pavan [120] are $Q_1 = -31.5$ G and $Q_2 = -7$ G. Bloor, Gilson, and Daykin [121] compared (9.77) and (9.84) using spin densities calculated from restricted Hartree-Fock functions with configuration interactions (RHF/CI). They reported no significant statistical improvement in the fitting of the experimental data by the second terms of either relation.

Negative Spin Densities in Odd Alternant Systems

The odd alternant systems have an odd number of carbons, and it is not possible to write valence-bond structures with a double bond to every carbon. There is always one carbon to which there are single bonds only. In free radicals formed from the odd-alternant molecules, the unpaired electron is in a non-bonding orbital on the odd carbon that can, in the valence-bond structures, switch between the alternant carbons, as can be simply illustrated with the equivalent structures **39** and **40** for the allyl radical:

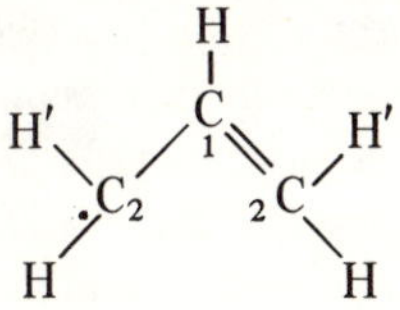

39

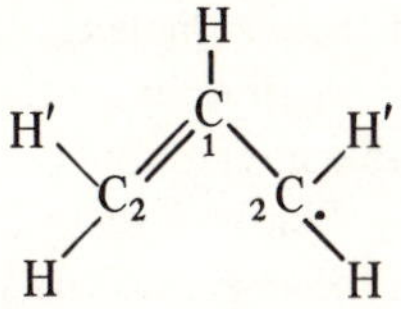

40

From symmetry, one expects the proton couplings of the two CH_2 groups to be equal, as is observed. Because the radical is planar, no hyperconjugation is expected to occur with the central CH; without configuration interaction, no spin density is expected on C_1, and no isotropic coupling is expected for the central H. The HMO approximation, which neglects configuration interaction, predicts zero spin density on the central carbon of the allyl radical and on alternate carbons of other odd-alternant radicals. Thus these radicals provide a good test of the more complete theories that include configuration interactions. The negative π spin density in odd alternant radicals may be explained with valence-bond theory [122,123] or with various molecular-orbital treatments that include π-π configuration interactions [124]. McLachlan [103] uses self-consistent field theory (SCF) for prediction of spin densities in the π-type radicals. His paper, which is a basic one for calculation of the distribution of spin density in aromatic hydrocarbon compounds, compares the experimental spin densities obtained from esr coupling data via the McConnell relation with those calculated from valence-bond theory and from self-consistent molecular-orbital wave functions. It also gives a lucid summary of the theoretical methods for calculation of spin density.

Fessenden and Schuler [22] measured the esr of the allyl radical in solution and found the proton on the central carbon C_1 to have a coupling of $a_1^H = 4.06$ G. Although the couplings of the two end CH_2 groups were found to be equivalent, the individual protons of each group differ slightly, having the values $a_2 =$ 13.93 G and $a_2' = 14.83$ G. This inequivalence is expected because of the steric factors of the triangular structure of the $C_2C_1C_2$. The coupling a_1^H had been detected earlier in solids and successfully explained by π-σ and π-π configuration interaction, which puts negative spin density on C_1. The spin density on each end-carbon may be calculated from the average $[(a_2^H + a_2^{H'})/2] = 14.38$ G with $Q_\alpha^H = -24.35$ G for the ethyl radical. Equation 9.74 gives $\rho_\pi(C_2) = 0.59$, and thus the combined π spin density on the two end carbons is 1.18. Normalization of the π spin density to unity indicates that $\rho_\pi(C_1) = 1 - 1.18 = -0.18$ on the central C_1. This density can be used with (9.74) to predict the αH coupling of the central CH. With this Q_α^H, the predicted value is $a_1^H = +4.38$ G, in good agreement with the observed value, $a_1^H = 4.06$. As theoretically predicted, the experimentally derived spin density on C_1 is negative, and the coupling a_1^H is positive.

Description of the complex molecular-orbital theories, which include π-π configuration interaction, would require too much digression here. The basic mechanism for the interaction is similar to that already described for the spin polarization of σ bonds (π-σ interaction) (see Chapter VI, Section 1.b), also similar to the one that causes the parallel alignment of spin in different orbitals of atoms (Hund's rule). The π-π interaction leads to spin polarization of the π bond, which puts negative π spin density on the alternate carbons, as can be seen from consideration of one of the valence-bond structures of the allyl radical. A paired π-orbital spin alignment on C_1, opposite to that of the unpaired π spin on C_2, would be favored because the antiparallel alignment would increase the probability of the switching of the double bond, that is, exchange between structures **39** and **40**, which would lower the energy of the system. This is likewise true for the alternate carbons in the odd-alternant aromatic radicals. Negative π spin density on a CH carbon induces positive spin density on the H, which gives rise to proton couplings opposite in sign to the usual proton coupling induced by positive π spin density. However, the signs of the couplings are not distinguished in normal observation of hyperfine structure. Since the total π-orbital spin density of the radical must sum to unity for a single unpaired electron, it is evident that negative spin density on some of the carbons increases the total spread of the proton hyperfine structure of hydrocarbon radicals.

Theoretical Spin Densities from Various Methods of Calculation

Harriman and Sando [125] used spin-extended SCF (SESCF) wave functions to calculate spin densities for both odd and even alternant radicals and correlated their results with experimental coupling by use of (9.74), (9.77), and (9.84). They obtained rather good agreement with experiment, except for positions 1 and 2 in anthracene and naphthalene. However, the parameter values used in these equations differ somewhat from those found by others. The values for these parameters derived in this manner depend on the theoretical spin densities, which, in turn, depend on the nature of the wave functions and the approximations used in their calculation. A comparison of spin densities calculated with various theoretical methods, from Harriman and Sando [125], is reproduced in Table 9.13.

5.c. β-Hydrogen Coupling in Hydrocarbon Groups

The theory of βH coupling in solution spectra is the same as that described in

Chapter IV and extensively applied to the isotropic components of βH coupling in single crystals. In solution spectra, the interatomic dipole-dipole components are averaged out, and only the isotropic component is observed. Nevertheless, this isotropic component is still dependent on the orientation θ of the $C_\alpha C_\beta H$ plane relative to the p_π^α-C_β plane, as indicated in Fig. 6.1. Thus considerable structural or conformational information has been gained from application of the formula [see (6.24)]

$$a_\beta^H = B_0 + B_2 \cos^2 \theta \tag{9.86}$$

to β coupling in solution spectra. For convenience in description of groups or in indication of the dihedral angle θ, it is customary to show the relative orientations of the C_α and C_β groups by a clock-like diagram similar to that shown in Fig. 9.30. The $C_\alpha C_\beta$ bond is perpendicular to the page and C_α and C_β, not shown, are both at the center of the circle. The C_α is assumed to be in front of

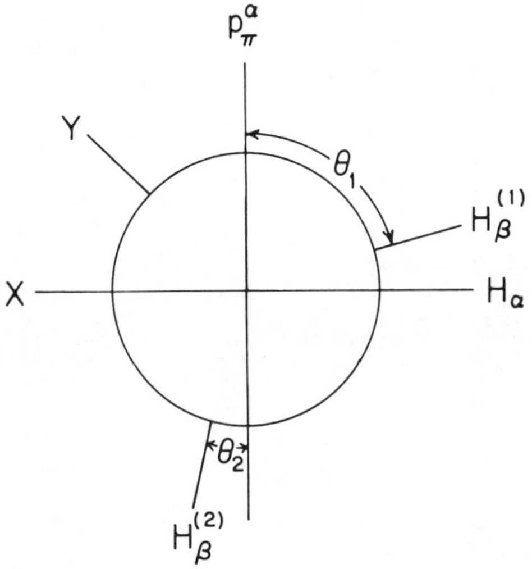

Fig. 9.30 Diagram for indication of the dihedral orientation θ of the β-coupling group relative to the p_π^α orbital. The $C_\alpha C_\beta$ bond is perpendicular to the page and at the center of the circle. The lines indicate perpendicular projections of the bonds in the plane of the page.

C_β so that the directions of the p_π^α orbital and bonds to C_α are indicated by the lines extending from the center of the circle. The bonds extending from C_β, assumed to be behind the "clock face," are shown only outside the circle. In the diagram the three bonds to C_α are assumed to be in a common plane per-

Table 9.13 Spin Densities Calculated by Various Methods[a]

System	*Atom*	*Hückel*[b]	*McLachlan*[c]	*VB*	*USCF*[d]	*PUSCF*[e]	*SESCF*
Allyl	1	0.500	0.594	0.588[f]	0.651	0.547	0.584
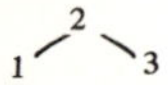	2	0	-0.187	-0.177	-0.302	-0.093	-0.167
Pentadienyl	1	0.333		0.433[f]	0.545	0.383	0.452
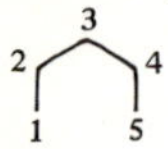	2	0		-0.131	-0.307	-0.094	-0.159
	3	0.333		0.398	0.524	0.422	0.415
Benzyl	1	0	-0.102		-0.189	-0.060	-0.134
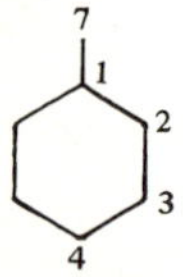	2	0.143	0.161		0.254	0.157	0.279
	3	0	-0.063		-0.158	-0.050	-0.143
	4	0.143	0.137		0.225	0.128	0.260
	7	0.571	0.770		0.771	0.718	0.602
Naphthalene$^{\pm}$	1	0.181	0.222	0.198	0.262	0.215	0.252
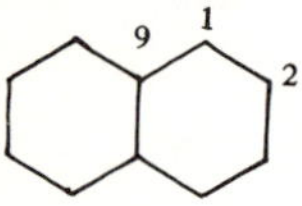	2	0.069	0.047	0.061	0.026	0.048	0.027
	9	0	-0.037	-0.019	-0.076	-0.024	-0.059
Anthracene$^{\pm}$	1	0.096	0.118		0.138	0.105	0.138
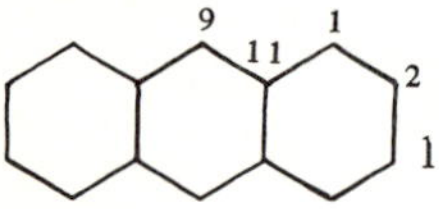	2	0.048	0.032		0.014	0.028	0.014
	9	0.193	0.256		0.319	0.260	0.293
	11	0.004	-0.028		-0.061	-0.014	-0.048
Azulene^{+}	1	0.284		0.210[g]	0.432		0.342
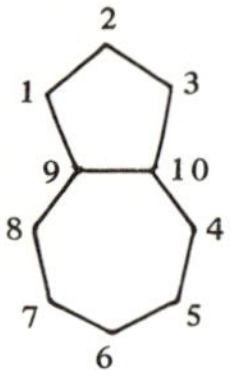	2	0		-0.011	-0.195		-0.097
	4	0.029		0.333	-0.068		-0.074
	5	0.119		-0.186	0.185		0.228
	6	0		0.374	-0.111		-0.128
	9	0.068		-0.039	0.104		0.117
Azulene^{-}	1	0.005	-0.027	0.130[g]	-0.011	-0.001	0.001
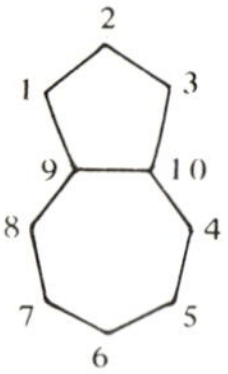	2	0.107	0.120	0.146	0.118	0.080	0.102
	4	0.213	0.292	0.306	0.313	0.236	0.317
	5	0.011	-0.081	-0.158	-0.178	-0.046	-0.135
	6	0.374	0.368	0.351	0.434	0.356	0.387
	9	0.089	0.071	-0.030	0.099	0.093	0.072

[a] From a tabulation by Harriman and Sando [125]. The abbrevaited methods are: valence-bond theory (VB); unrestricted, self-consistent-field theory (USCF); projected-unrestricted, self-consistent-field theory (PUSCF); and spin-extended, self-consistent-field approximation (SESCF).

[b] All values except those for azulene are from Snyder and Amos [118]. Azulene values are from T. H. Brown and M. Karplus, *J. Chem. Phys.*, 46, 870 (1967).

Table 9.13 (Continued)

---

[c]A. D. McLachlan [112,103], self-consistent-field method (SCF).
[d]All values except those for azulene(+) are from Snyder and Amos [118]; those for azulene(+) are from T. Amos, *Molec. Phys.*, 5, 91 (1962).
[e]Values for allyl, pentadienyl, and naphthalene are for the full-spin projections of functions reported by T. Amos and L. C. Snyder, *J. Chem. Phys.*, 41, 1773 (1964). The others are single-annihilation results from Snyder and Amos [118].
[f]From M. Hanna, A. D. McLachlan, H. H. Dearman, and H. M. McConnell, *J. Chem. Phsy.*, 37, 361 (1962).
[g]From T. H. Brown, *J. Chem. Phys.*, 41, 2223 (1964); T. H. Brown and M. Karplus, *J. Chem. Phys.*, 46, 870 (1967).

pendicular to p_π^α (π-type radical) with one bond to an H. Two hydrogens and an unspecified Y group are assumed to be bonded to C_β, but the diagram applies as well when Y is replaced by a third H or when one of the hydrogens is replaced by a second Y group. All lines represent projections of the bonds in the common plane of the page so that the angles measured from p_π^α are the dihedral angles θ of (9.86). When the bonding orbitals C_β are sp^3 hybrids, it can be assumed that the projections of the three bonds to C_β in a common plane normal to the $C_\alpha C_\beta$ direction, as indicated in Fig. 9.30, will be 120° apart, or approximately so. In this common case, $\theta_2 = \theta_1 \pm 120°$, and there is only one unknown θ with two observable coupling parameters, $a_\beta(H_1)$ and $a_\beta(H_2)$. By omission of the small orientation-independent constant B_0, one can derive

$$a_\beta^{(1)} = B_2 \cos^2 \theta_1 \tag{9.87}$$

$$a_\beta^{(2)} = B_2 \cos^2 (\theta_1 + 120°) \tag{9.88}$$

from which B_2 and θ_1 can be obtained. Applications of this approximation are given in Table 6.8. If there is a nonrotating methyl group attached to C_α, there will be three observable coupling constants and three observable βH couplings. With the assumption that the $C_\beta H_3$ bonds are equal, one can solve (9.86) for B_0 as well as for B_2 and θ_1. An illustration of this case is given in Chapter VI, Section 2.b. When there is only one coupling C_βH, it is still possible to get an approximate orientation of this group by assumption that $B_0 \approx 0$ and that B_2 can be derived from a probable Q_β^H with spin densities ρ_π^α obtained from a_α^H.

Numerous conformational studies of radicals in solution have been made with the observed β couplings by use of various assumptions and approximations similar to those described in the previous paragraph. Those interested in results on particular radicals should consult the comprehensive review of configurations and conformations of transient alkyls by Kochi [126].

The βH coupling (see Chapter VI, Section 2.b) may also be expressed as

$$a_\beta^{H} = (Q_0^\beta + Q_2^\beta \cos^2 \theta)\rho_\pi^\alpha \tag{9.89}$$

For $C_\beta H$ bonds executing rapid internal rotation about the $C_\alpha C_\beta$, it becomes

$$a_\beta^{H} = (Q_0^\beta + \frac{1}{2} Q_2^\beta)\rho_\pi^\alpha = Q_\beta^{H}\rho_\pi^\alpha \tag{9.90}$$

When there are methyl groups attached to the C_α, these usually undergo rapid internal rotation in the nonviscous solutions ordinarily employed. For reasons already given, the Q_β^{H} is expected to vary less from radical to radical than does Q_α^{H}; for the rotating methyl group in the alkyl radicals (see Table 9.9), Q_β^{H} is found to have the probable value

$$Q_\beta^{H}(CH_3) = Q_0^\beta + \frac{1}{2} Q_2^\beta = 29.25 \text{ G} \tag{9.91}$$

In Table 9.10 this value has been used with experimental $a_\beta^{H}(CH_3)$ couplings to derive the ρ_π^α values from which the more variable Q_α^{H} values are obtained. With the assumption that $Q_0^\beta \approx 0$, one can use the value

$$Q_2^\beta \approx 2\, Q_\beta^{H}(CH_3) = 58.5 \text{ G} \tag{9.92}$$

for calculating approximate θ and ρ_π^α values for alkyl radicals.

The application of (9.86) and (9.89) to β-coupling hydrogens on conjugated rings or on hydrocarbon groups attached to conjugated rings is less certain than to those attached to aliphatic radicals. The methyl protons for the hexamethylbenzene cation, $C_6(CH_3)_6^{+}$, have the equivalent value $a_\beta^{H} = 6.53 \pm 0.01$ G for all 18 protons [127]. If one assumes, as was done for $C_6H_6^{+}$ (Table 9.11), that the unpaired electron is entirely in a bonding π orbital on the ring, one would expect a spin denstiy $\rho_\pi^i = 1/6$ on each ringed carbon. This would indicate the rather large effective $Q_\beta(CH_3) \approx 6 \times 6.53 = 39$ G. However, the ρ_π^i value will clearly be less than 1/6 because of the unpaired electron density on the methyl group, which is caused by hyperconjugation. The reduction suggested by (9.75) leads to $\rho_\pi^i = (1/6)(1 - 0.081) = 0.153$. With this value, the indicated $Q_\beta(CH_3)$ is still higher (~43 G). Generally the couplings of the attached CH_3 group for the anion radicals of similar aromatic ringed systems are appreciably less than those of the corresponding cation radicals. For example, the $a_\beta(CH_3)$ for ion radicals of 9-methylanthracene is 7.79 G for the cation and only 4.27 G for the anion [128,129]. Vincow [130] estimates that $Q_\beta^{CH_3}$(cation) $\approx$ 40 G and that Q^{CH_3} (anion) = 20 ± 5 G for a series of methyl-substituted naphthalene anions. Because of the uncertainties in the effective Q_β for highly conjugated systems, efforts toward direct calculation of the βH spin densities and couplings with the various molecular-orbital theories would seem more likely to be fruitful than attempts to apply (9.89). Several such direct calculations with molecular-orbital theories have been made [105,127,129,131].

$X_\alpha C_\beta H$ Coupling.

Experimentally derived Q_β values for the coupling of $C_\beta H$ groups with π spin density on noncarbon X_α atoms are given in Chapter VI. Values for Q_β^{NCH} are derived in Chapter VI, Section 5.d and listed in Table 6.20 (in MHz); those for Q_β^{SCH} are derived in Chapter VI, Section 10.c and listed in Table 6.31 (in MHz). These parameters should be usable for obtaining approximate π spin densities on N and S atoms from solution spectra.

5.d. Coupling by γ and δ Hydrogens

The sequence of coupling alkyl groups is conventionally designated as

$$-\underset{|\alpha}{\overset{|}{\dot{C}}}-\underset{|\beta}{\overset{H}{\underset{|}{C}}}-\underset{|\gamma}{\overset{H}{\underset{|}{C}}}-\underset{|\delta}{\overset{H}{\underset{|}{C}}}-$$
$$\text{H} \quad \text{H} \quad \text{H} \quad \text{H}$$

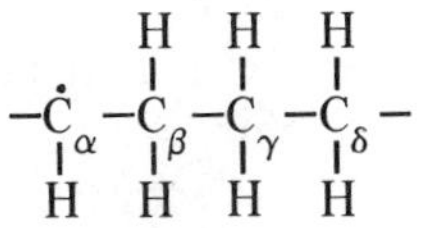

41

where the radical center or unpaired spin density is assumed to be on C_α. The exceptional resolution of esr spectroscopy of dilute solutions often makes possible the resolution and measurements of splitting by hydrogens bonded to C_γ and, in favorable cases, those to C_δ. The γ splittings are usually of the order of 1 G or less (see examples in Tables 9.4 and 9.5), and the δ splittings are much less. The accuracy of theoretical predictions of such small splittings does not justify attempts at their analysis here.

5.e. Coupling by Protons Bonded to Nitrogen

Experimental values of Q_{NH}^H derived from oriented radicals in single crystals are summarized in Table 6.10. They were obtained from the a_{NH}^H splitting with ρ_π^N derived from the measured anisotropic ^{14}N coupling. These magnitudes are reasonably consistent, with a mean value of 72 MHz or 25.7 G. They are in good agreement with the values of 25.9 G for the planar NH_3^+ radical and

23.9 G for NH_2 for which the ρ_π^N spin density can be taken as unity (see Chapter VIII, Sections 4.e and 5.a). The Q_{NH}^H values derived from solution spectra depend on theoretically predicted spin densities and are believed to be less accurate. Hence it seems desirable to use the $Q_{NH}^H = 25.7$ G, from single-crystal studies to obtain ρ_π^N spin densities from the a_{NH}^H values of solution spectra. There is an exception in the hydrazine radical ion $N_2H_4^+$, for which the HMO $\rho_\pi^N = 1/2$ on each N may be assumed from the symmetry. The isotropic couplings for this transient radical, $a_{NH}^H = 11.0 \pm 0.1$ G and $a(^{14}N) = 11.5 \pm 0.1$ G, have been measured with the Dixon-Norman method by Adams and Thomas [132]. The isotropic spin density obtained from the ^{14}N coupling, $(11.5/550) = 0.021$, is that expected from the spin polarization of the σ bonds. Thus it may be reasonably assumed that the radical is planar with the unpaired electron in an antibonding π orbital between the nitrogens. The indicated HMO unpaired π-electron density on each nitrogen is 1/2. With the measured a_{NH}^H, this density gives $Q_{NH}^H = 22$ G. The fact that this value is lower than 25.9 G, that for the NH_3^+, possibly results from the spin density in the overlap region between the nitrogens in $N_2H_4^+$, which is not taken into account in the normalization of the Hückel molecular orbitals.

Although many biologically significant, transient free radicals having NH groups are now observable with high resoltuion in aqueous solution [74], the unpaired π-electron densities are found to be mostly on carbons, and the $N_\alpha H$ couplings are usually small (~1 G) or not observed.

Proton coupling by $N_\beta H$ groups is not frequently observed because any conjugation ρ_π^α with $N_\beta H$ groups is more likely to occur with the unshared pair on N_β. However, β coupling may occur with $N_\beta H_3^+$ groups, as demonstrated in Table 6.14 for the $CO_2^-CHNH_3^+$ radical in an irradiated single crystal of glycine. The value $Q_\beta(CNH_3^+) = 52$ MHz $= 18.6$ G for the rotating $N_\beta H_3^+$ group should be usable for similar radicals in solution if such radicals are observed.

6. COUPLING BY NONHYDROGEN ELEMENTS IN FREE RADICALS

Isotropic nuclear coupling by elements other than hydrogen is often observed for rapidly tumbling radicals in solution. Although the isotropic hyperfine structure is usually easy to assign and the values obtained for the coupling constants are accurate, interpretation of these constants is more difficult and uncertain than it is for the hydrogen constants. These difficulties come primarily from two sources. The first is the presence of subvalence s electrons that interact with the unpaired electron-spin density of the valence shell and produce a spin density in the subvalence shell through exchange interactions. This spin density in the subvalence shells is slight, but it can exert a measurable effect on the coupling because of the large Fermi contact interactions of the inner electrons. The second complicating factor comes from the multiple orbitals of the valence shell and the multiple bonds often formed to the heavier elements other than the alkaline metals. These multiple bonds, like the CH bond, become spin polarized and produce spin density (of opposite sign) on both the bonded atoms. This polarization may be induced by spin density on the coupling atom or on the atoms to which it is bonded. From the observed isotropic coupling, there is, unfortunately, no way to ascertain the source of the polarization nor to distinguish between subvalence couplings and valence-shell couplings. Even the monovalent halogens, particularly fluorine, often form bonds that have a π-bonding component, in addition to their normal σ bonds.

Despite all these complexities, the pioneering theoretical work of Karplus and Fraenkel [133] has provided a means for semiquantitative interpretation of the isotropic couplings of ^{13}C and related atoms in π-type organic free radicals. Their formula, (6.37), is widely applied to ^{13}C couplings obtained from solution spectra. Since this formula is described and applied in Chapter VI, Section 4.a, it need not be further elaborated here. This formula with different parameters may be applied to other Group IV elements. Similar formulas with obvious alterations are applicable to other bonded atoms. The applications to isotropic coupling of ^{14}N are described in Chapter VI, Section 5.a. The mechanism of isotropic coupling of bonded ^{19}F is described in Chapter VI, Section 7.a. A fundamental difficulty in application of (6.37) and similar expressions for other heavier elements is the correct choice of the values for the various Q parameters involved. These parameters are probably known best for ^{13}C. Nevertheless, the number of coupling parameters and spin densities required for calculation of the one observed coupling constant makes the calculation uncertain, even for ^{13}C (see discussion in Chapter VI, Sections 4.a and 5.a).

Without complete theoretical analysis, isotropic hyperfine structure of the second- and higher-row elements can be of much help in identification of the chemical form of transient free radicals formed by rapid mixing of reacting chemicals, electrochemical oxidation and reduction, photolysis of liquids, or other means. Furthermore, the accurately measured couplings of the various nuclei can give valuable qualitative information about the structure and properties of large organic free radicals. Examples of this information are given in Sections 6.a and 6.b.

Couplings by ^{13}C, ^{14}N, and the halogens in particular, have been recorded in many hundreds of radicals in solution. Tabulation of the hyperfine coupling constants may be found in various survey papers and reviews. Bowers [134] gives an apparently complete tabulation of H and ^{14}N hyperfine splitting constants of organic radical ions measured in solution up to midyear 1964, with reference to literature sources. The 75 pages required for the tabulation is an indication of the enormous amount of esr work on liquid solutions. In a 1971 review of halogen hyperfine interactions, Hudson and Root [135] provide tabulations of isotropic couplings of F and Cl in organic free radicals measured

Table 9.14 Isotropic Coupling and Spin Density of ^{13}C and ^{29}Si in Selected Free Radicals in Liquid Solution except when Indicated.

Radical	*Coupling (G)*	*Spin Density*	*Structure of Radical Center*[a]	*Source of Coupling Data*
	$a_\alpha(^{13}C)$	$\rho_{2s}(^{13}C)$		
$\dot{C}H_3$	(+)38.34	0.0345	Planar	*b*
$CH_3\dot{C}H_2$	(+)39.07	0.0352	Planar	*b*
$(CH_3)_2\dot{C}H$	(+)41.3	0.0372	Planar	*c*
$(CH_3)_3\dot{C}$	(+)45.2	0.0405	Planar	*c*
$HO\dot{C}H_2$	(+)45.89	0.0413	Planar	*c*
$(COOH)_2\dot{C}H$ (solid)	(+)33.1	0.030	Planar	*d*
$F_2\dot{C}H$	(+)148.8	0.134	Pyramidal	*e*
CF_3	(+)271.6	0.245	~Tetrahedral	*e*
	$a_\alpha(^{29}Si)$	$\rho_{3s}(^{29}Si)$		
$\dot{Si}H_3$ (in Xe matrix)	(−)191	0.158	Pyramidal	*f*
$CH_3\dot{Si}H_2$ (solid)	(−)181	0.150	Pyramidal	*g*
$(CH_3)_2\dot{Si}H$	(−)183.05	0.152	Pyramidal	*h,i*
$(CH_3)_3\dot{Si}$	(−)181.14	0.150	Pyramidal	*h,i*
$(CH_3)_3\dot{Si}(CH_3)_2$	(−)137	0.114	Pyramidal	*i,j*
$(CH_3\dot{Si})_2SiCH_3$	(−)71	0.059	~Planar	*j*
$(CH_3Si)_3\dot{Si}$	(−)65	0.054	~Planar	*j*
$(CH_3Si)_2\dot{Si}Cl$ (solid)	(−)290	0.240	~Tetrahedral	*g*

[a] Refers to bond orientations of coupling carbon or silicon.
[b] R. W. Fessenden, *J. Phys. Chem.*, **71**, 74 (1967).
[c] H. Paul and H. Fischer, *Chem. Commun.*, 1038 (1971); H. Fischer, *Free Radicals*, J. K. Kochi, Ed., Vol. 2, Wiley, New York, 1973, Chapter 19.
[d] T. Cole and A. Heller, *J. Chem. Phys.*, **34**, 1085 (1961).
[e] Fessenden and Schuler [84].
[f] G. S. Jackel and W. Gordy, *Phys. Rev.*, **176**, 433 (1968).
[g] J. H. Sharp and M. C. R. Symons, *J. Chem. Soc. A*, 3084 (1970).
[h] P. J. Krusic and J. K. Kochi, *J. Am. Chem. Soc.*, **91**, 3938 (1969).
[i] S. W. Bennett, C. Eaborn, A. Hudson, R. A. Jackson and K. D. J. Root, *J. Chem. Soc. A*, 348 (1970).
[j] J. Cooper, A. Hudson, and R. A. Jackson, *Molec. Phys.*, **23**, 209 (1972).

in solution. Their tabulations also include couplings for ^{14}N and some for ^{13}C in halogenated organic radicals.

6.a. Isotropic Coupling and Structure of Radical Centers

From the hyperfine splitting of Group IV elements in simple XH_3 radicals trapped in inert matrices, it has been found that, unlike $\dot{C}H_3$ (which is a planar, π-type radical), SiH_3, GeH_3, and SnH_3 are σ-type radicals with pyramidal structure. Evidence for this is described in Chapter VIII, Section 4.d, and the derived pyramidal angles are given in Table 8.9. Although $\dot{C}H_3$ is planar, $\dot{C}F_3$ is approximately tetrahedral; CF_2CH is between tetrahedral and planar; and $CFCH_2$ is almost planar, as described in Chapter VIII, Section 4.f with comparisons given in Table 8.15. The trends of these simple radicals are followed by other, more complex, organic free radicals in which the unpaired electron density is concentrated on a Group IV element.

The isotropic ^{13}C and ^{29}Si hyperfine splitting constants of several such radicals measured in solution spectra are shown in Table 9.14. The constants for $\dot{C}H_3$ and $\dot{C}F_3$ are repeated for comparison. Effects of spin polarization in the carbon-centered radicals are expected to cause spin density $\rho_{2s}(C)$ or ~0.03. The constants measured for a single crystal of $(COOH)_2CH$ are also repeated in Table 9.14 for comparison. In Chapter VI, Section 4.a it is shown from theoretical considerations that the $\rho_{2s}(C)$ for this radical is expected to be due entirely to the effects of spin polarization and that the radical is a planar, π-type one. Similar analyses led to the conclusion that carbon-centered radicals with $a_\alpha(^{13}C) \geqslant 126$ MHz = 45 G are not strictly π-type; that is, the unpaired electron density on the C has some 2*s* character. It thus appears that carbon-centered radicals with $\rho_{2s}(C) \leqslant 0.04$ or $a_\alpha(^{13}C) < 45$ G are probably the planar π-type, whereas those with isotropic coupling constants greater than these are nonplanar, σ-type radicals. We have used this criterion in classification of the radicals in Table 9.14. The CH_2F radical with $a(^{13}C)$ = 154 MHz = 55 G is found to be slightly pyramidal (Table 8.15).

The ^{29}Si isotropic coupling $a_\alpha(^{29}Si)$ and density $\rho_{3s}(^{29}Si)$ of radicals for which the Si is bonded to methyl groups or to hydrogens are very close to those for SiH_3 in the xenon matrix. Therefore, the hybridization of the Si orbitals and pyramidal structure can be presumed to be about the same as those listed in Table 8.9 for SiH_3: the pyramidal angle is 74°, the bond angles 113.5°, and the 3*s* character of the Si bonding orbitals 28.5%. The bond angles are calculated from the *s* character of the orbitals with (7.50), as described in Chapter VIII, Section 4.d. It is of interest that the bond angle of SiH_3 in the Kr matrix [134], which is believed to be measured more accurately than it is in the Xe matrix, is only 110.6°. Presumably, this strong matrix effect is due to steric factors relating to the size of the radical compared with that of the matrix atom, but it is tempting to speculate that it may result from a slight exchange interaction between the unpaired electron orbitals and the atoms of the matrix.

The qualitative explanation originally given for the nonplanarity of the SiH_3 radical [136] has been borne out by later study of silicon-centered radicals in solution. The nonplanarity is related to the charge density in the bonding orbital relative to that of the orbital of the unpaired electron. Because the s orbitals have lower energy than the p orbitals of the same shell, the hybridization is favored that tends to put maximum charge density in the s orbital. An additional factor to be considered is the lowering of the covalent bond energy by the hybridization, which tends to increase the s character of the bonding orbitals. In CH_3 both of these factors favor the planar structure with the unpaired electron in a pure p orbital. Because the electronegativity of C, ($x_C = 2.5$), is greater than that of hydrogen ($x_H = 2.1$), the electron pair of the CH σ bond is unequally shared, with an average of more than one electron unit being in the C orbital. Thus more charge density will occur in the $2s$ orbital of the valence shell if this orbital is completely used by the bonding orbitals, with the result that the unpaired electron is left in a pure $2p$ orbital. In SiH_3 (also in GeH_3 and SnH_3) the polarity of the bonding is reversed because $x_{Si} < x_H$ and the charge density in each bonding orbital of the Si is less than one unit. For this reason, the total charge density in the $3s$ orbital would be greatest if the unpaired electron were completely in this orbital with the bonding-orbitals pure $3p$. However, this extreme is opposed by the lowered bond energy caused by bond-orbital hybridization. The result of these competing factors is a compromise in which all four of the orbitals have some degree of $3s$ character.

These qualitative considerations based on relative electronegativities may be applied (with caution) to other, more complex radicals in which the spin density is concentrated on a single Group IV element. For example, the value of $\rho_{3s}(Si)$ in Table 9.14 is lowest for $(CH_3Si)_3Si$, in which the Si_α is completely bonded to other silicons and the ionic character, which decreases the electronic charge density in the bonding orbitals of the Si_α, is lowest. In contrast, the $\rho_{3s}(Si)$ is greatest for the last radical shown, that for which the Si_α is bonded to the electronegative chlorine, $x_{Cl} = 3.0$. Even the carbon-centered radicals become nonplanar when C_α is bonded to the highly electronegative F (see Table 8.15) and also, but to a lesser degree, when the C is bonded to O or Cl. In Table 9.14, $\rho_{2s}(^{13}C) = 0.245$ for CF_3. For highly ionic bonds such as CF, considerations of orbital overlap are less significant, and those of relative electron charge density in the bonding orbitals of the central atom are more significant in the determination of the nature of the hybridization.

Radicals of the AB_3 type are treated with molecular-orbital theory by Walsh [137], who predicted the nonplanar structure for fluorinated methyl radicals long before the predictions were verified by the esr measurements of Fessenden and Schuler [84]. Also, the structures of the carbon-centered radicals are predicted by the molecular-orbital theory of Karplus and Fraenkel [133].

6.b. ESR Measurement of Electron-attracting or -repelling Properties of Chemical Groups

An esr adaptation of Hammett's well-known relation for comparing the relative electron-withdrawing and -donating powers of various substituent groups on benzene rings has been made by Bowers [138]. The parameter for comparing this property for substituent groups X, which Hammett [139] designated as σ, has become known as Hammett's sigma. Originally, he defined the σ for any substituent X on benzoic acid by

$$\sigma = \log_{10} \frac{K_i^{\mathrm{X}}}{K_i^0} \tag{9.93}$$

where K_i^0 is the ionization constant of the unsubstituted benzoic acid and K_i^{X} is that for the substituted acid XC_2H_4COOH. Obviously, σ is positive when $K_i^{\mathrm{X}} > K_i^0$ and negative when $K_i^{\mathrm{X}} < K_i^0$. Hammett found that reaction rates or equilibrium constants for other reactions of the benzoic acid derivatives could be related to σ by

$$\log_{10} \frac{k_x}{k_0} = \rho\, \sigma \tag{9.94}$$

where ρ depends on the specific type of reaction or reactant involved and where k_0 is the reaction constant for the unsubstituted acid and k_x is that for the X-substituted acid XC_6H_4COOH. Thus ρ is a constant that depends only on the type of reaction, and σ is a constant that depends only on the substituent group X.

Since σ evidently measures the ability of the X group to withdraw or supply electron-charge density through the ring to the COOH group, this property is measurable by reactions other than the ionization constant of the acid. For example, a close correlation should exist between Hammett's σ and the ^{17}O magnetic coupling constant of the OH of the benzoic acid derivatives and also between σ and the ^{17}O nuclear resonance frequencies. To the best of my knowledge, these comparisons with ^{17}O have not been made. However, a linear relationship has been found to exist between Hammett's σ and the chemical shifts of the ^{19}F nmr frequencies of XC_6H_4F [140,141]. This indicates that other groups can be substituted for the COOH in comparisons of the electron withdrawing (attraction) or supplying (repulsion) of the X substituents.

An esr measure of σ has been developed by Bowers [138] from anion radicals of the *p*-nitrobenzene derivatives $XC_6H_4NO_2$ in various solutions. The method is based on measurements of the effects of the various X-group substituents on the ^{14}N isotropic splitting constant. His equation relating σ to the measured coupling has the same form as Hammett's relation, (9.94), and is expressed as

$$\log_{10} \frac{a_N^X}{a_N^H} = \rho\,\sigma \tag{9.95}$$

where a_N^H is the ^{14}N coupling constant for the unsubstituted nitrobenzene $C_6H_5NO_2$ and a_N^X is that for the X-substituted derivatives. The ρ value, which in (9.94) depends on the particular reaction, depends here on the particular solvent used for the radicals. Since a_N^X depends directly on the unpaired electron density on the ^{14}N, it is evident that its value should vary with the electron-attracting or electron-repelling power of the X group, in a manner similar to that of the ionization constant of the substituted benzoic acids. However, it is likewise clear that the σ values for the nitrobenzene radical ions would not likely be equivalent to those for the benzoic acid derivatives. Thus Bowers derived a new σ scale for the substituted nitrobenzenes, which might be called "esr sigmas" or "Bowers' sigmas."

The method Bowers used for choosing these σ values was similar to that

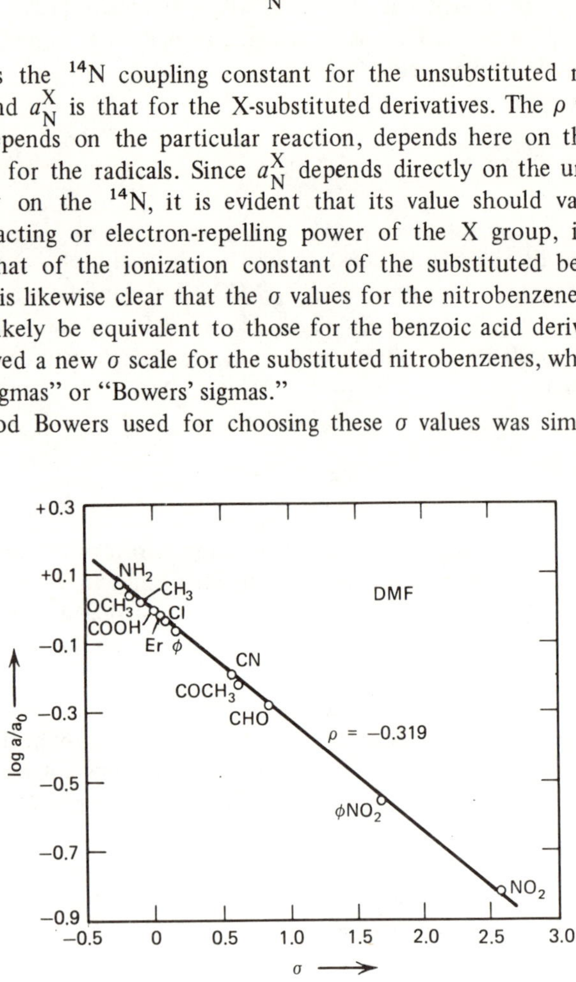

Fig. 9.31 Plots of $\log_{10}(a_N^X/a_N^H)$ versus σ for various substituted *p*-nitrobenzenes $XC_6H_4NO_2$ in the solvent dimethylformamide (DMF). The slope of the straight-line plot gives the ρ value for the particular solvent. From Bowers [138].

Table 9.15 σ Constants Derived from ^{14}N Isotropic Coupling of Anion Radicals of the *p*-Nitrobenzene Derivatives $XC_6H_4NO_2$ in Various Solutions[a]

Substituent X	Solvent Employed DMF	DMSO	AcCN	Final σ Value
CN	0.57	0.57	0.54	0.56
Br	0.07	0.10	0.10	0.09
$COCH_3$	0.66	0.63	0.57	0.62
OCH_3	-0.16	-0.16	-0.16	-0.16
CH_3	-0.09	-0.06	-0.06	-0.08
Cl	0.05	0.09	0.08	0.07
C_6H_5	0.17	0.16	0.17	0.17
OH	-0.51	-0.46	-0.42	-0.44
NH_2	-0.25	–	-0.23	-0.24
NO_2	2.55	2.59	2.60	2.58
CHO	0.85	0.83	0.84	0.84
COOH	-0.01	0.02	0.05	0.02
$C_6H_4NO_2$	1.72	1.68	1.63	1.68

[a]From Bowers [138].

originally employed by Hammett, but Bowers had the advantage of a high-speed computer to make the data correlations. With the high resolution characteristic of esr solution spectroscopy, accurate evaluation of the ^{14}N coupling presented no problem. The measured values of $\log_{10}(a_N^X/a_N^H)$ were plotted against the σ selected to give the best straight line for each solvent with an iterative computer program initiated with assumed σ values based on those available in the literature. Figure 9.31 illustrates the correlation finally achieved for the indicated X substituents in the dimethylformamide (DMF) solvent. The slope of the resulting straight line gives the value $\rho = -0.319$ for the solvent DMF. Plots for other solvents show equally close correlation, but with different slopes; $\rho = -0.312$ for dimethylsulfoxide (DMSO) and $\rho = -0.295$ for acetronitrile (AcCN). The resulting sets of σ values derived from these three solvents are quite close, as may be seen from examination of Table 9.15. The final σ values in the last column of the table are the averages of the three sets.

The esr σ values of Table 9.15 should correlate well with other properties of the nitrobenzene derivatives that depend on interactions with the nitrogroup. Furthermore, the method demonstrated by Bowers should be applicable to other chemical families derived from different parent molecules.

References

1. Harry G. Hecht, *Magnetic Resonance Spectroscopy,* Wiley, New York, 1967.

2. H. M. McConnell, *J. Chem. Phys.,* **25**, 709 (1956).
3. N. Bloembergen, E. M. Purcell, and R. V. Pound, *Phys. Rev.,* **73**, 679 (1948).
4. D. Kivelson, *J. Chem. Phys.,* **27**, 1087 (1957); **33**, 1094 (1960).
5. J. H. Freed and G. K. Fraenkel, *J. Chem. Phys.,* **39**, 326 (1963).
6. J. H. Freed and G. K. Fraenkel, *J. Chem. Phys.,* **41**, 699 (1964).
7. George E. Pake, *Paramagnetic Resonance,* Benjamin, New York, 1962.
8. R. Livingston and H. Zeldes, *J. Chem. Phys.,* **44**, 1245 (1966).
9. A. Horsfield, J. R. Morton, and D. H. Whiffen, *Molec. Phys.,* **4**, 327 (1961).
10. H. S. Gutowsky, D. M. McCall and C. P. Slichter, *J. Chem. Phys.,* **21**, 279 (1953).
11. H. M. McConnell, *J. Chem. Phys.,* **28**, 430 (1958).
12. C. S. Johnson, J., *Adv. Mag. Reson.,* **1**, 33 (1965).
13. P. D. Sullivan and J. R. Bolton, *Adv. Mag. Reson.,* **4**, 39 (1970).
14. N. M. Atherton, *Electron Spin Resonance: Theory and Applications,* Halsted Press, Wiley, Chichester, England, 1973.
15. R. K. Wangness and F. Bloch, *Phys. Rev.,* **89**, 728 (1953).
16. R. Kubo and K. Tomita, *J. Phys. Soc. Jap.,* **9**, 888 (1954).
17. A. G. Redfield, *Adv. Mag. Reson.*, **1**, 1 (1965).
18. C. P. Slichter, *Principles of Magnetic Resonance,* Harper, New York, 1963.
19. C. P. Poole, Jr. and H. A. Farach, *Relaxation in Magnetic Resonance: Dielectric and Mossbauer Applications,* Academic, New York, 1971.
20. G. K. Fraenkel, *J. Phys. Chem.,* **71**, 139 (1967).
21. F. J. Adrian, E. L. Cochran, and V. A. Bowers, *Free Radicals in Inorganic Chemistry,* American Chemical Society, Washington, D. C., 1962, p. 50.
22. R. W. Fessenden and R. H. Schuler, *J. Chem. Phys.,* **39**, 2147 (1963).
23. J. R. Bolton and A. Carrington, *Molec. Phys.,* **5**, 161 (1962).
24. J. H. Freed and G. K. Fraenkel, *J. Chem. Phys.,* **37**, 1156 (1962).
25. R. L. Ward and S. I. Weissman, *J. Am. Chem. Soc.,* **79**, 2086 (1957).
26. N. M. Atherton and S. I. Weissman, *J. Am. Chem. Soc.,* **83**, 1330 (1961).
27. P. J. Zandstra and S. I. Weissman, *J. Am. Chem. Soc.,* **84**, 4408 (1962).
28. J. P. Lloyd and G. E. Pake, *Phys. Rev.,* **94**, 579 (1954).
29. H. Fischer, *Molec. Phys.,* **9**, 149 (1965).
30. H. S. Gutowsky and C. H. Holm, *J. Chem. Phys.,* **25**, 1228 (1956).
31. H. Zeldes and R. Livingston, *J. Chem. Phys.,* **45**, 1946 (1966).
32. H. Zeldes and R. Livingston, *J. Chem. Phys.,* **47**, 1465 (1967).
33. S. I. Weissman, J. Townsend, D. E. Paul, and G. E. Pake, *J. Chem. Phys.,* **21**, 2227 (1953).
34. T. L. Chu, G. E. Pake, D. E. Paul, J. Townsend, and S. I. Weissman, *J. Phys. Chem.,* **57**, 504 (1953).

35. D. Lipkin, D. E. Paul, J. Townsend, and S. I. Weissman, *Science,* **117**, 534 (1953).
36. T. R. Tuttle and S. I. Weissman, *J. Am. Chem. Soc.,* **80**, 5342 (1958).
37. R. L. Ward and S. I. Weissman, *J. Am. Chem. Soc.,* **76**, 3612 (1954).
38. S. I. Weissman, E. de Boer, and J. J. Conradi, *J. Chem. Phys.,* **26**, 963 (1957).
39. S. I. Weissman, *J. Chem. Phys.,* 22, 1135 (1954).
40. I. C. Lewis and L. S. Singer, *J. Chem. Phys.,* **43**, 2712 (1965).
41. H. Nishiguchi, Y. Nakai, K. Nakamura, K. Ishuzi, Y. Deguchi, and H. Takaki, *J. Chem. Phys.,* **40**, 241 (1964).
42. E. de Boer and E. L. Makor, *Proc. Chem. Soc. (Lond.),* 23 (1963).
43. L. Michaelis, M. P. Schubert, R. K. Reber, J. A. Kuck, and S. Granick, *J. Am. Chem. Soc.,* **60**, 1678 (1938).
44. B. Venkataraman and G. K. Fraenkel, *J. Am. Chem. Soc.,* **77**, 2707 (1955).
45. B. Venkataraman and G. K. Fraenkel, *J. Chem. Phys.,* **23**, 588 (1955).
46. G. K. Fraenkel, *Ann. N. Y. Acad. Sci.,* **67**, 553 (1957).
47. J. E. Wertz and J. L. Vivo, *J. Chem. Phys.,* **23**, 2441 (1955).
48. D. H. Geske and A. H. Maki, *J. Am. Chem. Soc.,* **82**, 2671 (1960).
49. A. H. Maki and D. H. Geske, *J. Chem. Phys.,* **33**, 825 (1960).
50. A. H. Maki and D. H. Geske, *J. Am. Chem. Soc.,* **83**, 1852 (1961).
51. E. W. Stone and A. H. Maki, *J. Chem. Phys.,* **39**, 1635 (1963).
52. S. H. Glarum and L. C. Snyder, *J. Chem. Phys.,* **36**, 2989 (1962).
53. P. H. Rieger, I. Bernal, W. H. Reinmuth, and G. K. Fraenkel, *J. Am. Chem. Soc.,* **85**, 683 (1963).
54. N. Steinberger and G. K. Fraenkel, *J. Chem. Phys.,* **40**, 723 (1964).
55. J. R. Bolton and G. K. Fraenkel, *J. Chem. Phys.,* **40**, 3307 (1964).
56. B. L. Barton and G. K. Fraenkel, *J. Chem. Phys.,* **41**, 1455 (1964).
57. Z. Galus and R. N. Adams, *J. Chem. Phys.,* **36**, 2814 (1962).
58. W. M. Tolles and D. W. Moore, *J. Chem. Phys.,* **46**, 2102 (1967).
59. D. H. Levy and R. J. Myers, *J. Chem. Phys.,* **41**, 1062 (1964); 44, 4177 (1966).
60. E. Saito and B. H. J. Bielski, *J. Am. Chem. Soc.,* **83**, 4467 (1961).
61. W. T. Dixon and R. O. C. Norman, *Nature (Lond.),* **196**, 891 (1962).
62. W. T. Dixon and R. O. C. Norman, *J. Chem. Soc.,* 3119 (1963).
63. W. T. Dixon, R. O. C. Norman and A. L. Buley, *J. Chem. Soc.,* 3625 (1964).
64. W. T. Dixon and R. O. C. Norman, *J. Chem. Soc.,* 4850, 6857 (1964).
65. P. Smith, J. T. Pearson, P. B. Wood, and T. C. Smith, *J. Chem. Phys.,* **43**, 1535 (1965).
66. P. Smith, J. T. Pearson, and R. V. Tsina, *Can. J. Chem.,* **44**, 753 (1966).

67. J. T. Pearson, P. Smith, and T. C. Smith, *Can. J. Chem.*, **42**, 2022 (1965).
68. H. Fischer, *Z. Naturforsch.*, **A20**, 488 (1965).
69. D. J. Edge and R. O. C. Norman, *J. Chem. Soc. (B)*, 182 (1969).
70. R. O. C. Norman and P. R. West, *J. Chem. Soc. (B)*, 389 (1969).
71. A. L. J. Beckwith and R. O. C. Norman, *J. Chem. Soc. (B)*, 400 (1969).
72. A. L. J. Beckwith and R. O. C. Norman, *J. Chem. Soc. (B)*, 403 (1969).
73. W. A. Waters, *Science Progr.*, **53**, 413 (1965).
74. C. Nicolai, in *Physico-Chemical Properties of Nucleic Acids*, J. Duchesne, Ed., Academic, London, 1973, Chapter 6.
75. R. Poupko, A. Loenstein, and B. L. Silver, in *Magnetic Resonances in Biological Research*, C. Franconi, Ed., Gordon and Breach, New York, 1971, paper No. 19.
76. J. Q. Adams and J. R. Thomas, *J. Chem. Phys.*, **39**, 1904 (1963).
77. L. Burlamacchi and E. Tiezzi, in *Magnetic Resonances in Biological Research*, C. Franconi, Ed., Gordon and Breach, New York, 1971, paper No. 18.
78. J. F. Gibson, D. J. E. Ingram, M. C. R. Symons, and M. G. Townsend, *Trans. Faraday Soc.*, **53**, 914 (1957).
79. R. Livingston, J. K. Dohrmann, and H. Zeldes, *J. Chem. Phys.*, **53**, 2448 (1970).
80. P. Smith, R. A. Kaba, T. C. Smith J. T. Pearson, and P. B. Wood, *J. Magn. Reson.*, **18**, 254 (1975).
81. R. Livingston and H. Zeldes, *J. Magn. Reson.*, **1**, 169 (1969).
82. J. K. Dohrmann, R. Livingston, and H. Zeldes, *J. Am. Chem. Soc.*, **93**, 3343 (1971).
83. J. K. Dohrmann and R. Livingston, *J. Am. Chem. Soc.*, **93**, 5363 (1971).
84. R. W. Fessenden and R. H. Schuler, *J. Chem. Phys.*, **43**, 2704 (1965).
85. K. Eihen and R. W. Fessenden, *J. Phys. Chem.*, **75**, 1186 (1971).
86. G. P. Laroff and R. W. Fessenden, *J. Chem. Phys.*, **55**, 5000 (1971).
87. P. Neta, *Radiat. Res.*, **49**, 1 (1972).
88. R. W. Fessenden, *J. Phys. Chem.*, **68**, 1508 (1964).
89. B. Smaller, J. R. Remco, and E. C. Avery, *J. Chem. Phys.*, **48**, 5174 (1968).
90. G. Nucifora, B. Smaller, R. Remco, and E. C. Avery, *Radiat. Res.*, **49**, 96 (1972).
91. M. S. Blois, Jr., H. W. Brown, and J. E. Maling, in *Free Radicals in Biological Systems*, M. S. Blois, H. W. Brown, R. M. Lemmon, R. O. Lindblom, and M. Weissbluth, Eds., Academic, New York, 1961, Chapter 8; *Arch. Sci.*, **13**, 243 (1960).
92. A. J. Stone, *Molec. Phys.*, **6**, 509 (1963).

93. B. G. Segal, M. Kaplan, and G. K. Fraenkel, *J. Chem. Phys.*, **43**, 4191 (1965).
94. R. E. Moss and A. J. Perry, *Molec. Phys.*, 22, 789 (1971).
95. A. J. Stone, *Molec. Phys.*, 7, 311 (1964).
96. D. J. M. Fassaert and E. de Boer, *Molec. Phys.*, **21**, 485 (1971).
97. R. Biehl, M. Plato, and K. Möbius, *Molec. Phys.*, **35**, 985 (1978).
98. K. Möbius and M. Plato, *Z. Naturforsch.*, **A24**, 1078 (1969).
99. M. Plato and K. Möbius, *Z. Naturforsch.*, **A24**, 1084 (1969).
100. H. M. McConnell and D. B. Chesnut, *J. Chem. Phys.*, 28, 107 (1958).
101. E. de Boer and S. I. Weissman, *J. Am. Chem. Soc.*, **80**, 4549 (1958).
102. H. S. Jarrett, *J. Chem. Phys.*, **25**, 1289 (1956).
103. A. D. McLachlan, *Molec. Phys.*, **3**, 233 (1960).
104. A. D. McLachlan, H. H. Dearman, and R. Lefebvre, *J. Chem. Phys.*, **33**, 65 (1960).
105. J. P. Colpa and E. de Boer, *Molec. Phys.*, 7, 333 (1963).
106. P. Nordio, M. V. Pavan, and G. Giacometti, *Theor. Chim. Acta (Berlin)*, **1**, 302 (1963).
107. D. Lazdins and M. Karplus, *J. Chem. Phys.*, **44**, 1600 (1966).
108. K. D. Sales, *Advances in Free-Radical Chemistry*, Vol. 3, G. H. Williams, Ed., Academic, New York, 1969, Chapter 3.
109. D. B. Chesnut, *J. Chem. Phys.*, **29**, 43 (1958).
110. H. Fischer, *Z. Naturforsch.*, **A20**, 428 (1958).
111. A. J. Dobbs, B. C. Gilbert, and R. O. C. Norman, *J. Chem. Soc. (A)*, 124 (1971).
112. A. D. McLachlan, *Molec. Phys.*, 2, 271 (1959).
113. H. M. McConnell and R. E. Robertson, *J. Chem. Phys.*, **28**, 991 (1958).
114. J. R. Bolton and G. K. Fraenkel, *J. Chem. Phys.*, **40**, 3307 (1964).
115. J. P. Colpa and J. R. Bolton, *Molec. Phys.*, **6**, 273 (1963).
116. J. R. Bolton, *J. Chem. Phys.*, **43**, 309 (1965).
117. J. R. Bolton, in *Radical Ions*, E. T. Kaiser and L. Kevan, Eds., Interscience, New York, 1968, Chapter 1.
118. L. C. Snyder and T. Amos, *J. Chem. Phys.*, **42**, 3670 (1965).
119. I. C. Lewis and L. S. Singer, *J. Chem. Phys.*, **43**, 2712 (1965).
120. G. Giacometti, P. L. Nordio, and M. V. Pavan, *Theort. Chim. Acta*, **1**, 404 (1963).
121. J. E. Bloor, B. R. Gilson, and P. N. Daykin, *J. Phys. Chem.*, **70**, 1457 (1966).
122. P. Brovetto and S. Ferroni, *Il Nuovo Cim.*, **5**, 142 (1957).
123. H. M. McConnell and H. H. Dearman, *J. Chem. Phys.*, **28**, 51 (1958).

124. G. J. Hoijtink, *Molec. Phys.*, **1**, 157 (1958).

125. J. E. Harriman and K. M. Sando, *J. Chem. Phys.*, **48**, 5138 (1968).

126. J. K. Kochi, in *Advances in Free-Radical Chemistry*, Vol. 5, G. H. Williams, Ed., Academic, New York, 1975, Chapter 4.

127. M. K. Carter and G. Vincow, *J. Chem. Phys.*, **47**, 302 (1967).

128. J. A. Brivati, R. Hulme and M. C. R. Symons, *Proc. Chem. Soc.*, 384 (1961).

129. J. R. Bolton, A. Carrington, and A. D. McLachlan, *Molec. Phys.*, **5**, 31 (1962).

130. G. Vincow, in *Radical Ions,* E. T. Kaiser and L. Kevan, Eds., Interscience, New York, 1968, Chapter 4.

131. R. Bersohn, *J. Chem. Phys.*, **24**, 1066 (1956).

132. J. Q. Adams and J. R. Thomas, *J. Chem. Phys.*, **39**, 1904 (1963).

133. M. Karplus and G. K. Franekel, *J. Chem. Phys.*, **35**, 1312 (1961).

134. K. W. Bowers, *Adv. Mag. Reson.*, **1**, 317 (1965).

135. A. Hudson and K. D. T. Root, *Adv. Magn. Reson.*, **5**, 1 (1971).

136. R. L. Morehouse, J. J. Christiansen, and W. Gordy, *J. Chem. Phys.*, **45**, 1751 (1966).

137. A. D. Walsh, *J. Chem. Soc.*, 2301 (1953); 2306 (1953).

138. K. W. Bowers, in *Radical Ions,* E. T. Kaiser and L. Kevan, Eds., Interscience, New York, 1968, Chapter 5.

139. L. P. Hammett, *Physical Organic Chemistry,* McGraw Hill, New York, 1940.

140. H. S. Gutowsky, D. W. McCall, B. R. McGarvey, and L. E. Meyer, *J. Am. Chem. Soc.*, **74**, 4809 (1952).

141. H. S. Gutowsky and C. J. Hoffman, *J. Chem. Phys.*, **19**, 1259 (1951).

Chapter **X**

TRIPLE-STATE ESR

1. THE ESR OF MOLECULES IN TRIPLET STATES

Stable organic molecules usually have even numbers of electrons and singlet electronic ground states that are nonmagnetic. An electron can be lifted from one of the bonding orbitals to an antibonding orbital by the absorption of radiation having the proper frequency, but this process does not ordinarily reverse the spin, and the state thus excited is also a singlet. From the excited singlet state, the molecule can readily fall back to the original ground state by emission of radiation. Excited singlet states are very short lived (lifetimes $\sim 10^{-8}$ sec). If, however, a spin should be reversed during this process by some auxiliary mechanism, the Pauli exclusion principle would forbid the electron from falling back into the original bonding orbital. Before a molecule can return from the triplet state to the singlet ground state, it must await some event that will again cause a reversal of spin. For these reasons the triplet state is a metastable state. In pure organic substances, especially at low temperatures, the triplet state can be long lived (~ many seconds). The triplet state has a spin magnetic moment of two Bohr magnetons corresponding to $S = 1$, and under favorable conditions it is possible to observe electron-spin resonance of molecules in excited triplet states.

In the triplet state the orbital angular momentum of an organic molecule is effectively quenched by interaction with the strong electrical field of chemical bonds, just as it is for an organic free radical having a doublet spin state. The π orbitals of the triplet-state electrons are locked in fixed directions relative to the molecular frame and are not free to precess or become reoriented in an imposed magnetic field. For these reasons, organic molecules in triplet states have nearly isotropic g factors, very close to that of the free electron spin. There is, however, one major difference between the "spin resonance" of a triplet state and that of a free radical with a single unpaired electron. This difference arises from the dipole-dipole interaction of the two unpaired-electron spin moments. This interaction leads to a doublet splitting of the spin resonance of the triplet, the magnitude of which depends on the orientation of the molecule in the applied magnetic field.

Because of the magnetic dipole field of one unpaired spin on that of the other, the effective field on each differs from the applied field. As a result, there is a doublet splitting of the first-order ($\Delta M_S = \pm 1$) esr spectrum. This doublet splitting is large, approximately 1000 G in many organic molecules. Because the spin-spin interaction is strongly anisotropic, it causes the esr to be very broad for powdered or polycrystalline samples.

Electron-spin resonance of organic molecules in excited triplet states was first detected in 1958 by Hutchison and Mangum [1,2], with oriented naphthalene in a single crystal. The large anisotropy in the resonance caused by the electronic dipole-dipole interaction probably led to failure in the earlier attempts at detection of the resonance with randomly oriented molecules. Subsequently, however. the resonance was detected with phosphorescent molecules of powders and glasses.

1.a. Energies and Transition Frequencies

The spin Hamiltonian for the triplet state can be expressed adequately for most applications by the sum

$$\mathcal{H}_S = \mathcal{H}_{H,S} + \mathcal{H}_{S,S} + \mathcal{H}_{I,S} \tag{10.1}$$

where $\mathcal{H}_{H,S}$ signifies the interaction of the electron spin with the applied field,

$\mathcal{H}_{S,S}$, the dipole-dipole interaction of the electron spins, and $\mathcal{H}_{I,S}$, the interaction of the electronic and nuclear spin moments. In the most general form, all these interaction terms are tensor products, and (10.1) is

$$\mathcal{H}_S = \mathbf{S}\cdot\mathbf{g}\cdot\mathbf{H} + \mathbf{S}\cdot\mathbf{D}\cdot\mathbf{S} + \sum_i \mathbf{S}\cdot\mathbf{A}_i\cdot\mathbf{I}_i \tag{10.2}$$

where the summation is taken over all coupling nuclei. The first term on the right is the same as that applied to doublet states throughout this volume. The second term, the electron spin-spin interaction, is developed in Chapter II, Section 5 [see (2.78)]. The last tensor is like that applied to nuclear coupling in doublet spin states. Since the nuclear interactions in organic radicals are generally much smaller than the electron spin-spin interactions in triplet states, the last term can usually be treated as a perturbation on the energy states determined by the first two terms on the right. Hence we postpone consideration of the nuclear interactions and describe first the spin-spin splitting.

Expressed in terms of its principal axes x, y, z, the $\mathcal{H}_{S,S}$ of (2.78) becomes

$$\mathcal{H}_{S,S} = D_x S_x^2 + D_y S_y^2 + D_z S_z^2 \tag{10.3}$$

Because the trace of a dipole-dipole interaction tensor is zero, the diagonal elements are related by

$$D_x + D_y + D_z = 0 \tag{10.4}$$

The relationship between the coefficients, indicated by (10.4), makes it possible to express (10.3) with only two unknown parameters. By use of (10.4) and the additional relation, $S_x^2 + S_y^2 = S^2 - S_z^2 = S(S+1) - S_z^2$, the Hamiltonian can be put in the form

$$\mathcal{H}_{S,S} = D[(S_z^2 - \tfrac{1}{3}S(S+1)] + E(S_x^2 - S_y^2) \tag{10.5}$$

where

$$D = \tfrac{3}{2}D_z = -\tfrac{3}{2}(D_x + D_y) \tag{10.6}$$

and

$$E = \tfrac{1}{2}(D_x - D_y) \tag{10.7}$$

It should be mentioned that some spectroscopists use $-X$ for D_x, $-Y$ for D_y, and $-Z$ for D_z. However, the usage of D and E, as in (10.5), is rather universal. The term $-\frac{1}{3}DS(S + 1)$ in (10.5) shifts all triplet levels by the same amount, $-\frac{2}{3}D$, and hence has no effect on the transition frequencies. Consequently, it is omitted in many treatments, but we retain it for completeness.

Because of symmetry in most organic molecules, the principal axes of the **g** tensor and of the spin-spin coupling tensor are coincident. Thus we assume here that x, y, z are the principal axes for both tensors and express the spin Hamiltonian in this system. If the magnetic field is applied in some direction such that the direction cosines of **H** with the principal axes are l_x, l_y, and l_z, respectively ($H_x = Hl_x, H_y = Hl_y, H_z = Hl_z$), the Hamiltonian without the nuclear terms is

$$\mathcal{H}_S = \beta H(g_x l_x S_x + g_y l_y S_y + g_z l_z S_z) + D[S_z^2 - \frac{1}{3}S(S + 1)] + E(S_x^2 - S_y^2) \quad (10.8)$$

From the nonvanishing matrix elements of the spin operators for $S = 1$,

$$(\pm 1|S_z| \pm 1) = \pm 1$$
$$(0|S_x| \pm 1) = (\pm 1|S_x|0) = 1/\sqrt{2}$$
$$(0|S_y| \pm 1) = (\mp 1|S_y|0) = \pm i/\sqrt{2}$$
$$(\pm 1|S_z^2| \pm 1) = \pm 1$$
$$(\pm 1|S_x^2 - S_y^2| \mp 1) = \frac{1}{2}(\pm 1|S_+^2 + S_-^2| \mp 1) = 1$$

the matrix elements of $\mathcal{H}_S$ are readily found. These lead to the secular equation

$$\begin{array}{c|ccc|}
 & 1 & 0 & -1 \\
\hline
1 & \beta Hg_z l_z + \frac{D}{3} - W & \frac{\beta H}{\sqrt{2}}(g_x l_x - ig_y l_y) & E \\
0 & \frac{\beta H}{\sqrt{2}}(g_x l_x + ig_y l_y) & -\frac{2}{3}D - W & \frac{\beta H}{\sqrt{2}}(g_x l_x - ig_y l_y) \\
-1 & E & \frac{\beta H}{\sqrt{2}}(g_x l_x + ig_y l_y) & -\beta Hg_z l_z + \frac{D}{3} - W
\end{array} = 0 \quad (10.9)$$

The three roots W of this cubic equation correspond to the characteristic energies. The equation can be expressed in the form

$$(W + \frac{2}{3}D)(W - \frac{1}{3}D + E)(W - \frac{1}{3}D - E) - (\beta Hg_x l_x)^2(W - \frac{1}{3}D + E)$$

$$- (\beta H g_y l_y)^2 (W - \frac{1}{3}D - E) - (\beta H g_z l_z)^2 (W + \frac{2}{3}D) = 0 \qquad (10.10)$$

For arbitrary orientations of the crystal in the magnetic field, the solution of (10.10) is rather complicated and can be found most conveniently by numerical methods with the aid of a computer. However, for **H** along the three principal axes, it reduces to a form that can be readily solved to give closed formulas for the energies.

For **H** along z, $l_x = 0$, $l_y = 0$, and $l_z = 1$, and the solution yields

$$W_0 = -\frac{2D}{3} \qquad (10.11)$$

$$W_{\pm} = \frac{D}{3} \pm [(g_z \beta H)^2 + E^2]^{1/2} \qquad (10.12)$$

These formulas give the frequencies for the first-order transitions for the magnetic field **H** along z:

$$W_0 - W_- = h\nu_1 = [(g_z \beta H)^2 + E^2]^{1/2} - D \qquad (10.13)$$

$$W_+ - W_0 = h\nu_2 = [(g_z \beta H)^2 + E^2]^{1/2} + D \qquad (10.14)$$

These correspond to the usual $\Delta M_S = \pm 1$ transitions. The doublet separation for this case,

$$\Delta\nu = \nu_2 - \nu_1 = \frac{2D}{h} \qquad (10.15)$$

provides a direct measure of the constant D. Since a magnetic field is usually measured at a constant frequency, it is desirable to express (10.13) and (10.14) in the form

$$H_1 \text{ (along } z) = (g_z \beta)^{-1} [(h\nu_0 + D)^2 - E^2]^{1/2} \qquad (10.16)$$

$$H_2 \text{ (along } z) = (g_z \beta)^{-1} [(h\nu_0 - D)^2 - E^2]^{1/2} \qquad (10.17)$$

where H_1 and H_2 are the resonant field values at the same ν_0 frequency.

Solution of (10.10) for H along x ($l_y = 0, l_z = 0, l_x = 1$) yields the corresponding energies and first-order ($\Delta M_S = \pm 1$) transition frequencies:

$$W_0 = \frac{D}{3} - \text{E} \qquad (10.18)$$

$$W_{\pm} = -\frac{(D-3E)}{6} \pm \left[(g_x\beta H)^2 + \frac{(D+E)^2}{4}\right]^{1/2} \tag{10.19}$$

$$h\nu_1 = \frac{D-3E}{2} + \left[(g_x\beta H)^2 + \frac{(D+E)^2}{4}\right]^{1/2} \tag{10.20}$$

$$h\nu_2 = -\frac{D-3E}{2} + \left[(g_x\beta H)^2 + \frac{(D+E)^2}{4}\right]^{1/2} \tag{10.21}$$

$$\Delta\nu = \nu_1 - \nu_2 = \frac{D-3E}{h} \tag{10.22}$$

When the measurements are made at a constant frequency ν_0, the resonant field values are

$$H_1 \text{ (along } x) = (g_x\beta)^{-1}\left\{\left[h\nu_0 - \frac{D-3E}{2}\right]^2 - \frac{(D+E)^2}{4}\right\}^{1/2} \tag{10.23}$$

$$H_2 \text{ (along } x) = (g_x\beta)^{-1}\left\{\left[h\nu_0 + \frac{D-3E}{2}\right]^2 - \frac{(D+E)^2}{4}\right\}^{1/2} \tag{10.24}$$

For **H** along y, similar solutions yield the values

$$W_0 = \frac{D}{3} + \mathrm{E} \tag{10.25}$$

$$W_{\pm} = \frac{-(D+3E)}{6} \pm \left[(g_y\beta H)^2 + \frac{(D-E)^2}{4}\right]^{1/2} \tag{10.26}$$

$$h\nu_1 = \frac{D+3E}{2} + \left[(g_y\beta H)^2 + \frac{(D-E)^2}{4}\right]^{1/2} \tag{10.27}$$

$$h\nu_2 = \frac{-(D+3E)}{2} + \left[(g_y\beta H)^2 + \frac{(D-E)^2}{4}\right]^{1/2} \tag{10.28}$$

$$\Delta\nu = \nu_1 - \nu_2 = \frac{D+3E}{h} \tag{10.29}$$

The resonant field values at constant frequency ν_0 are

$$H_1 \text{ (along } y) = (g_y\beta)^{-1}\left\{\left[h\nu_0 - \frac{D+3E}{2}\right]^2 - \frac{(D-E)^2}{4}\right\}^{1/2} \tag{10.30}$$

$$H_2 \text{ (along } y) = (g_y\beta)^{-1}\left\{\left[h\nu_0 + \frac{D+3E}{2}\right]^2 - \frac{(D-E)^2}{4}\right\}^{1/2} \tag{10.31}$$

In each case it is assumed that the principal axes of $\mathcal{H}_{H,S}$ and $\mathcal{H}_{S,S}$ have the same direction.

The diagram of Fig. 10.1 illustrates energy levels as functions of field strength at constant frequency when the field is imposed along the z axis. Transitions between W_0 and either W_+ or W_- give the doublet resonance, as indicated.

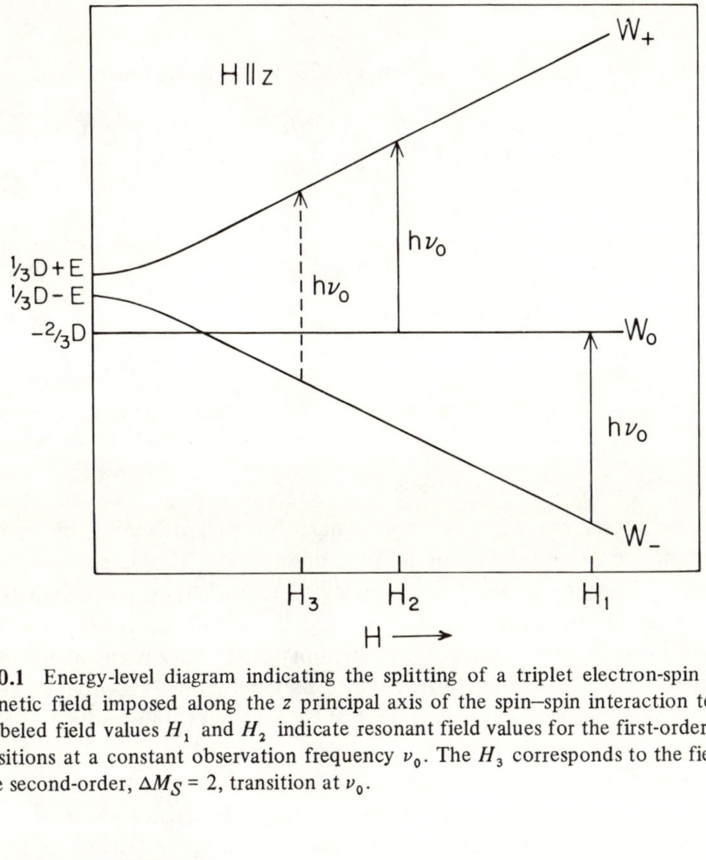

Fig. 10.1 Energy-level diagram indicating the splitting of a triplet electron-spin state by a magnetic field imposed along the z principal axis of the spin–spin interaction tensor **D**. The labeled field values H_1 and H_2 indicate resonant field values for the first-order, $\Delta M_S = 1$ transitions at a constant observation frequency ν_0. The H_3 corresponds to the field value for the second-order, $\Delta M_S = 2$, transition at ν_0.

Axially Symmetric Tensors

It has been found that the values of E are an order of magnitude smaller than those of D for many aromatic hydrocarbons, and also that the **g** anisotropy is very small (see tabulations in Section 1.b). Thus in approximate treatments one might set $E = 0$ and neglect the $\mathbf{g}$ anisotropy. This approximation is especially convenient in the study of randomly oriented molecules in glasses or powders.

To obtain the resonant field values at constant observation frequency ν_0 for axially symmetric $\mathbf{g}$ and $\mathbf{D}$ tensors (with common symmetry axes), one sets $g_z = g_\parallel, g_x = g_y = g_\perp$, and $E = 0$ into (10.16) and (10.17) and obtains

$$(H_1)_\parallel = H_0 + \frac{D}{g_\parallel \beta} \tag{10.32}$$

$$(H_2)_\parallel = H_0 - \frac{D}{g_\parallel \beta} \tag{10.33}$$

where $H_0 = (h\nu_0/g_\parallel\beta)$. Similarly, from (10.23) and (10.24) one obtains

$$(H_1)_\perp = H_0 \left(1 - \frac{D}{g_\perp \beta H_0}\right)^{1/2} \approx H_0 - \frac{1}{2}\left(\frac{D}{g_\perp \beta}\right) \tag{10.34}$$

$$(H_2)_\perp = H_0 \left(1 + \frac{D}{g_\perp \beta H_0}\right)^{1/2} \approx H_0 + \frac{1}{2}\left(\frac{D}{g_\perp \beta}\right) \tag{10.35}$$

where $H_0 = (h\nu_0/g_\perp\beta)$. The doublet splittings are

$$\Delta H_\parallel = \frac{2D}{g_\parallel \beta} \qquad \Delta H_\perp \approx \frac{D}{g_\perp \beta} \tag{10.36}$$

Second-order $\Delta M_S = \pm 2$ Transitions

These transitions are of great importance for measurements on randomly oriented molecules in glasses or powdered solutions. The anisotropies due to spin-spin interactions partly cancel, leaving a much sharper, more detectable resonance than the first-order transitions, even though the transition probabilities are much lower for the $\Delta M_S = \pm 2$ transitions. Also, only a single second-order transition occurs, $W_+ \longleftrightarrow W_-$, in contrast to the doublet of first-order transitions. Applications are described in Section 1.c. Here we give only the transition frequencies and resonant field values.

The second-order transition frequencies or resonant field values are easily found from the preceding energy values W_+ and W_- for **H** along each of the principal axes. Since M_S must change by 2, no second-order transitions involving $W_0(M_S = 0)$ can occur, only those for $W_+ \longleftrightarrow W_-(M_S = 1 \longleftrightarrow -1)$. The resonant conditions give the required field values for resonance with **H** along the respective principal axes. For **H** along z,

$$h\nu_3 = W_+ - W_- = 2[(g_z\beta H)^2 + E^2]^{1/2} \tag{10.37}$$

At constant observation frequency ν_0, with $h\nu_0 = g_z\beta H_0$, this formula may be expressed as

$$H_3 \text{ (along } z) = \frac{1}{2} H_0 \left[1 - 4 \left(\frac{E}{g_z \beta H_0} \right)^2 \right]^{1/2} \tag{10.38}$$

Similarly, for **H** along x,

$$h\nu_3 = 2 \left[(g_x \beta H)^2 + \frac{(D+E)^2}{4} \right]^{1/2} \tag{10.39}$$

and

$$H_3 \text{ (along } x) = \frac{1}{2} H_0 \left[1 - \left(\frac{D+E}{g_x \beta H_0} \right)^2 \right]^{1/2} \tag{10.40}$$

where $H_0 = (h\beta_0/g_x\beta)$.

For **H** along y,

$$h\nu_3 = 2 \left[(g_y \beta H)^2 + \frac{(D-E)^2}{4} \right]^{1/2} \tag{10.41}$$

and

$$H_3 \text{ (along } y) = \frac{1}{2} H_0 \left[1 - \left(\frac{D-E}{g_y \beta H_0} \right)^2 \right]^{1/2} \tag{10.42}$$

where $H_0 = (h\nu_0/g_y\beta)$.

In all these formulas the resonance field value H_3 occurs at approximately one-half the value $H_0 = (h\nu_0/g\beta)$ required for the first-order resonance. For this reason, the $\Delta M_S = 2$ transitions are often referred to as *half-field resonances*. Relative probabilities for these second-order transitions are given in Section 2.a. They are induced by an ac magnetic-field component that is parallel to H_0.

1.b. Fine Structure of Oriented Molecules in Single Crystals

A classic example of the triplet-state esr of oriented molecules in single crystals is that of phosphorescent naphthalene in substitutional sites in a single crystal of durene, originally observed by Hutchison and Mangum [2]. The naphthalene molecules are coplanar with the durene molecules that they replace; their long and short, twofold axes in this plane are parallel to the corresponding axes of the durene molecules as is indicated in Fig. 10.2. It is thus possible to use the structural data of the durene crystal to determine the orientation of the principal

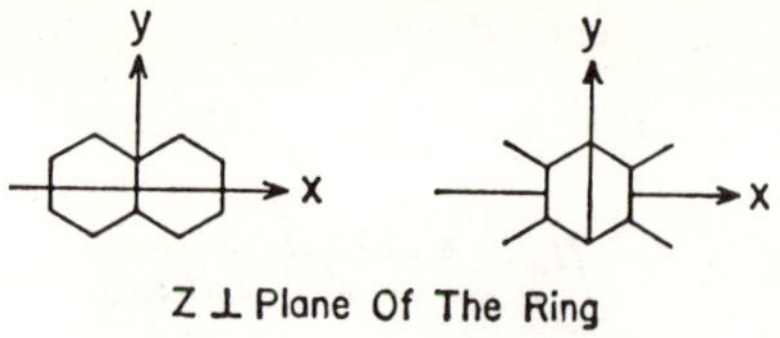

Fig. 10.2 Principal axes of the **D** tensor of substituted naphtahalene molecules (left) relative to the symmetry planes of the replaced durene molecules (right) of the host durene crystal. The z axis, not indicated, is perpendicular to the molecular planes of both molecules. From Hutchison and Mangum [2].

axes of naphthalene. The durene is monoclinic with two molecules in the unit cell [3], with relative orientations indicated by Fig. 10.3. Consequently, two sets of doublets are expected, one for each of the two distinct orientations of the naphthalene molecules that replace the durene molecules. Figure 10.4 shows plots of the observed resonant field values as functions of field orientation in the xz and yz principal magnetic planes of one set of the molecules having orientations in common. The plots for the xy plane are similar to those for the yz plane. The field values corresponding to the maxima and minima of the cosine-like curves are the resonant field values along the respective principal axes. These values can be used to give the spin-spin splitting constants D and E from (10.16),

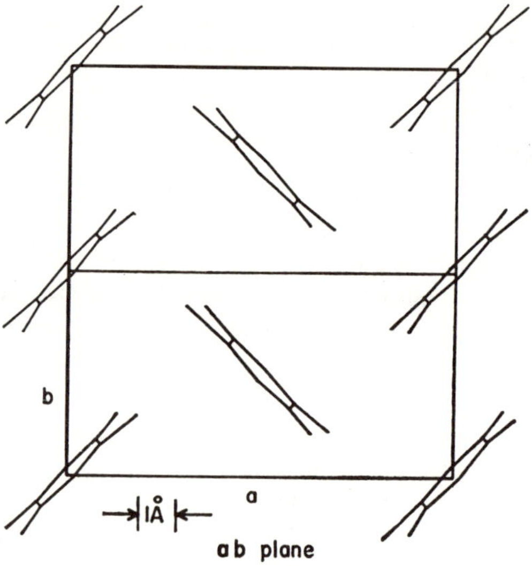

Fig. 10.3 Diagram indicating the orientations of the molecules (host durene or guest naphthalene) in the ab plane of the durene crystal. The molecular planes are perpendicular to the ab plane. From Hutchison and Mangum [2].

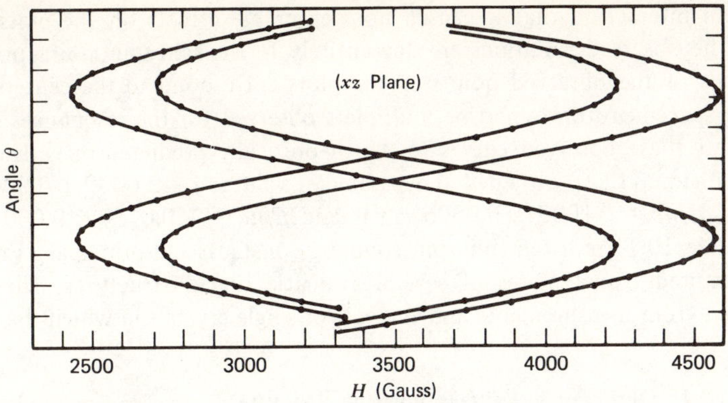

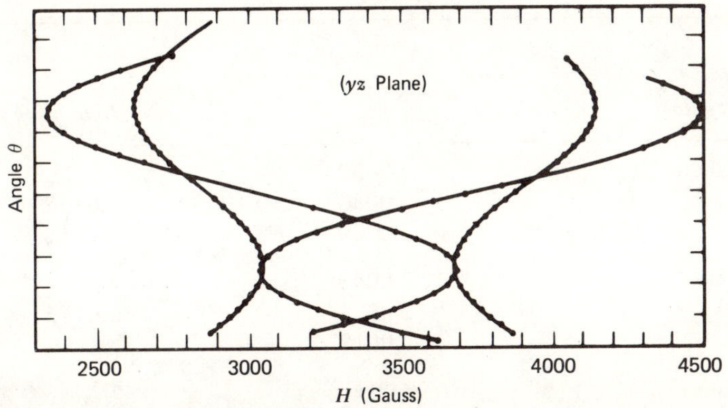

Fig. 10.4 The resonant field values (at constant ν_0) for the naphthalene triplet state plotted as functions of the orientation of the applied field in the *xz* and *yz* planes of one set of these molecules in the host durene crystal as indicated by Figs. 10.2 and 10.3. From Hutchison and Mangum [2].

(10.17), (10.23), (10.24), (10.30), and (10.31). The planes of the two molecules in the unit cell have approximate orthogonal orientation, as indicated in Fig. 10.3, with the y principal axis of each having the same direction. Consequently, the xz principal plane of one coincides approximately with the $z'x'$ plane for the other; rotation of the field about the y direction causes the simultaneous orientation change $x \to z$ for one molecule and $z' \to x'$ for the other. The similarity of the two doublet patterns for the xz plane in Fig. 10.4 results from the simultaneous 90°-out-of-phase rotation of the magnetic field in the xz principal plane for the molecules of both orientations in the unit cell. Note that for rotations in the yz plane, the plots for the two molecules coincide at one point. This coincidence occurs when the field is along their common y axis. The **g** anisotropy in the triplet-state naphthalene is so slight in comparison with the anisotropy in the spin-spin interaction that it caused no measurable effects on the plots in Fig. 10.4. The observed variations are due entirely to the spin-spin interactions. The positions of the observed points on the plots correspond to the centers of the closely spaced proton hyperfine multiplets observed for the resonances. From a fitting of these observed curves to the theoretically predicted ones, Hutchison and Mangum [1, 2] obtained the parameter values; $g = 2.0030 \pm 0.0004$ (isotropic), $D/hc = 0.1003 \pm 0.0006\ \text{cm}^{-1}$, and $E/hc = \mp 0.0137 \pm 0.0002\ \text{cm}^{-1}$.

In Table 10.1 are listed spin-spin coupling constants and principal g values for photoexcited triplet states of several aromatic ringed structures. These were obtained from measurements and analyses of single crystals in which the excited

Table 10.1 Observed g Values and Fine-structure Constants for Triplet States of Selected Molecules in Single-crystal Matrices

Molecule (Host Crystal)	*Temp. T (°K)*	g_{xx} g_{yy} g_{zz}	*D/hc* (*cm*$^{-1}$)	*E/hc* (*cm*$^{-1}$)	*Source*
Naphthalene (durene)	77	2.0030 2.0030 2.0029	+0.1003	−0.0137	*a*
Quinoline (durene)	77	2.0040 2.0029 2.0019	±0.1030	∓0.0162	*b*
Isoquinoline (durene)	77	2.003 2.003 2.003	±0.1004	∓0.0117	*b*
Quinoxaline (durene)	77	2.0047 2.0030 2.0019	±0.1007	∓0.0182	*c*
1,5-Naphthyridine (durene)	80	2.0043 2.0032 2.0018	+0.1030	−0.0167	*d*
Diphenylmethylene (benzophenone)	77	2.00451 2.00432 2.00251	±0.40505	∓0.01918	*e*
Diphenylmethylene (benzophenone)	2	2.00605 1.99891 2.00306	+0.407783	−0.020625	*f*

Table 10.1 (Continued)

Molecule (Host Crystal)	*Temp. T (°K)*	g_{xx} g_{yy} g_{zz}	*D/hc (cm⁻¹)*	*E/hc (cm⁻¹)*	*Source*
Diphenylmethylene (1,1-diphenylethylene)	77	2.0030 2.0028 2.0010	+0.3964	−0.01516	*g*
Tetramethylpyrazine (durene)	77	2.003 2.003 2.003	±0.0990	±0.0043	*h*
Anthracene (phenazine)	~1.5	2.0030 2.0033 2.0025	±0.0755	∓0.00791	*i*
Anthracene (diphenyl)	77	2.0029 2.0032 2.0024	+0.07156	−0.00844	*j*
Phenanthrene (diphenyl)	77	2.0041 2.00279 2.00209	±0.100430	∓0.046576	*k*
Pyrene (fluorene)	100	2.0033 2.0026 2.0033	±0.0678	∓0.0314	*l*
Fluorenylidene (diazafluorene)	77	2.00234 2.00272 2.00512	±0.40923	∓0.02828	*e*

[a]Hutchison and Mangum [1,2].
[b]J. S. Vincent and A. H. Maki, *J. Chem. Phys.*, **42**, 865 (1965).
[c]J. S. Vincent and A. H. Maki, *J. Chem. Phys.*, **39**, 3088 (1963).
[d]Vincent [19].
[e]Brandon, Closs, Davoust, Hutchison, Kohler, and Silbey [14].
[f]R. J. M. Anderson and B. E. Kohler, *J. Chem. Phys.*, **63**, 5081 (1975).
[g]Hutchison and Kohler [15].
[h]J. S. Vincent, *J. Chem. Phys.*, **47**, 1830 (1967).
[i]R. H. Clarke and C. A. Hutchison, *J. Chem. Phys.*, **54**, 2962 (1971).
[j]J. Grivet, *Chem. Phys. Lett.*, **4**, 104 (1969).
[k]Brandon, Gerkin, and Hutchison [61].
[l]O. H. Griffith, *J. Phys. Chem.*, **69**, 1429 (1965).

molecules were in substitutional sites of the crystals. Observations on photo-excited triplets of molecules of pure crystals have not proved feasible, apparently because of the rapid migration of the triplet excitation from molecule to molecule through resonance exchange. Although differing in details, the derivations of these characteristic parameters were obtained by application to the observed data of a Hamiltonian like that of (10.8), in a manner similar to that described previously for naphthalene in durene. Those who require specific information about methods and procedures should consult the original sources cited in Table 10.1. Following general usage, we have listed the D and E values in units of cm^{-1}. Multiplication by 29,979 ($\sim 3 \times 10^4$) will convert them to megahertz units. Since all the molecules in Table 10.1 are conjugated ring structures, it is not surprising that their spin-spin coupling constants are comparable in magnitude. For all the coupling constants, D is an order of magnitude larger than the asymmetry parameter E.

From the development of the spin-dipole Hamiltonian, (2.78) in Chapter II, Section 5, it is seen that the principal elements of D may be expressed as

$$D_x = \frac{1}{2} (g_e\beta)^2 \left\langle \frac{r_{12}^2 - 3x_{12}^2}{r_{12}^5} \right\rangle_{\text{av}} \tag{10.43}$$

$$D_y = \frac{1}{2} (g_e\beta)^2 \left\langle \frac{r_{12}^2 - 3y_{12}^2}{r_{12}^5} \right\rangle_{\text{av}} \tag{10.44}$$

$$D_z = \frac{1}{2} (g_e\beta)^2 \left\langle \frac{r_{12}^2 - 3z_{12}^2}{r_{12}^5} \right\rangle_{\text{av}} \tag{10.45}$$

where $r_{12} = r_1 - r_2$, $x_{12} = x_1 - x_2$, $y_{12} = y_1 - y_2$, and $z_{12} = z_1 - z_2$ are the coordinates separating the two interacting spins at specific points in their orbitals and where the average signifies that these quantities must be averaged over the orbitals of both electrons. With these expressions, (10.6) and (10.7) may be written as

$$D = \frac{3}{4} (g_e\beta)^2 \left\langle \frac{r_{12}^2 - 3z_{12}^2}{r_{12}^5} \right\rangle_{\text{av}} \tag{10.46}$$

$$E = -\frac{3}{4} (g_e\beta)^2 \left\langle \frac{x_{12}^2 - y_{12}^2}{r_{12}^5} \right\rangle_{\text{av}} \tag{10.47}$$

The theoretical calculations of the averages in (10.46) and (10.47) are quite complex. Theoretical methods using high-speed computer techniques with various assumptions or approximations of the molecular orbitals have been developed for handling the problem. The computer can be programmed to adjust the orbital parameters until the best fit of the observed spectral constants is achieved. In large, complex molecules there are more orbital parameters than the two observed constants D and E; however, other experimental data such as nuclear coupling can be used to reduce the unknowns. Various theoretical approaches to this problem may be found in the literature [4-12].

The two orbitals of unpaired triplet-state electrons of most conjugated planar molecules are both π-type orbitals perpendicular to the molecular plane. In this case the two electrons having parallel spins cannot occur on the same atom because of the Pauli exclusion principle, which prevents them from occupying the same p_π atomic orbital. However, the parallel spins can occur on the same atom if they have separate p_x, p_y, or p_z orbitals. As we see in Section 1.d, the ^{13}C nuclear coupling shows that significant unpaired p_x and p_y electron-spin density occurs on the central carbon in diphenylmethylene and fluorenylidene, a result indicating that the π orbitals of the two unpaired electrons of the triplet are in orthogonal planes. This finding is consistent with the abnormally large spin-spin coupling constant of these two molecules (Table 10.1). The relatively large p_x and p_y spin densities on the same carbon decrease the average separation of the unpaired electrons and increase their spin-spin coupling.

It is instructive to calculate the mean, effective separation of the two unpaired electrons in the planar molecules in Table 10.1 that have $D \approx 0.10$ cm^{-1} ≈ 3000 MHz = 1070 G. With the omission of the small E term and the assumption that the two electrons are aligned point dipoles separated by a distance R, the interaction constant D (in gauss) equals $3\beta/R^3$, where R is in centimeters. Thus the D = 1070 G indicates the $\langle R\rangle_{\text{eff}} \approx 2.96$ Å. This is approximately two CC-bond lengths. A similar calculation for fluorenylidene, which has $D = 0.40923$ cm^{-1} = 4537 G, indicates an $\langle R\rangle_{\text{eff}}$ of only 1.83 Å. It is of interest to compare these estimates with those for the triplet O_2 molecule, for which the spin-spin coupling constant in the gaseous state is 59,501 MHz [13]. The effective separation of the spin centers calculated in a similar way is 1.09 Å, compared with the measured internuclear distance of 1.2078 Å. If the p_π orbitals of the two spins are in the same plane, the exclusion principle would require them to be on the opposite oxygens. The spin-spin coupling of SO (158 209 MHz) is significantly greater than that for O_2, even though its internuclear distance is larger. This would suggest that the two interacting spins are in π_x and π_y orbitals, with much of the interacting spin density in p_x and p_y orbitals on the O. This is a likely result of the greater electronegativity of oxygen (3.5) than that for sulfur (2.5). Many elaborate molecular-orbital calculations of the spin-spin coupling in O_2 have been made, but the agreement among them is not close. Descriptions of these calculations and summaries of the results are given by Mizushima [13].

1.c. Theory of Hyperfine Structure in Oriented Molecules

The complete triplet-state spin Hamiltonian, including the direct interaction of the nuclei with the applied field **H**, is

$$\mathcal{H} = \beta \mathbf{S}\cdot\mathbf{g}\cdot\mathbf{H} + \mathbf{S}\cdot\mathbf{D}\cdot\mathbf{S} + \sum_k \mathbf{S}\cdot\mathbf{A}_k\cdot\mathbf{I}_k - \gamma\sum_k \mathbf{H}\cdot\mathbf{I}_k \tag{10.48}$$

where $\mathbf{A}_k$ is the coupling tensor, $\mathbf{I}_k$ is the spin of the kth coupling nucleus, and $\gamma = g_I\beta_I$ is the nuclear gyromagnetic ratio. The sum is taken over all the coupling nuclei. We assume, as is true for such triplet state molecules as those in Table 10.1, that the nuclear terms are small in comparison with the $\mathbf{S}\cdot\mathbf{D}\cdot\mathbf{S}$ term and that the anisotropy is **g** has negligible effects on the coupling. The nuclear interactions is considered as a perturbation on the eigenstates of the first two terms. For this purpose, it is convenient to express the nuclear interactions with reference to the principal axes x, y, z of the **D** tensor. In this system the interaction $\mathcal{H}_N$ for the kth nucleus, with k subscripts omitted for convenience is

$$\begin{aligned}\mathcal{H}_N &= \mathbf{S}\cdot\mathbf{A}\cdot\mathbf{I} - \gamma\mathbf{H}\cdot\mathbf{I}\\ &= (A_{xx}I_x + A_{xy}I_y + A_{xz}I_z)\,S_x - \gamma H_x I_x\\ &+ (A_{yx}I_x + A_{yy}I_y + A_{yz}I_z)\,S_y - \gamma H_y I_y\\ &+ (A_{zx}I_x + A_{zy}I_y + A_{zz}I_z)\,S_z - \gamma H_z I_z\end{aligned} \tag{10.49}$$

Special simplifying conditions are important. When the principal axes of $\mathbf{A}_k$ of the kth nucleus are the same as those of **D**, the cross terms vanish and $\mathcal{H}_N$ for this nucleus becomes

$$\mathcal{H}_N = A_x I_x S_x + A_y I_y S_y + A_z I_z S_z - \gamma(H_x I_x + H_y I_y + H_z I_z) \tag{10.50}$$

If the field is applied along either axis, x, y, or z, the energies are

$$E_N = A_i M_I M_S - \gamma H_i M_I \tag{10.51}$$

where $i = x, y$, or z. In the observation of normal first-order esr transitions, the nuclear quantum number does not change, and the last term is of no consequence. In ENDOR experiments, however, the nuclear spin is flipped by a radiofrequency field, and the last term must be retained. The hyperfine displacements observed for the esr transitions $M_S = 0 \longleftrightarrow \pm 1$, with **H** along x, y, or z, are

$$\Delta\nu_x = \frac{A_x M_I}{h}, \qquad \Delta\nu_y = \frac{A_y M_I}{h}, \qquad \Delta\nu_z = \frac{A_z M_I}{h} \tag{10.52}$$

Thus by making measurements of the splitting with **H** alternately imposed along x, y, or z, one can obtain the principal elements of the coupling tensors of any nuclei having principal coupling axes in common with the **D** tensor. For distinguishing the splittings of these nuclei from those of others, ENDOR frequently offers an advantage. The ENDOR frequencies in this case are simple:

$$\Delta\nu_{\text{ENDOR}} = \frac{A_i}{h} - \frac{\gamma H_i}{h} \tag{10.53}$$

where $i = x, y$, or z and $\gamma H/h$ is the nmr frequency.

When the field is applied off-axis, complications arise from the anisotropic **D** tensor that prevent the theory developed in Chapter III, Section 4.c or 4.e-g from applying with accuracy at the frequencies below 30 GHz (or kMc), those most often employed for esr measurements. In those developments in Chapter III it was assumed that the electron-spin vector **S** was aligned by the interaction $\beta\mathbf{H}\cdot\mathbf{g}\cdot\mathbf{S}$ along the effective field $\mathbf{H}_e$ (3.37) and that the eigenvalue of **S** is given by (3.89). When **g** is isotropic, $\mathbf{H} = \mathbf{H}_e$, **S** is aligned with the applied field, **H**, and $\langle S\rangle = (\mathbf{H}/|H|)M_S$. In the present case this relation holds to a good approximation only for very strong fields (high observational frequencies) for which $\beta\mathbf{S}\cdot\mathbf{g}\cdot\mathbf{H} \gg \mathbf{S}\cdot\mathbf{D}\cdot\mathbf{S}$. In effect, these two terms compete for the alignment of the spin vectors, and the applied-field interaction wins only when $g\beta H \gg D$. When the field is applied along a principal axis x, y, or z and when **g** has the same principal axes or is nearly isotropic, there is no competition; the formulas developed in Chapter III, Section 4 may then be applied.

Let us consider the $\mathcal{H}_N$ of (10.49) when **H** is applied along z, which is assumed to be a principal axis of **D** but not of $\mathbf{A}_k$. The spin **S** is aligned with the field along z, and the first-order reduced $\mathcal{H}_N$ is

$$\mathcal{H}_N = (A_{zz}I_z + A_{zx}I_x + A_{zy}I_y)M_S - \gamma H I_z \tag{10.54}$$

The eigenvalues for each of the spin states $M_S = 0 \pm 1$ can be readily found by solution of the secular equation formed from the matrix elements of this $\mathcal{H}_N$. Obviously, the first term vanishes for $M_S = 0$, and only the nmr term $-\gamma HM_I$ remains. For spins $I_k = 1/2$, as for H, ^{13}C, or ^{19}F, the secular equation is a quadratic and may be solved easily [cf. (5.75), (5.80), (5.81), and (5.82)]. For $M_S = 0, \pm 1$, with **H** along z, the result is

$$E_N\ (M_S = 0) = -\gamma HM_I = \pm\frac{1}{2}\gamma H \tag{10.55}$$

$$E_N\ (M_S = 1) = \pm\frac{1}{2}\left[(A_{zz} - \gamma H)^2 + A_{zx}^2 + A_{zy}^2\right]^{1/2} \tag{10.56}$$

$$E_N\ (M_S = -1) = \pm\frac{1}{2}\left[(A_{zz} + \gamma H)^2 + A_{zx}^2 + A_{zy}^2\right]^{1/2} \tag{10.57}$$

The energies for imposition of the field along x or y may be obtained from these equations by obvious changes in subscripts.

Now let us consider the more general case for which the field H is applied in an arbitrary direction. We designate the total $\mathcal{H}_S$ by

$$\mathcal{H}_S = \mathcal{H}_0 + \mathcal{H}_N \tag{10.58}$$

where $\mathcal{H}_0 = \beta \mathbf{S} \cdot \mathbf{g} \cdot \mathbf{H} + \mathbf{S} \cdot \mathbf{D} \cdot \mathbf{S}$ and $\mathcal{H}_N$ is given by (10.49) The condition $\mathcal{H}_0 >> \mathcal{H}_N$ applies for most triplet states; with this condition, $\mathcal{H}_N$ can be treated as a perturbation on the eigenstates of $\mathcal{H}_0$. The problem is more difficult than that for the doublet-state radicals (Chapter III), where the $\mathcal{H}_N$ may be treated as a perturbation of the $\beta \mathbf{S} \cdot \mathbf{g} \cdot \mathbf{H}$ term only. One must first solve the secular equation (10.9) for the eigenvalues and find the corresponding spin functions (eigenvectors) of $\mathcal{H}_0$ and then must use these spin functions for obtaining the characteristic energies of $\mathcal{H}_N$. In section 1.a we have derived the matrix elements and associated secular equation for $\mathcal{H}_0$ and have obtained the specific solutions for the eigenvalues for $\mathbf{H}$ along the principal axes x, y, and z. For off-axis directions of $\mathbf{H}$, the cubic secular equation (10.10) is best solved by numerical methods with a computer. Since the eigenvectors of $\mathcal{H}_0$ differ for each orientation of $\mathbf{H}$, we follow the customary procedure of simply indicating the expectation values of the spin functions of $\mathcal{H}_0$ by $\langle S \rangle$ in the matrix of $\mathcal{H}_N$. Thus we substitute the expectation values $\langle S_x \rangle$, $\langle S_y \rangle$, and $\langle S_z \rangle$ for the components of $\mathbf{S}$ in the $\mathcal{H}_N$ of (10.49). These expectation values must be computed as indicated for each orientation of $\mathbf{H}$; but since the component operators of $\mathbf{S}$ commute with those of $\mathbf{I}$, the secular equation of $\mathcal{H}_N$ for each electron-spin state can be solved with indicated expectation values $\langle S_x \rangle$, $\langle S_y \rangle$, and $\langle S_z \rangle$, which can be evaluated later by computer for each orientation of $\mathbf{H}$ as indicated.

For reasons given, we substitute $\langle S_x \rangle$, $\langle S_y \rangle$, and $\langle S_z \rangle$ for the components of $\mathbf{S}$ in (10.49) and also put $H_x = l_x H, H_y = l_y H$, and $H_z = l_z H$, where l_x, l_y, and l_z are the direction cosines of $\mathbf{H}$ with the respective axes. As an example, we describe the solution for $I = 1/2$, which is the simplest but perhaps also most important case for organic triplets. With these conditions and with z assumed to be the axis of quantization, $(\pm 1/2|I_z|\pm 1/2) = \pm 1/2$, $(\pm 1/2|I_x|\mp 1/2) = 1/2$, $(\pm 1/2|I_y|\mp 1/2) = \mp i/2$, the nonvanishing matrix elements of $\mathcal{H}_N$ of (10.49) are

$$\left(\pm\frac{1}{2}\middle|\mathcal{H}_N\middle|\pm\frac{1}{2}\right) = \pm\frac{1}{2}(A_{zz}\ \langle S_z\rangle + A_{xz}\ \langle S_x\rangle + A_{yz}\ \langle S_y\rangle - \gamma H l_z) = \pm c \tag{10.59}$$

$$\left(\mp\frac{1}{2}\middle|\mathcal{H}_N\middle|\pm\frac{1}{2}\right) = \frac{1}{2}(A_{zx}\ \langle S_z\rangle + A_{xx}\ \langle S_x\rangle + A_{yx}\ \langle S_y\rangle - \gamma H l_x)$$

$$\pm \frac{i}{2}(A_{zy} \langle S_z \rangle + A_{xy} \langle S_x \rangle + A_{yy} \langle S_y \rangle - \gamma H l_y)$$
$$= a \pm ib \tag{10.60}$$

The corresponding secular equation is

$$\begin{vmatrix} c - W & a + ib \\ a - ib & -c - W \end{vmatrix} = 0 \tag{10.61}$$

The solution is

$$W_\pm = \pm(a^2 + b^2 + c^2)^{1/2}$$
$$= \pm \frac{1}{2} [(A_{xx} \langle S_x \rangle + A_{yx} \langle S_y \rangle + A_{zx} \langle S_z \rangle - \gamma H l_x)^2$$
$$+ (A_{xy} \langle S_x \rangle + A_{yy} \langle S_y \rangle + A_{zy} \langle S_z \rangle - \gamma H l_y)^2$$
$$+ (A_{xz} \langle S_x \rangle + A_{yz} \langle S_y \rangle + A_{zz} \langle S_z \rangle - \gamma H l_z)^2]^{1/2} \tag{10.62}$$

If **H** is applied along z, then $l_z = 1$, $l_x = 0$, $l_y = 0$, $\langle S_z \rangle = M_S = 0, \pm 1$, $\langle S_y \rangle = 0$, $\langle S_x \rangle = 0$, and (10.62) reduces to (10.55) – (10.57), which were derived directly by solution of the secular equation for this specialized case.

For planar aromatic hydrocarbons, the axis perpendicular to the plane is, from symmetry, a principal coupling axis for the nuclei in the plane of the ring as well as a principal axis for the electron-spin-dipole tensor **D**. If x is an axis of the hyperfine structure as well as of the fine structure, the off-diagonal elements involving x vanish, and (10.62) reduces to

$$W_\pm = \pm \frac{1}{2} [(A_{xx} \langle S_x \rangle - \gamma H l_x)^2 + (A_{yy} \langle S_y \rangle + A_{yz} \langle S_z \rangle - \gamma H l_y)^2$$
$$+ (A_{yz} \langle S_y \rangle + A_{zz} \langle S_z \rangle - \gamma H l_z)^2]^{1/2} \tag{10.63}$$

For most of the molecules listed in Table 10.1, $D \approx 0.10\ \text{cm}^{-1} \approx 3000$ MHz. If the esr measurements are made at K-band frequencies $\approx 30{,}000$ MHz, where the $g_e \beta H$ term is an order of magnitude larger than D, the spin vector **S** will be quantized approximately along the applied field **H**. In approximate treatments, one can then set

$$\langle S \rangle = \left(\frac{H_0}{|H|}\right) M_I \tag{10.64}$$

and substitute

$$\langle S_x \rangle = l_x M_S, \ \langle S_y \rangle = l_y M_S, \ \langle S_z \rangle = l_z M_S \tag{10.65}$$

into (10.62), with $M_S = 0$ and ± 1 for the three triplet levels.

For finding the coupling matrix elements of the kth nucleus, one usually measures its splitting along the x, y, z axes and at various orientations in the three orthogonal planes of the x, y, z reference system. Procedures for finding the required elements and the diagonalization of the resulting **A** tensor are the same as those described for oriented doublet-state radicals in Chapter V, Section 2.

Because of the large number of closely spaced hyperfine components in the aromatic ringed systems with the lines not as sharp as those observed in liquid solutions, the ENDOR technique offers great advantage over esr for measurement of the nuclear couplings of these species in single crystals. The ENDOR frequencies correspond to the level spacing of the hyperfine splitting for the $M_S = \pm 1$ levels plus the nmr frequency corresponding to the interaction of the nuclear spin with the applied magnetic field. Thus for the common conditions under which (10.63) holds, the ENDOR displacements by the kth nucleus with $I_k = 1/2$, corresponding to $M_{I_k} = 1/2 \longleftrightarrow -1/2$, are given by

$$\Delta\nu_{\text{ENDOR}} = h^{-1} \left[(A_{xy}\langle S_x \rangle - \gamma H l_x)^2 + (A_{yy}\langle S_y \rangle + A_{yz}\langle S_z \rangle - \gamma H l_y)^2 + (A_{yz}\langle S_y \rangle + A_{zz}\langle S_z \rangle - \gamma H l_z)^2 \right]^{1/2} - \frac{\gamma H}{h} \tag{10.66}$$

Applications of (10.66) are described in the following section.

1.d. Some Measurements of Hyperfine Structure

Among the more important hyperfine measurements of the triplet state are those on the connecting carbon C_1 in diphenylmethylene and fluorenylidene.

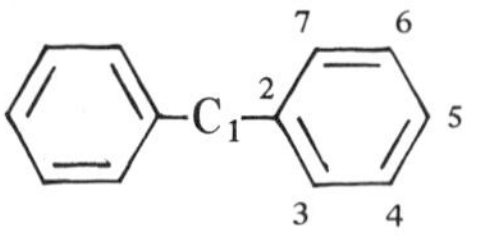

Diphenylmethylene
42

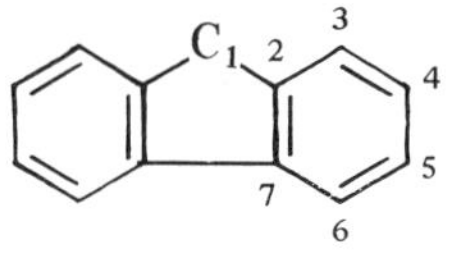

Fluorenylidene
43

As indicated in these structures, the C_1 is divalent, and the esr data indicate divalent carbon in this position [14]. These structures have abnormally large fine-structure constants in comparison with the other aromatic molecules shown in Table 10.1. Hyperfine coupling constants of $^{13}C_1$ measured by Hutchison and his associates [14,15] are listed in Table 10.2 with the derived spin densities. The large orthogonal $2p_x$ and $2p_y$ spin densities observed on the same atom, C_1, account for the abnormally large spin-spin interaction in these two molecules. Higuchi [16] has theoretically derived the value $D = 0.9055\ cm^{-1}$ for two parallel spins in orthogonal $2p$ orbitals of the same carbon. By scaling this predicted interaction to conform to these measured $2p_x$ and $2p_y$ spin densities on the C_1, Brandon, Closs, et al. [14] obtained the values $D = 0.374\ cm^{-1}$ for diphenylmethylene and $D = 0.405$ for fluorenylidene, which are close to the observed values. Evidently, most of the electron-spin-spin interaction in these molecules occurs on the C_1.

From the orientations of principal axes of the **D** tensor relative to the molecules of the host crystal as determined by X-ray crystallography, the probable directions of these principal axes x, y, z relative to the molecular structure of diphenylmethylene were found [15] to be as indicated in Fig. 10.5. These axes are approximately those of the ^{13}C coupling tensor.

Although the proton couplings for the triplet state of diphenylmethylene and fluorenylidene were not measured in the earlier esr work [14], Hutchison and

Table 10.2 Nuclear Couplings of ^{13}C and $2p$ Spin Densities on the Central Carbon C_1 of Triplet-state Diphenylmethylene and Fluorenylidene in Host Single Crystals

Molecule (Host Crystal)	*Coupling Element (MHz)*	*Isotropic Component (MHz)*	*Anisotropic Component (MHz)*	*C_1 Spin Density* ρ_{2p}	ρ_{2s}	*Ref. and Method*
Diphenylmethylene	A_{xx} 189.6		16.3	0.56 $(2p_x)$		*a*
(benzophenone)	A_{yy} 214.8	173.3	41.5	0.74 $(2p_y)$		esr
	A_{zz} 115.4		−57.8			
Diphenylmethylene	A_{xx} 166.6		21.6	0.565 $(2p_x)$		*b*
(1,1-diphenyl-	A_{yy} 177.2	144.9	32.3	0.646 $(2p_y)$	0.0867	
ethylene)	A_{zz} 91.0		−56.9			ENDOR
Fluorenylidene	A_{xx} 280.0		16.4	0.57 $(2p_x)$		*a*
(diazofluorene)	A_{yy} 307.3	263.6	43.8	0.78 $(2p_y)$	0.079	esr
	A_{zz} 203.4		−60.2			

[a] Brandon, Closs, Davoust, Hutchison, Kohler, and Silbey [14].
[b] Hutchison and Kohler [15].

Pearson [17] later measured the proton coupling with the ENDOR technique for the fluorenylidene triplet in host single crystals of diazofluorene. Also, Hutchison and Kohler [15] measured the ENDOR transitions of the protons of diphenylmethylene in the 1,1-diphenylethylene host crystal. Spin densities on each carbon of these molecules were derived from the proton couplings with the McConnell relation. Table 10.3 lists spin densities for one of the rings of each molecule together with that on C_1, as derived from the ENDOR measurements. The numbers designating the carbon positions of diphenylmethylene correspond to the unprimed numbers in Fig. 10.5. Descriptions of the measurements and

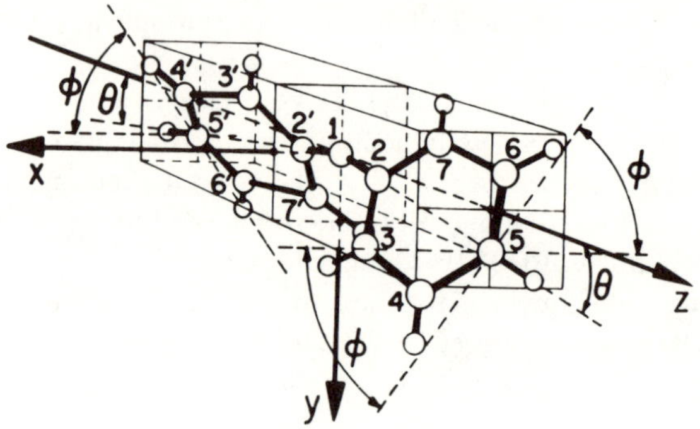

Fig. 10.5 Structural diagram of the diphenylmethylene molecule. From Hutchison and Kohler [15].

analyses of them may be found in the original papers [15,17]. Anderson and Kohler [18] have measured the ENDOR transitions as a function of crystal orientation in the applied field for all the protons in diphenylmethylene in a host single crystal of benzophenone in its $P2_1$ phase at about 2°K. The tensors for the hyperfine structure were derived for each proton in the molecule. The orientation of these hyperfine tensors was found to be consistent with the geometry of the molecule as found by Hutchison and Kohler [15] and shown in Fig. 10.5. Spin densities on all the carbons were derived from the proton-coupling tensors.

Vincent [19] has measured the esr hyperfine structure of ^{14}N and protons on carbons 2, 3, and 4 in 1,5-naphthyridene in the durene host crystal and has identified the coupling proton from a prediction of spin densities on the carbons by INDO theory. Second-order CH proton transitions like those described in Chapter V, Section 4 were observed for some orientations of the crystal.

Frosch, Goncalves, and Hutchinson [20] measured the esr hyperfine structure for several protons in phenanthrene-h_{10} in the host crystal of diphenyl-d_{10}. This was followed by ENDOR measurements of the deuterium level splitting in phenanthrene-d_{10} by Hutchison and McCann in a diphenyl host crystal

Table 10.3 Spin Densities Derived from ENDOR Measurements for Fluorenylidene in Diazofluorene Crystal and for Diphenylmethylene in 1,1Diphenylethylene

Carbon Number[a]	*Fluorenylidene*[b]		*Diphenylmethylene*[c]	
i	ρ_i[d]	$\sigma_i\{\rho_i\}$[e]	ρ_i[d]	$\sigma_i\{\rho_i\}$[e]
1	+0.710	0.106	+0.590	0.062
2	+0.020	0.064	-0.022	0.037
3	+0.072	0.004	+0.101	0.012
4	-0.051	0.014	-0.0392	0.0051
5	+0.093	0.010	+0.1108	0.0044
6	-0.022	0.004	-0.0381	0.0043
7	+0.021	0.027	+0.1215	0.0090

[a] See structural diagrams in text for number system, also Fig. 10.5.
[b] Hutchison and Pearson [17].
[c] Hutchison and Kohler [15].
[d] ρ_i = Spin density.
[e] $\sigma_i\{\rho_i\}$ = Standard deviation.

[21]. A rather complete mapping of triplet-state spin densities on the different carbons was obtained from these measurements.

2. TRIPLET-STATE RESONANCE IN RANDOMLY ORIENTED MOLECULES

2.a. Second-order $\Delta M_S = 2$ Transitions

Although accurate values of the parameters can be obtained from molecules in host single crystals, the number of species that can be investigated in this way is rather limited. Early attempts to detect esr of the triplet state for randomly oriented molecules in glassy solutions met with failure because of the wide spread of the spectrum caused by anisotropy in the **D** tensor. However, van der Waals and de Groot [22] noticed that the "half-field" transition, $\Delta M_S = 2$, showed much less anisotropy and should thus be more favorable for observations on randomly oriented molecules than are the normal $\Delta M_S = \pm 1$ transitions. This condition results from reduced effects of the D terms in the difference, $h\nu_3 = W_+ - W_-$, as shown in Section 1.a. Consequently, they calculated the transition probability for the so-called "forbidden transitions" and found them to be favorable when the applied field interaction $\beta\mathbf{S}\cdot\mathbf{g}\cdot\mathbf{H}$ is not appreciably larger than the spin-dipole term $\mathbf{S}\cdot\mathbf{D}\cdot\mathbf{S}$. If transitions are to be induced between the W_+ and W_- levels, the ac resonance-field component H_1 must be along the static field $\mathbf{H}_0$, whereas the ac field H_1 must be applied in the plane normal to $\mathbf{H}_0$ for induction of the first-order transition ($\Delta M_S \pm 1$). If the $\mathbf{H}_0$ field is applied along the z principal axis of **D**, then $l_z = 0$, $l_y = 0$, and the secular equation (10.9) with $g_z = g$ becomes

$$\begin{array}{c|ccc|}
 & 1 & 0 & -1 \\
1 & g\beta H_z + \frac{D}{z} - W & 0 & E \\
0 & 0 & -\frac{2}{3}D - W & 0 \\
-1 & E & 0 & -g\beta H_z + \frac{D}{z} - W
\end{array} = 0 \qquad (10.67)$$

which yields the solutions already obtained, (10.11) and (10.12). It is evident that the $g\beta H_z$ term intermixes the spin functions $|1\rangle$ and $|-1\rangle$, which can be expressed by the linear combination

$$\begin{aligned} \phi_+ &= a_+ |1\rangle + b_+ |-1\rangle \\ \phi_- &= a_- |1\rangle + b_- |-1\rangle \end{aligned} \qquad (10.68)$$

where ϕ_+ is the state function corresponding to W_+ and ϕ_- is that corresponding to W_-. The transition probabilities for an ac resonance field of amplitude H_{1z} imposed along z are proportional to

$$H_{1z}^2 |(\phi_+|S_z|\phi_-)|^2 = H_{1z}^2 |a_+ a_-^* - b_+ b_-^*|^2 \qquad (10.69)$$

The a and b coefficients are obtained by multiplication of the functions ϕ by the first or last row in the secular equation (10.67) with the respective roots W_+ and W_- and by an equating of the result to zero. For example, from the first row,

$$\begin{aligned} a_+ \left(g\beta H_z + \frac{D}{z} - W_+\right) + b_+ E &= 0 \\ a_- \left(g\beta H_z + \frac{D}{z} - W_-\right) + b_- E &= 0 \end{aligned} \qquad (10.70)$$

With the normalizing relations

$$a_+^2 + b_+^2 = 1, \qquad a_-^2 + b_-^2 = 1 \tag{10.71}$$

and with the values of W_+ and W_- from (10.67), one can solve (10.70) for specific coefficients and evaluate the squared transition moment of (10.69). The resulting moment is definitely not zero, as was first shown by van der Waals and de Groot [22], who evaluated the matrix elements with a different formulation $\mathcal{H}_{zz}$ in which $-X = D_x$, $-Y = D_y$, and $-Z = D_z$. With these substitutions, their results for the squared transition-matrix elements for the static and the radio-frequency magnetic fields imposed along the three respective principal axes of the fine-structure tensor are

$$|(\phi_+|S_z|\phi_-)|^2 = \left(\frac{D_x - D_y}{h\nu_0}\right)^2 \tag{10.72}$$

$$|(\phi_+|S_y|\phi_-)|^2 = \left(\frac{D_x - D_z}{h\nu_0}\right)^2 \tag{10.73}$$

$$|(\phi_+|S_x|\phi_-)|^2 = \left(\frac{D_y - D_z}{h\nu_0}\right)^2 \tag{10.74}$$

Using the values of D_x, D_y, and D_z evaluated for naphthalene by Hutchison and Mangum [1,2] at an observation frequency $\nu_0 = 0.3221\ \text{cm}^{-1}$, van der Waals and de Groot calculated that $|(\phi_+|S_i|\phi_-)|^2 = 0.126, 0.072$, and 0.008 for the field imposed along the x, y, z axes, respectively, with these axes labeled as indicated in Fig. 10.2. When compared with the normal $\Delta M_S = \pm 1$ transition, these quantities indicate that the probabilities of the "forbidden transitions" for naphthalene at this ν_0 range from 2% for **H** perpendicular to the molecular plane (z axis) to 25% for **H** along the x axis, of the normal transitions observed by Hutchison and Mangum [1,2].

It should be noted that when $D_x - D_y = E = 0$, the transition moment for the second-order transition for the field along z vanishes. Similarly, the second-order transition moment vanishes for the field along the other axes when $D_y - D_z = 0$. When the molecule has a threefold or sixfold axis of symmetry, as does benzene, the transition probability is strictly zero when the dc and ac field are both perpendicular to the molecular plane.

In the more general case for which the field is applied away from the principal axes x, y, z, one must find the eigenvectors of the Hamiltonian of (10.8). The secular equation for this $\mathcal{H}_S$ given by (10.9) can be used for this purpose with the three roots determined from (10.10). The function ϕ will be linear combinations of the three functions corresponding to $M_S = 0$ and ± 1, with the coefficients determined from the secular equation (10.8) with the three roots W_0, W_-, and W_+

obtained from solution of this equation, as already indicated in Section 1.a. For off-axis orientations, the solution of the cubic equation for the eigenvectors and eigenvalues can be achieved most easily by a numerical computer program. McGlynn, Azuni, and Kinoshito [23] give algebraic expressions for the eigenvectors C_{j,M_S}, where $M_S = 0,-1,+1$ and where j corresponds to the three roots that we label 0,−,+ and they label 1,2,3. They also give tabulations of the first-order ($\Delta M_S = \pm 1$) transition moments for **H** along each of the x, y, z axes.

Following their theoretical prediction of the line strengths of the "forbidden transitions," van der Waals and de Groot [22] proceeded to detect these transitions for oriented naphthalene in a single crystal of durene. Although they made no quantitative comparison of predicted and measured line strengths, the observed signals were strong, many times the noise level. They then made up a rigid glass solution of 0.012 mole% in glycerol and observed the signal of the half-field "forbidden transition" for randomly oriented molecules of naphthalene, shown

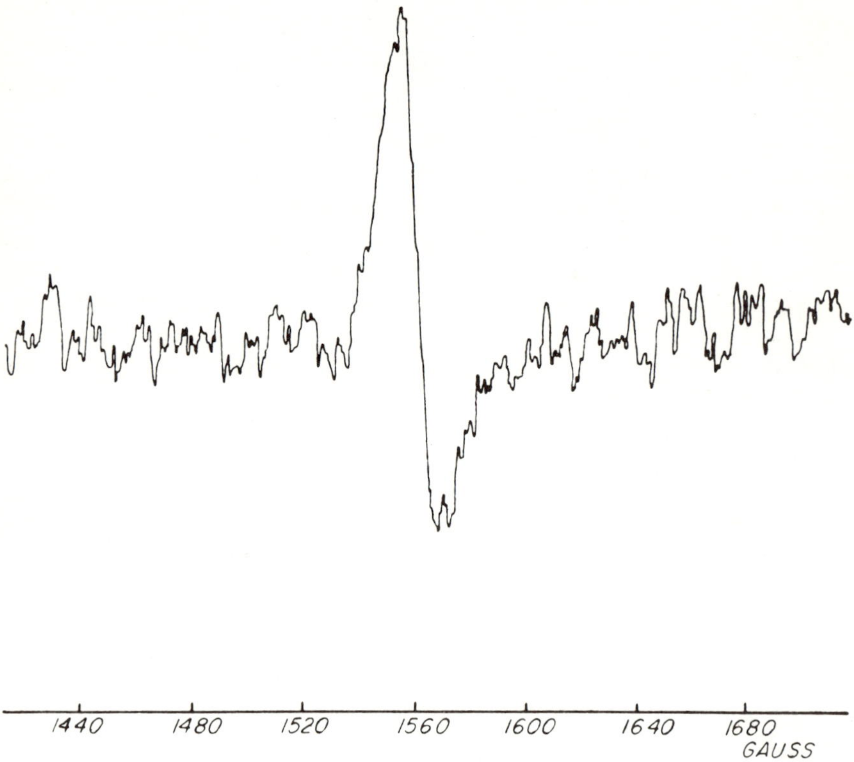

Fig. 10.6 The "forbidden" $\Delta M_S = 2$ transition of phosphorescent naphthalene in a glassy solution, as originally observed by van der Waals and de Groot [22].

in Fig. 10.6. Since this initial observation, many such second-order transitions have been observed for randomly oriented molecules in triplet states. From the squared transition-matrix elements of (10.72) to (10.74), it was found that for naphthalene the strongest $\Delta M = 2$ component for **H** along a principal axis occurs for the x axis. Thus one might assume the peak intensity in Fig. 10.6 occurs for **H** along x and might substitute the field value corresponding to this peak into (10.40) for an approximate evaluation of $D + E$. However, accurate line-shape calculations, reveal that the strongest absorption does not occur for this orientation but rather for an off-axis orientation corresponding to the minimum field value for the transition.

Line-shape Functions and Critical Field Values

Following the work by van der Waals and de Groot [22] on $\Delta M_S = 2$ transitions, Kottis and Lefebvre [24] designed a computer program for calculation of line shapes of these resonances for randomly oriented, static molecules and showed how it is possible to derive rather accurate values of the spin-spin coupling parameters from the peaks or inflection points in these shape functions. Their treatment is summarized here with the notation changed to conform to that used in previous section, that is D_x, D_y, and D_z for $-X$, $-Y$, and $-Z$ and $h\nu_0$ for δ.

If the small anisotropies in **g** are neglected and the direction of the field relative to the principal axes x, y, z of the fine-structure tensor are expressed in polar angles θ and ϕ, the spin Hamiltonian $\mathcal{H}_S$ can be written as

$$\mathcal{H}_S = (g\beta H)(S_x \sin\theta\cos\phi + S_y \sin\theta\sin\phi + S_z\cos\theta) + D_x S_x^2 + D_y S_y^2 + D_z S_z^2 \tag{10.75}$$

The secular equation of the $\mathcal{H}_S$ can then be expressed in the form [24]:

$$W^3 - W\,(g\beta H)^2 - [(D_x D_y + D_x D_z + D_y D_z)] - (g\beta H)^2\,(D_x \sin^2\theta\cos^2\phi + D_y \sin^2\theta\sin^2\phi + D_z\cos^2\theta) + D_x D_y D_z = 0 \tag{10.76}$$

For calculation of line shapes of randomly oriented samples, it is advantageous to express the cubic equation in terms of a constant interval $\delta = h\nu_0$ corresponding to the constant observing frequency. This is done by substitution of the roots $W = W_0$ and $W = W_0 + h\nu_0$ or $W = W_0 - h\nu_0$ alternately into (10.76) and by elimination of W_0 between the two equations. These algebraic manipulations, which are rather tedious, were carried out by Kottis and Lefebvre [24], who obtained the secular equations in the form

$$D_x \sin^2\theta\cos^2\phi + D_y \sin^2\theta\sin^2\phi + D_z\cos^2\theta$$
$$= D_xD_yD_z\ (g\beta H)^{-2} \mp 3^{-3/2}\ \{(g\beta H)^{-2}\ [(h\nu_0)^2 + D_xD_y + D_xD_z$$
$$+ D_yD_z] - 1\}\ [(4g\beta H)^2 - (h\nu_0)^2 - 4(D_xD_y + D_xD_z + D_yD_z)]^{1/2} \quad (10.77)$$

This form has an advantage in that the angular coordinates are separated from the magnetic-field values and frequency intervals. It is convenient for numerical, computer calculation of the line shapes of the randomly oriented molecules in glassy media. Functionally, (10.77) is expressed as

$$f(\theta,\phi) = F(H,h\nu_0) \quad (10.78)$$

The program for computer calculation of the line shapes of glassy solutions is like that already described in Chapter VIII, Sections 1 and 2 for calculation of line shapes resulting from **g** or **A** anisotropy. From (8.11), the fractional number of the randomly oriented molecules having orientations within the element of the solid angle $d\Omega = \sin\theta\, d\theta\, d\phi$ is $(1/4\pi)d(\cos\theta)d\phi = (1/4\pi)\sin\theta\, d\theta\, d\phi$, and the intensity of the resonance for any field value, from (8.16), is

$$I(H) \propto \int_0^{4\pi} f\,[H - H_r(\theta,\phi)]\, d\Omega$$

$$\propto \int_{\phi=0}^{2\pi}\int_{\theta=0}^{\pi} f\,[H - H_r(\theta,\phi)]\sin\theta\, d\theta\, d\phi \quad (10.79)$$

in which $f[(H - H_r(\theta,\phi)]$ is the line-shape function that in (8.17) is assumed to be the Lorentzian function. For their calculations of line shapes for triplet states, Kottis and Lefebvre [24] assumed a Gaussian line shape. No function is included to correct for the angular variation of the transition probability. This extra complication is hardly justified since the primary concern is with establishment of the line shape in the vicinity of the critical field values from which the information about the spin-spin coupling constant is gained. A computer program for evaluation of (10.79) with the functions in (10.78) is described by Kottis and Lefebvre, who used the program to evaluate typical esr line-shape functions for triplet-state molecules in glassy media.

The line-shape distribution curves for the $\Delta M_S = 2$ transition of randomly oriented molecules shown in Fig. 10.7 were calculated by Kottis and Lefebvre with their computer program. Although calculated for a specific frequency (X band) with the particular parameters for naphthalene, this figure is exceedingly useful for analysis of $\Delta M_S = 2$ transitions of other triplet-state molecules in glassy solution. A particularly important feature is the occurrence of the strongest and sharpest absorption peak not along any principal axis but at the minimum field for the transition. This is evidenced by the beautiful confluence

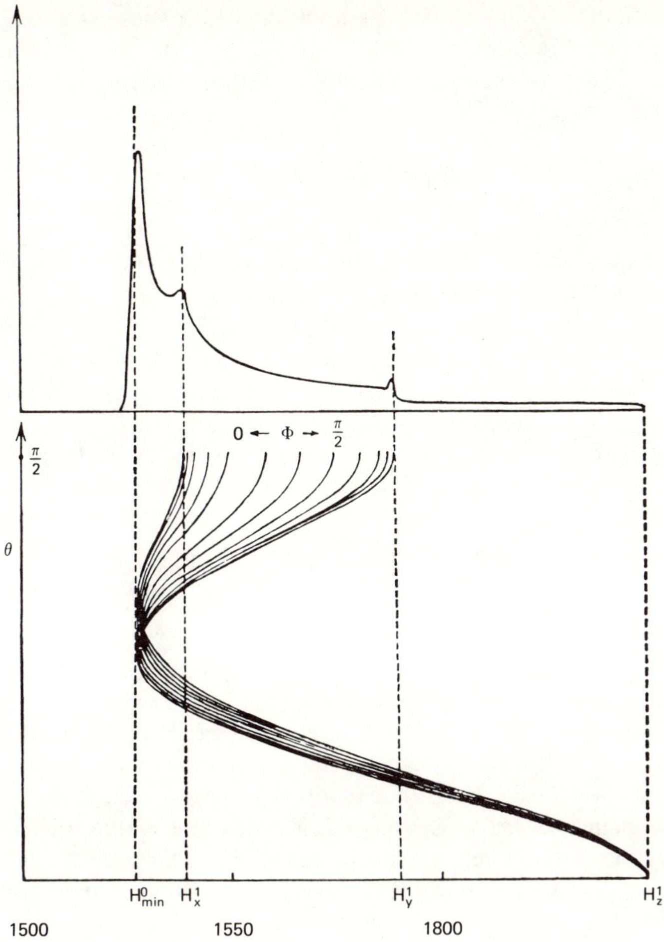

Fig. 10.7 Calculated esr absorption line-shape function for $\Delta M_S = 2$ of randomly oriented molecules in triplet states. From Kottis and Lefebvre [24].

of all the distribution curves in the lower figure at H_{min}. Because of the natural width of the resonance for any given orientation, the H_{min} is not, of course, the absolutely lowest field value at which absorption occurs, but is rather the lowest at which a peak absorption occurs. Although absorption peaks or critical points also occur for field values corresponding to **H** orientation along the principal axes of the spin-spin dipole tensor, these are much weaker than that for H_{min}. Consequently, observers usually detect or measure only the strong peaks corresponding to H_{min}. The lower graphs give evidence that this peak occurs at field values for which the slopes $d\theta/d$H of the distribution curves (lower graphs) become infinite. This lowest possible resonant field is also that at which the $F(H, h\nu_0)$ ceases to be a real function. Therefore, H_{min} can be found most easily if the term under the radical on the right of (10.77) is set equal to zero. It is then seen that

$$H_{min} = (2g\beta)^{-1} \; [(h\nu_0)^2 + 4(D_xD_y + D_xD_z + D_yD_z)]^{1/2} \qquad (10.80)$$

With the relationships of (10.6) and (10.7) and with $h\nu_0 = g\beta H_0$, this equation can also be expressed in the form

$$H_{min} = (2g\beta)^{-1} \; [(g\beta H_0)^2 - \frac{4}{3}(D^2 + 3E^2)]^{1/2} \qquad (10.81)$$

It is evident that for H_{min} to be real, the condition $(g\beta H_0) = (h\nu_0)^2 > 4/3\ (D^2 + 3E^2)$ must be satisfied. For convenience, (10.81) is usually expressed as

$$D^* = (D^2 + 3E^2)^{1/2} = \frac{\sqrt{3}}{2}[(g\beta H_0)^2 - (2g\beta H_{min})^2]^{1/2} \qquad (10.82)$$

From (10.82) it is apparent that the magnitude of $D^* = [(D^2/3) + E^2]^{1/2}$ can be obtained from measurement of H_{min}. Many such measurements have been made. Representative values of the parameters obtained are listed in Table 10.4. Whenever the weaker peaks in Fig. 10.7 corresponding to **H** along the principal axes x, y, and z can be detected, the magnitudes of both D and E can be obtained by application of (10.38), (10.40), and (10.42). These weaker peaks, as well as that corresponding to H_{min}, are most easily detected with a spectrometer that records the derivative of the absorption contour. This is obviously true because of the rapid change of slope of the absorption contour in these critical regions. To date, most of the measurements have been made on the peak corresponding to H_{min}.

The strong, sharp peak easily detected at H_{min} provides a ready means for detection of the presence of triplet excitation in numerous substances and for measurement of the lifetimes of these states in glassy solutions of polycrystalline solids. It should also be useful in the study of chemical reactions involving triplet-state intermediaries.

2.b. The $\Delta M_S = \pm 1$ Transitions

Following their success in detecting the $\Delta M_S = 2$ transitions, de Groot and van der Waals [25] and, independently, Wasserman, Snyder, and Yager [26], were able to detect the normal $\Delta M_S = \pm 1$ transitions in triplet states of randomly oriented molecules. To detect the $\Delta M_S = 1$ transition, which spreads over a few

Table 10.4 Triplet-state Parameters Derived from Observed $\Delta M_S = 2$ Transitions of Randomly Oriented Molecules in Glassy Solvents

Molecule	*Solvent*	$D^* = (D^2 + 3E^2)^{1/2}$ (cm^{-1})	*Source*
Diphenylacetylene	Ethanol	0.152	*a*
Phenylacetylene	Ethanol	0.150	*a*
Phenylmethylacetylene	Ethanol	0.145	*a*
Triphenylene	EPA	0.1353	*b*
	Lucite	0.1360	*c*
3,4-Benzpyrene	EPA	0.0758	*b*
Coronene	EPA	0.0971	*b*
	Lucite	0.0983	*c*
1,12-Benzperylene	Lucite	0.0718	*c*
Terphenyl	EPA	0.0961	*b*
Acenaphthene	Lucite	0.1029	*c*
Hydroquinone	EPA	0.1321	*b*
Benzoic acid	EPA	0.1385	*b*
1-Chloronaphthalene	Lucite	0.1056	*c*
1-Methylphenanthrene	Lucite	0.1288	*c*
1-Bromophenanthrene	Lucite	0.1327	*c*
Aniline	EPA	0.1317	*b*
Triphenylamine	Lucite	0.0801	*c*

[a] C. G. Wade and S. E. Webber, *J. Chem. Phys.*, **56**, 1919 (1972).
[b] B. Smaller, *J. Chem. Phys.*, **37**, 1578 (1962).
[c] C. Thomson, *J. Chem. Phys.*, **41**, 1 (1964).

thousand gauss, these observers took advantage of the rapid change of the absorption contour at field values in the vicinity of the canonical resonances, that is, fields giving rise to resonance absorptions by the small fractional number of molecules that have one of their principal axes oriented approximately in the direction of the applied field. The absorption contour changes rapidly as the field is swept through these regions. By use of the signal modulation and derivative detection methods, the resulting signals can be detected with exceptional sensitivity. The problem is like that of detecting similar inflection points along the principal axes of the **g** or **A** tensor for doublet radicals, described in Chapter VIII, Sections 1 and 2. However, the signals of the triplet state are weaker because of the much greater anisotropic spread of their absorption band, caused by the spin-spin interaction. In comparison to this spread, that caused by **g** anisotropy, and even by hyperfine splitting in organic molecules, is usually negligible. Figure 10.8 illustrates the type of absorption contour produced by the electron-

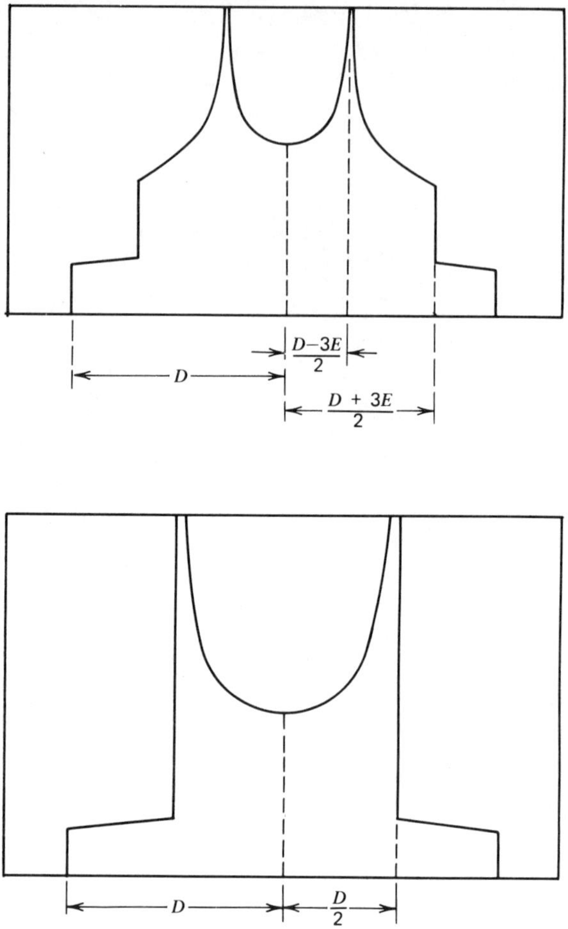

Fig. 10.8 The first-order, $\Delta M_S = 1$, triplet-state esr absorption contours due to anisotropy in the electron-spin–spin interaction alone. The upper diagram corresponds to the general case for which $E \neq 0$ and the lower diagram, to that for axial symmetry in **D** ($E = 0$). From Wasserman, Snyder, and Yager [26].

spin, dipole-dipole interaction when these other broadening factors are neglected. It is evident that one can obtain values of both D and E from measurements of the discrete changes in the contour with derivative detection. This curve was calculated by Wasserman, Snyder, and Yager [26] with a computer program similar to that of Kottis and Lefebvre [24] but designed by Snyder and Kornegay [27] to simulate the $\Delta M_S = 1$ spectra.

Figure 10.9 shows the first derivative, triplet-state esr that Wasserman, Snyder, and Yager et al. [26] observed for randomly oriented molecules of naphthalene-

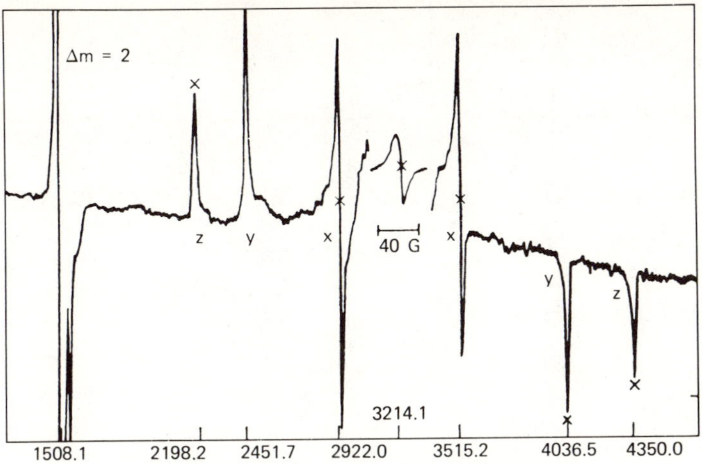

Fig. 10.9 First-derivative absorption peaks in the esr spectrum for randomly oriented naphthalene molecules in glassy solutions. The strongest signal is due to $\Delta M_S = 2$ transitions as indicated. The peaks marked x, y, and z are the $\Delta M_S = 1$ transitions occurring for molecules having their respective principal axes of **D** along the applied field (cf. with Fig. 10.8). The 40-G center inset line is attributed to double quantum absorptions. From Wasserman, Snyder, and Yager [26].

d_8 in tetrahydrofuran at 77°K. The three signals marked x, y, z for each $\Delta M_S = 1$ doublet component correspond to the absorption by molecules along the respective principal axes of the **D** tensor. The weak inset line in the 40-G span between the doublet components does not fit into the normal pattern. This interesting extra line was shown by the observers to arise from a double quantum transition. The strongest observed signal, that at the lowest field value, is the half-field, $\Delta M_S = 2$ component described in Section 2.a. With their computer program, Wasserman and co-workers calculated a derivative pattern that closely simulated the $\Delta M_S = 1$ and $\Delta M_S = 2$ components of this observed spectrum with the parameters $D = 0.10046$ cm^{-1}, $E = 0.01536$ cm^{-1}, and $g = 2.0028$. The probable error limits on D and E are stated as ± 0.00004 cm^{-1}. These parameters are rather close to the corresponding values $D = 0.10119$ cm^{-1} and $E = 0.01411$ cm^{-1} for naphthelene-d_8 measured in the host single crystal of durene [28].

Kottis and Lefebvre [29] also used their computer program to calculate the $\Delta M_S = 1$ spectrum for randomly oriented naphthalene. As in their earlier studies of the $\Delta M_S = 2$ transitions described in Section 2.a, they calculated resonance contours corresponding to all orientations of θ and ϕ. These are displayed in the lower graphs of Fig. 10.10. In the region of the principal axes their program simulated the $\Delta M_S = 1$ spectrum observed by de Groot and van der Waals [25], which is reproduced in the upper part of Fig. 10.10. The families of curves in the

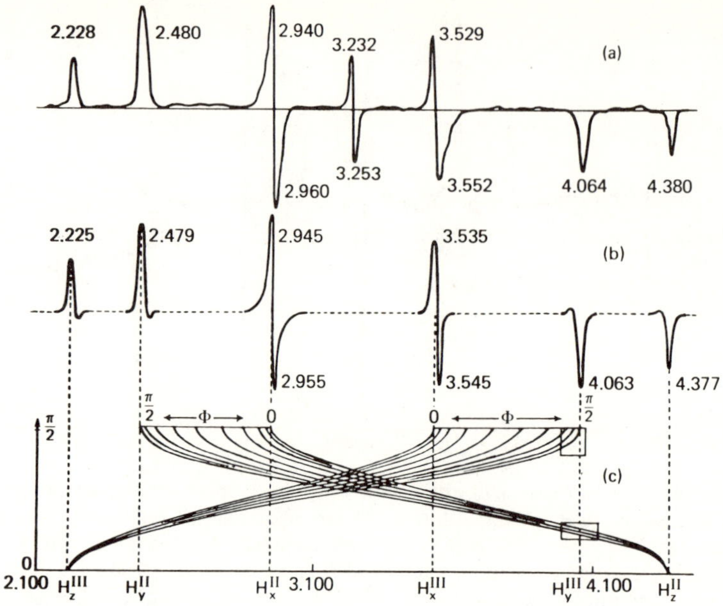

Fig. 10.10 Theoretical $\Delta M_S = 1$ esr spectrum for randomly oriented triplet-state molecules with zero-field parameters adjusted to those for naphthalene. The top diagram is the $\Delta M_S = 1$ derivative spectrum for phosphorescent naphthalene in glassy solution, observed by de Groot and van der Waals [25]. The theoretical curves were calculated by Kottis and Lefebvre [29].

lower part of the diagram help one to grasp quickly the nature of these complex spectra and to see at a glance the abrupt changes that give rise to the detectable inflection points corresponding to resonances for **H** along the principal axes.

The six $\Delta M_S = 1$ lines shown in Figs. 10.9 and 10.10 correspond to those for the doublet components calculated in Section 1.a for the single-crystal resonance when the crystal is oriented along the three principal axes of the **D** tensor. The formulas given there can be applied with the field values for these six lines for calculation of the magnitudes of both D and E. Although the resulting values are not quite as precise as those obtained with single crystals, the method is more widely applicable. It can be used to obtain spin-spin coupling constants for many substances, including large polymers, which cannot be studied in single-crystal hosts.

2.c. Applications

Although it may be advantageous in improving the accuracy of the derived parameters to have computer simulation of the spectra, useful work in esr of the triplet state fortunately does not require availability of a computer. Most of the information can be gained without a computer program now that the basic nature of the detected signals has been established with the aid of a computer. It is evident from Figs. 10.7 and 10.10 that the peaks corresponding to H_{min} and to the resonant fields along the principal axes of the **D** tensor are well defined. By careful measurements with optimum adjustment of modulation amplitudes of the spectrometer, one can obtain accurate values of D and E, with the relatively simple solution for **H** along the principal axes x, y, z given in Section 1.a or of D^* from H_{min} with the application of (10.82).

Examples of D^* values obtained from H_{min} of the $\Delta M_S = 2$ transitions are listed in Table 10.4. Examples of D and E values obtained from measurements of the H_x, H_y, and H_z peaks of the $\Delta M_S = 1$ transitions appear in Table 10.5. These examples are selected from the many such parameters that have been obtained from triplet-state esr of glassy solutions.

Among the more important simple molecules to be observed in glassy solution is CH_2. Bernheim, Bernard, et al. [30] confirmed the triplet ground state $^3\Sigma_g^-$ for CH_2 and obtained its spin dipole-dipole coupling parameters by observations of its $\Delta M_S = 1$ esr spectrum in a xenon matrix at 4.2°K. The molecules were produced in the matrix by photolysis of a mixture that contained 1 part in 10^4 of diazirine. The parameter values obtained were $D = 0.69$ cm^{-1} and $E = 0.003$ cm^{-1}. This large D value results from the fact that the two interacting parallel spins are in orthogonal p orbitals on the same carbon.

Biological Phenomena

Studies of biochemical and biological phenomena such as photosynthesis and the biochemical processes associated with vision are potentially important applications of triplet-state esr of randomly oriented molecules. Innovative work on biological polymers has already begun. The $\Delta M_S = 2$ triplet-state transitions like those of tyrosine and tryptophan in frozen solutions [31-36] have been observed in photoexcited glassy solutions of the proteins [31,33,36]. Zuchlich [36] also observed a $\Delta M_S = 1$ signal arising from the tryptophan constituent in ovalbumin. Shulman and his associates [37,38] observed $\Delta M_S = 2$ signals in phosphorescent DNA and in poly-dAT and showed by comparison with signals of the purines and pyramidines that they arise from the thymine constituent.

Active Dye-Laser Media

The organic dyes have increasing usage as the active media in tunable lasers. This function has greatly augmented the interest in the metastable triplet states of these substances. Important information on the zero-field separation of the triplet spin levels, on the decay rate of the triplet state, and on energy transfer can be gained from measurements of the triplet-state esr spectra of these dyes. Yamashita and his associates [39,40] have used both the $\Delta M_S = 2$ and the $\Delta M_S = 1$ esr transitions to study the properties of dye laser media during laser excita-

Table 10.5 Triplet-state Parameters Derived from Observed $\Delta M_S = 1$ Transitions of Randomly Oriented Molecules in Glassy Solvents

Molecule	*Glassy Solvent*	*D* (cm^{-1})	*E* (cm^{-1})	*Source*
	Ground-state Triplets			
$\mathrm{H-\overset{\bullet}{\underset{\bullet}{C}}-H}$	Argon (4.2°K)	0.69	0.003	*a*
$\mathrm{H-\overset{\bullet}{\underset{\bullet}{C}}-C{\equiv}C-H}$	Polychlorotrifluoro-ethylene (77°K)	0.6276	0	*b*
$\mathrm{H-\overset{\bullet}{\underset{\bullet}{C}}-C{\equiv}C-CH_3}$	Polychlorotrifluoro-ethylene (77°K)	0.6263	0	*b*
$\mathrm{H-\overset{\bullet}{\underset{\bullet}{C}}-C{\equiv}C-C_6H_5}$	Polychlorotrifluoro-ethylene (77°K)	0.5413	0.0035	*b*
$\mathrm{H-\overset{\bullet}{\underset{\bullet}{C}}-C{\equiv}C-C{\equiv}C-CH_3}$	Polychlorotrifluoro-ethylene (77°K)	0.6087	0	*b*
$\mathrm{H-\overset{\bullet}{\underset{\bullet}{C}}-C{\equiv}C-C{\equiv}C-C_6H_5}$	Polychlorotrifluoro-ethylene (77°K)	0.5530	0	*b*
$\mathrm{H-\overset{\bullet}{\underset{\bullet}{C}}-C{\equiv}N}$	Polychlorotrifluoro-ethylene (77°K)	0.8629	0	*b*
$\mathrm{N{\equiv}C-\overset{\bullet}{\underset{\bullet}{C}}-C{\equiv}N}$	Fluorolube (77°K)	1.002	<0.002	*c*
$\mathrm{\overset{\bullet}{\underset{\bullet}{C}}{=}N^{+}{=}N^{-}}$	Fluorolube (77°K)	1.153	<0.002	*c*
$\mathrm{H-\overset{\bullet}{\underset{\bullet}{C}}-CF_3}$	? (4.2°K)	0.712	0.021	*d*
$\mathrm{CF_3-\overset{\bullet}{\underset{\bullet}{C}}-CF_3}$	? (4.2°K)	0.7444	0.0437	*d*
	Excited-state Triplets			
Biphenyl	Diethylether	0.1092	0.0036	*e*
Carbazole	Diethylether	0.1022	0.0066	*e*
Dibenzofuran	Diethylether	0.1071	0.0092	*e*
Dibenzothiophene	Diethylether	0.1130	0.0021	*e*
Chrysene	2-Methyltetra-hydrofuran	0.095	0.025	*f*
1,2-Benzanthracene	2-Methyltetra-hydrofuran	0.079	0.014	*f*
1,2-Benzpyrene	2-Methyltetra-hydrofuran	0.090	0.023	*f*
Tryptophan	Aromatic amino acids	0.0984	0.0410	*g*
Tyrosine	Aromatic amino acids	0.1301	0.0558	*g*
Phenylalanine	Aromatic amino acids	0.1475	0.0439	*g*

Table 10.5 (Continued)

Molecule	*Glassy Solvent*	*D* (cm^{-1})	*E* (cm^{-1})	*Source*
Thymine mono-phosphate	Deuterated glass	0.203	0.010	*h*
Ortic acid	Polyethyleneglycol: H_2O	0.176	0.016	*i*
Guanine mono-phosphate	Polyethyleneglycol: H_2O	0.141	0.0172	*i*
Adenine mono-phosphate	Polyethyleneglycol: H_2O	0.119	0.027	*j*
Poly A	Polyethyleneglycol: H_2O	0.116	0.027	*j*

[a] Bernheim, Bernard, Wang, Wood, and Skell [30].
[b] R. A. Bernheim, R. J. Kempf, J. V. Gramas, and P. S. Skell, *J. Chem. Phys.*, **43**, 196 (1965).
[c] E. Wasserman, L. Barash, and W. A. Yager, *J. Am. Chem. Soc.*, **87**, 2075 (1965).
[d] E. Wasserman, L. Barash, and W. A. Yager, *J. Am. Chem. Soc.*, **87**, 4974 (1965).
[e] S. Siegel and H. S. Judeikis, *J. Phys. Chem.*, **70**, 2201 (1966).
[f] J. S. Brinen and N. K. Orloff, *J. Chem. Phys.*, **45**, 4747 (1966).
[g] Zuclich [36].
[h] Lamola, Gueron, Yamane, Eisinger, and Shulman [38].
[i] R. G. Shulman and R. O. Rahn, *J. Chem. Phys.*, **45**, 2940 (1966).
[j] R. O. Rahn, T. Yamane, J. Eisinger, J. W. Longworth, and R. G. Shulman, *J. Chem. Phys.*, **45**, 2947 (1966).

tion. They measured the zero-field splitting constants and the lifetimes in the triplet state for several types of dyes in ethanol at 77°K. Typical parameters obtained for rhodamine dyes are: $D = 0.057$ cm^{-1}, $E = 0.017$ cm^{-1}, and τ_T = 1.6 sec; for disodium fluorescein they obtained: D - 0.065 cm^{-1}, $E = 0.019$ cm^{-1}, and τ_T= 0.29 sec; for the acridine dye they obtained: $D = 0.061$ cm^{-1}, $E = 0.015$ cm^{-1}, and τ_T = 2.0 sec. The relatively small values for D in these molecules result from the spreading of the unpaired electron orbitals over several rings of these large systems.

Several dyes have been studied by Antonucci and Tolley [41] with the ΔM_S = 2 transition during mercury-arc pumping. The triplet-state esr of the thiacyanine and cyanine dyes have been studied by Pierce and Berg [42], who obtained rather accurate values of D, E, and D^* from observations of both $\Delta M_S = 1$ and $\Delta M_S = 2$ transitions.

3. ENERGY MIGRATION: TRIPLET EXCITONS

3.a. Charge-transfer Complexes: TCNQ

The pioneering esr work of Chesnut and his co-workers [43-45] on the low-lying, thermally populated, triplet states of the TCNQ (tetracyanoquinodimethane) charge-transfer complexes is essential to the understanding of the unusual electrical and magnetic behavior of these important systems. Great interest in these complexes has been generated among solid-state physicists by the discovery that in particular low-temperature ranges certain ones have an almost vanishing resistance to electrical conduction. This interest is indicated by the no less than three sessions devoted to TCNQ complexes at a recent meeting of the American Physical Society [46].

The original esr observations by Chesnut and Phillips [43] on several TCNQ charge-transfer complexes established that spin correlation in these complexes gives rise to a singlet ground state and a low-lying, thermally accessible triplet state. The triplet state is confirmed by the zero-field, dipole-dipole splitting of the resonance at reduced temperatures (−140°C) and by the presence of the half-

Table 10.6 Triplet-state ESR Parameters for TCNQ Charge-transfer Complexes

Complex[a]	g_x g_y g_z	$\|D/hc\|$ (cm^{-1})	$\|E/hc\|$ (cm^{-1})	*Source*
$(\phi_3XCH_3)^+(TCNQ)_2^-$	2.0040			
(X = P or As)	2.0031	0.0062	0.00098	*b*
	2.0027			
$(Cs^+)_2(TCNQ)_3^{=}$	2.0025			
	2.0015	0.00937	0.00151	*c*
	2.0003			
$Rb^+(TCNQ)^-$		0.0131	0.00159	*d*
$(TMB)^+(TCNQ)^-$		0.0140	0.00212	
$M^+(TCNQ)^-$		0.0150	0.00180	*e,f*
$(M^+)_2(TCNQ)_3^{=}$		0.00942	0.00151	*f*
$(M^+)_3(TCNQ)_4^{\equiv}$		0.0122	0.00125	*f*

[a] Abbreviations: TCNQ = tetracyanoquinodimethane, TMB = trimethylbenzimidazolium, M = morpholinium.
[b] Chesnut and Phillips [43].
[c] Chesnut and Arthur [44].
[d] Hibma, Sawatzky, and Kommandeur [58].
[e] M. A. Marechal and H. M. McConnell, *J. Chem. Phys.*, **43**, 497 (1965).
[f] Bailey and Chesnut [50].

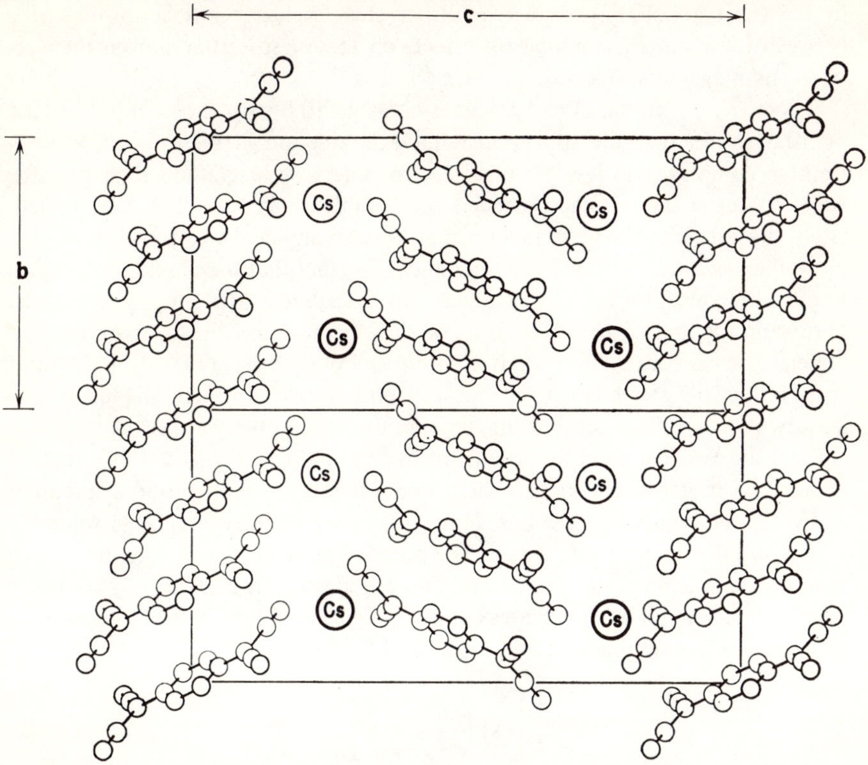

Fig. 10.11 A proposed structural arrangement of the components of a $(Cs^+)_2(TCNQ)_3^=$ crystal as viewed along the *a* axis. From Chesnut and Arthur [44].

field transitions ($\Delta M_S = 2$). At the higher temperatures, these features are averaged out by the exchange interaction.

A detailed analysis was made of single crystals of the $(\phi_3XCH_3)^+(TCNQ)_2^-$ salts (where X = P and As) at −150°C. The *g* values and zero-field splitting parameters are listed in Table 10.6 along with those for the complex $(Cs^+)_2(TCNQ)_3^=$, studied later by Chesnut and Arthur [44]. The single crystal for $(\phi_3XCH_3)^+$ $(TCNQ)_2^-$ is triclinic or pseudomonoclinic, and that for $(Cs^+)_2(TCNQ)_3^=$ is monoclinic. Figure 10.11 shows the proposed schematic arrangement of the TCNQ units

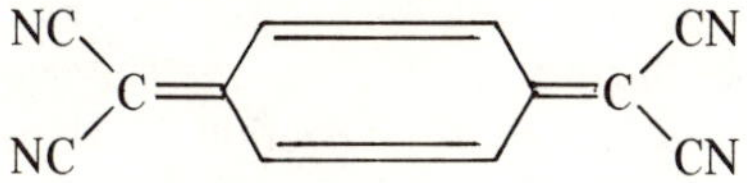

TCNQ
44

of the $(Cs^+)_2(TCNQ)_3^=$ complex in the crystal. Exchange interaction occuring between these units has important effects on the esr spectrum and on the electrical and magnetic properties.

A comparison of the D and E values in Table 10.6 with those of Table 10.1 or 10.5 shows that the spin dipole-dipole interaction in these TCNQ charge-transfer complexes is less by one to two orders of magnitude than that for triplet states in the normal molecules comparable in size to TCNQ. This immediately suggests that the wave functions of the two unpaired electrons spread over more than one unit in the TCNQ complex. The fact that the dipole-dipole splitting is observable, however, shows that this spreading at the temperature of the observation is limited to a few units, or that the movement of the two interacting electrons is coordinated (exciton motion). For $(\phi XCH_3)^+(TCNQ)_2^-$, Chesnut and Phillips [43] estimate that the electron pair of the triplet spreads over at least two of the TCNQ units at the temperature of the observation.

There are two temperature effects on the esr that give significant information about these systems. The most obvious one is that on the population of the thermally excited triplet state, which depends on the energy difference between the triplet and the singlet ground state. The population of the triplet state relative to that of the ground state (which gives no esr signal) can be found by measurements of the integrated intensities of the esr signal at the different temperatures. These intensities are related to temperature [47] by

$$I \propto \frac{1}{T}\left[\exp\left(\frac{J}{kT}\right) + 3\right]^{-1} \tag{10.83}$$

where J is the energy difference between the triplet and the singlet states. This energy is labeled J because it is essentially the isotropic spin-exchange energy J of the triplet that is positive for these complexes, for which the triplet is the upper state. Some singlet-triplet activation energies obtained from the measured intensities with this relation are listed in Table 10.7.

The effects of triplet exchange on the esr of the $(\phi_3XCH_3)^+(TCNQ)_2^-$ salts at different temperatures are illustrated in Fig. 10.12. These various curves are for a fixed orientation of the single crystal in the magnetic field. At the lowest temperature (−140°C) the lines are sharp, and the dipole-dipole doublet splitting characteristic of the triplet state is well resolved. As the temperature is raised, the lines begin to broaden, and the doublet-component separation decreases until the doublet fuses into a singlet line at −55°K. This singlet sharpens as the temperature is increased further, to −25°K. The behavior is exactly the same as that already encountered (Chapter IX, Section 2) where periodic exchange between two electrons, two protons, or two configurations of a free radical in a doublet state modulates the esr spectrum. For example, compare Fig. 10.12 with Fig. 9.8, where the exchange is between a proton of a free radical and the

Table 10.7 Exchange Parameters in TCNQ Salts

Salt	*Singlet-Triplet Splitting, J (eV)*	*Fast-exchange Line Narrowing,[c] E_{ex} (eV)*	*Slow-exchange Line Broadening,[c] E_{ex} (eV)*	*Slow-exchange Line Separation[c] E_{ex} (eV)*	*Source*
$(\phi_3AsCH_3)^+(TCNQ)_2^-$	0.065	0.16	0.11	0.050	*a*
$(\phi_3PCH_3)^+(TCNQ)_2^-$	0.065	0.19	0.13	0.042	*a*
$(Cs^+)_2(TCNQ)_3^=$	0.16	0.22	0.22	0.054	*a*
$(M^+)(TCNQ)^-$	0.36	–	0.65	0.11	*b*
$(M^+)_2(TCNQ)_3^=$	0.31	0.38	0.35	0.069	*b*
$(M^+)_3(TCNQ)_4^=$	0.33	–	0.38	–	*b*

[a]Jones and Chesnut [45].
[b]Bailey and Chesnut [50].
[c]Method for derivation of E_{ex}.

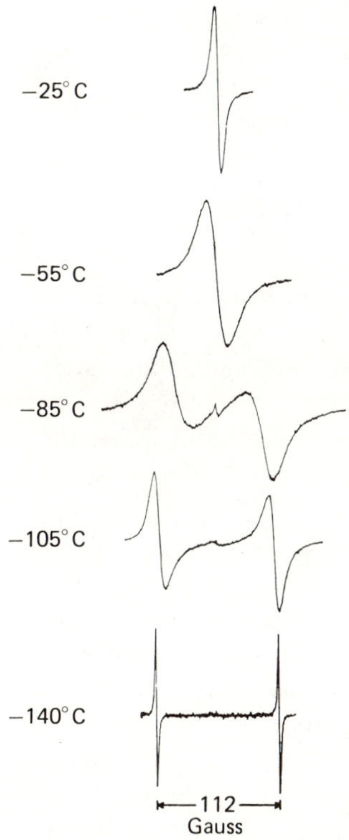

Fig. 10.12 The esr spectrum of $(\phi_3\ PCH_3)^+(TCNQ)_2^-$ at various temperatures, showing effects of increasing triplet–triplet exchange rates as the temperature of the sample is increased. The curves were recorded at a fixed orientation of the crystal in the magnetic field. From Chesnut and Phillips [43].

acid protons of a solution and where the frequency of the exchange is increased by the acid concentration rather than by a rise in temperature. The exchange that causes the changes in Fig. 10.12 occurs between two triplet entities. Estimates of the mean exchange frequency $2\pi\nu_{ex} = (1/\tau)$ can be calculated for the slow-exchange case (lowest temperature) from measurements of line widths with (9.31) or (9.32); for the rapid-exchange case (highest temperature), from (9.43) with $2\pi\nu_{ex} = [1/(T_2)_{ex}]$. The exchange frequency $\nu_{ex} = [1/(2\pi\tau)]$ at a given temperature can also be approximated with (9.52) from a measurement of the doublet separation $\Delta\omega = (\gamma_e/\sqrt{2})(\Delta H)$ at intermediate temperatures. Note, however, that this relation holds only in the region where $\Delta H \ll \Delta H_0$. The activation energy E_{ex} for the triplet exchange can be calculated from the exchange frequencies at particular temperatures from

$$\nu_{ex} = \nu_0 \exp\left(\frac{-E_{ex}}{kT}\right) \tag{10.84}$$

where $\omega_0 = 2\pi\nu_0 = (\gamma_e/\sqrt{2})\ \Delta H_0$, and ΔH_0 is the doublet splitting in gauss at the lowest temperature before exchange effects occurred. The theory on which these relations are based was apparently first developed by Anderson [48]. The activation energies E_{ex} derived by Chesnut and his associates [45,50] from line-width measurements in the fast and slow approximations and from line separations are given in Table 10.7. Those derived from the line separations (last column) agree poorly with these values and are believed to be less accurate. From line-width measurement as a function temperature Hibma and Kommandeur [57] derived the activation energies E_{ex} of 0.08 eV for $Rb^+(TCNQ)^-$ and 0.16 eV for $(TMB)^+(TCNQ)^-$. The values they obtained from the temperature dependence of the line separations were likewise found to be in poor agreement with those from line widths. They concluded that the temperature dependence of the line splittings probably results from processes differing from those affecting line widths.

The Exciton Model

Despite the resolution of electron dipole-dipole splitting with very sharp lines in single crystals, no hyperfine structure has been reported for the TCNQ charge-transfer triplets. This circumstance prompted the suggestion [45,49] that the observed triplets are really excitons moving through the specimens in the form of a coordinated electron pair with parallel spins, making a migrating unit with effective spin $S = 1$. The migration of the triplet would prevent resolution of any hyperfine splitting because the effective spin density would be spread about many nuclei, but would not eliminate their mutual dipole-dipole interaction because the movement of the two parallel spins is coordinated. In the idealized model the paired units with $S = 1$ would be in a triplet energy band with an energy gap between it and the ground-state singlet level. The energy gap would be determined primarily by the positive, isotropic spin-exchange interaction $J(s_1 \cdot s_2)$ of the coordinated electrons. In such a model, the line broadening, which depends on the triplet-state lifetime, results primarily from collisions between the "excitons" [45]. This model helps to coordinate the esr data with other unusual properties–electric, magnetic, and optical–exhibited by these charge-transfer complexes. It is evident that this idealized model of an excitation band applies to certain of these systems more than to others.

The exciton-band model has been discussed by many researchers, among the earliest of whom are Hardin McConnell [49,51] and Chesnut [52,53]. A theoretical model, the modified Hubbard Hamiltonian which coordinates many of the diverse properties of the TCNQ salts was developed by Soos and Klein [54], who give references to other useful papers on these salts. The later paper on esr of TCNQ salts by Bailey and Chesnut [50] gives extensive references, as do those on triplet excitons in TCNQ salts by Kommandeur and his associates [55-58].

Fractional Charge Transfer Complexes: TCNB

Hayashi, Iwata, and Nagakura [59] have studied a series of photoexcited complexes of TCNB (1,2,4,5-tetracyanobenzene) with organic molecules such as benzene (BE), toluene (TO), mesitylene (ME), durene (DU), pentamethylbenzene (PMB), and hexamethyl benzene (HMB). They obtained values of $D^* = (D + 2E)^{1/2}$ and triplet-state lifetimes from measurements of the esr spectra, the $\Delta M_S = 2$ transitions in glassy solvents. From theoretical considerations, they derived a relation between D^* and x, the fractional electron charge transfer from the complex molecule to the TCNB. In this way they found that rather larger fractional electronic charges were being transferred to the TCNB in certain of these molecular complexes. For example, x values found are: 0.95 for TCNB-HMB, 0.77 for TCNB-PMB, 0.72 for TCNB-DU, 0.42 for TCNB-ME, 0.11 for TCNB-TO, and 0.07 for TCNB-BE. Values of D^*, τ, and x for these and other similar complexes may be found in their paper.

3.b. Triplet Energy Transfer between Unlike Molecules

Farmer, Gardner, and McDowell [60] demonstrated that esr of the triplet state can be used for study of triplet-triplet energy transfer between similar molecules in a glassy matrix. They used the second-order $\Delta M_S = 2$ transitions that van der Waals and de Groot [22] had shown to be effective for detection of esr for randomly oriented molecules in the triplet state. Solutions of benzophenone and naphthalene (10^{-2} to 10^{-1} molar) in EPA were observed at 77°K. When light of wavelength 366 mμ (which is absorbed by benzophenone but is outside the absorption band of naphthalene) was directed on the mixture, esr of naphthalene in the triplet state was observed. When naphthalene alone (5×10^{-2} molar) was present in the EPA, no resonance was produced by the 366-mμ irradiation. This proved that the energy absorbed by the benzophenone was being transferred to excite the triplet state of naphthalene. Following these experiments, Hutchison and his collaborators [28,61] used the first-order $\Delta M_S = 1$ transition for quantative measurement of triplet energy transfer between phenanthrene and naphthalene at low concentration in single crystals of diphenyl.

In one energy-transfer mechanism, the donor molecule D is lifted to an excited singlet state S_D (of benzophenone or phenanthrene in the preceding experiments) by radiation that is too low in frequency to reach the excited singlet state S_A of the acceptor (naphthalene). Through internal conversion, a transition is made from S_D to the lower excited triplet state T_D of the donor. The energy T_D then migrates through the exciton band T_H of the host crystal to excite the triplet-state T_A of the acceptor, which has energy near, or slightly lower than, that of T_D and of T_H. However, detailed studies by Brenner and Hutchison [62] at low temperatures (~1.7°K) of phenanthrene (donor) and naphthalene (acceptor), both in dilute concentrations in diphenyl as the host single crystal, showed that there is, in addition to this process, a guest-host, intermolecular, intersystem crossing in which the S_D decays directly to the exciton band T_H of the host crystal. As before, the T_H excitation is then transferred to the acceptor guest to excite the triplet T_A. These studies also provide intensities of the esr signals and their rates of growth and decay under different light excitations at 1.7°K.

Hirota and Hutchison [63] studied rates of energy transfer as a function of temperature from phenanthrene to naphthalene in single crystals of diphenyl and observed different rate constants corresponding to different mechanisms of transfer. These studies were extended by Hirota [64]. Smaller, Avery, and Remco [65] used esr to study the triplet energy transfer between guest components in organic glassy solutions. The rates of energy transfer and signal decay were found to depend strongly on the viscosity of the glassy solutions.

In the study of triplet-triplet energy transfer, esr offers an advantage over optical spectroscopy in that ordinarily the esr of the triplet states of the donor and acceptor are resolvable and separately measurable, whereas the luminescence bands of the optical region are broad and overlapping.

4. PAIRED RADICAL TRIPLETS

Exchange-coupled radical pairs are now commonly observed in irradiated molecular powders and single crystals in which the trapped radicals are produced in pairs by bond breakage or in strongly irradiated samples where a detectable concentration of close-neighbor radicals is produced. When each of the separated radicals has only one unpaired electron, the two electrons of the exchange-coupled pair will have a triplet state. The interaction that causes the parallel alignment of the spins is the same as that giving rise to the parallel alignment of the spins of electrons in different orbitals of an atom (Hund's rule). This parallel alignment of their spins prevents the electrons from occurring together in the same orbital of one component of the radical pair where their mutual Coulomb repulsive interaction would be greater than if they were in separate orbitals on opposite components of the radical pair. However, the triplet is not necessarily the ground state of the complex. If the two radical centers are so close that significant overlapping of their orbitals occurs, a "bonded" state in which the two electrons have antiparallel spins may be the ground state. This is true for the copper acetate pair. Because the orbital overlap of the two components is never large, the exchange interaction is small compared with that in intramolecular bonds and is the order of a hundred microwave quanta or less. Because the exchange interaction of coupled molecular free radicals is isotropic or nearly so, the parallel-spin alignment is not broken down by the applied magnetic field, but it may be broken by thermal motions that cause flipping of the individual spins. Whether the ground state of the paired complex is the triplet or is a singlet determines whether the esr signal will increase or decrease with an increase of temperature. This assumes that the radical pairs are not displaced or dissociated by the increase of temperature. The integrated intensity of the esr lines is related to temperature and to the exchange energy by (10.83), in which the sign of J changes when the triplet is the ground state. By measurement of the line intensities as a function of temperature, an approximate value of J can be obtained, provided the radical pairs are stable in the temperature range covered. Most radical pairs are unstable above 100°K.

4.a. A Strongly Coupled Pair: Copper Acetate

In 1951 evidence for an exchange-coupled radical pair in copper acetate was obtained by Lancaster and Gordy [66], who observed from a powdered sample of copper acetate "half-field" esr signals corresponding to $g \simeq 4$ values similar to the $\Delta M_S = 2$ transitions described in Section 2.a. Shortly thereafter, Bleaney and Bowers [67] made detailed observations of the first-order $\Delta M_S = 1$ transitions of these coupled radical pairs in a single crystal of $Cu(C_2H_3O_2)\cdot H_2O$. For analysis of the data they used a triplet-state Hamiltonian like that of (10.8). By a fitting of the observed data to the predicted transition frequencies (or resonant field values), they obtained the spin dipole-dipole parameter values, $|D| = 0.34 \pm 0.03$ cm^{-1} and $|E| = 0.01 \pm 0.005$ cm^{-1}. Later, Abe and Shimoda [68] obtained the more precise values, $|D| = 0.345$ cm^{-1} and $|E| = 0.005$ cm^{-1}. The very small value of E relative to D proves that the spin-spin interaction is very nearly axially symmetric. From measurements of esr signal strength as a function of temperature, Bleaney and Bowers [67] derived the value of the exchange-coupling parameter as $J \approx 370°K = 257$ cm^{-1}, with an estimated error of about 20%. The more accurate value, $J = (310 \pm 15)$ cm^{-1}, was obtained from the magnetic susceptibility measurements (see Abragam and Bleaney [69]). This J corresponds to the triplet-singlet separation, and its sign proves that the singlet is the lower or ground-state level.

There are a number of significant differences between the esr parameters of the copper acetate radical pairs and those of molecular free-radical pairs produced by irradiation. The most important of these is the much larger value of D for the copper acetate pair (0.345 cm^{-1} or 10,340 MHz) compared with those (200 to 500 MHz) for the radical pairs commonly produced by irradiation of organic molecular solids. Another important difference is that the triplet is usually the ground state for the latter complexes, whereas the copper acetate pair has the singlet ground state.

Determination of the crystal structure with X-ray diffraction [70] shows that the Cu^{++} ions of the two molecules in the unit cell of the $Cu(C_2H_3O_2)_2 \cdot H_2O$ crystal are only 2.64 Å apart. Nevertheless, the dipole-dipole interaction of two electron spins at this distance cannot account for the observed $|D| = 0.345\ cm^{-1}$. With this R, Abragam and Bleaney [69] calculated the interaction $D_{dip} = -0.19\ cm^{-1}$, well below the measured magnitude of D. It is thus evident that anisotropy in the exchange interaction contributes to the measured D. Some anisotropy in J is expected from the residual spin-orbit coupling indicated by the relatively large **g** anisotropy, $g_x = 2.053$, $g_y = 2.093$, $g_z = 2.344$ for the radical pair [68]. The contribution to D by the anisotropic exchange interaction induced by the residual spin-orbit coupling is theoretically predicted [69] to be

$$D_{ex} = \frac{1}{8} J \left[\frac{1}{4} (g_z - 2)^2 - (g_\perp - 2)^2 \right] \tag{10.85}$$

With $J = 310\ cm^{-1}$ obtained from magnetic susceptibilities and the measured $g_z = 2.344$ and $g_\perp = (g_x + g_y)/2 = 2.073$, the value $D_{ex} = +0.95$ was obtained. The predicted $D = D_{ex} + D_{dip} = 0.76\ cm^{-1}$ is appreciably higher than the observed value. Reasons for this discrepancy are discussed by Abragam and Bleaney.

Since we are primarily concerned with triplet radical pairs induced by irradiation of molecular solids for which the **g** anisotropy and exchange coupling are much smaller than those of the copper complex, we do not pursue this problem further. The preceding summary is included to make one aware that residual spin orbital coupling can cause a sizeable contribution to D when, as evidenced by **g** anisotropy, the orbital momentum is not effectively quenched. A second-order anisotropic contribution to the $\mathbf{S} \cdot \mathbf{D} \cdot \mathbf{S}$ interaction induced by spin-orbit coupling is derived in Chapter II, Section 2. This contribution is given by the third term on the right of (2.13). Note that the tensor components Λ_{ij} given by (2.12) have the same form as those in the **g** tensor in (2.15). This relationship makes possible the evaluation of the anisotropy in **J** and $\mathbf{D}_{ex}$ from the measured anisotropy in **g** with calculation of Λ_{ij}.

4.b. Paired Radicals in Irradiated Single Crystals

The theory described in Section 1.a can be used for analysis of the esr of the triplet states of radical pairs in single crystals. However, the analysis is simpler than that of a molecule for two principal reasons. First, the zero-field doublet splitting for the radical pair is generally so much smaller than the Zeeman splitting at the conventional spectrometer frequency (X or K band) that the spin vector **S** can be considered as aligned with the applied field and the expectation values $\langle S \rangle$, $\langle S_x \rangle$, $\langle S_y \rangle$, and $\langle S_z \rangle$ can be expressed by (10.64) and (10.65). A second significant simplification of the analysis accrues from the axial symmetry that results from the rather large separation (several angstroms) of the spin centers on the two component monoradicals of the pair. The small D values and axial symmetry ($E \approx 0$) are demonstrated by the parameters for several radical pairs that are listed in Table 10.8. Calculation of the R values in the table is based on the assumption that the two aligned spins giving rise to the doublet splitting are separated by a fixed distance R. It is then easy to show that [71]

Table 10.8 Spin-spin Coupling Parameters and Separations of Radical-pair Centers in Irradiated Single Crystals

Crystal	$\lvert D \rvert$ (G)	$\lvert E \rvert$ (G)	R^a (Å)	*Source*
Dimethylglyoxime	155	~0	5.6	*c*
Hydroxyurea	107	~0	6.38	*d*
Carbazide	334	~0	4.37	*e*
ABNO[b]	773	47	3.30	*f*
Monofluoroacetamide	80	~0	7.03	*g*
Oxalic acid (dihydrate)	110	~0	6.33	*h*
Oxalic acid (anhydrous)	51	~0	9.93	*h*
Potassium persulfate	6.9	~0	15.8	*i*

[a]The value R is the effective separation of the two electron spins of the radical pair as determined by (10.86).
[b]9-Aza-bicyclo (3,3,1)nonan-3-one-9-oxyl.
[c]Y. Kurita, *J. Chem. Phys.*, **41**, 3926 (1964).
[d]K. Reiss and H. Shields, *J. Chem. Phys.*, **50**, 4368 (1969).
[e]K. Reiss and W. Gordy, *J. Chem. Phys.*, **55**, 5329 (1971).
[f]F. Genoud and M. Decorps, *Molec. Phys.*, **34**, 1583 (1977).
[g]M. Iwasaki and K. Toriyama, *J. Chem. Phys.*, **46**, 4693 (1967).
[h]G. C. Moulton, M. P. Cernansky, and D. C. Straw, *J. Chem. Phys.*, **46**, 4292 (1967).
[i]S. B. Barnes and M. C. R. Symons, *J. Chem. Soc. (A)*, 66, (1966).

$$D\,(\mathrm{G}) = \frac{3\beta}{R^3} = \frac{27{,}820}{R^3} \quad \text{where } R \text{ is in Angstrom units.} \tag{10.86}$$

An observational advantage is that the electron spin-spin doublet splitting of the radical dimer is often sufficiently large to cause the $\Delta M_S = 1$ doublets of the radical pair to avoid the central region where the $\Delta M_S = 1$ lines of the monoradicals occur. Although there is some overlapping of these two signals when there is widely spaced hyperfine structure, as for F coupling in the radicals of irradiated fluorocarbons, the patterns of the pairs and monoradicals are usually separated sufficiently that their hyperfine components can be correctly assigned. An example of the triplet esr of a radical pair formed from the unlike monoradicals $\dot{C}H_2CONH_2$ and $\dot{C}HFCONH_2$ is given in Fig. 10.13 for one orientation of the single crystal. Lines of both monoradicals as well as those of the radical pair are clearly evident. The zero-field parameters and the spin-center separations R of this radical pair are included in Table 10.8.

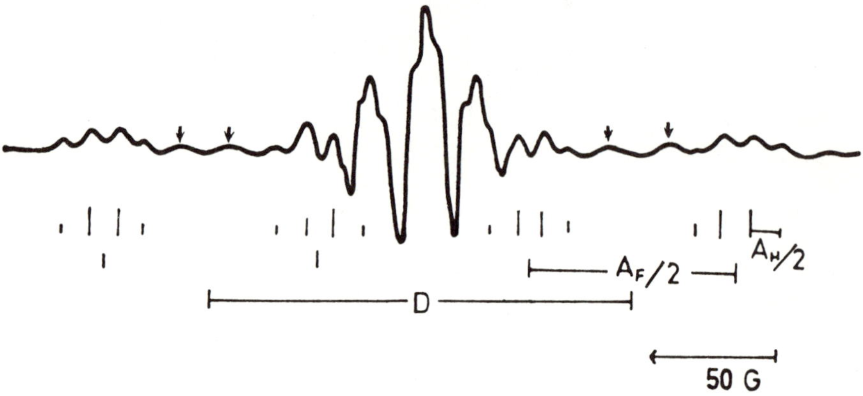

Fig. 10.13 Second-derivative esr spectrum ($\Delta M_S = 1$) of a radical pair between $\dot{C}H_2CONH_2$ and $\dot{C}HFCONH_2$ formed in a single crystal of CH_2FCONH_2 γ-irradiated at 77°K. The strong triplet at the center is due to the monoradical $\dot{C}H_2CONH_2$. Peaks indicated by small arrows are due to $\dot{C}HFCONH_2$ monoradicals. The stick diagram indicates the theoretical pattern for the radical pair at the particular orientation of the crystal when observations were made at 77°K. From M. Iwasaki and K. Toriyama, *J. Chem. Phys.*, **46**, 4693 (1967).

When it is not possible to unscramble the signals of the radical pair and those of the monoradicals, or when there is some question as to whether the signals detected are those of triplet state esr or simply signals of unknown monoradicals, it is advisable to search for the half-field $\Delta M_S = 2$ signals of the radical pair. These signals can arise only from triplet-state esr. For this purpose, a component magnetic vector of the ac microwave field must be along the static field $\mathbf{H}_0$, in contrast to the perpendicular arrangement for observation of the $\Delta M_S = 1$ transitions (Section 2.a).

Examples of the zero-field parameters and effective spin-center separations for paired radicals in irradiated single crystals are given in Table 10.8. These observations were made at or near 77°K. Generally, these radical pairs are unstable at room temperature, and most of them decay at temperatures well below 200°K. Descriptions of the probable structures of the monoradicals forming the pairs will be found in the literature cited in Table 10.8. The calculations of the R values are made with (10.86) and the assumption that the observed D value arises entirely from dipole-dipole coupling of the two parallel spins. This assumption is justified by the nearly isotropic g observed in the experiments.

4.c. Randomly Oriented Pairs in Matrix Isolation

Hydrogen Atom and Methyl Radical Complex

A radical pair formed by irradiation of methane at 4.2°K has been identified as an exchange-coupled H atom and CH_3 radical, separated by one methane molecule [71]. Because of the basic simplicity of this small pair and the possible detection of related species in other hydrocarbons at low temperature, its spectrum is described in some detail. However, one can expect to observe radical pairs involving H atoms only in a critical, low-temperature range. If the temperature is too high (much above 4°K), the dislocated H atoms are likely to escape through the lattice; and if too low, their displacement by the irradiation may be so small that re-formation of the dissociated CH bond will quickly occur.

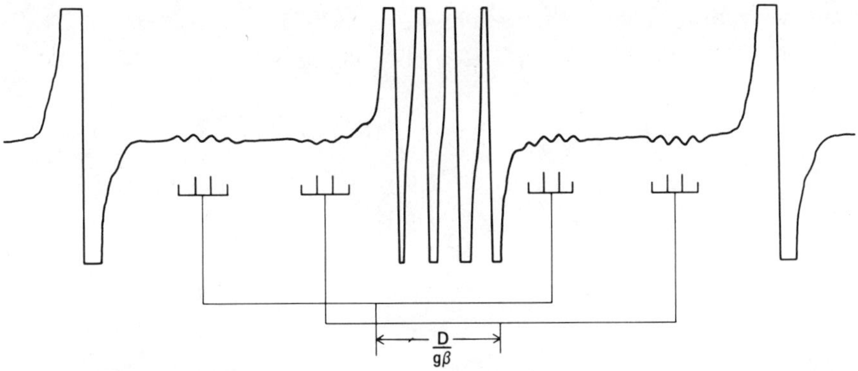

Fig. 10.14 The $\Delta M_S = 1$ spectrum of a coupled H atom–methyl radical pair (H–CH_3) in γ-irradiated methane at 4.2°K. The weak lines directly above the theoretical stick diagram are those for the radical pair. The two strong outside lines separated by 510 G are those of isolated H atoms. The strong central quartet is the spectrum of isolated CH_3 radicals. From Gordy and Morehouse [71].

Figure 10.14 shows the esr produced by γ-irradiation of methane at 4.2°K. It consists of the doublet and quartet hyperfine patterns of the H and CH_3 monoradicals (strong signals) and the much weaker triplet-state esr of the H---CH_3 pair indicated by the stick diagram. Because the coupled spin centers on the H and C atoms are separated by more than 6 Å, it can be assumed that the **D** tensor for the pair has axial symmetry about the line of separation. According to the statistical distribution, the strongest critical peak signal in the derivative curve for the $\Delta M_S = 1$ transition occurs for **H** perpendicular to the symmetry axis. Thus the observed peaks must correspond to $H_\perp$ values of the field at the frequency of the observation. Except for the hyperfine splitting, these critical fields are given by (10.34) and (10.35) and the doublet splitting, by $\Delta H = (D/g\beta)$ [see (10.36)]. This splitting, indicated in Fig. 10.14, was measured to be 91 G. With (10.86), this value yields $R = 6.76$ Å as the separation of the spin centers. This distance indicates that the coupled H and CH_3 are probably on opposite sides of a CH_4 molecule, with the H trapped in an octahedral site and the CH_3 in a substitutional site, most likely that originally occupied by the dislocated parent CH_4 molecule.

The hyperfine coupling of the H atom of the pair is isotropic and produces a splitting of the triplet levels of $M_S A_t^{\mathrm{H}} M_I$, where $M_I = 1/2$ or $-1/2$. The CH_3 of the pair evidently rotates about the symmetry axis in such a way as to equalize the perpendicular coupling of the three protons. The CH_3 splitting is small and can be treated with first-order perturbation theory to give the level splitting of $M_S A_t^{\mathrm{CH_3}} \sum_i M_{I_i}$, where the summation is taken over all combinations of the three proton quantum numbers. There is obviously no hyperfine splitting of the $M_S = 0$ level, and the splitting of the $M_S = 1$ and $M_S = -1$ levels is $\pm(A_t^{\mathrm{H}} M_I + A_t^{\mathrm{CH_3}} \sum_i M_{I_i})$. The energy-level diagram of Fig. 10.15 indicates the hyperfine splitting of the levels with the observed transitions shown by arrows. At constant frequencies, these transitions correspond to the following magnetic field positions:

$$(H_1)_\perp = H_0 + \frac{1}{2}\frac{D}{g\beta} + \frac{1}{g\beta}\left(A_t^{\mathrm{H}} M_I + A_t^{\mathrm{CH_3}} \sum_i M_{I_i}\right) \tag{10.87}$$

$$(H_2)_\perp = H_0 - \frac{1}{2}\frac{D}{g\beta} + \frac{1}{g\beta}\left(A_t^{\mathrm{H}} M_I + A_t^{\mathrm{CH_3}} \sum_i M_{I_i}\right) \tag{10.88}$$

The distinct combinations of the methyl proton spins, $\sum_i M_{I_i} = 3/2, 1/2, -1/2, -3/2$, produce a CH_3 quartet substructure on each proton doublet component ($M_I = 1/2$ and $-1/2$) of each of the electron spin-spin doublet components $(H_1)_\perp$ and $(H_2)_\perp$. The resulting esr pattern is indicated by the stick diagram under the observed pattern in Fig. 10.14. It should be noted that the observed hyperfine

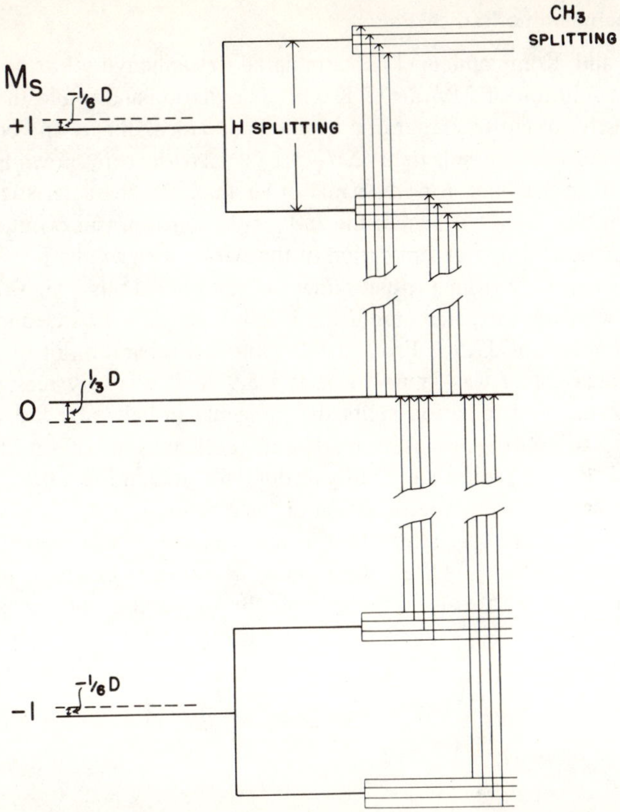

Fig. 10.15 Energy-level diagram for the exchange-coupled H···CH_3 radical pair with the observed transitions indicated by arrows. (see Fig. 10.14). From Gordy and Morehouse [71].

splitting and corresponding A_t^{H} and $A_t^{\mathrm{CH}_3}$ for the radical pair are only half those for the corresponding monoradicals, even though the splittings for the $M_S = \pm 1$ energy levels are the same as those for the $M_S = \pm 1/2$ levels in the monoradicals. This reduction in splitting is characteristic of the esr of the triplet state. It results from the zero splitting of the $M_S = 0$ level involved in each $\Delta M_S = 1$ transition.

The fact that the esr pattern of only one radical pair is observed, and that it has rather sharp components, indicates that the trapping sites for formation of detectable triplet pairs in the irradiated CH_4 at 4.2°K are unique. Triplet-state esr patterns with sharp components that can be assigned to only one pair-species with a single R value (see Table 10.8) are those most often observed in irradiated single crystals. In large hydrocarbon chains, however, molecules with multiple-pair signals have been observed (Section 4.d).

The Diphenylamino Radical Pair

Wiersma and Kommandeur [72] irradiated tetraphenylhydrazine (TPH) in a rigid glass solution of EPA at 77°K with uv radiation and found the induced esr to be mostly, or entirely, that of a triplet-state radical dimer. The observed $\Delta M_S = 1$ lines and also the half-field, $\Delta M_S = 2$ signal (inset) are shown in Fig. 10.16. There is little evidence for any signal other than the characteristic triplet-state doublet in the $\Delta M_S = 1$ signal; the $\Delta M_S = 2$ transition can occur only for the triplet. A close computer simulation of the $\Delta M_S = 1$ signal, with the assumption that it arose entirely from a triplet dimer radical, proved possible. This fitting was achieved with the esr parameter $|D| = 132 \pm 1.5$G, $E = 0 \pm 6$G, and with a line-width parameter of 37.5G. From the D value, the separation of the spin centers of the radical dimer was found to be $R = 5.9$ Å. It was very reasonably postulated that the uv irradiation splits the molecule in halves by breaking the NN bond. The two nitrogens move apart, and reorientations of the ringed groups occur to form the radical-pair configuration indicated in Fig. 10.17. This change in conformation prevents reformation of the NN bond.

The preceding example illustrates the power of the esr technique for the study of radiation chemistry. Often the primary reaction is breakage of a covalent bond to produce a short-lived radical pair. By observation of the esr of the rad-

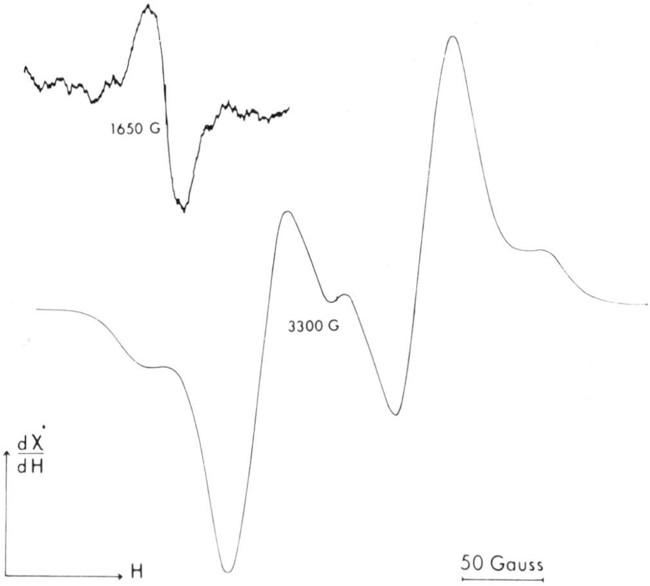

Fig. 10.16 The esr spectrum of the diphenylamino radical dimer in rigid EPA glass at 77°K. The lower spectrum is that for the $\Delta M_S = 1$ transition; the inset shows the half-field, $\Delta M_S = 2$ signal. From Wiersma and Kommandeur [72].

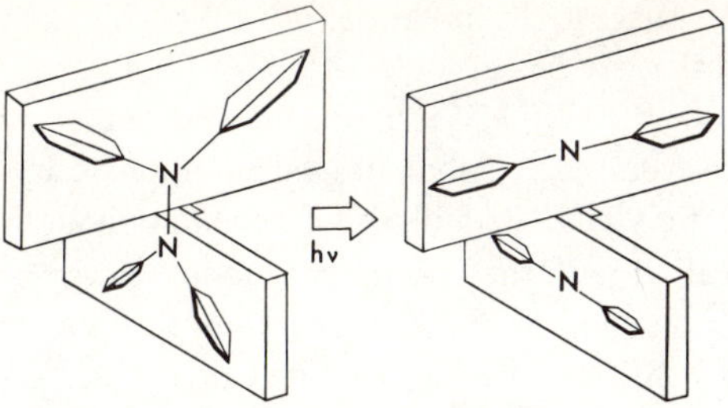

Fig. 10.17 Configurations for the undamaged tetraphenylhydrazine molecule and for the diphenylamino radical dimer produced by breakage of the NN bond. From Wiersma and Kommandeur [72].

ical pair, one should be able to learn which bond is broken. Clearly, observation of these transient broken-bond pairs is not always easy. The sample must be irradiated at a temperature such that thermal motions cause structural changes that prevent re-formation of the bond but one not so high as to cause separation of the parts sufficient to destroy the dimer radical. These conditions may be mutually exclusive in most substances. The dimer of the diphenylamino radical is sufficiently stable in the EPA glassy solution at 77°K to produce the strong dimer signal of Fig. 10.16 with little evidence of monomer signals; yet this signal was found to disappear completely at 90°K and to be replaced by signals of secondary monoradicals. The latter monoradicals were also observed and identified with esr [72].

Secondary or tertiary monoradicals are those usually observed in radiation studies with esr spectroscopy. Too little emphasis has been placed on observation of the primary radical pairs formed by the initial breakage of the bond. This has been caused most probably by the critical irradiation, temperature, viscosity, and other conditions that are required to stabilize these dimer radicals sufficiently for their observation. It would appear that the additional information to be gained would justify special efforts for observation of the primary radical pairs formed by dissociation of covalent bonds. The $\Delta M_S = 2$, $g = 4$ signals offer the simplest, most unambiguous method for this purpose. To detect these "forbidden" transitions, one should be aware that the inducing ac field component must be along the static field $\mathbf{H}_0$, in contrast to the requirement for the $\Delta M_S = 1$ transitions, the H_1 must be applied perpendicular to $\mathbf{H}_0$. In most microwave observational cavities, components of H_1 are both parallel and perpendicular to the $\mathbf{H}_0$, but certainly not in equal strengths.

4.d. Radical Pairs in Chain Hydrocarbons and High Polymers

Triplet-state esr of radical pairs has been observed in many large aliphatic hydrocarbons that were irradiated and observed at subroom temperatures. Samples with ordered as well as randomly oriented molecules have been studied. A few examples are given to suggest the potentialities of triplet-state esr in polymer chemistry and in radiation studies of hydrocarbons.

Iwasaki and his co-workers [73-75] have observed triplet-state esr of interchain radical pairs in several high polymers, including polyethylene, polypropylene, polyisobutylene, polyvinyl alcohol, cellulose, and polymethylmethacryrate. The radical pairs, together with the constituent monoradicals, were produced by γ-irradiation of the polymers at 77°K. The observed species were exchange-coupled pairs, as was confirmed by measurement of the $\Delta M_S = 2, g = 4$ transitions. Well-resolved proton hyperfine structure was observed for the γ-irradiated polyethylene and polypropylene, which provided a means for identification of the chemical form of the radical pairs. Figure 10.18 shows their observed spectrum for the $\Delta M_S = 2$ transition of polyethylene, together with a computer simulation of the spectrum. The simulated spectrum was computed for coupled $-CH_2\dot{C}HCH_2-$ pairs with proton couplings equivalent to one-half those observed for the constituent monoradicals. The reasons for the half-value splitting of the triplet relative to those of the monomer are explained in Section 4.c. In addition to this difference, the dimer radical formed from two identical mono-

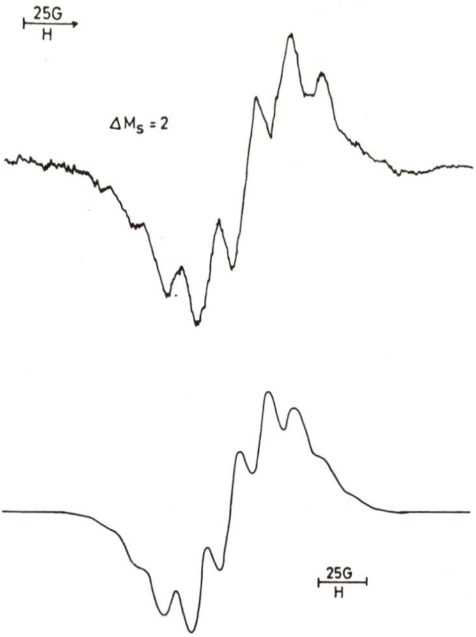

Fig. 10.18 Observed (upper curve) and computer-simulated (lower curve) hyperfine structure for the half-field, $\Delta M_S = 2$ transition of the interchain radical pairs in γ-irradiated polyethylene at 77° K. From Iwasaki, Ichikawa, and Ohmori [74].

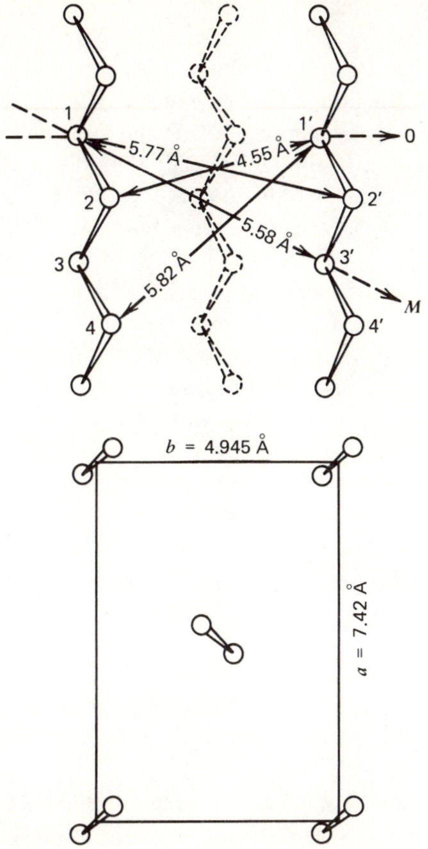

Fig. 10.19 Crystal structure of *n*-hydrocarbons with some interchain carbon separations indicated. The dashed arrows denoted by O and M are orthorhombic and monoclinic axes, respectively. From Iwasaki, Ichikawa, and T. Ohmori [75].

mers has twice the number of coupling nuclei as have the constituent monomers.

The good agreement between the observed and simulated patterns in Fig. 10.18 verifies the postulate that the triplet-state esr is that of two spin-coupled $-CH_2\dot{C}HCH_2-$ radicals. Similar comparisons for polypropylene proved that the observed radical pairs in this irradiated polymer were two coupled $-CH_2$-$\dot{C}(CH_3)$ -CH_2- monomers along neighboring strands in the polymer. The $|D| = 146$ G value observed for *n*-hydrocarbon chains [75] yields $R = 5.75$ Å for the separation of the spin centers of pair radicals in these hydrocarbon chains. From comparison with intercarbon distances of neighboring chains as measured by crystallographers in single crystals of the *n*-hydrocarbons, it was found that the distance of 5.75 Å corresponds closely to the 5.77-Å distance between carbons

1 and 2 on hydrocarbon chains having one intervening chain, as shown in Fig. 10.19.

Although Iwasaki, Ichikawa, and Ohmori [74] reported only interchain pair radicals, Fujimura and Tamura [76] have reported esr signals of intermolecular radical pairs between near-by spin centers on the same polymer molecule. It is evident that the formation of the radical pairs should depend on the condition of the specimens, the type of radiation used to induce the radicals, and the physical conditions, especially temperature, under which the observations are made.

Formation of radical pairs in γ-irradiated single crystals of n-decane ($C_{10}H_{22}$), n-decane d_{22}, and n-octane (C_8H_{18}) have been observed by Gillbro and Lund with esr spectroscopy [77]. In the n-decane, two types of monoradicals were found, $CH_3\dot{C}H(CH_2)_7CH_3$ and $-CH_2\dot{C}HCH_2-$, but only the latter type was observed to form radical pairs at 77°K. Only the $-CD_2\dot{C}DCD_2-$ monoradicals were detected in n-decane d_{22}, and these were likewise found to produce radical pairs. These pairs are intramolecular or intrachain radicals, of the same form as those described above for irradiated polymers. The hyperfine splitting, although somewhat better resolved in the single-crystal spectra, is in agreement with that observed in polyethylene (Fig. 10.18). Thus the single-crystal measurements provide confirmation of this type of cross-chain, spin-spin interaction in hydrocarbons.

The most interesting new feature discovered by Gillbro and Lund [77] is the formation in n-decane of cross-linked radical pairs of $-CD_2CDCD_2-$ having different separations R with different decay temperatures, as shown in Table 10.9.

Table 10.9 ESR Parameters of Radical Pairs in γ-Irradiated $C_{10}D_{22}$ Single Crystal[a]

Radical Pair	*Decay Temperature T (°C)*	*Coupling |D| (G)*	*Distance R (Å)*	*Concentration (%)*
I	-145	266	4.7	~0.1
II	-145	231	4.9	~0.1
III	-130	174	5.4	~0.2
IV	-90	122	6.1	~0.4
V	-30 (m.p.)	77.5	7.1	~0.4
VI	-30	49	8.3	~0.3

[a]From Gillbro and Lund [77].

Note that those having the greater separation R have the greater stability, as indicated by the higher decay temperature and also the higher concentration.

Those of type I have a separation of 4.7 Å within error limits of being the same as the repeat length, 4.75 Å, along the crystallographic *b* axis of *n*-decane (see Fig. 10.19). It was thus concluded that the exchange-coupled electrons for this type are located on the crystallographically equivalent carbon atoms on neighboring molecules, and that the radical pairs II, III, and IV are also comprised of monoradicals ($-CD_2\text{-}\dot{C}D\text{-}CD_2-$) located on adjacent molecules of the crystal. The most stable pairs, V and VI, were thought to be formed from monoradicals on molecules separated by one decane molecule. However, it would seem from comparisons with the work of Iwasaki and associates discussed earlier that the pairs III and IV having separations of 5.4 Å and 6.1 Å may also be formed from alternate as well as from adjacent molecules in the crystal.

References

1. C. A. Hutchison and B. W. Mangum, *J. Chem. Phys.*, **29**, 952 (1958).
2. C. A. Hutchison and B. W. Mangum, *J. Chem Phys.*, **34**, 908 (1961).
3. J. M Robertson, *Proc. Roy. Soc. (Lond.)*, **A141**, 594 (1933); **A142**, 659 (1933).
4. M. Gouterman and W. Moffitt, *J. Chem. Phys.*, **30**, 1107 (1959).
5. M. Gouterman, *J. Chem. Phys.*, **30**, 1369 (1959).
6. H. F. Hameka, *J. Chem. Phys.*, **31**, 315 (1959).
7. R. McWeeny, *J. Chem. Phys.*, **34**, 399, 1065 (1961).
8. R. McWeeny and Y. Minzuno, *Proc. Roy. Soc. (Lond.)*, **A259**, 554 (1961).
9. A. D. McLachlan, *Molec. Phys.*, **5**, 51 (1962).
10. S. A. Boorstein and M. Gouterman, *J. Chem. Phys.*, **39**, 2443 (1963); **41**, 2776 (1964).
11. S. A. Boorstein and M. Gouterman, *J. Chem. Phys.*, **42**, 3070 (1965).
12. Y. Gondo and A. H. Maki, *J. Chem. Phys.*, **50**, 3270 (1969).
13. M. Mizushima, *The Theory of Rotating Diatomic Molecules*, Wiley, New York, 1975.
14. R. W. Brandon, G. L. Closs, C. E. Davoust, C. A. Hutchison, B. E. Kohler, and R. Silbey, *J. Chem. Phys.*, **43**, 2006 (1965).
15. C. A. Hutchison and B. E. Kohler, *J. Chem. Phys.*, **51**, 3327 (1969).
16. J. Higuchi, *J. Chem. Phys.*, **38**, 1237 (1963).
17. C. A. Hutchison and G. A. Pearson, *J. Chem. Phys.*, **47**, 520 (1967).
18. R. J. M. Anderson and B. E. Kohler, *J. Chem. Phys.*, **65**, 2451 (1976).
19. J. S. Vincent, *J. Chem. Phys.*, **54**, 2237 (1971).
20. R. P. Frosch, A. M. P. Goncalves, and C. A. Hutchison, *J. Chem. Phys.*, **58**, 5209 (1973).
21. C. A. Hutchison and V H. McCann, *J. Chem. Phys.*, **61**, 820 (1974).
22. J. H. van der Waals and M. S. de Groot, *Molec. Phys.*, **2**, 333 (1959).

23. S. P. McGlynn, T. Azumi, and M. Kinoshito, *Molecular Spectroscopy of the Triplet State*, Prentice-Hall, Englewood Cliffs, N. J., 1969.
24. P. Kottis and R. Lefebvre, *J. Chem. Phys.*, **39**, 393 (1963).
25. M. S de Groot and J. H. van der Waals, *Molec. Phys.*, **6**, 545 (1963).
26. E. Wasserman, L. C. Snyder, and W. A. Yager, *J. Chem. Phys.*, **41**, 1763 (1964).
27. L. C. Snyder and R. L. Kornegay, *Bull. Am. Phys. Soc.*, **9**, 101 (1964).
28. R. W. Brandon, R. E. Gerkin, and C. A. Hutchison, *J. Chem. Phys.*, **37**, 447 (1962).
29. P Kottis and R. Lefebvre, *J. Chem. Phys.*, **41**, 379 (1964).
30. R. A. Bernheim, H. W. Bernard, P. S. Wang, L. S. Wood, and P. S. Skell, *J. Chem. Phys.*, **53**, 1280 (1970).
31. T. Shiga and L. H. Piette, *Photochem. Photobiol.*, **3**, 223 (1964).
32. J. E. Maling, K. Rosenkeck, and M. Weissbluth, *Photochem. Photobiol.*, **4**, 241 (1965).
33. T. Shiga, H S. Mason, and C. Simo, *Biochem.*, **5**, 1877 (1966).
34. B. Rabinovitch, *Arch. Biochem. Biophys.*, **124**, 258 (1968).
35. J. J. ten Bosch, R. O. Rahn, J. W. Longworth, and R. G. Shulman, *Proc. Nat. Acad. Sci. (USA)*, **59**, 1003 (1968).
36. J. Zuchlich, *J. Chem. Phys.*, **52**, 3586 (1970).
37. R. O. Rahn, R. G. Shulman, and J. W. Longworth, *J. Chem. Phys.*, **45**, 2955 (1966).
38. A. A. Lamola, M. Gueron, T. Yamane, J. Eisinger, and R. G. Shulman, *J. Chem. Phys.*, **47**, 2210 (1967).
39. M. Yamashita and H. Kashiwagi, *J. Chem. Phys.*, **59**, 2156 (1973).
40. M. Yamashita, H. Ikeda, and H. Kashiwagi, *J. Chem. Phys.*, **63**, 1127 (1975).
41. F. R. Antonucci and L. G. Tolley, *J. Phys. Chem.*, **77**, 2712 (1973).
42. R. A. Pierce and R. A. Berg, *J. Chem. Phys.*, **56**, 5087 (1972).
43. D. B. Chesnut and W. D. Phillips, *J. Chem. Phys.*, **35**, 1002 (1961).
44. D. B. Chesnut and P. Arthur, *J. Chem. Phys.*, **36**, 2969 (1962).
45. M. T. Jones and D. B. Chesnut, *J. Chem. Phys.*, **38**, 1311 (1963).
46. *Bull. Am. Phys. Soc.* 20 (3) (1975), Sessions EL, GI, and HK.
47. D. Bijl, H. Kainer, and A. C. Rose-Innes, *J. Chem. Phys.*, **30**, 765 (1959).
48. P. W. Anderson, *J. Phys. Soc. Jap.*, **9**, 316 (1954).
49. H. M. McConnell and R. Lynden-Bell, *J. Chem. Phys.*, **36**, 2393 (1962).
50. J. C. Bailey and D. B. Chesnut, *J. Chem. Phys.*, **51**, 5118 (1969).
51. H. M. McConnell, in *Molecular Biophysics*, B. Pullman and M. Weissbluth, Eds., Academic, New York, 1965, pp. 311-324.
52. D. B. Chesnut, *J. Chem. Phys.*, **40**, 405 (1964).
53. D. B. Chesnut, *J. Chem. Phys.*, **41**, 472 (1964).

54. Z. G. Soos and D. J. Klein, *J. Chem. Phys.*, **55**, 3284 (1971).
55. J. G. Vegter and J. Kommandeur, *Phys. Rev. B*, **7**, 2929 (1973).
56. T. Hibma, G. A. Sawatzky, and J. Kommandeur, *Chem. Phys. Lett.*, **23**, 21 (1973).
57. T. Hibma and J. Kommandeur, *Phys. Rev. B*, **12**, 2608 (1975).
58. T. Hibma, G. A. Sawatzky, and J. Kommandeur, *Phys. Rev. B*, **15**, 3959 (1977).
59. H. Hayashi, S. Iwata, and S. Nagakura, *J. Chem. Phys.*, **50**, 993 (1969).
60. J. B. Farmer, C. L. Gardner, and C. A. McDowell, *J. Chem. Phys.*, **34**, 1058 (1961).
61. R. W. Brandon, R. E. Gerkin, and C. A. Hutchison, *J. Chem. Phys.*, **41**, 3717 (1964).
62. H. C. Brenner and C. A. Hutchison, *J. Chem. Phys.*, **58**, 1328 (1973).
63. N. Hirota and C. A. Hutchison, *J. Chem. Phys.*, **42**, 2869 (1965).
64. N. Hirota, *J. Chem. Phys.*, **43**, 3354 (1965).
65. B. Smaller, E. C. Avery, and J. R. Remco, *J. Chem. Phys.*, **43**, 922 (1965).
66. F. W. Lancaster and W. Gordy, *J. Chem. Phys.*, **19**, 1181 (1951).
67. B. Bleaney and K. D. Bowers, *Proc. Royal Soc.*, **A214**, 451 (1952).
68. H. Abe and J. Shimoda, *J. Phys. Soc. Jap.*, **12**, 1255 (1957).
69. A. Abragam and B. Bleaney, *Electron Paramagnetic Resonance of Transition Ions*, Clarendon, Oxford, 1970.
70. J. N. van Niekerk and F. R. L. Schoening, *Acta Crystallogr.*, **6**, 227 (1953).
71. W. Gordy and R. Morehouse, *Phys. Rev.*, **151**, 207 (1966).
72. D. A. Wiersma and J. Kommandeur, *Molec. Phys.*, **13**, 241 (1967).
73. M. Iwasaki and T. Ichikawa, *J. Chem. Phys.*, **46**, 2851 (1967).
74. M. Iwasaki, T. Ichikawa, and T. Ohmori, *J. Chem. Phys.*, **50**, 1984 (1969).
75. M. Iwasaki, T. Ichikawa, and T. Ohmori, *J. Chem. Phys.*, **50**, 1991 (1969).
76. T. Fujimura and N. Tamura, *J. Polym. Sci.*, **B10**, 469 (1972).
77. T. Gillbro and A. Lund, *J. Chem. Phys.*, **61**, 1469 (1974).

APPENDIX

Table A.1 Relevant Nuclear Moments and Valence-shell Orbital Couplings

Isotope	Natural Abundance (%)	Nuclear Moments: I	Nuclear Moments: μ_I *(nm)*	Atomic-orbital Couplings: s *Orbitals* A_{ns} *(MHz)*	Atomic-orbital Couplings: p *Orbitals* B_{np} *(MHz)*	Source of Coupling Values
^{1}H	99.98	1/2	2.79255	1420.4		*a*
^{2}H	0.015	1	0.857348	218.26		*a*
^{6}Li	7.42	1	0.82189	152.14		*a*
^{7}Li	92.58	3/2	3.25586	401.76		*a*
^{9}Be	100	3/2	-1.1776	-358		*b*
^{10}B	19.61	3	1.8004	676	18	*c*
^{11}B	80.39	3/2	2.68858	2020	53	*c*
^{13}C	1.11	1/2	0.70225	3110	91	*c,d*
^{14}N	99.63	1	0.40365	1540	48	*c,d*
^{15}N	0.366	1/2	-0.28299	-2159	-67	*c,d*
^{17}O	0.037	5/2	-1.8928	-4628	-144	*c*
^{19}F	100	1/2	2.6285	47910	1515	*c,e*
^{23}Na	100	3/2	2.21711	885.8		*a*
^{27}Al	100	5/2	3.6414	2746	58.7	*b*
^{29}Si	4.70	1/2	-0.55492	-3381	-87	*c*
^{31}P	100	1/2	1.13165	10178	287	*c,d*
^{33}S	0.760	3/2	0.64292	2715	78	*c,d*
^{35}Cl	75.53	3/2	0.82191	4664	140	*b*

Table A.1 (Continued)

Isotope	Natural Abundance (%)	Nuclear Moments I	Nuclear Moments μ_I (nm)	Atomic-orbital Couplings s Orbitals A_{ns} (MHz)	Atomic-orbital Couplings p Orbitals B_{np} (MHz)	Source of Coupling Values
^{37}Cl	24.47	3/2	0.68414	3882	117	b
^{39}K	93.10	3/2	0.391	230.86		a
^{67}Zn	4.11	5/2	0.8757	1251		b
^{69}Ga	60.4	3/2	2.016	7454	149	b
^{73}Ge	7.76	9/2	-0.8792	1492	36	b
^{75}As	100	3/2	1.4347	9582	255	c
^{77}Se	7.58	1/2	0.534	13,468	383	b
^{79}Br	50.54	3/2	2.10576	21,738	646	b
^{81}Br	49.46	3/2	2.2696	23,429	696	b
^{85}Rb	72.15	5/2	1.3527	1,011.9		a
^{127}I	100	5/2	2.808		662	f
^{129}Xe	26.44	1/2	-0.7768	-33,030	-1052	c
^{133}Cs	100	7/2	2.579	2,298.19		a

[a]N. F. Ramsey, *Nuclear Moments*, Wiley, New York, 1953.
[b]J. R. Morton, J. R. Rowlands, and D. H. Whiffen, *National Physical Laboratory* (U. K.), Circular No. BPR 1.3, 1962.
[c]J. R. Morton, "Electron Spin Resonance Spectra of Oriented Radicals," *Chem. Rev.*, **64**, 453-471 (1964).
[d]G. W. Chantry, A. Horsfield, J. R. Morton, J. R. Rowlands, and D. H. Whiffen, *Molec. Phys.*, **5**, 233 (1962).
[e]R. J. Cook, J. R. Rowlands, and D. H. Whiffen, *Proc. Chem. Soc.*, 252 (1962).
[f]Derived from atomic-beam resonance constants, *a* and *b*.

Table A.2 Relevant Spin Orbital Couplings of Valence *p* Electrons

Element	ζ *(cm⁻¹)*	*Source*	*Element*	ζ *(cm⁻¹)*	*Source*
B	11	*a*	Cl	587	*b*
C	29	*a*	Zn	386	*a*
N	76	*a*	Ga	551	*a*
O	151	*a*	Ge	940	*a*
F	272	*b*	As	1500	*b*
Al	75	*a*	Se	1688	*a*
Si	142	*b*	Br	2460	*b*
P	299	*a*	Sn	2097	*b*
S	382	*a*	I	5060	*b*

[a] J. R. Morton, J. R. Rowlands, and D. H. Whiffen, *National Physical Laboratory* (U. K.), Circular No. BPR 1.3 1962.
[b] D. S. McClure, *J. Chem. Phys.*, **17**, 905 (1949).

Table A.3 Fundamental Physical Constants[a]

Speed of light	c	$(2.997925 \pm 0.000001) \times 10^{10}$ cm/sec
Electronic charge	e	$(4.80298 \pm 0.00007) \times 10^{-10}$ esu
		$(1.60210 \pm 0.00002) \times 10^{-20}$ emu
Avogadro's constant	N	$(6.02252 \pm 0.00009) \times 10^{23}$ molecules/mole
Atomic mass unit	$M(^{12}C)/12$	$(1.66043 \pm 0.00002) \times 10^{-24}$ g
Electron rest mass	m_e	$(9.10908 \pm 0.00013) \times 10^{-28}$ g
Proton rest mass	m_p	$(1.67252 \pm 0.00003) \times 10^{-24}$ g
Neutron rest mass	m_n	$(1.67482 \pm 0.00003) \times 10^{-24}$ g
Planck's constant	h	$(6.62559 \pm 0.00016) \times 10^{-27}$ erg sec
Fine-structure constant ($e^2/\hbar c$)	α	$(7.29720 \pm 0.00003) \times 10^{-3}$
Bohr magneton ($e\hbar/2m_e c$)	β	$(9.2732 \pm 0.0002) \times 10^{-21}$ erg/G
Nuclear magneton ($e\hbar/2m_p c$)	β_I	$(5.05050 \pm 0.00013) \times 10^{-24}$ erg/G
Boltzmann's constant	k	$(1.38054 \pm 0.00006) \times 10^{-16}$ erg/deg
Rydberg constant	R_∞	$(1.0973731 \pm 0.0000001) \times 10^{5}$ cm^{-1}
Bohr radius	a_0	$(5.29167 \pm 0.00002) \times 10^{-9}$ cm

[a] From E. R. Cohen and J. W. M. DuMond, *Rev. Mod. Phys.*, **37**, 537 (1965). The unified scale of atomic masses is used throughout ($^{12}C = 12$).

AUTHOR INDEX

Numbers in parentheses are reference numbers and show that an author's work is referred to although his name is not mentioned in the text. *Italics* indicate pages on which author is mentioned.

SUBJECT INDEX